UNDERSTANDING STATISTICS

IN THE BEHAVIORAL SCIENCES ▓ **TENTH EDITION**

UNDERSTANDING
STATISTICS
IN THE BEHAVIORAL SCIENCES ■ TENTH EDITION

ROBERT R. PAGANO

WADSWORTH
CENGAGE Learning™

Australia • Brazil • Japan • Korea • Mexico • Singapore • Spain • United Kingdom • United States

Understanding Statistics in the Behavioral Sciences, **Tenth Edition**
Robert R. Pagano

Publisher: Jon-David Hague
Psychology Editor: Tim Matray
Developmental Editor: Robert Jucha
Assistant Editor: Paige Leeds
Editorial Assistant: Lauren Moody
Media Editor: Mary Noel
Marketing Program Manager: Sean Foy
Content Project Manager: Charlene M.
    Carpentier
Design Director: Rob Hugel
Art Director: Pamela Galbreath
Print Buyer: Rebecca Cross
Rights Acquisitions Specialist: Dean
    Dauphinais
Production Service: Graphic World Inc.
Text Designer: Lisa Henry
Photo Researcher: PreMedia Global
Text Researcher: Sue Howard
Copy Editor: Graphic World Inc.
Illustrator: Graphic World Inc.
Cover Designer: Lisa Henry
Cover Image: School of Red Sea Bannerfish:
    © Strmko/Dreamstime.com
Compositor: Graphic World Inc.

For product information and technology assistance, contact us at
Cengage Learning Customer & Sales Support, 1-800-354-9706.
For permission to use material from this text or product, submit all requests online at **www.cengage.com/permissions**.
Further permissions questions can be e-mailed to
permissionrequest@cengage.com.

Library of Congress Control Number: 2011934938

Student Edition:
ISBN-13: 978-1-111-83726-6
ISBN-10: 1-111-83726-0

Loose-leaf Edition:
ISBN-13: 978-1-111-83938-3
ISBN-10: 1-111-83938-7

Wadsworth
20 Davis Drive
Belmont, CA 94002-3098
USA

Cengage Learning is a leading provider of customized learning solutions with office locations around the globe, including Singapore, the United Kingdom, Australia, Mexico, Brazil, and Japan. Locate your local office at **www.cengage.com/global**.

Cengage Learning products are represented in Canada by Nelson Education, Ltd.

For your course and learning solutions, visit **www.cengage.com.**

Purchase any of our products at your local college store or at our preferred online store **www.CengageBrain.com.**

Printed in the United States of America
1 2 3 4 5 6 7 15 14 13 12 11

I dedicate this tenth edition to all truth-seekers. May this textbook aid you in forming an objective understanding of reality. May the data-based, objective approach taught here help inform your decisions and beliefs to help improve your life and the lives of the rest of us.

ABOUT THE AUTHOR

ROBERT R. PAGANO received a Bachelor of Electrical Engineering degree from Rensselaer Polytechnic Institute in 1956 and a Ph.D. in Biological Psychology from Yale University in 1965. He was Assistant Professor and Associate Professor in the Department of Psychology at the University of Washington, Seattle, Washington, from 1965 to 1989. He was Associate Chairman of the Department of Neuroscience at the University of Pittsburgh, Pittsburgh, Pennsylvania, from 1990 to June 2000. While at the Department of Neuroscience, in addition to his other duties, he served as Director of Undergraduate Studies, was the departmental adviser for undergraduate majors, taught both undergraduate and graduate statistics courses, and served as a statistical consultant for departmental faculty. Bob was also Director of the Statistical Cores for two NIH center grants in schizophrenia and Parkinson's disease. He retired from the University of Pittsburgh in June 2000. Bob's current interests are in the physiology of consciousness, the physiology and psychology of meditation and in Positive Psychology. He has taught courses in introductory statistics at the University of Washington and at the University of Pittsburgh for over thirty years. He has been a finalist for the outstanding teaching award at the University of Washington for his teaching of introductory statistics.

Bob is married to Carol A. Eikleberry and they have a 21-year-old son, Robby. In addition, Bob has five grown daughters, Renee, Laura, Maria, Elizabeth, and Christina, one granddaughter, Mikaela, and a yellow lab. In his undergraduate years, Bob was an athlete, winning varsity letters in basketball, baseball and soccer. He loves tennis, but arthritis has temporarily caused a shift in retirement ambitions from winning the singles title at Wimbledon to watching the U.S. Open and getting in shape for doubles play sometime in the future. He also loves the outdoors, especially hiking, and his morning coffee. He especially values his daily meditation practice. His favorite cities to visit are Boulder, Estes Park, New York, Aspen, Santa Fe, and Santa Barbara.

CONTENTS

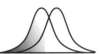

PART TWO DESCRIPTIVE STATISTICS 23

4 Measures of Central Tendency and Variability   79

5 The Normal Curve and Standard Scores   102

6 Correlation   122

7

Linear Regression   159

PART THREE   **INFERENTIAL STATISTICS   187**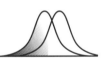

8

Random Sampling and Probability   189

9   Binomial Distribution   225

10   Introduction to Hypothesis Testing Using the Sign Test   248

11 Power 277

12 Sampling Distributions, Sampling Distribution of the Mean, the Normal Deviate (*z*) Test 298

13 Student's *t* Test for Single Samples 327

14

Student's *t* Test for Correlated and Independent Groups **356**

15 Introduction to the Analysis of Variance   401

16 Introduction to Two-Way Analysis of Variance   445

17 Chi-Square and Other Nonparametric Tests 482

18 Review of Inferential Statistics 527

I have been teaching a course in introductory statistics for more than 30 years, first within the Department of Psychology at the University of Washington, and most recently within the Department of Neuroscience at the University of Pittsburgh. Most of my students have been psychology majors pursuing the Bachelor of Arts degree, but many have also come from biology, business, education, neuroscience, nursing, the health sciences, and other fields. My introductory statistics course has been rated quite highly. While at the University of Washington, I was a finalist for the university's "Outstanding Teaching" award for teaching this course.

This textbook has been the mainstay of my teaching. Because most of my students have neither high aptitude nor strong interest in mathematics and are not well grounded in mathematical skills, I have used an informal, intuitive approach rather than a strictly mathematical one. My approach assumes only high-school algebra for background knowledge, and depends very little on equation derivation. It attempts to teach the introductory statistics material in a deep way, in a manner that facilitates conceptual understanding and critical thinking rather than mechanical, by-the-numbers problem solving.

My statistics course has been quite successful. Students are able to grasp the material, even the more complicated topics like "power," and at the same time they often report that they enjoy learning it. Student ratings of this course have been high. Their ratings of this textbook are even higher; among other things students say that the text is very clear. that they like the touches of humor, and that it helps them to have the material presented in such great detail. Some students have even commented that "this is the best textbook I have ever had." Admittedly, this kind of comment is not the most frequent one offered, but for an introductory statistics textbook, coming from psychology majors, I take it as high praise indeed.

I believe the factors that make my textbook successful are the following:

- It promotes understanding rather than mechanical problem solving.
- It is student-friendly and informally written, with touches of humor that connect with students and help lower anxiety.
- It is very clearly worded and written at the right level.

- It presents the material in great detail.
- It has good visuals.
- It uses a more extended treatment of sampling distributions and a particularly effective sequencing of the inferential material, beginning with the sign test instead of the conventional approach of beginning with the z test.
- It has interesting illustrative examples, and many ideally solved and end-of-chapter problems for students to practice with.

Rationale for Introducing Inferential Statistics with the Sign Test

Understanding the use of sampling distributions is critical to understanding inferential statistics. The first sampling distribution discussed by most texts is the sampling distribution of the mean, used in conjunction with the z test. The problem with this approach is that the sampling distribution of the mean is hard for students to understand. It cannot be generated from simple probability considerations, and its definition is very abstract and difficult to make concrete. Moreover, it is hard to relate the sampling distribution of the mean to its use in the z test. The situation is further complicated because at the same time as they are being asked to understand sampling distributions, students are being asked to understand a lot of other complicated concepts such as null hypothesis, alternative hypothesis, alpha level, Type I and Type II error, and so forth. As a result, many students do not develop an understanding of sampling distributions and why they are important in inferential statistics. I believe this lack of understanding persists throughout the rest of inferential statistics and undermines their understanding of this important material.

What appears to happen is that since students do not understand the use of sampling distributions, when they are asked to solve an inferential problem, they resort to mechanically going through the steps of (1) determining the appropriate statistic for the problem, (2) solving its equation by rote, (3) looking up the probability value in an appendix table, and (4) concluding regarding the null and alternative hypotheses. Many students follow this procedure without any insight as to why they are doing it, except that they know doing so will lead to the correct answer. Thus students are often able to solve problems without understanding what they are doing, all because they fail to develop a conceptual understanding of what a sampling distribution is and why it is important in inferential statistics.

To impart a basic understanding of sampling distributions, I believe it is much better to present an extended treatment of sampling distributions, beginning with the sign test rather than the z test. The sign test is a simple inference test for which the binomial distribution is the appropriate sampling distribution. The binomial distribution is very easy to understand and it can be derived from basic probability considerations. Moreover, its application to the inference process is clear and obvious. This combination greatly facilitates understanding inference and bolsters student confidence in their ability to successfully handle the inferential material. In my view, the appropriate pedagogical sequence is to present basic probability first, followed by the binomial distribution, which is then followed by the sign test. This is the sequence followed in this textbook (Chapters 8, 9, and 10, respectively).

Since the binomial distribution is entirely dependent on simple probability considerations, students can easily understand its generation and application. Moreover, the binomial distribution can also be generated by an empirical process that is used later in the text beginning with the sampling distribution of the mean in Chapter 12 and continuing for all of the remaining inference tests. Generating sampling distributions via an empirical approach helps make the concept of sampling distribution concrete and facilitates student understanding and application of sampling distributions. Since the sampling distribution of the sign test has been generated both by basic probability considerations and empirically, it serves as an important bridge to understanding all the sampling distributions discussed later in the textbook.

Introducing inferential statistics with the sign test has other advantages. All of the important concepts involving hypothesis testing can be illustrated; for example, null hypothesis, alternative hypothesis, alpha level, Type I and Type II errors, size of effect, and power. All of these concepts are learned before the formal discussion of sampling distributions and the z test in Chapter 12. Hence, they don't compete for the student's attention when the student is trying to understand sampling distributions. The sign test also provides an illustration of the before–after (repeated measures) experimental design. I believe this is a superior way to begin inference testing, because the before–after design is familiar to most students, is more intuitive, and is easier to understand than the single-sample design used with the z test.

After hypothesis testing is introduced using the sign test in Chapter 10, power is discussed using the sign test in Chapter 11. Many texts do not discuss power at all, or if they do, they give it abbreviated treatment. Power is a complicated topic. Using the sign test as the vehicle for a power analysis simplifies matters. Understanding power is necessary if one is to grasp the methodology of scientific investigation itself. When students gain insight into power, they can see why we bother discussing Type II errors. Furthermore, they see for the first time why we conclude by "retaining H_0" as a reasonable explanation of the data rather than by "accepting H_0 as true" (a most important and often unappreciated distinction). In this same vein, students also understand the error involved when one concludes that two conditions are equal from data that are not statistically significant. Thus power is a topic that brings the whole hypothesis-testing methodology into sharp focus.

At this state of the exposition, a diligent student can grasp the idea that data analysis basically involves two steps: (1) calculating the appropriate statistic, and (2) evaluating the statistic based on its sampling distribution. The time is ripe for a formal discussion of sampling distributions and how they can be generated. This is done at the beginning of Chapter 12. Then the sampling distribution of the mean is introduced. Rather than depending on an abstract theoretical definition of the sampling distribution of the mean, the text discusses how this sampling distribution can be generated empirically. This gives a much more concrete understanding of the sampling distribution of the mean and facilitates understanding its use with the z test.

Due to previous experience with the sign test and its easily understood sampling distribution, and using the empirical approach for generating the sampling distribution of the mean, most conscientious students have a good grasp of what sampling distributions are and why they are essential for inferential statistics. With this background, students comprehend that all of the concepts of hypothesis testing are the same as we go from inference test to inference test. What vary from experiment to experiment are the statistics used, and the accompanying sampling distribution. The stage is then set for moving through the remaining inference tests with understanding.

Other Important Textbook Features

There are other important features that are worth noting. Among them are the following:

- Chapter 1 discusses approaches for determining truth and establishes statistics as part of the scientific method, which is unusual for an introductory statistics textbook.

- Chapter 8 covers probability. It does not delve deeply into probability theory. I view this as a plus, because probability can be a very difficult topic and can cause students much unnecessary malaise unless treated at the right level. In my view the proper mathematical foundation for all of the inference tests contained in this textbook can be built simply by the use of basic probability definitions in conjunction with the addition and multiplication rules, as has been done in Chapter 8.

- In Chapter 14, the t test for correlated groups is introduced directly after the t test for single samples and is developed as a special case of the t test for single samples, only this time using difference scores rather than raw scores. This makes the t test for correlated groups quite easy to teach and easy for students to understand.

- In Chapter 14, understanding of power is deepened and the important principle of using the most powerful inference test is illustrated by analyzing the same data set with the t test for correlated groups and the sign test.

- In Chapter 14, the correlated and independent groups designs are compared with regard to power and utility.

- There is a discussion of the factors influencing the power of the t test in Chapter 14 and one-way ANOVA in Chapter 15.

- In Chapter 14, the confidence interval approach for evaluating the effect of the independent variable is presented along with the conventional hypothesis-testing approach.

- Chapter 18 is a summary chapter of all of the inferential statistics material. This chapter gives students the opportunity to choose among inference tests in solving problems. Students particularly like the decision tree presented here.

- *What Is the Truth?* sections: At the end of various chapters throughout the textbook, there are sections titled *What Is the Truth?* along with end-of chapter questions on these sections. These sections and questions are intended to illustrate real-world applications of statistics and to sharpen applied critical thinking.

Tenth Edition Changes

Textbook   The following changes have been made in the textbook.

- **SPSS material has been greatly expanded.** Because of increased use of statistical software in recent years and in response to reviewer advice, I have greatly expended the SPSS material. In the tenth edition, there is SPSS coverage at the end of Chapters 2, 3, 4, 5, 6, 7, 13, 14, 15, 16, and 17. For each chapter, this material is comprised of a detailed illustrative SPSS example and solution along with at least two new SPSS problems to practice on. In addition, a new Appendix E contains a general introduction to SPSS. Students can now learn SPSS and practice on chapter-relevant problems without recourse to additional outside sources. The SPSS material at the end of Chapters 4 and 6 that was contained in the ninth edition has been dropped. The old Appendix E, *Symbols*, has been moved to the inside cover of the textbook.

- **ANOVA symbols throughout Chapters 15 and 16 have been changed.** The symbols used in the previous editions of the textbook in the ANOVA chapters have been changed to more conventionally used symbols. The specific changes are as follows. In Chapter 15, s_W^2, s_B^2, SS_T, SS_W, SS_B, df_T, df_W, and df_B have been changed to MS_{within}, $MS_{between}$, SS_{total}, SS_{within}, $SS_{between}$, df_{total}, df_{within}, and $df_{between}$, respectively. In Chapter 16, s_W^2, s_R^2, s_C^2, s_{RC}^2, SS_T, SS_W, SS_R, SS_C,

SS_{RC}, df_T, df_W, df_R, df_C, and df_{RC} have been changed to $MS_{within-cells}$, MS_{rows}, $MS_{columns}$, $MS_{interaction}$, SS_{total}, $SS_{within-cells}$, SS_{rows}, $SS_{columns}$, $SS_{interaction}$, df_{total}, $df_{within-cells}$, df_{rows}, $df_{columns}$, and $df_{interaction}$, respectively.

I made these changes because I believe students will have an easier time transitioning to advanced statistical textbooks and using statistical software—including SPSS—and because of reviewer recommendations. I have some regrets with moving to the new symbols, because I believe the old symbols provide a better transition from the *t* test to ANOVA, and because of the extra time and effort it may require of instructors who are used to the old symbols (my apologies to these instructors for the inconvenience).

- **In Chapter 15, the Newman-Keuls test has been replaced with the Scheffé test.** The Newman-Keuls test has been dropped because of recent criticism from statistical experts that the Newman-Keuls procedure of adjusting r can result in an experimentwise or familywise Type I error rate that exceeds the specified level. I have replaced the Newman-Keuls test with the Scheffé test. The Scheffé test has the advantages that (1) it uses a modified ANOVA technique that is relatively easy to understand and compute; (2) it is very commonly used in the research literature; (3) it is the most flexible and conservative *post hoc* test available; and (4) it provides a good contrast to the Tukey HSD test.

- *What Is the Truth?* **questions have been added at the end of the chapters that contain** *What Is the Truth?* **sections (Chapters 1, 3, 6, 8, 10, 11, 15, and 17).** These questions have been added to provide closer integration of the *What Is the Truth?* sections with the rest of the textbook content and to promote applied critical thinking.

- **In Chapter 7, the section titled** *Regression of X on Y* **has been dropped.** This section has been dropped because students can compute the regression of Y on X or of X on Y by just designating the predicted variable as the Y variable. Therefore there is little practical gain in devoting a separate section to the regression of X on Y. Separate treatment of the regression of X on Y does contribute additional theoretical insight into the topic of regression, but was judged not important enough to justify precious introductory textbook space.

- **The** *To the Student* **section has been amplified to include a discussion of anxiety reduction.** This material has been added to help students who experience excessive anxiety when dealing with the statistics material. Five options for reducing anxiety have been presented: (1) seeking help at the university counseling center, (2) taking up the practice of meditation, (3) learning and practicing autogenic techniques, (4) increasing bodily relaxation via progressive muscle relaxation, and (5) practicing the techniques advocated by positive psychology.

- **The index has been revised.** I favor a detailed index. In previous editions, the index has only been partially revised, finally resulting in an index in the ninth edition that has become unwieldy and redundant. In the tenth edition the index has been completely revised. The result is a streamlined index that I believe retains the detail necessary for a good index.

- **Minor wording changes have been made throughout the textbook to increase clarity.**

Ancillaries

Aplia™ has replaced Enhanced WebAssign, which was used in the ninth edition. Aplia is an online interactive learning solution that improves comprehension and outcomes by increasing student effort and engagement. Founded by a professor to enhance his own courses, Aplia provides automatically graded assignments that were written to make the most of the Web medium and contain detailed, immediate explanations on every question. Our easy-to-use system has been used by more than 2,000,000 students at over 1,800 institutions.

Aplia for Pagano's *Understanding Statistics in the Behavioral Sciences* also includes end-of-chapter questions directly from the text.

WebTutor™   Jump-start your course with customizable, rich, text-specific content within your Course Management System. Whether you want to Web-enable your class or put an entire course online, WebTutor delivers. WebTutor offers a wide array of resources, including integrated eBook, quizzing, and more! Visit webtutor.cengage.com to learn more.

Instructor's Manual with Test Bank   The instructor's manual includes the textbook rationale, general teaching advice, advice to the student, and, for each chapter, a detailed chapter outline, learning objectives, a detailed chapter summary, teaching suggestions, discussion questions, and test questions and answers. Test questions are organized into multiple-choice, true/false, definitions, and short answer sections, and answers are also provided. Over 600 new test questions have been added to the tenth edition. The overall test bank has over 2,200 true/false, multiple-choice, and short-answer questions. The instructor's manual also includes answers to the end-of-chapter problems contained in the textbook for which no answers are given in the textbook.

Book Companion Website   Available for use by all students, the book companion website offers chapter-specific learning tools including Know and Be Able to Do, practice quizzes, flash cards, glossaries, a link to Statistics and Research Methods Workshops, and more. Go to www.cengagebrain.com for access.

Acknowledgments

I have received a great deal of help in the development and production of this edition. First, I would like to thank Timothy C. Matray, my Sponsoring Editor. He has been a continuing pillar of support throughout the development and production of this edition. I am especially grateful for his input in deciding on revision items, finding appropriate experts to review the revised material, for the role he has played in facilitating the transition from Enhanced WebAssign to Aplia and the ideas he has contributed to advertising. Next, I would like to thank Bob Jucha, the Developmental Editor for this edition. I am grateful for his conduct of surveys and evaluations, his role in coordinating with Production, his advice, and his hard work. I am indebted to Vernon Boes, the Senior Art Director and his team. I think they have produced an outstanding cover and interior design for the tenth edition. I believe he and his team have created a peaceful and esthetic cover that continues our animal theme, as well as a very clean, attractive interior textbook design. I am particularly pleased to have had the opportunity to collaborate on this edition with my daughter, Maria E. Pagano, who is an Associate Professor in the Department of Psychiatry at Case Western University. It was a lot of fun, and she helped greatly in reviewing parts of the textbook, especially the SPSS material. I am also grateful to Dr. Lynn Johnson for reviewing the positive psychology material presented in the *To the Student* section.

The remaining Cengage Learning/Wadsworth staff that I would like to thank are Content Project Manager Charlene M. Carpentier, Assistant Editor Paige Leeds, Editorial Assistant Lauren Moody, Media Editor Mary Noel, and Marketing Manager Sean Foy.

I wish to thank the following individuals who reviewed the ninth edition and made valuable suggestions for this revision.

Erin Buchanan, University of Mississippi
Ronald A. Craig. Edinboro University of Pennsylvania
David R. Dunaetz, Azusa Pacific University
Christine Ferri, Richard Stockton College of New Jersey
Carrie E. Hall, Miami University
Deborah J. Hendricks, West Virginia University
Mollie Herman, Towson University
Alisha Janowsky, University of Central Florida
Barry Kulhe, University of Scranton
Wanda C. McCarty, University of Cincinnati
Cora Lou Sherburne, Indiana University of Pennsylvania
Cheryl Terrance, University of North Dakota
Brigitte Vittrup, Texas Woman's University
Gary Welton, Grove City College

I am grateful to the Literary Executor of the Late Sir Ronald A. Fisher, F.R.S.; to Dr. Frank Yates, F.R.S.; and to the Longman Group Ltd., London, for permission to reprint Tables III, IV, and VII from their book *Statistical Tables for Biological, Agricultural and Medical Research* (sixth edition, 1974).

The material covered in this textbook, instructor's manual, and on the Web is appropriate for undergraduate students with a major in psychology or related behavioral science disciplines. I believe the approach I have followed helps considerably to impart this subject matter with understanding. I am grateful to receive any comments that will improve the quality of these materials.

Robert R. Pagano

TO THE STUDENT

Statistics uses probability, logic, and mathematics as ways of determining whether or not observations made in the real world or laboratory are due to random happenstance or due to an orderly effect one variable has on another. Separating happenstance, or chance, from cause and effect is the task of science, and statistics is a tool to accomplish that end. Occasionally, data will be so clear that the use of statistical analysis isn't necessary. Occasionally, data will be so garbled that no statistical analysis can meaningfully be applied to answer any reasonable question. However, most often, when analyzing the data from an experiment or study, statistics is useful in determining whether it is legitimate to conclude that an orderly effect has occurred. When this is the case, statistical analysis can also provide an estimate of the size of the effect.

It is useful to try to think of statistics as a means of learning a new set of problem-solving skills. You will learn new ways to ask questions, new ways to answer them, and a more sophisticated way of interpreting the data you read about in texts, journals, and newspapers.

In writing this textbook and creating the Web material, I have tried to make the material as clear, interesting, and easy to understand as I can. I have used a relaxed style, introduced humor, avoided equation derivation when possible, and chosen examples and problems that I believe will be interesting to students in the behavioral sciences. I have listed the objectives for each chapter so that you can see what is in store for you and guide your studying accordingly. I have also introduced "mentoring tips" throughout the textbook to help highlight important aspects of the material. While I was teaching at the University of Washington and the University of Pittsburgh, my statistics course was evaluated by each class of students that I taught. I found the suggestions of students invaluable in improving my teaching. Many of these suggestions have been incorporated into this textbook. I take quite a lot of pride in having been a finalist for the University of Washington Outstanding Teaching Award for teaching this statistics course, and in the fact that students have praised this textbook so highly. I believe much of my success derives from student feedback and the quality of this textbook.

Study Hints

- **Memorize symbols.** A lot of symbols are used in statistics. Don't make the material more difficult than necessary by failing to memorize what the symbols stand for. Treat them as though they were foreign vocabulary. Be able to go quickly from the symbol to what it stands for, and vice versa. The *Flash Cards* section in the accompanying Web material will help you accomplish this goal.

- **Learn the definitions for new terms.** Many new terms are introduced in this course. Part of learning statistics is learning the definitions of these new terms. If you don't know what the new terms mean, it will be impossible to do well in this course. Like the symbols, the new terms should be treated like foreign vocabulary. Be able to instantly associate each new term with its definition and vice versa. The *Flash Cards* section in the accompanying Web material will also help you accomplish this goal.

- **Work as many problems as needed for you to understand the material and produce correct answers.** In my experience there is a direct, positive relationship between working problems and doing well on this material. Be sure you try to understand the solutions. When using calculators and computers, there can be a tendency to press the keys and read the answer without really understanding the solution. I hope you won't fall into this trap. Also, work the problem from beginning to end, rather than just following someone else's solution and telling yourself that you could solve the problem if called upon to do so. Solving a problem from scratch is very different and often more difficult than "understanding" someone else's solution.

- **Don't fall behind.** The material in this course is cumulative. Do not let yourself fall behind. If you do, you will not understand the current material either.

- **Study several times each week, rather than just cramming.** A lot of research has shown that you will learn better and remember more material if you space your learning rather than just cramming for the test.

- **Read the material in the textbook prior to the lecture/discussion covering it.** You can learn a lot just by reading this textbook. Moreover, by reading the appropriate material just prior to when it is covered in class, you can determine the parts that you have difficulty with and ask appropriate questions when that material is covered by your instructor.

- **Pay attention and think about the material being covered in class.** This advice may seem obvious, but for whatever reason, it is frequently not followed by students. Often times I've had to stop my lecture or discussions to remind students about the importance of paying attention and thinking in class. I don't require students to attend my classes, but if they do, I assume they want to learn the material, and of course, attention and thinking are prerequisites for learning.

- **Ask the questions you need to ask.** Many of us feel our question is a "dumb" one, and we will be embarrassed because the question will reveal our ignorance to the instructor and the rest of the class. Almost always, the "dumb" question helps others sitting in the class because they have the same question. Even when this is not true, it is very often the case that if you don't ask the question, your learning is blocked and stops there, because the answer is necessary for you to continue learning the material. Don't let possible embarrassment hinder your learning. If it doesn't work for you to ask in class, then ask the question via email, or make an appointment with the instructor and ask then.

- **Compare your answers to mine.** For most of the problems I have used a hand calculator or computer to find the solutions. Depending on how many decimal places you carry your intermediate calculations, you may get slightly different answers than I do. In most cases I have used full calculator or computer accuracy for intermediate calculations (at least five decimal places). In general, you should carry all intermediate calculations to at least two more decimal places than the number of decimal places in the rounded final answer. For example, if you intend to round the final answer to two decimal places, than you should carry all intermediate calculations to at least four decimal places. If you follow this policy and your answer does not agree with ours, then you have probably made a calculation error.

■ **One final topic: Dealing with anxiety.** Anyone who has taught an introductory statistics class for psychology majors is aware that many students have a great deal of anxiety about taking a statistics course. Actually, a small or even moderate level of anxiety can facilitate learning, but too much anxiety can be an impediment. Fortunately, we know a fair amount about anxiety and techniques that help reduce it. If you think the level of fear or anxiety that you experience associated with statistics is causing you a problem, I suggest you avail yourself of one or more of the following options.

   ■ *Visit the counseling center at your college or university.* Many students experience anxiety associated with courses they are taking, especially math courses. Counseling centers have lots of experience helping these students overcome their anxiety. The services provided are confidential and usually free. If I were a student having a problem with course-related anxiety, this is the first place I would go for help.

   ■ *Meditation.* There has been a lot of research looking at the physiological and psychological effects of meditation. The research shows that meditation results in a more relaxed individual who generally experiences an increased sense of well-being. Being more relaxed can have the beneficial effect of lowering one's anxiety level in the day-to-day learning of statistics material. Meditation can have additional benefits, including developing mindfulness and equanimity. Equanimity is defined as calmness in the face of stress. Developing the trait of equanimity can be especially useful in dealing with anxiety. I have included a short reference section below for those interested in pursuing this topic further.

   ■ *Autogenic training and progressive muscle relaxation.* These are well-established techniques for promoting general relaxation. Autogenic training is a relaxation technique first developed by the German psychiatrist Johannes Schultz in the early 1930s. It involves repeating a series of autohypnotic sentences like "my right arm is heavy," or "my heartbeat is calm and regular," designed to calm one's autonomic nervous system. Progressive muscle relaxation is another well-established relaxation technique. It was developed by American physician Edmund Jacobson in the early 1920s. It is a technique developed to reduce anxiety by sequentially tensing and relaxing various muscle groups and focusing on the accompanying sensations. I have included a short reference section below for those interested in pursuing either of these techniques further

   ■ *Positive psychology.* Positive psychology is an area within psychology that was initiated about 15 years ago in the American Psychological Association presidential address of Martin Seligman. Its focus of interest is happiness or well-being. It studies normal and above-normal functioning individuals. Rather than attempting to help clients lead more positive lives by tracing the etiology of negative emotional states such as depression or anxiety, as does traditional clinical psychology, positive psychology attempts to promote greater happiness by incorporating techniques into one's life that research has revealed promote positive emotions. Some of the techniques recommended include learning and practicing optimism; learning to be grateful and to express gratitude; spending more time in areas involving personal strengths; sleeping well, getting proper nutrition, increasing exercise; savoring positive experiences, reframing (finding the good in the bad), skill training in detachment from negative thoughts, decreasing fear and anxiety by confronting or doing feared things or activities, doing good deeds and random acts of kindness, increasing connections with others, and increasing compassion and forgiveness. As with the other options, I have included below a short reference section for pursuing positive psychology further. I have also included a website for the Positive Psychology Center located at the University of Pennsylvania that I have found useful.

References

Meditation

S. Salzberg, *Real Happiness: The Power of Meditation*, Workman Publishing Company, New York, 2011.
W. Hart, *The Art of Living: Vipassana Meditation as taught by S.N. Goenka*, HarperCollins, New York, 1987.
B. Boyce (Ed), *The Mindfulness Revolution*, Shambala Publications Inc., Boston, 2011.
C. Beck, *Everyday Zen*, HarperCollins, New York, 1989

Autogenic Training

F. Stetter and S. Kupper, Autogenic training: a meta-analysis of clinical outcome studies, *Applied Psychophysiology and Biofeedback* 27(1):45:98, 2002.
W. Luthe and J. Schultz, *Autogenic Therapy*, The British Autogenic Society, 2001.
K. Kermani. *Autogenic Training: Effective Holistic Way to Better Health*, Souvenir Press, London, 1996.

Progressive Muscle Relaxation

E. Jacobson, E. *Progressive Relaxation*, University of Chicago Press, Chicago, 1938 (classic book).
M. Davis and E. R. Eshelman, *The Relaxation and Stress Reduction Workshop*, New Harbinger Publications, Inc., Oakland, 2008
F. McGuigan, *Progressive Relaxation: Origins, Principles and Clinical Applications*. In *Principles and Practice of Stress Management*, 2nd ed., Edited by P. Lehrer and R. Woolfold, New York, Guilford Press, 1993.

Positive Psychology

L. Johnson, *Enjoy Life: Healing with Happiness*, Head Acre Press, Salt Lake City, 2008.
M. Seligman, *Authentic Happiness: Using the New Positive Psychology to Realize Your Potential for Lasting Fulfillment*, The Free Press, New York, 2004.
C. Peterson, *A Primer in Positive Psychology*, Oxford University Press, Oxford, 2006.
Center for Positive Psychology, www.ppc.sas.upenn.edu.

I wish you great success in understanding the material contained in this textbook.

Robert R. Pagano

 PART
ONE

OVERVIEW

1 Statistics and Scientific Method

© Strmko / Dreamstime.com

Statistics and Scientific Method

LEARNING OBJECTIVES

After completing this chapter, you should be able to:

▪ Describe the four methods of establishing truth.
▪ Contrast observational and experimental research.
▪ Contrast descriptive and inferential statistics.
▪ Define the following terms: population, sample, variable, independent variable, dependent variable, constant, data, statistic, and parameter.
▪ Identify the population, sample, independent and dependent variables, data, statistic, and parameter from the description of a research study.
▪ Specify the difference between a statistic and a parameter.
▪ Give two reasons why random sampling is important.
▪ Understand the illustrative example, do the practice problem, and understand the solution.

INTRODUCTION

Have you ever wondered how we come to know truth? Most college students would agree that finding out what is true about the world, ourselves, and others constitutes a very important activity. A little reflection reveals that much of our time is spent in precisely this way. If we are studying geography, we want to know what is *true* about the geography of a particular region. Is the region mountainous or flat, agricultural or industrial? If our interest is in studying human beings, we want to know what is *true* about humans. Do we *truly* possess a spiritual nature, or are we *truly* reducible solely to atoms and molecules, as the reductionists would have it? How do humans think? What happens in the body to produce a sensation or a movement? When I get angry, is it *true* that there is a unique underlying physiological pattern? What is the pattern? Is my *true* purpose in life to become a teacher? Is it *true* that animals think? We could go on indefinitely with examples because so much of our lives is spent seeking and acquiring truth.

METHODS OF KNOWING

Historically, humankind has employed four methods to acquire knowledge. They are authority, rationalism, intuition, and the scientific method.

Authority

MENTORING TIP

Which of the four methods do you use most often?

When using the method of *authority*, we consider something true because of tradition or because some person of distinction says it is true. Thus, we may believe in the theory of evolution because our distinguished professors tell us it is true, or we may believe that God truly exists because our parents say so. Although this method of knowing is currently in disfavor and does sometimes lead to error, we use it a lot in living our daily lives. We frequently accept a large amount of information on the basis of authority, if for no other reason than we do not have the time or the expertise to check it out firsthand. For example, I believe, on the basis of physics authorities, that electrons exist, but I have never seen one; or perhaps closer to home, if the surgeon general tells me that smoking causes cancer, I stop smoking because I have faith in the surgeon general and do not have the time or means to investigate the matter personally.

Rationalism

The method of *rationalism* uses reasoning alone to arrive at knowledge. It assumes that if the premises are sound and the reasoning is carried out correctly according to the rules of logic, then the conclusions will yield truth. We are very familiar with reason because we use it so much. As an example, consider the following syllogism:

> All statistics professors are interesting people.
> Mr. X is a statistics professor.
> Therefore, Mr. X is an interesting person.

Assuming the first statement is true (who could doubt it?), then it follows that if the second statement is true, the conclusion must be true. Joking aside, hardly anyone would question the importance of the reasoning process in yielding truth. However, there are a great number of situations in which reason alone is inadequate in determining the truth.

To illustrate, let's suppose you notice that John, a friend of yours, has been depressed for a couple of months. As a psychology major, you know that psychological problems can produce depression. Therefore, it is reasonable to believe John may have psychological problems that are producing his depression. On the other hand, you also know that an inadequate diet can result in depression, and it is reasonable to believe that this may be at the root of his trouble. In this situation, there are two reasonable explanations of the phenomenon. Hence, reason alone is inadequate in distinguishing between them. We must resort to experience. Is John's diet in fact deficient? Will improved eating habits correct the situation? Or does John have serious psychological problems that, when worked through, will lift the depression? Reason alone, then, may be sufficient to yield truth in some situations, but it is clearly inadequate in others. As we shall see, the scientific method also uses reason to arrive at truth, but reasoning alone is only part of the process. Thus, the scientific method incorporates reason but is not synonymous with it.

Intuition

Knowledge is also acquired through *intuition*. By intuition, we mean that sudden insight, the clarifying idea that springs into consciousness all at once as a whole. It is not arrived at by reason. On the contrary, the idea often seems to occur after conscious reasoning has failed. Beveridge* gives numerous occurrences taken from prominent individuals. Here are a couple of examples:

> Here is Metchnikoff's own account of the origin of the idea of phagocytosis: "One day when the whole family had gone to the circus to see some extraordinary performing apes, I remained alone with my microscope, observing the life in the mobile cells of a transparent starfish larva, when a new thought suddenly flashed across my brain. It struck me that similar cells might serve in the defense of the organism against intruders. Feeling that there was in this something of surpassing interest, I felt so excited that I began striding up and down the room and even went to the seashore to collect my thoughts."

> Hadamard cites an experience of the mathematician Gauss, who wrote concerning a problem he had tried unsuccessfully to prove for years: "Finally two days ago I succeeded … like a sudden flash of lightning the riddle happened to be solved. I cannot myself say what was the conducting thread which connected what I previously knew with what made my success possible."

It is interesting to note that the intuitive idea often occurs after conscious reasoning has failed and the individual has put the problem aside for a while. Thus, Beveridge† quotes two scientists as follows:

> Freeing my mind of all thoughts of the problem I walked briskly down the street, when suddenly at a definite spot which I could locate today—as if from the clear sky above me—an idea popped into my head as emphatically as if a voice had shouted it.

> I decided to abandon the work and all thoughts relative to it, and then, on the following day, when occupied in work of an entirely different type, an idea came to my mind as suddenly as a flash of lightning and it was the solution … the utter simplicity made me wonder why I hadn't thought of it before.

Despite the fact that intuition has probably been used as a source of knowledge for as long as humans have existed, it is still a very mysterious process about which we have only the most rudimentary understanding.

---

*W. I. B. Beveridge, *The Art of Scientific Investigation*, Vintage Books/Random House, New York, 1957, pp. 94–95.
†Ibid., p. 92.

Scientific Method

Although the *scientific method* uses both reasoning and intuition for establishing truth, its reliance on objective assessment is what differentiates this method from the others. At the heart of science lies the *scientific experiment*. The method of science is rather straightforward. By some means, usually by reasoning deductively from existing theory or inductively from existing facts or through intuition, the scientist arrives at a hypothesis about some feature of reality. He or she then designs an experiment to objectively test the hypothesis. The data from the experiment are then analyzed statistically, and the hypothesis is either supported or rejected. The feature of overriding importance in this methodology is that no matter what the scientist believes is true regarding the hypothesis under study, the experiment provides the basis for an *objective* evaluation of the hypothesis. The data from the experiment force a conclusion consonant with reality. Thus, scientific methodology has a built-in safeguard for ensuring that truth assertions of any sort about reality must conform to what is demonstrated to be objectively true about the phenomena before the assertions are given the status of scientific truth.

An important aspect of this methodology is that the experimenter can hold incorrect hunches, and the data will expose them. The hunches can then be revised in light of the data and retested. This methodology, although sometimes painstakingly slow, has a self-correcting feature that, over the long run, has a high probability of yielding truth. Since in this textbook we emphasize statistical analysis rather than experimental design, we cannot spend a great deal of time discussing the design of experiments. Nevertheless, some experimental design will be covered because it is so intertwined with statistical analysis.

DEFINITIONS

In discussing this and other material throughout the book, we shall be using certain technical terms. The terms and their definitions follow:

- **Population** *A population is the complete set of individuals, objects, or scores that the investigator is interested in studying.* In an actual experiment, the population is the larger group of individuals from which the subjects run in the experiment have been taken.
- **Sample** *A sample is a subset of the population.* In an experiment, for economical reasons, the investigator usually collects data on a smaller group of subjects than the entire population. This smaller group is called the sample.
- **Variable** *A variable is any property or characteristic of some event, object, or person that may have different values at different times depending on the conditions.* Height, weight, reaction time, and drug dosage are examples of variables. A variable should be contrasted with a *constant*, which, of course, does not have different values at different times. An example is the mathematical constant π; it always has the same value (3.14 to two-decimal-place accuracy).
- **Independent variable (IV)** *The independent variable in an experiment is the variable that is systematically manipulated by the investigator.* In most experiments, the investigator is interested in determining the effect that one variable, say, variable *A*, has on one or more other variables. To do so, the investigator manipulates the levels of variable *A* and measures the effect on the other variables. Variable *A* is called the *independent* variable because its levels are controlled by the experimenter, independent of any change in the other variables. To illustrate, an investigator might be interested in the effect of alcohol on social behavior. To investigate this, he or she would probably vary the amount of alcohol

consumed by the subjects and measure its effect on their social behavior. In this example, the experimenter is manipulating the amount of alcohol and measuring its consequences on social behavior. Alcohol amount is the independent variable. In another experiment, the effect of sleep deprivation on aggressive behavior is studied. Subjects are deprived of various amounts of sleep, and the consequences on aggressiveness are observed. Here, the amount of sleep deprivation is being manipulated. Hence, it is the independent variable.

♦ **Dependent variable (DV)** *The dependent variable in an experiment is the variable that the investigator measures to determine the effect of the independent variable.* For example, in the experiment studying the effects of alcohol on social behavior, the amount of alcohol is the independent variable. The social behavior of the subjects is measured to see whether it is affected by the amount of alcohol consumed. Thus, social behavior is the dependent variable. It is called *dependent* because it may depend on the amount of alcohol consumed. In the investigation of sleep deprivation and aggressive behavior, the amount of sleep deprivation is being manipulated and the subjects' aggressive behavior is being measured. The amount of sleep deprivation is the independent variable, and aggressive behavior is the dependent variable.

♦ **Data** *The measurements that are made on the subjects of an experiment are called data.* Usually data consist of the measurements of the dependent variable or of other subject characteristics, such as age, gender, number of subjects, and so on. The data as originally measured are often referred to as *raw* or *original* scores.

♦ **Statistic** *A statistic is a number calculated on sample data that quantifies a characteristic of the sample.* Thus, the average value of a sample set of scores would be called a statistic.

♦ **Parameter** *A parameter is a number calculated on population data that quantifies a characteristic of the population.* For example, the average value of a population set of scores is called a parameter. It should be noted that a statistic and a parameter are very similar concepts. The only difference is that a statistic is calculated on a sample and a parameter is calculated on a population.

e x p e r i m e n t

Mode of Presentation and Retention

Let's now consider an illustrative experiment and apply the previously discussed terms.

MENTORING TIP

Very often parameters are unspecified. Is a parameter specified in this experiment?

An educator conducts an experiment to determine whether the mode of presentation affects how well prose material is remembered. For this experiment, the educator uses several prose passages that are presented visually or auditorily. Fifty students are selected from the undergraduates attending the university at which the educator works. The students are divided into two groups of 25 students per group. The first group receives a visual presentation of the prose passages, and the second group hears the passages through an auditory presentation. At the end of their respective presentations, the subjects are asked to write down as much of the material as they can remember. The average number of words remembered by each group is calculated, and the two group averages are compared to see whether the mode of presentation had an effect.

In this experiment, the independent variable is the mode of presentation of the prose passages (i.e., auditory or visual). The dependent variable is the number of words remembered. The sample is the 50 students who participated in the experiment. The population is the larger group of individuals from which the sample was taken, namely, the undergraduates attending the university. The data are the number of words recalled by each student in the sample. The average number of words recalled by each group is a statistic because it quantifies a characteristic of the sample scores. Since there was no measurement made of any population characteristic, there was no parameter calculated in this experiment. However, for illustrative purposes, suppose the entire population had been given a visual presentation of the passages. If we calculate the average number of words remembered by the population, the average number would be called a parameter because it quantifies a characteristic of the population scores.

Now, let's do a problem to practice identifying these terms.

Practice Problem 1.1

For the experiment described below, specify the following: the independent variable, the dependent variable(s), the sample, the population, the data, the statistic(s), and the parameter(s).

A professor of gynecology at a prominent medical school wants to determine whether an experimental birth control implant has side effects on body weight and depression. A group of 5000 adult women living in a nearby city volunteers for the experiment. The gynecologist selects 100 of these women to participate in the study. Fifty of the women are assigned to group 1 and the other fifty to group 2 such that the mean body weight and the mean depression scores of each group are equal at the beginning of the experiment. Treatment conditions are the same for both groups, except that the women in group 1 are surgically implanted with the experimental birth control device, whereas the women in group 2 receive a placebo implant. Body weight and depressed mood state are measured at the beginning and end of the experiment. A standardized questionnaire designed to measure degree of depression is used for the mood state measurement. The higher the score on this questionnaire is, the more depressed the individual is. The mean body weight and the mean depression scores of each group at the end of the experiment are compared to determine whether the experimental birth control implant had an effect on these variables. To safeguard the women from unwanted pregnancy, another method of birth control that does not interact with the implant is used for the duration of the experiment.

SOLUTION

Independent variable: The experimental birth control implant versus the placebo.

Dependent variables: Body weight and depressed mood state.

Sample: 100 women who participated in the experiment.

Population: 5000 women who volunteered for the experiment.

Data: The individual body weight and depression scores of the 100 women at the beginning and end of the experiment.

Statistics: Mean body weight of group 1 at the beginning of the experiment, mean body weight of group 1 at the end of the experiment, mean depression score of group 1 at the beginning of the experiment, mean depression score of group 1 at the end of the experiment, plus the same four statistics for group 2.

Parameter: No parameters were given or computed in this experiment. If the gynecologist had measured the body weights of all 5000 volunteers at the beginning of the experiment, the mean of these 5000 weights would be a parameter.

SCIENTIFIC RESEARCH AND STATISTICS

Scientific research may be divided into two categories: *observational studies* and *true experiments*. Statistical techniques are important in both kinds of research.

Observational Studies

In this type of research, no variables are actively manipulated by the investigator, and hence observational studies cannot determine causality. Included within this category of research are (1) naturalistic observation, (2) parameter estimation, and (3) correlational studies. With *naturalistic observation research,* a major goal is to obtain an accurate description of the situation being studied. Much anthropological and etiological research is of this type. *Parameter estimation research* is conducted on samples to estimate the level of one or more population characteristics (e.g., the population average or percentage). Surveys, public opinion polls, and much market research fall into this category. In *correlational research,* the investigator focuses attention on two or more variables to determine whether they are related. For example, to determine whether obesity and high blood pressure are related in adults older than 30 years, an investigator might measure the fat level and blood pressure of individuals in a sample of adults older than 30. The investigator would then analyze the results to see whether a relationship exists between these variables; that is, do individuals with low fat levels also have low blood pressure, do individuals with moderate fat levels have moderate blood pressure, and do individuals with high fat levels have high blood pressure?

True Experiments

MENTORING TIP

Only true experiments can determine causality.

In this type of research, an attempt is made to determine whether changes in one variable cause* changes in another variable. In a true experiment, an independent variable is manipulated and its effect on some dependent variable is studied. If desired, there can be more than one independent variable and more than one dependent variable. In the simplest case, there is only one independent and one dependent variable. One example of this case is the experiment mentioned previously that investigated the effect of alcohol on social behavior. In this experiment, you will recall, alcohol level was manipulated by the experimenter and its effect on social behavior was measured.

RANDOM SAMPLING

In all of the research described previously, data are usually collected on a sample of subjects rather than on the entire population to which the results are intended to apply. Ideally, of course, the experiment would be performed on the whole population,

---

*We recognize that the topic of cause and effect has engendered much philosophical debate. However, we cannot consider the intricacies of this topic here. When we use the term *cause*, we mean it in the common-sense way it is used by non-philosophers. That is, when we say that *A* caused *B*, we mean that a change in *A* produced a change in *B* with all other variables appropriately controlled.

but usually it is far too costly, so a sample is taken. Note that not just any sample will do. The sample should be a *random* sample. Random sampling is discussed in Chapter 8. For now, it is sufficient to know that random sampling allows the laws of probability, also discussed in Chapter 8, to apply to the data and at the same time helps achieve a sample that is representative of the population. Thus, the results obtained from the sample should also apply to the population. Once the data are collected, they are statistically analyzed and the appropriate conclusions about the population are drawn.

DESCRIPTIVE AND INFERENTIAL STATISTICS

Statistical analysis, of course, is the main theme of this textbook. It has been divided into two areas: (1) *descriptive statistics* and (2) *inferential statistics*. Both involve analyzing data. If an analysis is done for the purpose of describing or characterizing the data, then we are in the area of descriptive statistics. To illustrate, suppose your biology professor has just recorded the scores from an exam he has recently given you. He hands back the tests and now wants to describe the scores. He might decide to calculate the average of the distribution to describe its *central tendency*. Perhaps he will also determine its range to characterize its *variability*. He might also plot the scores on a graph to show the *shape* of the distribution. Since all of these procedures are for the purpose of describing or characterizing the data already collected, they fall within the realm of descriptive statistics.

Inferential statistics, on the other hand, is not concerned with just describing the obtained data. Rather, it embraces techniques that allow one to use obtained sample data to make inferences or draw conclusions about populations. This is the more complicated part of statistical analysis. It involves probability and various inference tests, such as *Student's* t *test* and the *analysis of variance*.

To illustrate the difference between descriptive and inferential statistics, suppose we were interested in determining the average IQ of the entire freshman class at your university. It would be too costly and time-consuming to measure the IQ of every student in the population, so we would take a random sample of, say, 200 students and give each an IQ test. We would then have 200 *sample* IQ scores, which we want to use to determine the average IQ in the *population*. Although we can't determine the exact value of the population average, we can estimate it using the sample data in conjunction with an inference test called Student's *t* test. The results would allow us to make a statement such as, "We are 95% confident that the interval of 115–120 contains the mean IQ of the population." Here, we are not just describing the obtained scores, as was the case with the biology exam. Rather, we are using the sample scores to infer to a population value. We are therefore in the domain of inferential statistics. *Descriptive* and *inferential statistics* can be defined as follows:

definitions ■ **Descriptive statistics** *is concerned with techniques that are used to describe or characterize the obtained data.*

■ **Inferential statistics** *involves techniques that use the obtained sample data to infer to populations.*

USING COMPUTERS IN STATISTICS

MENTORING TIP

It is easy to use this textbook with SPSS. Relevant chapters contain SPSS material specific to those chapters, and Appendix E presents a general introduction to SPSS.

The use of computers in statistics has increased greatly over the past decade. In fact, today almost all research data in the behavioral sciences are analyzed by statistical computer programs rather than "by hand" with a calculator. This is good news for students, who often like the ideas, concepts, and results of statistics but hate the drudgery of hand computation. The fact is that researchers hate computational drudgery too, and therefore almost always use a computer to analyze data sets of any appreciable size. Computers have the advantages of saving time and labor, minimizing the chances of computational error, allowing easy graphical display of the data, and providing better management of large data sets. However, as useful as computers are, there is often not enough time in a basic statistics course to include them. Therefore, this textbook is written so that you can learn the statistical content with or without the use of computers.

Several computer programs are available to do statistical analysis. The most popular are the Statistical Package for the Social Sciences (SPSS), Statistical Analysis System (SAS), SYSTAT, MINITAB, and Excel. Versions of these programs are available for both mainframes and microcomputers. I believe it is worth taking the extra time to learn one or more of them.

Although, as mentioned previously, I have written this textbook for use with or without computers, this edition makes it easy for you to use SPSS with the textbook. Of course, you or your instructor will have to supply the SPSS software. Chapters 2, 3, 4, 5, 6, 7, 13, 14, 15, 16, and 17 each contain chapter-specific SPSS material near the end. Appendix E, *Introduction to SPSS,* provides general instruction in SPSS. SPSS is probably the most popular statistical software program used in psychology. The software varies somewhat for different versions of SPSS; however, the changes are usually small. The SPSS material in this edition has been written using *SPSS, an IBM company, Windows Version 19.*

As you begin solving problems using computers, I believe you will begin to experience the fun and power that statistical software can bring to your study and use of statistics. In fact, once you have used software like SPSS to analyze data, you will probably wonder, "Why do I have to do any of these complicated calculations by hand?" Unfortunately, when you are using statistical software to calculate the value of a statistic, it does not help you understand that statistic. Understanding the statistic and its proper use is best achieved by doing hand calculations or step-by-step calculations using Excel. Of course, once you have learned everything you can from these calculations, using statistical software like SPSS to grind out correct values of the statistic seems eminently reasonable.

STATISTICS AND THE "REAL WORLD"

As I mentioned previously, one major purpose of statistics is to aid in the scientific evaluation of truth assertions. Although you may view this as rather esoteric and far removed from everyday life, I believe you will be convinced, by the time you have finished this textbook, that understanding statistics has very important practical aspects that can contribute to your satisfaction with and success in life. As you go through this textbook, I hope you will become increasingly aware of how frequently in ordinary life we are bombarded with "authorities" telling us, based on "truth assertions," what we should do, how we should live, what we should buy, what we should value, and so on. In areas of real importance to you, I hope you will begin to ask questions such as: "Are these truth assertions supported by data?" "How good are the data?" "Is chance a reasonable explanation of the data?" If there are no data presented, or if the data presented are of the form "My experience is that …" rather than from well-controlled experiments, I hope that you will begin to question how seriously you should take the authority's advice.

To help develop this aspect of your statistical decision making, I have included, at the end of certain chapters, applications taken from everyday life. These are titled, "What Is the Truth?" To begin, let's consider the following material.

WHAT IS THE TRUTH?   Data, Data, Where Are the Data?

 The accompanying advertisement was printed in an issue of *Psychology Today*. From a scientific point of view, what's missing?

Answer This ad is similar to a great many that appear these days. It promises a lot, but offers no experimental data to back up its claims. The ad puts forth a truth assertion: "**Think And Be Thin.**" It further claims "Here's a tape program that really works … and permanently!" The program consists of listening to a tape with subliminal messages that is supposed to program your mind to produce thinness. The glaring lack is that there are no controlled experiments, no data offered to substantiate the claim. This is the kind of claim that cries out for empirical verification. Apparently, the authors of the ad do not believe the readers of *Psychology Today* are very sophisticated, statistically. I certainly hope the readers of this textbook would ask for the data before they spend 6 months of their time listening to a tape, the message of which they can't even hear! ∎

Authorities Are Nice, but ...

An advertisement promoting Anacin-3 appeared in an issue of *Cosmopolitan*. The heading of the advertisement was "3 Good Reasons to Try Anacin-3." The advertisement pictured a doctor, a nurse, and a pharmacist making the following three statements:

1. "*Doctors* are recommending acetaminophen, the aspirin-free pain reliever in Anacin-3, more than any other aspirin-free pain reliever."

2. "*Hospitals* use acetaminophen, the aspirin-free pain reliever in Anacin-3, more than any other aspirin-free pain reliever."

3. "*Pharmacists* recommend acetaminophen, the aspirin-free pain reliever in Anacin-3, more than any other aspirin-free pain reliever."

From a scientific point of view, is anything missing?

Answer This is somewhat better than the previous ad. At least relevant authorities are invoked in support of the product. However,

the ad is misleading and again fails to present the appropriate data. Much better than the "3 Good Reasons to Try Anacin-3" given in the ad would be reason 4, data from well-conducted experiments showing that (a) acetaminophen is a better pain reliever than any other aspirin-free pain reliever and (b) Anacin-3 relieves pain better than any competitor. Any guesses about why these data haven't been presented? As a budding statistician, are you satisfied with the case made by this ad? ∎

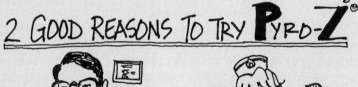

WHAT IS THE TRUTH?

Data, Data, What Are the Data?—1

In the previous "What Is the Truth?" sections, no data were presented to justify the authors' claims, a cardinal error in science. A little less grievous, but nonetheless questionable, is when data are presented in a study or experiment but the conclusions drawn by the authors seem to stay far from the actual data. A study reported in a recent newspaper article illustrates this point. The article is shown below.

CAVEMEN WERE PRETTY NICE GUYS

By VIRGINIA FENTON
The Associated Press

PARIS—Neanderthals might not have been as savage as we think. A 200,000-year-old jawbone discovered in France suggests the primitive hominids took care of each other, in this case feeding a toothless peer, an international team of experts said Friday.

A damaged jawbone, discovered last year in southern France, shows that its owner survived without teeth for up to several years— impossible without a helping hand from his or her peers, said Canadian paleontologist Serge Lebel.

Lebel directed an international team of experts who discovered the fossil in July 2000. Also on the team was the noted specialist of the Neanderthal period Erik Trinkaus of Washington University in St. Louis

and colleagues from Germany, Portugal and France.

"This individual must have been quite weak and needed preparation of his or her food, and the social group probably took care of him or her," Lebel said at a news conference.

"We mustn't dehumanize these beings. They show an entirely human kind of behavior," said Lebel, who works at the University of Quebec.

"Others in the group may have gone as far as chewing the food for their sick peer, as well as cutting and cooking it," he said.

"The discovery may push back ideas of the beginning of social care by 150,000 years," Lebel said. "A similar infection that caused a hominid to lose his teeth had previously only been found in fossils dating back 50,000 years," he said.

The team's findings were published in the Sept. 25 issue of the U.S. periodical Proceedings of the National Academy of Sciences.

However, University of Pittsburgh anthropologist Jeffrey Schwartz was skeptical.

"You can eat a lot without your teeth. There is no reason to think the individual couldn't have been chewing soft food—snails, mollusks, even worms."

Reprinted with permission of The Associated Press.

Do you think the finding of a 200,000-year-old, damaged jawbone by the international team of experts is an adequate database for the conclusions put forth by Dr. Lebel? Or, like Dr. Schwartz, are you skeptical? You can probably guess into which camp I fall. ∎

WHAT IS THE TRUTH?   Data, Data, What Are the Data?—2

As discussed in this chapter, using data as the basis for truth assertions is an important part of the scientific process. Sometimes, however, data are reported in a manner that is distorted, leading to false conclusions rather than truth. This point is illustrated in the following article that recently appeared in a local newspaper.

PUNDITS TWIST FACTS INTO URBAN LEGEND
By TIM RUTTEN

Urban legends are anecdotes so engaging, entertaining, anxiety-affirming or prejudice-confirming that people repeat them as true, even when they are not.

Their journalistic equivalent is the pithy "poll result" or "study finding" that so neatly encapsulates a social ill or point of partisan contention that no talk-show host or stump orator can resist it.

Ronald Reagan, for example, was so notoriously fond of dubious anecdotes that some of his close associates speculated that his years as an actor had rendered him utterly incapable of passing up a good line.

That provenance-be-damned attitude hardly ended with the Great Communicator's retirement.

CNN political analyst William Schneider, an authority on opinion polling, believes that highly charged political issues or campaigns are particularly prone to spin off fictional study results, which then take on a life of their own.

"Social Security is a classic case," Schneider said. "There is a political legend that a 'recent' poll, which is never specified, showed that more young Americans believe Elvis is alive than expect to collect a Social Security check when they retire. Over the past few years, every conservative on Earth and every Republican politician in Washington, including President Bush, has referred to this finding. They all just say, 'Polls or studies show this,' but none ever has. Republicans go on citing this bogus finding because—whether it's true or not—it has great value as political currency."

Recently, it was possible to watch another such journalistic legend being born. The Wall Street Journal's *Leisure & Arts page published an article excerpted from a speech delivered a week before by Bruce* Cole, chairman of the National Endowment for the Humanities. In it, he decried the fact "that Americans do not know their history" and reflected on this collective amnesia's particular perils in the wake of Sept. 11, whose memory he fears might even now be fading.

All fair points. However, in their support, Cole cited "a nationwide survey recently commissioned by Columbia Law School," which, he said, "found that almost two-thirds of all Americans think Karl Marx's socialist dogma, 'From each according to his ability, to each according to his needs,' was or may have been written by the Founding Fathers and was included in the Constitution."

Clear; helpful; alarming. Plausible, even, given the popular taste for altruistic banality. Everything, in other words, that a good rhetorical point should be. The problem is that is not what the Columbia poll found.

(continued)

WHAT IS THE TRUTH? *(continued)*

It is, however, what the May 29 press release announcing the poll's findings said. In fact, its first paragraph read: "Almost two-thirds of Americans think Karl Marx's maxim, 'From each according to his ability, to each according to his needs' was or could have been written by the framers and included in the Constitution."

Note the subtle shift from the release's "could have been written" to Cole's—or his speech writer's—"may have been written."

The announcement's second paragraph quotes Columbia Law School professor Michael Dorf on the troubling implications of the Marx "finding" and points the reader to a column the professor has written for a legal Web site, Findlaw.com. In that column, Dorf wrote that "The survey found that 69 percent of respondents either thought that the United States Constitution contained Marx's maxim, or did not know whether or not it did."

That is further still from Cole's recitation of the finding.

What of the actual poll, which asked a national sample of 1,000 respondents five yes-or-no questions about the Constitution?

In response to the query about whether the Marxist maxim is contained in the document, 35 percent said yes; 31 percent said no; and 34 percent said "they did not know."

Was Dorf entitled to conflate the yes and don't-know respondents

into a single, alarming two-thirds? No, Schneider said. "That is a total misinterpretation of what people mean when they tell a pollster they don't know. When people say that, they mean 'I have no idea.' Choosing to interpret it as meaning 'I think it may be true or could be true' simply is misleading on the writer's part."

What it does do is add pungency to a finding Schneider described as "wholly unremarkable. It's not at all surprising that about a third of respondents thought that is in the Constitution. Most people, for example, think the right to privacy exists in the Constitution, when no such right is enumerated anywhere in it."

In fact, what is remarkable—and far more relevant at the moment—about the Columbia poll is what it shows Americans do know about their Constitution: Fully 83 percent of the respondents recognized the first sentence of the 14th Amendment as part of the national charter, while 60 percent correctly understood that the president may not suspend the Bill of Rights in time of war.

More than two-thirds of those asked were aware that Supreme Court justices serve for life.

The only other question the respondents flubbed was whether the court's overturning of Roe vs. Wade would make abortion "illegal throughout the United States."

Fewer than a third of those polled correctly said it would not,

because state statutes guaranteeing choice would remain on the books.

These facts notwithstanding, it is a safe bet that radio and television talk-show hosts, editorial writers and all the rest of the usually suspect soon will be sighing, groaning, sneering and raging about Americans' inability to distinguish the thought of James Madison from that of Karl Marx. How long will it be before public education, the liberal media, humanism and single-parent families are blamed?

Commentary pages are the soft underbelly of American journalism. Their writers, however self-interested, are held to a different, which is to say lower, standard of proof because of their presumed expertise.

In fact, they are responsible for regularly injecting false information of this sort into our public discourse. In the marketplace of ideas, as on the used car lot, caveat emptor still is the best policy.

Source: "Regarding Media: Surveying a Problem with Polls" by Tim Rutten, *Los Angeles Times*, June 14, 2002. Copyright © 2002 Los Angeles Times. Reprinted with permission.

Do you think William Schneider is too cynical, or do you agree with the article's conclusion? "In fact, they are responsible for regularly injecting false information of this sort into our public discourse. In the marketplace of ideas, as on the used car lot, caveat emptor still is the best policy." ■

■ SUMMARY

In this chapter, I have discussed how truth is established. Traditionally, four methods have been used: authority, reason, intuition, and science. At the heart of science is the scientific experiment. By reasoning or through intuition, the scientist forms a hypothesis about some feature of reality. He or she designs an experiment to objectively test the hypothesis. The data from the experiment are then analyzed statistically, and the hypothesis is either confirmed or rejected.

Most scientific research falls into two categories: observational studies and true experiments. Natural observation, parameter estimation, and correlational studies are included within the observational category. Their major goal is to give an accurate description of the situation, estimate population parameters, or determine whether two or more of the variables are related. Since there is no systematic manipulation of any variable by the experimenter when doing an observational study, this type of research cannot determine whether changes in one variable will cause changes in another variable. Causal relationships can be determined only from true experiments.

In true experiments, the investigator systematically manipulates the independent variable and observes its effect on one or more dependent variables. Due to practical considerations, data are collected on only a sample of subjects rather than on the whole population. It is important that the sample be a random sample. The obtained data are then analyzed statistically.

The statistical analysis may be descriptive or inferential. If the analysis just describes or characterizes the obtained data, we are in the domain of descriptive statistics. If the analysis uses the obtained data to infer to populations, we are in the domain of inferential statistics. Understanding statistical analysis has important practical consequences in life.

■ IMPORTANT NEW TERMS

Constant (p. 6)
Correlational studies (p. 9)
Data (p. 7)
Dependent variable (p. 7)
Descriptive statistics (p. 10)
Independent variable (p. 6)
Inferential statistics (p. 10)
Method of authority (p. 4)

Method of intuition (p. 5)
Method of rationalism (p. 4)
Naturalistic observation
    research (p. 9)
Observational studies (p. 9)
Parameter (p. 7)
Parameter estimation research (p. 9)
Population (p. 6)

Sample (p. 6)
Scientific method (p. 6)
SPSS (p. 11)
Statistic (p. 7)
True experiment (p. 9)
Variable (p. 6)

■ QUESTIONS AND PROBLEMS

Note to the student: You will notice that at the end of specific problems in this and all other chapters except Chapter 2, I have identified, in color, a specific area within psychology and related fields to which the problems apply. For example, Problem 6, part **b,** is a problem in the area of biological psychology. It has been labeled ("biological" at the end of the problem ("psychology" is left off of each label for brevity). The specific areas identified are cognitive psychology, social psychology, developmental psychology, biological psychology, clinical psychology, industrial/organizational (I/O) psychology, health psychology, education, and other. I hope this labeling will be useful to your instructor in selecting assigned homework problems and to you in seeing the broad application of this material as well as in helping you select additional problems you might enjoy solving beyond the assigned ones.

1. Define each of the following terms:

   Population   Dependent variable
   Sample   Constant
   Data   Statistic
   Variable   Parameter
   Independent variable

2. What are four methods of acquiring knowledge? Write a short paragraph describing the essential characteristics of each.

3. How does the scientific method differ from each of the methods listed here?
   a. Method of authority
   b. Method of rationalism
   c. Method of intuition

4. Write a short paragraph comparing naturalistic observation and true experiments.

5. Distinguish between descriptive and inferential statistics. Use examples to illustrate the points you make.

6. In each of the experiments described here, specify (1) the independent variable, (2) the dependent variable, (3) the sample, (4) the population, (5) the data, and (6) the statistic:

   a. A health psychologist is interested in whether fear motivation is effective in reducing the incidence of smoking. Forty adult smokers are selected from individuals residing in the city in which the psychologist works. Twenty are asked to smoke a cigarette, after which they see a gruesome film about how smoking causes cancer. Vivid pictures of the diseased lungs and other internal organs of deceased smokers are shown in an effort to instill fear of smoking in these subjects. The other group receives the same treatment, except they see a neutral film that is unrelated to smoking. For 2 months after showing the film, the experimenter keeps records on the number of cigarettes smoked daily by the participants. A mean for each group is then computed, of the number of cigarettes smoked daily since seeing the film, and these means are compared to determine whether the fear-inducing film had an effect on smoking. health

   b. A physiologist wants to know whether a particular region of the brain (the hypothalamus) is involved in the regulation of eating. An experiment is performed in which 30 rats are selected from the university vivarium and divided into two groups. One of the groups receives lesions in the hypothalamus, whereas the other group gets lesions produced in a neutral area. After recovery from the operations, all animals are given free access to food for 2 weeks, and a record is kept of the daily food intake of each animal. At the end of the 2-week period, the mean daily food intake for each group is determined. Finally, these means are compared to see whether the lesions in the hypothalamus have affected the amount eaten. biological

   c. A clinical psychologist is interested in evaluating three methods of treating depression: medication, cognitive restructuring, and exercise. A fourth treatment condition, a waiting-only treatment group, is included to provide a baseline control group. Sixty depressed students are recruited from the undergraduate student body at a large state university, and fifteen are assigned to each treatment method. Treatments are administered for 6 months, after which each student is given a questionnaire designed to measure the degree of depression. The questionnaire is scaled from 0 to 100, with higher scores indicating a higher degree of depression. The mean depression values are then computed for the four treatments and compared to determine the relative effectiveness of each treatment. clinical, health

   d. A social psychologist is interested in determining whether individuals who graduate from high school but get no further education earn more money than high school dropouts. A national survey is conducted in a large Midwestern city, sampling 100 individuals from each category and asking each their annual salary. The results are tabulated, and mean salary values are calculated for each group. social

   e. A cognitive psychologist is interested in how retention is affected by the spacing of practice sessions. A sample of 30 seventh graders is selected from a local junior high school and divided into three groups of 10 students in each group. All students are asked to memorize a list of 15 words and are given three practice sessions, each 5 minutes long, in which to do so. Practice sessions for group 1 subjects are spaced 10 minutes apart; for group 2, 20 minutes apart; and for group 3, 30 minutes apart. All groups are given a retention test 1 hour after the last practice session. Results are recorded as the number of words correctly recalled in the test period. Mean values are computed for each group and compared. cognitive

f. A sport psychologist uses visualization in promoting enhanced performance in college athletes. She is interested in evaluating the relative effectiveness of visualization alone versus visualization plus appropriate self-talk. An experiment is conducted with a college basketball team. Ten members of the team are selected. Five are assigned to a visualization alone group, and five are assigned to a visualization plus self-talk group. Both techniques are designed to increase foul shooting accuracy. Each group practices its technique for 1 month. The foul shooting accuracy of each player is measured before and 1 month after beginning practice of the technique. Difference scores are computed for each player, and the means of the difference scores for each group are compared to determine the relative effectiveness of the two techniques. I/O, other

g. A typing teacher believes that a different arrangement of the typing keys will promote faster typing. Twenty secretarial trainees, selected from a large business school, participate in an experiment designed to test this belief. Ten of the trainees learn to type on the conventional keyboard. The other ten are trained using the new arrangement of keys. At the end of the training period, the typing speed in words per minute of each trainee is measured. The mean typing speeds are then calculated for both groups and compared to determine whether the new arrangement has had an effect. education

7. Indicate which of the following represent a variable and which a constant:
   a. The number of letters in the alphabet
   b. The number of hours in a day
   c. The time at which you eat dinner
   d. The number of students who major in psychology at your university each year
   e. The number of centimeters in a meter
   f. The amount of sleep you get each night
   g. The amount you weigh
   h. The volume of a liter

8. Indicate which of the following situations involve descriptive statistics and which involve inferential statistics:
   a. An annual stockholders' report details the assets of the corporation.
   b. A history instructor tells his class the number of students who received an A on a recent exam.
   c. The mean of a sample set of scores is calculated to characterize the sample.
   d. The sample data from a poll are used to estimate the opinion of the population.
   e. A correlational study is conducted on a sample to determine whether educational level and income in the population are related.
   f. A newspaper article reports the average salaries of federal employees from data collected on all federal employees.

9. For each of the following, identify the sample and population scores:
   a. A social psychologist interested in drinking behavior investigates the number of drinks served in bars in a particular city on a Friday during "happy hour." In the city, there are 213 bars. There are too many bars to monitor all of them, so she selects 20 and records the number of drinks served in them. The following are the data:

50	82	47	65
40	76	61	72
35	43	65	76
63	66	83	82
57	72	71	58

social

   b. To make a profit from a restaurant that specializes in low-cost quarter-pound hamburgers, it is necessary that each hamburger served weigh very close to 0.25 pound. Accordingly, the manager of the restaurant is interested in the variability among the weights of the hamburgers served each day. On a particular day, there are 450 hamburgers served. It would take too much time to weigh all 450, so the manager decides instead to weigh just 15. The following weights in pounds were obtained:

0.25	0.27	0.25
0.26	0.35	0.27
0.22	0.32	0.38
0.29	0.22	0.28
0.27	0.40	0.31

other

c. A machine that cuts steel blanks (used for making bolts) to their proper length is suspected of being unreliable. The shop supervisor decides to check the output of the machine. On the day of checking, the machine is set to produce 2-centimeter blanks. The acceptable tolerance is ±0.05 centimeter. It would take too much time to measure all 600 blanks produced in 1 day, so a representative group of 25 is selected. The following lengths in centimeters were obtained:

2.01	1.99	2.05	1.94	2.05
2.01	2.02	2.04	1.93	1.95
2.03	1.97	2.00	1.98	1.96
2.05	1.96	2.00	2.01	1.99
1.98	1.95	1.97	2.04	2.02

I/O

d. A physiological psychologist, working at Tacoma University, is interested in the resting, diastolic heart rates of all the female students attending the university. She randomly samples 30 females from the student body and records the following diastolic heart rates while the students are lying on a cot. Scores are in beats/min.

62	85	92	85	88	71
73	82	84	89	93	75
81	72	97	78	90	87
78	74	61	66	83	68
67	83	75	70	86	72

biological

What Is the Truth Questions

1. *Data, Data, Where Are the Data?*

   a. Find two magazine advertisements that make truth claims about the product(s) being advertised. From a scientific point of view, is anything missing? Discuss.

   b. If anyone, even an authority, asserts that "X" is true, e.g., "global warming is mainly due to human activity," what is necessary for the truth assertion to be adequately justified from a scientific point of view? In your opinion, based on your reading, TV viewing or other source, is the truth assertion that global warming is mainly due to human activity adequately justified from a scientific point of view? Discuss.

2. *Authorities Are Nice, but …*

   You are attending a lecture at your university. The lecturer is a prominent physician with an international reputation. Her field of expertise is Preventive Medicine. She asserts that the main cause of heart attack and stroke is eating red meat. To justify her view, she makes several logical arguments and claims all her colleagues agree with her. If you accept her view on the basis of logic or her national prominence or because her colleagues agree with her, what method(s) of determining truth are you using? From a scientific point of view, is this sufficient basis to conclude the claim is true? If not, what is missing? Discuss.

3. *Data, Data, What Are the Data?—1*

   You are reading a book written by a social psychologist on the benefits of developing conversation skills on relationships. There are five chapters, with each chapter describing a different conversation skill. Each chapter begins with a case study showing that when the individual learned the particular skill, his or her relationships improved. From a scientific point of view, are the case studies sufficient to justify the conclusion that developing the conversation skills presented in the book will improve relationships? Discuss.

■ ONLINE STUDY RESOURCES

CENGAGE**brain**.com

Login to CengageBrain.com to access the resources your instructor has assigned. For this book, you can access the book's companion website for chapter-specific learning tools including Know and Be Able to Do, practice quizzes, flash cards, and glossaries and a link to Statistics and Research Methods Workshops.

aplia

If your professor has assigned Aplia homework:

1. Sign in to your account
2. Complete the corresponding homework exercises as required by your professor
3. When finished, click "Grade It Now" to see which areas you have mastered and which need more work, and for detailed explanations of every answer.

Visit **www.cengagebrain.com** to access your account and to purchase materials.

DESCRIPTIVE STATISTICS

© Strmko / Dreamstime.com

© Strmko / Dreamstime.com

2

Basic Mathematical and Measurement Concepts

LEARNING OBJECTIVES

After completing this chapter, you should be able to:

- Assign subscripts using the X variable to a set of numbers.
- Do the operations called for by the summation sign for various values of i and N.
- Specify the differences in mathematical operations between $(\Sigma\, X)^2$ and $\Sigma\, X^2$ and compute each.
- Define and recognize the four measurement scales, give an example of each, and state the mathematical operations that are permissible with each scale.
- Define continuous and discrete variables and give an example of each.
- Define the real limits of a continuous variable and determine the real limits of values obtained when measuring a continuous variable.
- Round numbers with decimal remainders.
- Understand the illustrative examples, do the practice problems, and understand the solutions.

STUDY HINTS FOR THE STUDENT

Statistics is not an easy subject. It requires learning difficult concepts as well as doing mathematics. There is, however, some advice that I would like to pass on, which I believe will help you greatly in learning this material. This advice is based on many years of teaching the subject; I hope you will take it seriously.

Most students in the behavioral sciences have a great deal of anxiety about taking a course on mathematics or statistics. Without minimizing the difficulty of the subject, a good deal of this anxiety is unnecessary. To learn the material contained in this textbook, you do not have to be a whiz in calculus or differential equations. I have tried hard to present the material so that non–mathematically inclined students can understand it. I cannot, however, totally do away with mathematics. To be successful, you must be able to do elementary algebra and a few other mathematical operations. To help you review, I have included Appendix A, which covers prerequisite mathematics. You should study that material and be sure you can do the problems it contains. If you have difficulty with these problems, it will help to review the topic in a basic textbook on elementary algebra or use the website provided in Appendix A.

Another factor of which you should be aware is that a lot of symbols are used in statistics. For example, to designate the mean of a sample set of scores, we shall use the symbol $\overline{X}$ (read "X bar"). Students often make the material more difficult than necessary by failing to thoroughly learn what the symbols stand for. You can save yourself much grief by taking the symbols seriously. Treat them as though they are foreign vocabulary. Memorize them and be able to deal with them conceptually. For example, if the text says $\overline{X}$, the concept "the mean of the sample" should immediately come to mind.

MENTORING TIP

If you memorize the symbols and don't fall behind in assignments, you will find this material much easier to learn.

It is also important to realize that the material in statistics is cumulative. Do not let yourself fall behind. If you do, you will not understand the current material either. The situation can then snowball, and before you know it, you may seem hopelessly behind. Remember, do all you can to keep up with the material.

Finally, my experience indicates that a good deal of the understanding of statistics comes from working lots of problems. Very often, one problem is worth a thousand words. Frequently, although the text is clearly worded, the material won't come into focus until you have worked the problems associated with the topic. Therefore, do lots of problems, and afterward, reread the textual material to be sure you understand it.

In sum, I believe that if you can handle elementary algebra, work diligently on learning the symbols and studying the text, keep up with the material, and work lots of problems, you will do quite well. Believe it or not, as you begin to experience the elegance and fun that are inherent in statistics, you may even come to enjoy it.

MATHEMATICAL NOTATION

In statistics, we usually deal with group data that result from measuring one or more variables. The data are most often derived from samples, occasionally from populations. For mathematical purposes, it is useful to let symbols stand for the variables measured in the study. Throughout this text, we shall use the Roman capital letter X, and sometimes Y, to stand for the variable(s) measured. Thus, if we were measuring the age of subjects, we would let X stand for the variable "age." When there are many values of

MENTORING TIP

Be sure to distinguish between the *score value* and the *subject number*. For example, X_1 is the score value of the first subject or the first score in the distribution; $X_1 = 8$.

table 2.1   Age of six subjects

Subject Number	Score Symbol	Score Value, Age (yr)
1	X_1	8
2	X_2	10
3	X_3	7
4	X_4	6
5	X_5	10
6	X_6	12

the variable, it is important to distinguish among them. We do this by subscripting the symbol *X*. This process is illustrated in Table 2.1.

In this example, we are letting the variable "age" be represented by the symbol *X*. We shall also let *N* represent the number of scores in the distribution. In this example, $N = 6$. Each of the six scores represents a specific value of *X*. We distinguish among the six scores by assigning a subscript to *X* that corresponds to the number of the subject that had the specific value. Thus, the score symbol X_1 corresponds to the score value 8, X_2 to the score value 10, X_3 to the value 7, X_4 to 6, X_5 to 10, and X_6 to 12. In general, we can refer to a single score in the *X* distribution as X_i, where *i* can take on any value from 1 to *N*, depending on which score we wish to designate. To summarize,

- *X* or *Y* stands for the variable measured.
- *N* stands for the total number of subjects or scores.
- X_i is the *i*th score, where *i* can vary from 1 to *N*.

SUMMATION

One of the most frequent operations performed in statistics is to sum all or part of the scores in the distribution. Since it is awkward to write out "sum of all the scores" each time this operation is required, particularly in equations, a symbolic abbreviation is used instead. The capital Greek letter sigma (Σ) indicates the operation of summation. The algebraic phrase employed for summation is

$$\sum_{i=1}^{N} X_i$$

This is read as "sum of the *X* variable from $i = 1$ to *N*." The notations above and below the summation sign designate which scores to include in the summation. The term below the summation sign tells us the first score in the summation, and the term above the summation sign designates the last score. This phrase, then, indicates that we are to add the *X* scores, beginning with the first score and ending with the *N*th score. Thus,

$$\sum_{i=1}^{N} X_i = X_1 + X_2 + X_3 + \cdots + X_N \qquad \textit{summation equation}$$

Applied to the age data of the previous table,

$$\sum_{i=1}^{N} X_i = X_1 + X_2 + X_3 + X_4 + X_5 + X_6$$

$$= 8 + 10 + 7 + 6 + 10 + 12 = 53$$

When the summation is over all the scores (from 1 to N), the summation phrase itself is often abbreviated by omitting the notations above and below the summation sign and by omitting the subscript i. Thus,

$$\sum_{i=1}^{N} X_i \text{ is often written as } \Sigma X.$$

In the previous example,

$$\Sigma X = 53$$

This says that the sum of all the X scores is 53.

Note that it is not necessary for the summation to be from 1 to N. For example, we might desire to sum only the second, third, fourth, and fifth scores. Remember, the notation below the summation sign tells us where to begin the summation, and the term above the sign tells us where to stop. Thus, to indicate the operation of summing the second, third, fourth, and fifth scores, we would use the symbol $\sum_{i=2}^{5} X_i$. For the preceding age data,

$$\sum_{i=2}^{5} X_i = X_2 + X_3 + X_4 + X_5 = 10 + 7 + 6 + 10 = 33$$

Let's do some practice problems in summation.

Practice Problem 2.1

a. For the following scores, find $\sum_{i=1}^{N} X_i$:

X: 6, 8, 13, 15 $\Sigma X = 6 + 8 + 13 + 15 = 42$

X: 4, −10, −2, 20, 25, 8 $\Sigma X = 4 - 10 - 2 + 20 + 25 + 8 = 45$

X: 1.2, 3.5, 0.8, 4.5, 6.1 $\Sigma X = 1.2 + 3.5 + 0.8 + 4.5 + 6.1 = 16.1$

b. For the following scores, find $\sum_{i=1}^{3} X_i$:

$$X_1 = 10, X_2 = 12, X_3 = 13, X_4 = 18$$

$$\sum_{i=1}^{3} X_i = 10 + 12 + 13 = 35$$

c. For the following scores, find $\displaystyle\sum_{i=2}^{4} X_i + 3$:

$$X_1 = 20, X_2 = 24, X_3 = 25, X_4 = 28, X_5 = 30, X_6 = 31$$

$$\sum_{i=2}^{4} X_i + 3 = (24 + 25 + 28) + 3 = 80$$

d. For the following scores, find $\displaystyle\sum_{i=2}^{4} (X_i + 3)$:

$$X_1 = 20, X_2 = 24, X_3 = 25, X_4 = 28, X_5 = 30, X_6 = 31$$

$$\sum_{i=2}^{4} (X_i + 3) = (24 + 3) + (25 + 3) + (28 + 3) = 86$$

There are two more summations that we shall frequently encounter later in the textbook. They are ΣX^2 and $(\Sigma X)^2$. Although they look alike, they are different and generally will yield different answers. The symbol ΣX^2 (sum of the squared X scores) indicates that we are first to square the X scores and then sum them. Thus,

$$\Sigma X^2 = X_1^2 + X_2^2 + X_3^2 + \cdots + X_N^2$$

Given the scores $X_1 = 3, X_2 = 5, X_3 = 8,$ and $X_4 = 9,$

$$\Sigma X^2 = 3^2 + 5^2 + 8^2 + 9^2 = 179$$

The symbol $(\Sigma X)^2$ (sum of the X scores, quantity squared) indicates that we are first to sum the X scores and then square the resulting sum. Thus,

$$(\Sigma X)^2 = (X_1 + X_2 + X_3 + \cdots + X_N)^2$$

For the previous scores, namely, $X_1 = 3, X_2 = 5\ X_3 = 8,$ and $X_4 = 9,$

$$(\Sigma X)^2 = (3 + 5 + 8 + 9)^2 = (25)^2 = 625$$

MENTORING TIP

Caution: be sure you know the difference between ΣX^2 and $(\Sigma X)^2$ and can compute each.

Note that $\Sigma X^2 \neq (\Sigma X)^2$ [$179 \neq 625$]. Confusing ΣX^2 and $(\Sigma X)^2$ is a common error made by students, particularly when calculating the standard deviation. We shall return to this point when we take up the standard deviation in Chapter 4.*

Order of Mathematical Operations

As you no doubt have noticed in understanding the difference between ΣX^2 and $(\Sigma X)^2$, the order in which you perform mathematical operations can make a great difference in the result. Of course, you should follow the order indicated by the symbols in the mathematical phrase or equation. This is something that is taught in elementary algebra. However, since many students either did not learn this when taking elementary algebra or have forgotten it in the ensuing years, I have decided to include a quick review here.

---

*See Note 2.1 at the end of this chapter for additional summation rules, if desired.

Mathematical operations should be done in the following order:

1. Always do what is in parentheses first. For example, $(\Sigma\ X)^2$ indicates that you are to sum the X scores first and then square the result. Another example showing the priority given to parentheses is the following: $2(5 + 8) = 2(13) = 26$

MENTORING TIP

If your algebra is somewhat rusty, see Appendix A, *Review of Prerequisite Mathematics,* p. 553.

2. If the mathematical operation is summation (Σ), perform the summation last, unless parentheses indicate otherwise. For example, $\Sigma\ X^2$ indicates that you should square each X value first, and then sum the squared values. $(\Sigma\ X)^2$ indicates that you should sum the X scores first and then square the result. This is because of the order imposed by the parentheses.

3. If multiplication and addition or subtraction are specified, the multiplication should be performed first, unless parentheses indicate otherwise. For example,

$$4 \times 5 + 2 = 20 + 2 = 22$$

$$6 \times (4 + 3) \times 2 = 6 \times 7 \times 2 = 84$$

$$6 \times (14 - 12) \times 3 = 6 \times 2 \times 3 = 36$$

4. If division and addition or subtraction are specified, the division should be performed first, unless parentheses indicate otherwise. For example,

$$12 \div 4 + 2 = 3 + 2 = 5$$

$$12 \div (4 + 2) = 12 \div 6 = 2$$

$$12 \div 4 - 2 = 3 - 2 = 1$$

$$12 \div (4 - 2) = 12 \div 2 = 6$$

5. The order in which numbers are added does not change the result. For example,

$$6 + 4 + 11 = 4 + 6 + 11 = 11 + 6 + 4 = 21$$

$$6 + (-3) + 2 = -3 + 6 + 2 = 2 + 6 + (-3) = 5$$

6. The order in which numbers are multiplied does not change the result. For example,

$$3 \times 5 \times 8 = 8 \times 5 \times 3 = 5 \times 8 \times 3 = 120$$

MEASUREMENT SCALES

Since statistics deals with data and data are the result of measurement, we need to spend some time discussing measuring scales. This subject is particularly important because the type of measuring scale employed in collecting the data helps determine which statistical inference test is used to analyze the data. Theoretically, a measuring scale can have one or more of the following mathematical attributes: magnitude, an equal interval between adjacent units, and an absolute zero point. Four types of scales are commonly encountered in the behavioral sciences: *nominal, ordinal, interval*, and *ratio*. They differ in the number of mathematical attributes that they possess.

Nominal Scales

MENTORING TIP

When using a nominal scale, you cannot do operations of addition, subtraction, multiplication, division, or ratios.

A *nominal scale* is the lowest level of measurement and is most often used with variables that are qualitative in nature rather than quantitative. Examples of qualitative variables are brands of jogging shoes, kinds of fruit, types of music, days of the month, nationality, religious preference, and eye color. When a nominal scale is used, the variable is divided into its several categories. These categories comprise the "units" of the scale, and objects are "measured" by determining the category to which they belong.

definition ■ *A **nominal scale** is one that has categories for the units.*

Thus, measurement with a nominal scale really amounts to classifying the objects and giving them the name (hence, *nominal* scale) of the category to which they belong.

To illustrate, if you are a jogger, you are probably interested in the different brands of jogging shoes available for use, such as Brooks, Nike, Adidas, Saucony, and New Balance, to name a few. Jogging shoes are important because, in jogging, each shoe hits the ground about 800 times a mile. In a 5-mile run, that's 4000 times. If you weigh 125 pounds, you have a total impact of 300 tons on each foot during a 5-mile jog. That's quite a pounding. No wonder joggers are extremely careful about selecting shoes.

The variable "brand of jogging shoes" is a qualitative variable. It is measured on a nominal scale. The different brands mentioned here represent some of the possible categories (units) of this scale. If we had a group of jogging shoes and wished to measure them using this scale, we would take each one and determine to which brand it belonged. It is important to note that because the units of a nominal scale are categories, there is no magnitude relationship between the units of a nominal scale. Thus, there is no quantitative relationship between the categories of Nike and Brooks. The Nike is no more or less a brand of jogging shoe than is the Brooks. They are just different brands. The point becomes even clearer if we were to call the categories jogging shoes 1 and jogging shoes 2 instead of Nike and Brooks. Here, the numbers 1 and 2 are really just names and bear no magnitude relationship to each other.

A fundamental property of nominal scales is that of *equivalence*. By this we mean that all members of a given class are the same from the standpoint of the classification variable. Thus, all pairs of Nike jogging shoes are considered the same from the standpoint of "brand of jogging shoes"—this despite the fact that there may be different types of Nike jogging shoes present.

An operation often performed in conjunction with nominal measurement is that of counting the instances within each class. For example, if we had a bunch of jogging shoes and we determined the brand of each shoe, we would be doing nominal measurement. In addition, we might want to count the number of shoes in each category. Thus, we might find that there are 20 Nike, 19 Saucony, and 6 New Balance shoes in the bunch. These frequencies allow us to compare the number of shoes within each category. This quantitative comparison of numbers within each category should not be confused with the statement made earlier that there is no magnitude relationship between the units of a nominal scale. We can compare quantitatively the numbers of Nike with the numbers of Saucony shoes, but Nike is no more or less a brand of jogging shoe than is Saucony. Thus, a nominal scale does not possess any of the mathematical attributes of magnitude, equal interval, or absolute zero point. It merely allows categorization of objects into mutually exclusive categories. The main mathematical operation done with nominal scales is to count the number of objects that occur within each category on the scale.

Ordinal Scales

MENTORING TIP

When using an ordinal scale, you cannot do operations of addition, subtraction, multiplication, division, or ratios.

An *ordinal scale* represents the next higher level of measurement. It possesses a relatively low level of the property of magnitude. With an ordinal scale, we rank-order the objects being measured according to whether they possess more, less, or the same amount of the variable being measured. Thus, an ordinal scale allows determination of whether $A > B$, $A = B$, or $A < B$.

definition   ■ *An **ordinal scale** is one in which the numbers on the scale represent rank orderings, rather than raw score magnitudes.*

An example of an ordinal scale is the rank ordering of the top five contestants in a speech contest according to speaking ability. Among these speakers, the individual ranked 1 was judged a better speaker than the individual ranked 2, who in turn was judged better than the individual ranked 3. The individual ranked 3 was judged a better speaker than the individual ranked 4, who in turn was judged better than the individual ranked 5. It is important to note that although this scale allows better than, equal to, or less than comparisons, it says nothing about the magnitude of the difference between adjacent units on the scale. In the present example, the difference in speaking ability between the individuals ranked 1 and 2 might be large, and the difference between individuals ranked 2 and 3 might be small. Thus, an ordinal scale does not have the property of equal intervals between adjacent units. Furthermore, since all we have is relative rankings, the scale doesn't tell the absolute level of the variable. Thus, all five of the top-ranked speakers could have a very high level of speaking ability or a low level. This information can't be obtained from an ordinal scale.

Other examples of ordinal scaling are the ranking of runners in the Boston Marathon according to their finishing order, the rank ordering of college football teams according to merit by the Associated Press, the rank ordering of teachers according to teaching ability, and the rank ordering of students according to motivation level.

Interval Scales

MENTORING TIP

When using an interval scale, you can do operations of addition and subtraction. You cannot do multiplication, division, or ratios.

The *interval scale* represents a higher level of measurement than the ordinal scale. It possesses the properties of magnitude and equal interval between adjacent units but doesn't have an absolute zero point.

definition   ■ *An **interval scale** is one in which the units represent raw score magnitudes, there are equal intervals between adjacent units on the scale, and there is no absolute zero point.*

Thus, the interval scale possesses the properties of the ordinal scale and has equal intervals between adjacent units. The phrase "equal intervals between adjacent units" means that there are equal amounts of the variable being measured between adjacent units on the scale.

The Celsius scale of temperature measurement is a good example of the interval scale. It has the property of equal intervals between adjacent units but does not have an absolute zero point. The property of equal intervals is shown by the fact that a given change of heat will cause the same change in temperature reading on the scale no matter where on the scale the change occurs. Thus, the additional amount of heat that will cause a temperature reading to change from 2° to 3° Celsius will also cause a reading to change from 51° to 52° or from 105° to 106° Celsius. This illustrates the fact that equal amounts of heat are indicated between adjacent units throughout the scale.

Since with an interval scale there are equal amounts of the variable between adjacent units on the scale, equal differences between the numbers on the scale represent equal differences in the magnitude of the variable. Thus, we can say the difference in heat is the same between 78° and 75° Celsius as between 24° and 21° Celsius. It also follows logically that greater differences between the numbers on the scale represent greater differences in the magnitude of the variable being measured, and smaller differences between the numbers on the scale represent smaller differences in the magnitude of the variable being measured. Thus, the difference in heat between 80° and 65° Celsius is greater than between 18° and 15° Celsius, and the difference in heat between 93° and 91° Celsius is less than between 48° and 40° Celsius. In light of the preceding discussion, we can see that in addition to being able to determine whether $A = B$, $A > B$, or $A < B$, an interval scale allows us to determine whether $A - B = C - D$, $A - B > C - D$, or $A - B < C - D$.

Ratio Scales

MENTORING TIP

When using a ratio scale, you can perform all mathematical operations.

The next, and highest, level of measurement is called a *ratio scale*. It has all the properties of an interval scale and, in addition, has an absolute zero point.

definition ■ *A **ratio scale** is one in which the units represent raw score magnitudes, there are equal intervals between adjacent units on the scale, and there is an absolute zero point.*

Without an absolute zero point, it is not legitimate to compute ratios with the scale readings. Since the ratio scale has an absolute zero point, ratios are permissible (hence the name *ratio* scale).

A good example to illustrate the difference between interval and ratio scales is to compare the Celsius scale of temperature with the Kelvin scale. Zero on the Kelvin scale is an absolute zero; it signifies the complete absence of heat. Zero degrees on the Celsius scale does not signify the complete absence of heat. Rather, it signifies the temperature at which water freezes. It is an arbitrary zero point that actually occurs at a reading of 273.15 kelvins (K).

The Celsius scale is an interval scale, and the Kelvin scale is a ratio scale. The difference in heat between adjacent units measured on the Celsius scale is the same throughout the Celsius scale. The same is true for the Kelvin scale. A heat change that causes the Celsius reading to change from 6° to 7° Celsius will also cause a change from 99° to 100° Celsius. Similarly, a heat change that causes the Kelvin scale to change from 10K to 11K will also cause the Kelvin scale to change from 60K to 61K. However, we cannot compute ratios with the Celsius scale. A reading of 20° Celsius is not twice as hot as 10° Celsius. This can be seen by converting the Celsius readings to the actual heat they represent. In terms of actual heat, 20° Celsius is really 293.15K (273.15 + 20), and 10° Celsius is really 283.15K (273.15 + 10). It is obvious that 293.15K is not twice 283.15K. Since the Kelvin scale has an absolute zero, a reading of 20K on it is twice as hot as 10K. Thus, ratios are permissible with the Kelvin scale.

Other examples of variables measured with ratio scales include reaction time, length, weight, age, and frequency of any event, such as the number of Nike shoes contained in the bunch of jogging shoes discussed previously. With a ratio scale, you can construct ratios and perform all the other mathematical operations usually associated with numbers (e.g., addition, subtraction, multiplication, and division). The four scales of measurement and their characteristics are summarized in Figure 2.1.

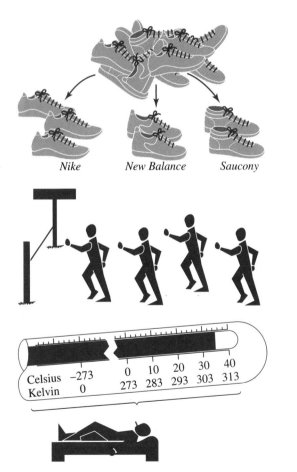

Nominal
Units of scale are categories. Objects are measured by determining the category to which they belong. There is no magnitude relationship between the categories.

Nike   *New Balance*   *Saucony*

Ordinal
Possesses the property of magnitude. Can rank-order the objects according to whether they possess more, less, or the same amount of the variable being measured. Thus, can determine whether $A > B$, $A = B$, or $A < B$.

Interval and Ratio
Interval: Possesses the properties of magnitude and equal intervals between adjacent units. Can do same determinations as with an ordinal scale, plus can determine whether $A - B = C - D$, $A - B > C - D$, or $A - B < C - D$.
Ratio: Possesses the properties of magnitude, equal interval between adjacent units, and an absolute zero point. Can do all the mathematical operations usually associated with numbers, including ratios.

| Celsius | −273 | 0 | 10 | 20 | 30 | 40 |
| Kelvin | 0 | 273 | 283 | 293 | 303 | 313 |

figure 2.1 Scales of measurement and their characteristics.

MEASUREMENT SCALES IN THE BEHAVIORAL SCIENCES

In the behavioral sciences, many of the scales used are often treated as though they are of interval scaling without clearly establishing that the scale really does possess equal intervals between adjacent units. Measurement of IQ, emotional variables such as anxiety and depression, personality variables (e.g., self-sufficiency, introversion, extroversion, and dominance), end-of-course proficiency or achievement variables, attitudinal variables, and so forth fall into this category. With all of these variables, it is clear that the scales are not ratio. For example, with IQ, if an individual actually scored a zero on the Wechsler Adult Intelligence Scale (WAIS), we would not say that he had zero intelligence. Presumably, some questions could be found that he could answer, which would indicate an IQ greater than zero. Thus, the WAIS does not have an absolute zero point, and ratios are not appropriate. Hence, it is not correct to say that a person with an IQ of 140 is twice as smart as someone with an IQ of 70.

On the other hand, it seems that we can do more than just specify a rank ordering of individuals. An individual with an IQ of 100 is closer in intelligence to someone with an IQ of 110 than to someone with an IQ of 60. This appears to be interval scaling, but it is difficult to establish that the scale actually possesses *equal* intervals between adjacent units. Many researchers treat those variables as though they were measured on interval scales, particularly when the measuring instrument is well standardized, as is the WAIS. It is more controversial to treat poorly standardized scales measuring psychological variables as interval scales. This issue arises particularly in inferential statistics, where the level of scaling can influence the selection of the test to be used for data analysis. There are two schools of thought. The first claims that certain tests, such as Student's *t* test and the analysis of variance, should be limited in use to data that are interval or ratio in scaling. The second disagrees, claiming that these tests can also be used with nominal and ordinal data. The issue, however, is too complex to be treated here.*

CONTINUOUS AND DISCRETE VARIABLES

In Chapter 1, we defined a variable as a property or characteristic of something that can take on more than one value. We also distinguished between independent and dependent variables. In addition, variables may be continuous or discrete:

definitions
■ *A **continuous variable** is one that theoretically can have an infinite number of values between adjacent units on the scale.*

■ *A **discrete variable** is one in which there are no possible values between adjacent units on the scale.*

---

*The interested reader should consult N. H. Anderson, "Scales and Statistics: Parametric and Nonparametric," *Psychological Bulletin, 58* (1961), 305–316; F. M. Lord, "On the Statistical Treatment of Football Numbers," *American Psychologist, 8* (1953), 750–751; W. L. Hays, *Statistics for the Social Sciences*, 2nd ed., Holt, Rinehart and Winston, New York, 1973, pp. 87–90; S. Siegel, *Nonparametric Statistics for the Behavioral Sciences*, McGraw-Hill, New York, 1956, pp. 18–20; and S. S. Stevens, "Mathematics, Measurement, and Psychophysics," in *Handbook of Experimental Psychology*, S. S. Stevens, ed., Wiley, New York, 1951, pp. 23–30.

Weight, height, and time are examples of continuous variables. With each of these variables, there are potentially an infinite number of values between adjacent units. If we are measuring time and the smallest unit on the clock that we are using is 1 second, between 1 and 2 seconds there are an infinite number of possible values: 1.1 seconds, 1.01 seconds, 1.001 seconds, and so forth. The same argument could be made for weight and height.

This is not the case with a discrete variable. "Number of children in a family" is an example of a discrete variable. Here the smallest unit is one child, and there are no possible values between one or two children, two or three children, and so on. The characteristic of a discrete variable is that the variable changes in fixed amounts, with no intermediate values possible. Other examples include "number of students in your class," "number of professors at your university," and "number of dates you had last month."

Real Limits of a Continuous Variable

Since a continuous variable may have an infinite number of values between adjacent units on the scale, all measurements made on a continuous variable are *approximate*. Let's use weight to illustrate this point. Suppose you began dieting yesterday. Assume it is spring, heading into summer, and bathing suit weather is just around the corner. Anyway, you weighed yourself yesterday morning, and your weight was shown by the solid needle in Figure 2.2. Assume that the scale shown in the figure has accuracy only to the nearest pound. The weight you record is 180 pounds. This morning, when you weighed yourself after a day of starvation, the pointer was shown by the dashed needle. What weight do you report this time? We know as a humanist that you would like to record 179 pounds, but as a budding scientist, it is *truth* at all costs. Therefore, you again record 180 pounds. When would you be justified in reporting 179 pounds? When the needle is below the halfway point between 179 and 180 pounds. Similarly, you would still record 180 pounds if the needle was above 180 but below the halfway point between 180 and 181 pounds. Thus, any time the weight 180 pounds is recorded, we don't necessarily

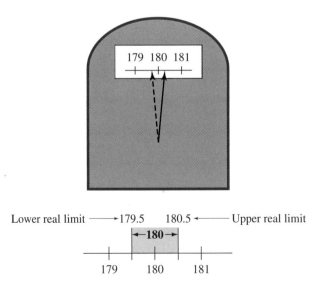

figure 2.2 Real limits of a continuous variable.

mean exactly 180 pounds, but rather that the weight was 180 ± 0.5 pounds. We don't know the exact value of the weight, but we are sure it is in the range 179.5 to 180.5. This range specifies the *real limits* of the weight 180 pounds. The value 179.5 is called the *lower real limit*, and 180.5 is the *upper real limit*.

definition ■ *The* **real limits of a continuous variable** *are those values that are above and below the recorded value by one-half of the smallest measuring unit of the scale.*

To illustrate, if the variable is weight, the smallest unit is 1 pound, and we record 180 pounds, the real limits are above and below 180 pounds by $\frac{1}{2}$ pound. Thus, the real limits are 179.5 and 180.5 pounds.* If the smallest unit were 0.1 pound rather than 1 pound and we recorded 180.0 pounds, then the real limits would be $180 \pm \frac{1}{2}(0.1)$, or 179.95 and 180.05.

Significant Figures

In statistics, we analyze data, and data analysis involves performing mathematical calculations. Very often, we wind up with a decimal remainder (e.g., after doing a division). When this happens, we need to decide to how many decimal places we should carry the remainder.

In the physical sciences, we usually follow the practice of carrying the same number of significant figures as are in the raw data. For example, if we measured the weights of five subjects to three significant figures (173, 156, 162, 165, and 175 pounds) and we were calculating the average of these weights, our answer should also contain only three significant figures. Thus,

$$\overline{X} = \frac{\Sigma X}{N} = \frac{173 + 156 + 162 + 165 + 175}{5} = \frac{831}{5} = 166.2 = 166$$

The answer of 166.2 would be rounded to three significant figures, giving a final answer of 166 pounds. For various reasons, this procedure has not been followed in the behavioral sciences. Instead, a tradition has evolved in which most final values are reported to two or three decimal places, regardless of the number of significant figures in the raw data. Since this is a text for use in the behavioral sciences, we have chosen to follow this tradition. Thus, in this text, we shall report most of our final answers to two decimal places. Occasionally there will be exceptions. For example, correlation and regression coefficients have three decimal places, and probability values are often given to four places, as is consistent with tradition. *It is standard practice to carry all intermediate calculations to two or more decimal places further than will be reported in the final answer.* Thus, when the final answer is required to have two decimal places, you should carry intermediate calculations to at least four decimal places and round the final answer to two places.

---

*Actually, the real limits are 179.500000 … and 180.499999 …, but it is not necessary to be that accurate here.

Rounding

Given that we shall be reporting our final answers to two and sometimes three or four decimal places, we need to decide how we determine what value the last digit should have. Happily, the rules to be followed are rather simple and straightforward:

1. Divide the number you wish to round into two parts: the potential answer and the remainder. The potential answer is the original number extending through the desired number of decimal places. The remainder is the rest of the number.
2. Place a decimal point in front of the first digit of the remainder, creating a decimal remainder.
3. If the decimal remainder is greater than $\frac{1}{2}$, add 1 to the last digit of the answer.
4. If the decimal remainder is less than $\frac{1}{2}$, leave the last digit of the answer unchanged.
5. If the decimal remainder is equal to $\frac{1}{2}$, add 1 to the last digit of the answer if it is an odd digit, but if it is even, leave it unchanged.

MENTORING TIP

Caution: students often have trouble when the remainder is equal to $\frac{1}{2}$. Be sure you can do problems of this type (see rule 5 and the last two rows of Table 2.2).

Let's try a few examples. Round the numbers in the left-hand column of Table 2.2 to two decimal places.

To accomplish the rounding, the number is divided into two parts: the potential answer and the remainder. Since we are rounding to two decimal places, the potential answer ends at the second decimal place. The rest of the number constitutes the remainder. For the first number, 34.01350, 34.01 constitutes the potential answer and .350 the remainder. Since .350 is below $\frac{1}{2}$, the last digit of the potential answer remains unchanged and the final answer is 34.01. For the second number, 34.01761, the decimal remainder (.761) is above $\frac{1}{2}$. Therefore, we must add 1 to the last digit, making the correct answer 34.02. For the next two numbers, the decimal remainder equals $\frac{1}{2}$. The number 45.04500 becomes 45.04 because the last digit in the potential answer is even. The number 45.05500 becomes 45.06 because the last digit is odd.

table 2.2 Rounding

Number	Answer; Remainder	Decimal Remainder	Final Answer	Reason
34.01350	34.01;350	.350	34.01	Decimal remainder is below $\frac{1}{2}$.
34.01761	34.01;761	.761	34.02	Decimal remainder is above $\frac{1}{2}$.
45.04500	45.04;500	.500	45.04	Decimal remainder equals $\frac{1}{2}$, and last digit is even.
45.05500	45.05;500	.500	45.06	Decimal remainder equals $\frac{1}{2}$, and last digit is odd.

■ SUMMARY

In this chapter, I have discussed basic mathematical and measurement concepts. The topics covered were notation, summation, measuring scales, discrete and continuous variables, and rounding. In addition, I pointed out that to do well in statistics, you do not need to be a mathematical whiz. If you have a sound knowledge of elementary algebra, do lots of problems, pay special attention to the symbols, and keep up, you should achieve a thorough understanding of the material.

■ IMPORTANT NEW TERMS

Continuous variable (p. 35)
Discrete variable (p. 35)
Interval scale (p. 32)

Nominal scale (p. 31)
Ordinal scale (p. 32)
Ratio scale (p. 33)

Real limits of a continuous
    variable (p. 36)
Summation (p. 27)

■ QUESTIONS AND PROBLEMS

1. Define and give an example of each of the terms in the Important New Terms section.
2. Identify which of the following represent continuous variables and which are discrete variables:
   a. Time of day
   b. Number of females in your class
   c. Number of bar presses by a rat in a Skinner box
   d. Age of subjects in an experiment
   e. Number of words remembered
   f. Weight of food eaten
   g. Percentage of students in your class who are females
   h. Speed of runners in a race
3. Identify the scaling of each of the following variables:
   a. Number of bicycles ridden by students in the freshman class
   b. Types of bicycles ridden by students in the freshman class
   c. The IQ of your teachers (assume equal interval scaling) — interval — $0 \neq$ no intelligence
   d. Proficiency in mathematics, graded in the categories of poor, fair, and good
   e. Anxiety over public speaking, scored on a scale of 0–100 (Assume the difference in anxiety between adjacent units throughout the scale is not the same.)
   f. The weight of a group of dieters
   g. The time it takes to react to the sound of a tone
   h. Proficiency in mathematics, scored on a scale of 0–100 (The scale is well standardized and can be thought of as having equal intervals between adjacent units.) interval $(0 \neq$ no proficiency$)$
   i. Ratings of professors by students on a 50-point scale (There is an insufficient basis for assuming equal intervals between adjacent units.)
4. A student is measuring assertiveness with an interval scale. Is it correct to say that a score of 30 on the scale represents half as much assertiveness as a score of 60? Explain.
5. For each of the following sets of scores, find $\sum_{i=1}^{N} X_i$:
   a. 2, 4, 5, 7
   b. 2.1, 3.2, 3.6, 5.0, 7.2
   c. 11, 14, 18, 22, 25, 28, 30
   d. 110, 112, 115, 120, 133

6. Round the following numbers to two decimal places:
   a. 14.53670
   b. 25.26231
   c. 37.83500
   d. 46.50499
   e. 52.46500
   f. 25.48501
7. Determine the real limits of the following values:
   a. 10 pounds (assume the smallest unit of measurement is 1 pound)
   b. 2.5 seconds (assume the smallest unit of measurement is 0.1 second)
   c. 100 grams (assume the smallest unit of measurement is 10 grams)
   d. 2.01 centimeters (assume the smallest unit of measurement is 0.01 centimeter)
   e. 5.232 seconds (assume the smallest unit of measurement is 1 millisecond)
8. Find the values of the expressions listed here:
   a. Find $\sum_{i=1}^{4} X_i$ for the scores $X_1 = 3, X_2 = 5, X_3 = 7, X_4 = 10$.
   b. Find $\sum_{i=1}^{6} X_i$ for the scores $X_1 = 2, X_2 = 3, X_3 = 4, X_4 = 6, X_5 = 9, X_6 = 11, X_7 = 14$.
   c. Find $\sum_{i=2}^{N} X_i$ for the scores $X_1 = 10, X_2 = 12, X_3 = 13, X_4 = 15, X_5 = 18$.
   d. Find $\sum_{i=3}^{N-1} X_i$ for the scores $X_1 = 22, X_2 = 24, X_3 = 28, X_4 = 35, X_5 = 38, X_6 = 40$.
9. In an experiment measuring the reaction times of eight subjects, the following scores in milliseconds were obtained:

Subject	Reaction Time
1	250
2	378
3	451
4	275

5	225
6	430
7	325
8	334

a. If X represents the variable of reaction time, assign each of the scores its appropriate X_i symbol.
b. Compute ΣX for these data.

10. Represent each of the following with summation notation. Assume the total number of scores is 10.
   a. $X_1 + X_2 + X_3 + X_4 + \ldots + X_{10}$
   b. $X_1 + X_2 + X_3$
   c. $X_2 + X_3 + X_4$
   d. $X_2^2 + X_3^2 + X_4^2 + X_5^2$

11. Round the following numbers to one decimal place:
   a. 1.423
   b. 23.250
   c. 100.750
   d. 41.652
   e. 35.348

12. For each of the sets of scores given in Problems 5b and 5c, show that $\Sigma X^2 \neq (\Sigma X)^2$.

13. Given the scores $X_1 = 3$, $X_2 = 4$, $X_3 = 7$, and $X_4 = 12$, find the values of the following expressions. (This question pertains to Note 2.1.)

   a. $\displaystyle\sum_{i=1}^{N} (X_i + 2)$

   b. $\displaystyle\sum_{i=1}^{N} (X_i - 3)$

   c. $\displaystyle\sum_{i=1}^{N} (2X_i)$

   d. $\displaystyle\sum_{i=1}^{N} (X_i/4)$

14. Round each of the following numbers to one decimal place and two decimal places.
   a. 4.1482
   b. 4.1501
   c. 4.1650
   d. 4.1950

■ SPSS ILLUSTRATIVE EXAMPLE 2.1

Before beginning the SPSS instruction contained in the various chapters, I suggest you read Appendix E, *Introduction to SPSS*. It describes the general operation of SPSS and its procedures for data entry. Reading Appendix E will help you understand the material that follows.

   In this illustrative example, I will present a detailed description of how to use SPSS to compute the sum of a set of scores. Included are screen shots of the computer display you will see if you are actually performing the steps of data entry and data analysis using the SPSS software instead of just reading about them. The screen shots will vary somewhat depending on the version of SPSS that you are using. In the remaining chapters, due to space limitations, I am not able to include screen shots and hence will use a different formatting. It is true that we don't need SPSS to compute the sum of a small set of scores like in the example below. However, this simple example is a good place to begin learning how SPSS works.

example

Use SPSS to find the sum of the following set of scores. As part of the solution, name the set of scores *Height*.

12, 16, 14, 19, 25, 13, 18.

SOLUTION

STEP 1: Enter the Data.

Assume that you are in the Data Editor, Data View with a blank table displayed and the cursor located in the first cell of the first column. To enter the scores,

1. Type 12 and then **press** Enter.

12.00 is entered into the first cell of the first column; after entry, the cursor moves down one cell, and at the heading of the column SPSS changes the column name from **var** to **VAR00001**.

2. Type 16 and then **press** Enter.
3. Type 14 and then **press** Enter.
4. Type 19 and then **press** Enter.
5. Type 25 and then **press** Enter.
6. Type 13 and then **press** Enter.
7. Type 18 and then **press** Enter.

16.00 is entered into the cell and the cursor moves down one cell.
14.00 is entered into the cell and the cursor moves down one cell.
19.00 is entered into the cell and the cursor moves down one cell.
25.00 is entered into the cell and the cursor moves down one cell.
13.00 is entered into the cell and the cursor moves down one cell.
18.00 is entered into the cell and the cursor moves down one cell.

The Data Editor, Data View with the entered scores is shown below.

STEP 2: Name the Variables. In using SPSS, you don't have to give the variables names. When you enter the data, SPSS automatically gives the scores entered into each column a default name (**VAR00001**, **VAR00002**, etc.). Since we entered the scores in the first column, SPSS automatically gave these scores the default name **VAR00001**. However, rather than use the name **VAR00001**, this example asks us to give the set of scores the name Height. Let's see how to do this.

1. Click the **Variable View** tab in the lower left corner of the Data Editor.

This displays the **Variable View** on screen with the first cell in the **Name** column containing **VAR00001**.

2. Click VAR00001: type Height in the highlighted cell and then **press Enter**.

Height is entered as the variable name, replacing **VAR00001**.

STEP 3:   **Analyze the Data.** The next step is to analyze the data. This example asks that we use SPSS to compute the sum of the Height scores. Although you can analyze the data from the Data Editor–Variable View, I recommend as a general rule that you switch to the Data Editor–Data View for analyzing data. This allows you to better conceptually link the data to the analysis results. If you are in the Data Editor–Variable View and want to switch to the Data Editor–Data View, **click** the **Variable View** tab on the lower left corner of the Data Editor–Variable View. In this and the remaining SPSS material in the remaining chapters, I will assume that you are in the Data Editor–Data View prior to doing the analysis. Assuming you are ready to do the analysis, let's do it!!

1. **Click** **Analyze** on the menu bar at the top of the screen; then **select** **Descriptive Statistics**; then, **click** **Descriptives…**.

This tells SPSS that you want to compute some descriptive statistics, and SPSS responds by displaying the **Descriptives** dialog box, which is shown below. The large box on the left displays the variable names of all the scores that are included in the Data Editor. In this example, **Height** is the only variable; it is displayed in the large box on the left, highlighted.

2. **Click** the **arrow** in the middle of the dialog box.

This moves **Height** from the large box on the left into the **Variable(s)** box on the right. If the data had multiple variables, they would be listed in this box and you would select the variable(s) that you wanted analyzed before clicking the **arrow**. SPSS would then move all selected variables to the **Variable(s)** box on the right. SPSS analyses are carried out only on the variables located in the designated box. In this example, the designated box is the **Variable(s)** box.

3. **Click** **Options …** in the right hand corner of the dialog box.

This produces the **Descriptives: Options** dialog box, which is shown below. This dialog box allows you to select among the various descriptive statistics that SPSS can compute. A couple of other options are included that need not concern us here. The checked entries are the default descriptive statistics that SPSS computes.

4. Click all statistics with **checks** to remove the default **check** entries.

The **checked** entries tell SPSS which statistics to compute. If a box is already **checked**, **clicking** it removes the check. If the box is **unchecked**, **clicking** it puts a **check** in the box. We removed all default checks because for this example we want to compute only the sum of the scores, and none of the default statistics do this.

5. Click Sum.

This produces a **check** in the **Sum** box and tells SPSS to compute the sum of the **Height** scores when given the **OK** to do so. SPSS is now ready to compute the sum of the **Height** scores when it gets the **OK** command, which is located on the **Descriptives** dialog box.

6. Click Continue.

This returns you to the **Descriptives** dialog box so you can give the **OK** command.

7. Click OK.

After getting the **OK** command, SPSS then analyzes the data and displays the results in the **Descriptive Statistics** table displayed on the following page.

Analysis Results

Descriptive Statistics

	N	Sum
Height	7	117.00
Valid N (listwise)	7	

The results for this problem show that the sum of the **Height** scores is **117.00**. The table also shows that **N** = **7**. When N is small, using SPSS to obtain this result may not seem worth the work. However, suppose that you also want to compute the mean, standard deviation, and variance of the scores. To make the case even more compelling, suppose instead of having only one variable, you had five variables you wanted to analyze. In addition to checking the **Sum** box, all you need do is check the boxes for these statistics as well, move all of the variable names into the designated box, and voila, the sum, mean, standard deviation, and variance of all five variables would be computed and displayed. I hope you are beginning to see what a wonderful help SPSS can be in doing data analysis.

■ SPSS ADDITIONAL PROBLEMS

1. Use SPSS to find the sum for each of the set of scores shown below. In solving this problem, do not give the scores a new name.
   a. 12, 15, 18, 14, 13, 18, 17, 23, 22, 14, 10.
   b. 1.3, 0.9, 2.2, 2.4, 3.2, 5.6, 7.8, 3.3, 2.6.
   c. 214, 113, 115, 314, 215, 423, 500, 125, 224, 873.

2. For the following set of scores, use SPSS to find $\sum_{i=1}^{N} X_i$. As part of the solution, name the variable X.
   a. 2.44, 3.57, 6.43, 3.21, 8.45, 6.37, 8.25, 3.98.
   b. 23, 65, 43, 87, 89, 64, 59, 67, 53, 34, 21, 18, 28.

■ NOTES

2.1 Many textbooks present a discussion of additional summation rules, such as the summation of a variable plus a constant, summation of a variable times a constant, and so forth. Since this knowledge is not necessary for understanding any of the material in this textbook, I have not included it in the main body but have presented the material here. Knowledge of summation rules may come in handy as background for statistics courses taught at the graduate level.

Rule 1 The sum of the values of a variable plus a constant is equal to the sum of the values of the variable plus N times the constant. In equation form,

$$\sum_{i=1}^{N} (X_i + a) = \sum_{i=1}^{N} X_i + Na$$

The validity of this equation can be seen from the following simple algebraic proof:

$$\sum_{i=1}^{N} (X_i + a) = (X_1 + a) + (X_2 + a) + (X_3 + a)$$

$$+ \cdots + (X_N + a)$$

$$= (X_1 + X_2 + X_3 + \cdots + X_N)$$

$$+ (a + a + a + \cdots + a)$$

$$= \sum_{i=1}^{N} X_i + Na$$

To illustrate the use of this equation, suppose we wish to find the sum of the following scores with a constant of 3 added to each score:

X: 4, 6, 8, 9

$$\sum_{i=1}^{N} (X_i + 3) = \sum_{i=1}^{N} X_i + Na = 27 + 4(3) = 39$$

Rule 2 The sum of the values of a variable minus a constant is equal to the sum of the values of the variable minus N times the constant. In equation form,

$$\sum_{i=1}^{N} (X_i - a) = \sum_{i=1}^{N} X_i - Na$$

The algebraic proof of this equation is as follows:

$$\sum_{i=1}^{N} (X_i - a) = (X_1 - a) + (X_2 - a) + (X_3 - a)$$

$$+ \cdots + (X_N - a)$$

$$= (X_1 + X_2 + X_3 + \cdots + X_N)$$

$$+ (-a - a - a - a - \cdots - a)$$

$$= \sum_{i=1}^{N} X_i - Na$$

To illustrate the use of this equation, suppose we wish to find the sum of the following scores with a constant of 2 subtracted from each score:

X: 3, 5, 6, 10

$$\sum_{i=1}^{N} (X_i - 2) = \sum_{i=1}^{N} X_i - Na = 24 - 4(2) = 16$$

Rule 3 The sum of a constant multiplied by the value of a variable is equal to the constant multiplied by the sum of the values of the variable. In equation form,

$$\sum_{i=1}^{N} aX_i = a \sum_{i=1}^{N} X_i$$

The validity of this equation is shown here:

$$\sum_{i=1}^{N} aX_i = aX_1 + aX_2 + aX_3 + \cdots + aX_N$$

$$= a(X_1 + X_2 + X_3 + \cdots + X_N)$$

$$= a \sum_{i=1}^{N} X_i$$

To illustrate the use of this equation, suppose we wish to determine the sum of 4 times each of the following scores:

X: 2, 5, 7, 8, 12

$$\sum_{i=1}^{N} 4X_i = 4 \sum_{i=1}^{N} X_i = 4(34) = 136$$

Rule 4 The sum of a constant divided into the values of a variable is equal to the constant divided into the sum of the values of the variable. In equation form,

$$\sum_{i=1}^{N} \frac{X_i}{a} = \frac{\sum_{i=1}^{N} X_i}{a}$$

The validity of this equation is shown here:

$$\sum_{i=1}^{N} \frac{X_i}{a} = \frac{X_1}{a} + \frac{X_2}{a} + \frac{X_3}{a} + \cdots + \frac{X_N}{a}$$

$$= \frac{X_1 + X_2 + X_3 + \cdots + X_N}{a}$$

$$= \frac{\sum_{i=1}^{N} X_i}{a}$$

Again, let's do an example to illustrate the use of this equation. Suppose we want to find the sum of 4 divided into the following scores:

X: 3, 4, 7, 10, 11

$$\sum_{i=1}^{N} \frac{X_i}{4} = \frac{\sum_{i=1}^{N} X_i}{4} = \frac{35}{4} = 8.75$$

■ ONLINE STUDY RESOURCES

CENGAGE**brain**.com

Login to CengageBrain.com to access the resources your instructor has assigned. For this book, you can access the book's companion website for chapter-specific learning tools including Know and Be Able to Do, practice quizzes, flash cards, and glossaries and a link to Statistics and Research Methods Workshops.

aplia

If your professor has assigned Aplia homework:

1. Sign in to your account
2. Complete the corresponding homework exercises as required by your professor
3. When finished, click "Grade It Now" to see which areas you have mastered and which need more work, and for detailed explanations of every answer.

Visit **www.cengagebrain.com** to access your account and to purchase materials.

3

Frequency Distributions

LEARNING OBJECTIVES

After completing this chapter, you should be able to:

- Define a frequency distribution and explain why it is a useful type of descriptive statistic.
- Contrast ungrouped and grouped frequency distributions.
- Construct a frequency distribution of grouped scores.
- Define and construct relative frequency, cumulative frequency, and cumulative percentage distributions.
- Define and compute percentile point and percentile rank.
- Describe bar graph, histogram, frequency polygon, and cumulative percentage curve, and recognize instances of each.
- Define symmetrical curve, skewed curve, and positive and negative skew, and recognize instances of each.
- Construct stem and leaf diagrams, and state their advantage over histograms.
- Understand the illustrative examples, do the practice problems, and understand the solutions.

© Strmko / Dreamstime.com

INTRODUCTION: UNGROUPED FREQUENCY DISTRIBUTIONS

Let's suppose you have just been handed back your first exam in statistics. You received an 86. Naturally, you are interested in how well you did relative to the other students. You have lots of questions: How many other students received an 86? Were there many scores higher than yours? How many scores were lower? The raw scores from the exam are presented haphazardly in Table 3.1. Although all the scores are shown, it is difficult to make much sense out of them the way they are arranged in the table. A more efficient arrangement, and one that conveys more meaning, is to list the scores with their frequency of occurrence. This listing is called a *frequency distribution*.

definition ■ *A* **frequency distribution** *presents the score values and their frequency of occurrence.* When presented in a table, the score values are listed in rank order, with the lowest score value usually at the bottom of the table.

The scores in Table 3.1 have been arranged into a frequency distribution that is shown in Table 3.2. The data now are more meaningful. First, it is easy to see that there are 2 scores of 86. Furthermore, by summing the appropriate frequencies (*f*), we can determine the number of scores higher and lower than 86. It turns out that there are 15 scores higher and 53 scores lower than your score. It is also quite easy to determine the range of the scores when they are displayed as a frequency distribution. For the statistics test, the scores ranged from 46 to 99. From this illustration, it can be seen that the major purpose of a frequency distribution is to present the scores in such a way to facilitate ease of understanding and interpretation.

table 3.1   Scores from statistics exam (*N* = 70)

95	57	76	93	86	80	89
76	76	63	74	94	96	77
65	79	60	56	72	82	70
67	79	71	77	52	76	68
72	88	84	70	83	93	76
82	96	87	69	89	77	81
87	65	77	72	56	78	78
58	54	82	82	66	73	79
86	81	63	46	62	99	93
82	92	75	76	90	74	67

table 3.2   Scores from Table 3.1 organized into a frequency distribution

Score	f	Score	f	Score	f	Score	f
99	1	85	0	71	1	57	1
98	0	84	1	70	2	56	2
97	0	83	1	69	1	55	0
96	2	82	5	68	1	54	1
95	1	81	2	67	2	53	0
94	1	80	1	66	1	52	1
93	3	79	3	65	2	51	0
92	1	78	2	64	0	50	0
91	0	77	4	63	2	49	0
90	1	76	6	62	1	48	0
89	2	75	1	61	0	47	0
88	1	74	2	60	1	46	1
87	2	73	1	59	0		
86	2	72	3	58	1		

GROUPING SCORES

When there are many scores and the scores range widely, as they do on the statistics exam we have been considering, listing individual scores results in many values with a frequency of zero and a display from which it is difficult to visualize the shape of the distribution and its central tendency. Under these conditions, the individual scores are usually grouped into class intervals and presented as a *frequency distribution of grouped scores*. Table 3.3 shows the statistics exam scores grouped into two frequency distributions, one with each interval being 2 units wide and the other having intervals 19 units wide.

When you are grouping data, one of the important issues is how wide each interval should be. Whenever data are grouped, some information is lost. The wider the interval, the more information lost. For example, consider the distribution shown in Table 3.3 with intervals 19 units wide. Although an interval this large does result in a smooth display (there are no zero frequencies), a lot of information has been lost. For instance, how are the 38 scores distributed in the interval from 76 to 94? Do they fall at 94? Or at 76? Or are they evenly distributed throughout the interval? The point is that we do not know how they are distributed in the interval. We have lost that information by the grouping. Note that the larger the interval, the greater the ambiguity.

It should be obvious that the narrower the interval, the more faithfully the original data are preserved. The extreme case is where the interval is reduced to 1 unit wide and we are back to the individual scores. Unfortunately, when the interval is made too narrow, we encounter the same problems as with individual scores—namely, values with zero frequency and an unclear display of the shape of the distribution and its central tendency. The frequency distribution with intervals 2 units wide, shown in Table 3.3, is an example in which the intervals are too narrow.

table 3.3 Scores from Table 3.1 grouped into class intervals of different widths

Class Interval (width = 2)	f	Class Interval (width = 19)	f
98–99	1	95–113	4
96–97	2	76–94	38
94–95	2	57–75	23
92–93	4	38–56	5
90–91	1		N = 70
88–89	3		
86–87	4		
84–85	1		
82–83	6		
80–81	3		
78–79	5		
76–77	10		
74–75	3		
72–73	4		
70–71	3		
68–69	2		
66–67	3		
64–65	2		
62–63	3		
60–61	1		
58–59	1		
56–57	3		
54–55	1		
52–53	1		
50–51	0		
48–49	0		
46–47	1		
	N = 70		

MENTORING TIP

Using 10 to 20 intervals works well for most distributions.

From the preceding discussion, we can see that in grouping scores there is a trade-off between losing information and presenting a meaningful visual display. To have the best of both worlds, we must choose an interval width neither too wide nor too narrow. In practice, we usually determine interval width by dividing the distribution into 10 to 20 intervals. Over the years, this range of intervals has been shown to work well with most distributions. Within this range, the specific number of intervals used depends on the number and range of the raw scores. Note that the more intervals used, the narrower each interval becomes.

Constructing a Frequency Distribution of Grouped Scores

The steps for constructing a frequency distribution of grouped scores are as follows:

1. Find the range of the scores.
2. Determine the width of each class interval (i).
3. List the limits of each class interval, placing the interval containing the lowest score value at the bottom.
4. Tally the raw scores into the appropriate class intervals.
5. Add the tallies for each interval to obtain the interval frequency.

Let's apply these steps to the data of Table 3.1.

1. *Finding the range.*

$$\text{Range} = \text{Highest score minus lowest score} = 99 - 46 = 53$$

2. *Determining interval width* (i). Let's assume we wish to group the data into approximately 10 class intervals.

$$i = \frac{\text{Range}}{\text{Number of class intervals}} = \frac{53}{10} = 5.3 \quad \textit{(round to 5)}$$

When i has a decimal remainder, we'll follow the rule of rounding i to the same number of decimal places as in the raw scores. Thus, i rounds to 5.

3. *Listing the intervals.* We begin with the lowest interval. The first step is to determine the lower limit of this interval. There are two requirements:

   a. The lower limit of this interval must be such that the interval contains the lowest score.

   b. It is customary to make the lower limit of this interval evenly divisible by i.

   Given these two requirements, the lower limit is assigned the value of the lowest score in the distribution if it is evenly divisible by i. If not, then the lower limit is assigned the next lower value that is evenly divisible by i. In the present example, the lower limit of the lowest interval begins with 45 because the lowest score (46) is not evenly divisible by 5.

   Once the lower limit of the lowest interval has been found, we can list all of the intervals. Since each interval is 5 units wide, the lowest interval ranges from 45 to 49. Although it may seem as though this interval is only 4 units wide, it really is 5. If in doubt, count the units (45, 46, 47, 48, 49). In listing the other intervals, we must be sure that the intervals are continuous and mutually exclusive. By mutually exclusive, we mean that the intervals must be such that no score can be legitimately included in more than one interval. Following these rules, we wind up with the intervals shown in Table 3.4. Note that, consistent with our discussion of real limits in Chapter 2, the class intervals shown in the first column represent apparent limits. The real limits are shown in the second column. The usual practice is to list just the apparent limits of each interval and omit the real limits. We have followed this practice in the remaining examples.

4. *Tallying the scores.* Next, the raw scores are tallied into the appropriate class intervals. Tallying is a procedure whereby one systematically goes through the distribution and for each raw score enters a tally mark next to the interval that contains the score. Thus, for 95 (the first score in Table 3.1), a tally mark is placed in the interval 95–99. This procedure has been followed for all the scores, and the results are shown in Table 3.4.

MENTORING TIP

After completing Step 3, the resulting number of intervals often slightly exceeds the number of intervals specified in Step 2, because the lowest interval and the highest interval usually extend beyond the lowest and highest scores.

table 3.4   Construction of frequency distribution for grouped scores

Class Interval	Real Limits	Tally	f
95–99	94.5–99.5	(score of 95) → ////	4
90–94	89.5–94.5	/N/ /	6
85–89	84.5–89.5	/N/ //	7
80–84	79.5–84.5	/N/ /N/	10
75–79	74.5–79.5	/N/ /N/ /N/ /	16
70–74	69.5–74.5	/N/ ////	9
65–69	64.5–69.5	/N/ //	7
60–64	59.5–64.5	////	4
55–59	54.5–59.5	////	4
50–54	49.5–54.5	//	2
45–49	44.5–49.5	/	1
			N = 70

5. *Summing into frequencies.* Finally, the tally marks are converted into frequencies by adding the tallies within each interval. These frequencies are also shown in Table 3.4.

Practice Problem 3.1

Let's try a practice problem. Given the following 90 scores, construct a frequency distribution of grouped scores having approximately 12 intervals.

112	68	55	33	72	80	35	55	62
102	65	104	51	100	74	45	60	58
92	44	122	73	65	78	49	61	65
83	76	95	55	50	82	51	138	73
83	72	89	37	63	95	109	93	65
75	24	60	43	130	107	72	86	71
128	90	48	22	67	76	57	86	114
33	54	64	82	47	81	28	79	85
42	62	86	94	52	106	30	117	98
58	32	68	77	28	69	46	53	38

SOLUTION

1. *Find the range.* Range = Highest score − Lowest score = 138 − 22 = 116.
2. *Determine the interval width (i):*

$$i = \frac{\text{Range}}{\text{Number of intervals}} = \frac{116}{12} = 9.7 \quad \textit{i rounds to 10}$$

3. *List the limits of each class interval.* Because the lowest score in the distribution (22) is not evenly divisible by i, the lower limit of the lowest interval is 20. Why 20? Because it is the next lowest scale value evenly divisible by 10. The limits of each class interval have been listed in Table 3.5.

4. *Tally the raw scores into the appropriate class intervals.* This has been done in Table 3.5.

5. *Add the tallies for each interval to obtain the interval frequency.* This has been done in Table 3.5.

MENTORING TIP

Note that if tallying is done correctly, the sum of the tallies ($\sum f$) should equal N.

table 3.5 Frequency distribution of grouped scores for Practice Problem 3.1

Class Interval	Tally	f	Class Interval	Tally	f
130–139	//	2	70–79	/// /// ///	13
120–129	//	2	60–69	/// /// ///	15
110–119	///	3	50–59	/// /// //	12
100–109	/// /	6	40–49	/// ///	8
90–99	/// //	7	30–39	/// //	7
80–89	/// /// /	11	20–29	////	4
					$N = 90$

Practice Problem 3.2

Given the 130 scores shown here, construct a frequency distribution of grouped scores having approximately 15 intervals.

1.4	2.9	3.1	3.2	2.8	3.2	3.8	1.9	2.5	4.7
1.8	3.5	2.7	2.9	3.4	1.9	3.2	2.4	1.5	1.6
2.5	3.5	1.8	2.2	4.2	2.4	4.0	1.3	3.9	2.7
2.5	3.1	3.1	4.6	3.4	2.6	4.4	1.7	4.0	3.3
1.9	0.6	1.7	5.0	4.0	1.0	1.5	2.8	3.7	4.2
2.8	1.3	3.6	2.2	3.5	3.5	3.1	3.2	3.5	2.7
3.8	2.9	3.4	0.9	0.8	1.8	2.6	3.7	1.6	4.8
3.5	1.9	2.2	2.8	3.8	3.7	1.8	1.1	2.5	1.4
3.7	3.5	4.0	1.9	3.3	2.2	4.6	2.5	2.1	3.4
1.7	4.6	3.1	2.1	4.2	4.2	1.2	4.7	4.3	3.7
1.6	2.8	2.8	2.8	3.5	3.7	2.9	3.5	1.0	4.1
3.0	3.1	2.7	2.2	3.1	1.4	3.0	4.4	3.3	2.9
3.2	0.8	3.2	3.2	2.9	2.6	2.2	3.6	4.4	2.2

(continued)

SOLUTION

1. *Find the range.* Range = Highest score − Lowest score = 5.0 − 0.6 = 4.4.
2. *Determine the interval width (i):*

$$i = \frac{\text{Range}}{\text{Number of intervals}} = \frac{4.4}{15} = 0.29 \quad i \text{ rounds to } 0.3$$

3. *List the limits of each class interval.* Since the lowest score in the distribution (0.6) is evenly divisible by i, it becomes the lower limit of the lowest interval. The limits of each class interval are listed in Table 3.6.
4. *Tally the raw scores into the appropriate class intervals.* This has been done in Table 3.6.
5. *Add the tallies for each interval to obtain the interval frequency.* This has been done in Table 3.6. Note that since the smallest unit of measurement in the raw scores is 0.1, the real limits for any score are ± 0.05 away from the score. Thus, the real limits for the interval 4.8–5.0 are 4.75–5.05.

table 3.6 Frequency distribution of grouped scores for Practice Problem 3.2

Class Interval	Tally	f	Class Interval	Tally	f
4.8–5.0	//	2	2.4–2.6	/N/ /N/	10
4.5–4.7	/N/	5	2.1–2.3	/N/ ////	9
4.2–4.4	/N/ ///	8	1.8–2.0	/N/ ////	9
3.9–4.1	/N/ /	6	1.5–1.7	/N/ ///	8
3.6–3.8	/N/ /N/ /	11	1.2–1.4	/N/ /	6
3.3–3.5	/N/ /N/ /N/ /	16	0.9–1.1	////	4
3.0–3.2	/N/ /N/ /N/ /	16	0.6–0.8	///	3
2.7–2.9	/N/ /N/ /N/ //	17			$N = 130$

Relative Frequency, Cumulative Frequency, and Cumulative Percentage Distributions

It is often desirable to express the data from a frequency distribution as a relative frequency, a cumulative frequency, or a cumulative percentage distribution.

definitions ■ A **relative frequency distribution** *indicates the proportion of the total number of scores that occurs in each interval.*

■ A **cumulative frequency distribution** *indicates the number of scores that fall below the upper real limit of each interval.*

■ A **cumulative percentage distribution** *indicates the percentage of scores that fall below the upper real limit of each interval.*

table 3.7 Relative frequency, cumulative frequency, and cumulative percentage distributions for the grouped scores in Table 3.4

Class Interval	f	Relative f	Cumulative f	Cumulative %
95–99	4	0.06	70	100
90–94	6	0.09	66	94.29
85–89	7	0.10	60	85.71
80–84	10	0.14	53	75.71
75–79	16	0.23	43	61.43
70–74	9	0.13	27	38.57
65–69	7	0.10	18	25.71
60–64	4	0.06	11	15.71
55–59	4	0.06	7	10.00
50–54	2	0.03	3	4.29
45–49	1	0.01	1	1.43
	70	1.00		

Table 3.7 shows the frequency distribution of statistics exam scores expressed as relative frequency, cumulative frequency, and cumulative percentage distributions. To convert a frequency distribution into a relative frequency distribution, the frequency for each interval is divided by the total number of scores. Thus,

$$\text{Relative } f = \frac{f}{N}$$

For example, the relative frequency for the interval 45–49 is found by dividing its frequency (1) by the total number of scores (70). Thus, the relative frequency for this interval $= \frac{1}{70} = 0.01$. The relative frequency is useful because it tells us the proportion of scores contained in the interval.

The cumulative frequency for each interval is found by adding the frequency of that interval to the frequencies of all the class intervals below it. Thus, the cumulative frequency for the interval 60–64 $= 4 + 4 + 2 + 1 = 11$.

The cumulative percentage for each interval is found by converting cumulative frequencies to cumulative percentages. The equation for doing this is:

$$\text{cum\%} = \frac{\text{cum } f}{N} \times 100$$

For the interval 60–64,

$$\text{cum\%} = \frac{\text{cum } f}{N} \times 100 = \frac{11}{70} \times 100 = 15.71\%$$

Cumulative frequency and cumulative percentage distributions are especially useful for finding percentiles and percentile ranks.

PERCENTILES

Percentiles are measures of relative standing. They are used extensively in education to compare the performance of an individual to that of a reference group. Thus, the

60th percentile point is the value on the measurement scale below which 60% of the scores in the distribution fall.

definition ■ A **percentile** or **percentile point** is the value on the measurement scale below which a specified percentage of the scores in the distribution fall.

Computation of Percentile Points

MENTORING TIP

Caution: many students find this section and the one following on Percentile Rank more difficult than the other sections. Be prepared to expand more effort on these sections if needed.

Let's assume we are interested in computing the 50th percentile point for the statistics exam scores. The scores have been presented in Table 3.8 as cumulative frequency and cumulative percentage distributions. We shall use the symbol P_{50} to stand for the 50th percentile point. What do we mean by the 50th percentile point? From the definition of percentile point, P_{50} is the scale value below which 50% of the scores fall. Since there are 70 scores in the distribution, P_{50} must be the value below which 35 scores fall (50% of 70 = 35). Looking at the cumulative frequency column and moving up from the bottom, we see that P_{50} falls in the interval 75–79. At this stage, however, we do not know what scale value to assign P_{50}. All we know is that it falls somewhere between the real limits of the interval 75–79, which are 74.5 to 79.5. To find where in the interval P_{50} falls, we make the assumption that all the scores in the interval are equally distributed throughout the interval.

Since 27 of the scores fall below a value of 74.5, we need to move into the interval until we acquire 8 more scores (Figure 3.1). Because there are 16 scores in the interval and the interval is 5 scale units wide, each score in the interval takes up $\frac{5}{16}$ of a unit. To acquire 8 more scores, we need to move into the interval $\frac{5}{16} \times 8 = 2.5$ units. Adding 2.5 to the lower limit of 74.5, we arrive at P_{50}. Thus,

$$P_{50} = 74.5 + 2.5 = 77.0$$

table 3.8 Computation of percentile points for the scores of Table 3.1

Class Interval	f	Cum f	Cum %	Percentile Computation
95–99	4	70	100	Percentile point $= X_L + (i/f_i)(\text{cum} f_P - \text{cum} f_L)$
90–94	6	66	94.29	
85–89	7	60	85.71	
80–84	10	53	75.71	
75–79	16	43	61.43	$P_{50} = 74.5 + \left(\frac{5}{16}\right)(35 - 27) = 77.00$
70–74	9	27	38.57	
65–69	7	18	25.71	$P_{20} = 64.5 + \left(\frac{5}{7}\right)(14 - 11) = 66.64$
60–64	4	11	15.71	
55–59	4	7	10.00	
50–54	2	3	4.29	
45–49	1	1	1.43	

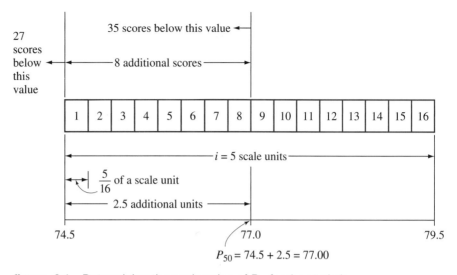

figure 3.1   Determining the scale value of P_{50} for the statistics exam scores.

From *Statistical Reasoning in Psychology and Education* by E.W. Minium. Copyright © 1978 John Wiley & Sons, Inc. Adapted by permission.

To find any percentile point, follow these steps:

1. *Determine the frequency of scores below the percentile point.* We will symbolize this frequency as "cum f_P."

$$\text{cum } f_P = (\% \text{ of scores below the percentile point}) \times N$$
$$\text{cum } f_P \text{ for } P_{50} = 50\% \times N = (0.50) \times 70 = 35$$

2. *Determine the lower real limit of the interval containing the percentile point.* We will call the real lower limit X_L. Knowing the number of scores below the percentile point, we can locate the interval containing the percentile point by comparing cum f_P with the cumulative frequency for each interval. Once the interval containing the percentile point is located, we can immediately ascertain its lower real limit, X_L. For this example, the interval containing P_{50} is 75–79 and its real lower limit, $X_L = 74.5$.

3. *Determine the number of additional scores we must acquire in the interval to reach the percentile point.*

$$\text{Number of additional scores} = \text{cum } f_P - \text{cum } f_L$$

where   cum f_L = frequency of scores below the lower real limit of the interval containing the percentile point.

For the preceding example,

$$
\begin{aligned}
\text{Number of additional scores} &= \text{cum } f_P - \text{cum } f_L \\
&= 35 - 27 \\
&= 8
\end{aligned}
$$

4. *Determine the number of additional units into the interval we must go to acquire the additional number of scores.*

$$
\begin{aligned}
\text{Additional units} &= (\text{Number of units per score}) \times \text{Number of additional scores} \\
&= (i/f_i) \times \text{Number of additional scores} \\
&= \left(\tfrac{5}{16}\right) \times 8 \\
&= 2.5
\end{aligned}
$$

Note that

f_i is the number of scores in the interval and
i/f_i gives us the number of units per score for the interval

5. *Determine the percentile point.* This is accomplished by adding the additional units to the lower real limit of the interval containing the percentile point.

$$\text{Percentile point} = X_L + \text{Additional units}$$
$$P_{50} = 74.5 + 2.5 = 77.00$$

These steps can be put into equation form. Thus,

$$\text{Percentile point} = X_L + (i/f_i)(\text{cum } f_P - \text{cum } f_L)\text{*}$$

*equation for computing
percentile point*

where

X_L = value of the lower real limit of the interval containing the percentile point
cum f_p = frequency of scores below the percentile point
cum f_L = frequency of scores below the lower real limit of the interval containing the percentile point
f_i = frequency of the interval containing the percentile point
i = width of the interval

Using this equation to calculate P_{50}, we obtain

$$\text{Percentile point} = X_L + (i/f_i)(\text{cum } f_P - \text{cum } f_L)$$

$$P_{50} = 74.5 + (\tfrac{5}{16})(35 - 27)$$
$$= 74.5 + 2.5 = 77.00$$

Practice Problem 3.3

Let's try another problem. This time we'll calculate P_{20}, the value below which 20% of the scores fall.

In terms of cumulative frequency, P_{20} is the value below which 14 scores fall (20% of 70 = 14). From Table 3.8 (p. 56), we see that P_{20} lies in the interval 65–69. Since 11 scores fall below a value of 64.5, we need 3 more scores. Given there are 7 scores in the interval and the interval is 5 units wide, we must move $\frac{5}{7} \times 3 = 2.14$ units into the interval. Thus,

$$P_{20} = 64.5 + 2.14 = 66.64$$

P_{20} could also have been found directly by using the equation for percentile point. Thus,

$$\text{Percentile point} = X_L + (i/f_i)(\text{cum } f_P - \text{cum } f_L)$$

$$P_{20} = 64.5 + (\tfrac{5}{7})(14 - 11)$$

$$= 64.5 + 2.14 = 66.64$$

---

*I am indebted to LeAnn Wilson for suggesting this form of the equation.

Practice Problem 3.4

Let's try one more problem. This time let's compute P_{75}. P_{75} is the scale value below which 75% of the scores fall.

In terms of cumulative frequency, P_{75} is the scale value below which 52.5 scores fall (cum f_P = 75% of 70 = 52.5). From Table 3.8 (p. 56), we see that P_{75} falls in the interval 80–84. Since 43 scores fall below this interval's lower limit of 79.5, we need to add to 79.5 the number of scale units appropriate for 52.5 – 43 = 9.5 more scores. Since there are 10 scores in this interval and the interval is 5 units wide, we need to move into the interval $\frac{5}{10} \times 9.5 = 4.75$ units. Thus,

$$P_{75} = 79.5 + 4.75 = 84.25$$

P_{75} also could have been found directly by using the equation for percentile point. Thus,

$$\text{Percentile point} = X_L + (i/f_i)(\text{cum } f_P - \text{cum } f_L)$$

$$P_{75} = 79.5 + (\tfrac{5}{10})(52.5 - 43)$$

$$= 79.5 + 4.75$$

$$= 84.25$$

PERCENTILE RANK

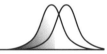

Sometimes we want to know the percentile rank of a raw score. For example, since your score on the statistics exam was 86, it would be useful to you to know the percentile rank of 86.

definition ■ *The **percentile rank** of a score is the percentage of scores with values lower than the score in question.*

Computation of Percentile Rank

This situation is just the reverse of calculating a percentile point. Now, we are given the score and must calculate the percentage of scores below it. Again, we must assume that the scores within any interval are evenly distributed throughout the interval. From the class interval column of Table 3.9, we see that the score of 86 falls in the interval 85–89. There are 53 scores below 84.5, the lower limit of this interval. Since there are 7 scores in the interval and the interval is 5 scale units wide, there are $\frac{7}{5}$ scores per scale unit. Between a score of 86 and 84.5, there are $(\frac{7}{5})(86 - 84.5) = 2.1$ additional scores. There are, therefore, a total of 53 + 2.1 = 55.1 scores below 86. Since there are 70 scores in the distribution, the percentile rank of 86 = $(\frac{55.1}{70}) \times 100 = 78.71$.

These operations are summarized in the following equation:

$$\text{Percentile rank} = \frac{\text{cum } f_L + (f_i/i)(X - X_L)}{N} \times 100 \qquad \textit{equation for computing percentile rank}$$

where   cum f_L = frequency of scores below the lower real limit of the interval containing the score X

X = score whose percentile rank is being determined

X_L = scale value of the lower real limit of the interval containing the score X

i = interval width

f_i = frequency of the interval containing the score X

N = total number of raw scores

Using this equation to find the percentile rank of 86, we obtain

$$\text{Percentile rank} = \frac{\text{cum}\, f_L + (f_i/i)(X - X_L)}{N} \times 100$$

$$= \frac{53 + (\frac{7}{5})(86 - 84.5)}{70} \times 100$$

$$= \frac{53 + 2.1}{70} \times 100$$

$$= \frac{55.1}{70} \times 100$$

$$= 78.71$$

Practice Problem 3.5

Let's do another problem for practice. Find the percentile rank of 59.

The score of 59 falls in the interval 55–59. There are 3 scores below 54.5. Since there are 4 scores within the interval, there are $(\frac{4}{5})(59 - 54.5) = 3.6$ scores within the interval below 59. In all, there are $3 + 3.6 = 6.6$ scores below 59. Thus, the percentile rank of 59 = $(\frac{6.6}{70}) \times 100 = 9.43$.

The solution is presented in equation form in Table 3.9.

table 3.9   Computation of percentile rank for the scores of Table 3.1

Class Interval	f	Cum f	Cum %	Percentile Rank Computation
95–99	4	70	100	Percentile rank = $\dfrac{\text{cum}\, f_L + (f_i/i)(X - X_L)}{N} \times 100$
90–94	6	66	94.29	
85–89	7	60	85.71	Percentile rank of 86 = $\dfrac{53 + (\frac{7}{5})(86 - 84.5)}{70} \times 100$
80–84	10	53	75.71	= 78.71
75–79	16	43	61.43	
70–74	9	27	38.57	
65–69	7	18	25.71	
60–64	4	11	15.71	
55–59	4	7	10.00	Percentile rank of 59 = $\dfrac{3 + (\frac{4}{5})(59 - 54.5)}{70} \times 100$
50–54	2	3	4.29	= 9.43
45–49	1	1	1.43	

84.5

Practice Problem 3.6

Let's do one more practice problem. Using the frequency distribution of grouped scores shown in Table 3.5 (p. 53), determine the percentile rank of a score of 117.

The score of 117 falls in the interval 110–119. The lower limit of this interval is 109.5. There are $6 + 7 + 11 + 13 + 15 + 12 + 8 + 7 + 4 = 83$ scores below 109.5. Since there are 3 scores within the interval and the interval is 10 units wide, there are $(\frac{3}{10})(117 - 109.5) = 2.25$ scores within the interval that are below a score of 117. In all, there are $83 + 2.25 = 85.25$ scores below a score of 117. Thus, the percentile rank of $117 = (\frac{85.25}{90}) \times 100 = 94.72$.

This problem could also have been solved by using the equation for percentile rank. Thus,

$$\text{Percentile rank} = \frac{\text{cum} f_L + (f_i/i)(X - X_L)}{N} \times 100$$

$$= \frac{83 + (\frac{3}{10})(117 - 109.5)}{90} \times 100$$

$$= 94.72$$

GRAPHING FREQUENCY DISTRIBUTIONS

Frequency distributions are often displayed as graphs rather than tables. Since a graph is based completely on the tabled scores, the graph does not contain any new information. However, a graph presents the data pictorially, which often makes it easier to see important features of the data. I have assumed, in writing this section, that you are already familiar with constructing graphs. Even so, it is worthwhile to review a few of the important points.

1. A graph has two axes: vertical and horizontal. The vertical axis is called the *ordinate*, or Y axis, and the horizontal axis is the *abscissa*, or X axis.
2. Very often the independent variable is plotted on the X axis and the dependent variable on the Y axis. In graphing a frequency distribution, the score values are usually plotted on the X axis and the frequency of the score values is plotted on the Y axis.
3. Suitable units for plotting scores should be chosen along the axes.
4. To avoid distorting the data, it is customary to set the intersection of the two axes at zero and then choose scales for the axes such that the height of the graphed data is about three-fourths of the width. Figure 3.2 shows how violation of this rule can bias the impression conveyed by the graph. The figure shows two graphs plotted from the same data, namely, enrollment at a large university during the years 1998–2010. Part (a) follows the rule we have just elaborated. In part (b), the scale on the ordinate does not begin at zero and is greatly expanded from that of part (a). The impressions conveyed by the two graphs are very different. Part (a) gives the correct impression of a very stable enrollment, whereas part (b) greatly distorts the data, making them seem as though there were large enrollment fluctuations.

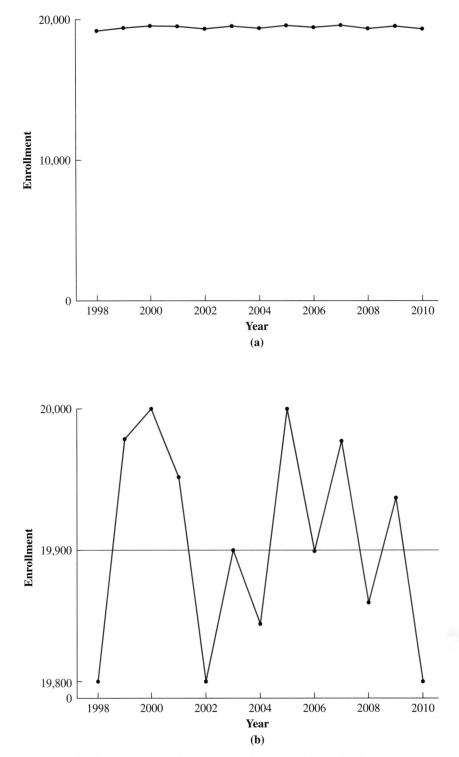

figure 3.2 Enrollment at a large university from 1998 to 2010.

5. Ordinarily, the intersection of the two axes is at zero for both scales. When it is not, this is indicated by breaking the relevant axis near the intersection. For example, in Figure 3.4, the horizontal axis is broken to indicate that a part of the scale has been left off.
6. Each axis should be labeled, and the title of the graph should be both short and explicit.

Four main types of graphs are used to graph frequency distributions: the *bar graph*, the *histogram*, the *frequency polygon*, and the *cumulative percentage* curve.

The Bar Graph

Frequency distributions of nominal or ordinal data are customarily plotted using a bar graph. This type of graph is shown in Figure 3.3. A bar is drawn for each category, where the height of the bar represents the frequency or number of members of that category. Since there is no numerical relationship between the categories in nominal data, the various groups can be arranged along the horizontal axis in any order. In Figure 3.3, they are arranged from left to right according to the magnitude of frequency in each category. Note that the bars for each category in a bar graph do not touch each other. This further emphasizes the lack of a quantitative relationship between the categories.

The Histogram

The histogram is used to represent frequency distributions composed of interval or ratio data. It resembles the bar graph, but with the histogram, a bar is drawn for each class interval. The class intervals are plotted on the horizontal axis such that each class bar begins and terminates at the real limits of the interval. The height of the bar corresponds to the frequency of the class interval. Since the intervals are continuous, the vertical bars must touch each other rather than be spaced apart as is done with the bar graph. Figure 3.4 shows the statistics exam scores (Table 3.4, p. 52) displayed as a histogram. Note that it is customary to plot the midpoint of each class interval on the abscissa. The grouped scores have been presented again in the figure for your convenience.

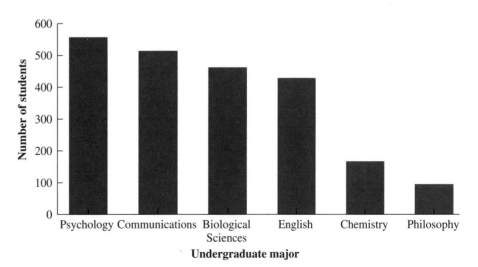

figure 3.3 Bar graph: Students enrolled in various undergraduate majors in a college of arts and sciences.

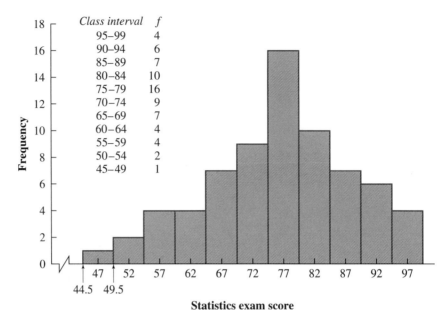

Class interval	f
95–99	4
90–94	6
85–89	7
80–84	10
75–79	16
70–74	9
65–69	7
60–64	4
55–59	4
50–54	2
45–49	1

figure 3.4 Histogram: Statistics exam scores of Table 3.4.

The Frequency Polygon

The frequency polygon is also used to represent interval or ratio data. The horizontal axis is identical to that of the histogram. However, for this type of graph, instead of using bars, a point is plotted over the midpoint of each interval at a height corresponding to the frequency of the interval. The points are then joined with straight lines. Finally, the line joining the points is extended to meet the horizontal axis at the midpoint of the two class intervals falling immediately beyond the end class intervals containing scores. This closing of the line with the horizontal axis forms a polygon, from which the name of this graph is taken.

Figure 3.5 displays the scores listed in Table 3.4 as a frequency polygon. The major difference between a histogram and a frequency polygon is the following: The histogram displays the scores as though they were equally distributed over the interval, whereas the frequency polygon displays the scores as though they were all concentrated at the midpoint of the interval. Some investigators prefer to use the frequency polygon when they are comparing the shapes of two or more distributions. The frequency polygon also has the effect of displaying the scores as though they were continuously distributed, which in many instances is actually the case.

The Cumulative Percentage Curve

Cumulative frequency and cumulative percentage distributions may also be presented in graphical form. We shall illustrate only the latter because the graphs are basically the same and cumulative percentage distributions are more often encountered. You will recall that the cumulative percentage for a class interval indicates the percentage of scores that fall below the upper real limit of the interval. Thus, the vertical axis for the cumulative percentage curve is plotted in cumulative percentage units. On the horizontal axis, instead of plotting points at the midpoint of each class interval, we plot them

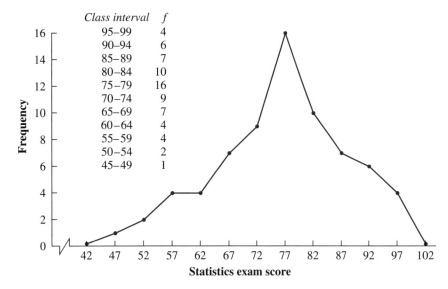

figure 3.5 Frequency polygon: Statistics exam scores of Table 3.4.

at the upper real limit of the interval. Figure 3.6 shows the scores of Table 3.7 (p. 55) displayed as a cumulative percentage curve. It should be obvious that the cumulative frequency curve would have the same shape, the only difference being that the vertical axis would be plotted in cumulative frequency rather than in cumulative percentage units. Both percentiles and percentile ranks can be read directly off the cumulative percentage curve. The cumulative percentage curve is also called an *ogive*, implying an *S* shape.

Shapes of Frequency Curves

Frequency distributions can take many different shapes. Some of the more commonly encountered shapes are shown in Figure 3.7. Curves are generally classified as *symmetrical* or *skewed*.

definitions ■ *A curve is* **symmetrical** *if when folded in half the two sides coincide. If a curve is not symmetrical, it is* **skewed**.

The curves shown in Figure 3.7(a), (b), and (c) are symmetrical. The curves shown in parts (d), (e), and (f) are skewed. If a curve is skewed, it may be *positively* or *negatively skewed*.

definitions ■ *When a curve is* **positively skewed**, *most of the scores occur at the lower values of the horizontal axis and the curve tails off toward the higher end. When a curve is* **negatively skewed**, *most of the scores occur at the higher values of the horizontal axis and the curve tails off toward the lower end.*

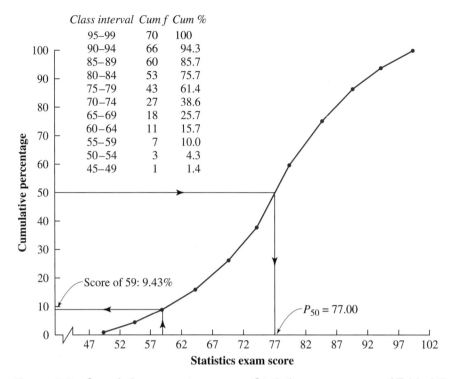

Class interval	Cum f	Cum %
95–99	70	100
90–94	66	94.3
85–89	60	85.7
80–84	53	75.7
75–79	43	61.4
70–74	27	38.6
65–69	18	25.7
60–64	11	15.7
55–59	7	10.0
50–54	3	4.3
45–49	1	1.4

figure 3.6 Cumulative percentage curve: Statistics exam scores of Table 3.7.

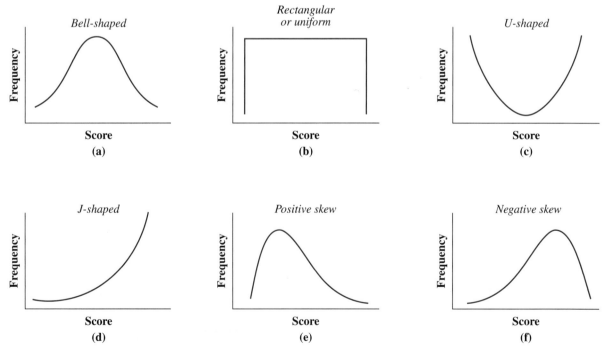

figure 3.7 Shapes of frequency curves.

The curve in part (e) is positively skewed, and the curve in part (f) is negatively skewed.

Frequency curves are often referred to according to their shape. Thus, the curves shown in parts (a), (b), (c), and (d) are, respectively, called *bell-shaped*, rectangular or uniform, *U-shaped*, and *J-shaped* curves.

EXPLORATORY DATA ANALYSIS

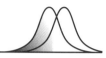

Exploratory data analysis is a recently developed procedure. It employs easy-to-construct diagrams that are quite useful in summarizing and describing sample data. One of the most popular of these is the *stem and leaf diagram*.

Stem and Leaf Diagrams

Stem and leaf diagrams were first developed in 1977 by John Tukey, working at Princeton University. They are a simple alternative to the histogram and are most useful for summarizing and describing data when the data set includes less than 100 scores. Unlike what happens with a histogram, however, a stem and leaf diagram does not lose any of the original data. A stem and leaf diagram for the statistics exam scores of Table 3.1 is shown in Figure 3.8.

In constructing a stem and leaf diagram, each score is represented by a *stem* and a *leaf*. The stem is placed to the left of the vertical line and the leaf to the right. For example, the stems and leafs for the first and last original scores are:

stem	leaf		stem	leaf
9	5		6	7

In a stem and leaf diagram, stems are placed in order vertically down the page, and the leafs are placed in order horizontally across the page. The leaf for each score is usually the last digit, and the stem is the remaining digits. Occasionally, the leaf is the last two digits, depending on the range of the scores.

Note that in stem and leaf diagrams, stem values can be repeated. In Figure 3.8, the stem values are repeated twice. This has the effect of stretching the stem—that is, creating more intervals and spreading the scores out. A stem and leaf diagram for the statistics scores with stem values listed only once is shown here.

4	6
5	2 4 6 6 7 8
6	0 2 3 3 5 5 6 7 7 8 9
7	0 0 1 2 2 2 3 4 4 5 6 6 6 6 6 6 7 7 7 7 8 8 9 9 9
8	0 1 1 2 2 2 2 2 3 4 6 6 7 7 8 9 9
9	0 2 3 3 3 4 5 6 6 9

Listing stem values only once results in fewer, wider intervals, with each interval generally containing more scores. This makes the display appear more crowded. Whether stem values should be listed once, twice, or even more than twice depends on the range of the scores.

Original Scores

95	57	76	93	86	80	89
76	76	63	74	94	96	77
65	79	60	56	72	82	70
67	79	71	77	52	76	68
72	88	84	70	83	93	76
82	96	87	69	89	77	81
87	65	77	72	56	78	78
58	54	82	82	66	73	79
86	81	63	46	62	99	93
82	92	75	76	90	74	67

Stem and Leaf Diagram

4	6
5	2 4
5	6 6 7 8
6	0 2 3 3
6	5 5 6 7 7 8 9
7	0 0 1 2 2 2 3 4 4
7	5 6 6 6 6 6 7 7 7 7 8 8 9 9 9
8	0 1 1 2 2 2 2 2 3 4
8	6 6 7 7 8 9 9
9	0 2 3 3 3 4
9	5 6 6 9

figure 3.8 Stem and leaf diagram: Statistics exam scores of Table 3.1.

You should observe that rotating the stem and leaf diagram of Figure 3.8 counterclockwise 90°, such that the stems are at the bottom, results in a diagram very similar to the histogram shown in Figure 3.4. With the histogram, however, we have lost the original scores; with the stem and leaf diagram, the original scores are preserved.

WHAT IS THE TRUTH?   Stretch the Scale, Change the Tale

An article appeared in the business section of a newspaper, discussing the rate increases of Puget Power & Light Company. The company was in poor financial condition and had proposed still another rate increase in 1984 to try to help it get out of trouble. The issue was particularly sensitive because rate increases had plagued the region recently to pay for huge losses in nuclear power plant construction. The graph at right appeared in the article along with the caption "Puget Power rates have climbed steadily during the past 14 years." Do you notice anything peculiar about the graph?

Answer Take a close look at the *X* axis. From 1970 to 1980, the scale is calibrated in 2-year intervals.  After 1980, the same distance on the *X* axis represents 1 year rather than 2 years. Given the data, stretching this part of the scale gives the false impression that costs have risen "steadily." When plotted properly, as is done in the bottom graph, you can see that rates have not risen steadily, but instead have greatly accelerated over the last 3 years (including the proposed rate increase). Labeling the

rise as "steady" rather than a greatly accelerating increase obviously is in the company's interest. It is unclear whether the company furnished the graph or whether the newspaper constructed its own. In any case, when the axes of graphs are not uniform, reader beware! ■

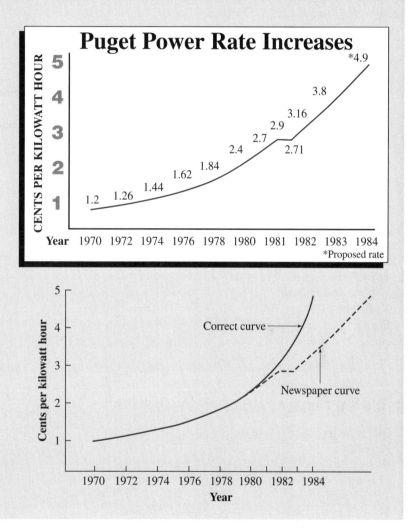

■ SUMMARY

In this chapter, I have discussed frequency distributions and how to present them in tables and graphs. In descriptive statistics, we are interested in characterizing a set of scores in the most meaningful manner. When faced with a large number of scores, it is easier to understand, interpret, and discuss the scores when they are presented as a frequency distribution. A frequency distribution is a listing of the score values in rank order along with their frequency of occurrence. If there are many scores existing over a wide range, the scores are usually grouped together in equal intervals to allow a more meaningful interpretation. The scores can be presented as an ordinary frequency distribution, a relative frequency distribution, a cumulative frequency distribution, or a cumulative percentage distribution. I discussed each of these and how

to construct them. I also presented the concepts of percentile point and percentile rank and discussed how to compute each.

When frequency distributions are graphed, frequency is plotted on the vertical axis and the score value on the horizontal axis. Four main types of graphs are used: the bar graph, the histogram, the frequency polygon, and the cumulative percentage curve. I discussed the use of each type and how to construct them.

Frequency curves can also take on various shapes. I illustrated some of the common shapes encountered (e.g., bell-shaped, U-shaped, and J-shaped) and discussed the difference between symmetrical and skewed curves. Finally, I discussed the use of an exploratory data analysis technique: stem and leaf diagrams.

■ IMPORTANT NEW TERMS

Bar graph (p. 63)
Bell-shaped curve (p. 67)
Cumulative frequency
    distribution (p. 54)
Cumulative percentage curve (p. 64)
Cumulative percentage
    distribution (p. 54)
Exploratory data analysis (p. 67)
Frequency distribution (p. 48)

Frequency distribution of grouped
    scores (p. 49)
Frequency polygon (p. 63)
Histogram (p. 63)
J-shaped curve (p. 67)
Negatively skewed curve (p. 65)
Percentile point (p. 56)
Percentile rank (p. 59)
Positively skewed curve (p. 65)

Relative frequency
    distribution (p. 54)
Skewed curve (p. 65)
Stem and leaf diagrams (p. 67)
Symmetrical curve (p. 65)
U-shaped curve (p. 67)
X axis (abscissa) (p. 61)
Y axis (ordinate) (p. 61)

■ QUESTIONS AND PROBLEMS

1. Define each of the terms in the Important New Terms section.
2. How do bar graphs, histograms, and frequency polygons differ in construction? What type of scaling is appropriate for each?
3. The following table gives the 2002 median annual salaries of various categories of scientists in the United States holding PhDs. Construct a bar graph for these data with "Annual Salary" on the Y axis and "Category of Scientist" on the X axis. Arrange the categories so that the salaries decrease from left to right.

Category of Scientist	Annual Salary ($)
Biological and Health Sciences	70,100
Chemistry	79,100
Computer and Math Sciences	75,000
Psychology	66,700
Sociology and Anthropology	63,100

4. A graduate student has collected data involving 66 scores. Based on these data, he has constructed two

frequency distributions of grouped scores. These are shown here. Do you see anything wrong with these distributions? Explain.

a.

Class Interval	f
48–63	17
29–47	28
10–28	21

b.

Class Interval	f	Class Interval	f
62–63	2	34–35	2
60–61	4	32–33	0
58–59	3	30–31	5
56–57	1	28–29	3
54–55	0	26–27	0
52–53	4	24–25	4
50–51	5	22–23	5
48–49	2	20–21	2
46–47	0	18–19	0
44–45	5	16–17	3
42–43	4	14–15	1
40–41	3	12–13	0
38–39	0	10–11	2
36–37	6		

5. The following scores were obtained by a college sophomore class on an English exam:

60	94	75	82	72	57	92	75	85	77	91
72	85	64	78	75	62	49	70	94	72	84
55	90	88	81	64	91	79	66	68	67	74
45	76	73	68	85	73	83	85	71	87	57
82	78	68	70	71	78	69	98	65	61	83
84	69	77	81	87	79	64	72	55	76	68
93	56	67	71	83	72	82	78	62	82	49
63	73	89	78	81	93	72	76	73	90	76

a. Construct a frequency distribution of the ungrouped scores ($i = 1$).
b. Construct a frequency distribution of grouped scores having approximately 15 intervals. List both the apparent and real limits of each interval.
c. Construct a histogram of the frequency distribution constructed in part **b**.

d. Is the distribution skewed or symmetrical? If it is skewed, is it skewed positively or negatively?
e. Construct a stem and leaf diagram with the last digit being a leaf and the first digit a stem. Repeat stem values twice.
f. Which diagram do you like better, the histogram of part **c** or the stem and leaf diagram of part **e?** Explain. education
6. Express the grouped frequency distribution of part **b** of Problem 5 as a relative frequency, a cumulative frequency, and a cumulative percentage distribution. education
7. Using the cumulative frequency arrived at in Problem 6, determine
   a. P_{75}
   b. P_{40} education
8. Again, using the cumulative distribution and grouped scores arrived at in Problem 6, determine
   a. The percentile rank of a score of 81
   b. The percentile rank of a score of 66
   c. The percentile rank of a score of 87 education
9. Construct a histogram of the distribution of grouped English exam scores determined in Problem 5, part **b.** education
10. The following scores show the amount of weight lost (in pounds) by each client of a weight control clinic during the last year:

10	13	22	26	16	23	35	53	17	32
41	35	24	23	27	16	20	60	48	43
52	31	17	20	33	18	23	8	24	15
26	46	30	19	22	13	22	14	21	39
28	43	37	15	20	11	25	9	15	21
21	25	34	10	23	29	28	18	17	24
16	26	7	12	28	20	36	16	14	
18	16	57	31	34	28	42	19	26	

a. Construct a frequency distribution of grouped scores with approximately 10 intervals.
b. Construct a histogram of the frequency distribution constructed in part **a.**
c. Is the distribution skewed or symmetrical? If it is skewed, is it skewed positively or negatively?
d. Construct a stem and leaf diagram with the last digit being a leaf and the first digit a stem. Repeat stem values twice.
e. Which diagram do you like better, the histogram of part **b** or the stem and leaf diagram of part **d?** Explain. clinical, health
11. Convert the grouped frequency distribution of weight losses determined in Problem 10 to a relative

frequency and a cumulative frequency distribution. clinical, health

12. Using the cumulative frequency distribution arrived at in Problem 11, determine
    a. P_{50}
    b. P_{25} clinical, health

13. Again using the cumulative frequency distribution of Problem 11, determine
    a. The percentile rank of a score of 41
    b. The percentile rank of a score of 28 clinical, health

14. Construct a frequency polygon using the grouped frequency distribution determined in Problem 10. Is the curve symmetrical? If not, is it positively or negatively skewed? clinical, health

15. A small eastern college uses the grading system of 0–4.0, with 4.0 being the highest possible grade. The scores shown here are the grade point averages of the students currently enrolled as psychology majors at the college.

2.7	1.9	1.0	3.3	1.3	1.8	2.6	3.7
3.1	2.2	3.0	3.4	3.1	2.2	1.9	3.1
3.4	3.0	3.5	3.0	2.4	3.0	3.4	2.4
2.4	3.2	3.3	2.7	3.5	3.2	3.1	3.3
2.1	1.5	2.7	2.4	3.4	3.3	3.0	3.8
1.4	2.6	2.9	2.1	2.6	1.5	2.8	2.3
3.3	3.1	1.6	2.8	2.3	2.8	3.2	2.8
2.8	3.8	1.4	1.9	3.3	2.9	2.0	3.2

    a. Construct a frequency distribution of grouped scores with approximately 10 intervals.
    b. Construct a histogram of the frequency distribution constructed in part **a**.
    c. Is the distribution skewed or symmetrical? If skewed, is it skewed positively or negatively?
    d. Construct a stem and leaf diagram with the last digit being a leaf and the first digit a stem. Repeat stem values five times.
    e. Which diagram do you like better, the histogram of part **b** or the stem and leaf diagram of part **d**? Explain. education

16. For the grouped scores in Problem 15, determine
    a. P_{80}
    b. P_{20} education

17. Sarah's grade point average is 3.1. Based on the frequency distribution of grouped scores constructed in Problem 15, part **a**, what is the percentile rank of Sarah's grade point average? education

18. The policy of the school in Problem 15 is that to graduate with a major in psychology, a student must have a grade point average of 2.5 or higher.

    a. Based on the ungrouped scores shown in Problem 15, what percentage of current psychology majors needs to raise its grades?
    b. Based on the frequency distribution of grouped scores, what percentage needs to raise its grades?
    c. Explain the difference between the answers to parts **a** and **b**. education

19. Construct a frequency polygon using the distribution of grouped scores constructed in Problem 15. Is the curve symmetrical or positively or negatively skewed?

20. The psychology department of a large university maintains its own vivarium of rats for research purposes. A recent sampling of 50 rats from the vivarium revealed the following rat weights (grams):

320	282	341	324	340	302	336	265	313	317
310	335	353	318	296	309	308	310	277	288
314	298	315	360	275	315	297	330	296	274
250	274	318	287	284	267	292	348	302	297
270	263	269	292	298	343	284	352	345	325

    a. Construct a frequency distribution of grouped scores with approximately 11 intervals.
    b. Construct a histogram of the frequency distribution constructed in part **a**.
    c. Is the distribution symmetrical or skewed?
    d. Construct a stem and leaf diagram with the last digit being a leaf and the first two digits a stem. Do not repeat stem values.
    e. Which diagram do you like better, the histogram or the stem and leaf diagram? Why? biological

21. Convert the grouped frequency distribution of rat weights determined in Problem 20 to a relative frequency, cumulative frequency, and cumulative percentage distribution. biological

22. Using the cumulative frequency distribution arrived at in Problem 21, determine
    a. P_{50}
    b. P_{75} biological

23. Again using the cumulative frequency distribution arrived at in Problem 21, determine
    a. The percentile rank of a score of 275
    b. The percentile rank of a score of 318 biological

24. A professor is doing research on individual differences in the ability of students to become hypnotized. As part of the experiment, she administers a portion of the Stanford Hypnotic Susceptibility Scale to 85 students who volunteered for the experiment. The results are scored from 0 to 12, with 12 indicating the highest degree of hypnotic susceptibility and 0 the lowest. The scores are shown here.

9	7	11	4	9	7	8	8	10	6
6	4	3	5	5	4	6	2	6	8
10	8	6	7	3	7	1	6	5	3
2	7	6	2	6	9	4	7	9	6
5	9	5	0	5	6	3	6	7	9
7	5	4	2	9	8	11	7	12	3
8	6	5	4	10	7	4	10	8	7
6	2	7	5	3	4	8	6	4	5
4	6	5	8	7					

a. Construct a frequency distribution of the scores.
b. Construct a histogram of the frequency distribution constructed in part **a.**
c. Is the distribution symmetrical or skewed?
d. Determine the percentile rank of a score of 5 and a score of 10. Hint: Use the frequency distribution constructed in part **a** to determine percentile rank.
clinical, cognitive, health

What Is the Truth? Questions

1. *Stretch the Scale, Change the Tale.*
   a. What is the purpose of graphing data? Be sure to include the concept of accuracy in your answer.
   b. In reading articles that present graphs, why is it important to check the X and Y values at the origin of each graph?
   c. Why is it important for the scales of the X and Y axis to be uniform throughout each scale? Is it ever appropriate to use scales that are not uniform? Explain.

■ SPSS ILLUSTRATIVE EXAMPLE 3.1

The general operation of SPSS and its procedures for data entry are described in Appendix E, *Introduction to SPSS*. Chapter 3 of the textbook discusses frequency distributions. SPSS can be a great help in constructing and graphing frequency distributions.

example

For this example, let's use the statistics exam scores given in Table 3.1, p. 48 of the textbook. For convenience the scores are repeated below.

95	57	76	93	86	80	89
76	76	63	74	94	96	77
65	79	60	56	72	82	70
67	79	71	77	52	76	68
72	88	84	70	83	93	76
82	96	87	69	89	77	81
87	65	77	72	56	78	78
58	54	82	82	66	73	79
86	81	63	46	62	99	93
82	92	75	76	90	74	67

a. Use SPSS to construct an ungrouped frequency distribution of the scores.
b. Use SPSS to construct a histogram of the scores. Is the distribution symmetrical?

SOLUTION

STEP 1: **Enter the Data.** Enter the statistics exam scores in the first column (**VAR00001**) of the Data Editor, beginning with the first score in the first cell of the first column. The heading of this column changes from **var** to **VAR00001** after you enter the first score. Data entry can be a bit tedious, especially when *N* is large. Be sure to check the accuracy of your entries. No sense analyzing erroneous data.

STEP 2: **Name the Variables.** This step is optional. If we chose to skip this step, SPSS will use the default variable name that heads the column in which the scores are entered. Since we are just beginning with SPSS instruction, let's skip this step. We will give variables new names in subsequent chapters. Since we are not giving the scores a new name and the scores are entered in the first column of the Data Editor, SPSS will use the default heading of the first column (**VAR00001**) as the name of the scores.

STEP 3: **Analyze the Data.**

Part a. **Construct an Ungrouped Frequency Distribution of the Scores.** If you are currently not displaying the Data Editor–Data View, I suggest you do so before going on. To switch to the Data Editor–Data View from the Data Editor–Variable View, **click** the **Variable View** tab on the lower left corner of the Data Editor–Variable View. Let's now go on with the analysis.

1. Click **A**nalyze on the tool bar at the top of the screen; then **select D**escriptive **Statistics**; then **click F**requencies....	This produces the **Frequencies** dialog box with **VAR00001** located in the large box on the left; **VAR00001** is highlighted. Be sure the box for **Display Frequency Tables** has a **check** in it. One of the functions of the **Frequencies** dialog box is to produce ungrouped frequency distributions.
2. **Click** the **arrow** in the middle of the dialog box.	Since **VAR00001** is already highlighted, clicking the arrow moves **VAR00001** from the large box on the left into the **V**ariable(s) box on the right.
3. **Click OK**.	SPSS then analyzes the data and displays the two tables in the output window that are shown below. One is titled **Statistics** and tells us that *N* = **70**. The other is titled **VAR00001**. It displays ungrouped **Frequency, Percent,** and **Cumulative Percent** distributions for the **VAR00001** data.

Analysis Results

Statistics

VAR00001

N	Valid	70
	Missing	0

VAR00001

		Frequency	Percent	Valid Percent	Cumulative Percent
Valid	46.00	1	1.4	1.4	1.4
	52.00	1	1.4	1.4	2.9
	54.00	1	1.4	1.4	4.3
	56.00	2	2.9	2.9	7.1
	57.00	1	1.4	1.4	8.6
	58.00	1	1.4	1.4	10.0
	60.00	1	1.4	1.4	11.4
	62.00	1	1.4	1.4	12.9
	63.00	2	2.9	2.9	15.7
	65.00	2	2.9	2.9	18.6
	66.00	1	1.4	1.4	20.0
	67.00	2	2.9	2.9	22.9
	68.00	1	1.4	1.4	24.3
	69.00	1	1.4	1.4	25.7
	70.00	2	2.9	2.9	28.6
	71.00	1	1.4	1.4	30.0
	72.00	3	4.3	4.3	34.3
	73.00	1	1.4	1.4	35.7
	74.00	2	2.9	2.9	38.6
	75.00	1	1.4	1.4	40.0
	76.00	6	8.6	8.6	48.6
	77.00	4	5.7	5.7	54.3
	78.00	2	2.9	2.9	57.1
	79.00	3	4.3	4.3	61.4
	80.00	1	1.4	1.4	62.9
	81.00	2	2.9	2.9	65.7
	82.00	5	7.1	7.1	72.9
	83.00	1	1.4	1.4	74.3
	84.00	1	1.4	1.4	75.7
	86.00	2	2.9	2.9	78.6
	87.00	2	2.9	2.9	81.4
	88.00	1	1.4	1.4	82.9
	89.00	2	2.9	2.9	85.7
	90.00	1	1.4	1.4	87.1
	92.00	1	1.4	1.4	88.6
	93.00	3	4.3	4.3	92.9
	94.00	1	1.4	1.4	94.3
	95.00	1	1.4	1.4	95.7
	96.00	2	2.9	2.9	98.6
	99.00	1	1.4	1.4	100.0
	Total	70	100.0	100.0	

It is worth comparing the ungrouped frequency distribution, shown in the second column, with the one shown in Table 3.2, p. 49 of the textbook. They are essentially the same except SPSS does not display any **0** frequency scores, and the scale is inverted.

Part b. Construct a Histogram of the Scores. To construct a histogram of the **VAR00001** scores,

1. **Click <u>A</u>nalyze** on the tool bar at the top of the screen; then **Select D<u>e</u>scriptive Statistics**; then **Click <u>F</u>requencies...**.

This produces the **Frequencies** dialog box which is also used to construct histograms. **VAR00001** is in the large box on the right, because you moved it there in part **a**. If for some reason it is not there, please do so before moving on.

2. **Click Display Frequency Tables**.

This removes the **check** in the **Display Frequency Tables** box to prevent unwanted frequency tables from being displayed as output. If there is no check in this box, skip this step. Don't worry about the warning message if it comes up. Just click **OK** on it.

4. **Click <u>C</u>harts...**.

This produces the **Frequencies: Charts** dialog box that is used to construct histograms and a few other graphs.

5. **Click <u>H</u>istograms:**

This produces a blue dot in the **<u>H</u>istograms:** box, telling SPSS to construct a histogram when it gets the OK.

6. **Click Continue**.

This returns you to the **Frequencies** dialog box.

7. **Click OK**.

SPSS constructs a table and a histogram, and displays them. The histogram is shown below. We will ignore the table.

Analysis Results

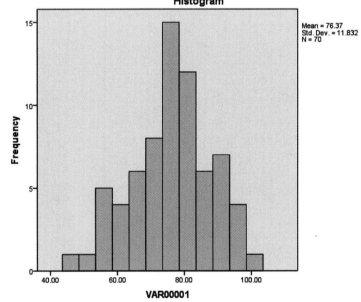

Comparing the SPSS histogram with the one presented in Figure 3.4, p. 52 of the textbook, you can see they are different. This is because SPSS grouped the scores into 12 intervals, whereas the textbook used 11.

As a fun extra exercise, let's do the following. SPSS allows the option of superimposing a normal curve on the histogram. Let's ask SPSS to do this. We will assume that the entries you made to construct the histogram haven't changed. To superimpose the normal curve on the histogram, go through the steps you have just completed to construct a histogram. Stop when you have displayed the **Frequencies: Charts** dialog box. The **Histograms:** box should already have a blue dot in it. Then,

---

INSTRUCTIONS	EXPLANATION
1. Click <u>S</u>how normal curve on histogram located under **Histograms:** on the **Frequencies: Charts** dialog box.	This produces a **check** in the **Show normal curve** box. When you give the OK to plot the histogram, this tells SPSS to superimpose a normal curve on the histogram. Seeing both on the same plot can help you tell how closely the data approximates the normal distribution.
2. Click **Continue**.	This returns you to the **Frequencies** dialog box.
3. Click **OK**.	SPSS constructs and displays the histogram with a superimposed normal curve that is shown below. Pretty neat, eh?

Analysis Results

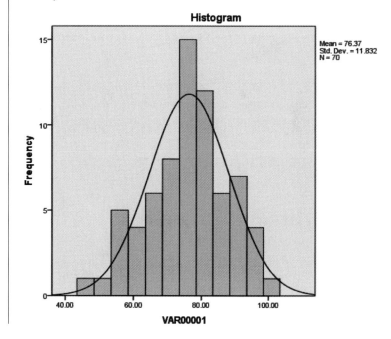

■ SPSS ADDITIONAL PROBLEMS

1. For this problem, use the weight loss scores of Chapter 3, End-of-Chapter Problem 10, p. 71 of the textbook. Do not give the scores a new name. For convenience the scores are repeated here.

10	13	22	26	16	23	35	53	17	32
41	35	24	23	27	16	20	60	48	43
52	31	17	20	33	18	23	8	27	15
26	46	30	19	22	13	22	14	21	39
28	43	37	15	20	11	25	9	15	21
21	25	34	10	23	29	28	18	17	24
16	26	7	12	28	20	36	16	14	
18	16	57	31	34	28	42	19	26	

   a. Use SPSS to construct an ungrouped frequency distribution of the scores.
   b. Use SPSS to construct a histogram of the scores. Is the distribution symmetrical? If not, is it positively or negatively skewed?

2. For this problem, use the grade point averages given in the textbook, Chapter 3, end-of-chapter Problem 15, p. 72. Give the scores the new name, *GPA*. For convenience the scores are repeated here.

2.7	1.9	1.0	3.3	1.3	1.8	2.6	3.7
3.1	2.2	3.0	3.4	3.1	2.2	1.9	3.1
3.4	3.0	3.5	3.0	2.4	3.0	3.4	2.4
2.4	3.2	3.3	2.7	3.5	3.2	3.1	3.3
2.1	1.5	2.7	2.4	3.4	3.3	3.0	3.8
1.4	2.6	2.9	2.1	2.6	1.5	2.8	2.3
3.3	3.1	1.6	2.8	2.3	2.8	3.2	2.8
2.8	3.8	1.4	1.9	3.3	2.9	2.0	3.2

   Use SPSS to construct a histogram of the scores, with a normal curve superimposed on the histogram. Is the distribution symmetrical? If not, is it positively or negatively skewed? Is the distribution normally distributed?

■ ONLINE STUDY RESOURCES

CENGAGE**brain**.com

Login to CengageBrain.com to access the resources your instructor has assigned. For this book, you can access the book's companion website for chapter-specific learning tools including Know and Be Able to Do, practice quizzes, flash cards, and glossaries and a link to Statistics and Research Methods Workshops

aplia™

If your professor has assigned Aplia homework:

1. Sign in to your account
2. Complete the corresponding homework exercises as required by your professor
3. When finished, click "Grade It Now" to see which areas you have mastered and which need more work, and for detailed explanations of every answer.

Visit **www.cengagebrain.com** to access your account and to purchase materials.

4

Measures of Central Tendency and Variability

© Strmko / Dreamstime.com

LEARNING OBJECTIVES

After completing this chapter, you should be able to:

- Contrast central tendency and variability.
- Define arithmetic mean, deviation score, median, mode, overall mean, range, standard deviation, sum of squares, and variance.
- Specify how the arithmetic mean, median, and mode differ conceptually; specify the properties of the mean, median, and mode.
- Compute the following: arithmetic mean, overall mean, median, mode, range, deviation scores, sum of squares, standard deviation, and variance.
- Specify how the mean, median, and mode are affected by skew in unimodal distributions.
- Explain how the standard deviation of a sample, as calculated in the textbook, differs from the standard deviation of a population, and why they differ.
- Understand the illustrative examples, do the practice problems, and understand the solutions.

INTRODUCTION

In Chapter 3, we discussed how to organize and present data in meaningful ways. The frequency distribution and its many derivatives are useful in this regard, but by themselves, they do not allow quantitative statements that characterize the distribution as a whole to be made, nor do they allow quantitative comparisons to be made between two or more distributions. It is often desirable to describe the characteristics of distributions quantitatively. For example, suppose a psychologist has conducted an experiment to determine whether men and women differ in mathematical aptitude. She has two sets of scores, one from the men and one from the women in the experiment. How can she compare the distributions? To do so, she needs to quantify them. This is most often done by computing the average score for each group and then comparing the averages. The measure computed is a measure of the *central tendency* of each distribution.

A second characteristic of distributions that is very useful to quantify is the *variability* of the distribution. Variability specifies the extent to which scores are different from each other, are dispersed, or are spread out. It is important for two reasons. First, determining the variability of the data is required by many of the statistical inference tests that we shall be discussing later in this book. In addition, the variability of a distribution can be useful in its own right. For example, suppose you were hired to design and evaluate an educational program for disadvantaged youngsters. When evaluating the program, you would be interested not only in the average value of the end-of-program scores but also in how variable the scores were. The variability of the scores is important because you need to know whether the effect of the program is uniform or varies over the youngsters. If it varies, as it almost assuredly will, how large is the variability? Is the program doing a good job with some students and a poor job with others? If so, the program may need to be redesigned to do a better job with those youngsters who have not been adequately benefiting from it.

Central tendency and variability are the two characteristics of distributions that are most often quantified. In this chapter, we shall discuss the most important measures of these two characteristics.

MEASURES OF CENTRAL TENDENCY

The three most often used measures of central tendency are the arithmetic mean, the median, and the mode.

The Arithmetic Mean

You are probably already familiar with the arithmetic mean. It is the value you ordinarily calculate when you average something. For example, if you wanted to know the average number of hours you studied per day for the past 5 days, you would add the hours you studied each day and divide by 5. In so doing, you would be calculating the arithmetic mean.

definition   ■   *The **arithmetic mean** is defined as the sum of the scores divided by the number of scores. In equation form,*

$$\overline{X} = \frac{\Sigma X_i}{N} = \frac{X_1 + X_2 + X_3 + \cdots + X_N}{N} \qquad \textit{mean of sample}$$

or

$$\mu = \frac{\Sigma X_i}{N} = \frac{X_1 + X_2 + X_3 + \cdots + X_N}{N} \qquad \begin{array}{l}\textit{mean of population}\\ \textit{set of scores}\end{array}$$

where

$$X_1 \ldots X_N = \text{raw scores}$$
$$\overline{X} \text{ (read "X bar")} = \text{mean of a sample set of scores}$$
$$\mu \text{ (read "mew")} = \text{mean of a population set of scores}$$
$$\Sigma \text{ (read "sigma")} = \text{summation sign}$$
$$N = \text{number of scores}$$

Note that we use two symbols for the mean: $\overline{X}$ if the scores are sample scores and μ (the Greek letter mu) if the scores are population scores. The computations, however, are the same regardless of whether the scores are sample or population scores. We shall use μ without any subscript to indicate that this is the mean of a population of raw scores. Later on in the text, we shall calculate population means of other kinds of scores for which we shall add the appropriate subscript.

Let's try a few problems for practice.

Practice Problem 4.1

Calculate the mean for each of the following sample sets of scores:

a. X: 3, 5, 6, 8, 14

$$\overline{X} = \frac{\Sigma X_i}{N} = \frac{3 + 5 + 6 + 8 + 14}{5}$$
$$= \frac{36}{5} = 7.20$$

b. X: 20, 22, 28, 30, 37, 38

$$\overline{X} = \frac{\Sigma X_i}{N} = \frac{20 + 22 + 28 + 30 + 37 + 38}{6}$$
$$= \frac{175}{6} = 29.17$$

c. X: 2.2, 2.4, 3.1, 3.1

$$\overline{X} = \frac{\Sigma X_i}{N} = \frac{2.2 + 2.4 + 3.1 + 3.1}{4}$$
$$= \frac{10.8}{4} = 2.70$$

table 4.1 Demonstration that $\Sigma (X_i - \overline{X}) = 0$

Subject Number	X_i	$X_i - \overline{X}$	Calculation of $\overline{X}$
1	2	−4	
2	4	−2	$\overline{X} = \dfrac{\Sigma X_i}{N} = \dfrac{30}{5}$
3	6	−0	
4	8	+2	$= 6.00$
5	$\underline{10}$	$\underline{+4}$	
	$\Sigma X_i = 30$	$\Sigma (X_i - \overline{X}) = 0$	

Properties of the mean   The mean has many important properties or characteristics. First,

The mean is sensitive to the exact value of all the scores in the distribution.

To calculate the mean you have to add *all* the scores, so a change in any of the scores will cause a change in the mean. This is not true of the median or the mode.

A second property is the following:

The sum of the deviations about the mean equals zero. Written algebraically, this property becomes $\Sigma (X_i - \overline{X}) = 0$.

This property says that if the mean is subtracted from each score, the sum of the differences will equal zero. The algebraic proof is presented in Note 4.1 at the end of this chapter. A demonstration of its validity is shown in Table 4.1. This property results from the fact that the mean is the balance point of the distribution. The mean can be thought of as the fulcrum of a seesaw, to use a mechanical analogy. The analogy is shown in Figure 4.1, using the scores of Table 4.1. When the scores are distributed along the seesaw according to their values, the mean of the distribution occupies the position where the scores are in balance.

A third property of the mean also derives from the fact that the mean is the balance point of the distribution:

MENTORING TIP

An *extreme* score is one that is far from the mean.

The mean is very sensitive to extreme scores.

A glance at Figure 4.1 should convince you that, if we added an extreme score (one far from the mean), it would greatly disrupt the balance. The mean would have to shift a considerable distance to reestablish balance. The mean is more sensitive to extreme scores than is the median or the mode. We shall discuss this more fully when we take up the median.

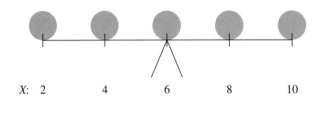

$$X: \quad 2 \qquad 4 \qquad 6 \qquad 8 \qquad 10$$

$$\overline{X} = 6.00$$

figure 4.1   The mean as the balance point in the distribution.

table 4.2 Demonstration that $\Sigma (X_i - \overline{X})^2$ is a minimum

(1) X_i	(2) $(X_i - 3.00)^2$	(3) $(X_i - 4.00)^2$	(4) $(X_i - 5.00)^2$	(5) $(X_i - 6.00)^2$	(6) $(X_i - 7.00)^2$
2	1	4	9	6	25
4	1	0	1	4	9
6	9	4	1	0	1
8	25	16	9	4	1
$\overline{X} = 5.00$	36	24	20	24	36

A fourth property of the mean has to do with the variability of the scores about the mean. This property states the following:

The sum of the squared deviations of all the scores about their mean is a minimum. Stated algebraically, $\Sigma (X_i - \overline{X})^2$ is a minimum.

MENTORING TIP

At this point, just concentrate on understanding this property; don't worry about its application.

This is an important characteristic used in many areas of statistics, particularly in regression. Elaborated a little more fully, this property states that although the sum of the squared deviations about the mean does not usually equal zero, this sum is smaller than if the squared deviations were taken about any other value. The validity of this property is demonstrated in Table 4.2. The scores of the distribution are given in the first column. Their mean equals 5.00. The fourth column shows the squared deviations of the X_i scores about their mean $(X_i - 5.00)^2$. The sum of these squared deviations is 20. The other columns show the squared deviations of the X_i scores about values other than the mean. In the second column, the value is 3.00 $(X_i - 3.00)^2$; in the third column, 4.00; in the fifth column, 6.00; and in the sixth column, 7.00. Note that the sum of the squared deviations about each of these values is larger than the sum of the squared deviations about the mean of the distribution. Not only is the sum larger, but the farther the value gets from the mean, the larger the sum becomes. This implies that although we've compared only four other values, it holds true for all other values. Thus, although the sum of the squared deviations about the mean does not usually equal zero, it is smaller than if the squared deviations are taken about any other value.

The last property has to do with the use of the mean for statistical inference. This property states the following:

Under most circumstances, of the measures used for central tendency, the mean is least subject to sampling variation.

If we were repeatedly to take samples from a population on a random basis, the mean would vary from sample to sample. The same is true for the median and the mode. However, the mean varies less than these other measures of central tendency. This is very important in inferential statistics and is a major reason why the mean is used in inferential statistics whenever possible.

The Overall Mean

Occasionally, the situation arises in which we know the mean of several groups of scores and we want to calculate the mean of all the scores combined. Of course, we could start from the beginning again and just sum all the raw scores and divide by the total number of scores. However, there is a shortcut available if we already know the mean of the groups and the number of scores in each group. The equation for this method derives from the basic definition of the mean. Suppose we have several groups of scores that we wish to combine to calculate the overall mean. We'll let k equal the number of groups. Then,

$$\overline{X}_{\text{overall}} = \frac{\text{Sum of all scores}}{N}$$

$$= \frac{\Sigma X_i(\text{first group}) + \Sigma X_i(\text{second group}) + \cdots + \Sigma X_i(\text{last group})}{n_1 + n_2 + \cdots + n_k}$$

where N = total number of scores
n_1 = number of scores in the first group
n_2 = number of scores in the second group
n_k = number of scores in the last group

Since $\overline{X}_1 = \Sigma X_i$ (first group)$/n_1$, multiplying by n_1, we have ΣX_i (first group) $= n_1\overline{X}_1$. Similarly, ΣX_i (second group) $= n_2 \overline{X}_2$, and ΣX_i (last group) $= n_k\overline{X}_k$, where $\overline{X}_k$ is the mean of the last group. Substituting these values in the numerator of the preceding equation, we arrive at

$$\overline{X}_{\text{overall}} = \frac{n_1\overline{X}_1 + n_2\overline{X}_2 + \ldots + n_k\overline{X}_k}{n_1 + n_2 + \ldots + n_k} \qquad \textit{overall mean of several groups}$$

In words, this equation states that the *overall mean* is equal to the sum of the mean of each group times the number of scores in the group, divided by the sum of the number of scores in each group.

To illustrate how this equation is used, suppose a sociology professor gave a final exam to two classes. The mean of one of the classes was 90, and the number of scores was 20. The mean of the other class was 70, and 40 students took the exam. Calculate the mean of the two classes combined.

The solution is as follows: Given that $\overline{X}_1 = 90$ and $n_1 = 20$ and that $\overline{X}_2 = 70$ and $n_2 = 40$,

$$\overline{X}_{\text{overall}} = \frac{n_1\overline{X}_1 + n_2\overline{X}_2}{n_2 + n_2} = \frac{20(90) + 40(70)}{20 + 40} = 76.67$$

MENTORING TIP

The overall mean is often called the *weighted* mean.

The overall mean is much closer to the average of the class with 40 scores than the class with 20 scores. In this context, we can see that each of the means is being *weighted* by its number of scores. We are counting the mean of 70 forty times and the mean of 90 only twenty times. Thus, the overall mean really is a weighted mean, where the weights are the number of scores used in determining each mean. Let's do one more problem for practice.

Practice Problem 4.2

A researcher conducted an experiment involving three groups of subjects. The mean of the first group was 75, and there were 50 subjects in the group. The mean of the second group was 80, and there were 40 subjects. The third group had a mean of 70 and 25 subjects. Calculate the overall mean of the three groups combined.

SOLUTION

The solution is as follows: Given that $\overline{X}_1 = 75$, $n_1 = 50$; $\overline{X}_2 = 80$, $n_2 = 40$; and $\overline{X}_3 = 70$, $n_3 = 25$,

$$\overline{X}_{\text{overall}} = \frac{n_1\overline{X}_1 + n_2\overline{X}_2 + n_3\overline{X}_3}{n_1 + n_2 + n_3} = \frac{50(75) + 40(80) + 25(70)}{50 + 40 + 25}$$

$$= \frac{8700}{115} = 75.65$$

The Median

The second most frequently encountered measure of central tendency is the median.

definition ■ *The **median** (symbol Mdn) is defined as the scale value below which 50% of the scores fall.* It is therefore the same thing as P_{50}.

In Chapter 3, we discussed how to calculate P_{50}; therefore, you already know how to calculate the median for grouped scores. For practice, however, Practice Problem 4.3 contains another problem and its solution. You should try this problem and be sure you can solve it before going on.

Practice Problem 4.3

Calculate the median of the grouped scores listed in Table 4.3.

table 4.3 Calculating the median from grouped scores

Class Interval	f	Cum f	Cum %	Calculation of Median
3.6–4.0	4	52	100.00	
3.1–3.5	6	48	92.31	
2.6–3.0	8	42	80.77	
2.1–2.5	10	34	65.38	Mdn $= P_{50}$
1.6–2.0	9	24	46.15	$= X_L + (i/f_i)(\text{cum} f_P - \text{cum} f_L)$
1.1–1.5	7	15	28.85	$= 2.05 + (0.5/10)(26 - 24)$
0.6–1.0	5	8	15.38	$= 2.05 + 0.10 = 2.15$
0.1–0.5	3	3	5.77	

(continued)

SOLUTION

The median is the value below which 50% of the scores fall. Since $N = 52$, the median is the value below which 26 of the scores fall (50% of 52 = 26). From Table 4.3, we see that the median lies in the interval 2.1–2.5. Since 24 scores fall below a value of 2.05, we need two more scores to make up the 26. Given that there are 10 scores in the interval and the interval is 0.5 unit wide, we must move 0.5/10 × 2 = 0.10 unit into the interval. Thus,

$$\text{Median} = 2.05 + 0.10 = 2.15$$

The median could also have been found by using the equation for percentile point. This solution is shown in Table 4.3.

When dealing with raw (ungrouped) scores, it is quite easy to find the median. First, arrange the scores in rank order.

MENTORING TIP

To help you remember that the median is the centermost score, think of the median of a road (the center line) that divides the road in half.

The median is the centermost score if the number of scores is odd. If the number is even, the median is taken as the average of the two centermost scores.

To illustrate, suppose we have the scores 5, 2, 3, 7, and 8 and want to determine their median. First, we rank-order the scores: 2, 3, 5, 7, 8. Since the number of scores is odd, the median is the centermost score. In this example, the median is 5. It may seem that 5 is not really P_{50} for the set of scores. However, consider the score of 5 to be evenly distributed over the interval 4.5–5.5. Now it becomes obvious that half of the scores fall below 5.0. Thus, 5.0 is P_{50}.

Let's try another example, this time with an even number of scores. Given the scores 2, 8, 6, 4, 12, and 10, determine their median. First, we rank-order the scores: 2, 4, 6, 8, 10, 12. Since the number of scores is even, the median is the average of the two centermost scores. The median for this example is (6 + 8)/2 = 7. For additional practice, Practice Problem 4.4 presents a few problems dealing with raw scores.

Practice Problem 4.4

Calculate the median for the following sets of scores

a. 8, 10, 4, 3, 1, 15	Rank order: 1, 3, 4, 8, 10, 15	Mdn = (4 + 8)/2 = 6
b. 100, 102, 108, 104, 112	Rank order: 100, 102, 104, 108, 112	Mdn = 104
c. 2.5, 1.8, 1.2, 2.4, 2.0	Rank order: 1.2, 1.8, 2.0, 2.4, 2.5	Mdn = 2.0
d. 10, 11, 14, 14, 16, 14, 12	Rank order: 10, 11, 12, 14, 14, 14, 16	Mdn = 14

In the last set of scores in Practice Problem 4.4, the median occurs at 14, where there are three scores. Technically, we should consider the three scores equally spread out over the interval 13.5–14.5. Then we would find the median by using the equation shown in Table 4.3 (p. 85), with $i = 1$ (Mdn = 13.67). However, when raw scores are being used, this refinement is often not made. Rather, the median is taken at 14. We shall follow this procedure. Thus, if the median occurs at a value where there are tied scores, we shall use the tied score as the median.

table 4.4 Effect of extreme scores
on the mean and median

Scores	Mean	Median
3, 4, 6, 7, 10	6	6
3, 4, 6, 7, 100	24	6
3, 4, 6, 7, 1000	204	6

Properties of the median   There are two properties of the median worth noting. First,

The median is less sensitive than the mean to extreme scores.

To illustrate this property, consider the scores shown in the first column of Table 4.4. The three distributions shown are the same except for the last score. In the second distribution, the score of 100 is very different in value from the other scores. In the third distribution, the score of 1000 is even more extreme. Note what happens to the mean in the second and third distributions. Since the mean is sensitive to extreme scores, it changes considerably with the extreme scores. How about the median? Does it change too? As we see from the third column, the answer is no! The median stays the same. Since the median is not responsive to each individual score but rather divides the distribution in half, it is not as sensitive to extreme scores as is the mean. For this reason, when the distribution is strongly skewed, it is probably better to represent the central tendency with the median rather than the mean. Certainly, in the third distribution of Table 4.4, the median of 6 does a better job representing most of the scores than does the mean of 204.

The second property of the median involves its sampling stability. It states that,

Under usual circumstances, the median is more subject to sampling variability than the mean but less subject to sampling variability than the mode.

Because the median is usually less stable than the mean from sample to sample, it is not as useful in inferential statistics.

The Mode

The third and last measure of central tendency that we shall discuss is the mode.

definition   ■   *The **mode** is defined as the most frequent score in the distribution.**

Clearly, this is the easiest of the three measures to determine. The mode is found by inspection of the scores; there isn't any calculation necessary. For instance, to find the mode of the data in Table 3.2 (p. 49), all we need to do is search the frequency column. The mode for these data is 76. With grouped scores, the mode is designated as the midpoint of the interval with the highest frequency. The mode of the grouped scores in Table 3.4 (p. 52) is 77.

---

*When all the scores in the distribution have the same frequency, it is customary to say that the distribution has no mode.

Usually, distributions are *unimodal*; that is, they have only one mode. However, it is possible for a distribution to have many modes. When a distribution has two modes, as is the case with the scores 1, 2, 3, 3, 3, 3, 4, 5, 7, 7, 7, 7, 8, 9, the distribution is called *bimodal*. Histograms of a unimodal and bimodal distribution are shown in Figure 4.2. Although the mode is the easiest measure of central tendency to determine, it is not used very much in the behavioral sciences because it is not very stable from sample to sample and often there is more than one mode for a given set of scores.

Measures of Central Tendency and Symmetry

If the distribution is unimodal and symmetrical, the mean, median, and mode will all be equal. An example of this is the bell-shaped curve mentioned in Chapter 3 and shown in Figure 4.3. When the distribution is skewed, the mean and median will not be equal. Since the mean is most affected by extreme scores, it will have a value closer to the extreme scores than will the median. Thus, with a negatively skewed distribution, the mean will be lower than the median. With a positively skewed curve, the mean will be larger than the median. Figure 4.3 shows these relationships.

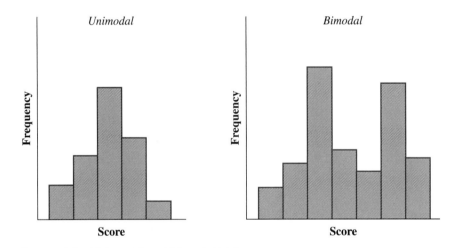

figure 4.2 Unimodal and bimodal histograms.

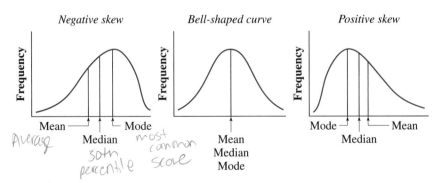

figure 4.3 Symmetry and measures of central tendency.

From *Statistical Reasoning in Psychology and Education* by E.W. Minium. Copyright © 1978 John Wiley & Sons, Inc. Adapted by permission of John Wiley & Sons, Inc.

MEASURES OF VARIABILITY

Previously in this chapter, we pointed out that variability specifies how far apart the scores are spread. Whereas measures of central tendency are a quantification of the average value of the distribution, measures of variability quantify the extent of *dispersion*. Three measures of variability are commonly used in the behavioral sciences: the range, the standard deviation, and the variance.

The Range

We have already used the range when we were constructing frequency distributions of grouped scores.

definition   ■   *The* **range** *is defined as the difference between the highest and lowest scores in the distribution. In equation form,*

Range = Highest score – Lowest score

The range is easy to calculate but gives us only a relatively crude measure of dispersion, because the range really measures the spread of only the extreme scores and not the spread of any of the scores in between. Although the range is easy to calculate, we've included some problems for you to practice on. Better to be sure than sorry.

Practice Problem 4.5

Calculate the range for the following distributions:

a. 2, 3, 5, 8, 10	Range = 10 – 2 = 8
b. 18, 12, 28, 15, 20	Range = 28 – 12 = 16
c. 115, 107, 105, 109, 101	Range = 115 – 101 = 14
d. 1.2, 1.3, 1.5, 1.8, 2.3	Range = 2.3 – 1.2 = 1.1

The Standard Deviation

Before discussing the standard deviation, it is necessary to introduce the concept of a deviation score.

Deviation scores   So far, we've been dealing mainly with raw scores. You will recall that a raw score is the score as originally measured. For example, if we are interested in IQ and we measure an IQ of 126, then 126 is a raw score.

definition   ■ A **deviation score** *tells how far away the raw score is from the mean of its distribution.*

In equation form, a deviation score is defined as

$$X - \overline{X} \qquad \textit{deviation score for sample data}$$
$$X - \mu \qquad \textit{deviation score for population data}$$

As an illustration, consider the sample scores in Table 4.5. The raw scores are shown in the first column, and their transformed deviation scores are in the second column. The deviation score tells how far the raw score lies above or below the mean. Thus, the raw score of $2(X = 2)$ lies 4 units below the mean $(X - \overline{X} = -4)$. The raw scores and their deviation scores are also shown pictorially in Figure 4.4.

Let's suppose that you are a budding mathematician (use your imagination if necessary). You have been assigned the task of deriving a measure of dispersion that gives the average deviation of the scores about the mean. After some reflection, you say, "That's easy. Just calculate the deviation from the mean of each score and average the deviation scores." Your logic is impeccable. There is only one stumbling block. Consider the scores in Table 4.6. For the sake of this example, we will assume this is a population set of scores. The first column contains the population raw scores and the second column the deviation scores. We want to calculate the average deviation of the raw scores about their mean. According to your method, we would first compute the deviation scores (second column) and average them by dividing the sum of the deviation scores $[\Sigma (X - \mu)]$ by N. The stumbling block is that $\Sigma (X - \mu) = 0$. Remember, this is a general property of the mean. The sum of the deviations about the mean always equals zero. Thus, if we follow your suggestion, the average of the deviations would always equal zero, no matter how dispersed the scores were $[\Sigma (X - \mu)/N = 0/N = 0]$.

You are momentarily stunned by this unexpected low blow. However, you don't give up. You look at the deviation scores and you see that the negative scores are canceling the positive ones. Suddenly, you have a flash of insight. Why not square each deviation score? Then all the scores would be positive, and their sum would no longer be zero. Eureka! You have solved the problem. Now you can divide the sum of the squared scores by N to get the average value $[\Sigma (X - \mu)^2/N]$, and the average won't equal zero. You should note that the numerator of this formula $[\Sigma (X - \mu)^2]$ is called the *sum of squares* or, more accurately, *sum of squared deviations,* and is symbolized as SS_{pop}. The only trouble at this point is that you have now calculated the average *squared deviation*, not the average deviation. What you need to do is "unsquare" the answer. This is done by taking the square root of SS_{pop}/N.

table 4.5   Calculating deviation scores

X	$X - \overline{X}$	Calculation of $\overline{X}$
2	$2 - 6 = -4$	$\overline{X} = \dfrac{\Sigma X}{N} = \dfrac{30}{5}$
4	$4 - 6 = -2$	
6	$6 - 6 = \ \ 0$	$= 6.00$
8	$8 - 6 = +2$	
10	$10 - 6 = +4$	

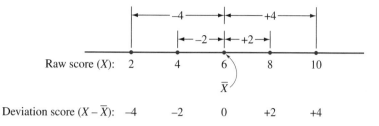

Deviation score $(X - \bar{X})$:   −4   −2   0   +2   +4

figure 4.4   Raw scores and their corresponding deviation scores.

Your reputation as a mathematician is vindicated! You have come up with the equation for standard deviation used by many statisticians. The symbol for the standard deviation of population scores is σ (the lowercase Greek letter sigma), and for samples it is s. Your derived equation for population scores is as follows:

$$\sigma = \sqrt{\frac{SS_{pop}}{N}} = \sqrt{\frac{\Sigma\,(X - \mu)^2}{N}}$$ *standard devition of a population set of raw scores—deviation method*

where $SS_{pop} = \Sigma\,(X - \mu)^2$ *sum of squares—population data*

Calculation of the standard deviation of a population set of scores using the deviation method is shown in Table 4.6.

Technically, the equation is the same for calculating the standard deviation of sample scores. However, when we calculate the standard deviation of sample data, we usually want to use our calculation to estimate the population standard deviation. It can be shown algebraically that the equation with N in the denominator gives an estimate that on the average is too small. Dividing by $N - 1$, instead of N, gives a more accurate estimate of σ. Since estimation of the population standard deviation is an important use of the sample standard deviation and since it saves confusion later on in this textbook (when we cover Student's t test and the F test), we have chosen to adopt the equation with $N - 1$ in the denominator for calculating the standard deviation of sample scores. Thus,

MENTORING TIP

Caution: be sure you understand why we compute s with $N - 1$ in the denominator.

$$s = \text{Estimated } \sigma = \sqrt{\frac{SS}{N - 1}} = \sqrt{\frac{\Sigma\,(X - \bar{X})^2}{N - 1}}$$ *standard deviation of a sample set of raw scores— deviation method*

where $SS = \Sigma\,(X - \bar{X})^2$ *sum of squares—sample data*

table 4.6   Calculation of the standard deviation of a population set of scores by the deviation method

X	$X - \mu$	$(X - \mu)^2$	Calculation of μ and σ
3	−2	4	$\mu = \dfrac{\Sigma X}{N} = \dfrac{25}{5} = 5.00$
4	−1	1	
5	0	0	
6	+1	1	$\sigma = \sqrt{\dfrac{SS_{pop}}{N}} = \sqrt{\dfrac{\Sigma\,(X - \mu)^2}{N}} = \sqrt{\dfrac{10}{5}} = 1.41$
7	+2	4	
	$\Sigma\,(X - \mu) = 0$	$\Sigma\,(X - \mu)^2 = 10$	

table 4.7 Calculation of the standard deviation of sample scores by the deviation method

X	$X - \overline{X}$	$(X - \overline{X})^2$	Calculation of $\overline{X}$ and s
2	−4	16	$\overline{X} = \dfrac{\Sigma X}{N} = \dfrac{30}{5} = 6.00$
4	−2	4	
6	0	0	
8	+2	4	$s = \sqrt{\dfrac{SS}{N-1}} = \sqrt{\dfrac{\Sigma (X - \overline{X})^2}{N-1}} = \sqrt{\dfrac{40}{5-1}}$
10	+4	16	
	$\Sigma (X - \overline{X}) = 0$	$SS = 40$	$= \sqrt{10} = 3.16$

In most practical situations, the data are from samples rather than populations. Calculation of the standard deviation of a sample using the preceding equation for samples is shown in Table 4.7. Although this equation gives the best conceptual understanding of the standard deviation and it does yield the correct answer, it is quite cumbersome to use in practice. This is especially true if the mean is not a whole number. Table 4.8 shows an illustration using the previous equation with a mean that has a decimal remainder. Note that each deviation score has a decimal remainder that must be squared to get $(X - \overline{X})^2$. A great deal of rounding is necessary, which may contribute to inaccuracy. In addition, we are dealing with adding five-digit numbers, which increases the possibility of error. You can see how cumbersome using this equation becomes when the mean is not an integer, and in most practical problems, the mean is not an integer!

Calculating the standard deviation of a sample by the raw scores method   It can be shown algebraically that

$$SS = \Sigma X^2 - \frac{(\Sigma X)^2}{N} \qquad \textit{sum of squares}$$

table 4.8 Calculation of the standard deviation with use of deviation scores when the mean is not a whole number

X	$X - \overline{X}$	$(X - \overline{X})^2$	Calculation of $\overline{X}$ and s
10	−6.875	47.2656	$\overline{X} = \dfrac{\Sigma X}{N} = \dfrac{135}{8} = 16.875$
12	−4.875	23.7656	
13	−3.875	15.0156	
15	−1.875	3.5156	$s = \sqrt{\dfrac{SS}{N-1}} = \sqrt{\dfrac{\Sigma (X - \overline{X})^2}{N-1}}$
18	1.125	1.2656	
20	3.125	9.7656	
22	5.125	26.2656	$= \sqrt{\dfrac{192.8748}{7}}$
25	8.125	66.0156	
$\Sigma X = 135$	$\Sigma (X - \overline{X}) = 0.000$	$SS = 192.8748$	$= \sqrt{27.5535}$
$N = 8$			$= 5.25$

The derivation is presented in Note 4.2. Using this equation to find *SS* allows us to use the raw scores without the necessity of calculating deviation scores. This, in turn, avoids the decimal remainder difficulties described previously. We shall call this method of computing *SS*, "the raw score method" to distinguish it from the "deviation method." Since the raw score method is generally easier to use and avoids potential errors, it is the method of choice in computing *SS* and will be used throughout the remainder of this text. When using the raw score method, *you must be sure not to confuse* ΣX^2 *and* $(\Sigma X)^2$. ΣX^2 *is read "sum X square," or "sum of the squared X scores," and* $(\Sigma X)^2$ *is read "sum X quantity squared," or "sum of the X scores, squared."* To find ΣX^2, we square each score and then sum the squares. To find $(\Sigma X)^2$, we sum the scores and then square the sum. The result is different for the two procedures. In addition, *SS* must be positive. If your calculation turns out negative, you have probably confused ΣX^2 and $(\Sigma X)^2$.

Table 4.9 shows the calculation of the standard deviation, using the raw score method, of the data presented in Table 4.8. When using this method, we first calculate *SS* from the raw score equation and then substitute the obtained value in the equation for the standard deviation.

Properties of the standard deviation   The standard deviation has many important characteristics. First, the *standard deviation gives us a measure of dispersion relative to the mean.* This differs from the range, which gives us an absolute measure of the spread between the two most extreme scores. Second, *the standard deviation is sensitive to each score in the distribution.* If a score is moved closer to the mean, then the standard deviation will become smaller. Conversely, if a score shifts away from the mean, then the standard deviation will increase. Third, *like the mean, the standard deviation is stable with regard to sampling fluctuations.* If samples were taken repeatedly from populations of the type usually encountered in the behavioral sciences, the standard deviation of the samples would vary much less from sample to sample than the range. This property is one of the main reasons why the standard deviation is used so much more often than the range for reporting variability. Finally, both the mean and the standard deviation can be manipulated algebraically. This allows mathematics to be done with them for use in inferential statistics.

Now let's do Practice Problems 4.6 and 4.7.

table **4.9**   Calculation of the standard deviation by the raw score method

X	X^2	Calculation of *SS*	Calculation of *s*
10	100	$SS = \Sigma X^2 - \dfrac{(\Sigma X)^2}{N}$	$s = \sqrt{\dfrac{SS}{N-1}}$
12	144		
13	169		
15	225	$= 2471 - \dfrac{(135)^2}{8}$	$= \sqrt{\dfrac{192.875}{7}}$
18	324		
20	400	$= 2471 - 2278.125$	$= \sqrt{27.5536}$
22	484		
25	625	$= 192.875$	$= 5.25$
$\Sigma X = 135$	$\Sigma X^2 = 2471$		
$N = 8$			

Practice Problem 4.6

Calculate the standard deviation of the sample scores contained in the first column of the following table:

X	X^2	Calculation of SS	Calculation of s
25	625	$SS = \Sigma X^2 - \dfrac{(\Sigma X)^2}{N}$	$s = \sqrt{\dfrac{SS}{N-1}}$
28	784		
35	1,225	$= 15{,}545 - \dfrac{(387)^2}{10}$	$= \sqrt{\dfrac{568.1}{9}}$
37	1,369		
38	1,444		
40	1,600	$= 15{,}545 - 14{,}976.9$	$= \sqrt{63.1222}$
42	1,764		
45	2,025	$= 568.1$	$= 7.94$
47	2,209		
50	2,500		
$\Sigma X = 387$	$\Sigma X^2 = 15{,}545$		
$N = 10$			

Practice Problem 4.7

Calculate the standard deviation of the sample scores contained in the first column of the following table:

X	X^2	Calculation of SS	Calculation of s
1.2	1.44	$SS = \Sigma X^2 - \dfrac{(\Sigma X)^2}{N}$	$s = \sqrt{\dfrac{SS}{N-1}}$
1.4	1.96		
1.5	2.25	$= 60.73 - \dfrac{(25.9)^2}{12}$	$= \sqrt{\dfrac{4.8292}{11}}$
1.7	2.89		
1.9	3.61		
2.0	4.00	$= 60.73 - 55.9008$	$= \sqrt{0.4390}$
2.2	4.84		
2.4	5.76	$= 4.8292$	$= 0.66$
2.5	6.25		
2.8	7.84		

X	X^2	Calculation of *SS*	Calculation of *s*
3.0	9.00		
3.3	10.89		
$\Sigma X = 25.9$	$\Sigma X^2 = 60.73$		
$N = 12$			

The Variance

The *variance* of a set of scores is just the square of the standard deviation. For sample scores, the variance equals

$$s^2 = \text{Estimated } \sigma^2 = \frac{SS}{N - 1} \qquad \textit{variance of a sample}$$

For population scores, the variance equals

$$\sigma^2 = \frac{SS_{\text{pop}}}{N} \qquad \textit{variance of a sample}$$

The variance is not used much in descriptive statistics because it gives us squared units of measurement. However, it is used quite frequently in inferential statistics.

■ SUMMARY

In this chapter, I have discussed the central tendency and variability of distributions. The most common measures of central tendency are the arithmetic mean, the median, and the mode. The arithmetic mean gives the average of the scores and is computed by summing the scores and dividing by N. The median divides the distribution in half and, hence, is the scale value that is at the 50th percentile point of the distribution. The mode is the most frequent score in the distribution. The mean possesses special properties that make it by far the most commonly used measure of central tendency. However, if the distribution is quite skewed, the median should be used instead of the mean because it is less affected by extreme scores. In addition to presenting these measures, I showed how to calculate each and elaborated their most important properties. I also showed how to obtain the overall mean when the average of several means is desired. Finally, we discussed the relationship between the mean, median, and mode of a distribution and its symmetry.

The most common measures of variability are the range, the standard deviation, and the variance. The range is a crude measure that tells the dispersion between the two most extreme scores. The standard deviation is the most frequently encountered measure of variability. It gives the average dispersion about the mean of the distribution. The variance is just the square of the standard deviation. As with the measures of central tendency, our discussion of variability included how to calculate each measure. Finally, since the standard deviation is the most important measure of variability, I also presented its properties.

■ IMPORTANT NEW TERMS

Arithmetic mean (p. 80)
Central tendency (p. 80)
Deviation score (p. 89)
Dispersion (p. 89)

Median (p. 85)
Mode (p. 87)
Overall mean (p. 83)
Range (p. 89)

Standard deviation (p. 89)
Sum of squares (p. 91, 92)
Variability (p. 80)
Variance (p. 95)

■ QUESTIONS AND PROBLEMS

1. Define or identify the terms in the Important New Terms section.
2. State four properties of the mean and illustrate each with an example.
3. Under what condition might you prefer to use the median rather than the mean as the best measure of central tendency? Explain why.
4. Why is the mode not used very much as a measure of central tendency?
5. The overall mean ($\overline{X}_{overall}$) is a weighted mean. Is this statement correct? Explain.
6. Discuss the relationship between the mean and median for distributions that are symmetrical and skewed.
7. Why is the range not as useful a measure of dispersion as the standard deviation?
8. The standard deviation is a relative measure of average dispersion. Is this statement correct? Explain.
9. Why do we use $N - 1$ in the denominator for computing s but use N in the denominator for determining σ?
10. What is the raw score equation for SS? When is it useful?
11. Give three properties of the standard deviation.
12. How are the variance and standard deviation related?
13. If $s = 0$, what must be true about the scores in the distribution? Verify your answer, using an example.
14. Can the value of the range, standard deviation, or variance of a set of scores be negative? Explain.
15. Give the symbol for each of the following:
    a. Mean of a sample
    b. Mean of a population
    c. Standard deviation of a sample
    d. Standard deviation of a population
    e. A raw score
    f. Variance of a sample
    g. Variance of a population
16. Calculate the mean, median, and mode for the following scores:
    a. 5, 2, 8, 2, 3, 2, 4, 0, 6
    b. 30, 20, 17, 12, 30, 30, 14, 29
    c. 1.5, 4.5, 3.2, 1.8, 5.0, 2.2

17. Calculate the mean of the following set of sample scores: 1, 3, 4, 6, 6.
    a. Add a constant of 2 to each score. Calculate the mean for the new values. Generalize to answer the question, "What is the effect on the mean of adding a constant to each score?"
    b. Subtract a constant of 2 from each score. Calculate the mean for the new values. Generalize to answer the question, "What is the effect on the mean of subtracting a constant from each score?"
    c. Multiply each score by a constant of 2. Calculate the mean for the new values. Generalize to answer the question, "What is the effect on the mean of multiplying each score by a constant?"
    d. Divide each score by a constant of 2. Calculate the mean for the new values. Generalize to answer the question, "What is the effect on the mean of dividing each score by a constant?"
18. The following scores resulted from a biology exam:

Scores	f	Scores	f
95–99	3	65–69	7
90–94	3	60–64	6
85–89	5	55–59	5
80–84	6	50–54	3
75–79	6	45–49	2
70–74	8		

    a. What is the median for this exam?
    b. What is the mode? education
19. Using the scores shown in Table 3.5 (p. 53),
    a. Determine the median.
    b. Determine the mode.
20. Using the scores shown in Table 3.6 (p. 54),
    a. Determine the median.
    b. Determine the mode.

21. For the following distributions, state whether you would use the mean or the median to represent the central tendency of the distribution. Explain why.
    a. 2, 3, 8, 5, 7, 8
    b. 10, 12, 15, 13, 19, 22
    c. 1.2, 0.8, 1.1, 0.6, 25

22. Given the following values of central tendency for each distribution, determine whether the distribution is symmetrical, positively skewed, or negatively skewed:
    a. Mean = 14, median = 12, mode = 10
    b. Mean = 14, median = 16, mode = 18
    c. Mean = 14, median = 14, mode = 14

23. A student kept track of the number of hours she studied each day for a 2-week period. The following daily scores were recorded (scores are in hours): 2.5, 3.2, 3.8, 1.3, 1.4, 0, 0, 2.6, 5.2, 4.8, 0, 4.6, 2.8, 3.3. Calculate
    a. The mean number of hours studied per day
    b. The median number of hours studied per day
    c. The modal number of hours studied per day education

24. Two salesmen working for the same company are having an argument. Each claims that the average number of items he sold, averaged over the last month, was the highest in the company. Can they both be right? Explain. I/O, other

25. An ornithologist studying the glaucous-winged gull on Puget Sound counts the number of aggressive interactions per minute among a group of sea gulls during 9 consecutive minutes. The following scores resulted: 24, 9, 12, 15, 10, 13, 22, 20, 14. Calculate
    a. The mean number of aggressive interactions per minute
    b. The median number of aggressive interactions per minute
    c. The modal number of aggressive interactions per minute biological

26. A reading specialist tests the reading speed of children in four ninth-grade English classes. There are 42 students in class A, 35 in class B, 33 in class C, and 39 in class D. The mean reading speed in words per minute for the classes were as follows: class A, 220; class B, 185; class C, 212; and class D, 172. What is the mean reading speed for all classes combined? education

27. For the following sample sets of scores, calculate the range, the standard deviation, and the variance:
    a. 6, 2, 8, 5, 4, 4, 7
    b. 24, 32, 27, 45, 48
    c. 2.1, 2.5, 6.6, 0.2, 7.8, 9.3

28. In a particular statistics course, three exams were given. Each student's grade was based on a weighted average of his or her exam scores. The first test had a weight of 1, the second test had a weight of 2, and the third test had a weight of 2. The exam scores for one student are listed here. What was the student's overall average?

Exam	1	2	3
Score	83	97	92

education

29. The timekeeper for a particular mile race uses a stopwatch to determine the finishing times of the racers. He then calculates that the mean time for the first three finishers was 4.25 minutes. After checking his stopwatch, he notices to his horror that the stopwatch begins timing at 15 seconds rather than at 0, resulting in scores each of which is 15 seconds too long. What is the correct mean time for the first three finishers? I/O, other

30. The manufacturer of brand A jogging shoes wants to determine how long the shoes last before resoling is necessary. She randomly samples from users in Chicago, New York, and Seattle. In Chicago, the sample size was 28, and the mean duration before resoling was 7.2 months. In New York, the sample size was 35, and the mean duration before resoling was 6.3 months. In Seattle, the sample size was 22, and the mean duration before resoling was 8.5 months. What is the overall mean duration before resoling is necessary for brand A jogging shoes? I/O, other

31. Calculate the standard deviation of the following set of sample scores: 1, 3, 4, 6, 6.
    a. Add a constant of 2 to each score. Calculate the standard deviation for the new values. Generalize to answer the question, "What is the effect on the standard deviation of adding a constant to each score?"
    b. Subtract a constant of 2 from each score. Calculate the standard deviation for the new values. Generalize to answer the question, "What is the effect on the standard deviation of subtracting a constant from each score?"
    c. Multiply each score by a constant of 2. Calculate the standard deviation for the new values. Generalize to answer the question, "What is the effect on the standard deviation of multiplying each score by a constant?"
    d. Divide each score by a constant of 2. Calculate the standard deviation for the new values. Generalize to answer the question, "What is the effect

on the standard deviation of dividing each score by a constant?"

32. An industrial psychologist observed eight drill press operators for 3 working days. She recorded the number of times each operator pressed the "faster" button instead of the "stop" button to determine whether the design of the control panel was contributing to the high rate of accidents in the plant. Given the scores 4, 7, 0, 2, 7, 3, 6, 7, compute the following:
    a. Mean
    b. Median
    c. Mode
    d. Range
    e. Standard deviation
    f. Variance I/O

33. Without actually calculating the variability, study the following sample distributions:
    Distribution a: 21, 24, 28, 22, 20
    Distribution b: 21, 32, 38, 15, 11
    Distribution c: 22, 22, 22, 22, 22
    a. Rank-order them according to your best guess of their relative variability.
    b. Calculate the standard deviation of each to verify your rank ordering.

34. Compute the standard deviation for the following sample scores. Why is s so high in part **b**, relative to part **a**?
    a. 6, 8, 7, 3, 6, 4
    b. 6, 8, 7, 3, 6, 35

35. A social psychologist interested in the dating habits of college undergraduates samples 10 students and determines the number of dates they have had in the last month. Given the scores 1, 8, 12, 3, 8, 14, 4, 5, 8, 16, compute the following:
    a. Mean
    b. Median
    c. Mode
    d. Range
    e. Standard deviation
    f. Variance social

36. A cognitive psychologist measures the reaction times of 6 subjects to emotionally laden words. The following scores in milliseconds are recorded: 250, 310, 360, 470, 425, 270. Compute the following:
    a. Mean
    b. Median
    c. Mode
    d. Range
    e. Standard deviation
    f. Variance cognitive

37. A biological psychologist records the number of cells in a particular brain region of cats that respond to a tactile stimulus. Nine cats are used. The following cell counts/animal are recorded: 15, 28, 33, 19, 24, 17, 21, 34, 12. Compute the following:
    a. Mean
    b. Median
    c. Mode
    d. Range
    e. Standard deviation
    f. Variance biological

38. What happens to the mean of a set of scores if
    a. A constant a is added to each score in the set?
    b. A constant a is subtracted from each score in the set?
    c. Each score is multiplied by a constant a?
    d. Each score is divided by a constant a?
    Illustrate each of these with a numerical example.

39. What happens to the standard deviation of a set of scores if
    a. A constant a is added to each score in the set?
    b. A constant a is subtracted from each score in the set?
    c. Each score is multiplied by a constant a?
    d. Each score is divided by a constant a?
    Illustrate each of these with a numerical example.

40. Suppose that, as is done in some lotteries, we sample balls from a big vessel. The vessel contains a large number of balls, each labeled with a single number, 0–9. There are an equal number of balls for each number, and the balls are continually being mixed. For this example, let's collect 10 samples of three balls each. Each sample is formed by selecting balls one at a time and replacing each ball back in the vessel before selecting the next ball. The selection process used ensures that every ball in the vessel has an equal chance of being chosen on each selection. Assume the following samples are collected.

| 1, 3, 4 | 2, 2, 6 | 3, 8, 8 | 1, 6, 7 | 5, 6, 9 |
| 3, 4, 7 | 1, 2, 6 | 2, 3, 7 | 6, 8, 9 | 4, 7, 9 |

    a. Calculate the mean of each sample.
    b. Calculate the median of each sample.
    Based on the properties of the mean and median discussed previously in the chapter, do you expect more variability in the means or medians? Verify this by calculating the standard deviation of the means and medians. other

■ SPSS ILLUSTRATIVE EXAMPLE 4.1

The general operation of SPSS and data entry are described in Appendix E, *Introduction to SPSS*. SPSS is very useful for computing statistics used to quantify central tendency and variability. The illustrative example will show you how to compute some of them.

example

Use SPSS to compute the mean, standard deviation, variance, and range for the following set of mathematics exam scores. Label the scores *Mathexam*.

Mathexam: 78, 65, 47, 38, 86, 57, 88, 66, 43, 95, 73, 82, 61

SOLUTION

STEP 1:   Enter the Data. Enter the statistics exam scores in the first column (**VAR00001**) of the Data Editor, Data View, beginning with the first score in the first cell of the first column.

STEP 2:   Name the Variables. For this example, we will name the scores, *Mathexam*.

Click the **Variable View** tab in the lower left corner of the Data Editor.	This displays the **Variable View** on screen with first cell of the **Name** column containing **VAR00001**.
Click VAR00001; then **type Mathexam** in the highlighted cell and **press Enter.**	**Mathexam** is entered as the variable name, replacing **VAR00001**.

STEP 3:   Analyze the Data. We will compute the mean, standard deviation, variance, and range for the *Mathexam* scores. Before doing so, **click** on the **Data View** tab of the Data Editor to display the **Data Editor–Data View** screen. Displaying this screen helps to relate the data to the results shown in the output table.

Click Analyze on the menu bar at the top of the screen; then **select Descriptive Statistics;** then **click Descriptives....**	This produces the **Descriptives** dialog box which SPSS uses to do descriptive statistics. **Mathexam** is displayed, highlighted in the large box on the left.
Click the **arrow** in the middle of the dialog box.	This moves **Mathexam** from the large box on the left into the **Variable(s):** box on the right.
Click Options... at the top right of the dialog box.	This produces the **Descriptions: Options** dialog box which allows you to select the descriptive statistics that you wish to compute. The checked boxes indicate the default statistics that SPSS computes.
Click Minimum and **Maximum;** then **Click Variance;** then **Click Range.**	This removes the default **checked** entries for **Minimum** and **Maximum**, and produces a **check** in the **Variance** and **Range** boxes. Since the **Mean** and **Std. deviation** boxes were already checked, the boxes **for Mean, Std. deviation, Variance**, and **Range** should now be the only boxes checked. SPSS will compute these statistics when given the **OK** command from the **Descriptions** dialog box.

Click Continue.

This returns you to the **Descriptions** dialog box where you can give the **OK** command.

Click OK.

SPSS then analyzes the data and displays the results shown below.

Analysis Results

Descriptive Statistics

	N	Range	Mean	Std. Deviation	Variance
Mathexam	13	57.00	67.6154	18.07640	326.756
Valid N (listwise)	13				

Sure beats computing these statistics by hand!!

■ SPSS ADDITIONAL PROBLEMS

1. Use SPSS to compute the mean, standard deviation, variance, and range for the following sets of scores. Name the scores, *Scores*.
   a. 8, 2, 6, 12, 4, 7, 4, 10, 13, 15, 11, 12, 5
   b. 7.2, 2.3, 5.4, 2.3, 3.4, 9.2, 7.6, 4.7, 2.8, 6.5
   c. 23, 65, 47, 38, 86, 57, 32, 66, 43, 85, 29, 40, 42
   d. 212, 334, 250, 436, 425, 531, 600, 487, 529, 234, 515
   e. Does SPSS use N or $N - 1$ in the denominator when computing the standard deviation? How could you determine the correct answer without looking it up or asking someone?

2. Use SPSS to demonstrate that the mean of a set of scores can vary without changing the standard deviation.

3. Use SPSS to demonstrate that the standard deviation of a set of scores can vary without changing the mean.

■ NOTES

4.1 To show that $\Sigma (X_i - \overline{X}) = 0$,

$$\Sigma (X_i - \overline{X}) = \Sigma X_i - \Sigma \overline{X}$$
$$= \Sigma X_i - N \overline{X}$$
$$= \Sigma X_i - N\left(\frac{\Sigma X_i}{N}\right)$$
$$= \Sigma X_i - \Sigma X_i$$
$$= 0$$

4.2 To show that $SS = \Sigma X^2 - [(\Sigma X)^2/N]$,

$$SS = \Sigma (X - \overline{X})^2$$
$$= \Sigma (X^2 - 2X\overline{X} + \overline{X}^2)$$
$$= \Sigma X^2 - \Sigma 2X\overline{X} + \Sigma \overline{X}^2$$
$$= \Sigma X^2 - 2\overline{X} \Sigma X + N\overline{X}^2$$
$$= \Sigma X^2 - 2\left(\frac{\Sigma X}{N}\right)\Sigma X + \frac{N(\Sigma X)^2}{N^2}$$
$$= \Sigma X^2 - \frac{2(\Sigma X)^2}{N} + \frac{(\Sigma X)^2}{N}$$
$$= \Sigma X^2 - \frac{(\Sigma X)^2}{N}$$

■ ONLINE STUDY RESOURCES

CENGAGE **brain**.com

Login to CengageBrain.com to access the resources your instructor has assigned. For this book, you can access the book's companion website for chapter-specific learning tools including Know and Be Able to Do, practice quizzes, flash cards, and glossaries and a link to Statistics and Research Methods Workshops

aplia

If your professor has assigned Aplia homework:

1. Sign in to your account
2. Complete the corresponding homework exercises as required by your professor
3. When finished, click "Grade It Now" to see which areas you have mastered and which need more work, and for detailed explanations of every answer.

Visit **www.cengagebrain.com** to access your account and to purchase materials.

5

The Normal Curve and Standard Scores

© Strmko / Dreamstime.com

LEARNING OBJECTIVES

After completing this chapter, you should be able to:

- Describe the typical characteristics of a normal curve.
- Define a z score.
- Compute the z score for a raw score, given the raw score, the mean, and standard deviation of the distribution.
- Compute the z score for a raw score, given the raw score and the distribution of raw scores.
- Explain the three main features of z distributions.
- Use z scores with a normal curve to find: (a) the percentage of scores falling below any raw score in the distribution, (b) the percentage of scores falling above any raw score in the distribution, and (c) the percentage of scores falling between any two raw scores in the distribution.
- Understand the illustrative examples, do the practice problems, and understand the solutions.

INTRODUCTION

The normal curve is a very important distribution in the behavioral sciences. There are three principal reasons why. First, many of the variables measured in behavioral science research have distributions that quite closely approximate the normal curve. Height, weight, intelligence, and achievement are a few examples. Second, many of the inference tests used in analyzing experiments have sampling distributions that become normally distributed with increasing sample size. The sign test and Mann–Whitney U test are two such tests, which we shall cover later in the text. Finally, many inference tests require sampling distributions that are normally distributed (we shall discuss sampling distributions in Chapter 12). The z test, Student's t test, and the F test are examples of inference tests that depend on this point. Thus, much of the importance of the normal curve occurs in conjunction with inferential statistics.

THE NORMAL CURVE

The *normal curve* is a theoretical distribution of population scores. It is a bell-shaped curve that is described by the following equation:

$$Y = \frac{N}{\sqrt{2\pi}\,\sigma}\, e^{-(X-\mu)^2/2\sigma^2} \quad \textit{equation of the normal curve}$$

where
Y = frequency of a given value of X*
X = any score in the distribution
μ = mean of the distribution
σ = standard deviation of the distribution
N = total frequency of the distribution
π = a constant of 3.1416
e = a constant of 2.7183

MENTORING TIP

Note that the normal curve is a theoretical curve and is only approximated by real data.

Most of us will never need to know the exact equation for the normal curve. It has been given here primarily to make the point that the normal curve is a theoretical curve that is mathematically generated. An example of the normal curve is shown in Figure 5.1.

Note that the curve has two inflection points, one on each side of the mean. Inflection points are located where the curvature changes direction. In Figure 5.1, the inflection points are located where the curve changes from being convex downward to being convex upward. If the bell-shaped curve is a normal curve, the inflection points are at 1 standard deviation from the mean ($\mu + 1\sigma$ and $\mu - 1\sigma$). Note also that as the curve approaches the horizontal axis, it is slowly changing its Y value. Theoretically, the curve never quite reaches the axis. It approaches the horizontal axis and gets closer and closer to it, but it never quite touches it. The curve is said to be *asymptotic* to the horizontal axis.

---

*The labeling of Y as "frequency" is a slight simplification. I believe this simplification aids considerably in understanding and applying the material that follows. Strictly speaking, it is the area under the curve, between any two X values, that is properly referred to as "frequency." For a discussion of this point, see E. Minium and B. King, *Statistical Reasoning in Psychology and Education*, 4th ed., John Wiley and Sons, New York, 2008, p. 119.

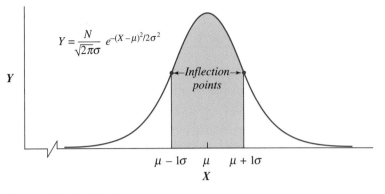

figure 5.1   Normal curve.

Area Contained Under the Normal Curve

In distributions that are normally shaped, there is a special relationship between the mean and the standard deviation with regard to the area contained under the curve. When a set of scores is normally distributed, 34.13% of the area under the curve is contained between the mean (μ) and a score that is equal to $\mu + 1\sigma$; 13.59% of the area is contained between a score equal to $\mu + 1\sigma$ and a score of $\mu + 2\sigma$; 2.15% of the area is contained between scores of $\mu + 2\sigma$ and $\mu + 3\sigma$; and 0.13% of the area exists beyond $\mu + 3\sigma$. This accounts for 50% of the area. Since the curve is symmetrical, the same percentages hold for scores below the mean. These relationships are shown in Figure 5.2. Since frequency is plotted on the vertical axis, these percentages represent the *percentage of scores* contained within the area.

To illustrate, suppose we have a population of 10,000 IQ scores. The distribution is normally shaped, with $\mu = 100$ and $\sigma = 16$. Since the scores are normally distributed, 34.13% of the scores are contained between scores of 100 and 116 ($\mu + 1\sigma = 100 + 16 = 116$), 13.59% between 116 and 132 ($\mu + 2\sigma = 100 + 32 = 132$), 2.15% between 132 and 148, and 0.13% above 148. Similarly, 34.13% of the scores fall between 84 and 100, 13.59% between 68 and 84, 2.15% between 52 and 68, and 0.13% below 52. These relationships are also shown in Figure 5.2.

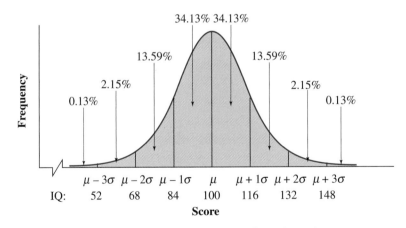

figure 5.2   Areas under the normal curve for selected scores.

To calculate the number of scores in each area, all we need to do is multiply the relevant percentage by the total number of scores. Thus, there are 34.13% × 10,000 = 3413 scores between 100 and 116, 13.59% × 10,000 = 1359 scores between 116 and 132, and 215 scores between 132 and 148; 13 scores are greater than 148. For the other half of the distribution, there are 3413 scores between 84 and 100, 1359 scores between 68 and 84, and 215 scores between 52 and 68; there are 13 scores below 52. Note that these frequencies would be true only if the distribution is exactly normally distributed. In actual practice, the frequencies would vary slightly depending on how close the distribution is to this theoretical model.

STANDARD SCORES (z SCORES)

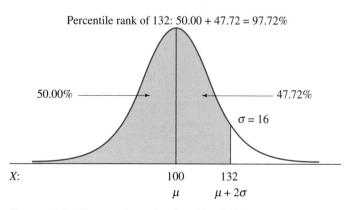

Suppose someone told you your IQ is 132. Would you be happy or sad? In the absence of additional information, it is difficult to say. An IQ of 132 is meaningless unless you have a reference group to compare against. Without such a group, you can't tell whether the score is high, average, or low. For the sake of this illustration, let's assume your score is one of the 10,000 scores of the distribution just described. Now we can begin to give your IQ score of 132 some meaning. For example, we can determine the percentage of scores in the distribution that are lower than 132. You will recognize this as determining the percentile rank of the score of 132. (As you no doubt recall, the percentile rank of a score is defined as the percentage of scores that are below the score in question.) Referring to Figure 5.2, we can see that 132 is 2 standard deviations above the mean. In a normal curve, there are 34.13 + 13.59 = 47.72% of the scores between the mean and a score that is 2 standard deviations above the mean. To find the percentile rank of 132, we need to add to this percentage the 50.00% that lie below the mean. Thus, 97.72% (47.72 + 50.00) of the scores fall below your IQ score of 132. You should be quite happy to be so intelligent. The solution is shown in Figure 5.3.

To solve this problem, we had to determine how many standard deviations the raw score of 132 was above or below the mean. In so doing, we transformed the raw score into a *standard score*, also called a *z score*.

figure 5.3   Percentile rank of an IQ of 132.

definition ■ *A z score is a transformed score that designates how many standard deviation units the corresponding raw score is above or below the mean.*

In equation form,

$$z = \frac{X - \mu}{\sigma} \quad \text{z score for population data}$$

$$z = \frac{X - \overline{X}}{s} \quad \text{z score for sample data}$$

For the previous example,

$$z = \frac{X - \mu}{\sigma} = \frac{132 - 100}{16} = 2.00$$

The process by which the raw score is altered is called a *score transformation*. We shall see later that the *z* transformation results in a distribution having a mean of 0 and a standard deviation of 1. The reason *z* scores are called standard scores is that they are expressed relative to a distribution mean of 0 and a standard deviation of 1.

In conjunction with a normal curve, *z* scores allow us to determine the number or percentage of scores that fall above or below any score in the distribution. In addition, *z* scores allow comparison between scores in different distributions, even when the units of the distributions are different. To illustrate this point, let's consider another population set of scores that are normally distributed. Suppose that the weights of all the rats housed in a university vivarium are normally distributed, with $\mu = 300$ and $\sigma = 20$ grams. What is the percentile rank of a rat weighing 340 grams?

The solution is shown in Figure 5.4. First, we need to convert the raw score of 340 grams to its corresponding *z* score:

$$z = \frac{X - \mu}{\sigma} = \frac{340 - 300}{20} = 2.00$$

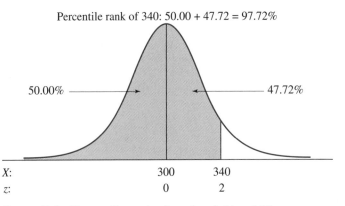

Percentile rank of 340: 50.00 + 47.72 = 97.72%

50.00% ⟶   ⟵ 47.72%

X: 300   340
z: 0   2

figure 5.4 Percentile rank of a rat weighing 340 grams.

Since the scores are normally distributed, 34.13 + 13.59 = 47.72% of the scores are between the score and the mean. Adding the remaining 50.00% that lie below the mean, we arrive at a percentile rank of 47.72 + 50.00 = 97.72% for the weight of 340 grams. Thus, the IQ score of 132 and the rat's weight of 340 grams have something in common. They both occupy the same relative position in their respective distributions. The rat is as heavy as you are smart.

This example, although somewhat facetious, illustrates an important use of z scores—namely, to compare scores that are not otherwise directly comparable. Ordinarily, we would not be able to compare intelligence and weight. They are measured on different scales and have different units. But by converting the scores to their z-transformed scores, we eliminate the original units and replace them with a universal unit, the standard deviation. Thus, your score of 132 IQ units becomes a score of 2 standard deviation units above the mean, and the rat's weight of 340 grams also becomes a score of 2 standard deviation units above the mean. In this way, it is possible to compare "anything with anything" as long as the measuring scales allow computation of the mean and standard deviation. The ability to compare scores that are measured on different scales is of fundamental importance to the topic of correlation. We shall discuss this in more detail when we take up that topic in Chapter 6.

So far, the examples we've been considering have dealt with populations. It might be useful to practice computing z scores using sample data. Let's do this in the next practice problem.

Practice Problem 5.1

For the set of sample raw scores $X = 1, 4, 5, 7, 8$ determine the z score for each raw score.

STEP 1: Determine the mean of the raw scores.

$$\overline{X} = \frac{\Sigma X_i}{N} = \frac{25}{5} = 5.00$$

STEP 2: Determine the standard deviation of the scores.

$$s = \sqrt{\frac{SS}{N-1}} \qquad SS = \Sigma X^2 - \frac{(\Sigma X)^2}{N}$$

$$= \sqrt{\frac{30}{4}} \qquad = 155 - \frac{(25)^2}{5}$$

$$= 2.7386 \qquad = 30$$

(continued)

STEP 3: Compute the z score for each raw score.

X	z
1	$z = \dfrac{X - \bar{X}}{s} = \dfrac{1 - 5}{2.7386} = -1.46$
4	$z = \dfrac{X - \bar{X}}{s} = \dfrac{4 - 5}{2.7386} = -0.37$
5	$z = \dfrac{X - \bar{X}}{s} = \dfrac{5 - 5}{2.7386} = 0.00$
7	$z = \dfrac{X - \bar{X}}{s} = \dfrac{7 - 5}{2.7386} = 0.73$
8	$z = \dfrac{X - \bar{X}}{s} = \dfrac{8 - 5}{2.7386} = 1.10$

Characteristics of z Scores

There are three characteristics of z scores worth noting. First, *the z scores have the same shape as the set of raw scores.* Transforming the raw scores into their corresponding z scores does not change the shape of the distribution. Nor do the scores change their relative positions. All that is changed are the score values. Figure 5.5 illustrates this point by showing the IQ scores and their corresponding z scores. You should note that although we have used z scores in conjunction with the normal distribution, all z distributions are not normally shaped. If we use the z equation given previously, z scores can be calculated for distributions of any shape. The resulting z scores will take on the shape of the raw scores.

Second, *the mean of the z scores always equals zero* ($\mu_z = 0$). This follows from the observation that the scores located at the mean of the raw scores will also be at the mean of the z scores (see Figure 5.5). The z value for raw scores at the mean equals zero. For example, the z transformation for a score at the mean of the IQ distribution is given by $z = (X - \mu)/\sigma = (100 - 100)/16 = 0$. Thus, the mean of the z distribution equals zero.

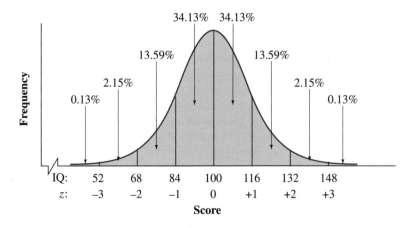

figure 5.5 Raw IQ scores and corresponding z scores.

The last characteristic of importance is that the *standard deviation of z scores always equals 1* ($\sigma_z = 1$). This follows because a raw score that is 1 standard deviation above the mean has a *z* score of +1:

$$z = \frac{(\mu + 1\sigma) - \mu}{\sigma} = 1$$

Finding the Area, Given the Raw Score

In the previous examples with IQ and weight, the *z* score was carefully chosen so that the solution could be found from Figure 5.2. However, suppose instead of an IQ of 132, we desire to find the percentile rank of an IQ of 142. Assume the same population parameters. The solution is shown in Figure 5.6. First, draw a curve showing the population and locate the relevant area by entering the score 142 on the horizontal axis. Then shade in the area desired. Next, calculate *z*:

$$z = \frac{X - \mu}{\sigma} = \frac{142 - 100}{16} = \frac{42}{16} = 2.62$$

Since neither Figure 5.2 nor Figure 5.5 shows a percentage corresponding to a *z* score of 2.62, we cannot use these figures to solve the problem. Fortunately, the areas under the normal curve for various *z* scores have been computed, and the resulting values are shown in Table A of Appendix D.

The first column of the table (column A) contains the z score. Column B lists the proportion of the total area between a given z score and the mean. Column C lists the proportion of the total area that exists beyond the z score.

We can use Table A to find the percentile rank of 142. First, we locate the *z* score of 2.62 in column A. Next, we determine from column B the proportion of the total area between the *z* score and the mean. For a *z* score of 2.62, this area equals 0.4956. To this value we must add 0.5000 to take into account the scores lying below the mean (the picture helps remind us to do this). Thus, the proportion of scores that lie below an IQ of 142 is 0.4956 + 0.5000 = 0.9956. To convert this proportion to a percentage, we must multiply by 100. Thus, the percentile rank of 142 is 99.56. Table A can be used to find the area for any *z* score, provided the scores are normally distributed. When using Table A, it is usually sufficient to round *z* values to two-decimal-place accuracy. Let's do a few more illustrative problems for practice.

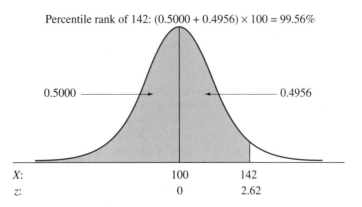

Percentile rank of 142: $(0.5000 + 0.4956) \times 100 = 99.56\%$

0.5000 — 0.4956

| X: | 100 | 142 |
| z: | 0 | 2.62 |

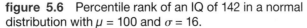

figure 5.6   Percentile rank of an IQ of 142 in a normal distribution with $\mu = 100$ and $\sigma = 16$.

Practice Problem 5.2

The scores on a nationwide mathematics aptitude exam are normally distributed, with $\mu = 80$ and $\sigma = 12$. What is the percentile rank of a score of 84?

SOLUTION

MENTORING TIP

Always draw the picture first.

In solving problems involving areas under the normal curve, it is wise, at the outset, to draw a picture of the curve and locate the relevant areas on it. The accompanying figure shows such a picture. The shaded area contains all the scores lower than 84. To find the percentile rank of 84, we must first convert 84 to its corresponding z score:

$$z = \frac{X - \mu}{\sigma} = \frac{84 - 80}{12} = \frac{4}{12} = 0.33$$

To find the area between the mean and a z score of 0.33, we enter Table A, locate the z value in column A, and read off the corresponding entry in column B. This value is 0.1293. Thus, the proportion of the total area between the mean and a z score of 0.33 is 0.1293. From the accompanying figure, we can see that the remaining scores below the mean occupy 0.5000 proportion of the total area. If we add these two areas together, we shall have the proportion of scores lower than 84. Thus, the proportion of scores lower than 84 is $0.1293 + 0.5000 = 0.6293$. The percentile rank of 84 is then $0.6293 \times 100 = 62.93$.

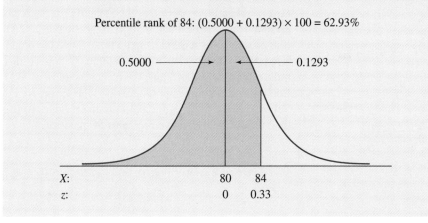

Percentile rank of 84: $(0.5000 + 0.1293) \times 100 = 62.93\%$

0.5000 ⟶   ⟵ 0.1293

X:	80	84
z:	0	0.33

Practice Problem 5.3

What percentage of aptitude scores are below a score of 66?

SOLUTION

Again, the first step is to draw the appropriate diagram. This is shown in the accompanying figure. From this diagram, we can see that the relevant area (shaded) lies beyond the score of 66. To find the percentage of scores contained in this area, we must first convert 66 to its corresponding z score. Thus,

$$z = \frac{X - \mu}{\sigma} = \frac{66 - 80}{12} = \frac{-14}{12} = -1.17$$

From Table A, column C, we find that the area beyond a z score of 1.17 is 0.1210. Thus, the percentage of scores below 66 is 0.1210 × 100 = 12.10%. Table A does not show any negative z scores. However, this does not cause a problem because the normal curve is symmetrical and negative z scores have the same proportion of area as positive z scores of the same magnitude. Thus, the proportion of total area lying beyond a z score of +1.17 is the same as the proportion lying beyond a z score of −1.17.

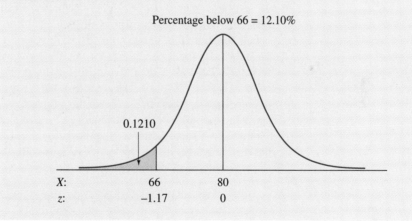

Practice Problem 5.4

Using the same population as in Practice Problem 5.3, what percentage of scores fall between 64 and 90?

SOLUTION

The relevant diagram is shown at the end of the practice problem. This time, the shaded areas are on either side of the mean. To solve this problem, we must find the area between 64 and 80 and add it to the area between 80 and 90. As before, to determine area, we must calculate the appropriate z score. This time, however, we must compute two z scores. For the area to the left of the mean,

$$z = \frac{64 - 80}{12} = \frac{-16}{12} = -1.33$$

For the area to the right of the mean,

$$z = \frac{90 - 80}{12} = \frac{10}{12} = 0.83$$

Since the areas we want to determine are between the mean and the z score, we shall use column B of Table A. The area corresponding to a z score of -1.33 is 0.4082, and the area corresponding to a z score of 0.83 is 0.2967. The total area equals the sum of these two areas. Thus, the proportion of scores falling between 64 and 90 is $0.4082 + 0.2967 = 0.7049$. The percentage of scores between 64 and 90 is $0.7049 \times 100 = 70.49\%$. Note that in this problem we cannot just subtract 64 from 90 and divide by 12. The areas in Table A are designated with the mean as a reference point. Therefore, to solve this problem, we must relate the scores of 64 and 90 to the mean of the distribution. You should also note that you cannot just subtract one z value from the other because the curve is not rectangular; rather, it has differing amounts of area under various points of the curve.

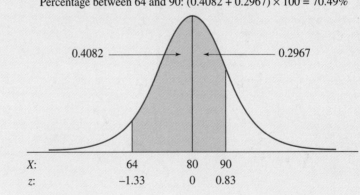

Percentage between 64 and 90: $(0.4082 + 0.2967) \times 100 = 70.49\%$

0.4082 — 0.2967

| X: | 64 | 80 | 90 |
| z: | −1.33 | 0 | 0.83 |

Practice Problem 5.5

Another type of problem arises when we want to determine the area between two scores and both scores are either above or below the mean. Let's try a problem of this sort. Find the percentage of aptitude scores falling between the scores of 95 and 110.

SOLUTION

The accompanying figure shows the distribution and the relevant area. As in Practice Problem 5.4, we can't just subtract 95 from 110 and divide by 12 to find the appropriate z score. Rather, we must use the mean as our reference point. In this problem, we must find (1) the area between 110 and the mean and (2) the area between 95 and the mean. By subtracting these two areas, we shall arrive at the area between 95 and 110. As before, we must calculate two z scores:

$$z = \frac{110 - 80}{12} = \frac{30}{12} = 2.50 \quad \textit{z transformation of 110}$$

$$z = \frac{95 - 80}{12} = \frac{15}{12} = 1.25 \quad \textit{z transformation of 95}$$

From column B of Table A,

$$\text{Area } (z = 2.50) = 0.4938$$

and

$$\text{Area } (z = 1.25) = 0.3944$$

Thus, the proportion of scores falling between 95 and 110 is 0.4938 − 0.3944 = 0.0994. The percentage of scores is 0.0994 × 100 = 9.94%.

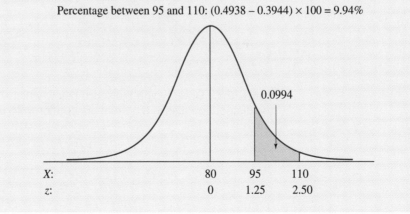

Percentage between 95 and 110: (0.4938 − 0.3944) × 100 = 9.94%

Finding the Raw Score, Given the Area

Sometimes we know the area and want to determine the corresponding score. The following problem is of this kind. Find the raw score that divides the distribution of aptitude scores such that 70% of the scores are below it.

This problem is just the reverse of the previous one. Here, we are given the area and need to determine the score. Figure 5.7 shows the appropriate diagram. Although we don't know what the raw score value is, we can determine its corresponding z score from Table A. Once we know the z score, we can solve for the raw score using the z equation. If 70% of the scores lie below the raw score, then 30% must lie above it. We can find the z score by searching in Table A, column C, until we locate the area closest to 0.3000 (30%) and then noting that the z score corresponding to this area is 0.52. To find the raw score, all we need to do is substitute the relevant values in the z equation and solve for X. Thus,

$$z = \frac{X - \mu}{\sigma}$$

Substituting and solving for X,

$$0.52 = \frac{X - 80}{12}$$

$$X = 80 + 12(0.52) = 86.24$$

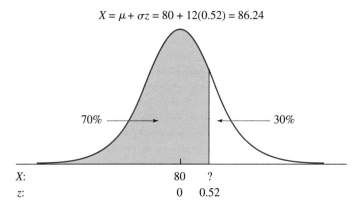

$$X = \mu + \sigma z = 80 + 12(0.52) = 86.24$$

70% ⟶   ⟵ 30%

X: 80   ?
z: 0   0.52

figure 5.7 Determining the score below which 70% of the distribution falls in a normal distribution with $\mu = 80$ and $\sigma = 12$.

Practice Problem 5.6

Let's try another problem of this type. What is the score that divides the distribution such that 99% of the area is below it?

SOLUTION

The diagram is shown below. If 99% of the area is below the score, 1% must be above it. To solve this problem, we locate the area in column C of Table A that is closest to 0.0100 (1%) and note that $z = 2.33$. We convert the z score to its corresponding raw score by substituting the relevant values in the z equation and solving for X. Thus,

$$z = \frac{X - \mu}{\sigma}$$

$$2.33 = \frac{X - 80}{12}$$

$$X = 80 + 12(2.33) = 107.96$$

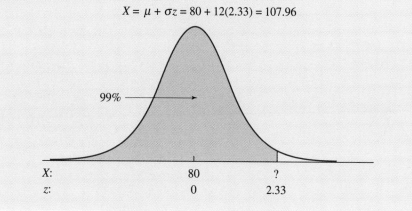

$$X = \mu + \sigma z = 80 + 12(2.33) = 107.96$$

X:	80	?
z:	0	2.33

99%

Practice Problem 5.7

Let's do one more problem. What are the scores that bound the middle 95% of the distribution?

SOLUTION

The diagram is shown below. There is an area of 2.5% above and below the middle 95%. To determine the scores that bound the middle 95% of the distribution, we must first find the z values and then convert these values to raw scores. The z scores are found in Table A by locating the area in column C closest to 0.0250 (2.5%) and reading the associated z score in column A. In this case, $z = \pm 1.96$. The raw scores are found by substituting the relevant values in the z equation and solving for X. Thus,

$$z = \frac{X - \mu}{\sigma}$$

$$-1.96 = \frac{X - 80}{12}$$
$$X = 80 + 12(-1.96)$$
$$= 56.48$$

$$+1.96 = \frac{X - 80}{12}$$
$$X = 80 + 12(1.96)$$
$$= 103.52$$

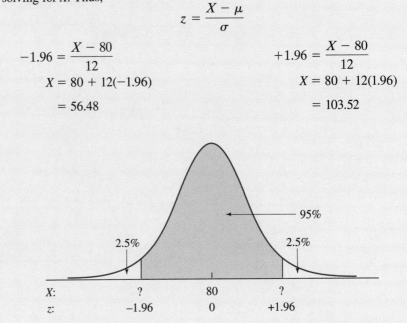

■ SUMMARY

In this chapter, I have discussed the normal curve and standard scores. I pointed out that the normal curve is a bell-shaped curve and gave the equation describing it. Next, I discussed the area contained under the normal curve and its relation to z scores. A z score is a transformation of a raw score. It designates how many standard deviation units the corresponding raw score is above or below the mean. A z distribution has the following characteristics: (1) the

z scores have the same shape as the set of raw scores, (2) the mean of z scores always equals 0, and (3) the standard deviation of z scores always equals 1. Finally, I showed how to use z scores in conjunction with a normal distribution to find (1) the percentage or frequency of scores corresponding to any raw score in the distribution and (2) the raw score corresponding to any frequency or percentage of scores in the distribution.

■ IMPORTANT NEW TERMS

Asymptotic (p. 103)
Normal curve (p. 103)

Standard scores (z scores) (p. 105)

■ QUESTIONS AND PROBLEMS

1. Define
   a. Asymptotic
   b. The normal curve
   c. z scores
   d. Standard scores
2. What is a score transformation? Provide an example.
3. What are the values of the mean and standard deviation of the z distribution?
4. Must the shape of a z distribution be normal? Explain.
5. Are all bell-shaped distributions normal distributions? Explain.
6. If a set of scores is normally distributed, what information does the area under the curve give us?
7. What proportion of scores in a normal distribution will have values lower than $z = 0$? What proportion will have values greater than $z = 0$?
8. Given the set of sample raw scores 10, 12, 16, 18, 19, 21,
   a. Convert each raw score to its z-transformed value.
   b. Compute the mean and standard deviation of the z scores.
9. Assume the raw scores in Problem 8 are population scores and perform the calculations called for in parts **a** and **b**.
10. A population of raw scores is normally distributed with $\mu = 60$ and $\sigma = 14$. Determine the z scores for the following raw scores taken from that population:
    a. 76
    b. 48
    c. 86
    d. 60
    e. 74
    f. 46

11. For the following z scores, determine the percentage of scores that lie beyond z:
    a. 0
    b. 1
    c. 1.54
    d. –2.05
    e. 3.21
    f. –0.45
12. For the following z scores, determine the percentage of scores that lie between the mean and the z score:
    a. 1
    b. –1
    c. 2.34
    d. –3.01
    e. 0
    f. 0.68
    g. –0.73
13. For each of the following, determine the z score that divides the distribution such that the given percentage of scores lies above the z score (round to two decimal places):
    a. 50%
    b. 2.50%
    c. 5%
    d. 30%
    e. 80%
    f. 90%
14. Given that a population of scores is normally distributed with $\mu = 110$ and $\sigma = 8$, determine the following:
    a. The percentile rank of a score of 120
    b. The percentage of scores that are below a score of 99

c. The percentage of scores that are between a score of 101 and 122
d. The percentage of scores that are between a score of 114 and 124
e. The score in the population above which 5% of the scores lie

15. At the end of a particular quarter, Carol took four final exams. The mean and standard deviation for each exam along with Carol's grade on each exam are listed here. Assume that the grades on each exam are normally distributed.

Exam	Mean	Standard Deviation	Carol's Grade
French	75.4	6.3	78.2
History	85.6	4.1	83.4
Psychology	88.2	3.5	89.2
Statistics	70.4	8.6	82.5

a. On which exam did Carol do best, relative to the other students taking the exam?
b. What was her percentile rank on this exam? education

16. A hospital in a large city records the weight of every infant born at the hospital. The distribution of weights is normally shaped, with a mean $\mu = 2.9$ kilograms and a standard deviation $\sigma = 0.45$. Determine the following:
a. The percentage of infants who weighed less than 2.1 kilograms
b. The percentile rank of a weight of 4.2 kilograms
c. The percentage of infants who weighed between 1.8 and 4.0 kilograms
d. The percentage of infants who weighed between 3.4 and 4.1 kilograms
e. The weight that divides the distribution such that 1% of the weights are above it
f. Beyond what weights do the most extreme 5% of the scores lie?
g. If 15,000 infants have been born at the hospital, how many weighed less than 3.5 kilograms? health, I/O

17. A statistician studied the records of monthly rainfall for a particular geographic locale. She found that the average monthly rainfall was normally distributed with a mean $\mu = 8.2$ centimeters and a standard deviation $\sigma = 2.4$. What is the percentile rank of the following scores?
a. 12.4
b. 14.3
c. 5.8

d. 4.1
e. 8.2 I/O, other

18. Using the same population parameters as in Problem 17, find what percentage of scores are above the following scores:
a. 10.5
b. 13.8
c. 7.6
d. 3.5
e. 8.2 I/O, other

19. Using the same population parameters as in Problem 17, find what percentage of scores are between the following scores:
a. 6.8 and 10.2
b. 5.4 and 8.0
c. 8.8 and 10.5 I/O, other

20. A jogging enthusiast keeps track of how many miles he jogs each week. The following scores are sampled from his year 2007 records:

Week	Distance*	Week	Distance
5	32	30	36
8	35	32	38
10	30	38	35
14	38	43	31
15	37	48	33
19	36	49	34
24	38	52	37

*Scores are miles run.

a. Determine the z scores for the distances shown in the table. Note that the distances are sample scores.
b. Plot a frequency polygon for the raw scores.
c. On the same graph, plot a frequency polygon for the z scores.
d. Is the z distribution normally shaped? If not, explain why.
e. Compute the mean and standard deviation of the z distribution. I/O, other

21. A stock market analyst has kept records for the past several years of the daily selling price of a particular blue-chip stock. The resulting distribution of scores is normally shaped with a mean $\mu = \$84.10$ and a standard deviation $\sigma = \$7.62$.
a. Determine the percentage of selling prices that were below a price of $95.00.
b. What percentage of selling prices were between $76.00 and $88.00?
c. What percentage of selling prices were above $70.00?

d. What selling price divides the distribution such that 2.5% of the scores are above it? I/O

22. Anthony is deciding whether to go to graduate school in business or law. He has taken nationally administered aptitude tests for both fields. Anthony's scores along with the national norms are shown here. Based solely on Anthony's relative standing on these tests, which field should he enter? Assume that the scores on both tests are normally distributed.

	National Norms		
Field	μ	σ	**Anthony's Scores**
Business	68	4.2	80.4
Law	85	3.6	89.8

education

23. On which of her two exams did Rebecca do better? How about Maurice? Assume the scores on each exam are normally distributed.

	μ	σ	**Rebecca's Scores**	**Maurice's Scores**
Exam 1	120	6.8	130	132
Exam 2	50	2.4	56	55

education

24. A psychologist interested in the intelligence of children develops a standardized test for selecting "gifted" children. The test scores are normally distributed, with $\mu = 75$ and $\sigma = 8$. Assume a gifted child is defined as one who scores in the upper 1% of the distribution. What is the minimum score needed to be selected as gifted? cognitive, developmental

■ SPSS ILLUSTRATIVE EXAMPLE 5.1

The general operation of SPSS and data entry are described in Appendix E, *Introduction to SPSS*. SPSS can be quite useful for transforming sample raw scores into z scores. Let's see how to do this.

example

Let's use SPSS to solve Practice Problem 5.1 on p. 107 of the textbook. For convenience, the practice problem is repeated here.

For the set of raw scores $X = 1, 4, 5, 7, 8$, determine the z score for each raw score.

In solving the problem, name the variable X.

SOLUTION

STEP 1: **Enter the Data.** Enter the statistics exam scores in the first column (**VAR00001**) of the SPSS Data Editor, beginning with the first score in the first cell of the first column.

STEP 2: **Name the Variables.** For this example, we will name the scores X.

Click the **Variable View** tab in the lower left corner of the Data Editor.	This displays the **Variable View** on screen with first cell of the **Name** column containing **VAR00001**.
Click VAR00001; then **type X** in the highlighted cell and **press Enter**.	**X** is entered as the variable name, replacing **VAR00001**.

STEP 3:   **Analyze the Data.** Next, we will use SPSS to compute the *z* score for each raw score. I suggest you switch to the Data Editor, Data View if you haven't done so already; now, on with the analysis.

Click **Analyze** on the menu bar at the top of the screen; then select **Descriptive Statistics;** then **click Descriptives....**

Click the **arrow** in the middle of the dialog box.

Click Save **standardized values as variables**.

Click **OK**.

This produces the **Descriptives** dialog box, which SPSS uses to do descriptive statistics. **X** is displayed, highlighted in the large box on the left.

This moves **X** from the large box on the left into the **Variable(s):** box on the right, ready for analysis.

This produces a **check** in the **Save standardized values as variables** box.

SPSS then analyzes the data, computes and displays the default or selected statistics as discussed in Chapter 4. It also computes the *z* values of the raw scores and displays the *z* scores as a new variable **ZX** in the Data Editor, Data View as shown below. Note: to view the Data Editor from the Output screen, click on **Window** on the menu bar at the top; then click on the file that contains your data set. Once in the Data Editor, you may have to switch from the Variable View to the Data View.

Analysis Results

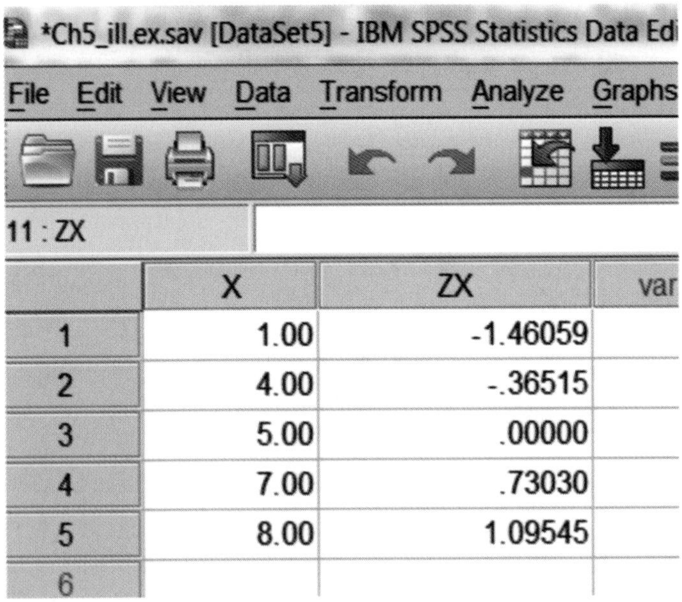

*Ch5_ill.ex.sav [DataSet5] - IBM SPSS Statistics Data Edi

File   Edit   View   Data   Transform   Analyze   Graphs

11 : ZX

	X	ZX	var
1	1.00	-1.46059	
2	4.00	-.36515	
3	5.00	.00000	
4	7.00	.73030	
5	8.00	1.09545	
6			

If you compare the SPSS-generated *z* scores with those in Practice Problem 5.1, you can see that the SPSS *z* scores when rounded to two decimal places are the same as the *z* scores shown in the textbook. A note of caution: SPSS computes z scores for sample, not population raw scores. If you want to get z scores for population raw scores, multiply each sample z score by $\sqrt{\dfrac{N}{N-1}}$.

■ SPSS ADDITIONAL PROBLEMS

1. Use SPSS to compute z scores for the following data set. In solving the problem, name the scores Y.
   10, 13, 15, 16, 18, 20, 21, 24
2. Use SPSS to compute z scores for the distance scores given in Chapter 5, Problem 20, p. 118 in the textbook.

Compare your answer with that given in Appendix C for this problem. Name the scores *Distance*.
3. Use SPSS to demonstrate that the mean of a z distribution of scores equals 0 and the standard deviation equals 1.

■ ONLINE STUDY RESOURCES

CENGAGE **brain**.com

Login to CengageBrain.com to access the resources your instructor has assigned. For this book, you can access the book's companion website for chapter-specific learning tools including Know and Be Able to Do, practice quizzes, flash cards, and glossaries and a link to Statistics and Research Methods Workshops.

aplia

If your professor has assigned Aplia homework:

1. Sign in to your account
2. Complete the corresponding homework exercises as required by your professor
3. When finished, click "Grade It Now" to see which areas you have mastered and which need more work, and for detailed explanations of every answer.

Visit **www.cengagebrain.com** to access your account and to purchase materials.

© Strmko / Dreamstime.com

6

Correlation

LEARNING OBJECTIVES

After completing this chapter, you should be able to:

▪ Define, recognize graphs of, and distinguish between the following: linear and curvilinear relationships, positive and negative relationships, direct and inverse relationships, and perfect and imperfect relationships.

▪ Specify the equation of a straight line and understand the concepts of slope and intercept.

▪ Define scatter plot, correlation coefficient, and Pearson *r*.

▪ Compute the value of Pearson *r*, and state the assumptions underlying Pearson *r*.

▪ Define the coefficient of determination (r^2); specify and explain an important use of r^2.

▪ List three correlation coefficients other than Pearson *r* and specify the factors that determine which correlation coefficient to use; specify the effects on correlation of range and of an extreme score.

▪ Compute the value of Spearman rho (r_s) and specify the scaling of the variables appropriate for its use.

▪ Explain why correlation does not imply causation.

▪ Understand the illustrative examples, do the practice problems, and understand the solutions.

INTRODUCTION

In the previous chapters, we were mainly concerned with single distributions and how to best characterize them. In addition to describing individual distributions, it is often desirable to determine whether the scores of one distribution are related to the scores of another distribution. For example, the person in charge of hiring employees for a large corporation might be very interested in knowing whether there was a relationship between the college grades that were earned by their employees and their success in the company. If a strong relationship between these two variables did exist, college grades could be used to predict success in the company and hence would be very useful in screening prospective employees.

Aside from the practical utility of using a relationship for prediction, why would anyone be interested in determining whether two variables are related? One important reason is that if the variables are related, it is possible that one of them is the cause of the other. As we shall see later in this chapter, the fact that two variables are related is not sufficient basis for proving causality. Nevertheless, because correlational studies are among the easiest to carry out, showing that a correlation exists between the variables is often the first step toward proving that they are causally related. Conversely, if a correlation does not exist between the two variables, a causal relationship can be ruled out.

Another very important use of correlation is to assess the "test–retest reliability" of testing instruments. Test–retest reliability means consistency in scores over repeated administrations of the test. For example, assuming an individual's IQ is stable from month to month, we would expect a good test of IQ to show a strong relationship between the scores of two administrations of the test 1 month apart to the same people. Correlational techniques allow us to measure the relationship between the scores derived on the two administrations and, hence, to measure the test–retest reliability of the instrument.

Correlation and regression are very much related. They both involve the relationship between two or more variables. Correlation is primarily concerned with finding out whether a relationship exists and with determining its magnitude and direction, whereas regression is primarily concerned with using the relationship for prediction. In this chapter, we discuss correlation, and in Chapter 7, we will take up the topic of linear regression.

RELATIONSHIPS

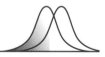

Correlation is a topic that deals primarily with the magnitude and direction of relationships. Before delving into these special aspects of relationships, we will discuss some general features of relationships. With these in hand, we can better understand the material specific to correlation.

Linear Relationships

To begin our discussion of relationships, let's illustrate a linear relationship between two variables. Table 6.1 shows one month's salary for five salespeople and the dollar value of the merchandise each sold that month.

table 6.1 Salary and merchandise sold

Salesperson	X Variable Merchandise Sold ($)	Y Variable Salary ($)
1	0	500
2	1000	900
3	2000	1300
4	3000	1700
5	4000	2100

The relationship between these variables can best be seen by plotting a graph using the paired *X* and *Y* values for each salesman as the points on the graph. Such a graph is called a *scatter plot.*

definition ■ *A **scatter plot** is a graph of paired X and Y values.*

The scatter plot for the salesperson data is shown in Figure 6.1. Referring to this figure, we see that all of the points fall on a straight line. When a straight line describes the relationship between two variables, the relationship is called *linear.*

definition ■ *A **linear relationship** between two variables is one in which the relationship can be most accurately represented by a straight line.*

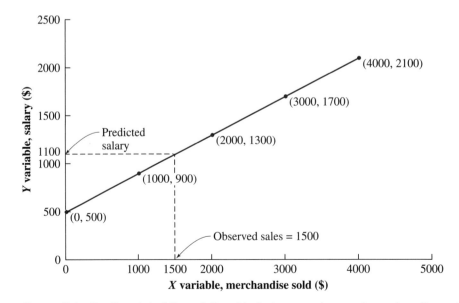

figure 6.1 Scatter plot of the relationship between salary and merchandise sold.

Note that not all relationships are linear. Some relationships are *curvilinear*. In these cases, when a scatter plot of the X and Y variables is drawn, a *curved* line fits the points better than a straight line.

Deriving the equation of the straight line   The relationship between "salary" and "merchandise sold" shown in Figure 6.1 can be described with an equation. Of course, this equation is the equation of the line joining all of the points. The general form of the equation is given by

$$Y = bX + a \qquad equation\ of\ a\ straight\ line$$

where $a = Y$ intercept (value of Y when $X = 0$)
$b = $ slope of the line

Finding the *Y* intercept *a*   The *Y intercept* is the value of Y where the line intersects the Y axis. Thus, it is the Y value when $X = 0$. In this problem, we can see from Figure 6.1 that

$$a = Y\ intercept = 500$$

Finding the slope *b*   The *slope* of a line is a measure of its rate of change. It tells us how much the Y score changes for each unit change in the X score. In equation form,

$$b = \text{slope} = \frac{\Delta Y}{\Delta X} = \frac{Y_2 - Y_1}{X_2 - X_1} \qquad slope\ of\ a\ straight\ line$$

Since we are dealing with a straight line, its slope is constant. This means it doesn't matter what values we pick for X_2 and X_1; the corresponding Y_2 and Y_1 scores will yield the same value of slope. To calculate the slope, let's vary X from 2000 to 3000. If $X_1 = 2000$, then $Y_1 = 1300$. If $X_2 = 3000$, then $Y_2 = 1700$. Substituting these values into the slope equation,

$$b = \text{slope} = \frac{\Delta Y}{\Delta X} = \frac{Y_2 - Y_1}{X_2 - X_1} = \frac{1700 - 1300}{3000 - 2000} = \frac{400}{1000} = 0.40$$

Thus, the slope is 0.40. This means that the Y value increases 0.40 units for every 1-unit increase in X. The slope and Y intercept determinations are also shown in Figure 6.2. Note that the same slope would occur if we had chosen other values for X_1 and X_2. For example, if $X_1 = 1000$ and $X_2 = 4000$, then $Y_1 = 900$ and $Y_2 = 2100$. Solving for the slope,

$$b = \text{slope} = \frac{\Delta Y}{\Delta X} = \frac{Y_2 - Y_1}{X_2 - X_1} = \frac{2100 - 900}{4000 - 1000} = \frac{1200}{3000} = 0.40$$

Again, the slope is 0.40.

The full equation for the linear relationship that exists between salary and merchandise sold can now be written:

$$Y = bX + a$$

Substituting for a and b,

$$Y = 0.40X + 500$$

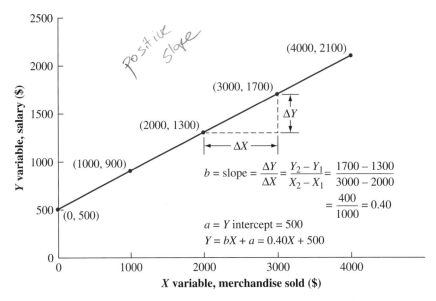

figure 6.2 Graph of salary and amount of merchandise sold.

The equation $Y = 0.40X + 500$ describes the relationship between the Y variable (salary) and the X variable (merchandise sold). It tells us that Y increases by 1 unit for every 0.40 increase in X. Moreover, as long as the relationship holds, this equation lets us compute an appropriate value for Y, given any value of X. That makes the equation very useful for prediction.

Predicting Y, given X   When used for prediction, the equation becomes

$$Y' = 0.40X + 500$$

where Y' = the predicted value of the Y variable

With this equation, we can predict any Y value just by knowing the corresponding X value. For example, if $X = 1500$ as in our previous problem, then

$$
\begin{aligned}
Y' &= 0.40X + 500 \\
&= 0.40(1500) + 500 \\
&= 600 + 500 \\
&= 1100
\end{aligned}
$$

Thus, if a salesperson sells $1500 worth of merchandise, his or her salary would equal $1100.

Of course, prediction could also have been done graphically, as shown in Figure 6.1. By vertically projecting the X value of $1500 until it intersects with the straight line, we can read the predicted Y value from the Y axis. The predicted value is $1100, which is the same value we arrived at using the equation.

Positive and Negative Relationships

In addition to being linear or curvilinear, the relationship between two variables may be *positive* or *negative*.

definitions ■ *A **positive relationship** indicates that there is a direct relationship between the variables. A **negative relationship** indicates that there is an inverse relationship between X and Y.*

The slope of the line tells us whether the relationship is positive or negative. When the relationship is positive, the slope is positive. The previous example had a positive slope; that is, higher values of X were associated with higher values of Y, and lower values of X were associated with lower values of Y. When the slope is positive, the line runs upward from left to right, indicating that as X increases, Y increases. Thus, a *direct relationship* exists between the two variables.

When the relationship is negative, there is an inverse relationship between the variables, making the slope negative. An example of a negative relationship is shown in Figure 6.3. Note that with a negative slope, the curve runs downward from left to right. Low values of X are associated with high values of Y, and high values of X are associated with low values of Y. Another way of saying this is that as X increases, Y decreases.

Perfect and Imperfect Relationships

In the relationships we have graphed so far, all of the points have fallen on the straight line. When this is the case, the relationship is a *perfect* one (see "definition" on p. 128). Unfortunately, in the behavioral sciences, perfect relationships are rare. It is much more common to find imperfect relationships.

As an example, Table 6.2 shows the IQ scores and grade point averages of a sample of 12 college students. Suppose we wanted to determine the relationship between these

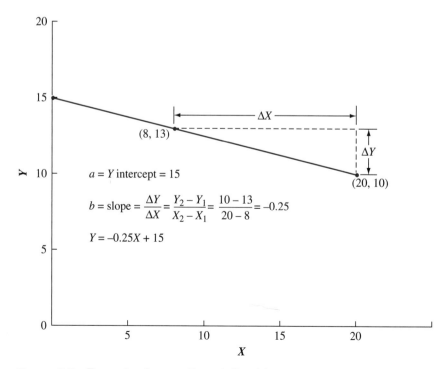

$a = Y \text{ intercept} = 15$

$b = \text{slope} = \dfrac{\Delta Y}{\Delta X} = \dfrac{Y_2 - Y_1}{X_2 - X_1} = \dfrac{10 - 13}{20 - 8} = -0.25$

$Y = -0.25X + 15$

figure 6.3 Example of a negative relationship.

table 6.2 IQ and grade point average of 12 college students

Student No.	IQ	Grade Point Average
1	110	1.0
2	112	1.6
3	118	1.2
4	119	2.1
5	122	2.6
6	125	1.8
7	127	2.6
8	130	2.0
9	132	3.2
10	134	2.6
11	136	3.0
12	138	3.6

hypothetical data. The scatter plot is shown in Figure 6.4. From the scatter plot, it is obvious that the relationship between IQ and college grades is imperfect. The imperfect relationship is positive because lower values of IQ are associated with lower values of grade point average, and higher values of IQ are associated with higher values of grade point average. In addition, the relationship appears linear.

definitions ■ *A* **perfect relationship** *is one in which a positive or negative relationship exists and all of the points fall on the line. An* **imperfect relationship** *is one in which a relationship exists, but all of the points do not fall on the line.*

To describe this relationship with a straight line, the best we can do is to draw the line that *best* fits the data. Another way of saying this is that, when the relationship is imperfect, we cannot draw a single straight line through all of the points. We can, however, construct a straight line that most accurately fits the data. This line has been drawn in Figure 6.4. This best-fitting line is often used for prediction; when so used, it is called a *regression* line.*

A *USA Today* article reported that there is an inverse relationship between the amount of television watched by primary school students and their reading skills. Suppose the sixth-grade data for the article appeared as shown in Figure 6.5. This is an example of a negative, imperfect, linear relationship. The relationship is negative because higher values of television watching are associated with lower values of reading skill, and lower values of television watching are associated with higher values of reading skill. The linear relationship is imperfect because not all of the points fall on a single straight line. The regression line for these data is also shown in Figure 6.5.

Having completed our background discussion of relationships, we can now move on to the topic of correlation.

---

*The details on how to construct this line will be discussed in Chapter 7.

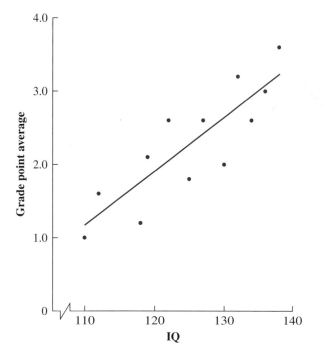

figure 6.4   Scatter plot of IQ and grade point average.

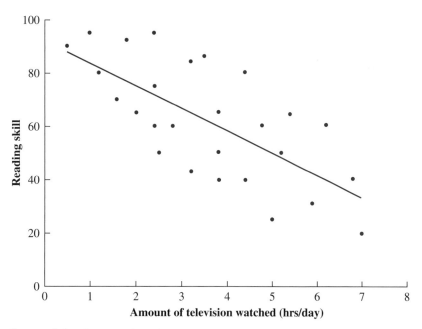

figure 6.5   Scatter plot of reading skill and amount of television watched by sixth graders.

CORRELATION

Correlation is a topic that focuses on the *direction* and *degree* of the relationship. The direction of the relationship refers to whether the relationship is positive or negative. The degree of relationship refers to the magnitude or strength of the relationship. The degree of relationship can vary from nonexistent to perfect. When the relationship is perfect, correlation is at its highest and we can exactly predict from one variable to the other. In this situation, as *X* changes, so does *Y*. Moreover, the same value of *X* always leads to the same value of *Y*. Alternatively, the same value of *Y* always leads to the same value of *X*. The points all fall on a straight line, assuming the relationship is linear. When the relationship is nonexistent, correlation is at its lowest and knowing the value of one of the variables doesn't help at all in predicting the other. Imperfect relationships have intermediate levels of correlation, and prediction is approximate. Here, the same value of *X* doesn't always lead to the same value of *Y*. Nevertheless, on the average, *Y* changes systematically with *X*, and we can do a better job of predicting *Y* with knowledge of *X* than without it.

Although it suffices for some purposes to talk rather loosely about "high" or "low" correlations, it is much more often desirable to know the exact magnitude and direction of the correlation. A *correlation coefficient* gives us this information.

definition
■ *A **correlation coefficient** expresses quantitatively the magnitude and direction of the relationship.*

A correlation coefficient can vary from +1 to –1. The *sign* of the coefficient tells us whether the relationship is *positive* or *negative*. The numerical part of the correlation coefficient describes the magnitude of the correlation. The higher the number, the greater is the correlation. Since 1 is the highest number possible, it represents a perfect correlation. A correlation coefficient of +1 means the correlation is perfect and the relationship is positive. A correlation coefficient of –1 means the correlation is perfect and the relationship is negative. When the relationship is nonexistent, the correlation coefficient equals 0. Imperfect relationships have correlation coefficients varying in magnitude between 0 and 1. They will be plus or minus depending on the direction of the relationship.

Figure 6.6 shows scatter plots of several different linear relationships and the correlation coefficients for each. The Pearson *r* correlation coefficient has been used because the relationships are linear. We shall discuss Pearson *r* in the next section. Each scatter plot is made up of paired *X* and *Y* values. Note that the closer the points are to the regression line, the higher the magnitude of the correlation coefficient and the more accurate the prediction. Also, when the correlation is zero, there is no relationship between *X* and *Y*. This means that *Y* does not increase or decrease systematically with increases or decreases in *X*. Thus, with zero correlation, the regression line for predicting *Y* is horizontal and knowledge of *X* does not aid in predicting *Y*.

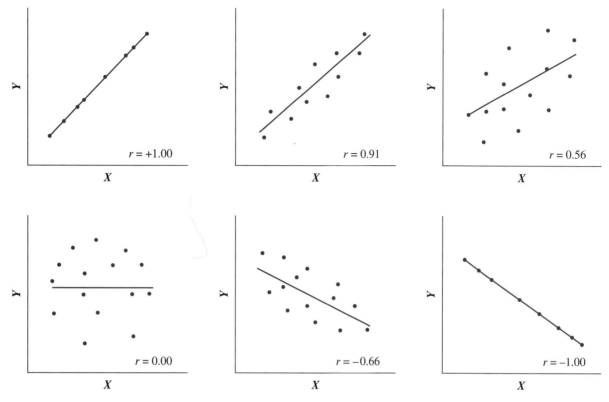

figure 6.6 Scatter plots of several linear relationships.

The Linear Correlation Coefficient Pearson *r*

You will recall from our discussion in Chapter 5 that a basic problem in measuring the relationship between two variables is that very often the variables are measured on different scales and in different units. For example, if we are interested in measuring the correlation between IQ and grade point average for the data presented in Table 6.2, we are faced with the problem that IQ and grade point average have very different scaling. As was mentioned in Chapter 5, this problem is resolved by converting each score to its z-transformed value, in effect putting both variables on the same scale, a z scale.

To appreciate how useful z scores are for determining correlation, consider the following example. Suppose your neighborhood supermarket is having a sale on oranges. The oranges are bagged, and each bag has the total price marked on it. You want to know whether there is a relationship between the weight of the oranges in each bag and their cost. Being a natural-born researcher, you randomly sample six bags and weigh each one. The cost and weight in pounds of the six bags are shown in Table 6.3. A scatter plot of the data is graphed in Figure 6.7. Are these two variables related? Yes; in fact, all the points fall on a straight line. There is a perfect positive correlation between the cost and weight of the oranges. Thus, the correlation coefficient must equal +1.

Next, let's see what happens when we convert these raw scores to z scores. The raw scores for weight (X) and cost (Y) have been expressed as standard scores in the fourth and fifth columns of Table 6.3. Something quite interesting has happened. *The paired raw scores for each bag of oranges have the same z value.* For example, the paired raw scores for bag A are 2.25 and 0.75. However, their respective z scores are both −1.34. The raw

table 6.3 Cost and weight in pounds of six bags of oranges

Bag	Weight (lb) X	Cost ($) Y	z_X	z_Y
A	2.25	0.75	−1.34	−1.34
B	3.00	1.00	−0.80	−0.80
C	3.75	1.25	−0.27	−0.27
D	4.50	1.50	0.27	0.27
E	5.25	1.75	0.80	0.80
F	6.00	2.00	1.34	1.34

score of 2.25 is as many standard deviation units below the mean of the X distribution as the raw score of 0.75 is below the mean of the Y distribution. The same is true for the other paired scores. All of the paired raw scores occupy the same relative position within their own distributions. That is, they have the same z values. When using raw scores, this relationship is obscured because of differences in scaling between the two variables. If the paired scores occupy the same relative position within their own distributions, then the correlation must be perfect ($r = 1$), because knowing one of the paired values will allow us to exactly predict the other value. If prediction is perfect, the relationship must be perfect.

This brings us to the definition of Pearson r.

definition ■ **Pearson r** *is a measure of the extent to which paired scores occupy the same or opposite positions within their own distributions.*

Note that this definition also includes the paired scores occupying opposite positions. If the paired z scores have the same magnitude but opposite signs, the correlation would again be perfect and r would equal −1.

This example highlights a very important point. Since correlation is concerned with the relationship between two variables and the variables are often measured in different units and scaling, the magnitude and direction of the correlation coefficient

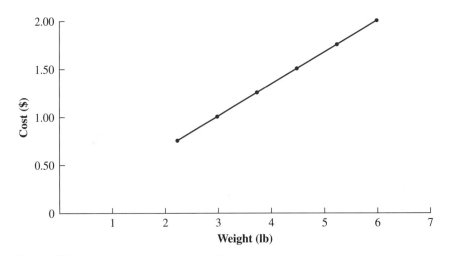

figure 6.7 Cost of oranges versus their weight in pounds.

must be independent of the differences in units and scaling that exist between the two variables. Pearson r achieves this by using z scores. Thus, we can correlate such diverse variables as time of day and position of the sun, percent body fat and caloric intake, test anxiety and examination grades, and so forth.

Since this is such an important point, we would like to illustrate it again by taking the previous example one more step. In the example involving the relationship between the cost of oranges and their weight, suppose you weighed the oranges in kilograms rather than in pounds. Should this change the degree of relationship between the cost and weight of the oranges? In light of what we have just presented, the answer is surely no. Correlation must be independent of the units used in measuring the two variables. If the correlation is 1 between the cost of the oranges and their weight in pounds, the correlation should also be 1 between the cost of the oranges and their weight in kilograms. We've converted the weight of each bag of oranges from pounds to kilograms. The data are presented in Table 6.4, and the raw scores are plotted in Figure 6.8. Again, all the scores fall on a straight line, so the correlation equals 1.00. Notice the values of the paired z scores in the fourth and fifth columns of Table 6.4. Once more, they have the same values, and these values are the same as when the oranges were weighed in pounds. Thus, using z scores allows a measurement of the relationship between the two variables that is independent of differences in scaling and of the units used in measuring the variables.

Calculating Pearson r The equation for calculating Pearson r using z scores is

$$r = \frac{\sum z_X z_Y}{N - 1} \quad \textit{conceptual equation}$$

where $\sum z_X z_Y$ = the sum of the product of each z score pair

To use this equation, you must first convert each raw score into its z-transformed value. This can take a considerable amount of time and possibly create rounding errors. With some algebra, this equation can be transformed into a calculation equation that uses the raw scores:

$$r = \frac{\sum XY - \dfrac{(\sum X)(\sum Y)}{N}}{\sqrt{\left[\sum X^2 - \dfrac{(\sum X)^2}{N} \right]\left[\sum Y^2 - \dfrac{(\sum Y)^2}{N} \right]}} \quad \begin{array}{l}\textit{computational equation for} \\ \textit{Pearson r}\end{array}$$

where $\sum XY$ = the sum of the product of each X and Y pair ($\sum XY$ is also called the sum of the cross products)

   N = the number of paired scores

table 6.4   Cost and weight in kilograms of six bags of oranges

Bag	Weight (kg) X	Cost ($) Y	z_X	z_Y
A	1.02	0.75	−1.34	−1.34
B	1.36	1.00	−0.80	−0.80
C	1.70	1.25	−0.27	−0.27
D	2.04	1.50	0.27	0.27
E	2.38	1.75	0.80	0.80
F	2.72	2.00	1.34	1.34

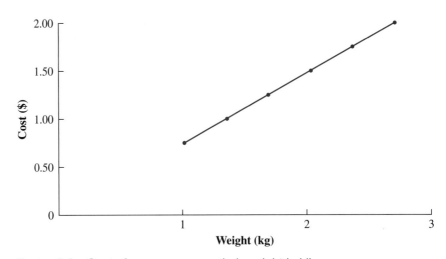

figure 6.8 Cost of oranges versus their weight in kilograms.

Table 6.5 contains some hypothetical data collected from five subjects. Let's use these data to calculate Pearson r:

$$r = \frac{\Sigma \, XY - \dfrac{(\Sigma \, X)(\Sigma \, Y)}{N}}{\sqrt{\left[\Sigma \, X^2 - \dfrac{(\Sigma \, X)^2}{N} \right]\left[\Sigma \, Y^2 - \dfrac{(\Sigma \, Y)^2}{N} \right]}}$$

table 6.5 Hypothetical data for computing Pearson r

Subject	X	Y	X^2	Y^2	XY
A	1	2	1	4	2
B	3	5	9	25	15
C	4	3	16	9	12
D	6	7	36	49	42
E	7	5	49	25	35
Total	21	22	111	112	106

$N = 5$

$$r = \frac{\Sigma XY - \dfrac{(\Sigma X)(\Sigma Y)}{N}}{\sqrt{\left[\Sigma X^2 - \dfrac{(\Sigma X)^2}{N} \right]\left[\Sigma Y^2 - \dfrac{(\Sigma Y)^2}{N} \right]}}$$

MENTORING TIP

Caution: remember that N is the number of paired scores; $N = 5$ in this example.

$$= \frac{106 - \dfrac{21(22)}{5}}{\sqrt{\left[111 - \dfrac{(21)^2}{5} \right]\left[112 - \dfrac{(22)^2}{5} \right]}}$$

$$= \frac{13.6}{18.616} = 0.731 = 0.73$$

ΣXY is called the sum of the cross products. It is found by multiplying the X and Y scores for each subject and then summing the resulting products. Calculation of ΣXY and the other terms is illustrated in Table 6.5. Substituting these values in the previous equation, we obtain

$$r = \frac{106 - \dfrac{21(22)}{5}}{\sqrt{\left[111 - \dfrac{(21)^2}{5}\right]\left[112 - \dfrac{(22)^2}{5}\right]}} = \frac{13.6}{\sqrt{22.8(15.2)}} = \frac{13.6}{18.616} = 0.731 = 0.73$$

Practice Problem 6.1

Let's try another problem. This time we shall use data given in Table 6.2. For your convenience, these data are reproduced in the first three columns of the accompanying table. In this example, we have an imperfect linear relationship, and we are interested in computing the magnitude and direction of the relationship using Pearson r. The solution is also shown in the following table.

SOLUTION

Student No.	IQ X	Grade Point Average Y	X^2	Y^2	XY
1	110	1.0	12,100	1.00	110.0
2	112	1.6	12,544	2.56	179.2
3	118	1.2	13,924	1.44	141.6
4	119	2.1	14,161	4.41	249.9
5	122	2.6	14,884	6.76	317.2
6	125	1.8	15,625	3.24	225.0
7	127	2.6	16,129	6.76	330.2
8	130	2.0	16,900	4.00	260.0
9	132	3.2	17,424	10.24	422.4
10	134	2.6	17,956	6.76	348.4
11	136	3.0	18,496	9.00	408.0
12	138	3.6	19,044	12.96	496.8
Total	1503	27.3	189,187	69.13	3488.7

$$r = \frac{\Sigma XY - \dfrac{(\Sigma X)(\Sigma Y)}{N}}{\sqrt{\left[\Sigma X^2 - \dfrac{(\Sigma X)^2}{N}\right]\left[\Sigma Y^2 - \dfrac{(\Sigma Y)^2}{N}\right]}}$$

$$= \frac{3488.7 - \dfrac{1503(27.3)}{12}}{\sqrt{\left[189,187 - \dfrac{(1503)^2}{12}\right]\left[69.13 - \dfrac{(27.3)^2}{12}\right]}} = \frac{69.375}{81.085} = 0.856 = 0.86$$

Practice Problem 6.2

Let's try one more problem. Have you ever wondered whether it is true that opposites attract? We've all been with couples in which the two individuals seem so different from each other. But is this the usual experience? Does similarity or dissimilarity foster attraction?

A social psychologist investigating this problem asked 15 college students to fill out a questionnaire concerning their attitudes toward a variety of topics. Some time later, they were shown the "attitudes" of a stranger to the same items and were asked to rate the stranger as to probable liking for the stranger and probable enjoyment of working with him. The "attitudes" of the stranger were really made up by the experimenter and varied over subjects regarding the proportion of attitudes held by the stranger that were similar to those held by the rater. Thus, for each subject, data were collected concerning his attitudes and the attraction of a stranger based on the stranger's attitudes to the same items. If similarities attract, then there should be a direct relationship between the attraction of the stranger and the proportion of his similar attitudes. The data are presented in the table at the end of this practice problem. The higher the attraction, the higher is the score. The maximum possible attraction score is 14. Compute the Pearson r correlation coefficient* to determine whether there is a direct relationship between similarity of attitudes and attraction.

SOLUTION

The solution is shown in the following table.

Student No.	Proportion of Similar Attitudes X	Attraction Y	X^2	Y^2	XY
1	0.30	8.9	0.090	79.21	2.670
2	0.44	9.3	0.194	86.49	4.092
3	0.67	9.6	0.449	92.16	6.432
4	0.00	6.2	0.000	38.44	0.000
5	0.50	8.8	0.250	77.44	4.400
6	0.15	8.1	0.022	65.61	1.215
7	0.58	9.5	0.336	90.25	5.510
8	0.32	7.1	0.102	50.41	2.272
9	0.72	11.0	0.518	121.00	7.920
10	1.00	11.7	1.000	136.89	11.700
11	0.87	11.5	0.757	132.25	10.005
12	0.09	7.3	0.008	53.29	0.657
13	0.82	10.0	0.672	100.00	8.200
14	0.64	10.0	0.410	100.00	6.400
15	0.24	7.5	0.058	56.25	1.800
Total	7.34	136.5	4.866	1279.69	73.273

*As will be pointed out later in the chapter, it is legitimate to calculate Pearson r only where the data are of interval or ratio scaling. Therefore, to calculate Pearson r for this problem, we must assume the data are at least of interval scaling.

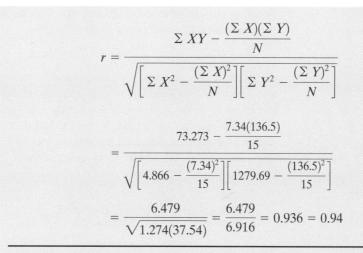

$$r = \frac{\Sigma\ XY - \dfrac{(\Sigma\ X)(\Sigma\ Y)}{N}}{\sqrt{\left[\Sigma\ X^2 - \dfrac{(\Sigma\ X)^2}{N}\right]\left[\Sigma\ Y^2 - \dfrac{(\Sigma\ Y)^2}{N}\right]}}$$

$$= \frac{73.273 - \dfrac{7.34(136.5)}{15}}{\sqrt{\left[4.866 - \dfrac{(7.34)^2}{15}\right]\left[1279.69 - \dfrac{(136.5)^2}{15}\right]}}$$

$$= \frac{6.479}{\sqrt{1.274(37.54)}} = \frac{6.479}{6.916} = 0.936 = 0.94$$

Therefore, based on these students, there is a very strong relationship between similarity and attractiveness.

MENTORING TIP

Caution: students often find this section difficult. Be prepared to spend additional time on it to achieve understanding.

A second interpretation for Pearson *r*   Pearson *r* can also be interpreted in terms of the variability of *Y* accounted for by *X*. This approach leads to important additional information about *r* and the relationship between *X* and *Y*. Consider Figure 6.9, in which an imperfect relationship is shown between *X* and *Y*. In this example, the *X* variable represents spelling competence and the *Y* variable is writing ability of six students in the third grade. Suppose we are interested in predicting the writing score for Maria, the student whose spelling score is 88. If there were no relationship between writing and spelling, we would predict a score of 50, which is the overall mean of all the writing scores. In the absence of a relationship between *X* and *Y*, the overall mean is the best predictor. When there is no relationship between *X* and *Y*, using the mean minimizes prediction errors because the sum of the squared deviations from it is a minimum. You will recognize this as

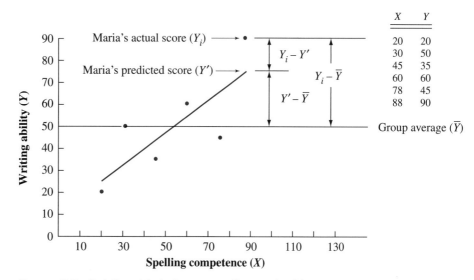

figure 6.9   Relationship between spelling and writing.

the fourth property of the mean, discussed in Chapter 4. Maria's actual writing score is 90, so our estimate of 50 is in error by 40 points. Thus,

$$\text{Maria's actual writing score} - \text{Group average} = Y_i - \overline{Y} = 90 - 50 = 40$$

However, in this example, the relationship between X and Y is not zero. Although it is not perfect, a relationship greater than zero exists between X and Y. Therefore, the overall mean of the writing scores is not the best predictor. Rather, as discussed previously in the chapter, we can use the regression line for these data as the basis of our prediction. The regression line for the writing and spelling scores is shown in Figure 6.9. Using this line, we would predict a writing score of 75 for Maria. Now the error is only 15 points. Thus,

$$\left(\begin{matrix}\text{Maria's actual}\\\text{score}\end{matrix}\right) - \left(\begin{matrix}\text{Maria's predicted}\\\text{score using } X\end{matrix}\right) = Y_i - Y' = 90 - 75 = 15$$

It can be observed in Figure 6.9 that the distance between Maria's score and the mean of the Y scores is divisible into two segments. Thus,

$$Y_i - \overline{Y} = (Y_i - Y') + (Y' - \overline{Y})$$

Deviation of Y_i	=	Error in prediction using the relationship between X and Y	+	Deviation of Y_i accounted for by the relationship between X and Y

The segment $Y_i - Y'$ represents the error in prediction. The remaining segment $Y' - \overline{Y}$ represents that part of the deviation of Y_i that is accounted for by the relationship between X and Y. You should note that "accounted for by the relationship between X and Y" is often abbreviated as "accounted for by X."

Suppose we now determine the predicted Y score (Y') for each X score using the regression line. We could then construct $Y_i - \overline{Y}$ for each score. If we squared each $Y_i - \overline{Y}$ and summed over all the scores, we would obtain

$$\Sigma (Y_i - \overline{Y})^2 = \Sigma (Y_i - Y')^2 + \Sigma (Y' - \overline{Y})^2$$

Total variability of Y	=	Variability of prediction errors	+	Variability of Y accounted for by X

Note that $\Sigma (Y_i - \overline{Y})^2$ is the sum of squares of the Y scores. It represents the total variability of the Y scores. Thus, this equation states that the total variability of the Y scores can be divided into two parts: the variability of the prediction errors and the variability of Y accounted for by X.

We know that, as the relationship gets stronger, the prediction gets more accurate. In the previous equation, as the relationship gets stronger, the prediction errors get smaller, also causing the variability of prediction errors $\Sigma (Y_i - Y')^2$ to decrease. Since the total variability $\Sigma (Y_i - \overline{Y})^2$ hasn't changed, *the variability of Y accounted for by X,* namely, $\Sigma (Y' - \overline{Y})^2$, must increase. Thus, the proportion of the total variability of the Y scores that is accounted for by X, namely, $\Sigma (Y' - \overline{Y})^2/\Sigma (Y_i - \overline{Y})^2$, is a measure of the strength of relationship. It turns out that if we take the square root of this ratio and substitute for Y' the appropriate values, we obtain the computational formula for Pearson r. We previously defined Pearson r as a measure of the extent to which paired scores occupy the same or opposite positions within their own distributions. From what we have just said, it is also the case that Pearson r equals the square root of the proportion of the variability of Y accounted for by X. In equation form,

$$r = \sqrt{\frac{\Sigma(Y' - \bar{Y})^2}{\Sigma(Y_i - \bar{Y})^2}} = \sqrt{\frac{\text{Variability of } Y \text{ that is accounted for by } X}{\text{Total variability of } Y}}$$

$$r = \sqrt{\text{Proportion of the total variability of } Y \text{ that is accounted for by } X}$$

It follows from this equation that the higher r is, the greater the proportion of the variability of Y that is accounted for by X.

Relationship of r^2 and explained variability

If we square the previous equation, we obtain

$$r^2 = \text{Proportion of the total variability of } Y \text{ that is accounted for by } X$$

Thus, r^2 is called the *coefficient of determination*. As shown in the equation, r^2 equals the proportion of the total variability of Y that is accounted for or explained by X. In the problem dealing with grade point average and IQ, the correlation was 0.86. If we square r, we obtain

$$r^2 = (0.86)^2 = 0.74$$

This means that 74% of the variability in Y can be accounted for by IQ. If it turns out that IQ is a causal factor in determining grade point average, then r^2 tells us that IQ accounts for 74% of the variability in grade point average. What about the remaining 26%? Other factors that can account for the remaining 26% must be influencing grade point average. The important point here is that one can be misled by using r into thinking that X may be a major cause of Y when really it is r^2 that tells us how much of the change in Y can be accounted for by X.* The error isn't so serious when you have a correlation coefficient as high as 0.86. However, in the behavioral sciences, such high correlations are rare. Correlation coefficients of $r = 0.50$ or 0.60 are considered fairly high, and yet correlations of this magnitude account for only 25% to 36% of the variability in Y ($r^2 = 0.25$ to 0.36). Table 6.6 shows the relationship between r and the explained variability expressed as a percentage.

MENTORING TIP

If one of the variables is causal, then r^2 is a measure of the size of its effect.

Other Correlation Coefficients

So far, we have discussed correlation and described in some detail the linear correlation coefficient Pearson r. We have chosen Pearson r because it is the most frequently encountered correlation coefficient in behavioral science research. However, you should be aware that there are many different correlation coefficients one might employ, each of which is appropriate under different conditions. In deciding which correlation coefficient to calculate, the *shape of the relationship* and the *measuring scale* of the data are the two most important considerations.

---

*Viewed in this manner, if IQ is a causal factor, then r^2 is a measure of the size of the IQ effect.

table 6.6 Relationship between r and explained variability

r	Explained Variability, r^2 (%)
0.10	1
0.20	4
0.30	9
0.40	16
0.50	25
0.60	36
0.70	49
0.80	64
0.90	81
1.00	100

Shape of the relationship The choice of which correlation coefficient to calculate depends on whether the relationship is linear or curvilinear. If the data are curvilinear, using a linear correlation coefficient such as Pearson r can seriously underestimate the degree of relationship that exists between X and Y. Accordingly, another correlation coefficient η (eta) is used for curvilinear relationships. An example is the relationship between motor skills and age. There is an inverted U-shaped relationship between motor skills and age. In early life, motor skills are low. They increase during the middle years, and then decrease in later life. However, since η is not frequently encountered in behavioral science research, we have not presented a detailed discussion of it.* This does, however, emphasize the importance of doing a scatter plot to determine whether the relationship is linear before just routinely going ahead and calculating a linear correlation coefficient. It is also worth noting here that, like r^2, if one of the variables is causal, η^2 is a measure of the size of effect. We discuss this aspect of η^2 in Chapter 15.

Measuring scale The choice of correlation coefficient also depends on the type of measuring scale underlying the data. We've already discussed the linear correlation coefficient Pearson r. It assumes the data are measured on an interval or ratio scale. Some examples of other linear correlation coefficients are the *Spearman rank order correlation coefficient rho* (r_s), the *biserial correlation coefficient* (r_b), and the *phi* (ϕ) *coefficient*. In actuality, each of these coefficients is the equation for Pearson r simplified to apply to the lower-order scaling. Rho is used when one or both of the variables are of ordinal scaling, r_b is used when one of the variables is at least interval and the other is dichotomous, and phi is used when each of the variables is dichotomous. Although it is beyond the scope of this textbook to present each of these correlation coefficients in detail, the Spearman rank order correlation coefficient rho occurs frequently enough to warrant discussion here.

---

*A discussion of η as well as the other coefficients presented in this section is contained in N. Downie and R. Heath, *Basic Statistical Methods,* 4th ed., Harper & Row, New York, 1974, pp. 102–114.

The Spearman rank order correlation coefficient rho (r_s)   As mentioned, the Spearman rank order correlation coefficient rho is used when one or both of the variables are only of ordinal scaling. *Spearman rho* is really the linear correlation coefficient Pearson *r* applied to data that meet the requirements of ordinal scaling. The easiest equation for calculating rho when there are no ties or just a few ties relative to the number of paired scores is

$$r_s = 1 - \frac{6 \Sigma D_i^2}{N^3 - N} \quad \textit{computational equation for rho}$$

where D_i = difference between the *i*th pair of ranks = $R(X_i) - R(Y_i)$
$R(X_i)$ = rank of the *i*th *X* score
$R(Y_i)$ = rank of the *i*th *Y* score
N = number of pairs of ranks

It can be shown that, with ordinal data having no ties, Pearson *r* reduces algebraically to the previous equation.

To illustrate the use of rho, let's consider an example. Assume that a large corporation is interested in rating a current class of 12 management trainees on their leadership ability. Two psychologists are hired to do the job. As a result of their tests and interviews, the psychologists each independently rank-order the students according to leadership ability. The rankings are from 1 to 12, with 1 representing the highest level of leadership. The data are given in Table 6.7. What is the correlation between the rankings of the two psychologists?

Since the data are of ordinal scaling, we should compute rho. The solution is shown in Table 6.7. Note that subjects 5 and 6 were tied in the rankings of psychologist A. When ties occur, the rule is to give each subject the average of the tied ranks. For example, subjects 5 and 6 were tied for ranks 2 and 3. Therefore, they each received a ranking of 2.5 [(2 + 3)/2 = 2.5]. In giving the two subjects a rank of 2.5, we have effectively used up ranks 2 and 3. The next rank is 4. D_i is the difference between the paired rankings for

table 6.7   Calculation of r_s for leadership example

Subject	Rank Order of Psychologist A $R(X_i)$	Rank Order of Psychologist B $R(Y_i)$	$D_i =$ $R(X_i) - R(Y_i)$	D_i^2
1	6	5	1	1
2	5	3	2	4
3	7	4	3	9
4	10	8	2	4
5	2.5	1	1.5	2.25
6	2.5	6	−3.5	12.25
7	9	10	−1	1
8	1	2	−1	1
9	11	9	2	4
10	4	7	−3	9
11	8	11	−3	9
12	12	12	0	0
$N = 12$				$\Sigma D_i^2 = 56.5$

$$r_s = 1 - \frac{6 \Sigma D_i^2}{N^3 - N} = 1 - \frac{6(56.5)}{(12)^3 - 12} = 1 - \frac{339}{1716} = 0.80$$

MENTORING TIP

Remember: when ties occur, give each tied score the average of the tied ranks and give the next highest score the next unused rank. For example, if three scores are tied at ranks 5, 6, and 7, they each would receive a rank of 6 and the next highest score would be assigned a rank of 8.

the ith subject. Thus, $D_i = 1$ for subject 1. It doesn't matter whether you subtract $R(X_i)$ from $R(Y_i)$ or $R(Y_i)$ from $R(X_i)$ to get D_i because we square each D_i value. The squared D_i values are then summed ($\Sigma D_i^2 = 56.5$). This value is then entered in the equation along with N ($N = 12$) and r_s is computed. For this problem,

$$r_s = 1 - \frac{6\Sigma D_i^2}{N^3 - N} = 1 - \frac{6(56.5)}{12^3 - 12} = 1 - \frac{339}{1716} = 0.80$$

Practice Problem 6.3

To illustrate computation of r_s, let's assume that the raters' attitude and attraction scores given in Practice Problem 6.2 were only of ordinal scaling. Given this assumption, determine the value of the linear correlation coefficient rho for these data and compare the value with the value of Pearson r determined in Practice Problem 6.2.

SOLUTION

The data and solution are shown in the following table.

Subject	Proportion of Similar Attitudes X_i	Attraction Y_i	Rank of X_i $R(X_i)$	Rank of Y_i $R(Y_i)$	$D_i =$ $R(X_i) - R(Y_i)$	D^2_i
1	0.30	8.9	5	7	−2	4
2	0.44	9.3	7	8	−1	1
3	0.67	9.6	11	10	1	1
4	0.00	6.2	1	1	0	0
5	0.50	8.8	8	6	2	4
6	0.15	8.1	3	5	−2	4
7	0.58	9.5	9	9	0	0
8	0.32	7.1	6	2	4	16
9	0.72	11.0	12	13	−1	1
10	1.00	11.7	15	15	0	0
11	0.87	11.5	14	14	0	0
12	0.09	7.3	2	3	−1	1
13	0.82	10.0	13	11.5	1.5	2.25
14	0.64	10.0	10	11.5	−1.5	2.25
15	0.24	7.5	4	4	0	0
$N = 15$						$\Sigma D_i^2 = 36.5$

$$r_s = 1 - \frac{6\Sigma D_i^2}{N^3 - N} = 1 - \frac{6(36.5)}{(15)^3 - 15} = 1 - \frac{219}{3360} = 0.93$$

Note that $r_s = 0.93$ and $r = 0.94$. The values are not identical but quite close. In general, when Pearson r is calculated using the interval or ratio properties of data, its values will be close but not exactly the same as when calculated on only the ordinal properties of those data.

Effect of Range on Correlation

If a correlation exists between X and Y, restricting the range of either of the variables will have the effect of *lowering* the correlation. This can be seen in Figure 6.10, where we have drawn a scatter plot of freshman grade point average and College Entrance Examination Board (CEEB) scores. The figure has been subdivided into low, medium, and high CEEB scores. Taking the figure as a whole (i.e., considering the full range of the CEEB scores), there is a high correlation between the two variables. However, if we were to consider the three sections separately, the correlation for each section would be much lower. Within each section, the points show much less systematic change in Y with changes in X. This, of course, indicates a lower correlation between X and Y. The effect of range restriction on correlation is often encountered in education or industry. For instance, suppose that on the basis of the high correlation between freshman grades and CEEB scores as shown in Figure 6.10, a college decided to admit only high school graduates who have scored in the high range of the CEEB scores. If the subsequent freshman grades of these students were correlated with their CEEB scores, we would expect a much lower correlation because of the range restriction of the CEEB scores for these freshmen. In a similar vein, if one is doing a correlational study and obtains a low correlation coefficient, one should check to be sure that range restriction is not responsible for the low value.

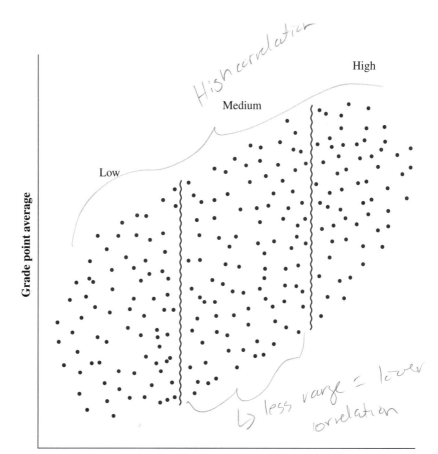

College Entrance Examination Board (CEEB) scores

figure 6.10 Freshman grades and CEEB scores.

Effect of Extreme Scores

Consider the effect of an extreme score on the magnitude of the correlation coefficient. Figure 6.11(a) shows a set of scores where all the scores cluster reasonably close together. The value of Pearson *r* for this set of scores is 0.11. Figure 6.11(b) shows the same set of scores with an extreme score added. The value of Pearson *r* for this set of scores is 0.94. The magnitude of Pearson *r* has changed from 0.11 to 0.94. This is a demonstration of the point that *an extreme score can drastically alter the magnitude of the correlation coefficient* and, hence, change the interpretation of the data. Therefore, it is a good idea to check the scatter plot of the data for extreme scores before computing the correlation coefficient. If an extreme score exists, caution must be exercised in interpreting the relationship. If the sample is a large random sample, an extreme value usually will not greatly alter the size of the correlation. However, if the sample is a small one, as in this example, an extreme score can have a large effect.

Correlation Does Not Imply Causation

MENTORING TIP

Remember: it takes a true experiment to determine causality.

When two variables (*X* and *Y*) are correlated, it is tempting to conclude that one of them is the cause of the other. However, to do so without further experimentation would be a serious error, because whenever two variables are correlated, there are four possible explanations of the correlation: (1) the correlation between *X* and *Y* is spurious, (2) *X* is the cause of *Y*, (3) *Y* is the cause of *X*, or (4) a third variable is the cause of the correlation between *X* and *Y*. The first possibility asserts that it was just due to accidents of sampling unusual people or unusual behavior that the sample showed a correlation; that is, if the experiment were repeated or more samples were taken, the correlation would disappear. If the correlation is really spurious, it is obviously wrong to conclude that there is a causal relationship between *X* and *Y*.

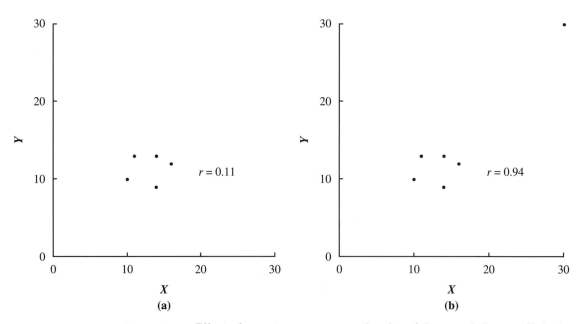

figure 6.11 Effect of an extreme score on the size of the correlation coefficient.

It is also erroneous to assume causality between X and Y if the fourth alternative is correct. Quite often, when X and Y are correlated, they are not causally related to each other but rather a third variable is responsible for the correlation. For example, do you know that there is a close relationship between the salaries of university professors and the price of a fifth of scotch whiskey? Which is the cause and which the effect? Do the salaries of university professors dominate the scotch whiskey market such that when the professors get a raise and thereby can afford to buy more scotch, the price of scotch is raised accordingly? Or perhaps the university professors are paid from the profits of scotch whiskey sales, so when the professors need a raise, the price of a fifth of scotch whiskey goes up? Actually, neither of these explanations is correct. Rather, a third factor is responsible for this correlation. What is that factor? Inflation! Recently, a newspaper article reported a positive correlation between obesity and female crime. Does this mean that if a woman gains 20 pounds, she will become a criminal? Or does it mean that if she is a criminal, she is doomed to being obese? Neither of these explanations seems satisfactory. Frankly, we are not sure how to interpret this correlation. One possibility is that it is a spurious correlation. If not, it could be due to a third factor, namely, socioeconomic status. Both obesity and crime are related to lower socioeconomic status.

The point is that a correlation between two variables is not sufficient to establish causality between them. There are other possible explanations. To establish that one variable is the cause of another, we must conduct an experiment in which we systematically vary *only* the suspected causal variable and measure its effect on the other variable.

WHAT IS THE TRUTH? "Good Principal = Good Elementary School," or Does It?

A major newspaper of a large city carried as a front page headline, printed in large bold letters, **"Equation for success: Good principal = good elementary school."** The article that followed described a study in which elementary school principals were rated by their teachers on a series of questions indicating whether the principals were strong, average, or weak leaders. Students in these schools were evaluated in reading and mathematics on the annual California Achievement Tests. As far as we can tell from the newspaper article, the ratings and test scores were obtained from ongoing principal assignments, with no attempt in the study to randomly assign principals to schools. The results showed that (1) in 11 elementary schools that had strong principals, students were making big academic strides; (2) in 11 schools where principals were weak leaders, students were showing less improvement than average or even falling behind; and (3) in 39 schools where principals were rated as average, the students' test scores were showing just average improvement.

The newspaper reporter interpreted these data as indicated by the headline, "Equation for success: Good principal = good elementary school." In the article, an elementary school principal was quoted, "I've always said 'Show me a good school, and I'll show you a good principal,' but now we have powerful, incontrovertible data that corroborates that." The article further quoted the president of the principals' association as saying: "It's exciting information that carries an enormous responsibility. It shows we can make a real difference in our students' lives." In your view, do the data warrant these conclusions?

Answer Although, personally, I believe that school principals are important to educational quality, the study seems to be strictly a correlational one: paired measurements on two variables, without random assignment to groups. From what we said previously, it is impossible to determine causality from such a study. The individuals quoted herein have taken a study that shows there is a *correlation* between "strong" leadership by elementary school principals and educational gain and concluded that the principals *caused* the educational gain. The conclusion is too strong. The correlation could be spurious or due to a third variable.

It is truly amazing how often this error is made in real life. Stay on the lookout, and I believe you will be surprised how frequently you encounter individuals concluding causation when the data are only correlational. Of course, now that you are so well informed on this point, you will never make this mistake yourself! ■

A recent article in the *New York Times* dealing with the topic "Does Money Buy Happiness?" shows a very interesting variant of the basic scatter plot that we discussed in this chapter. I thought you might like to see it. The variant is shown below. It graphs average life satisfaction as a function of Gross Domestic Product (GDP)/capita for a wide variety of countries. Pairing is by country, rather than by student as was the case for the IQ and grade point average data shown in Figure 6.4 on p. 129. Now that you are so good at interpreting scatter plots, I'm sure that you can tell from viewing the new scatter plot that there appears to be a positive, linear relationship between life satisfaction and GDP/capita. That is to say, as you go from countries with low GDP/capita to high GDP/capita, life satisfaction increases and the increase to a first approximation seems linear. In general, the positive correlation appears to be true within countries as well as across countries.

In addition to showing you this interesting scatter plot, I thought you might also be interested in the topic itself. Here is an excerpt from the article.

MONEY DOESN'T BUY HAPPINESS. WELL, ON SECOND THOUGHT...

In the aftermath of World War II, the Japanese economy went through one of the greatest booms the world has ever known. From 1950 to 1970, the economy's output per person grew more than sevenfold. Japan, in just a few decades, remade itself from a war-torn country into one of the richest nations on earth.

Yet, strangely, Japanese citizens didn't seem to become any more satisfied with their lives. According to one poll, the percentage of people who gave the most positive possible answer about their life satisfaction actually fell from the late 1950's to the early '70s. They were richer, but apparently no happier.

This contrast became the most famous example of a theory known as the Easterlin paradox. In 1974, Richard Easterlin, then an economist at the University of Pennsylvania, published a study in which he argued that economic growth didn't necessarily lead to more satisfaction.

(continued)

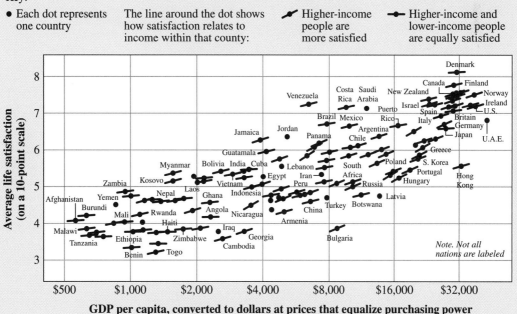

■ SUMMARY

In this chapter, I have discussed the topic of correlation. Correlation is a measure of the relationship that exists between two variables. The magnitude and direction of the relationship are given by a correlation coefficient. The correlation coefficient can vary from +1 to −1. The sign of the coefficient tells us whether the relationship is positive or negative. The numerical part describes the magnitude of the correlation. When the relationship is perfect, the magnitude is 1. If the relationship is nonexistent, the magnitude is 0. Magnitudes between 0 and 1 indicate imperfect relationships.

There are many correlation coefficients that can be computed, depending on the scaling of the data and the shape of the relationship. In this chapter, I emphasized Pearson r and Spearman rho. Pearson r is defined as a measure of the extent to which paired scores occupy the same or opposite positions within their own distributions. Using standard scores allows measurement of the relationship that is independent of the differences in scaling and of the units used in measuring the variables. Pearson r is also equal to the square root of the proportion of the total variability in Y that is accounted for by X. In addition to these concepts, I presented a computational equation for r and practiced calculating r.

Spearman rho is used for linear relationships when one or both of the variables are only of ordinal scaling.

The computational equation for rho was presented and several practice problems worked out. Next, I discussed the effect of an extreme score on the size of the correlation. After that, I discussed the effect of range on correlation and pointed out that truncated range will result in a lower correlation coefficient.

As the last topic of correlation, I discussed correlation and causation. I pointed out that if a correlation exists between two variables in an experiment, we cannot conclude they are causally related on the basis of the correlation alone because there are other possible explanations. The correlation may be spurious, or a third variable may be responsible for the correlation between the first two variables. To establish causation, one of the variables must be independently manipulated and its effect on the other variable measured. All other variables should be held constant or varied unsystematically. Even if the two variables are causally related, it is important to keep in mind that r^2, rather than r, indicates the size of the effect of one variable on the other.

■ IMPORTANT NEW TERMS

Biserial coefficient (p. 140)
Coefficient of determination
  (p. 139)
Correlation (p. 130)
Correlation coefficient (p. 130)
Curvilinear relationship (p. 125)
Direct relationship (p. 127)

Imperfect relationship (p. 128)
Inverse relationship (p. 127)
Linear relationship (p. 124)
Negative relationship (p. 127)
Pearson r (p. 131)
Perfect relationship (p. 128)
Phi coefficient (p. 140)

Positive relationship (p. 127)
Scatter plot (p. 124)
Slope (p. 125)
Spearman rho (p. 141)
Variability accounted for by
  X (p. 138)
Y intercept (p. 125)

■ QUESTIONS AND PROBLEMS

1. Define or identify each of the terms in the Important New Terms section.
2. Discuss the different kinds of relationships that are possible between two variables.
3. For each scatter plot in the accompanying figure (parts a–f, on p. 150), determine whether the relationship is
   a. Linear or curvilinear. If linear, further determine whether it is positive or negative.
   b. Perfect or imperfect
4. Professor Taylor does an experiment and establishes that a correlation exists between variables A and B. On the basis of this correlation, she asserts that A is the cause of B. Is this assertion correct? Explain.
5. Give two meanings of Pearson r.
6. Why are z scores used as the basis for determining Pearson r?
7. What is the range of values that a correlation coefficient may take?
8. A study has shown that the correlation between fatigue and irritability is 0.53. On the basis of this correlation, the author concludes that fatigue is an important factor in producing irritability. Is this conclusion justified? Explain.
9. What factors influence the choice of whether to use a particular correlation coefficient? Give some examples.
10. The Pearson r and Spearman rho correlation coefficients are related. Is this statement correct? Explain.
11. When two variables are correlated, there are four possible explanations of the correlation. What are they?
12. What effect might an extreme score have on the magnitude of relationship between two variables? Discuss.
13. What effect does decreasing the range of the paired scores have on the correlation coefficient?
14. Given the following sets of paired sample scores:

A		B		C	
X	Y	X	Y	X	Y
1	1	4	2	1	5
4	2	5	4	4	4
7	3	8	5	7	3
10	4	9	1	10	2
13	5	10	4	13	1

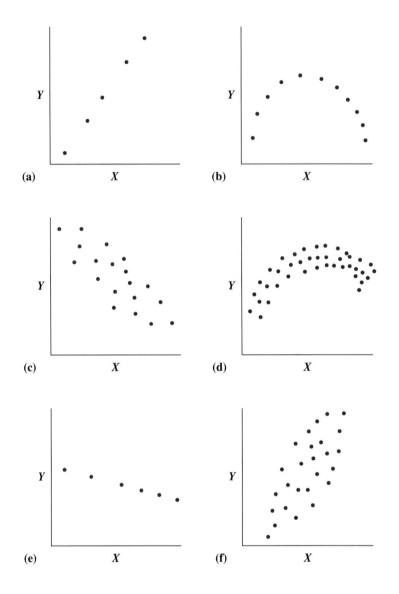

(a)   X

(b)   X

(c)   X

(d)   X

(e)   X

(f)   X

a. Use the equation

$$r = \Sigma \, z_X z_Y / (N - 1)$$

to compute the value of Pearson r for each set. Note that in set B, where the correlation is lowest, some of the $z_X z_Y$ values are positive and some are negative. These tend to cancel each other, causing r to have a low magnitude. However, in both sets A and C, all the products have the same sign, causing r to be large in magnitude. When the paired scores occupy the same or opposite positions within their own distributions, the $z_X z_Y$ products have the same sign, resulting in high magnitudes for r.

b. Compute r for set B, using the raw score equation. Which do you prefer: using the raw score or the z score equation?

c. Add the constant 5 to the X scores in set A and compute r again, using the raw score equation. Has the value changed?

d. Multiply the X scores in set A by 5 and compute r again. Has the value changed?

e. Generalize the results obtained in parts **c** and **d** to subtracting and dividing the scores by a constant. What does this tell you about r?

15. In a large introductory sociology course, a professor gives two exams. The professor wants to determine

whether the scores students receive on the second exam are correlated with their scores on the first exam. To make the calculations easier, a sample of eight students is selected. Their scores are shown in the accompanying table.

Student	Exam 1	Exam 2
1	60	60
2	75	100
3	70	80
4	72	68
5	54	73
6	83	97
7	80	85
8	65	90

a. Construct a scatter plot of the data, using exam 1 score as the X variable. Does the relationship look linear?
b. Assuming a linear relationship exists between scores on the two exams, compute the value for Pearson r.
c. How well does the relationship account for the scores on exam 2? education

16. A graduate student in developmental psychology believes there may be a relationship between birth weight and subsequent IQ. She randomly samples seven psychology majors at her university and gives them an IQ test. Next she obtains the weight at birth of the seven majors from the appropriate hospitals (after obtaining permission from the students, of course). The data are shown in the following table.

Student	Birth Weight (lbs)	IQ
1	5.8	122
2	6.5	120
3	8.0	129
4	5.9	112
5	8.5	127
6	7.2	116
7	9.0	130

a. Construct a scatter plot of the data, plotting birth weight on the X axis and IQ on the Y axis. Does the relationship appear to be linear?
b. Assume the relationship is linear and compute the value of Pearson r. developmental

17. A researcher conducts a study to investigate the relationship between cigarette smoking and illness. The number of cigarettes smoked daily and the number of days absent from work in the last year due to illness is determined for 12 individuals employed at the company where the researcher works. The scores are given in the following table.

Subject	Cigarettes Smoked	Days Absent
1	0	1
2	0	3
3	0	8
4	10	10
5	13	4
6	20	14
7	27	5
8	35	6
9	35	12
10	44	16
11	53	10
12	60	16

a. Construct a scatter plot for these data. Does the relationship look linear?
b. Calculate the value of Pearson r.
c. Eliminate the data from subjects 1, 2, 3, 10, 11, and 12. This decreases the range of both variables. Recalculate r for the remaining subjects. What effect does decreasing the range have on r?
d. If you use the full set of scores, what percentage of the variability in the number of days absent is accounted for by the number of cigarettes smoked daily? Of what use is this value?
clinical, health

18. An educator has constructed a test for mechanical aptitude. He wants to determine how reliable the test is over two administrations spaced by 1 month. A study is conducted in which 10 students are given two administrations of the test, with the second administration being 1 month after the first. The data are given in the following table.

Student	Administration 1	Administration 2
1	10	10
2	12	15
3	20	17
4	25	25
5	27	32
6	35	37
7	43	40
8	40	38
9	32	30
10	47	49

a. Construct a scatter plot of the paired scores.
b. Determine the value of *r*.
c. Would it be fair to say that this is a reliable test? Explain using r^2.
   education

19. A group of researchers has devised a stress questionnaire consisting of 15 life events. They are interested in determining whether there is cross-cultural agreement on the relative amount of adjustment each event entails. The questionnaire is given to 300 Americans and 300 Italians. Each individual is instructed to use the event of "marriage" as the standard and to judge each of the other life events in relation to the adjustment required in marriage. Marriage is arbitrarily given a value of 50 points. If an event is judged to require greater adjustment than marriage, the event should receive more than 50 points. How many more points depends on how much more adjustment is required. After each subject within each culture has assigned points to the 15 life events, the points for each event are averaged. The results are shown in the following table.

Life Event	Americans	Italians
Death of spouse	100	80
Divorce	73	95
Marital separation	65	85
Jail term	63	52
Personal injury	53	72
Marriage	50	50
Fired from work	47	40
Retirement	45	30
Pregnancy	40	28
Sex difficulties	39	42
Business readjustment	39	36
Trouble with in-laws	29	41
Trouble with boss	23	35
Vacation	13	16
Christmas	12	10

a. Assume the data are at least of interval scaling and compute the correlation between the American and Italian ratings.
b. Assume the data are only of ordinal scaling and compute the correlation between ratings of the two cultures. clinical, health

20. Given the following set of paired scores from five subjects,

Subject No.	1	2	3	4	5
Y	5	6	9	9	11
X	6	8	4	8	7

a. Construct a scatter plot of the data.
b. Compute the value of Pearson *r*.
c. Add the following paired scores from a sixth subject to the data: $Y = 26$, $X = 25$.
d. Construct another scatter plot, this time for the six paired scores.
e. Compute the value of Pearson *r* for the six paired scores.
f. Is there much of a difference between your answers for parts **b** and **e**? Explain the difference.

21. The director of an obesity clinic in a large northwestern city believes that drinking soft drinks

contributes to obesity in children. To determine whether a relationship exists between these two variables, she conducts the following pilot study. Eight 12-year-old volunteers are randomly selected from children attending a local junior high school. Parents of the children are asked to monitor the number of soft drinks consumed by their child over a 1-week period. The children are weighed at the end of the week and their weights converted into body mass index (BMI) values. The BMI is a common index used to measure obesity and takes into account both height and weight. An individual is considered obese if he or she has a BMI value $\geq$ 30. The following data are collected.

Child	Number of Soft Drinks Consumed	BMI
1	3	20
2	1	18
3	14	32
4	7	24
5	21	35
6	5	19
7	25	38
8	9	30

a. Graph a scatter plot of the data. Does the relationship appear linear?
b. Assume the relationship is linear and compute Pearson r. health

22. A social psychologist conducts a study to determine the relationship between religion and self-esteem. Ten eighth graders are randomly selected for the study. Each individual undergoes two tests, one measuring self-esteem and the other religious involvement. For the self-esteem test, the lower the score is, the higher self-esteem is; for the test measuring religious involvement, the higher the score is, the higher religious involvement is. The self-esteem test has a range from 1 to 10 and the religious involvement test ranges from 0 to 50. For the purposes of this question, assume both tests are well standardized and of interval scaling. The following data are collected.

Subject	Religious Involvement	Self-Esteem
1	5	8
2	25	3
3	45	2
4	20	7
5	30	5
6	40	5
7	1	4
8	15	4
9	10	7
10	35	3

a. If a relationship exists such that the more religiously involved one is, the higher actual self-esteem is, would you expect r computed on the provided values to be negative or positive? Explain.
b. Compute r. Were you correct in your answer to part **a**? social, developmental

23. A psychologist has constructed a paper & pencil test purported to measure depression. To see how the test compares with the ratings of experts, 12 "emotionally disturbed" individuals are given the paper & pencil test. The individuals are also independently rank-ordered by two psychiatrists according to the degree of depression each psychiatrist finds as a result of detailed interviews. The scores are given here. Higher scores represent greater depression.

Individual	Paper & Pencil Test	Psychiatrist A	Psychiatrist B
1	48	12	9
2	37	11	12
3	30	4	5
4	45	7	8
5	31	10	11
6	24	8	7
7	28	3	4
8	18	1	1
9	35	9	6
10	15	2	2
11	42	6	10
12	22	5	3

a. What is the correlation between the rankings of the two psychiatrists?

b. What is the correlation between the scores on the paper & pencil test and the rankings of each psychiatrist? clinical, health

24. For this problem, let's suppose that you are a psychologist employed in the human resources department of a large corporation. The corporation president has just finished talking with you about the importance of hiring productive personnel in the manufacturing section of the corporation and has asked you to help improve the corporation's ability to do so. There are 300 employees in this section, with each employee making the same item. Until now, the corporation has been depending solely on interviews for selecting these employees. You search the literature and discover two well-standardized paper & pencil performance tests that you think might be related to the performance requirements of this section. To determine whether either might be used as a screening device, you select 10 representative employees from the manufacturing section, making sure that a wide range of performance is represented in the sample, and administer the two tests to each employee. The data are shown in the table. The higher the score, the better the performance. The work performance scores are the actual number of items completed by each employee per week, averaged over the past 6 months.

a. Construct a scatter plot of work performance and test 1, using test 1 as the X variable. Does the relationship look linear?

b. Assuming it is linear, compute the value of Pearson r.

c. Construct a scatter plot of work performance and test 2, using test 2 as the X variable. Is the relationship linear?

d. Assuming it is linear, compute the value of Pearson r.

e. If you could use only one of the two tests for screening prospective employees, would you use either test? If yes, which one? Explain. I/O

	Employee									
	1	**2**	**3**	**4**	**5**	**6**	**7**	**8**	**9**	**10**
Work performance	50	74	62	90	98	52	68	80	88	76
Test 1	10	19	20	20	21	14	10	24	16	14
Test 2	25	35	40	49	50	29	32	44	46	35

What Is the Truth? Questions

1. *'Good Principal = Good Elementary School,' or Does It?*

   a. Give another explanation for the relationship, besides *"good principals produce good elementary schools."*

   b. Give one example that you have encountered where someone of authority has used correlational data to impute causation. Discuss the possible errors that are involved, using the specific variables of your example.

2. *Money Doesn't Buy Happiness, or Does It?*

   a. Do you like the scatter plot? Is it convincing? Did you read any of the references on happiness? So what do you think: does money buy happiness?

■ SPSS ILLUSTRATIVE EXAMPLE 6.1 _____

The general operation of SPSS and data entry are described in Appendix E, *Introduction to SPSS.* SPSS can be very helpful when dealing with correlation by graphing scatter plots and computing correlation coefficients.

example

For this example, let's use the IQ and grade point average (GPA) data shown in Table 6.2, p. 128, of the textbook. For your convenience the data are shown again below.

a. Use SPSS to construct a scatter plot of the data. In so doing, name the two variables *GPA* and *IQ*. Make *IQ* the *X* axis variable.
b. Assuming a linear relationship exists between *IQ* and *GPA,* use SPSS to compute the value of Pearson *r.*

Student No.	1	2	3	4	5	6	7	8	9	10	11	12
IQ	110	112	118	119	122	125	127	130	132	134	136	138
GPA	1.0	1.6	1.2	2.1	2.6	1.8	2.6	2.0	3.2	2.6	3.0	3.6

SOLUTION

STEP 1: Enter the Data.

1. Enter the *IQ* scores in the first column (**VAR00001**) of the SPSS Data Editor, beginning with the first score in the first cell of the first column of the Data Editor.
2. Enter the *GPA* scores in the second column (**VAR00002**) of the SPSS Data Editor, beginning with the first score in the first cell of the second column of the Data Editor.

STEP 2: Name the Variables. In this example, we will give the default variables **VAR00001** and **VAR00002** the new names of **IQ** and **GPA**, respectively. Here's how it is done.

1. Click the **Variable View** tab in the lower left corner of the Data Editor.	This displays the **Variable View** on screen, with **VAR00001** and **VAR00002** displayed in the first and second cells of the **Name** column.
2. Click VAR00001; then **type IQ** in the highlighted cell and **press Enter.**	**IQ** is entered as the variable name, replacing **VAR00001**. The cursor then moves to the next cell, highlighting **VAR00002**.
3. Replace VAR00002 with **GPA** and then **press Enter.**	**GPA** is entered as the variable name, replacing **VAR00002**.

STEP 3:   **Analyze the Data.**

Part a.   Construct a Scatter Plot of the Data. I suggest you switch to the Data Editor-Data View if you haven't already done so. Now, let's proceed with the analysis.

---

1. Click **Graphs** on the menu bar at the top of the screen; then select **Legacy Dialogs**; then click **Scatter/Dot....**

This produces the **Scatter/Dot** dialog box, which is used to produce scatter plots. The default is the **Simple Scatter** plot, which is what we want. Therefore, we don't have to click it.

2. Click **Define**.

This produces the **Simple Scatterplot** dialog box with **IQ** and **GPA** in the large box on the left. **IQ** is highlighted.

3. Click the **arrow** for the **X axis:**.

This moves **IQ** from the large box on the left into the **X axis:** box on the right. We have done this because we want to plot **IQ** on the X axis.

4. Click **GPA** in the large box on the left; then click the **arrow** for the **Y axis:**.

This moves **GPA** from the large box on the left into the **Y axis:** box on the right. We have done this because we want to plot **GPA** on the Y axis.

5. Click **OK**.

SPSS then constructs and displays in the output window a scatter plot of the two variables, with **IQ** plotted on the X axis and **GPA** plotted on the Y axis. The scatter plot is shown below.

Analysis Results

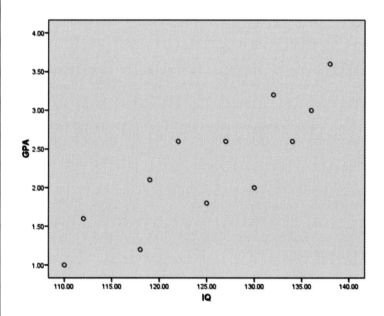

If you compare this scatter plot with that shown in Figure 6.4, p. 129 of the textbook, they appear identical, except for the axes labeling and the regression line.

Part b. **Compute the Value of Pearson *r* for IQ and GPA.** Ordinarily, the first step would be to enter the data. However, that has already been done in STEP 1. Therefore, we will go directly to computing Pearson *r*.

1. Click **Analyze** on the tool bar at the top of the screen; then **select Correlate;** then **click Bivariate…**

   This produces the **Bivariate Correlations** dialog box, with **IQ** and **GPA** displayed in the large box on the left. **IQ** is highlighted.

2. **Click** the **arrow** between the two large boxes.

   This moves **IQ** into the **Variables:** box on the right.

3. **Click GPA** in the large box on the left; then **click** the **arrow** between the two large boxes.

   This moves **GPA** into the **Variables:** box on the right. Notice that the **Pearson** box already has a **check** in it, telling SPSS to compute Pearson *r* when it gets the **OK**.

4. **Click OK**.

   SPSS computes Pearson *r* for **IQ** and **GPA** and displays the results in the Output window in the **Correlations** table. The **Correlations** table is shown below.

   Analysis Results

   Correlations

		IQ	GPA
IQ	Pearson Correlation	1	.856**
	Sig. (2-tailed)		.000
	N	12	12
GPA	Pearson Correlation	.856**	1
	Sig. (2-tailed)	.000	
	N	12	12

   ** Correlation is significant at the 0.01 level (2-tailed).

   Note, SPSS uses the term **Pearson Correlation** for *Pearson r*. The value of the **Pearson Correlation** between **IQ** and **GPA** given in the SPSS **Correlations** table is **.856**. This is the same value arrived at for these data in the textbook in Practice Problem 6.1, p. 135. The **Correlations** table also gives additional information that we do not need at this time, so we'll ignore it for now.

■ SPSS ADDITIONAL PROBLEMS

1. For this problem, we will use the data given in Chapter 6, Problem 19, p. 152.
   a. Use SPSS to construct a scatter plot of the data. Name the variables *American* and *Italian*. Use *American* as the *X* axis variable and *Italian* as the *Y* axis variable.
   b. Describe the relationship.
   c. Use SPSS to calculate the value of Pearson *r*.

2. A psychology professor is interested in whether there is a relationship between *reaction time* and *age*. The following data are collected on thirty individuals randomly sampled from the city in which the professor works. Random sampling is conducted in a manner to ensure that a wide age range is represented in the sample.

Subject No.	1	2	3	4	5	6	7	8	9	10
RT (msec)	300	538	832	747	610	318	822	693	461	460
Age (yrs)	24.6	28.5	45.2	31.1	22.0	14.4	57.9	55.0	37.3	14.0

Subject No.	11	12	13	14	15	16	17	18	19	20
RT (msec)	418	706	830	584	515	222	740	582	398	661
Age (yrs)	24.7	41.5	51.3	45.8	19.0	10.0	57.4	32.2	19.7	25.7

Subject No.	21	22	23	24	25	26	27	28	29	30
RT (msec)	605	890	717	610	803	702	241	360	890	424
Age (yrs)	52.1	55.4	46.0	38.6	39.7	34.4	15.0	29.2	59.5	32.0

a. Use SPSS to construct a scatter plot of the data. Name the variables *RT* and *Age*. Assign *RT* as the *Y* axis variable and *Age* as the *X* axis variable.

b. Describe the relationship.

c. Use SPSS to compute Pearson *r*.

d. How much of the variability of *RT* is accounted for by *Age*?

■ ONLINE STUDY RESOURCES

CENGAGE **brain**.com

Login to CengageBrain.com to access the resources your instructor has assigned. For this book, you can access the book's companion website for chapter-specific learning tools, including Know and Be Able to Do, practice quizzes, flash cards, and glossaries, and a link to Statistics and Research Methods Workshops.

aplia

If your professor has assigned Aplia homework:

1. Sign in to your account
2. Complete the corresponding homework exercises as required by your professor
3. When finished, click "Grade It Now" to see which areas you have mastered and which need more work, and for detailed explanations of every answer.

Visit **www.cengagebrain.com** to access your account and to purchase materials.

© Strmko / Dreamstime.com

7

Linear Regression

LEARNING OBJECTIVES

After completing this chapter, you should be able to:

- Define regression, regression line, and regression constant.
- Specify the relationship between strength of relationship and prediction accuracy.
- Construct the least-squares regression line for predicting Y given X, specify what the least-squares regression line minimizes, and specify the convention for assigning X and Y to the data variables.
- Explain what is meant by standard error of estimate, state the relationship between errors in prediction and the magnitude of $s_{Y|X}$, and define homoscedasticity and explain its use.
- Specify the condition(s) that must be met to use linear regression.
- Specify the relationship between regression constants and Pearson r.
- Explain the use of multiple variables and their relationship to prediction accuracy.
- Compute R^2 for two variables; specify what R^2 stands for and what it measures.
- Understand the illustrative examples, do the practice problems, and understand the solutions.

INTRODUCTION

Regression and correlation are closely related. At the most basic level, they both involve the relationship between two variables, and they both utilize the same set of basic data: paired scores taken from the same or matched subjects. As we saw in Chapter 6, correlation is concerned with the magnitude and direction of the relationship. Regression focuses on using the relationship for prediction. Prediction is quite easy when the relationship is perfect. If the relationship is perfect, all the points fall on a straight line and all we need do is derive the equation of the straight line and use it for prediction. As you might guess, when the relationship is perfect, so is prediction. All predicted values are exactly equal to the observed values and prediction error equals zero. The situation is more complicated when the relationship is imperfect.

definitions
- **Regression** *is a topic that considers using the relationship between two or more variables for prediction.*

- *A* **regression line** *is a best fitting line used for prediction.*

PREDICTION AND IMPERFECT RELATIONSHIPS

Let's return to the data involving grade point average and IQ that were presented in Chapter 6. For convenience, the data have been reproduced in Table 7.1. Figure 7.1 shows a scatter plot of the data. The relationship is imperfect, positive, and linear. The problem we face for prediction is how to determine the single straight line that best describes the data. The solution most often used is to construct the line that minimizes errors of prediction according to a *least-squares* criterion. Appropriately, this line is called the *least-squares* regression line.

table **7.1** IQ and grade point average of 12 college students

Student No.	IQ	Grade Point Average
1	110	1.0
2	112	1.6
3	118	1.2
4	119	2.1
5	122	2.6
6	125	1.8
7	127	2.6
8	130	2.0
9	132	3.2
10	134	2.6
11	136	3.0
12	138	3.6

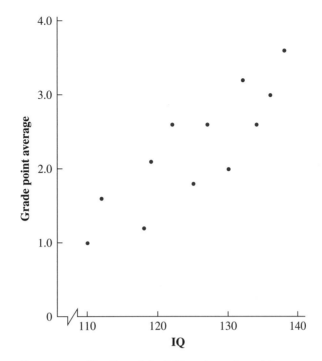

figure 7.1 Scatter plot of IQ and grade point average.

The least-squares regression line for the data in Table 7.1 is shown in Figure 7.2(a). The vertical distance between each point and the line represents the error in prediction. If we let Y' = the predicted Y value and Y = the actual value, then $Y - Y'$ equals the error for each point. It might seem that the total error in prediction should be the simple algebraic sum of $Y - Y'$ summed over all of the points. If this were true, since we are interested in minimizing the error, we would construct the line that minimizes $\Sigma (Y - Y')$. However, the total error in prediction does not equal $\Sigma (Y - Y')$ because some of the Y' values will be greater than Y and some will be less. Thus, there will be both positive and negative error scores, and the simple algebraic sums of these would cancel each other. We encountered a similar situation when considering measures of the average dispersion. In deriving the equation for the standard deviation, we squared $X - \overline{X}$ to overcome the fact that there were positive and negative deviation scores that canceled each other. The same solution works here, too. Instead of just summing $Y - Y'$, we first compute $(Y - Y')^2$ for each score. This removes the negative values and eliminates the cancellation problem. Now, if we minimize $\Sigma (Y - Y')^2$, we minimize the total error of prediction.

definition ■ *The* **least-squares regression line** *is the prediction line that minimizes the total error of prediction, according to the least-squares criterion of $\Sigma (Y - Y')^2$.*

For any linear relationship, there is only one line that will minimize $\Sigma (Y - Y')^2$. Thus, there is only one least-squares regression line for each linear relationship.

We said before that there are many "possible" prediction lines we could construct when the relationship is imperfect. Why should we use the least-squares regression line? We use the least-squares regression line because it gives the greatest

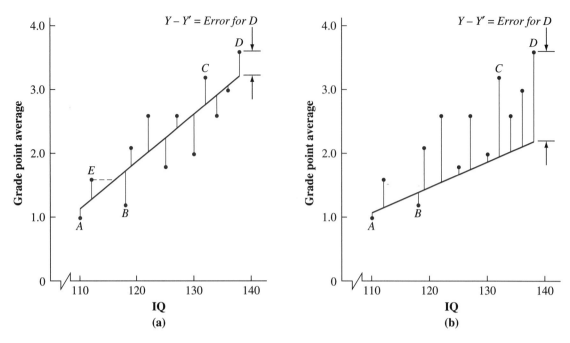

figure 7.2   Two regression lines and prediction error.

overall accuracy in prediction. To illustrate this point, another prediction line has been drawn in Figure 7.2(b). This line has been picked arbitrarily and is just one of an infinite number that could have been drawn. How does it compare in prediction accuracy with the least-squares regression line? We can see that it actually does better for some of the points (e.g., points A and B). However, it also misses badly on others (e.g., points C and D). If we consider all of the points, it is clear that the line of Figure 7.2(a) fits the points better than the line of Figure 7.2(b). The total error in prediction, represented by $\Sigma (Y - Y')^2$, is less for the least-squares regression line than for the line in Figure 7.2(b). In fact, the total error in prediction is less for the least-squares regression line than for any other possible prediction line. Thus, the least-squares regression line is used because it gives greater overall accuracy in prediction than any other possible regression line.

CONSTRUCTING THE LEAST-SQUARES REGRESSION LINE: REGRESSION OF *Y* ON *X*

The equation for the least-squares regression line for predicting Y given X is

$$Y' = b_Y X + a_Y \qquad \textit{linear regression equation for predicting Y given X}$$

where Y' = predicted or estimated value of Y
b_Y = slope of the line for minimizing errors in predicting Y
a_Y = Y axis intercept for minimizing errors in predicting Y

This is, of course, the general equation for a straight line that we have been using all along. In this context, however, a_Y and b_Y are called *regression constants*. This line is called the regression line of *Y* on *X*, or simply the *regression of Y on X*, because we are predicting *Y* given *X*.

The b_Y regression constant is equal to

$$b_Y = \frac{\Sigma\, XY - \dfrac{(\Sigma\, X)(\Sigma\, Y)}{N}}{SS_X}$$

where SS_X = sum of squares of *X* scores $= \Sigma\, X^2 - \dfrac{(\Sigma\, X)^2}{N}$
            N = number of *paired* scores
            $\Sigma\, XY$ = sum of the product of each *X* and *Y* pair (also called the sum of the cross products)

The equation for computing b_Y from the raw scores is

$$b_Y = \frac{\Sigma\, XY - \dfrac{(\Sigma\, X)(\Sigma\, Y)}{N}}{\Sigma\, X^2 - \dfrac{(\Sigma\, X)^2}{N}}$$

*computational equation for determining the **b** regression constant for predicting Y given X*

The a_Y regression constant is given by

$$a_Y = \overline{Y} - b_Y\overline{X}$$

*computational equation for determining the **a** regression constant for predicting Y given X*

Since we need the b_Y constant to determine the a_Y constant, the procedure is to first find b_Y and then a_Y. Once both are found, they are substituted into the regression equation. Let's construct the least-squares regression line for the IQ and grade point data presented previously. For convenience, the data have been presented again in Table 7.2.

$$b_Y = \frac{\Sigma\, XY - \dfrac{(\Sigma\, X)(\Sigma\, Y)}{N}}{\Sigma\, X^2 - \dfrac{(\Sigma\, X)^2}{N}}$$

$$= \frac{3488.7 - \dfrac{1503(27.3)}{12}}{189{,}187 - \dfrac{(1503)^2}{12}}$$

$$= \frac{69.375}{936.25} = 0.0741 = 0.074$$

$$a_Y = \overline{Y} - b_Y\overline{X}$$

$$= 2.275 - 0.0741(125.25)$$

$$= -7.006$$

and

$$Y' = 0.074X - 7.006$$

The full solution is also shown in Table 7.2. The regression line has been plotted in Figure 7.3. You can now use the equation for *Y'* to predict the grade point

table 7.2 IQ and grade point average of 12 college students: predicting Y from X

Student No.	IQ X	Grade Point Average Y	XY	X^2
1	110	1.0	110.0	12,100
2	112	1.6	179.2	12,544
3	118	1.2	141.6	13,924
4	119	2.1	249.9	14,161
5	122	2.6	317.2	14,884
6	125	1.8	225.0	15,625
7	127	2.6	330.2	16,129
8	130	2.0	260.0	16,900
9	132	3.2	422.4	17,424
10	134	2.6	348.4	17,956
11	136	3.0	408.0	18,496
12	138	3.6	496.8	19,044
Total	1503	27.3	3488.7	189,187

MENTORING TIP

Remember: N is the number of *paired* scores, not the total number of scores. In this example, $N = 12$.

$$b_Y = \frac{\Sigma XY - \dfrac{(\Sigma X)(\Sigma Y)}{N}}{\Sigma X^2 - \dfrac{(\Sigma X)^2}{N}} = \frac{3488.7 - \dfrac{1503(27.3)}{12}}{189,187 - \dfrac{(1503)^2}{12}} = \frac{69.375}{936.25} = 0.0741 = 0.074$$

$$a_Y = \overline{Y} - b_Y\overline{X} = 2.275 - 0.0741(125.25) = -7.006$$

$$Y' = b_Y X + a_Y = 0.074X - 7.006$$

MENTORING TIP

When plotting the regression line, a good procedure is to select the lowest and highest X values in the sample data, and compute Y' for these X values. Then locate these X, Y coordinates on the graph and draw the straight line between them.

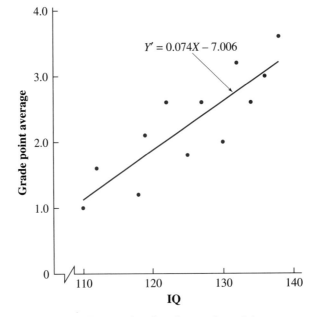

figure 7.3 Regression line for grade point average and IQ.

average, knowing only the student's IQ score. For example, suppose a student's IQ score is 124; using this regression line, what is the student's predicted grade point average?

$$Y' = 0.074X - 7.006$$
$$= 0.074(124) - 7.006$$
$$= 2.17$$

MENTORING TIP

Label the variable to which you are predicting as the Y variable, and the variable you are predicting from as the X variable.

Please note that it is customary to label the variable to which we are predicting as the Y variable, and the variable we are predicting from as the X variable. Accordingly, *Grade Point Average* was given the label Y and *IQ* was given the label X. If we were interested in predicting *IQ* from *Grade Point Average*, we would have labeled *IQ* as the Y variable and *Grade Point Average* as the X variable. Following this convention, whichever variable is labeled Y becomes the predicted variable, and the equations previously derived for Y', a_Y, and b_Y are the appropriate equations to use.

Let's try a couple of practice problems.

Practice Problem 7.1

A developmental psychologist is interested in determining whether it is possible to use the heights of young boys to predict their eventual height at maturity. To answer this question, she collects the data shown in the following table. Since we are interested in predicting *Height at Age 20* from *Height at Age 3*, we have labeled *Height at Age 20* as the Y variable and *Height at Age 3* as the X variable.

a. Draw a scatter plot of the data.
b. If the data are linearly related, derive the least-squares regression line.
c. Based on these data, what height would you predict for a 20-year-old if at 3 years his height were 42 inches?

Individual No.	Height at Age 3 (in.) X	Height at Age 20 (in.) Y	XY	X^2
1	30	59	1,770	900
2	30	63	1,890	900
3	32	62	1,984	1,024
4	33	67	2,211	1,089
5	34	65	2,210	1,156
6	35	61	2,135	1,225
7	36	69	2,484	1,296
8	38	66	2,508	1,444
9	40	68	2,720	1,600
10	41	65	2,665	1,681
11	41	73	2,993	1,681
12	43	68	2,924	1,849

(continued)

Individual No.	Height at Age 3 (in.) X	Height at Age 20 (in.) Y	XY	X^2
13	45	71	3,195	2,025
14	45	74	3,330	2,025
15	47	71	3,337	2,209
16	48	75	3,600	2,304
Total	618	1077	41,956	24,408

$$b_Y = \frac{\Sigma XY - \dfrac{(\Sigma X)(\Sigma Y)}{N}}{\Sigma X^2 - \dfrac{(\Sigma X)^2}{N}} = \frac{41,956 - \dfrac{618(1077)}{16}}{24,408 - \dfrac{(618)^2}{16}} = 0.6636 = 0.664$$

$$a_Y = \overline{Y} - b_Y \overline{X} - 67.3125 - 0.6636(38.625) = 41.679$$

$$Y' = b_Y X + a_Y = 0.664X + 41.679$$

SOLUTION

a. The scatter plot is shown in the following figure. It is clear that an imperfect relationship that is linear and positive exists between the heights at ages 3 and 20.

b. Derive the least-squares regression line.

$$Y' = b_Y X + a_Y$$

$$b_Y = \frac{\Sigma XY - \dfrac{(\Sigma X)(\Sigma Y)}{N}}{\Sigma X^2 - \dfrac{(\Sigma X)^2}{N}} = \frac{41,956 - \dfrac{618(1077)}{16}}{24,408 - \dfrac{(618)^2}{16}}$$

$$= 0.6636 = 0.664$$

$$a_Y = \overline{Y} - b_Y \overline{X} = 67.3125 - 0.6636(38.625)$$

$$= 41.679$$

Therefore,

$$Y' = b_Y X + a_Y = 0.664X + 41.679$$

This solution is also shown in the previous table. The least-squares regression line is shown on the scatter plot below.

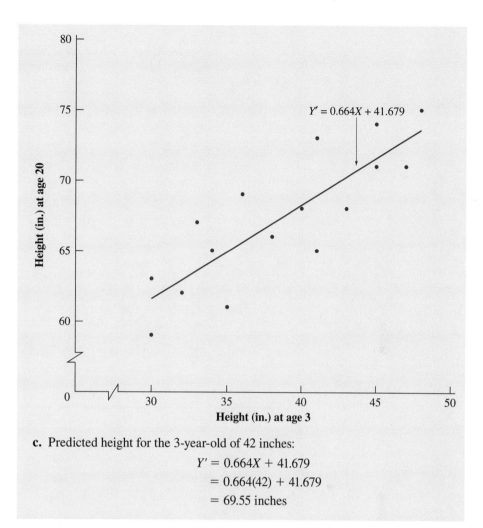

c. Predicted height for the 3-year-old of 42 inches:

$$Y' = 0.664X + 41.679$$
$$= 0.664(42) + 41.679$$
$$= 69.55 \text{ inches}$$

Practice Problem 7.2

A neuroscientist suspects that low levels of the brain neurotransmitter serotonin may be causally related to aggressive behavior. As a first step in investigating this hunch, she decides to do a correlative study involving nine rhesus monkeys. The monkeys are observed daily for 6 months, and the number of aggressive acts is recorded. Serotonin levels in the striatum (a brain region associated with aggressive behavior) are also measured once per day for each animal. The resulting data are shown in the following table. The number of aggressive acts for each animal is the average for the 6 months, given on a per-day basis. Serotonin levels are also average values over the 6-month period.

a. Draw a scatter plot of the data.
b. If the data are linearly related, derive the least-squares regression line for predicting the number of aggressive acts from serotonin level.
c. On the basis of these data, what is the number of aggressive acts per day you would predict if a rhesus monkey had a serotonin level of 0.46 microgm/gm?

(continued)

Subject No.	Serotonin Level (microgm/gm) X	Number of Aggressive Acts/day Y	XY	X²
1	0.32	6.0	1.920	0.1024
2	0.35	3.8	1.330	0.1225
3	0.38	3.0	1.140	0.1444
4	0.41	5.1	2.091	0.1681
5	0.43	3.0	1.290	0.1849
6	0.51	3.8	1.938	0.2601
7	0.53	2.4	1.272	0.2809
8	0.60	3.5	2.100	0.3600
9	0.63	2.2	1.386	0.3969
Total	4.16	32.8	14.467	2.0202

$$b_Y = \frac{\Sigma XY - \dfrac{(\Sigma X)(\Sigma Y)}{N}}{\Sigma X^2 - \dfrac{(\Sigma X)^2}{N}} = \frac{14.467 - \dfrac{(4.16)(32.8)}{9}}{2.0202 - \dfrac{(4.16)^2}{9}} = -7.127$$

$$a_Y = \overline{Y} - b_Y \overline{X} = 3.6444 - (-7.1274)(0.4622) = 6.939$$

$$Y' = b_Y X + a_Y = -7.127X + 6.939$$

SOLUTION

a. The scatter plot follows. It is clear that an imperfect, linear, negative relationship exists between the two variables.

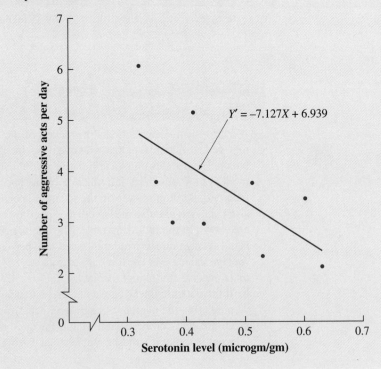

> **b.** Derive the least-squares regression line. The solution is shown at the bottom of the previous table and the regression line has been plotted on the scatter plot above.
>
> **c.** Predicted number of aggressive acts:
>
> $$Y' = -7.127\,X + 6.939$$
> $$= -7.127(0.46) + 6.939$$
> $$= 3.7 \text{ aggressive acts per day}$$

MEASURING PREDICTION ERRORS: THE STANDARD ERROR OF ESTIMATE

The regression line represents our best estimate of the Y scores, given their corresponding X values. However, unless the relationship between X and Y is perfect, most of the actual Y values will not fall on the regression line. Thus, when the relationship is imperfect, there will necessarily be prediction errors. It is useful to know the magnitude of the errors. For example, it sounds nice to say that, on the basis of the relationship between IQ and grade point average given previously, we predict that John's grade point average will be 3.2 when he is a senior. However, since the relationship is imperfect, it is unlikely that our prediction is exactly correct. Well, if it is not exactly correct, then how far off is it? If it is likely to be very far off, we can't put much reliance on the prediction. However, if the error is likely to be small, the prediction can be taken seriously and decisions made accordingly.

Quantifying prediction errors involves computing the *standard error of estimate*. The standard error of estimate is much like the standard deviation. You will recall that the standard deviation gave us a measure of the average deviation about the mean. The standard error of estimate gives us a measure of the average deviation of the prediction errors about the regression line. In this context, the regression line can be considered an estimate of the mean of the Y values for each of the X values. It is like a "floating" mean of the Y values, which changes with the X values. With the standard deviation, the sum of the deviations, $\Sigma\,(X - \overline{X})$, equaled 0. We had to square the deviations to obtain a meaningful average. The situation is the same with the standard error of estimate. Since the sum of the prediction errors, $\Sigma\,(Y - Y')$, equals 0, we must square them also. The average is then obtained by summing the squared values, dividing by $N - 2$, and taking the square root of the quotient (very much like with the standard deviation). The equation for the standard error of estimate for predicting Y given X is

$$s_{Y|X} = \sqrt{\dfrac{\Sigma\,(Y - Y')^2}{N - 2}} \qquad \textit{standard error of estimate when predicting Y given X}$$

Note that we have divided by $N - 2$ rather than $N - 1$, as was done with the sample standard deviation.* The calculations involved in using this equation are quite laborious. The computational equation, which is given here, is much easier to use. In

---

*We divide by $N - 2$ because calculation of the standard error of estimate involves fitting the data to a straight line. To do so requires estimation of two parameters, slope and intercept, leaving the deviations about the line with $N - 2$ degrees of freedom. We shall discuss degrees of freedom in Chapter 13.

determining the b_Y regression coefficient, we have already calculated the values for SS_X and SS_Y.

$$s_{Y|X} = \sqrt{\dfrac{SS_Y - \dfrac{[\Sigma XY - (\Sigma X)(\Sigma Y)/N]^2}{SS_X}}{N - 2}}$$

computational equation: standard error of estimate when predicting Y given X

To illustrate the use of these equations, let's calculate the standard error of estimate for the grade point and IQ data shown in Tables 7.1 and 7.2. As before, we shall let grade point average be the Y variable and IQ the X variable, and we shall calculate the standard error of estimate for predicting grade point average, given IQ. As computed in the tables, $SS_X = 936.25$, $SS_Y = 7.022$, $\Sigma XY - (\Sigma X)(\Sigma Y)/N = 69.375$, and $N = 12$. Substituting these values in the equation for the standard error of estimate for predicting Y given X, we obtain

$$s_{Y|X} = \sqrt{\dfrac{SS_Y - \dfrac{[\Sigma XY - (\Sigma X)(\Sigma Y)/N]^2}{SS_X}}{N - 2}}$$

$$= \sqrt{\dfrac{7.022 - \dfrac{(69.375)^2}{936.25}}{12 - 2}}$$

$$= \sqrt{0.188} = 0.43$$

Thus, the standard error of estimate $= 0.43$. This measure has been computed over all the Y scores. For it to be meaningful, we must assume that the variability of Y remains constant as we go from one X score to the next. This assumption is called the assumption of *homoscedasticity*. Figure 7.4(a) shows an illustration where the homoscedasticity assumption is met. Figure 7.4(b) shows an illustration where the assumption is violated. The homoscedasticity assumption implies that if we divided the X scores into columns, the variability of Y would not change from column to column. We can see how this is true for Figure 7.4(a) but not for 7.4(b).

What meaning does the standard error of estimate have? Certainly, it is a quantification of the errors of prediction. The larger its value, the less confidence we have in the prediction. Conversely, the smaller its value, the more likely the prediction will be accurate.

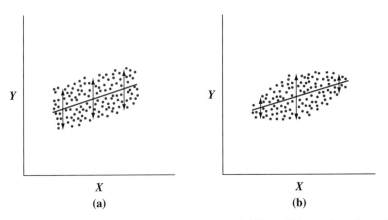

figure 7.4 Scatter plots showing the variability of Y as a function of X.

From E.W. Minium, *Statistical Reasoning in Psychology and Education.* Copyright © 1978 by John Wiley & Sons, Inc. Adapted with permission of John Wiley & Sons, Inc.

We can still be more quantitative. We can assume the points are normally distributed about the regression line (Figure 7.5). If the assumption is valid and we were to construct two lines parallel to the regression line at distances of $\pm 1s_{Y|X}$, $\pm 2s_{Y|X}$, and $\pm 3s_{Y|X}$, we would find that approximately 68% of the scores fall between the lines at $\pm 1s_{Y|X}$, approximately 95% lie between $\pm 2s_{Y|X}$, and approximately 99% lie between $\pm 3s_{Y|X}$. To illustrate this point, in Figure 7.6 we have drawn two dashed lines parallel to the regression line for the grade point and IQ data at a distance of $\pm 1s_{Y|X}$. We have also entered the scores in the figure. According to what we said previously, approximately 68% of the scores should lie between these lines. There are 12 points, so we would expect $0.68(12) = 8$ of the scores to be contained within the lines. In fact, there are 8. The agreement isn't always this good, particularly when there are only 12 scores in the sample. As N increases, the agreement usually increases as well.

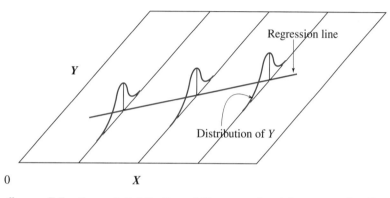

figure 7.5 Normal distribution of Y scores about the regression line.

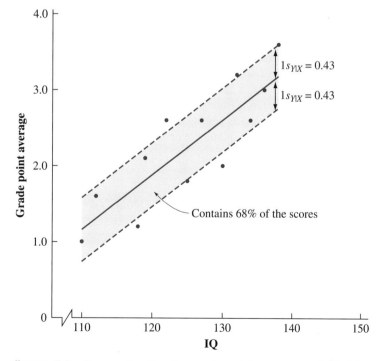

figure 7.6 Regression line for grade point average and IQ data with parallel lines $1s_{Y|X}$ above and below the regression line.

CONSIDERATIONS IN USING LINEAR REGRESSION FOR PREDICTION

The procedures we have described are appropriate for predicting scores based on presumption of a linear relationship existing between the *X* and *Y* variables. If the relationship is nonlinear, the prediction will not be very accurate. It follows, then, that the first assumption for successful use of this technique is that *the relationship between X and Y must be linear.* Second, we are not ordinarily interested in using the regression line to predict scores of the individuals who were in the group used for calculating the regression line. After all, why predict their scores when we already know them? Generally, a regression line is determined for use with subjects where one of the variables is unknown. For instance, in the IQ and grade point average problem, a university admissions officer might want to use the regression line to predict the grade point averages of prospective students, knowing their IQ scores. It doesn't make any sense to predict the grade point averages of the 12 students whose data were used in the problem. He already knows their grade point averages. If we are going to use data collected on one group to predict scores of another group, it is important that *the basic computation group be representative of the prediction group.* Often this requirement is handled by randomly sampling from the prediction population and using the sample for deriving the regression equation. Random sampling is discussed in Chapter 8. Finally, the linear regression equation is properly used *just for the range of the variable on which it is based.* For example, when we were predicting grade point average from IQ, we should have limited our predictions to IQ scores ranging from 110 to 138. Since we do not have any data beyond this range, we do not know whether the relationship continues to be linear for more extreme values of IQ.

To illustrate this point, consider Figure 7.7, where we have extended the regression line to include IQ values up to 165. At the university from which these data were sampled, the highest possible grade point average is 4.0. If we used the extended regression line to predict the grade point average for an IQ of 165, we would predict a grade point average of 5.2, a value that is obviously wrong. Prediction for IQs greater than 165 would be even worse. Looking at Figure 7.7, we can see that if the relationship does extend beyond an IQ of 138, it can't extend beyond an IQ of about 149 (the IQ value where the regression line meets a grade point average of 4.0). Of course, there is no reason to believe the relationship exists beyond the base data point of IQ = 138, and hence predictions using this relationship should not be made for IQ values greater than 138.

RELATION BETWEEN REGRESSION CONSTANTS AND PEARSON *r*

Although we haven't presented this aspect of Pearson *r* before, it can be shown that Pearson *r* is the slope of the least-squares regression line when the scores are plotted as *z* scores. As an example, let's use the data given in Table 6.3 on the weight and cost of six bags of oranges. For convenience, the data have been reproduced in Table 7.3. Figure 7.8(a) shows the scatter plot of the raw scores and the least-squares regression line for these raw scores. This is a perfect, linear relationship, so *r* = 1.00. Figure 7.8(b) shows the scatter plot of the paired *z* scores and the least-squares regression line for these *z* scores. The slope of the regression line for the raw scores is *b*, and

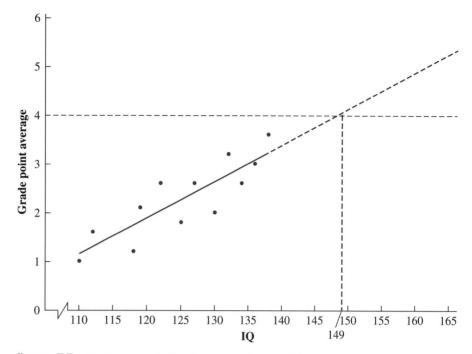

figure 7.7 Limiting prediction to range of base data.

table 7.3 Cost and weight in pounds of six bags of oranges

Bag	Weight (lb) X	Cost ($) Y	z_X	z_Y
A	2.25	0.75	−1.34	−1.34
B	3.00	1.00	−0.80	−0.80
C	3.75	1.25	−0.27	−0.27
D	4.50	1.50	0.27	0.27
E	5.25	1.75	0.80	0.80
F	6.00	2.00	1.34	1.34

the slope of the regression line for the z scores is r. Note that the slope of this latter regression line is 1.00, as it should be, because $r = 1.00$.

Since Pearson r is a slope, it is related to b_Y. It can be shown algebraically that

$$b_Y = r \frac{s_Y}{s_X}$$

MENTORING TIP

Note that if the standard deviations of the X and Y scores are the same, then $b_Y = r$.

This equation is useful if we have already calculated r, s_Y, and s_X and want to determine the least-squares regression line. For example, in the problem involving IQ and grade point average, $r = 0.8556$, $s_Y = 0.7990$, and $s_X = 9.2257$. Suppose we want to find b_Y and a_Y having already calculated r, s_Y, and s_X. The simplest way is to use the equation

$$b_Y = r \frac{s_Y}{s_X} = 0.8556 \left(\frac{0.7990}{9.2257} \right) = 0.074$$

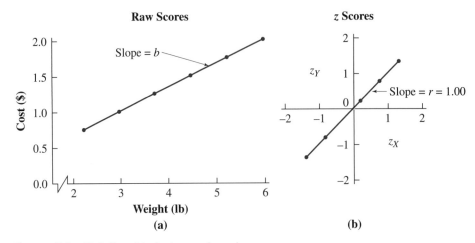

figure 7.8 Relationship between *b* and *r*.

MULTIPLE REGRESSION

Note that this is the same value arrived at previously in the chapter, on p. 163. Having found b_Y, we would calculate a_Y in the usual way.

Thus far, we have discussed regression and correlation using examples that have involved only two variables. When we were discussing the relationship between grade point average and IQ, we determined that $r = 0.856$ and that the equation of the regression line for predicting grade point average from IQ was

$$Y' = 0.74X - 7.006$$

where Y' = predicted value of grade point average
            X = IQ score

This equation gave us a reasonably accurate prediction. Although we didn't compute it, total prediction error squared $[\Sigma(Y - Y')^2]$ was 1.88, and the amount of variability accounted for was 73.2%. Of course, there are other variables besides IQ that might affect grade point average. The amount of time that students spend studying, motivation to achieve high grades, and interest in the courses taken are a few that come to mind. Even though we have reasonably good prediction accuracy using IQ alone, we might be able to do better if we also had data relating grade point average to one or more of these other variables.

Multiple regression is an extension of simple regression to situations that involve two or more predictor variables. To illustrate, let's assume we had data from the 12 college students that include a second predictor variable called "study time," as well as the original grade point average and IQ scores. The data for these three variables are shown in columns 2, 3, and 4 of Table 7.4. Now we can derive a regression equation for predicting grade point average using the two predictor variables, IQ and study time. The general form of the multiple regression equation for two predictor variables is

$$Y' = b_1X_1 + b_2X_2 + a$$

where Y' = predicted value of Y
b_1 = coefficient of the first predictor variable
X_1 = first predictor variable
b_2 = coefficient of the second predictor variable
X_2 = second predictor variable
a = prediction constant

This equation is very similar to the one we used in simple regression except that we have added another predictor variable and its coefficient. As before, the coefficient and constant values are determined according to the least-squares criterion that $\Sigma (Y - Y')^2$ is a minimum. However, this time the mathematics are rather formidable and the actual calculations are almost always done on a computer, using statistical software. For the data of our example, the multiple regression equation that minimizes errors in Y is given by

$$Y' = 0.049X_1 + 0.118X_2 - 5.249$$

where Y' = predicted value of grade point average
b_1 = 0.049
X_1 = IQ score
b_2 = 0.118
X_2 = study time score
a = 5.249

To determine whether prediction accuracy is increased by using the multiple regression equation, we have listed in column 5 of Table 7.4 the predicted grade

table 7.4 A comparison of prediction accuracy using one or two predictor variables

Student No.	IQ (X_1)	Study Time (hr/wk) (X_2)	Grade Point Average GPA (Y)	Predicted GPA Using IQ (Y')	Predicted GPA Using IQ + Study Time (Y')	Error Using Only IQ	Error Using IQ + Study Time
1	110	8	1.0	1.14	1.13	−0.14	−0.13
2	112	10	1.6	1.29	1.46	0.31	0.13
3	118	6	1.2	1.74	1.29	−0.54	−0.09
4	119	13	2.1	1.81	2.16	0.29	−0.06
5	122	14	2.6	2.03	2.43	0.57	0.17
6	125	6	1.8	2.26	1.63	−0.46	0.17
7	127	13	2.6	2.40	2.56	0.20	0.04
8	130	12	2.0	2.63	2.59	−0.63	−0.59
9	132	13	3.2	2.77	2.81	0.42	0.39
10	134	11	2.6	2.92	2.67	−0.32	−0.07
11	136	12	3.0	3.07	2.88	−0.07	0.12
12	138	18	3.6	3.21	3.69	0.38	0.09
					Total error squared = $\Sigma (Y - Y')^2$ = 1.88		= 0.63

point average scores using only IQ for prediction, in column 6 the predicted grade point average scores using both IQ and study time as predictor variables, and prediction errors from using each in columns 7 and 8, respectively. We have also plotted in Figure 7.9(a) the actual Y value and the two predicted Y' values for each student. Students have been ordered from left to right on the X axis according to the increased prediction accuracy that results for each by using the multiple regression equation. In Figure 7.9(b), we have plotted the percent improvement in prediction accuracy for each student that results from using IQ + study time rather than just IQ. It is clear from Table 7.4 and Figure 7.9 that using the multiple regression equation has greatly improved overall prediction accuracy. For example, prediction accuracy was increased for all students except student number 11, and for student number 3, accuracy was increased by almost 40%. We have also shown $\Sigma (Y - Y')^2$ for each regression line at the bottom of Table 7.4. Adding the second predictor variable reduced the total prediction error squared from 1.88 to 0.63, an improvement of more than 66%.

Since, in the present example, prediction accuracy was increased by using two predictors rather than one, it follows that the proportion of the variability of Y accounted for has also increased. In trying to determine this proportion, you might be tempted, through extension of the concept of r^2 from our previous discussion of correlation, to compute r^2 between grade point average and each predictor and then simply add the resulting values. Table 7.5 shows a Pearson r correlation matrix involving grade point average, IQ, and study time. If we followed this procedure, the proportion of variability accounted for would be greater than 1.00 [$(0.856)^2 + (0.829)^2 = 1.42$], which is clearly impossible. One cannot account for more than 100% of the variability. The error occurs because there is overlap in variability accounted for between IQ and study time. Students with higher IQs also tend to study more. Therefore, part of the variability in grade point average that is explained by IQ is also explained by study time. To correct for this, we must take the correlation between IQ and study time into account.

The correct equation for computing the proportion of variance accounted for when there are two predictor variables is

$$R^2 = \frac{r_{YX_1}^2 + r_{YX_2}^2 - 2r_{YX_1}r_{YX_2}r_{X_1X_2}}{1 - r_{X_1X_2}^2}$$

where R^2 = the multiple coefficient of determination
r_{YX_1} = the correlation between Y and predictor variable X_1
r_{YX_2} = the correlation between Y and predictor variable X_2
$r_{X_1X_2}$ = the correlation between predictor variables X_1 and X_2

R^2 is also often called the *squared multiple correlation*. Based on the data of the present study, r_{YX_1} = the correlation between grade point average and IQ = 0.856,

table 7.5 Pearson correlation matrix between IQ, study time, and grade point average

	IQ (X_1)	Study Time (X_2)	Grade Point Average (Y)
IQ (X_1)	1.000		
Study time (X_2)	0.560	1.000	
Grade point average (Y)	0.856	0.829	1.000

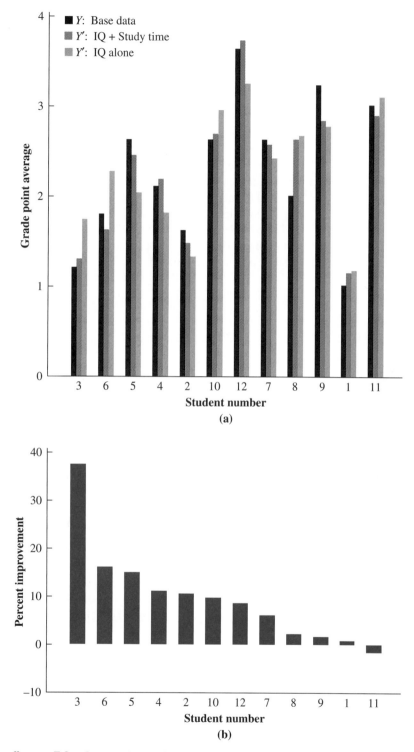

figure 7.9 Comparison of prediction accuracy using one or two predictor variables.

r_{YX_2} = the correlation between Y and study time = 0.829, and $r_{X_1X_2}$ = the correlation between IQ and study time = 0.560. For these data,

$$R^2 = \frac{(0.856)^2 + (0.829)^2 - 2(0.856)(0.829)(0.560)}{1 - (0.560)^2}$$

$$= 0.910$$

Thus, the proportion of variance accounted for has increased from 0.73 to 0.91 by using IQ and study time.

Of course, just adding another predictor variable per se will not necessarily increase prediction accuracy or the amount of variance accounted for. Whether prediction accuracy and the amount of variance accounted for are increased depends on the strength of the relationship between the variable being predicted and the additional predictor variable and also on the strength of the relationship between the predictor variables themselves. For example, notice what happens to R^2 when $r_{X_1X_2} = 0$. This topic is taken up in more detail in advanced textbooks.*

■ SUMMARY

In this chapter, I have discussed how to use the relationship between two variables for prediction. When the line that best fits the points is used for prediction, it is called a regression line. The regression line most used for linear imperfect relationships fits the points according to a least-squares criterion. Next, I presented the equations for determining the least-squares regression line when predicting Y given X and the convention that the predicted variable is symbolized by Y and the variable from which we predict by X. I then used these equations to construct regression lines for various sets of data and showed how to use these lines for prediction. Next, I discussed how to quantify the errors in prediction by computing the standard error of estimate. I presented the conditions under which the use of the linear regression line was appropriate: The relationship must be linear, the regression line must have been derived from data representative of the group to which prediction is desired, and prediction must be limited to the range of the base data. Next, I discussed the relationship between b and r. Finally, I introduced the topic of multiple regression and multiple correlation, discussed the multiple coefficient of determination, and showed how using two predictor variables can increase the accuracy of prediction.

■ IMPORTANT NEW TERMS

Homoscedasticity (p. 170)
Least-squares regression line (p. 161)
Multiple coefficient of
    determination (p. 176)

Multiple regression (p. 174)
Regression (p. 160)
Regression constant (p. 162)
Regression line (p. 160)

Regression of Y on X (p. 162)
Standard error of estimate (p. 169)

---

*For a more advanced treatment of multiple regression, see D. C. Howell, *Statistical Methods for Psychology,* 7th ed., Wadsworth Cengage Learning, Belmont, CA, 2010, pp. 515–577.

■ QUESTIONS AND PROBLEMS

1. Define or identify each of the terms in the Important New Terms section.
2. List some situations in which it would be useful to have accurate prediction.
3. The least-squares regression line minimizes $\Sigma (Y - Y')^2$ rather than $\Sigma (Y - Y')$. Is this statement correct? Explain.
4. The least-squares regression line is the prediction line that results in the most direct "hits." Is this statement correct? Explain.
5. State the convention that is used to assign X and Y to the *predicted to* and *predicted from* variables.
6. How are r and b_Y related? Explain.
7. Of what value is it to know the standard error of estimate for a set of paired X and Y scores?
8. What is R^2 called? Is it true that conceptually R^2 is analogous to r^2, except that R^2 applies to situations in which there are two or more predictor variables? Explain. Will using a second predictor variable always increase the precision of prediction? Explain.
9. Given the set of paired X and Y scores,

X	7	10	9	13	7	11	13
Y	1	2	4	3	3	4	5

   a. Construct a scatter plot of the paired scores. Does the relationship appear linear?
   b. Determine the least-squares regression line for predicting Y given X.
   c. Draw the regression line on the scatter plot.
   d. Using the relationship between X and Y, what value would you predict for Y if X = 12? (Round to two decimal places.)
10. A clinical psychologist is interested in the relationship between testosterone level in married males and the quality of their marital relationship. A study is conducted in which the testosterone levels of eight married men are measured. The eight men also fill out a standardized questionnaire assessing quality of marital relationship. The questionnaire scale is 0 – 25, with higher numbers indicating better relationships. Testosterone scores are in nanomoles/liter of serum. The data are shown below.

Subject Number	1	2	3	4	5	6	7	8
Relationship Score	24	15	15	10	19	11	20	19
Testosterone Level	12	13	19	25	9	16	15	21

   a. On a piece of graph paper, construct a scatter plot of the data. Use testosterone level as the X variable.
   b. Describe the relationship shown on the graph.
   c. Compute the value of Pearson r.
   d. Determine the least-squares regression line for predicting relationship score from testosterone level. Should b_Y be positive or negative? Why?
   e. Draw the least-squares regression line of part **d** on the scatter plot of part **a**.
   f. Based on the data of the eight men, what relationship score would you predict for a male who has a testosterone level of 23 nanomoles/liter of serum? clinical, health, biological
11. A popular attraction at a carnival recently arrived in town is the booth where Mr. Clairvoyant (a bright statistics student of somewhat questionable moral character) claims that he can guess the weight of females to within 1 kilogram by merely studying the lines in their hands and fingers. He offers a standing bet that if he guesses incorrectly the woman can pick out any stuffed animal in the booth. However, if he guesses correctly, as a reward for his special powers, she must pay him $2. Unknown to the women who make bets, Mr. Clairvoyant is able to surreptitiously measure the length of their left index fingers while "studying" their hands. Also unknown to the bettors, but known to Mr. Clairvoyant, is the following relationship between the weight of females and the length of their left index fingers:

Length of Left Index Finger (cm)	5.6	6.2	6.0	5.4
Weight (kg)	79.0	83.5	82.0	77.5

   a. If you were a prospective bettor, having all this information before you, would you make the bet with Mr. Clairvoyant? Explain.
   b. Using the data in the accompanying table, what is the least-squares regression line for predicting a woman's weight, given the length of her index finger?
   c. Using the least-squares regression line determined in part **b**, if a woman's index finger is 5.7 centimeters, what would be her predicted weight? (Round to two decimal places.) cognitive
12. A statistics professor conducts a study to investigate the relationship between the performance of his students on exams and their anxiety. Ten students from

his class are selected for the experiment. Just before taking the final exam, the 10 students are given an anxiety questionnaire. Here are final exam and anxiety scores for the 10 students:

Student No.	1	2	3	4	5	6	7	8	9	10
Anxiety	28	41	35	39	31	42	50	46	45	37
Final Exam	82	58	63	89	92	64	55	70	51	72

a. On a piece of graph paper, construct a scatter plot of the paired scores. Use anxiety as the X variable.

b. Describe the relationship shown in the graph.

c. Assuming the relationship is linear, compute the value of Pearson r.

d. Determine the least-squares regression line for predicting the final exam score, given the anxiety level. Should b_Y be positive or negative? Why?

e. Draw the least-squares regression line of part **d** on the scatter plot of part **a**.

f. Based on the data of the 10 students, if a student has an anxiety score of 38, what value would you predict for her final exam score? (Round to two decimal places.)

g. Calculate the standard error of estimate for predicting final exam scores from anxiety scores. clinical, health, education

13. The sales manager of a large sporting goods store has recently started a national advertising campaign. He has kept a record of the monthly costs of the advertising and the monthly profits. These are shown here. The entries are in thousands of dollars.

Month	Jan.	Feb.	Mar.	Apr.	May	Jun.	Jul.
Monthly Advertising Cost	10.0	14.0	11.4	15.6	16.8	11.2	13.2
Monthly Profit	125	200	160	150	210	110	125

a. Assuming a linear relationship exists, derive the least-squares regression line for predicting monthly profits from monthly advertising costs.

b. In August, the manager plans to spend $17,000 on advertising. Based on the data, how much profit should he expect that month? (Round to the nearest $1000.)

c. Given the relationship shown by the paired scores, can you think of a reason why the manager doesn't spend a lot more money on advertising? I/O

14. A newspaper article reported that "there is a strong correlation between continuity and success when it comes to NBA coaches." The article was based on the following data:

Coach, Team	Tenure as Coach with Same Team (yr)	1996–1997 Record (% games won)
Jerry Sloan, Utah	9	79
Phil Jackson, Chicago	8	84
Rudy Tomjanovich, Houston	6	70
George Karl, Seattle	6	70
Lenny Wilkens, Atlanta	4	68
Mike Fratello, Cleveland	4	51
Larry Brown, Indiana	4	48

a. Is the article correct in claiming that there is a strong correlation between continuity and success when it comes to NBA coaches?

b. Derive the least-squares regression line for predicting success (% games won) from tenure.

c. Based on your answer to part **b**, what "% games won" would you predict for an NBA coach who had 7 years' "tenure" with the same team? I/O, other

15. During inflationary times, Mr. Chevez has become budget conscious. Since his house is heated electrically, he has kept a record for the past year of his monthly electric bills and of the average monthly outdoor temperature. The data are shown in the following table. Temperature is in degrees Celsius, and the electric bills are in dollars.

a. Assuming there is a linear relationship between the average monthly temperature and the monthly electric bill, determine the least-squares regression line for predicting the monthly electric bill from the average monthly temperature.

b. On the basis of the almanac forecast for this year, Mr. Chevez expects a colder winter. If February is 8 degrees colder this year, how much should Mr. Chevez allow in his budget for February's electric bill? In calculating your answer, assume that the costs of electricity will rise 10% from last year's costs because of inflation.

c. Calculate the standard error of estimate for predicting the monthly electric bill from average monthly temperature.

Month	Average Temp.	Elec. Bill ($)
Jan.	10	120
Feb.	18	90
Mar.	35	118
Apr.	39	60
May	50	81
Jun.	65	64
Jul.	75	26
Aug.	84	38
Sep.	52	50
Oct.	40	80
Nov.	25	100
Dec.	21	124

other

16. In Chapter 6, Problem 16 (p. 151), data were presented on the relationship between birth weight and the subsequent IQ of seven randomly selected psychology majors from a particular university. The data are again presented below.

Student	Birth Weight (lbs)	IQ
1	5.8	122
2	6.5	120
3	8.0	129
4	5.9	112
5	8.5	127
6	7.2	116
7	9.0	130

a. Assuming there is a linear relationship, use these data and determine the least-squares regression line for predicting IQ, given birth weight.
b. Using this regression line, what IQ would you predict for a birth weight of 7.5?
developmental

17. In Chapter 6, Problem 21 (p. 152), data were given on the relationship between the number of soft drinks consumed in a week by eight 12-year-olds and their body mass index (BMI). The 12-year-olds were randomly selected from a junior high school in a large northwestern city. The data are again presented here.

Child	Number of Soft Drinks Consumed	BMI
1	3	20
2	1	18
3	14	32
4	7	24
5	21	35
6	5	19
7	25	38
8	9	30

a. Assuming the data show a linear relationship, derive the least-squares regression line for predicting BMI, given the number of soft drinks consumed.
b. Using this regression line, what BMI would you predict for a 12-year-old from this school who consumes a weekly average of 17 soft drinks?
health

18. In Chapter 6, Problem 22 (p. 153), data were presented from a study conducted to determine the relationship between religious involvement and self-esteem. The data are again presented below.

Subject	Religious Involvement	Self-Esteem
1	5	8
2	25	3
3	45	2
4	20	7
5	30	5
6	40	5
7	1	4
8	15	4
9	10	7
10	35	3

a. Assuming a linear relationship, derive the least-squares regression line for predicting self-esteem from religious involvement.
b. Using this regression line, what value of self-esteem would you predict for an eighth grader whose value of religious involvement is 43?
social, developmental

19. In Chapter 6, Problem 24 (p. 154), data were shown on the relationship between the work performance of 10 workers randomly chosen from the manufacturing section of a large corporation and two possible screening tests. The data are again shown below.

In that problem you were asked to recommend which of the two tests should be used as a screening device for prospective employees for that section of the company. On the basis of the data presented, you recommended using test 2. Now the question is: Would it be better to use both test 1 and test 2, rather than test 2 alone? Explain your answer, using R^2 and r^2. Use a computer and statistical software to solve this problem if you have access to them. I/O

					Employee					
	1	**2**	**3**	**4**	**5**	**6**	**7**	**8**	**9**	**10**
Work performance	50	74	62	90	98	52	68	80	88	76
Test 1	10	19	20	20	21	14	10	24	16	14
Test 2	25	35	40	49	50	29	32	44	46	35

■ SPSS ILLUSTRATIVE EXAMPLE 7.1 _____

The general operation of SPSS and data entry are described in Appendix E, *Introduction to SPSS*. Chapter 7 of the textbook discusses the topic of linear regression. Statistical software can be a great help in doing the graphs and computations contained in this section.

example

Let's use SPSS to do Chapter 7, problem 12, p. 179, parts a, b, d, and g of the textbook. For convenience, the problem is repeated below.

A statistics professor conducts a study to investigate the relationship between the performance of his students on exams and their anxiety. Ten students from his class are selected for the experiment. Just before taking the final exam, the 10 students are given an anxiety questionnaire. Here are final exam and anxiety scores for the 10 students.

Student Number	1	2	3	4	5	6	7	8	9	10
Anxiety	28	41	35	39	31	42	50	46	45	37
Final Exam	82	58	63	89	92	64	55	70	51	72

Use SPSS to do the following:

a. Construct a scatter plot of the paired scores. Let's assume that we want to predict *Final Exam* scores from *Anxiety* scores. Therefore, we will use *Anxiety* as the *X* variable.
b. Describe the relationship shown in the graph.
d. Determine the least-squares regression line for predicting the *Final Exam* score, given *Anxiety* level. Should b_Y be positive or negative? Why?
g. Calculate the standard error of estimate for predicting *Final Exam* scores from *Anxiety* scores.

SOLUTION

STEP 1:   Enter the Data.

1. Enter the *Anxiety* scores in the first column (**VAR00001**) of the SPSS Data Editor, beginning with the first score in the top cell of the first column.
2. Enter the *Final Exam* scores in the second column (**VAR00002**) of the SPSS Data Editor, beginning with the first score in the top cell of the second column.

STEP 2:   Name the Variables. In this example, we will give the default variables **VAR00001** and **VAR00002** the new names of **Anxiety** and **FinalExam**, respectively. Here's how it is done.

1. Click the **Variable View** tab in the lower left corner of the Data Editor.	This displays the **Variable View** on screen with **VAR00001** and **VAR00002** displayed in the first and second cells of the **Name** column.
2. Click VAR00001; then **type Anxiety** in the highlighted cell and then **press Enter**.	**Anxiety** is entered as the variable name, replacing **VAR00001**. The cursor then moves to the next cell, highlighting **VAR00002**.
3. Replace VAR00002 with **FinalExam** and then **press Enter**.	**FinalExam** is entered as the variable name, replacing **VAR00002**.

STEP 3:   Analyze the Data.

Part a.   Construct a Scatter Plot of the Data. I suggest you switch to the Data Editor-Data View if you haven't already done so. Now, let's proceed with the analysis.

1. Click on **Graphs** on the menu bar at the top of the screen; then **select Legacy Dialogs**; then **click Scatter/Dot…**.	This produces the **Scatter/Dot** dialog box, which is used to control scatters plots. The default is the **Simple Scatter** graph, which is what we want. Therefore, we don't have to click it.
2. Click Define.	This produces the **Simple Scatterplot** dialog box with **Anxiety** and **FinalExam** displayed in the large box on the left. **Anxiety** is highlighted.
3. Click the **arrow** for the **X axis**.	This moves **Anxiety** from the large box on the left into the **X axis** box on the right. We have done this because we want to plot **Anxiety** on the *X* axis.
4. Click FinalExam in the large box on the left; then **click** the **arrow** for the **Y axis**.	This moves **FinalExam** from the large box on the left into the **Y axis** box on the right. We have done this because we want to plot **FinalExam** on the *Y* axis.
5. Click on **OK**.	SPSS then constructs a scatter plot of the two variables, with **FinalExam** plotted on the *X* axis and **Anxiety** plotted on the *Y* axis. The scatter plot is shown below.

Analysis Results

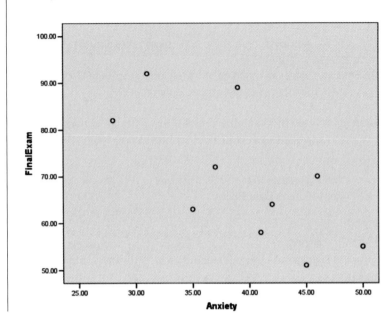

Part b.   The relationship is linear, imperfect, and negative.

Part d.   **Derive the Least-Squares Regression Line for Predicting "FinalExam," given "Anxiety."**

---

1. **Click Analyze** from the menu bar at the top of the screen; then **select Regression** from the drop-down menu; then **click** on **Linear...**.

   This produces the **Linear Regression** dialog box with **Anxiety** and **FinalExam** displayed in the large box on the left. **Anxiety** is highlighted.

2. **Click** the **arrow** for the **Independent(s):** box.

   This moves **Anxiety** into the **Independent(s):** box. This tells SPSS that **Anxiety** is the independent variable, i.e., the variable from which we are predicting.

3. **Click FinalExam** in the large box on the left; then **click** the **arrow** for the **Dependent:** box.

   This moves **FinalExam** into the **Dependent:** box. This tells SPSS that **FinalExam** is the dependent variable, i.e., the variable to which we are predicting.

4. **Click OK**.

   SPSS does its computations and outputs the results in four tables. For the purposes of this problem, we are interested in only two of the tables, the **Model Summary** table and the **Coefficients** table. These tables are shown below.

Analysis Results

Model Summary

Model	R	R Square	Adjusted R Square	Std. Error of the Estimate
1	.691[a]	.477	.412	10.86532

a. Predictors: (Constant), Anxiety

Coefficients[a]

Model		Unstandardized Coefficients		Standardized Coefficients	t	Sig.
		B	Std. Error	Beta		
1	(Constant)	125.883	21.111		5.963	.000
	Anxiety	-1.429	.529	-.691	-2.702	.027

a. Dependent Variable: FinalExam

The **Coefficients** table gives the a_Y and b_Y values for the regression equation. a_Y (the constant) = 125.883 and b_Y (the **Anxiety** Coefficient) = −1.429. Substituting these values into the general equation for predicting Y given X, we arrive at

$$Y' = -1.429X + 125.883$$

where Y' = predicted **FinalExam** score

X = **Anxiety**

This agrees with the equation given for these data in the textbook. b_Y should be negative because the relationship is negative.

Part g. **Calculate the Standard Error of Estimate for Predicting "Final Exam" Scores from "Anxiety" Scores.** This information is given in the last column of the **Model Summary** table shown above. From this table,

Std Error of the Estimate = 10.86532 or 10.87 (2 decimal places)

This also agrees with the answer given in the textbook. Yea for SPSS!!

■ SPSS ADDITIONAL PROBLEMS

1. For this example, let's use the IQ and GPA (Grade Point Average) data shown in Table 7.2, p. 164, of the textbook. For your convenience the data is shown again here.

Student No.	1	2	3	4	5	6	7	8	9	10	11	12
GPA	1.0	1.6	1.2	2.1	2.6	1.8	2.6	2.0	3.2	2.6	3.0	3.6
IQ	110	112	118	119	122	125	127	130	132	134	136	138

a. Use SPSS to construct a scatter plot of the data. In so doing, name the two variables, *IQ* and *GPA*. Plot *IQ* on the *X* axis and *GPA* on the *Y* axis. Compare your answer with Figure 7.1, p. 161.

b. Assuming a linear relationship exists between *IQ* and *GPA*, use SPSS to derive the least squares regression line for predicting *GPA* given *IQ*. Compare your answer with that shown in Table 7.2.

2. A psychology professor is interested in the relationship between grade point average (GPA) in graduate school and Graduate Record Exam (GRE) scores. A random sample of 20 graduate students is used for the study. The GPA and GRE score for each student is shown in the table that follows.

Student No.	1	2	3	4	5	6	7	8	9	10
GPA	3.70	3.18	2.90	2.93	3.02	2.65	3.70	3.77	3.41	2.38
GRE	637	562	520	624	500	500	700	680	655	525

Student No.	11	12	13	14	15	16	17	18	19	20
GPA	3.54	3.12	3.21	3.35	2.60	3.25	3.48	2.74	2.90	3.28
GRE	593	656	592	689	550	536	629	541	588	619

a. Use SPSS to construct a scatter plot of the paired scores. If you choose to use new variable names, name the variables *GPA* and *GRE*. Make the *Y* axis variable *GPA* and the *X* axis variable *GRE*.

b. Describe the relationship shown in the graph.

c. Use SPSS to construct the least-squares regression line for predicting *GPA* from *GRE* scores.

d. Compute the standard error of estimate for predicting *GPA* from *GRE* scores.

■ ONLINE STUDY RESOURCES

CENGAGE**brain**.com

Login to CengageBrain.com to access the resources your instructor has assigned. For this book, you can access the book's companion website for chapter-specific learning tools including Know and Be Able to Do, practice quizzes, flash cards, and glossaries, and a link to Statistics and Research Methods Workshops.

aplia

If your professor has assigned Aplia homework:

1. Sign in to your account.
2. Complete the corresponding homework exercises as required by your professor.
3. When finished, click "Grade It Now" to see which areas you have mastered and which need more work, and for detailed explanations of every answer.

Visit **www.cengagebrain.com** to access your account and to purchase materials.

INFERENTIAL STATISTICS

© Strmko / Dreamstime.com

© Strmko / Dreamstime.com

8

Random Sampling and Probability

LEARNING OBJECTIVES

After completing this chapter, you should be able to:

- Define a random sample; specify why the sample used in a study should be a random sample, and explain two methods of obtaining a random sample.
- Define sampling with replacement, sampling without replacement, *a priori* and *a posteriori* probability.
- List three basic points concerning probability values.
- Define the addition and multiplication rules, and solve problems involving their use.
- Define independent, mutually exclusive, and mutually exhaustive events.
- Define probability in conjunction with a continuous variable and solve problems when the variable is continuous and normally distributed.
- Understand the illustrative examples, do the practice problems, and understand the solutions.

INTRODUCTION

We have now completed our discussion of descriptive statistics and are ready to begin considering the fascinating area of inferential statistics. With descriptive statistics, we were concerned primarily with presenting and describing sets of scores in the most meaningful and efficient way. With inferential statistics, we go beyond mere description of the scores. A basic aim of inferential statistics is to use the sample scores to make a statement about a characteristic of the population. There are two kinds of statements made. One has to do with *hypothesis testing* and the other with *parameter estimation*.

In hypothesis testing, the experimenter is collecting data in an experiment on a sample set of subjects in an attempt to validate some hypothesis involving a population. For example, suppose an educational psychologist believes a new method of teaching mathematics to the third graders in her school district (population) is superior to the usual way of teaching the subject. In her experiment, she employs two samples of third graders, one of which is taught by the new teaching method and the other by the old one. Each group is tested on the same final exam. In doing this experiment, the psychologist is not satisfied with just reporting that the mean of the group that received the new method was higher than the mean of the other group. She wants to make a statement such as, "The improvement in final exam scores was due to the new teaching method and not chance factors. Furthermore, the improvement does not apply just to the particular sample tested. Rather, the improvement would be found in the whole population of third graders if they were taught by the new method." The techniques used in inferential statistics make these statements possible.

In parameter estimation experiments, the experimenter is interested in determining the magnitude of a population characteristic. For example, an economist might be interested in determining the average monthly amount of money spent last year on food by single college students. Using sample data, with the techniques of inferential statistics, he can estimate the mean amount spent by the population. He would conclude with a statement such as, "The probability is 0.95 that the interval of $300–$400 contains the population mean."

The topics of *random sampling* and *probability* are central to the methodology of inferential statistics. In the next section, we shall consider random sampling. In the remainder of the chapter, we shall be concerned with presenting the basic principles of probability.

RANDOM SAMPLING

To generalize validly from the sample to the population, both in hypothesis testing and in parameter estimation experiments, the sample cannot be *just any subset* of the population. Rather, it is crucial that the sample is a *random* sample.

definition ■ *A* **random sample** *is defined as a sample selected from the population by a process that ensures that (1) each possible sample of a given size has an equal chance of being selected and (2) all the members of the population have an equal chance of being selected into the sample.**

---

*See Note 8.1, p. 223.

To illustrate, consider the situation in which we have a population comprising the scores 2, 3, 4, 5, and 6 and we want to randomly draw a sample of size 2 from the population. Note that normally the population would have a great many more scores in it. We've restricted the population to five scores for ease in understanding the points we wish to make. Let's assume we shall be sampling from the population one score at a time and then placing it back into the population before drawing again. This is called *sampling with replacement* and is discussed later in this chapter. The following comprise all the samples of size 2 we could get from the population by using this method of sampling:

2, 2	3, 2	4, 2	5, 2	6, 2
2, 3	3, 3	4, 3	5, 3	6, 3
2, 4	3, 4	4, 4	5, 4	6, 4
2, 5	3, 5	4, 5	5, 5	6, 5
2, 6	3, 6	4, 6	5, 6	6, 6

There are 25 samples of size 2 we might get when sampling one score at a time with replacement. To achieve random sampling, the process must be such that (1) all of the 25 possible samples have an equally likely chance of being selected and (2) all of the population scores (2, 3, 4, 5, and 6) have an equal chance of being selected into the sample.

The sample should be a random sample for two reasons. First, to generalize from a sample to a population, it is necessary to apply the laws of probability to the sample. If the sample has not been generated by a process ensuring that each possible sample of that size has an equal chance of being selected, then we can't apply the laws of probability to the sample. The importance of this aspect of randomness and of probability to statistical inference will become apparent when we have covered the chapters on hypothesis testing and sampling distributions (see Chapters 10 and 12, respectively).

The second reason for random sampling is that, to generalize from a sample to a population, it is necessary that the sample be representative of the population. One way to achieve representativeness is to choose the sample by a process that ensures that all the members of the population have an equal chance of being selected into the sample. Thus, requiring the sample to be random allows the laws of probability to be used on the sample and at the same time results in a sample that should be representative of the population.

It is tempting to think that we can achieve representativeness by using methods other than random sampling. Very often, however, the selected procedure results in a biased (unrepresentative) sample. An example of this was the famous *Literary Digest* presidential poll of 1936, which predicted a landslide victory for Landon (57% to 43%). In fact, Roosevelt won, gaining 62% of the ballots. The *Literary Digest* prediction was grossly in error. Why? Later analysis showed that the error occurred because the sample was not representative of the voting population. It was a *biased* sample. The individuals selected were chosen from sources like the telephone book, club lists, and lists of registered automobile owners. These lists systematically excluded the poor, who were unlikely to have telephones or automobiles. It turned out that the poor voted overwhelmingly for Roosevelt. Even if other methods of sampling do on occasion result in a representative sample, the methods would not be useful for inference because we could not apply the laws of probability necessary to go from the sample to the population.

Techniques for Random Sampling

It is beyond the scope of this textbook to delve deeply into the ways of generating random samples. This topic can be complex, particularly when dealing with surveys. We shall, however, present a few of the more commonly used techniques in conjunction with some simple situations so that you can get a feel for what is involved. Suppose we have a population of 100 people and wish to randomly sample 20 for an experiment. One way to do this would be to number the individuals in the population from 1 to 100, then take 100 slips of paper and write one of the numbers on each slip, and put the slips into a hat, shake them around a lot, and pick out one. We would repeat the shaking and pick out another. Then we would continue this process until 20 slips have been picked. The numbers contained on the slips of paper would identify the individuals to be used in the sample. With this method of random sampling, it is crucial that the population be thoroughly mixed to ensure randomness.

A common way to produce random samples is to use a table of random numbers, such as Table J in Appendix D. These tables are most often constructed by a computer using a program that guarantees that all the digits (0–9) have an equal chance of occurring each time a digit is printed.

The table may be used as successive single digits, as successive two-digit numbers, as successive three-digit numbers, and so forth. For example, in Table J, p. 612, if we begin at row 1 and move horizontally across the page, the random order of single digits would be 3, 2, 9, 4, 2,.... If we wish to use two-digit numbers, the random order would be 32, 94, 29, 54, 16,....

Since the digits in the table are random, they may be used vertically in both directions and horizontally in both directions. The direction to be used should be specified before entering the table. To use the table properly, it should be entered randomly. One way would be to make cards with row and column numbers and place the cards in a box, mix them up, and then pick a row number and a column number. The intersection of the row and column would be the location of the first random number. The remaining numbers would be located by moving from the first number in the direction specified prior to entering the table. To illustrate, suppose we wanted to form a random sample of 3 subjects from a population of 10 subjects.* For this example, we have decided to move horizontally to the right in the table. To choose the sample, we would first assign each individual in the population a number from 0 to 9. Next, the table would be entered randomly to locate the first number. Let's assume the entry turns out to be the first number of row 7, p. 612, which is 3. This number designates the first subject in the sample. Thus, the first subject in the sample would be the subject bearing the number 3. We have already decided to move to the right in the table, so the next two numbers are 5 and 6. Thus, the individuals bearing the numbers 5 and 6 would complete the sample.

Next, let's do a problem in which there are more individuals in the population. For purposes of illustration, we shall assume that a random sample of 15 subjects is desired from a population of 100. To vary things a bit, we have decided to move vertically down in the table for this problem, rather than horizontally to the right. As before, we need to assign a number to each member of the population. This time, the numbers assigned are from 00 to 99 instead of from 0 to 9. Again the table is entered randomly. This time, let's assume the entry occurs at the intersection of the first two-digit number of column 3 with row 12. The two-digit number located at this intersection is 70. Thus, the first subject in the sample is the individual bearing the number 70. The next subject would be located by moving vertically down from 70. Thus, the second subject in the sample would be the individual bearing the number 33. This process would be continued until 15 subjects have been selected. The complete set of subject numbers would be 70, 33,

---

*Of course, in real experiments, the number of elements in the population is much greater than 10. We are using 10 in the first example to help you understand how to use the table.

82, 22, 96, 35, 14, 12, 13, 59, 97, 37, 54, 42, and 89. In arriving at this set of numbers, the number 82 appeared twice in the table. Since the same individual cannot be in the sample more than once, the repeated number was not included.

Sampling With or Without Replacement

So far, we have defined a random sample, discussed the importance of random sampling, and presented some techniques for producing random samples. To complete our discussion, we need to distinguish between sampling with replacement and sampling without replacement. To illustrate the difference between these two methods of sampling, let's assume we wish to form a sample of two scores from a population composed of the scores 4, 5, 8, and 10. One way would be to randomly draw one score from the population, record its value, and then place it back in the population before drawing the second score. Thus, the first score would be eligible for selection again on the second draw. This method of sampling is called *sampling with replacement*. A second method would be to randomly draw one score from the population and not replace it before drawing the second one. Thus, the same member of the population could appear in the sample only once. This method of sampling is called *sampling without replacement*.

definitions ■ **Sampling with replacement** *is defined as a method of sampling in which each member of the population selected for the sample is returned to the population before the next member is selected.*

■ **Sampling without replacement** *is defined as a method of sampling in which the members of the sample are not returned to the population before subsequent members are selected.*

When subjects are being selected to participate in an experiment, sampling without replacement must be used because the same individual can't be in the sample more than once. You will probably recognize this as the method we used in the preceding section. Sampling with replacement forms the mathematical basis for many of the inference tests discussed later in the textbook. Although the two methods do not yield identical results, when sample size is small relative to population size, the differences are negligible and "with-replacement" techniques are much easier to use in providing the mathematical basis for inference. Let's now move on to the topic of probability.

PROBABILITY

Probability may be approached in two ways: (1) from an *a priori*, or classical, viewpoint and (2) from an *a posteriori*, or empirical, viewpoint. *A priori* means that which can be deduced from reason alone, without experience. From the *a priori*, or classical, viewpoint, probability is defined as

$$p(A) = \frac{\text{Number of events classifiable as } A}{\text{Total number of possible events}} \quad a\ priori\ probability$$

The symbol $p(A)$ is read "the probability of occurrence of event A." Thus, the equation states that the probability of occurrence of event A is equal to the number of events classifiable as A divided by the number of possible events. To illustrate how this equation is used, let's look at an example involving dice. Figure 8.1 shows a pair of dice. Each die (the singular of dice is die) has six sides, with a different number of spots painted on each side. The spots vary from one to six. These innocent-looking cubes are used for gambling in a game called *craps*. They have been the basis of many tears and much happiness depending on the "luck" of the gambler.

Returning to *a priori* probability, suppose we are going to roll a die once. What is the probability it will come to rest with a 2 (the side with two spots on it) facing upward? Since there are six possible numbers that might occur and only one of these is 2, the probability of a 2, in one roll of one die, is

MENTORING TIP

Because of tradition, probability values in this chapter have been rounded to 4-decimal-place accuracy. Unless you are told otherwise, your answers to end-of-chapter problems for this chapter should also be rounded to 4 decimal places.

$$p(A) = p(2) = \frac{\text{Number of events classifiable as 2}}{\text{Total number of possible events}} = \frac{1}{6} = 0.1667*$$

Let's try one more problem using the *a priori* approach. What is the probability of getting a number greater than 4 in one roll of one die? This time there are two events classifiable as A (rolling 5 or 6). Thus,

$$p(A) = p(5 \text{ or } 6) = \frac{\text{Number of events classifiable as 5 or 6}}{\text{Total number of possible events}} = \frac{2}{6} = 0.3333$$

Note that the previous two problems were solved by reason alone, without recourse to any data collection. This approach is to be contrasted with the *a posteriori*, or empirical, approach to probability. *A posteriori* means "after the fact," and in the context of probability, it means after some data have been collected. From the *a posteriori*, or empirical, viewpoint, probability is defined as

$$p(A) = \frac{\text{Number of times } A \text{ has occurred}}{\text{Total number of occurrences}} \quad \textit{a posteriori probability}$$

To determine the probability of a 2 in one roll of one die by using the empirical approach, we would have to take the actual die, roll it many times, and count the number of times a 2 has occurred. The more times we roll the die, the better. Let's assume for this

figure 8.1   A pair of dice.

---

*In this and all other problems involving dice, we shall assume that the dice will not come to rest on any of their edges.

problem that we roll the die 100,000 times and that a 2 occurs 16,000 times. The probability of a 2 occurring in one roll of the die is found by

$$p(2) = \frac{\text{Number of times 2 has occurred}}{\text{Total number of occurrences}} = \frac{16{,}000}{100{,}000} = 0.1600$$

Note that, with this approach, it is necessary to have the actual die and to collect some data before determining the probability. The interesting thing is that if the die is evenly balanced (all numbers are equally likely), then when we roll the die many, many times, the *a posteriori* probability approaches the *a priori* probability. If we roll an infinite number of times, the two probabilities will equal each other. Note also that, if the die is loaded (weighted so that one side comes up more often than the others), then the *a posteriori* probability will differ from the *a priori* determination. For example, if the die is heavily weighted for a 6 to come up, a 2 might never appear. We can see now that the *a priori* equation assumes that each possible outcome has an equal chance of occurrence. For most of the problems in this chapter and the next, we shall use the *a priori* approach to probability.

Some Basic Points Concerning Probability Values

Since probability is fundamentally a proportion, it ranges in value from 0.00 to 1.00. If the probability of an event occurring equals 1.00, then the event is certain to occur. If the probability equals 0.00, then the event is certain *not* to occur. For example, an ordinary die does not have a side with 7 dots on it. Therefore, the probability of rolling a 7 with a single die equals 0.00. Rolling a 7 is certain not to occur. On the other hand, the probability that a number from 1 to 6 will occur equals 1.00. It is certain that one of the numbers 1, 2, 3, 4, 5, or 6 will occur.

The probability of occurrence of an event is expressed as a fraction or a decimal number. For example, the probability of randomly picking the ace of spades in one draw from a deck of ordinary playing cards is $\frac{1}{52}$, or 0.0192.* The answer may be left as a fraction $\left(\frac{1}{52}\right)$, but usually is converted to its decimal equivalent (0.0192).

Sometimes probability is expressed as "chances in 100." For example, someone might say the probability that event A will occur is 5 chances in 100. What he really means is $p(A) = 0.05$. Occasionally, probability is also expressed as the odds for or against an event occurring. For example, a betting person might say that the odds are 3 to 1 favoring Fred to win the race. In probability terms, $p(\text{Fred's winning}) = \frac{3}{4} = 0.75$. If the odds were 3 to 1 against Fred's winning, then $p(\text{Fred's winning}) = \frac{1}{4} = 0.25$.

Computing Probability

Determining the probability of events can be complex. In fact, whole courses are devoted to this topic, and they are quite difficult. Fortunately, for our purposes, there are only two major probability rules we need to learn: the addition rule and the multiplication rule. These rules provide the foundation necessary for understanding the statistical inference tests that follow in this textbook.

---

*For the uninitiated, a deck of ordinary playing cards is composed of 52 cards, 4 suits (spades, hearts, diamonds, and clubs), and 13 cards in each suit (Ace, 2, 3, 4, 5, 6, 7, 8, 9, 10, Jack, Queen, and King).

The Addition Rule

The *addition rule* is concerned with determining the probability of occurrence of any one of several possible events. To begin our discussion, let's assume there are only two possible events, *A* and *B*. When there are two events, the addition rule states the following.

definition ■ *The **probability of occurrence of A or B** is equal to the probability of occurrence of A plus the probability of occurrence of B minus the probability of occurrence of both A and B.*

In equation form, the addition rule states:

$$p(A \text{ or } B) = p(A) + p(B) - p(A \text{ and } B)$$

addition rule for two events—general equation

Let's illustrate how this rule is used. Suppose we want to determine the probability of picking an ace or a club in one draw from a deck of ordinary playing cards. The problem has been solved in two ways in Figure 8.2. Refer to the figure as you read this paragraph. The first way is by enumerating all the events classifiable as an ace or a club and using the basic equation for probability. There are 16 ways to get an ace or a club, so the probability of getting an ace or a club $= \frac{16}{52} = 0.3077$. The second method uses the addition rule. The probability of getting an ace $= \frac{4}{52}$, and the probability of getting a club $= \frac{13}{52}$. The probability of getting both an ace and a club $= \frac{1}{52}$. By the addition rule, the probability of getting an ace or a club $= \frac{4}{52} + \frac{13}{52} - \frac{1}{52} = \frac{16}{52} = 0.3077$. Why do we need to subtract the probability of getting both an ace and a club? Because we have already counted the ace of clubs twice. Without subtracting it, we would be misled into thinking there are 17 favorable events rather than just 16.

In this course, we shall be using the addition rule almost entirely in situations where the events are *mutually exclusive*.

definition ■ **Two events are mutually exclusive** *if both cannot occur together. Another way of saying this is that* **two events are mutually exclusive** *if the occurrence of one precludes the occurrence of the other.*

The events of rolling a 1 and of rolling a 2 in one roll of a die are mutually exclusive. If the roll ends with a 1, it cannot also be a 2. The events of picking an ace and a king in one draw from a deck of ordinary playing cards are mutually exclusive. If the card is an ace, it precludes the card also being a king. This can be contrasted with the events of picking an ace and a club in one draw from the deck. These events are not mutually exclusive because there is a card that is both an ace and a club (the ace of clubs).

When the events are mutually exclusive, the probability of both events occurring together is zero. Thus, $p(A \text{ and } B) = 0$ when *A* and *B* are mutually exclusive. Under these conditions, the addition rule simplifies to:

$$p(A \text{ or } B) = p(A) + p(B)$$

addition rule when A and B are mutually exclusive

Let's practice solving some problems involving situations in which A and B are mutually exclusive.

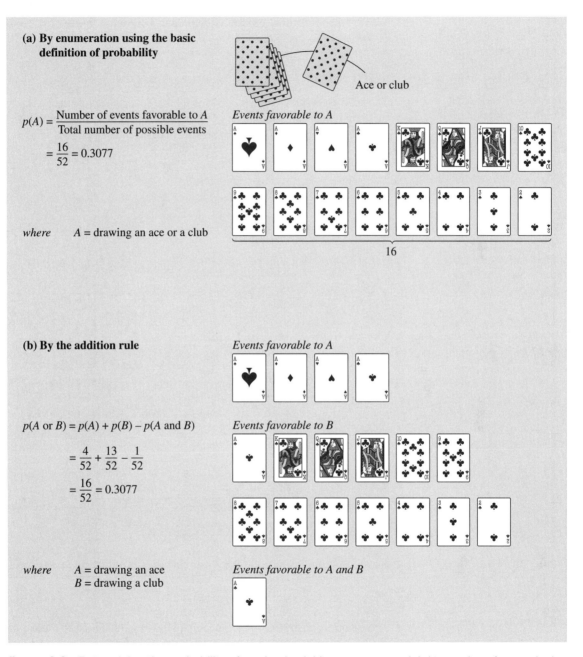

(a) By enumeration using the basic definition of probability

Ace or club

$p(A) = \dfrac{\text{Number of events favorable to } A}{\text{Total number of possible events}}$

$= \dfrac{16}{52} = 0.3077$

where A = drawing an ace or a club

Events favorable to A

16

(b) By the addition rule

Events favorable to A

$p(A \text{ or } B) = p(A) + p(B) - p(A \text{ and } B)$

$= \dfrac{4}{52} + \dfrac{13}{52} - \dfrac{1}{52}$

$= \dfrac{16}{52} = 0.3077$

Events favorable to B

where A = drawing an ace
B = drawing a club

Events favorable to A and B

figure 8.2 Determining the probability of randomly picking an ace or a club in one draw from a deck of ordinary playing cards.

Practice Problem 8.1

What is the probability of randomly picking a 10 or a 4 in one draw from a deck of ordinary playing cards?

SOLUTION

The solution is shown in the following figure. Since we want either a 10 *or* a 4 and these two events are mutually exclusive, the addition rule with mutually exclusive events is appropriate. Thus, $p(10 \text{ or } 4) = p(10) + p(4)$.

There are four 10s, four 4s, and 52 cards, so $p(10) = \frac{4}{52}$ and $p(4) = \frac{4}{52}$. Thus, $p(10 \text{ or } 4) = \frac{4}{52} + \frac{4}{52} = \frac{8}{52} = 0.1538$.

$$p(A \text{ or } B) = p(A) + p(B)$$

$$p(\text{a } 10 \text{ or a } 4) = p(10) + p(4)$$

$$= \frac{4}{52} + \frac{4}{52}$$

$$= \frac{8}{52} = 0.1538$$

where A = drawing a 10
         B = drawing a 4

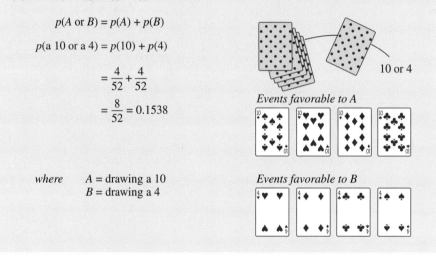

10 or 4

Events favorable to A

Events favorable to B

Practice Problem 8.2

In rolling a fair die once, what is the probability of rolling a 1 or an even number?

SOLUTION

The solution is shown in the accompanying figure. Since the events are mutually exclusive and the problem asks for either a 1 *or* an even number, the addition rule with mutually exclusive events applies. Thus, $p(1 \text{ or an even number}) = p(1) + p(\text{an even number})$. There is one way to roll a 1; there are three ways to roll an even number (2, 4, 6), and there are six possible outcomes. Thus, $p(1) = \frac{1}{6}$, $p(\text{an even number}) = \frac{3}{6}$, and $p(1 \text{ or an even number}) = \frac{1}{6} + \frac{3}{6} = \frac{4}{6} = 0.6667$

$$p(A \text{ or } B) = p(A) + p(B)$$

$$p(1 \text{ or an even number}) = p(1) + p(\text{an even number})$$

$$= \frac{1}{6} + \frac{3}{6} = \frac{4}{6}$$

$$= 0.6667$$

where $\quad A$ = rolling a 1
$\quad\quad\quad B$ = rolling an even number

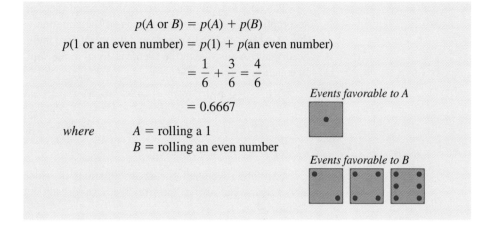

Events favorable to A

Events favorable to B

Practice Problem 8.3

Suppose you are going to randomly sample 1 individual from a population of 130 people. In the population, there are 40 children younger than 12, 60 teenagers, and 30 adults. What is the probability the individual you select will be a teenager or an adult?

SOLUTION

The solution is shown in the accompanying figure. Since the events are mutually exclusive and we want a teenager or an adult, the addition rule with mutually exclusive events is appropriate. Thus, $p(\text{teenager or adult}) = p(\text{teenager}) + p(\text{adult})$. Since there are 60 teenagers, 30 adults, and 130 people in the population, $p(\text{teenager}) = \frac{60}{130}$ and $p(\text{adult}) = \frac{30}{130}$. Thus, $p(\text{teenager or adult}) = \frac{60}{130} + \frac{30}{130} = \frac{90}{130} = 0.6923$.

$$p(A \text{ or } B) = p(A) + p(B)$$

$$p(\text{a teenager or an adult}) = p(\text{a teenager}) + p(\text{an adult})$$

$$= \frac{60}{130} + \frac{30}{130}$$

$$= \frac{90}{130} = 0.6923$$

where $\quad A$ = a teenager
$\quad\quad\quad B$ = an adult

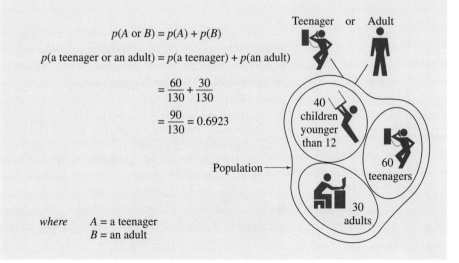

The addition rule may also be used when there are more than two events. This is accomplished by a simple extension of the equation used for two events. Thus, when there are more than two events and the events are mutually exclusive, the probability of occurrence of any one of the events is equal to the sum of the probability of each event. In equation form,

$$p(A \text{ or } B \text{ or } C \ldots \text{ or } Z) = p(A) + p(B) + p(C) + \ldots + p(Z)$$

*addition rule with more than
two mutually exclusive events*

where $Z =$ the last event

Very often we shall encounter situations in which the events are not only mutually exclusive but also exhaustive. We have already defined mutually exclusive but not *exhaustive*.

---

definition   ■   *A set of events is **exhaustive** if the set includes all of the possible events.*

For example, in rolling a die once, the set of events of getting a 1, 2, 3, 4, 5, or 6 is exhaustive because the set includes all of the possible events. When a set of events is both exhaustive and mutually exclusive, a very useful relationship exists. Under these conditions, the sum of the individual probabilities of each event in the set must equal 1. Thus,

$$p(A) + p(B) + p(C) + \ldots + p(Z) = 1.00$$

*when events are exhaustive and
mutually exclusive*

where $A, B, C \ldots Z =$ the events

To illustrate this relationship, let's consider the set of events of getting a 1, 2, 3, 4, 5, or 6 in rolling a fair die once. Since the events are exhaustive and mutually exclusive, the sum of their probabilities must equal 1. We can see this is true because $p(1) = \frac{1}{6}$, $p(2) = \frac{1}{6}$, $p(3) = \frac{1}{6}$, $p(4) = \frac{1}{6}$, $p(5) = \frac{1}{6}$, and $p(6) = \frac{1}{6}$. Thus,

$$p(1) + p(2) + p(3) + p(4) + p(5) + p(6) = 1.00$$

$$\tfrac{1}{6} + \tfrac{1}{6} + \tfrac{1}{6} + \tfrac{1}{6} + \tfrac{1}{6} + \tfrac{1}{6} = 1.00$$

When there are only two events and the events are mutually exclusive, it is customary to assign the symbol P to the probability of occurrence of one of the events and Q to the probability of occurrence of the other event. For example, if I were flipping a penny and only allowed it to come up heads or tails, this would be a situation in which there are only two possible events with each flip (a head or a tail), and the events are mutually exclusive (if it is a head, it can't be a tail, and vice versa). It is customary to let P equal the probability of occurrence of one of the events, say, a head, and Q equal the probability of occurrence of the other event, a tail.

When flipping coins, it is useful to distinguish between fair coins and biased coins.

---

definition   ■   *A fair coin or unbiased coin is one where if flipped once, the probability of a head = the probability of a tail = $\frac{1}{2}$. If the coin is biased, the probability of a head ≠ the probability of a tail ≠ $\frac{1}{2}$.*

Thus, if we are flipping a coin, if we let P equal the probability of a head and Q equal the probability of a tail, <u>and the coin is a fair coin,</u> then $P = \frac{1}{2}$ or 0.50 and $Q = \frac{1}{2}$ or 0.50. Since the events of getting a head or a tail in a single flip of a coin are exhaustive and mutually exclusive, their probabilities must equal 1. Thus,

$$P + Q = 1.00 \qquad \textit{when two events are exhaustive and mutually exclusive}$$

We shall be using the symbols P and Q extensively in Chapter 9 in conjunction with the binomial distribution.

The Multiplication Rule

Whereas the addition rule gives the probability of occurrence of any one of several events, the *multiplication rule* is concerned with the joint or successive occurrence of several events. Note that the multiplication rule often deals with what happens on more than one roll or draw, whereas the addition rule covers just one roll or one draw. If we are interested in the joint or successive occurrence of two events A and B, then the multiplication rule states the following:

definition   ■   *The* **probability of occurrence of both A and B** *is equal to the probability of occurrence of A times the probability of occurrence of B, given A has occurred.*

In equation form, the multiplication rule is

$$p(A \text{ and } B) = p(A)p(B|A) \qquad \textit{multiplication rule with two events—general equation}$$

Note that the symbol $p(B|A)$ is read "probability of occurrence of B given A has occurred." It does not mean B divided by A. Note also that the multiplication rule is concerned with the occurrence of *both A and B*, whereas the addition rule applies to the occurrence of *either A or B*.

In discussing the multiplication rule, it is useful to distinguish among three conditions: when the events are mutually exclusive, when the events are independent, and when the events are dependent.

Multiplication rule: mutually exclusive events   We have already discussed the joint occurrence of A and B when A and B are mutually exclusive. You will recall that if A and B are mutually exclusive, then

$$p(A \text{ and } B) = 0 \qquad \textit{multiplication rule with mutually exclusive events}$$

because when events are mutually exclusive, the occurrence of one precludes the occurrence of the other. The probability of their joint occurrence is zero.

Multiplication rule: independent events   To understand how the multiplication rule applies in this situation, we must first define *independent*.

definition   ■   **Two events are independent** *if the occurrence of one has no effect on the probability of occurrence of the other.*

Sampling with replacement illustrates this condition well. For example, suppose we are going to draw two cards, one at a time, with replacement, from a deck of ordinary playing cards. We can let *A* be the card drawn first and *B* be the card drawn second. Since *A* is replaced before drawing *B*, the occurrence of *A* on the first draw has no effect on the probability of occurrence of *B*. For instance, if *A* were an ace, because it is replaced in the deck before picking the second card, the occurrence of an ace on the first draw has no effect on the probability of occurrence of the card picked on the second draw. If *A* and *B* are independent, then the probability of *B* occurring is unaffected by *A*. Therefore, $p(B|A) = p(B)$. Under this condition, the multiplication rule becomes

$$p(A \text{ and } B) = p(A)p(B|A) = p(A)p(B)$$

*multiplication rule with
independent events*

Let's see how to use this equation. Suppose we are going to randomly draw two cards, one at a time, with replacement, from a deck of ordinary playing cards. What is the probability both cards will be aces?

The solution is shown in Figure 8.3. Since the problem requires an ace on the first draw *and* an ace on the second draw, the multiplication rule is appropriate. We can let *A* be an ace on the first draw and *B* be an ace on the second draw. Since sampling is with replacement, *A* and *B* are independent. Thus, p(an ace on first draw and an ace on second draw) = p(an ace on first draw)p(an ace on second draw). There are four aces possible on the first draw, four aces possible on the second draw (sampling is *with* replacement), and 52 cards in the deck, so p(an ace on first draw) = $\frac{4}{52}$ and p(an ace on second draw) = $\frac{4}{52}$. Thus, p(an ace on first draw and an ace on second draw) = $\frac{4}{52}(\frac{4}{52}) = \frac{16}{2704} = 0.0059$. Let's do a few more problems for practice.

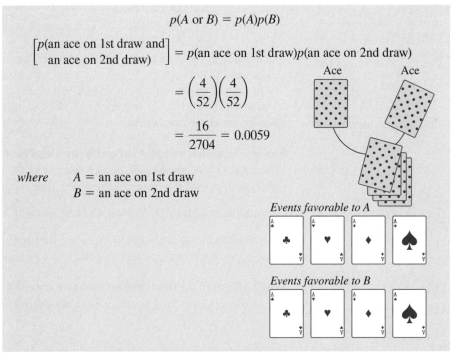

figure 8.3 Determining the probability of randomly sampling two aces in two draws from a deck of ordinary playing cards. Sampling is one at a time with replacement: multiplication rule with independent events.

Practice Problem 8.4

Suppose we roll a pair of fair dice once. What is the probability of obtaining a 2 on die 1 and a 4 on die 2?

SOLUTION

The solution is shown in the following figure. Since there is independence between the dice and the problem asks for a 2 *and* a 4, the multiplication rule with independent events applies. Thus, p(a 2 on die 1 and a 4 on die 2) = p(a 2 on die 1) p(a 4 on die 2). There is one way to get a 2 on die 1, one way to get a 4 on die 2, and six possible outcomes with each die. Therefore, p(2 on die 1) = $\frac{1}{6}$, p(4 on die 2) = $\frac{1}{6}$, and p(2 on die 1 and 4 on die 2) = $\frac{1}{6}(\frac{1}{6}) = \frac{1}{36} = 0.0278$.

$$p(2 \text{ on die 1 and 4 on die 2}) = p(2 \text{ on die 1})p(4 \text{ on die 2})$$

$$= \left(\frac{1}{6}\right)\left(\frac{1}{6}\right) = \frac{1}{36} = 0.0278$$

Events favorable to A
Die 1

Events favorable to B
Die 2

where A = a 2 on die 1
        B = a 4 on die 2

Practice Problem 8.5

If two pennies are flipped once, what is the probability both pennies will turn up heads? Assume that the pennies are fair coins and that a head or tail is the only possible outcome with each coin.

SOLUTION

The solution is shown in the accompanying figure. Since the outcome with the first coin has no effect on the outcome of the second coin, there is independence between events. The problem requires a head with the first coin *and* a head with the second coin, so the multiplication rule with independent events is appropriate. Thus, p(a head with the first penny and a head with the second penny) = p(a head with first penny)p(a head with second penny). Since

(continued)

there is only one way to get a head with each coin and two possibilities with each coin (a head or a tail), p(a head with first penny) $= \frac{1}{2}$, and p(a head with second penny) $= \frac{1}{2}$. Thus, p(head with first penny and head with second penny) $= \frac{1}{2}(\frac{1}{2}) = \frac{1}{4} = 0.2500$.

$$p(A \text{ and } B) = p(A)p(B)$$

$$\left[\begin{array}{c} p(\text{a head with 1st penny and} \\ \text{a head with 2nd penny}) \end{array}\right] = p(\text{a head with 1st penny})p(\text{a head with 2nd penny})$$

$$= \left(\frac{1}{2}\right)\left(\frac{1}{2}\right) = 0.2500$$

where A = a head with 1st penny   *Events favorable to A*
              B = a head with 2nd penny   *First penny*

Events favorable to B
Second penny

Practice Problem 8.6

Suppose you are randomly sampling from a bag of fruit. The bag contains four apples, six oranges, and five peaches. If you sample two fruits, one at a time, with replacement, what is the probability you will get an orange and an apple in that order?

SOLUTION

The solution is shown in the accompanying figure. Since there is independence between draws (sampling is with replacement) and we want an orange *and* an apple, the multiplication rule with independent events applies. Thus, p(an orange on first draw and an apple on second draw) = p(an orange on first draw) p(an apple on second draw). Since there are 6 oranges and 15 pieces of fruit in the bag, p(an orange on first draw) = $\frac{6}{15}$. Because the fruit selected on the first draw is replaced before the second draw, it has no effect on the fruit picked on the second draw. There are 4 apples and 15 pieces of fruit, so p(an apple on second draw) $= \frac{4}{15}$. Therefore, p(an orange on first draw and an apple on second draw) $= \frac{6}{15}(\frac{4}{15}) = 0.1067$.

$$p(A \text{ and } B) = p(A)p(B)$$

$$\begin{bmatrix} p(\text{an orange on 1st draw} \\ \text{and apple on 2nd draw}) \end{bmatrix} = p(\text{an orange on 1st draw})p(\text{an apple on 2nd draw})$$

$$= \left(\frac{6}{15}\right)\left(\frac{4}{15}\right)$$

$$= \frac{24}{225} = 0.1067$$

where $A =$ an orange on 1st draw
   $B =$ an apple on 2nd draw

Orange   Apple

Events favorable to A
Oranges

Events favorable to B
Apples

Practice Problem 8.7

Suppose you are randomly sampling 2 individuals from a population of 110 men and women. There are 50 men and 60 women in the population. Sampling is one at a time, with replacement. What is the probability the sample will contain all women?

SOLUTION

The solution is shown in the accompanying figure. Since the problem requires a woman on the first draw and a woman on the second draw and there is independence between these two events (sampling is with replacement), the multiplication

(continued)

rule with independent events is appropriate. Thus, p(a woman on first draw and a woman on second draw) = p(a woman on first draw)p(a woman on second draw). There are 60 women and 110 people in the population, so p(a woman on first draw) = $\frac{60}{110}$, and p(a woman on second draw) = $\frac{60}{110}$. Therefore, p(a woman on first draw and a woman on second draw) = $\frac{60}{110}\left(\frac{60}{110}\right) = \frac{3600}{12,100} = 0.2975$.

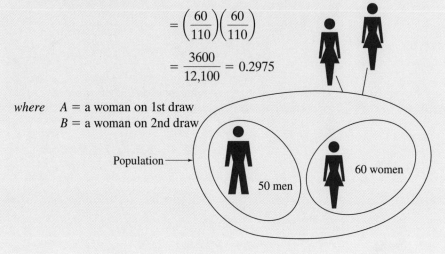

$$\begin{bmatrix} p\text{(a woman on 1st draw and} \\ \text{a woman on 2nd draw)} \end{bmatrix} = p\text{(a woman on 1st draw)}p\text{(a woman on 2nd draw)}$$

$$= \left(\frac{60}{110}\right)\left(\frac{60}{110}\right)$$

$$= \frac{3600}{12,100} = 0.2975$$

where A = a woman on 1st draw
B = a woman on 2nd draw

Population

50 men

60 women

The multiplication rule with independent events also applies to situations in which there are more than two events. In such cases, the probability of the joint occurrence of the events is equal to the product of the individual probabilities of each event. In equation form,

$$p(A \text{ and } B \text{ and } C \text{ and } \dots Z) = p(A)p(B)p(C) \dots p(Z)$$

multiplication rule with more than two independent events

To illustrate the use of this equation, let's suppose that instead of sampling 2 individuals from the population in Practice Problem 8.7, you are going to sample 4 persons. Otherwise the problem is the same. The population is composed of 50 men and 60 women. As before, sampling is one at a time, with replacement. What is the probability you will pick 3 women and 1 man in that order? The solution is shown in Figure 8.4. Since the problem requires a woman on the first *and* second *and* third draws *and* a man on the fourth draw and sampling is with replacement, the multiplication rule with more than two independent events is appropriate. This rule is just like the multiplication rule with two independent events, except there are more terms to multiply. Thus, p(a woman on first draw and a woman on second draw and a woman on third draw and a man on fourth draw) = p(a woman on first draw)p(a woman on second draw)p(a woman on third draw)p(a man on fourth draw). There are 60 women, 50 men, and 110 people in the population. Since sampling is with replacement, p(a woman on first draw) = $\frac{60}{110}$, p(a woman on second draw) = $\frac{60}{110}$, p(a woman on third draw) = $\frac{60}{110}$, and p(a man on fourth draw) = $\frac{50}{110}$. Thus, p(a woman on

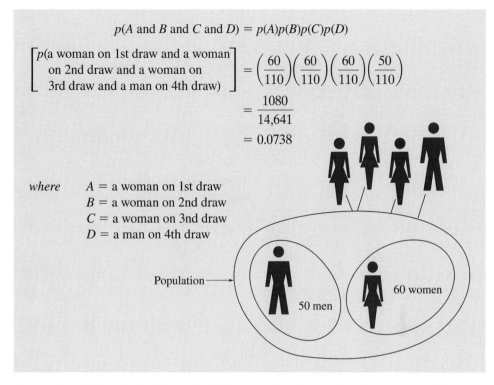

$$p(A \text{ and } B \text{ and } C \text{ and } D) = p(A)p(B)p(C)p(D)$$

$$\begin{bmatrix} p(\text{a woman on 1st draw and a woman} \\ \text{on 2nd draw and a woman on} \\ \text{3rd draw and a man on 4th draw)} \end{bmatrix} = \left(\frac{60}{110}\right)\left(\frac{60}{110}\right)\left(\frac{60}{110}\right)\left(\frac{50}{110}\right)$$

$$= \frac{1080}{14{,}641}$$

$$= 0.0738$$

where A = a woman on 1st draw
B = a woman on 2nd draw
C = a woman on 3nd draw
D = a man on 4th draw

Population → 50 men   60 women

figure 8.4 Determining the probability of randomly sampling 3 women and 1 man, in that order, in four draws from a population of 50 men and 60 women. Sampling is one at a time with replacement: multiplication rule with several independent events.

first draw and a woman on second draw and a woman on third draw and a man on fourth draw) $= \frac{60}{110}\left(\frac{60}{110}\right)\left(\frac{60}{110}\right)\left(\frac{50}{110}\right) = 1080/14{,}641 = 0.0738$.

Multiplication rule: dependent events   When A and B are dependent, the probability of occurrence of B is affected by the occurrence of A. In this case, we cannot simplify the equation for the probability of A and B. We must use it in its original form. Thus, if A and B are dependent,

$$p(A \text{ and } B) = p(A)p(B|A) \qquad \textit{multiplication rule with dependent events}$$

Sampling without replacement provides a good illustration for dependent events. Suppose you are going to draw two cards, one at a time, *without* replacement, from a deck of ordinary playing cards. What is the probability both cards will be aces?

The solution is shown in Figure 8.5. We can let A be an ace on the first draw and B be an ace on the second draw. Since sampling is without replacement (whatever card is picked the first time is kept out of the deck), the occurrence of A *does* affect the probability of B. A and B are dependent. Since the problem asks for an ace on the first draw *and* an ace on the second draw, and these events are dependent, the multiplication rule with dependent events is appropriate. Thus, p(an ace on first draw and an ace on second draw) = p(an ace on first draw)p(an ace on second draw, given an ace was obtained on first draw). For the first draw, there are 4 aces and 52 cards. Therefore, p(an ace on first draw) $= \frac{4}{52}$. Since sampling is without replacement, p(an ace on second draw given an ace on first draw) $= \frac{3}{51}$. Thus, p(an ace on first draw and an ace on second draw) $= \frac{4}{52}\left(\frac{3}{51}\right) = \frac{12}{2652} = 0.0045$.

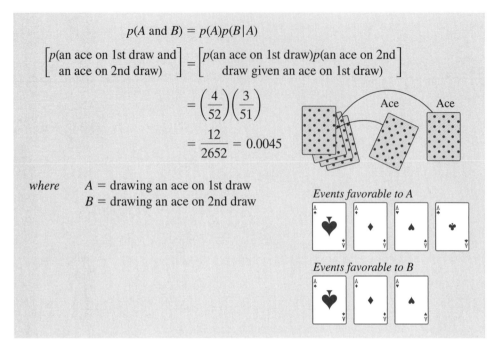

$$p(A \text{ and } B) = p(A)p(B|A)$$

$$\begin{bmatrix} p(\text{an ace on 1st draw and} \\ \text{an ace on 2nd draw}) \end{bmatrix} = \begin{bmatrix} p(\text{an ace on 1st draw})p(\text{an ace on 2nd} \\ \text{draw given an ace on 1st draw}) \end{bmatrix}$$

$$= \left(\frac{4}{52}\right)\left(\frac{3}{51}\right)$$

$$= \frac{12}{2652} = 0.0045$$

where A = drawing an ace on 1st draw
           B = drawing an ace on 2nd draw

Events favorable to A

Events favorable to B

figure 8.5 Determining the probability of randomly picking two aces in two draws from a deck of ordinary playing cards. Sampling is one at a time without replacement: multiplication rule with dependent events.

Practice Problem 8.8

Suppose you are randomly sampling two fruits, one at a time, from the bag of fruit in Practice Problem 8.6. As before, the bag contains four apples, six oranges, and five peaches. However, this time you are sampling *without* replacement. What is the probability you will get an orange and an apple in that order?

SOLUTION

The solution is shown in the accompanying figure. Since the problem requires an orange *and* an apple and sampling is without replacement, the multiplication rule with dependent events applies. Thus, p(an orange on first draw and an apple on second draw) = p(an orange on first draw)p(an apple on second draw given an orange was obtained on first draw). On the first draw, there are 6 oranges and 15 fruits. Therefore, p(an orange on first draw) = $\frac{6}{15}$. Since sampling is without replacement, p(an apple on second draw given an orange on first draw) = $\frac{4}{14}$. Therefore, p(an orange on first draw and an apple on second draw) = $\frac{6}{15}(\frac{4}{14})$ = $\frac{24}{210}$ = 0.1143.

$$p(A \text{ and } B) = p(A)p(B|A)$$

$$\begin{bmatrix} p(\text{an orange on 1st draw and} \\ \text{an apple on 2nd draw)} \end{bmatrix} = \begin{bmatrix} p(\text{an orange on 1st draw})p(\text{an apple on} \\ \text{2nd draw given an orange on 1st draw)} \end{bmatrix}$$

$$= \left(\frac{6}{15}\right)\left(\frac{4}{14}\right)$$

$$= \frac{24}{210} = 0.1143$$

where A = an orange on 1st draw
B = an apple on 2nd draw

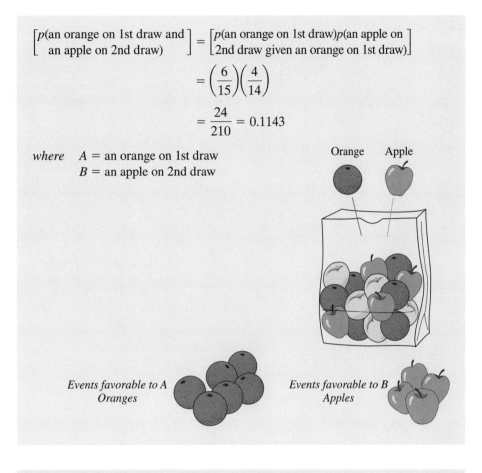

Orange   Apple

Events favorable to A
Oranges

Events favorable to B
Apples

Practice Problem 8.9

In a particular college class, there are 15 music majors, 24 history majors, and 46 psychology majors. If you randomly sample 2 students from the class, what is the probability they will both be history majors? Sampling is one at a time, without replacement.

SOLUTION

The solution is shown in the accompanying figure. Since the problem requires a history major on the first draw *and* a history major on the second draw and sampling is without replacement, the multiplication rule with dependent events is appropriate. Thus, p(a history major on first draw and a history major on second draw) = p(a history major on first draw)p(a history major on second draw given a history major was obtained on first draw). On the first draw, there were 24 history majors and 85 people in the population. Therefore, p(a history major on first draw) = $\frac{24}{85}$. Since sampling is without replacement, p(a history major on second draw given a history major on first draw) = $\frac{23}{84}$. Therefore, p(a history major on first draw and a history major on second draw) = $\frac{24}{85}(\frac{23}{84}) = \frac{552}{7140} = 0.0773$.

(continued)

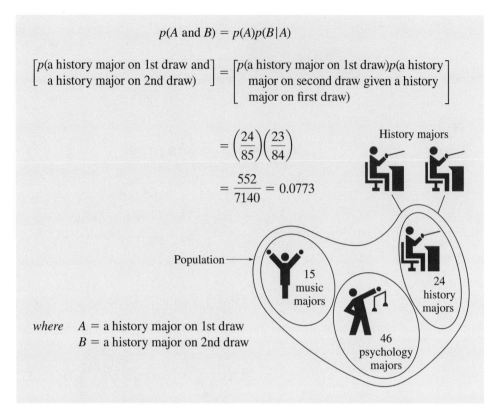

$$p(A \text{ and } B) = p(A)p(B|A)$$

$$\begin{bmatrix} p(\text{a history major on 1st draw and} \\ \text{a history major on 2nd draw}) \end{bmatrix} = \begin{bmatrix} p(\text{a history major on 1st draw})p(\text{a history} \\ \text{major on second draw given a history} \\ \text{major on first draw}) \end{bmatrix}$$

$$= \left(\frac{24}{85}\right)\left(\frac{23}{84}\right)$$

$$= \frac{552}{7140} = 0.0773$$

where A = a history major on 1st draw
B = a history major on 2nd draw

History majors

Population →

15 music majors

46 psychology majors

24 history majors

Like the multiplication rule with independent events, the multiplication rule with dependent events also applies to situations in which there are more than two events. In such cases, the equation becomes

$$p(A \text{ and } B \text{ and } C \text{ and } \ldots Z) = p(A)p(B|A)p(C|AB) \ldots p(Z|ABC \ldots)$$

*multiplication rule with more
than two dependent events*

where $p(A)$ = probability of A
$p(B|A)$ = probability of B given A has occurred
$p(C|AB)$ = probability of C given A and B have occurred
$p(Z|ABC \ldots)$ = probability of Z given A, B, C, and all other events
have occurred

To illustrate how to use this equation, let's do a problem that involves more than two dependent events. Suppose you are going to sample 4 students from the college class given in Practice Problem 8.9. In that class, there were 15 music majors, 24 history majors, and 46 psychology majors. If sampling is one at a time, without replacement, what is the probability you will obtain 4 history majors?

The solution is shown in Figure 8.6. Since the problem requires a history major on the first *and* second *and* third *and* fourth draws and sampling is without replacement, the multiplication rule with more than two dependent events is appropriate. This rule is very much like the multiplication rule with two dependent events, except more multiplying is required. Thus, for this problem, p(a history major on first draw and a history major on second draw and a history major on third draw and a history major on fourth draw) = p(a history major on first draw)p(a history major on second draw given a history major on first draw)p(a history major on third draw given a

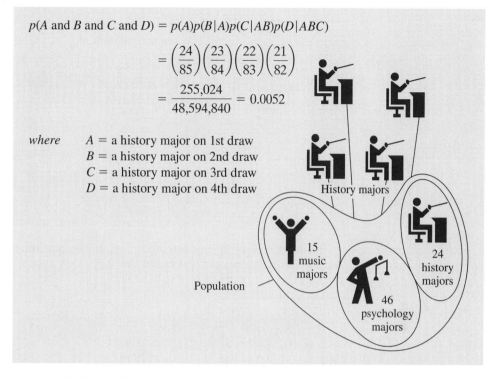

$$p(A \text{ and } B \text{ and } C \text{ and } D) = p(A)p(B|A)p(C|AB)p(D|ABC)$$

$$= \left(\frac{24}{85}\right)\left(\frac{23}{84}\right)\left(\frac{22}{83}\right)\left(\frac{21}{82}\right)$$

$$= \frac{255{,}024}{48{,}594{,}840} = 0.0052$$

where A = a history major on 1st draw
B = a history major on 2nd draw
C = a history major on 3rd draw
D = a history major on 4th draw

History majors

15 music majors

46 psychology majors

24 history majors

Population

figure 8.6   Determining the probability of randomly sampling 4 history majors on four draws from a population of 15 music majors, 24 history majors, and 46 psychology majors. Sampling is one at a time without replacement: multiplication rule ith several dependent events.

history major on first and second draws)p(a history major on fourth draw given a history major on first, second, and third draws). On the first draw, there are 24 history majors and 85 individuals in the population. Thus, p(a history major on first draw) $= \frac{24}{85}$. Since sampling is without replacement, p(a history major on second draw given a history major on first draw) $= \frac{23}{84}$, p(a history major on third draw given a history major on first and second draws) $= \frac{22}{83}$, and p(a history major on fourth draw given a history major on first, second, and third draws) $= \frac{21}{82}$. Therefore, p(a history major on first draw and a history major on second draw and a history major on third draw and a history major on fourth draw) $= \frac{24}{85}(\frac{23}{84})(\frac{22}{83})(\frac{21}{82}) = 255{,}024/48{,}594{,}840 = 0.0052$.

Multiplication and Addition Rules

Some situations require that we use both the multiplication and addition rules for their solutions. For example, suppose that I am going to roll two fair dice once. What is the probability the sum of the numbers showing on the dice will equal 11? The solution is shown in Figure 8.7. There are two possible outcomes that yield a sum of 11 (die 1 = 5 and die 2 = 6, which we shall call outcome A; and die 1 = 6 and die 2 = 5, which we shall call outcome B). Since the dice are independent, we can use the multiplication rule with independent events to find the probability of each outcome. By using this rule, $p(A) = \frac{1}{6}(\frac{1}{6}) = \frac{1}{36}$, and $p(B) = \frac{1}{6}(\frac{1}{6}) = \frac{1}{36}$. Since either of the outcomes yields a sum of 11, $p(\text{sum of } 11) = p(A \text{ or } B)$. These outcomes are mutually exclusive, so we can use the addition rule with mutually exclusive events to find $p(A \text{ or } B)$. Thus, $p(\text{sum of } 11) = p(A \text{ or } B) = p(A) + p(B) = \frac{1}{36} + \frac{1}{36} = \frac{2}{36} = 0.0556$.

Let's try one more problem that involves both the multiplication and addition rules.

$$p(A) = p(5 \text{ on die 1 and 6 on die 2})$$
$$= p(5 \text{ on die 1})p(6 \text{ on die 2})$$
$$= \left(\frac{1}{6}\right)\left(\frac{1}{6}\right) = \frac{1}{36}$$

$$p(B) = p(6 \text{ on die 1 and 5 on die 2})$$
$$= p(6 \text{ on die 1})p(5 \text{ on die 2})$$
$$= \left(\frac{1}{6}\right)\left(\frac{1}{6}\right) = \frac{1}{36}$$

$$p(\text{sum of 11}) = p(A \text{ or } B) = p(A) + p(B)$$
$$= \frac{1}{36} + \frac{1}{36} = \frac{2}{36} = 0.0556$$

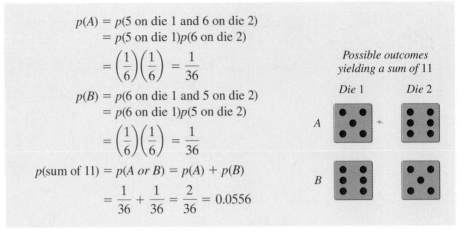

Possible outcomes yielding a sum of 11

Die 1   *Die 2*

A

B

figure 8.7 Determining the probability of rolling a sum of 11 in one roll of two fair dice: multiplication and addition rules.

Practice Problem 8.10

Suppose you have arrived in Las Vegas and you are going to try your "luck" on a one-armed bandit (slot machine). In case you are not familiar with slot machines, basically a slot machine has three wheels that rotate independently. Each wheel contains pictures of different objects. Let's assume the one you are playing has seven different fruits on wheel 1: a lemon, a plum, an apple, an orange, a pear, some cherries, and a banana. Wheels 2 and 3 have the same fruits as wheel 1. When the lever is pulled down, the three wheels rotate independently and then come to rest. On the slot machine, there is a window in front of each wheel. The pictures of the fruits pass under the window during rotation. When the wheel stops, one of the fruits from each wheel will be in view. We shall assume that each fruit on a wheel has an equal probability of appearing under the window at the end of rotation. You insert your silver dollar and pull down the lever. What is the probability that two lemons and a pear will appear? Order is not important; all you care about is getting two lemons and a pear, in any order.

SOLUTION

The solution is shown in the accompanying figure. There are three possible orders of two lemons and a pear: lemon, lemon, pear; lemon, pear, lemon; and pear, lemon, lemon. Since the wheels rotate independently, we can use the multiplication rule with independent events to determine the probability of each order. Since each fruit is equally likely, $p(\text{lemon and lemon and pear}) = p(\text{lemon})p(\text{lemon})p(\text{pear}) = \frac{1}{7}\left(\frac{1}{7}\right)\left(\frac{1}{7}\right) = \frac{1}{343}$. The same probability also applies to the other two orders. Since the three orders give two lemons and a pear, $p(\text{two lemons and a pear}) = p(\text{order 1, 2, or 3})$. By using the addition rule with independent events, $p(\text{order 1, 2, or 3}) = \frac{3}{343} = 0.0087$. Thus, the probability of getting two lemons and a pear, without regard to order, equals 0.0087.

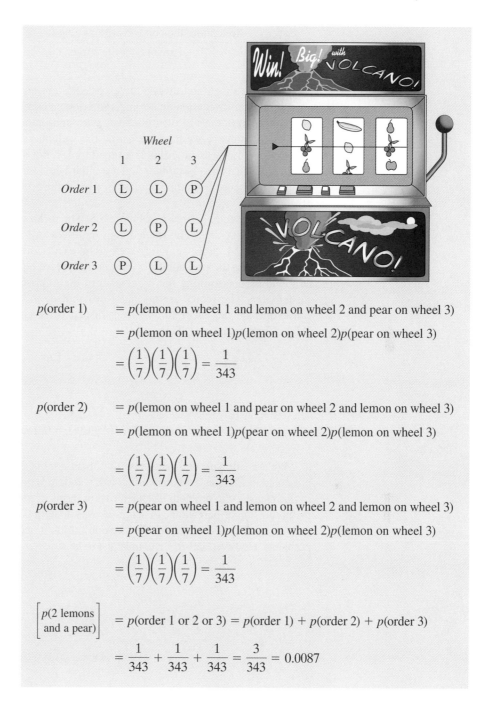

$p(\text{order 1})$ $= p(\text{lemon on wheel 1 and lemon on wheel 2 and pear on wheel 3})$

             $= p(\text{lemon on wheel 1})p(\text{lemon on wheel 2})p(\text{pear on wheel 3})$

$$= \left(\frac{1}{7}\right)\left(\frac{1}{7}\right)\left(\frac{1}{7}\right) = \frac{1}{343}$$

$p(\text{order 2})$ $= p(\text{lemon on wheel 1 and pear on wheel 2 and lemon on wheel 3})$

             $= p(\text{lemon on wheel 1})p(\text{pear on wheel 2})p(\text{lemon on wheel 3})$

$$= \left(\frac{1}{7}\right)\left(\frac{1}{7}\right)\left(\frac{1}{7}\right) = \frac{1}{343}$$

$p(\text{order 3})$ $= p(\text{pear on wheel 1 and lemon on wheel 2 and lemon on wheel 3})$

             $= p(\text{pear on wheel 1})p(\text{lemon on wheel 2})p(\text{lemon on wheel 3})$

$$= \left(\frac{1}{7}\right)\left(\frac{1}{7}\right)\left(\frac{1}{7}\right) = \frac{1}{343}$$

$\begin{bmatrix} p(2 \text{ lemons} \\ \text{and a pear}) \end{bmatrix}$ $= p(\text{order 1 or 2 or 3}) = p(\text{order 1}) + p(\text{order 2}) + p(\text{order 3})$

$$= \frac{1}{343} + \frac{1}{343} + \frac{1}{343} = \frac{3}{343} = 0.0087$$

Probability and Continuous Variables

So far in our discussion of probability, we have considered variables that have been discrete, such as sampling from a deck of cards or rolling a pair of dice. However, many of the dependent variables that are evaluated in experiments are continuous, not discrete. When a variable is continuous,

$$p(A) = \frac{\text{Area under the curve corresponding to } A}{\text{Total area under the curve}} \qquad \textit{probability of A with a continuous variable}$$

Often (although not always) these variables are normally distributed, so we shall concentrate our discussion on normally distributed continuous variables.

To illustrate the use of probability with continuous variables that are normally distributed, suppose we have measured the weights of all the sophomore women at your college. Let's assume this is a population set of scores that is normally distributed, with a mean of 120 pounds and a standard deviation of 8 pounds. If we randomly sampled one score from the population, what is the probability it would be equal to or greater than a score of 134?

The population is drawn in Figure 8.8. The mean of 120 and the score of 134 are located on the X axis. The shaded area represents all the scores that are equal to or greater than 134. Since sampling is random, each score has an equal chance of being selected. Thus, the probability of obtaining a score equal to or greater than 134 can be found by determining the proportion of the total scores that are contained in the shaded area. The scores are normally distributed, so we can find this proportion by converting the raw score to its z-transformed value and then looking up the area in Table A in Appendix D. Thus,

$$z = \frac{X - \mu}{\sigma} = \frac{134 - 120}{8} = \frac{14}{8} = 1.75$$

From Table A, column C,

$$p(X \geq 134) = 0.0401$$

We are sure you will recognize that this type of problem is quite similar to those presented in Chapter 5 when dealing with standard scores. The main difference is that, in this chapter, the problem has been cast in terms of probability rather than asking for the proportion or percentage of scores as was done in Chapter 5. Since you are already familiar with this kind of problem, we don't think it necessary to give a lot of practice problems. However, let's try a couple just to be sure.

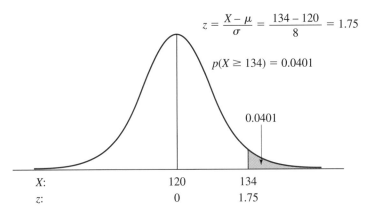

figure 8.8 Probability of obtaining $X \geq 134$ if randomly sampling one score from a normal population, with $\mu = 120$ and $\sigma = 8$.

Practice Problem 8.11

Consider the same population of sophomore women just discussed in the text. If one score is randomly sampled from the population, what is the probability it will be equal to or less than 110?

SOLUTION

MENTORING TIP

Remember: draw the picture first, as you did in Chapter 5.

The solution is presented in the accompanying figure. The shaded area represents all the scores that are equal to or less than 110. Since sampling is random, each score has an equal chance of being selected. To find $p(X \leq 110)$, first we must transform the raw score of 110 to its z score. Then we can find the proportion of the total scores that are contained in the shaded area by using Table A. Thus,

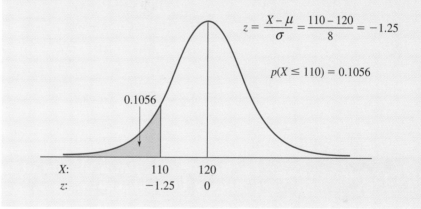

$$z = \frac{X - \mu}{\sigma} = \frac{110 - 120}{8} = -1.25$$

$$p(X \leq 110) = 0.1056$$

0.1056

X:	110	120
z:	-1.25	0

Practice Problem 8.12

Considering the same population again, what is the probability of randomly sampling a score that is as far or farther from the mean than a score of 138?

SOLUTION

The solution is shown in the accompanying figure. The score of 138 is 18 units above the mean. Since the problem asks for scores as far or farther from the mean, we must also consider scores that are 18 units or more below the mean. The shaded areas contain all of the scores that are 18 units or more away from the mean. Since sampling is random, each score has an equal chance of being selected. To find $p(X \leq 102 \text{ or } X \geq 138)$, first we must transform the raw scores of 102 and 138 to their z scores. Then we can find the proportion of the total scores that are

(continued)

contained in the shaded areas by using Table A. $p(X \le 102 \text{ or } X \ge 138)$ equals this proportion. Thus,

$$z = \frac{X - \mu}{\sigma} = \frac{102 - 120}{8} = -\frac{18}{8} \qquad z = \frac{X - \mu}{\sigma} = \frac{138 - 120}{8} = \frac{18}{8}$$

$$= -2.25 \qquad\qquad\qquad = 2.25$$

From Table A, column C,

$$p(X \le 102 \text{ or } X \ge 138) = 0.0122 + 0.0122 = 0.0244$$

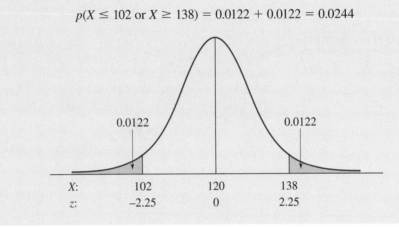

| X: | 102 | 120 | 138 |
| z: | −2.25 | 0 | 2.25 |

 Despite a tradition of qualitatively rather than quantitatively based decision making, the legal field is increasingly using statistics as a basis for decisions. The following case from Sweden is an example.

In a Swedish trial, the defendant was contesting a charge of overtime parking. An officer had marked the position of the valves of the front and rear tires of the accused driver's car, according to a clock representation (e.g., front valve to one o'clock and rear valve to six o'clock), in both cases to the nearest hour (see diagram). After the allowed time had elapsed, the car was still there, with the two valves pointing to one and six o'clock as before. The accused was given a parking ticket.

In court, however, he pleaded innocent, claiming that he had left the parking spot in time, but returned to it later, and the valves had just happened to come to rest in the same position as before. The judge, not having taken a basic course in statistics, called in a statistician to evaluate the defendant's claim of coincidence. Is the defendant's claim reasonable? Assume you are the statistician. What would you tell the judge? In formulating your answer, assume independence between the wheels, as did the statistician who advised the judge.

Answer As a statistician, your job is to determine how reasonable the plea of coincidence really is. If we assume the defendant's story is true about leaving and coming back to the parking spot, what is the probability of the two valves returning to their one and six o'clock positions? Since there are 12 possible positions for each valve, assuming independence between the wheels, using the multiplication rule,

$$p(\text{one and six}) = \left(\tfrac{1}{12}\right)\left(\tfrac{1}{12}\right) = \tfrac{1}{144}$$
$$= 0.0069$$

Thus, if coincidence (or chance alone) is at work, the probability of the valves returning to their original positions is about 7 times in 1000.

What do you think the judge did when given this information? Believe it or not, the judge acquitted the defendant, saying that if all four wheels had been checked and found to point in the same directions as before ($p = \frac{1}{12} \times \frac{1}{12} \times \frac{1}{12} \times \frac{1}{12} = \frac{1}{20,736} = 0.00005$), then the coincidence claim would have been rejected as too improbable and the defendant convicted. Thus, the judge considered the coincidence explanation as too probable to reject, even though the results would be obtained only 1 out of 144 times if coincidence was

at work. Actually, because the wheels do not rotate independently, the formulas used most likely understate somewhat the probability of a chance

return to the original position. (How did you do? Can we call on you in the future as a statistical expert to help mete out justice?) ◼

WHAT IS THE TRUTH?

Sperm Count Decline—Male or Sampling Inadequacy?

The headline of an article that appeared in 1995 in a leading metropolitan newspaper read, "20-year study shows sperm count decline among fertile men." Excerpts from the article are reproduced here.

A new study has found a marked decline in sperm counts among fertile men over the past 20 years …

The paper, published today in The New England Journal of Medicine, was based on data collected over a 20-year period at a Paris

sperm bank. Some experts in the United States took strong exception to the findings …

The new study, by Dr. Pierre Jouannet of the Center for the Study of the Conservation of Human Eggs and Sperm in Paris, examined semen collected by a sperm bank in Paris beginning in 1973. They report that sperm counts fell by an average of 2.1 percent a year, going from 89 million sperm per milliliter in 1973 to an average of 60 million per milliliter in 1992. At the same time they found the percentages of sperm that moved normally and were properly formed declined by 0.5 to 0.6 of 1 percent a year.

The paper is accompanied by an invited editorial by an expert on male infertility, Dr. Richard Sherins, director of the division of andrology at the Genetics and IVF Institute in Fairfax, Va., who said the current studies and several preceding it suffered from methodological flaws that made their data uninterpretable.

Sherins said that the studies did not look at sperm from randomly selected men and that sperm counts and sperm quality vary so much from week to week that it is hazardous to rely on single samples to measure sperm quality, as these studies did.

(continued)

WHAT IS THE TRUTH? *(continued)*

What do you think? Why might it be important to use samples from randomly selected men rather than from men who deposit their sperm at a sperm bank? Why might large week-to-week variability in sperm counts and sperm quality complicate interpretation of the data? ■

WHAT IS THE TRUTH?   **A Sample of a Sample**

The following article on polling appeared in a recent issue of *The New York Times* on the Web. I found the article very interesting and hope it is to you as well. The article is presented in its entirety.

HOW THE 'TYPICAL' RESPONDENT IS FOUND
By MICHAEL KAGAY

"What kind of people do you call for your polls? You must not be polling anyone around here in Texas."

Pollsters receive a lot of calls like this, usually from people who are genuinely puzzled to find that their own views on some controversial issue, or even the views of most people in their particular locality, are not in the majority nationwide—according to the polls.

But most pollsters take enormous care to ensure a representative sample of respondents to their polls. Good question writing may still be an art, and skillful interviewing may be a craft, but proper sampling is what puts the science into polling.

Random Digit Dialing, or RDD, is the standard sampling procedure in use throughout the polling profession today. The objective is to give every residential telephone number in the United States an equal chance of being called for an interview. Almost every polling organization uses some form of RDD.

The New York Times/CBS News Poll, for example, relies on the GENESYS system, developed and maintained by Marketing Systems Group of Philadelphia. That system consists of a database of over 42,000 residential telephone exchanges throughout the United States, updated every few months to include newly created area codes and exchanges.

That system also contains computer software to draw a random sample of those exchanges for each new poll, and to randomly make up the last four digits of each individual telephone number to be called.

That random choice from the universe of all possible telephone numbers is what guarantees the representativeness of the sample of households called for a poll.

In the case of the poll the caller was inquiring about, Texans constitute about seven percent of all Americans, but they formed six percent of the respondents in that particular poll. Pretty close to the mark, but a little bit short.

Pursuing Respondents Vigorously

Some types of people are harder to reach and to interview than others. How can bias in the results be reduced or avoided?

The first remedy is vigorous pursuit of the household and then of the designated individual respondent within the household.

Most polls make multiple calls to the household over the course of the poll, which often includes daytime calling, nighttime calling, weekdays and weekends. Interviewers leave messages on answering machines about why they are calling, make appointments when the designated individual is not at home, and offer to call back at a time more convenient to the respondent.

For respondents who refuse to be interviewed, many polls also employ a special squad of interviewers who make an additional callback to try to convert the refusal into a completed interview. That task takes special skills—the ability to convince the respondent of the

importance of the interview and the right personality—an ability to tolerate possible rejection.

Weighting or Balancing After the Interview

Still, just about every poll contains a few too many women, retired people, college graduates, and whites—compared to what the Census Bureau calculates the U.S. population contains at any given time.

This occurs because people vary in how frequently they are at home, how willing they are to talk to strangers on the phone, and how confident they feel in sharing their views about political and social issues.

Therefore, most polls mildly weight their respondents—slightly upping the proportion of men, younger people, those with less education, and racial minorities—to bring the proportions into proper balance. Some pollsters refer to this procedure as balancing.

At the same time the Times/CBS News Poll also takes the opportunity to make a slight adjustment for the

size of household and the number of telephone lines into the residence. This equalizes the probability of selection for individuals in situations where some households have multiple telephone lines and when some households have more adult residents than others.

All these minor adjustments serve to make a good sample even better, and more representative.

A Sample That Looks Like America

So what do the respondents in a typical New York Times/CBS News Poll look like? After all the random dialing, the callbacks, the refusal conversion, and the weighting, here is what a typical poll consists of:

* In terms of geography, respondents are 22 percent living in the northeast, 33 percent in the south, 24 percent in the mid-west, and 21 percent in the west.
* In terms of gender, they are 47 percent men and 53 percent women.
* In terms of race, they are 80 percent white, 11 percent black, 1 percent Asian, and 6 percent other.

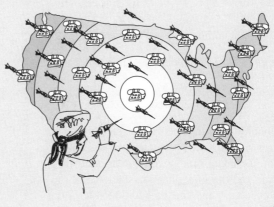

(continued)

■ SUMMARY

In this chapter, I have discussed the topics of random sampling and probability. A random sample is defined as a sample that has been selected from a population by a process that ensures that (1) each possible sample of a given size has an equal chance of being selected and (2) all members of the population have an equal chance of being selected into the sample. After defining and discussing the importance of random sampling, I described various methods for obtaining a random sample. In the last section on random sampling, I discussed sampling with and without replacement.

In presenting probability, I pointed out that probability may be approached from two viewpoints: *a priori* and *a posteriori*. According to the *a priori* view, $p(A)$ is defined as

$$p(A) = \frac{\text{Number of events classifiable as } A}{\text{Total number of possible events}}$$

From an *a posteriori* standpoint, $p(A)$ is defined as

$$p(A) = \frac{\text{Number of times } A \text{ has occurred}}{\text{Total number of occurrences}}$$

Since probability is fundamentally a proportion, it ranges from 0.00 to 1.00. Next, I presented two probability rules necessary for understanding inferential statistics: the addition rule and the multiplication rule. Assuming there are two events (A and B), the addition rule gives the probability of A or B, whereas the multiplication rule gives the probability of A and B. The addition rule states the following:

$$p(A \text{ or } B) = p(A) + p(B) - p(A \text{ and } B)$$

If the events are mutually exclusive,

$$p(A \text{ or } B) = p(A) + p(B)$$

If the events are mutually exclusive and exhaustive,

$$p(A) + p(B) = 1.00$$

The multiplication rule states the following:

$$p(A \text{ and } B) = p(A)p(B|A)$$

If the events are mutually exclusive,

$$p(A \text{ and } B) = 0$$

If the events are independent,

$$p(A \text{ and } B) = p(A)p(B)$$

If the events are dependent, we must use the general equation

$$p(A \text{ and } B) = p(A)p(B|A)$$

In addition, I discussed (1) the generalization of these equations to situations in which there were more than two events and (2) situations that required both the addition and multiplication rules for their solution. Finally, I discussed the probability of A with continuous variables and described how to find $p(A)$ when the variable was both normally distributed and continuous. The equation for determining the probability of A when the variable is continuous is

$$p(A) = \frac{\text{Area under the curve corresponding to } A}{\text{Total area under the curve}}$$

■ IMPORTANT NEW TERMS

Addition rule (p. 196)
A posteriori probability (p. 194)
A priori probability (p. 193)
Biased coin (p. 200)
Exhaustive set of events (p. 200)
Fair coin (p. 200)

Independence of two events (p. 201)
Multiplication rule (p. 201)
Mutually exclusive events (p. 196)
Probability (p. 193)
Probability of occurrence of *A* or *B* (p. 196)

Probability of occurrence of both *A* and *B* (p. 201)
Random sample (p. 190)
Sampling with replacement (p. 193)
Sampling without replacement (p. 193)

■ QUESTIONS AND PROBLEMS

1. Define or identify each term in the Important New Terms section.
2. What two purposes does random sampling serve?
3. Assume you want to form a random sample of 20 subjects from a population of 400 individuals. Sampling will be without replacement, and you plan to use Table J in Appendix D to accomplish the randomization. Explain how you would use the table to select the sample.
4. A developmental psychologist is interested in assessing the "emotional intelligence" of college students. The experimental design calls for administering a questionnaire that measures emotional intelligence to a sample of 100 undergraduate student volunteers who are enrolled in an introductory psychology course currently being taught at her university. Assume this is the only sample being used for this study and discuss the adequacy of the sample.
5. What is the difference between *a priori* and *a posteriori* probability?
6. The addition rule gives the probability of occurrence of any one of several events, whereas the multiplication rule gives the probability of the joint or successive occurrence of several events. Is this statement correct? Explain, using examples to illustrate your explanation.
7. When solving problems involving the multiplication rule, is it useful to distinguish among three conditions? What are these conditions? Why is it useful to distinguish among them?
8. What is the definition of probability when the variable is continuous?
9. Which of the following are examples of independent events?
   a. Obtaining a 3 and a 4 in one roll of two fair dice
   b. Obtaining an ace and a king in that order by drawing twice without replacement from a deck of cards
   c. Obtaining an ace and a king in that order by drawing twice with replacement from a deck of cards

d. A cloudy sky followed by rain
   e. A full moon and eating a hamburger
10. Which of the following are examples of mutually exclusive events?
   a. Obtaining a 4 and a 7 in one draw from a deck of ordinary playing cards
   b. Obtaining a 3 and a 4 in one roll of two fair dice
   c. Being male and becoming pregnant
   d. Obtaining a 1 and an even number in one roll of a fair die
   e. Getting married and remaining a bachelor
11. Which of the following are examples of exhaustive events?
   a. Flipping a coin and obtaining a head or a tail (edge not allowed)
   b. Rolling a die and obtaining a 2
   c. Taking an exam and either passing or failing
   d. Going out on a date and having a good time
12. At the beginning of the baseball season in a particular year, the odds that the New York Yankees will win the American League pennant are 3 to 2.
   a. What are the odds that the Yankees will lose the pennant?
   b. What is the probability that the Yankees will win the pennant? Express your answer as a decimal.
   c. What is the probability that the Yankees will lose the pennant? Express your answer as a decimal. other
13. If you draw a single card once from a deck of ordinary playing cards, what is the probability that it will be
   a. The ace of diamonds?
   b. A 10?
   c. A queen or a heart?
   d. A 3 or a black card? other
14. If you roll two fair dice once, what is the probability that you will obtain
   a. A 2 on die 1 and a 5 on die 2?
   b. A 2 and a 5 without regard to which die has the 2 or 5?

c. At least one 2 or one 5?

d. A sum of 7? other

15. If you are randomly sampling one at a time with replacement from a bag that contains eight blue marbles, seven red marbles, and five green marbles, what is the probability of obtaining

a. A blue marble in one draw from the bag?

b. Three blue marbles in three draws from the bag?

c. A red, a green, and a blue marble in that order in three draws from the bag?

d. At least two red marbles in three draws from the bag? other

16. Answer the same questions as in Problem 15, except sampling is one at a time without replacement. other

17. You are playing the one-armed bandit (slot machine) described in Practice Problem 8.10, p. 212. There are three wheels, and on each wheel there is a picture of a lemon, a plum, an apple, an orange, a pear, cherries, and a banana (seven different pictures). You insert your silver dollar and pull down the lever. What is the probability that

a. Three oranges will appear?

b. Two oranges and a banana will appear, without regard to order?

c. At least two oranges will appear? other

18. You want to call a friend on the telephone. You remember the first three digits of her phone number, but you have forgotten the last four digits. What is the probability that you will get the correct number merely by guessing once? other

19. You are planning to win big at the race track. In a particular race, there are seven horses entered. If the horses are all equally matched, what is the probability of your correctly picking the winner and runner-up? other

20. A gumball dispenser has 38 orange gumballs, 30 purple ones, and 18 yellow ones. The dispenser operates such that one quarter delivers 1 gumball.

a. Using three quarters, what is the probability of obtaining 3 gumballs in the order orange, purple, orange?

b. Using one quarter, what is the probability of obtaining 1 gumball that is either purple or yellow?

c. Using three quarters, what is the probability that of the 3 gumballs obtained, exactly 1 will be purple and 1 will be yellow? other

21. If two cards are randomly drawn from a deck of ordinary playing cards, one at a time, with replacement, what is the probability of obtaining at least one ace? other

22. A state lottery is paying $1 million to the holder of the ticket with the correct eight-digit number. Tickets cost $1 apiece. If you buy one ticket, what is the probability you will win? Assume there is only one ticket for each possible eight-digit number and the winning number is chosen by a random process (round to eight decimal places). other

23. Given a population comprising 30 bats, 15 gloves, and 60 balls, if sampling is random and one at a time without replacement,

a. What is the probability of obtaining a glove if one object is sampled from the population?

b. What is the probability of obtaining a bat and a ball in that order if two objects are sampled from the population?

c. What is the probability of obtaining a bat, a glove, and a bat in that order if three objects are sampled from the population? other

24. A distribution of scores is normally distributed with a mean $\mu = 85$ and a standard deviation $\sigma = 4.6$. If one score is randomly sampled from the distribution, what is the probability that it will be

a. Greater than 96?

b. Between 90 and 97?

c. Less than 88? other

25. Assume the IQ scores of the students at your university are normally distributed, with $\mu = 115$ and $\sigma = 8$. If you randomly sample one score from this distribution, what is the probability it will be

a. Higher than 130?

b. Between 110 and 125?

c. Lower than 100? cognitive

26. A standardized test measuring mathematics proficiency in sixth graders is administered nationally. The results show a normal distribution of scores, with $\mu = 50$ and $\sigma = 5.8$. If one score is randomly sampled from this population, what is the probability it will be

a. Higher than 62?

b. Between 40 and 65?

c. Lower than 45? education

27. Assume we are still dealing with the population of Problem 24. If, instead of randomly sampling from the population, the single score was sampled, using a nonrandom process, would that affect any of the answers to Problem 24 part **a**, **b**, or **c**? Explain. other

28. An ethologist is interested in how long it takes a certain species of water shrew to catch its prey. On 20 occasions each day, he lets a dragonfly loose inside the cage of a shrew and times how long it takes until the shrew catches the dragonfly. After months of research, the ethologist concludes that the mean prey-catching time was 30 seconds, the standard deviation was 5.5 seconds, and the scores were normally distributed. Based on the shrew's past record, what is the probability
    a. It will catch a dragonfly in less than 18 seconds?
    b. It will catch a dragonfly in between 22 and 45 seconds?
    c. It will take longer than 40 seconds to catch a dragonfly? biological
29. An instructor at the U.S. Navy's underwater demolition school believes he has developed a new technique for staying under water longer. The school commandant gives him permission to try his technique with a student who has been randomly selected from the current class. As part of their qualifying exam, all students are tested to see how long they can stay under water without an air tank. Past records show that the scores are normally distributed with a mean $\mu = 130$ seconds and a standard deviation $\sigma = 14$ seconds. If the new technique has no additional effect, what is the probability that the randomly selected student will stay under water for
    a. More than 150 seconds?
    b. Between 115 and 135 seconds?
    c. Less than 90 seconds? education
30. If you are randomly sampling two scores one at a time with replacement from a population comprising the scores 2, 3, 4, 5, and 6, what is the probability that
    a. The mean of the sample ($\overline{X}$) will equal 6.0?
    b. $\overline{X} \geq 5.5$?
    c. $\overline{X} \leq 2.0$?
    Hint: All of the possible samples of size 2 are listed on p. 191. other

What Is the Truth? Questions

1. *'Not Guilty, I'm a Victim of Coincidence': Gutsy Plea or Truth?*
    a. What do you think of using statistics to mete out justice? Is it appropriate? Discuss.
    b. Assuming the actual probability was in fact 0.0069, do you think the judge was right in acquitting the defendant? Discuss.
    c. Give one example of a recent use of statistics in the legal field.
2. *Sperm Count Decline: Male or Sampling Inadequacy?*
    a. Why might it be important to use samples from randomly selected men rather than from men who deposit their sperm at a sperm bank?
    b. Why might large week-to-week variability in sperm counts and sperm quality complicate interpretation of the data?
3. *A Sample of a Sample*
    a. Given that many of the polls taken on TV or Facebook or the Internet are admittedly not scientific, do they have any value? What are the limitations of such polls?
    b. Give an example of an unscientific poll that you believe was valuable. Discuss.
    c. In exit polls taken in conjunction with national elections, do you believe the results should be shown on television in areas where voting is still taking place? Discuss.

■ NOTES

8.1 I realize that if the process ensures that each possible sample of a given size has an equal chance of being selected, then it also ensures that all the members of the population have an equal chance of being selected into the sample. I included the latter statement in the definition because I believed it is sufficiently important to deserve this special emphasis.

■ ONLINE STUDY RESOURCES

CENGAGE**brain**.com

Login to CengageBrain.com to access the resources your instructor has assigned. For this book, you can access the book's companion website for chapter-specific learning tools including Know and Be Able to Do, practice quizzes, flash cards and glossaries, and a link to Statistics and Research Methods Workshops.

aplia™

If your professor has assigned Aplia homework:

1. Sign in to your account.
2. Complete the corresponding homework exercises as required by your professor.
3. When finished, click "Grade It Now" to see which areas you have mastered and which need more work, and for detailed explanations of every answer.

Visit **www.cengagebrain.com** to access your account and to purchase materials.

9

© Strmko / Dreamstime.com

Binomial Distribution

LEARNING OBJECTIVES

After completing this chapter, you should be able to:

- Specify the five conditions that should be met to result in a binomial distribution.
- Describe the relationship between binomial distribution and binomial expansion, and explain how the binomial table relates to the binomial expansion.
- Specify what each term in the expanded binomial expansion stands for in terms of P and Q events.
- Specify for what P and Q values the binomial distribution is symmetrical, for what values it is skewed, and what happens to the shape of the binomial distribution as N increases.
- Solve binomial problems for $N \leq 20$, using the binomial table.
- Solve binomial problems for $N > 20$, using the normal approximation.
- Understand the illustrative examples, do the practice problems, and understand the solutions.

INTRODUCTION

In Chapter 10, we'll discuss the topic of hypothesis testing. This topic is a very important one. It forms the basis for most of the material taken up in the remainder of the textbook. For reasons explained in Chapter 10, we've chosen to introduce the concepts of hypothesis testing by using a simple inference test called the sign test. However, to understand and use the sign test, we must first discuss a probability distribution called the binomial distribution.

DEFINITION AND ILLUSTRATION OF THE BINOMIAL DISTRIBUTION

The binomial distribution may be defined as follows:

definition ■ *The* **binomial distribution** *is a probability distribution that results when the following five conditions are met: (1) there is a series of N trials; (2) on each trial, there are only two possible outcomes; (3) on each trial, the two possible outcomes are mutually exclusive; (4) there is independence between the outcomes of each trial; and (5) the probability of each possible outcome on any trial stays the same from trial to trial. When these requirements are met, the binomial distribution tells us each possible outcome of the N trials and the probability of getting each of these outcomes.*

Let's use coin flipping as an illustration for generating the binomial distribution. Suppose we flip a fair, or unbiased, penny once. Suppose further that we restrict the possible outcomes at the end of the flip to either a head or a tail. You will recall from Chapter 8 that a *fair coin* means the probability of a head with the coin equals the probability of a tail. Since there are only two possible outcomes in one flip,

$$p(\text{head}) = p(H) = \frac{\text{Number of outcomes classifiable as heads}}{\text{Total number of outcomes}}$$

$$= \frac{1}{2} = 0.5000$$

$$p(\text{tail}) = p(T) = \frac{\text{Number of outcomes classifiable as tails}}{\text{Total number of outcomes}}$$

$$= \frac{1}{2} = 0.5000$$

MENTORING TIP

Again, probability values have been given to four-decimal-place accuracy. Answers to end-of-chapter problems should also be given to four decimal places, unless you are told otherwise.

Now suppose we flip two pennies that are unbiased. The flip of each penny is considered a trial. Thus, with two pennies, there are two trials ($N = 2$). The possible outcomes of flipping two pennies are given in Table 9.1. There are four possible outcomes: one in which there are 2 heads (row 1), two in which there are 1 head and 1 tail (rows 2 and 3), and one in which there are 2 tails (row 4).

table 9.1 All possible outcomes of flipping two coins once

Row No.	Penny 1	Penny 2	No. of Outcomes
1	H	H	1
2	H	T	
3	T	H	2
4	T	T	1
		Total outcomes	4

Next, let's determine the probability of getting each of these outcomes due to chance. If chance alone is operating, then each of the outcomes is equally likely. Thus,

$$p(2 \text{ heads}) = p(HH) = \frac{\text{Number of outcomes classifiable as 2 heads}}{\text{Total number of outcomes}}$$

$$= \frac{1}{4} = 0.2500$$

$$p(1 \text{ head}) = p(HT \text{ or } TH) = \frac{\text{Number of outcomes classifiable as 1 head}}{\text{Total number of outcomes}}$$

$$= \frac{2}{4} = 0.5000$$

$$p(0 \text{ head}) = p(TT) = \frac{\text{Number of outcomes classifiable as 0 heads}}{\text{Total number of outcomes}}$$

$$= \frac{1}{4} = 0.2500$$

You should note that we could have also found these probabilities from the multiplication and addition rules. For example, $p(1 \text{ head})$ could have been found from a combination of the addition and multiplication rules as follows:

$$p(1 \text{ head}) = p(HT \text{ or } TH)$$

Using the multiplication rule, we obtain

$$p(HT) = p(\text{head on coin 1 and tail on coin 2})$$
$$= p(\text{head on coin 1})p(\text{tail on coin 2})$$
$$= \frac{1}{2}\left(\frac{1}{2}\right) = \frac{1}{4}$$

$$p(TH) = p(\text{tail on coin 1 and head on coin 2})$$
$$= p(\text{tail on coin 1})p(\text{head on coin 2})$$
$$= \frac{1}{2}\left(\frac{1}{2}\right) = \frac{1}{4}$$

Using the addition rule, we obtain

$$p(1 \text{ head}) = p(HT \text{ or } TH)$$
$$= p(HT) + p(TH)$$
$$= \frac{1}{4} + \frac{1}{4} = \frac{2}{4} = 0.5000$$

Next, suppose we increase N from 2 to 3. The possible outcomes of flipping three unbiased pennies once are shown in Table 9.2. This time there are eight possible outcomes: one way to get 3 heads (row 1), three ways to get 2 heads and 1 tail (rows 2, 3, and 4), three ways to get 1 head and 2 tails (rows 5, 6, and 7), and one way to get 0 heads (row 8). Since each outcome is equally likely,

$$p(3 \text{ heads}) = \frac{1}{8} = 0.1250$$

$$p(2 \text{ heads}) = \frac{3}{8} = 0.3750$$

$$p(1 \text{ head}) = \frac{3}{8} = 0.3750$$

$$p(0 \text{ heads}) = \frac{1}{8} = 0.1250$$

table 9.2   All possible outcomes of flipping three pennies once

Row No.	Penny 1	Penny 2	Penny 3	No. of Outcomes
1	H	H	H	1
2	H	H	T	
3	H	T	H	3
4	T	H	H	
5	T	T	H	
6	T	H	T	3
7	H	T	T	
8	T	T	T	1
			Total outcomes	8

The distributions resulting from flipping one, two, or three fair pennies are shown in Table 9.3. These are binomial distributions because they are probability distributions that have been generated by a situation in which there is a series of

table 9.3   Binomial distribution for coin flipping when the number of coins equals 1, 2, or 3

N	Possible Outcomes	Probability
1	1 H	0.5000
	0 H	0.5000
2	2 H	0.2500
	1 H	0.5000
	0 H	0.2500
3	3 H	0.1250
	2 H	0.3750
	1 H	0.3750
	0 H	0.1250

trials ($N = 1$, 2, or 3), where on each trial there are only two possible outcomes (head or tail), on each trial the possible outcomes are mutually exclusive (if it's a head, it cannot be a tail), there is independence between trials (there is independence between the outcomes of each coin), and the probability of a head or tail on any trial stays the same from trial to trial. Note that each distribution gives two pieces of information: (1) all possible outcomes of the N trials and (2) the probability of getting each of the outcomes.

GENERATING THE BINOMIAL DISTRIBUTION
FROM THE BINOMIAL EXPANSION

We could continue this enumeration process for larger values of N, but it becomes too laborious. It would indeed be a dismal prospect if we had to use enumeration for every value of N. Think about what happens when N gets to 15. With 15 pennies, there are $(2)^{15} = 32,768$ different ways that the 15 coins could fall. Fortunately, there is a mathematical expression that allows us to generate in a simple way everything we've been considering. The expression is called the *binomial expansion*. The binomial expansion is given by

$$(P + Q)^N \qquad \textit{binomial expansion}$$

where P = probability of one of the two possible outcomes of a trial
Q = probability of the other possible outcome
N = number of trials

To generate the possible outcomes and associated probabilities we arrived at in the previous coin-flipping experiments, all we need to do is expand the expression $(P + Q)^N$ for the number of coins in the experiment and evaluate each term in the expansion. For example, if there are two coins, $N = 2$ and

$$2P \; events \quad 1P \; and \; 1Q \; event \quad 2Q \; events$$
$$(P + Q)^N = (P + Q)^2 = P^2 + 2P^1Q^1 + Q^2$$

The terms P^2, $2P^1Q^1$, and Q^2 represent all the possible outcomes of flipping two coins once.

*The letters of each term (**P** or **PQ** or **Q**) tell us the kinds of events that comprise the outcome, the exponent of each letter tells us how many of that kind of event there are in the outcome, and the coefficient of each term tells us how many ways there are of obtaining the outcome.*

Thus,

1. P^2 indicates that one possible outcome is composed of two P events. The lone P tells us this outcome is composed entirely of P events. The exponent 2 indicates there are two of this kind of event. If we associate P with heads, then P^2 tells us one possible outcome is two heads.
2. $2P^1Q^1$ indicates that another possible outcome is one P and one Q event, or one head and one tail. The coefficient 2 tells us there are two ways to obtain one P and one Q event.
3. Q^2 represents an outcome of two Q events, or two tails (zero heads).

The probability of getting each of these possible outcomes is found by evaluating their respective terms using the numerical values of P and Q. If the coins are fair, then $P = Q = 0.50$. Thus,

$$p(2 \text{ heads}) = P^2 = (0.50)^2 = 0.2500$$
$$p(1 \text{ head}) = 2P^1Q^1 = 2(0.50)(0.50) = 0.5000$$
$$p(0 \text{ heads}) = Q^2 = (0.50)^2 = 0.2500$$

These results are the same as those obtained by enumeration. Note that in using the binomial expansion to find the probability of each possible outcome, we do not add the terms but use them separately. At this point, it probably seems much easier to use enumeration than the binomial expansion. However, the situation reverses itself quickly as N gets larger.

Let's do one more example, this time with $N = 3$. As before, we need to expand $(P + Q)^N$ and evaluate each term in the expansion using $P = Q = 0.50$.* Thus,

$$(P + Q)^N = (P + Q)^3 = P^3 + 3P^2Q + 3PQ^2 + Q^3$$

The terms P^3, $3P^2Q$, $3PQ^2$, and Q^3 represent all of the possible outcomes of flipping three pennies once. P^3 tells us there are three P events, or 3 heads. The term $3P^2Q$ indicates that this outcome has two P events and one Q event, or 2 heads and 1 tail. The term $3PQ^2$ represents one P event and two Q events, or 1 head and 2 tails. Finally, the term Q^3 designates three Q events and zero P events, or 3 tails and 0 heads. We can find the probability of each of these outcomes by evaluation of their respective terms. Since each coin is a fair coin, $P = Q = 0.50$. Thus,

$$p(3 \text{ heads}) = P^3 = (0.50)^3 = 0.1250$$
$$p(2 \text{ heads}) = 3P^2Q = 3(0.50)^2(0.05) = 0.3750$$
$$p(1 \text{ head}) = 3PQ^2 = 3(0.50)(0.50)^2 = 0.3750$$
$$p(0 \text{ heads}) = Q^3 = (0.50)^3 = 0.1250$$

These are the same results we derived previously by enumeration.

The binomial distribution may be generated for any N, P, and Q by using the binomial expansion. We have graphed the binomial distributions for $N = 3, 8,$ and 15 in Figure 9.1. $P = Q = 0.50$ for each of these distributions.

MENTORING TIP

Caution: if $P \neq 0.50$, the binomial distribution is not symmetrical. This is important for some applications in Chapter 11.

Note that (1) with $P = 0.50$, the binomial distribution is symmetrical; (2) it has two tails (i.e., it tails off as we go from the center toward either end); (3) it involves a discrete variable (e.g., we can't have $2\frac{1}{2}$ heads); and (4) as N increases, the binomial distribution gets closer to the shape of a normal curve.

USING THE BINOMIAL TABLE

Although in principle any problem involving binomial data can be answered by directly substituting into the binomial expansion, mathematicians have saved us the work. They have solved the binomial expansion for many values of N and reported the results in tables. One such table is Table B in Appendix D. This table gives the binomial distribution for values of N up to 20. Glancing at Table B (p. 595), you observe that N (the number of trials) is given in the first column and the possible outcomes are given in the second column, which is headed by "No. of P or Q Events." The rest of the columns contain probability entries for various values of P or Q. The values of P or Q are given

*See Note 9.1 for the general equation to expand $(P + Q)^N$.

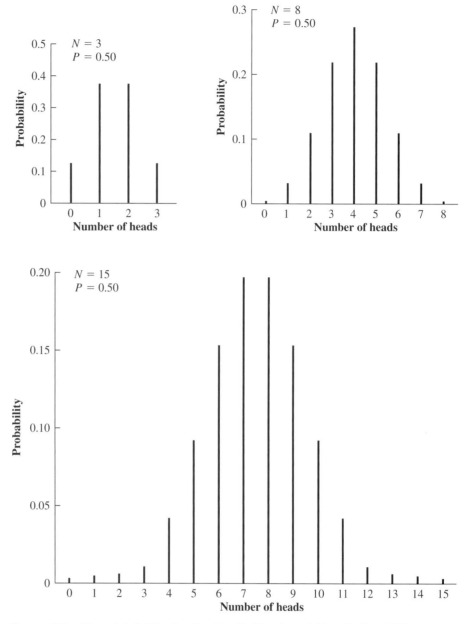

figure 9.1 Binomial distribution for *N* = 3, *N* = 8, and *N* = 15; *P* = 0.50.

at the top of each column. Thus, the second column contains probability values for *P* or *Q* = 0.10 and the last column has the values for *P* or *Q* = 0.50. In practice, any problem involving binomial data can be solved by looking up the appropriate probability in this table. This, of course, applies only for *N* ≤ 20 and the *P* or *Q* values given in the table.

The reader should note that Table B can be used to solve problems in terms of *P* or *Q*. Thus, with the exception of the first column, the column headings are given in terms of *P* or *Q*. To emphasize which we are using (*P* or *Q*) in a given problem, if we are entering the table under *P* and the number of *P* events, we shall refer to the second column as "number of *P* events" and the remaining column headings as "*P*" probability values. If we are entering Table B under *Q* and the number of events, we shall refer to the second column heading as "number of Q events" and the rest of the column headings as "Q" probability values. Let's now see how to use this table to solve problems involving binomial situations.

example

If I flip three unbiased coins once, what is the probability of getting 2 heads and 1 tail? Assume each coin can be only a head or tail.

SOLUTION

In this problem, N is the number of coins, which equals 3. We can let P equal the probability of a head in one flip of any coin. The coins are unbiased, so $P = 0.50$. Since we want to determine the probability of getting 2 heads, the number of P events equals 2. Having determined the foregoing, all we need do is enter Table B under $N = 3$. Next, we locate the 2 in the number of P events column. The answer is found where the row containing the 2 intersects the column headed by $P = 0.50$. This is shown in Table 9.4. Thus,

table 9.4 Table B entry

	No. of P	P
N	Events	**0.50**
3	2	0.3750

$p(\text{2 heads and 1 tail}) = 0.3750$

Note that this is the same answer we arrived at before. In fact, if you look at the remaining entries in that column ($P = 0.50$) for $N = 2$ and 3, you will see that they are the same as we arrived at earlier using the binomial expansion—and they ought to be because the table entries are taken from the binomial expansion. Let's try some practice problems using this table.

Practice Problem 9.1

If six unbiased coins are flipped once, what is the probability of getting
a. Exactly 6 heads?
b. 4, 5, or 6 heads?

SOLUTION

a. Given there are six coins, $N = 6$. Again, we can let P be the probability of a head in one flip of any coin. The coins are unbiased, so $P = 0.50$. Since we want to know the probability of getting exactly 6 heads, the number of P events = 6. Entering Table B under $N = 6$, number of P events = 6, and $P = 0.50$, we find

Table B entry

	No. of P	P
N	Events	**0.50**
6	6	0.0156

$p(\text{exactly 6 head}) = 0.0156$

b. Again, $N = 6$ and $P = 0.50$. We can find the probability of 4, 5, and 6 heads by entering Table B under number of P events = 4, 5, and 6, respectively. Thus,

Table B entry

	No. of P	P
N	Events	0.50
6	4	0.2344
	5	0.0938
	6	0.0156

$p(4 \text{ heads}) = 0.2344$
$p(5 \text{ heads}) = 0.0938$
$p(6 \text{ heads}) = 0.0156$

From the addition rule with mutually exclusive events,

$$p(4, 5, \text{ or } 6 \text{ heads}) = p(4) + p(5) + p(6)$$
$$= 0.2344 + 0.0938 + 0.0156$$
$$= 0.3438$$

Practice Problem 9.2

If 10 unbiased coins are flipped once, what is the probability of getting a result as extreme or more extreme than 9 heads?

SOLUTION

There are 10 coins, so $N = 10$. As before, we shall let $P =$ the probability of getting a head in one flip of any coin. The coins are unbiased, so $P = Q = 0.50$. The phrase "as extreme or more extreme than" means "as far from the center of the distribution or farther from the center of the distribution than." Thus, "as extreme or more extreme than 9 heads" means results that are as far from the center of the distribution or farther from the center of the distribution than 9 heads. Thus, the number of P events = 0, 1, 9, or 10. In Table B under $N = 10$, number of events = 0, 1, 9, or 10, and $P = 0.50$, we find

$$p\begin{pmatrix} \text{as extreme or} \\ \text{more extreme} \\ \text{than 9 heads} \end{pmatrix} = p(0, 1, 9, \text{ or } 10)$$
$$= p(0) + p(1) + p(9) + p(10)$$
$$= 0.0010 + 0.0098 + 0.0098$$
$$\quad + 0.0010$$
$$= 0.0216$$

Table B entry

	No. of P	P
N	Events	0.50
10	0	0.0010
	1	0.0098
	9	0.0098
	10	0.0010

The binomial expansion is very general. It is not limited to values where $P = 0.50$. Accordingly, Table B also lists probabilities for values of P other than 0.50. Let's try some problems where P is not equal to 0.50.

Practice Problem 9.3

Assume you have eight *biased coins*. You will recall from Chapter 8 that a biased coin is one where $P \neq Q$. Each coin is weighted such that the probability of a head with it is 0.30. If the eight biased coins are flipped once, then

a. What is the probability of getting 7 heads?

b. What is the probability of getting 7 or 8 heads?

c. The probability found in part **a** comes from evaluating which of the term(s) in the following binomial expansion?

$$P^8 + 8P^7Q^1 + 28P^6Q^2 + 56P^5Q^3 + 70P^4Q^4 + 56P^3Q^5 + 28P^2Q^6 + 8P^1Q^7 + Q^8$$

d. With your calculator, evaluate the term(s) selected in part **c** using $P = 0.30$. Compare your answer with the answer in part **a**. Explain.

SOLUTION

a. Given there are eight coins, $N = 8$. Let P = the probability of getting a head in one flip of any coin. Since the coins are biased such that the probability of a head on any coin is 0.30, $P = 0.30$. Since we want to determine the probability of getting exactly 7 heads, the number of P events = 7. In Table B under $N = 8$, number of P events = 7, and $P = 0.30$, we find the following:

p(exactly 7 heads) = 0.0012

	Table B entry	
N	**No. of P Events**	**P 0.30**
8	7	0.0012

b. Again, $N = 8$ and $P = 0.30$. We can find the probability of 7 and 8 heads in Table B under number of P events = 7 and 8, respectively. Thus,

p(7 heads) = 0.0012
p(8 heads) = 0.0001

	Table B entry	
N	**No. of P Events**	**P 0.30**
8	7	0.0012
	8	0.0001

From the addition rule with mutually exclusive events,

$$p(7 \text{ or } 8 \text{ heads}) = p(7) + p(8)$$
$$= 0.0012 + 0.0001$$
$$= 0.0013$$

c. $8P^7Q^1$

d. $8P^7Q^1 = 8(0.30)^7(0.7) = 0.0012$. As expected, the answers are the same. The table entry was computed using $8P^7Q^1$ with $P = 0.30$ and $Q = 0.70$.

Thus, using Table B when P is less than 0.50 is very similar to using it when $P = 0.50$. We just look in the table under the new P value rather than under $P = 0.50$.

Table B can also be used when $P > 0.50$. To illustrate, consider the following example.

example

$P > 0.50$

If five biased coins are flipped once, what is the probability of getting (**a**) 5 heads and (**b**) 4 or 5 heads? Each coin is weighted such that the probability of a head on any coin is 0.75.

SOLUTION

a. 5 heads. There are five coins, so $N = 5$. Again, let $P =$ the probability of getting a head in one flip of any coin. Since the bias is such that the probability of a head on any coin is 0.75, $P = 0.75$. Since we want to determine the probability of getting 5 heads, the number of P events equals 5. Following our usual procedure, we would enter Table B under $N = 5$, number of P events $= 5$, and $P = 0.75$. However, Table B does not have a column headed by 0.75. All of the column headings are equal to or less than 0.50. Nevertheless, we can use Table B to solve this problem.

When $P > 0.50$, all we need do is solve the problem in terms of Q and the number of Q events, rather than P and the number of P events. Since the probability values given in Table B are for either P or Q, once the problem is put in terms of Q, we can refer to Table B using Q rather than P. Translating the problem into Q terms involves two steps: determining Q and determining the number of Q events. Let's follow these steps using the present example:

1. *Determining Q.*

$$Q = 1 - P = 1 - 0.75 = 0.25$$

2. *Determining the number of Q events.*

$$\text{Number of Q events} = N - \text{Number of P events} = 5 - 5 = 0$$

Thus, to solve this example, we refer to Table B under $N = 5$, number of Q events $= 0$, and $Q = 0.25$. The Table B entry is shown in Table 9.5. Thus,

$p(5 \text{ heads}) = p(0 \text{ tails}) = 0.2373$

table 9.5   Table B entry

N	No. of Q Events	Q 0.25
5	0	0.2373

b. 4 or 5 heads. Again, $N = 5$ and $Q = 0.25$. This time, the number of Q events $= 0$ or 1. The Table B entry is shown in Table 9.6. Thus,

$p(4 \text{ or } 5 \text{ heads}) = p(0 \text{ or } 1 \text{ tail})$
$= 0.2373 + 0.3955$
$= 0.6328$

table 9.6   Table B entry

N	No. of Q Events	Q 0.25
5	0	0.2373
	1	0.3955

We are now ready to try a practice problem.

Practice Problem 9.4

If 12 biased coins are flipped once, what is the probability of getting
a. Exactly 10 heads?
b. 10 or more heads?
The coins are biased such that the probability of a head with any coin equals 0.65.

SOLUTION

a. Given there are 12 coins, $N = 12$. Let P = the probability of a head in one flip of any coin. Since the probability of a head with any coin equals 0.65, $P = 0.65$. Since $P > 0.50$, we shall enter Table B with Q rather than P. If there are 10 P events, there must be 2 Q events ($N = 12$). If $P = 0.65$, then $Q = 0.35$. Using Q in Table B, we obtain

$p(10 \text{ heads}) = 0.1088$

Table B entry

N	No. of Q Events	Q 0.35
12	2	0.1088

b. Again, $N = 12$ and $P = 0.65$. This time, the number of P events equals 10, 11, or 12. Since $P > 0.50$, we must use Q in Table B rather than P. With $N = 12$, the number of Q events equals 0, 1, or 2 and $Q = 0.35$. Using Q in Table B, we obtain

$$p(10, 11, \text{ or } 12 \text{ heads}) = 0.1088 + 0.0368 \\ + 0.0057 \\ = 0.1513$$

Table B entry

N	No. of Q Events	Q 0.35
12	0	0.0057
	1	0.0368
	2	0.1088

So far, we have dealt exclusively with coin flipping. However, the binomial distribution is not limited to just coin flipping. It applies to all situations involving a series of trials where on each trial there are only two possible outcomes, the possible outcomes on each trial are mutually exclusive, there is independence between the outcomes of each trial, and the probability of each possible outcome on any trial stays the same from trial to trial. There are many situations that fit these requirements. To illustrate, let's do a couple of practice problems.

Practice Problem 9.5

A student is taking a multiple-choice exam with 15 questions. Each question has five choices. If the student guesses on each question, what is the probability of passing the test? The lowest passing score is 60% of the questions answered correctly. Assume that the choices for each question are equally likely.

SOLUTION

This problem fits the binomial requirements. There is a series of trials (questions). On each trial, there are only two possible outcomes. The student is either right or wrong. The possible outcomes are mutually exclusive. If she is right on a question, she can't be wrong. There is independence between the outcomes of each trial. If she is right on question 1, it has no effect on the outcome of question 2. Finally, if we assume the student guesses on each trial, then the probability of being right and the probability of being wrong on any trial stay the same from trial to trial. Thus, the binomial distribution and Table B apply.

   We can consider each question a trial (no pun intended). Given there are 15 questions, $N = 15$. We can let $P =$ the probability that she will guess correctly on any question. Since there are five choices that are equally likely on each question, $P = 0.20$. A passing grade equals 60% correct answers or more. Therefore, the student will pass if she gets 9 or more answers correct (60% of 15 is 9). Thus, the number of P events equals 9, 10, 11, 12, 13, 14, and 15. Looking in Table B under $N = 15$, number of P events = 9, 10, 11, 12, 13, 14, and 15, and $P = 0.20$, we obtain

Table B entry

	No. of P	P
N	**Events**	**0.20**
15	9	0.0007
	10	0.0001
	11	0.0000
	12	0.0000
	13	0.0000
	14	0.0000
	15	0.0000

$$p(9, 10, 11, 12, 13, 14,$$
$$\text{or 15 correct guesses}) = 0.0007 + 0.0001$$
$$= 0.0008$$

Practice Problem 9.6

Your friend claims to be a coffee connoisseur. He always drinks Starbucks and claims no other coffee even comes close to tasting so good. You suspect he is being a little grandiose. In fact, you wonder whether he can even taste the difference between Starbucks and the local roaster's coffee. Your friend agrees to the following experiment. While blindfolded, he is given six opportunities to taste from two cups of coffee and tell you which of the two cups contains Starbucks. The cups are identical and contain the same type of coffee except that one contains coffee made from beans supplied and roasted by Starbucks and the other by the local roaster. After each tasting of the two cups, you remove any telltale signs and randomize which of the two cups he is given first for the next trial. Believe it or not, your friend correctly identifies Starbucks on all six trials! What do you conclude? Can you think of a way to increase your confidence in the conclusion?

SOLUTION

The logic of our analysis is as follows. We will assume that your friend really can't tell the difference between the two coffees. He must then be guessing on each trial. We will compute the probability of getting six out of six correct, assuming guessing on each trial. If this probability is very low, we will reject guessing as a reasonable explanation and conclude that your friend can really taste the difference.

This experiment fits the requirements for the binomial distribution. Each comparison of the two coffees can be considered a trial (again, no pun intended). On each trial, there are only two possible outcomes. Your friend is either right or wrong. The outcomes are mutually exclusive. There is independence between trials. If your friend is correct on trial 1, it has no effect on the outcome of trial 2. Finally, if we assume your friend guesses on any trial, then the probability of being correct and the probability of being wrong stay the same from trial to trial.

Given each comparison of coffees is a trial, $N = 6$. We can let $P =$ the probability your friend will guess correctly on any trial. There are only two coffees, so $P = 0.50$. Your friend was correct on all six trials. Therefore, the number of P events $= 6$. Thus,

Table B entry

$p(6 \text{ correct guesses}) = 0.0156$

N	No. of P Events	P 0.50
6	6	0.0156

Assuming your friend is guessing, the probability of his getting six out of six correct is 0.0156. Since this is a fairly low value, you would probably reject guessing as a reasonable explanation and conclude that your friend can really taste the difference. To increase your confidence in rejecting guessing, you could include more brands of coffee on each trial, or you could increase the number of trials. For example, even with only two coffees, the probability of guessing correctly on 12 out of 12 trials is 0.0002.

USING THE NORMAL APPROXIMATION

A limitation of using the binomial table is that when N gets large, the table gets huge. Imagine how big the table would be if it went up to $N = 200$, rather than to $N = 20$ as it does in this textbook. Not only that, but imagine solving the problem of determining the probability of getting 150 or more heads if we were flipping a fair coin 200 times. Not only would the table have to be very large, but we would wind up having to add 51 four-digit probability values to get our answer! Even statistics professors are not that sadistic. Not to worry!

Remember, I pointed out earlier that as N increases, the binomial distribution becomes more normally shaped. When the binomial distribution approximates the normal distribution closely enough, we can solve binomial problems using z scores and the normal curve, as we did in Chapter 8, rather than having to look up many discrete values in a table. I call this approach the *normal approximation approach*.

How close the binomial distribution is to the normal distribution depends on N, P, and Q. As N increases, the binomial distribution gets more normally shaped. As P and Q deviate from 0.50, the binomial distribution gets less normally shaped. A criterion that is commonly used, and one that we shall adopt, is that if $NP \geq 10$ and $NQ \geq 10$, then the binomial distribution is close enough to the normal distribution to use the normal approximation approach without unduly sacrificing accuracy. Table 9.7 shows the minimum value of N for several values of P and Q necessary to meet this criterion. Notice that as P and Q get further from 0.50, N must get larger to meet the criterion.

table 9.7 Minimum value of N for several values of P and Q

P	Q	N
0.50	0.50	20
0.30	0.70	34
0.10	0.90	100

The normal distribution that the binomial approximates has the following parameters.

1. The mean of the distribution equals NP. Thus,

$$\mu = NP \qquad \textit{Mean of the normal distribution approximated by the binomial distribution}$$

2. The standard deviation of the distribution equals $\sqrt{NPQ}$. Thus,

$$\sigma = \sqrt{NPQ} \qquad \textit{Standard deviation of the normal distribution approximated by the binomial distribution}$$

To use the normal approximation approach, we first compute the z score of the frequency given in the problem. Next we determine the appropriate probability by entering column B or C of Table A, using the computed z score. Table A, as you probably remember, gives us areas under the normal curve. Let's try an example to see how this works. For the first example, let's do one of the sort we are used to.

example

If I flip 20 unbiased coins once, what is the probability of getting 18 or more heads?

SOLUTION

To solve this example, let's follow these steps.

1. *Determine if the criterion is met for normal approximation approach.*

   Since the coins are unbiased, $P = Q = 0.50, N = 20$.

   $$NP = 20(0.50) = 10$$
   $$NQ = 20(0.50) = 10$$

   Since $NP = 10$ and $NQ = 10$, the criterion that both $NP \geq 10$ and $NQ \geq 10$ is met. Therefore we can assume the binomial distribution is close enough to a normal distribution to solve the example using the normal approximation, rather than the binomial table. Note that both the NP and the NQ criterion must be met to use the normal approximation approach.

2. *Determine the parameters of the approximated normal curve*

   $$\mu = NP = 20(0.50) = 10$$
   $$\sigma = \sqrt{NPQ} = \sqrt{20(0.50)(0.50)} = \sqrt{5.00} = 2.24$$

3. *Draw the picture and locate the important information on it.*

   Next, let's draw the picture of the distribution and locate the important information on it as we did in Chapter 8. This is shown in Figure 9.2. The figure shows the normal distribution with $\mu = 10$, $X = 18$. The shaded area corresponds to the probability of getting 18 or more heads. We can determine this probability by computing the z value of 18 and looking up the probability in Table A.

4. *Determining the probability of 18 or more heads*

   The z value of 18 is given by

   $$z = \frac{X - \mu}{\sigma} = \frac{18 - 10}{2.24} = 3.58$$

   Entering Table A, Column C, using the z score of 3.58, we obtain

   $$p(18 \text{ or more heads}) = p(X \geq 18) = 0.0002$$

   Thus, if I flip 20 unbiased coins once, the probability of getting 18 or more heads is 0.0002.

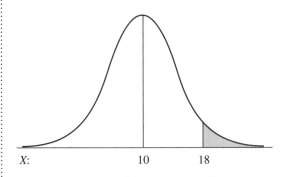

X: 10 18

figure 9.2 Determining the probability of 18 or more heads using the normal approximation approach

You might be wondering how close this value is to that which we would have obtained using the binomial table. Let's check it out. Looking in Table B, under $N = 20$, $P = 0.50$, and number of P events = 18, 19, and 20, we obtain

$$p(18, 19, \text{ or } 20 \text{ heads}) = 0.0002 + 0.0000 + 0.0000 + 0.0002$$

Not too shabby! The normal approximation yielded exactly the same value (four decimal-place accuracy) as the value given by the actual binomial distribution. Of course, the values given by the normal approximation are not always this accurate, but the accuracy is usually close enough for most statistical purposes. This is especially true if N is large and P is close to 0.50.*

Next, let's do an example in which $P \neq Q$.

example

Over the past 10 years, the football program at a large university graduated 70% of its varsity athletes. If the same probability applies to this year's group of 65 varsity football players,

a. What is the probability that 50 or more players of the group will graduate?
b. What is the probability that 48 or fewer players of the group will graduate?

SOLUTION

a. Probability that 50 or more players will graduate. For the solution, let's follow these steps:

1. *Determine if the criterion is met for normal approximation approach.*

   Let P = the probability any player in the group will graduate = 0.70.
   Let Q = the probability any player in the group will not graduate = 0.30.

   $$NP = 65(0.70) = 45.5$$
   $$NQ = 65(0.30) = 19.5$$

   Since $NP = 45.5$ and $NQ = 19.5$, the criterion that both $NP \geq 10$ and $NQ \geq 10$ is met. Therefore, we can use the normal approximation approach.

2. *Determine the parameters of the approximated normal curve*

   $$\mu = NP = 65(0.70) = 45.5$$
   $$\sigma = \sqrt{NPQ} = \sqrt{65(0.70)(0.30)} = \sqrt{13.65} = 3.69$$

3. *Draw the picture and locate the important information on it.*

   This is shown in Figure 9.3. The figure shows the normal distribution with $\mu = 45.5$, $X = 50$. The shaded area corresponds to the probability that 50 or more players of the group will graduate. We can determine this probability by computing the z value of 50 and looking up the probability in Table A, Column C.

4. *Determining the probability that 50 or more players will graduate*

   The z value of 50 is given by

   $$z = \frac{X - \mu}{\sigma} = \frac{50 - 45.5}{3.69} = 1.22$$

---

*There is a correction for continuity procedure available that increases accuracy. However, for the intended readers of this textbook, it introduces unnecessary complexity and so it has not been included. For a discussion of this correction, see D. S. Moore and G. P. McCabe, *Introduction to the Practice of Statistics,* W. H. Freeman and Company, New York, 1989, pp. 402–403.

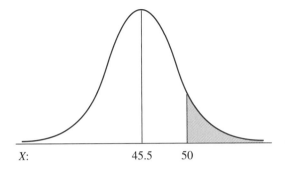

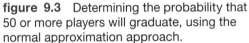

figure 9.3 Determining the probability that 50 or more players will graduate, using the normal approximation approach.

Entering Table A, Column C, using the z score of 1.22, we obtain

$$p(50 \text{ or more graduates}) = p(X \geq 50) = 0.1112$$

Thus, the probability that 50 or more players of the group will graduate is 0.1112.

b. Probability that 48 or fewer players will graduate.

Since we have already completed steps 1 and 2 in part **a,** we will begin with step 3.

3. *Draw the picture and locate the important information on it.*

This is shown in Figure 9.4. The figure shows the normal distribution with $\mu = 45.5$, $X = 48$. The shaded area corresponds to the probability that 48 or fewer players will graduate.

4. *Determining the probability that 48 or fewer players will graduate*

The probability that 48 or fewer players will graduate is found by computing the z value of 48, consulting Table A, Column B, for the probability of between 48 and 45.5 graduates, and then adding 0.5000 for the probability of graduates below 45.5. The z value of 48 is given by

$$z = \frac{X - \mu}{\sigma} = \frac{48 - 45.5}{3.69} = 0.68$$

Entering Table A, Column B, using the z score of 0.68, we obtain

$$p(\text{graduates between 48 and 45.5}) = 0.2517$$

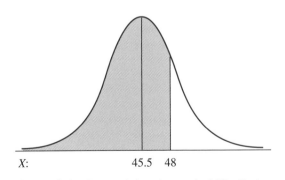

figure 9.4 Determining the probability that 48 or fewer players will graduate, using the normal approximation approach.

Next, we need to add 0.5000 to include the graduates below 45.5, making the total probability = 0.2517 + 0.5000 = 0.7517. Thus, the probability of 48 or fewer football players graduating is 0.7517.

Next, let's do a practice problem.

Practice Problem 9.7

A local union has 10,000 members, of which 20% are Hispanic. The union selects 150 representatives to vote in the coming national election for union president. Sixteen of the 150 selected representatives are Hispanics. Although you have been told that the selection was random and that there was no ethnic bias involved in the selection, you are not sure since the number of Hispanics seems low.

a. If the selection were really random, what is the probability that there would be 16 or fewer Hispanics selected as representatives? In answering, assume that P and Q do not change from selection to selection.

b. Given the answer obtained in part **a,** what is your tentative conclusion about random selection and possible ethnic bias?

SOLUTION

a. Probability of getting 16 or fewer Hispanic representatives. Let's follow these steps to solve this problem.

STEP 1. **Determine if criterion is met to use the normal approximation.**
Let P = probability of getting a Hispanic on any selection. Therefore, $P = 0.20$.

Let Q = probability of not getting a Hispanic on any selection. Therefore, $Q = 0.80$.

$$NP = 150(0.20) = 30$$
$$NQ = 150(0.80) = 120$$

Since $NP = 30$ and $NQ = 120$, the criterion that both $NP \geq 10$ and $NQ \geq 10$ is met. It's reasonable to use the normal approximation to solve the problem.

STEP 2. **Determine the parameters of the approximated normal curve.**

$$\mu = NP = 150(0.20) = 30$$
$$\sigma = \sqrt{NPQ} = \sqrt{150(0.20)(0.80)} = \sqrt{24.00} = 4.90$$

STEP 3. **Draw the picture and locate the important information on it.**
The picture is drawn in Figure 9.5. It shows the normal distribution with $\mu = 30$ and $X = 16$. The shaded area corresponds to the probability of getting 16 or fewer Hispanics as representatives.

(continued)

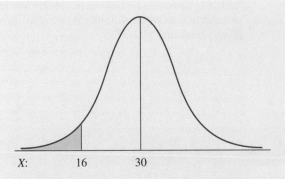

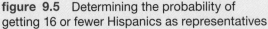

X: 16   30

figure 9.5   Determining the probability of
getting 16 or fewer Hispanics as representatives

> **STEP 4.   Determining the probability of 16 or fewer Hispanics.**
> We can determine this probability by computing the *z* value of 16
> and looking up the probability in Table A, Column C. The *z* value
> of 16 is given by
>
> $$z = \frac{X - \mu}{\sigma} = \frac{16 - 30}{4.90} = -2.86$$
>
> Entering Table A, Column C, using the *z* score of 2.86, we obtain
>
> $$p(16 \text{ or fewer Hispanics}) = p(X \leq 16) = 0.0021$$
>
> Thus, if sampling is random, the probability of getting 16 or fewer
> Hispanic representatives is 0.0021.
>
> **b.** Tentative conclusion, given the probability obtained in Part **a**:
>
> While random selection might have actually been the case, the probability
> obtained in part **a** is quite low and doesn't inspire much confidence in this
> possibility. A more reasonable explanation is that something systematic was
> going on in the selection process that resulted in fewer Hispanic representa-
> tives than would be expected via random selection. Of course, there may be
> reasons other than ethnic bias that could explain the data.

■ SUMMARY

In this chapter, I have discussed the binomial distribution.
The binomial distribution is a probability distribution that
results when the following conditions are met: (1) there
is a series of *N* trials; (2) on each trial, there are only two
possible outcomes; (3) the outcomes are mutually exclusive; (4) there is independence between trials; and (5) the
probability of each possible outcome on any trial stays the
same from trial to trial. When these conditions are met,
the binomial distribution tells us each possible outcome

of the *N* trials and the probability of getting each of these
outcomes.

I illustrated the binomial distribution through
coin-flipping experiments and then showed how the
binomial distribution could be generated through the
binomial expansion. The binomial expansion is given
by $(P + Q)^N$, where P = the probability of occurrence
of one of the events and Q = the probability of occurrence of the other event. Next, I showed how to use the

binomial table (Table B in Appendix D) to solve problems where $N \leq 20$. Finally, I showed how to use the normal approximation to solve problems where $N > 20$.

The binomial distribution is appropriate whenever the five conditions listed at the beginning of this summary are met.

■ IMPORTANT NEW TERMS

Binomial distribution (p. 226)
Binomial expansion (p. 229)

Binomial table (p. 230)
Normal approximation (p. 239)

Number of P events (p. 229)
Number of Q events (p. 229)

■ QUESTIONS AND PROBLEMS

1. Briefly define or explain each of the terms in the Important New Terms section.
2. What are the five conditions necessary for the binomial distribution to be appropriate?
3. In a binomial situation, if $P = 0.10$, $Q = $ ____.
4. Using Table B, if $N = 6$ and $P = 0.40$,
   a. The probability of getting exactly five events = ____.
   b. This probability comes from evaluating which of the terms in the following equation?

   $$P^6 + 6P^5Q + 15P^4Q^2 + 20P^3Q^3$$
   $$+ 15P^2Q^4 + 6PQ^5 + Q^6$$

   c. Evaluate the term(s) of your answer in part **b** using $P = 0.40$ and compare your answer with part **a.**
5. Using Table B, if $N = 12$ and $P = 0.50$,
   a. What is the probability of getting exactly 10 P events?
   b. What is the probability of getting 11 or 12 P events?
   c. What is the probability of getting at least 10 P events?
   d. What is the probability of getting a result as extreme as or more extreme than 10 P events?
6. Using Table B, if $N = 14$ and $P = 0.70$,
   a. What is the probability of getting exactly 13 P events?
   b. What is the probability of getting at least 13 P events?
   c. What is the probability of getting a result as extreme as or more extreme than 13 P events?
7. Using Table B, if $N = 20$ and $P = 0.20$,
   a. What is the probability of getting exactly two P events?
   b. What is the probability of getting two or fewer P events?

   c. What is the probability of getting a result as extreme as or more extreme than two P events?
8. An individual flips nine fair coins. If she allows only a head or a tail with each coin,
   a. What is the probability they all will fall heads?
   b. What is the probability there will be seven or more heads?
   c. What is the probability there will be a result as extreme as or more extreme than seven heads?
9. Someone flips 15 biased coins once. The coins are weighted such that the probability of a head with any coin is 0.85.
   a. What is the probability of getting exactly 14 heads?
   b. What is the probability of getting at least 14 heads?
   c. What is the probability of getting exactly 3 tails?
10. Thirty biased coins are flipped once. The coins are weighted so that the probability of a head with any coin is 0.40. What is the probability of getting at least 16 heads?
11. A key shop advertises that the keys made there have a $P = 0.90$ of working effectively. If you bought 10 keys from the shop, what is the probability that all of the keys would work effectively?
12. A student is taking a true/false exam with 15 questions. If he guesses on each question, what is the probability he will get at least 13 questions correct? education
13. A student is taking a multiple-choice exam with 16 questions. Each question has five alternatives. If the student guesses on 12 of the questions, what is the probability she will guess at least 8 correct? Assume all of the alternatives are equally likely for each question on which the student guesses. education
14. You are interested in determining whether a particular child can discriminate the color green from blue. Therefore, you show the child five wooden

blocks. The blocks are identical except that two are green and three are blue. You randomly arrange the blocks in a row and ask him to pick out a green block. After a block is picked, you replace it and randomize the order of the blocks once more. Then you again ask him to pick out a green block. This procedure is repeated until the child has made 14 selections. If he really can't discriminate green from blue, what is the probability he will pick a green block at least 11 times? cognitive

15. Let's assume you are an avid horse racing fan. You are at the track and there are eight races. On this day, the horses and their riders are so evenly matched that chance alone determines the finishing order for each race. There are 10 horses in every race. If, on each race, you bet on one horse to show (to finish first, second, or third),
    a. What is the probability that you will win your bet in all eight races?
    b. What is the probability that you will win in at least six of the races? other

16. A manufacturer of valves admits that its quality control has gone radically "downhill" such that currently the probability of producing a defective valve is 0.50. If it manufactures 1 million valves in a month and you randomly sample from these valves 10,000 samples, each composed of 15 valves,
    a. In how many samples would you expect to find exactly 13 good valves?
    b. In how many samples would you expect to find at least 13 good valves? I/O

17. Assume that 15% of the population is left-handed and the remainder is right-handed (there are no ambidextrous individuals). If you stop the next five people you meet, what is the probability that
    a. All will be left-handed?
    b. All will be right-handed?
    c. Exactly two will be left-handed?
    d. At least one will be left-handed?
    For the purposes of this problem, assume independence in the selection of the five individuals. other

18. In your voting district, 25% of the voters are against a particular bill and the rest favor it. If you randomly poll four voters from your district, what is the probability that
    a. None will favor the bill?
    b. All will favor the bill?
    c. At least one will be against the bill? I/O

19. At your university, 30% of the undergraduates are from out of state. If you randomly select eight of the undergraduates, what is the probability that
    a. All are from within the state?
    b. All are from out of state?
    c. Exactly two are from within the state?
    d. At least five are from within the state? education

20. Twenty students living in a college dormitory participated in a taste contest between the two leading colas.
    a. If there really is no preference, what is the probability that all 20 would prefer Brand X to Brand Y?
    b. If there really is no preference, what is the probability that at least 17 would prefer Brand X to Brand Y?
    c. How many of the 20 students would have to prefer Brand X before you would be willing to conclude that there really is a preference for Brand X? other

21. In your town, the number of individuals voting in the next election is 800. Of those voting, 600 are Republicans. If you randomly sample 60 individuals, one at a time, from the voting population, what is the probability there will be 42 or more Republicans in the sample? Assume the probability of getting a Republican on each sampling stays the same. social

22. A large bowl contains 1 million marbles. Half of the marbles have a plus (+) painted on them and the other half have a minus (–).
    a. If you randomly sample 10 marbles, one at a time with replacement from the bowl, what is the probability you will select 9 marbles with pluses and 1 with a minus?
    b. If you take 1000 random samples of 10 marbles, one at a time with replacement, how many of the samples would you expect to be all pluses? other

■ NOTES

9.1 The equation for expanding $(P + Q)^N$ is

$$(P + Q)^N = P^N + \frac{N}{1}P^{N-1}Q + \frac{N(N - 1)}{1(2)} P^{N-2} Q^2$$

$$+ \frac{N(N - 1)(N - 2)}{1(2)(3)} P^{N-3}Q^3$$

$$+ \ldots + Q^N$$

■ ONLINE STUDY RESOURCES

CENGAGE**brain**.com

Login to CengageBrain.com to access the resources your instructor has assigned. For this book, you can access the book's companion website for chapter-specific learning tools including Know and Be Able to Do, practice quizzes, flash cards, and glossaries, and a link to Statistics and Research Methods Workshops.

aplia™

If your professor has assigned Aplia homework:

1. Sign in to your account.
2. Complete the corresponding homework exercises as required by your professor.
3. When finished, click "Grade It Now" to see which areas you have mastered and which need more work, and for detailed explanations of every answer.

Visit **www.cengagebrain.com** to access your account and to purchase materials.

10

Introduction to Hypothesis Testing Using the Sign Test

© Strmko / Dreamstime.com

LEARNING OBJECTIVES

After completing this chapter, you should be able to:

▪ Specify the essential features of the repeated measures design.
▪ Define the alternative (H_1) and null hypotheses (H_0), and explain the relationship between them. Include a discussion of directional and nondirectional H_1s and the H_0s that go with them.
▪ Define alpha level, explain the purpose of the alpha level, and specify the decision rule for determining when to reject or retain the null hypothesis.
▪ Explain the difference between *significant* and *important*.
▪ Explain the process of evaluating the null hypothesis, beginning with H_1 and H_0, and ending with the possibility of making a Type I or Type II error.
▪ Explain why we evaluate H_0 first and then H_1 indirectly, rather than directly evaluate H_1; explain why we evaluate the tail result and not the exact result itself.
▪ Explain when it is appropriate to do one- and two-tailed evaluations.
▪ Define Type I and Type II errors and explain why it is important to discuss these possible errors; specify the relationship between Type I and Type II errors, and between the alpha level and Type I and Type II errors.
▪ Formulate H_1 and H_0 for the sign test and solve problems using the sign test.
▪ Understand the illustrative example, do the practice problems, and understand the solutions.

INTRODUCTION

We pointed out previously that inferential statistics has two main purposes: (1) hypothesis testing and (2) parameter estimation. By far, most of the applications of inferential statistics are in the area of hypothesis testing. As discussed in Chapter 1, scientific methodology depends on this application of inferential statistics. Without objective verification, science would cease to exist, and objective verification is often impossible without inferential statistics. You will recall that at the heart of scientific methodology is an experiment. Usually, the experiment has been designed to test a hypothesis, and the resulting data must be analyzed. Occasionally, the results are so clear-cut that statistical inference is not necessary. However, such experiments are rare. Because of the variability that is inherent from subject to subject in the variable being measured, it is often difficult to detect the effect of the independent variable without the help of inferential statistics. In this chapter, we shall begin the fascinating journey into how experimental design, in conjunction with mathematical analysis, can be used to verify truth assertions or hypotheses, as we have been calling them. We urge you to apply yourself to this chapter with special rigor. The material it contains applies to all of the inference tests we shall take up (which constitutes most of the remaining text).

LOGIC OF HYPOTHESIS TESTING

experiment

Marijuana and the Treatment of AIDS Patients

We begin with an experiment. Let's assume that you are a social scientist working in a metropolitan hospital that serves a very large population of AIDS patients. You are very concerned about the pain and suffering that afflict these patients. In particular, although you are not yet convinced, you think there may be an ethically proper place for using marijuana in the treatment of these patients, particularly in the more advanced stages of the illness. Of course, before seriously considering the other issues involved in advocating the use of marijuana for this purpose, you must be convinced that it does have important positive effects. Thus far, although there have been many anecdotal reports from AIDS patients that using marijuana decreases their nausea, increases their appetite, and increases their desire to socialize, there have not been any scientific experiments to shore up these reports.

As a scientist, you realize that although personal reports are suggestive, they are not conclusive. Experiments must be done before one can properly assess cause and effect—in this case, the effects claimed for marijuana. This is very important to you, so you decide to embark on a research program directed to this end. The first experiment you plan is to investigate the effect of marijuana on appetite in AIDS patients. Of course, if marijuana actually decreases appetite rather than increases it, you want to be able to detect this as well because it has important practical consequences. Therefore, this will be a basic fact-finding experiment in which you attempt to determine whether marijuana has any effect at all, either to increase or to decrease appetite. The first experiment will be a modest one. You plan to randomly sample 10 individuals from the population of AIDS patients who are being treated at your hospital. You realize that the generalization will be limited to this population, but for many reasons, you are willing to accept this limitation for this initial experiment. After getting permission from the appropriate authorities, you conduct the following experiment.

A random sample of 10 AIDS patients who agree to participate in the experiment is selected from a rather large population of AIDS patients being treated on an outpatient basis at your hospital. None of the patients in this population are being treated with marijuana. Each patient is admitted to the hospital for a week to participate in the experiment. The first 2 days are used to allow each patient to get used to the hospital. On the third day, half of the patients receive a pill containing a synthetic form of marijuana's active ingredient, THC, prior to eating each meal, and on the sixth day, they receive a placebo pill before each meal. The other half of the patients are treated the same as in the experimental condition, except that they receive the pills in the reverse order, that is, the placebo pills on the third day and the THC pills on the sixth day. The dependent variable is the amount of food eaten by each patient on day 3 and day 6.

In this experiment, each subject is tested under two conditions: an experimental condition and a control condition. We have labeled the condition in which the subject receives the THC pills as the experimental condition and the condition in which the subject receives the placebo pills as the control condition. Thus, there are two scores for each subject: the amount of food eaten (calories) in the experimental condition and the amount of food eaten in the control condition. If marijuana really does affect appetite, we would expect different scores for the two conditions. For example, if marijuana increases appetite, then more food should be eaten in the experimental condition. If the control score for each subject is subtracted from the experimental score, we would expect a predominance of positive difference scores. The results of the experiment are given in Table 10.1.

These data could be analyzed with several different statistical inference tests such as the *sign test*, Wilcoxon matched-pairs signed ranks test, and Student's *t* test for correlated groups. The choice of which test to use in an actual experiment is an important one. It depends on the sensitivity of the test and on whether the data of the experiment meet the assumptions of the test. We shall discuss each of these points in subsequent chapters. In this chapter, we shall analyze the data of your experiment with the sign test. We have chosen the sign test because (1) it is easy to understand and (2) all of the major concepts concerning hypothesis testing can be illustrated clearly and simply.

The sign test ignores the magnitude of the difference scores and considers only their direction or sign. This omits a lot of information, which makes the test

table 10.1 Results of the marijuana experiment

Patient No.	Experimental Condition THC Pill Food Eaten (calories)	Control Condition Placebo Pill Food Eaten (calories)	Difference Score (calories)
1	1325	1012	+313
2	1350	1275	+ 75
3	1248	950	+298
4	1087	840	+247
5	1047	942	+105
6	943	860	+ 83
7	1118	1154	− 36
8	908	763	+145
9	1084	920	+164
10	1088	876	+212

rather insensitive (but much easier to understand). If we consider only the signs of the difference scores, then your experiment produced 9 out of 10 pluses. The amount of food eaten in the experimental condition was greater after taking the THC pill in all but one of the patients. Are we therefore justified in concluding that marijuana produces an increase in appetite? Not necessarily.

Suppose that marijuana has absolutely no effect on appetite. Isn't it still possible to have obtained 9 out of 10 pluses in your experiment? Yes, it is. If marijuana has no effect on appetite, then each subject would have received two conditions that were identical except for chance factors. Perhaps when subject 1 was run in the THC condition, he had slept better the night before and his appetite was higher than when run in the control condition before any pills were taken. If so, we would expect him to eat more food in the THC condition even if THC has no effect on appetite. Perhaps subject 2 had a cold when run in the placebo condition, which blunted her appetite relative to when run in the experimental condition. Again we would expect more food to be eaten in the experimental condition even if THC has no effect.

We could go on giving examples for the other subjects. The point is that these explanations of the greater amount eaten in the THC condition are chance factors. They are different factors, independent of one another, and they could just as easily have occurred on either of the two test days. It seems unlikely to get 9 out of 10 pluses simply as a result of chance factors. The crucial question really is, "How unlikely is it?" Suppose we know that if chance alone is responsible, we shall get 9 out of 10 pluses only 1 time in 1 billion. This is such a rare occurrence, we would no doubt reject chance and, with it, the explanation that marijuana has no effect on appetite. We would then conclude by accepting the hypothesis that marijuana affects appetite because it is the only other possible explanation. Since the sample was a random one, we can assume it was representative of the AIDS patients being treated at your hospital, and we therefore would generalize the results to that population.

Suppose, however, that the probability of getting 9 out of 10 pluses due to chance alone is really 1 in 3, not 1 in 1 billion. Can we reject chance as a cause of the data? The decision is not as clear-cut this time. What we need is a rule for determining when the obtained probability is small enough to reject chance as an underlying cause. We shall see that this involves setting a critical probability level (called the alpha level) against which to compare the results.

Let's formalize some of the concepts we've been presenting.

Repeated Measures Design

The experimental design that we have been using is called the *repeated measures, replicated measures,* or *correlated groups design. The essential features are that there are paired scores in the conditions, and the differences between the paired scores are analyzed.* In the marijuana experiment, we used the same subjects in each condition. Thus, the subjects served as their own controls. Their scores were paired, and the differences between these pairs were analyzed. Instead of the same subjects, we could have used identical twins or subjects who were matched in some other way. In animal experimentation, littermates have often been used for pairing. The most basic form of this design employs just two conditions: an experimental and a control condition. The two conditions are kept as identical as possible except for values of the independent variable, which, of course, are intentionally made different. In our example, marijuana is the independent variable.

Alternative Hypothesis (H_1)

In any experiment, there are two hypotheses that compete for explaining the results: the *alternative hypothesis* and the *null hypothesis*. The alternative hypothesis is the one that claims the difference in results between conditions is due to the independent variable. In this case, it is the hypothesis that claims "marijuana affects appetite." The alternative hypothesis can be *directional* or *nondirectional*. The hypothesis "marijuana affects appetite" is nondirectional because it does not specify the direction of the effect. If the hypothesis specifies the direction of the effect, it is a directional hypothesis. "Marijuana increases appetite" is an example of a directional alternative hypothesis.

Null Hypothesis (H_0)

The null hypothesis is set up to be the logical counterpart of the alternative hypothesis such that if the null hypothesis is false, the alternative hypothesis must be true. Therefore, these two hypotheses must be mutually exclusive and exhaustive. If the alternative hypothesis is nondirectional, it specifies that the independent variable has an effect on the dependent variable. For this nondirectional alternative hypothesis, the null hypothesis asserts that the independent variable has *no* effect on the dependent variable. In the present example, since the alternative hypothesis is nondirectional, the null hypothesis specifies that "marijuana does not affect appetite." We pointed out previously that the alternative hypothesis specifies "marijuana affects appetite." You can see that these two hypotheses are mutually exclusive and exhaustive. If the null hypothesis is false, then the alternative hypothesis must be true. As you will see, we always first evaluate the null hypothesis and try to show that it is false. If we can show it to be false, then the alternative hypothesis must be true.*

MENTORING TIP

Be sure you understand that for a directional H_1, the null hypothesis doesn't predict just "no effect." Rather, it predicts "no effect or a real effect in the direction opposite to that predicted by H_1."

If the alternative hypothesis is directional, the null hypothesis asserts that the independent variable does not have an effect in the direction specified by the alternative hypothesis; it either has no effect or an effect in the direction opposite to H_1.† For example, for the alternative hypothesis "marijuana increases appetite," the null hypothesis asserts that "marijuana either has no effect on appetite, or it increases appetite." Again, note that the two hypotheses are mutually exclusive and exhaustive. If the null hypothesis is false, then the alternative hypothesis must be true.

Decision Rule (α Level)

We always evaluate the results of an experiment by assessing the null hypothesis. The reason we directly assess the null hypothesis instead of the alternative hypothesis is that we can calculate the probability of chance events, but there is no way to calculate the probability of the alternative hypothesis. We evaluate the null hypothesis by assuming it is true and testing the reasonableness of this assumption by calculating the probability of getting the results *if chance alone is operating*. If the obtained probability turns out to be equal to or less than a critical probability level called the *alpha (α) level*, we reject the null hypothesis. Rejecting the null hypothesis allows us, then, to accept indirectly the alternative hypothesis because, if the experiment is done properly, it is the only other possible explanation. When we reject H_0, we say

*See Note 10.1.
†See Note 10.2.

MENTORING TIP

Caution: if the obtained probability $> \alpha$, it is incorrect to conclude by "accepting H_0." The correct conclusion is "retain H_0" or "fail to reject H_0." You will learn why in Chapter 11.

the results are *significant* or *reliable*. If the obtained probability is greater than the alpha level, we conclude by *failing to reject H_0*. Since the experiment does not allow rejection of H_0, we *retain H_0*, as a reasonable explanation of the data. Throughout the text, we shall use the expressions "failure to reject H_0" and "retain H_0" interchangeably. When we retain H_0, we say the results are not significant or reliable. Of course, when the results are not significant, we cannot accept the alternative hypothesis. Thus, the decision rule states:

If the obtained probability $\leq \alpha$, reject H_0.
If the obtained probability $> \alpha$, fail to reject H_0, retain H_0.

The alpha level is set at the beginning of the experiment. Commonly used alpha levels are $\alpha = 0.05$ and $\alpha = 0.01$. Later in this chapter, we shall discuss the rationale underlying the use of these levels.

For now let's assume $\alpha = 0.05$ for the marijuana data. Thus, to evaluate the results of the marijuana experiment, we need to (1) determine the probability of getting 9 out of 10 pluses if chance alone is responsible and (2) compare this probability with alpha.

Evaluating the Marijuana Experiment

The data of this experiment fit the requirements for the binomial distribution. The experiment consists of a series of trials (the exposure of each patient to the experimental and control conditions is a trial). On each trial, there are only two possible outcomes: a plus and a minus. Note that this model does not allow ties. If any ties occur, they must be discarded and the N reduced accordingly. The outcomes are mutually exclusive (a plus and a minus cannot occur simultaneously), there is independence between trials (the score of patient 1 in no way influences the score of patient 2, etc.), and the probability of a plus and the probability of a minus stay the same from trial to trial. Since the binomial distribution is appropriate, we can use Table B in Appendix D (Table 10.2) to determine the probability of getting 9 pluses out of 10 trials when chance alone is responsible. We solve this problem in the same way we did with the coin-flipping problems in Chapter 9.

Given there are 10 patients, $N = 10$. We can let $P =$ the probability of getting a plus with any patient.* If chance alone is operating, the probability of a plus is equal to the probability of a minus. There are only two equally likely alternatives, so $P = 0.50$. Since we want to determine the probability of 9 pluses, the number of P events = 9. In Table B under $N = 10$, number of P events = 9, and $P = 0.50$, we obtain

$p(9 \text{ pluses}) = 0.0098$

table 10.2 Table B entry

N	No. of P Events	P 0.50
10	9	0.0098

*Throughout this chapter and the next, whenever using the sign test, we shall always let $P =$ the probability of a *plus* with any subject. This is arbitrary; we could have chosen Q. However, using the same letter (P or Q) to designate the probability of a plus for all problems does avoid unnecessary confusion.

Alpha has been set at 0.05. The analysis shows that only 98 times in 10,000 would we get 9 pluses if chance alone is the cause. Since 0.0098 is lower than alpha, we *reject the null hypothesis.** It does not seem to be a reasonable explanation of the data. Therefore, we conclude by accepting the alternative hypothesis that marijuana affects appetite. It appears to increase it. Since the sample was randomly selected, we assume the sample is representative of the population. Therefore, it is legitimate to assume that this conclusion applies to the *population* of AIDS patients being treated at your hospital.

It is worth noting that very often in practice the results of an experiment are generalized to groups that were not part of the population from which the sample was taken. For instance, on the basis of this experiment, we might be tempted to claim that marijuana would increase the appetites of AIDS patients being treated at other hospitals. Strictly speaking, the results of an experiment apply only to the population from which the sample was randomly selected. Therefore, generalization to other groups should be made with caution. This caution is necessary because the other groups may differ from the subjects in the original population in some way that would cause a different result. Of course, as the experiment is replicated in different hospitals with different patients, the legitimate generalization becomes much broader.

TYPE I AND TYPE II ERRORS

When making decisions regarding the null hypothesis, it is possible to make errors of two kinds. These are called *Type I* and *Type II errors.*

definitions   ■   *A* **Type I error** *is defined as a decision to reject the null hypothesis when the null hypothesis is true. A* **Type II error** *is defined as a decision to retain the null hypothesis when the null hypothesis is false.*

To illustrate these concepts, let's return to the marijuana example. Recall the logic of the decision process. First, we assume H_0 is true and evaluate the probability of getting the obtained score differences between conditions if chance alone is responsible. If the obtained probability $\leq \alpha$, we reject H_0. If the obtained probability $> \alpha$, we retain H_0. In the marijuana experiment, the obtained probability $[p(9 \text{ pluses})] = 0.0098$. Since this was lower than alpha, we rejected H_0 and concluded that marijuana was responsible for the results. Can we be certain that we made the correct decision? How do we know that chance wasn't really responsible? Perhaps the null hypothesis is really true. Isn't it possible that this was one of those 98 times in 10,000 we would get 9 pluses and 1 minus if chance alone was operating? *The answer is that we never know for sure that chance wasn't responsible.* It is possible that the 9 pluses and 1 minus were really due to chance.

---

*This is really a simplification made here for clarity. In actual practice, we evaluate the probability of getting the obtained result *or any more extreme.* The point is discussed in detail later in this chapter in the section titled "Evaluating the Tail of the Distribution."

table 10.3 Possible conclusions and the state of reality

	State of Reality	
Decision	H_0 *is true*	H_0 *is false*
Retain H_0	[1]Correct decision	[2]Type II error
Reject H_0	[3]Type I error	[4]Correct decision

If so, then we made an error by rejecting H_0. This is a Type I error—a rejection of the null hypothesis when it is true.

A Type II error occurs when we retain H_0 and it is false. Suppose that in the marijuana experiment p(9 pluses) = 0.2300 instead of 0.0098. In this case, 0.2300 > α, so we would retain H_0. If H_0 is false, we have made a Type II error, that is, retaining H_0 when it is false.

To help clarify the relationship between the decision process and possible error, we've summarized the possibilities in Table 10.3. The column heading is *State of Reality*. This means the correct state of affairs regarding the null hypothesis. There are only two possibilities. Either H_0 is true or it is false. The row heading is the decision made when analyzing the data. Again, there are only two possibilities. Either we reject H_0 or we retain H_0. If we retain H_0 and H_0 is true, we've made a correct decision (see the first cell in the table). If we reject H_0 and H_0 is true, we've made a Type I error. This is shown in cell 3. If we retain H_0 and it is false, we've made a Type II error (cell 2). Finally, if we reject H_0 and H_0 is false, we've made a correct decision (cell 4). Note that when we reject H_0, the only possible error is a Type I error. If we retain H_0, the only error we may make is a Type II error.

You may wonder why we've gone to the trouble of analyzing all the logical possibilities. We've done so because it is very important to know the possible errors we may be making when we draw conclusions from an experiment. From the preceding analysis, we know there are only two such possibilities, a Type I error or a Type II error. Knowing these are possible, we can design experiments before conducting them to help minimize the probability of making a Type I or a Type II error. By minimizing the probability of making these errors, we maximize the probability of concluding correctly, regardless of whether the null hypothesis is true or false. We shall see in the next section that alpha limits the probability of making a Type I error. Therefore, by controlling the alpha level we can minimize the probability of making a Type I error. *Beta* (read "bayta") is defined as the probability of making a Type II error. We shall discuss ways to minimize beta in the next chapter.

ALPHA LEVEL AND THE DECISION PROCESS

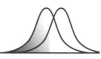

It should be clear that whenever we are using sample data to evaluate a hypothesis, we are never certain of our conclusion. When we reject H_0, we don't know for sure that it is false. We take the risk that we may be making a Type I error. Of course, the less reasonable it is that chance is the cause of the results, the more confident we are that we haven't made an error by rejecting the null hypothesis. For example, when the probability of getting the results is 1 in 1 million ($p = 0.000001$) under the assumption of chance, we are more confident that the null hypothesis is false than when the probability is 1 in 10 ($p = 0.10$).

definition   ■ *The **alpha level** that the scientist sets at the beginning of the experiment is the level to which he or she wishes to limit the probability of making a Type I error.*

Thus, when a scientist sets $\alpha = 0.05$, he is in effect saying that when he collects the data he will reject the null hypothesis if, under the assumption that chance alone is responsible, the obtained probability is equal to or less than 5 times in 100. In so doing, he is saying that he is willing to limit the probability of rejecting the null hypothesis when it is true to 5 times in 100. Thus, he limits the probability of making a Type I error to 0.05.

There is no magical formula that tells us what the alpha level should be to arrive at truth in each experiment. To determine a reasonable alpha level for an experiment, we must consider the consequences of making an error. In science, the effects of rejecting the null hypothesis when it is true (Type I error) are costly. When a scientist publishes an experiment in which he rejects the null hypothesis, other scientists either attempt to replicate the results or accept the conclusion as valid and design experiments based on the scientist having made a *correct decision*. Since many work-hours and dollars go into these follow-up experiments, scientists would like to minimize the possibility that they are pursuing a false path. Thus, they set rather conservative alpha levels: $\alpha = 0.05$ and $\alpha = 0.01$ are commonly used. You might ask, "Why not set even more stringent criteria, such as $\alpha = 0.001$?" Unfortunately, when alpha is made more stringent, the probability of making a Type II error increases.

We can see this by considering an example. This example is best understood in conjunction with Table 10.4. Suppose we do an experiment and set $\alpha = 0.05$ (top row of Table 10.4). We evaluate chance and get an obtained probability of 0.02. We reject H_0. If H_0 is true, we have made a Type I error (cell 1). Suppose, however, that alpha had been set at $\alpha = 0.01$ instead of 0.05 (bottom row of Table 10.4). In this case, we would retain H_0 and no longer would be making a Type I error (cell 3). Thus, the more stringent the alpha level, the lower the probability of making a Type I error.

On the other hand, what happens if H_0 is really false (last column of the table)? With $\alpha = 0.05$ and the obtained probability = 0.02, we would reject H_0 and thereby

table 10.4   Effect on beta of making alpha more stringent

Alpha Level	Obtained Probability	Decision	State of Reality	
			H_0 *is true*	H_0 *is false*
0.05	0.02	Reject H_0	[1]Type I error	[2]Correct decision
0.01	0.02	Retain H_0	[3]Correct decision	[4]Type II error

make a correct decision (cell 2). However, if we changed alpha to $\alpha = 0.01$, we would retain H_0 and we would make a Type II error (cell 4). Thus, making alpha more stringent decreases the probability of making a Type I error but increases the probability of making a Type II error. Because of this interaction between alpha and beta, the alpha level chosen for an experiment depends on the intended use of the experimental results. As mentioned previously, if the results are to communicate a new fact to the scientific community, the consequences of a Type I error are great, and therefore stringent alpha levels are used (0.05 and 0.01). If, however, the experiment is exploratory in nature and the results are to guide the researcher in deciding whether to do a full-fledged experiment, it would be foolish to use such stringent levels. In such cases, alpha levels as high as 0.10 or 0.20 are often used.

Let's consider one more example. Imagine you are the president of a drug company. One of your leading biochemists rushes into your office and tells you that she has discovered a drug that increases memory. You are of course elated, but you still ask to see the experimental results. Let's assume it will require a $30 million outlay to install the apparatus to manufacture the drug. This is quite an expense, but if the drug really does increase memory, the potential benefits and profits are well worth it. In this case, you would want to be very sure that the results are not due to chance. The consequences of a Type I error are great. You stand to lose $30 million. You will probably want to use an extremely stringent alpha level before deciding to reject H_0 and risk the $30 million.

We hasten to reassure you that *truth* is not dependent on the alpha level used in an experiment. Either marijuana affects appetite or it doesn't. Either the drug increases memory or it doesn't. Setting a stringent alpha level merely diminishes the possibility that we shall conclude for the alternative hypothesis when the null hypothesis is really true.

Since we never know for sure what the real truth is as a result of a single experiment, replication is a necessary and essential part of the scientific process. Before an "alleged fact" is accepted into the body of scientific knowledge, it must be demonstrated independently in several laboratories. The probability of making a Type I error decreases greatly with independent replication.

EVALUATING THE TAIL OF THE DISTRIBUTION

In the previous discussion, the obtained probability was found by using just the specific outcome of the experiment (i.e., 9 pluses and 1 minus). However, we did that to keep things simple for clarity when presenting the other major concepts. In fact, it is incorrect to use *just the specific outcome* when evaluating the results of an experiment. Instead, we must determine the probability of getting the obtained outcome or *any outcome even more extreme*. It is this probability that we compare with alpha to assess the reasonableness of the null hypothesis. In other words, we evaluate the *tail* of the distribution, beginning with the obtained result, rather than just the obtained result itself. If the alternative hypothesis is nondirectional, we evaluate the obtained result or any even more extreme in both directions (both tails). If the alternative hypothesis is directional, we evaluate only the tail of the distribution that is in the direction specified by H_1.

To illustrate, let's again evaluate the data in the present experiment, this time evaluating the tails rather than just the specific outcome. Figure 10.1 shows the

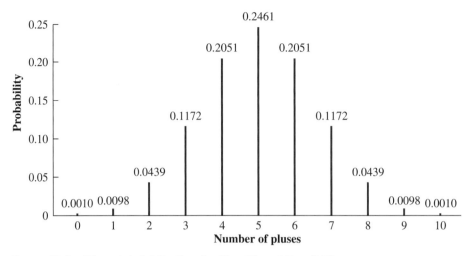

figure 10.1 Binomial distribution for N = 10 and P = 0.50.

binomial distribution for N = 10 and P = 0.50. The distribution has two tails, one containing few pluses and one containing many pluses. The alternative hypothesis is nondirectional, so to calculate the obtained probability, we must determine the probability of getting the obtained result or a result even more extreme *in both direc-tions.* Since the obtained result was 9 pluses, we must include outcomes as extreme as or more extreme than 9 pluses. From Figure 10.1, we can see that the outcome of 10 pluses is more extreme in one direction and the outcomes of 1 plus and 0 pluses are as extreme or more extreme in the other direction. Thus, the obtained probability is as follows:

$$p(0, 1, 9, \text{ or } 10 \text{ pluses}) = p(0) + p(1) + p(9) + p(10)$$
$$= 0.0010 + 0.0098 + 0.0098 + 0.0010$$
$$= 0.0216$$

It is this probability (0.0216, not 0.0098) that we compare with alpha to reject or retain the null hypothesis. This probability is called a *two-tailed probability* value because the outcomes we evaluate occur under both tails of the distribution. Thus, alternative hypotheses that are nondirectional are evaluated with two-tailed prob-ability values. If the alternative hypothesis is nondirectional, the alpha level must also be two-tailed. If $\alpha = 0.05_{2 \text{ tail}}$, this means that the two-tailed obtained prob-ability value must be equal to or less than 0.05 to reject H_0. In this example, 0.0216 is less than 0.05, so we reject H_0 and conclude as we did before that marijuana affects appetite.

If the alternative hypothesis is directional, we evaluate the tail of the distribution that is in the direction predicted by H_1. To illustrate this point, suppose the alternative hypothesis was that "marijuana increases appetite" and the obtained result was 9 pluses and 1 minus. Since H_1 specifies that marijuana increases appetite, we evaluate the tail with the higher number of pluses. Remember that a plus means more food eaten in the marijuana condition. Thus, if marijuana increases appetite, we expect mostly pluses.

The outcome of 10 pluses is the only possible result in this direction more extreme than 9 pluses. The obtained probability is

$$p(9 \text{ or } 10 \text{ pluses}) = 0.0098 + 0.0010$$
$$= 0.0108$$

This probability is called a *one-tailed probability* because all of the outcomes we are evaluating are under one tail of the distribution. Thus, alternative hypotheses that are directional are evaluated with one-tailed probabilities. If the alternative hypothesis is directional, the alpha level must be one-tailed. Thus, directional alternative hypotheses are evaluated against one-tailed alpha levels. In this example, if $\alpha = 0.05_{1 \text{ tail}}$, we would reject H_0 because 0.0108 is less than 0.05.

The reason we evaluate the tail has to do with the alpha level set at the beginning of the experiment. In the example we have been using, suppose the hypothesis is that "marijuana increases appetite." This is a directional hypothesis, so a one-tailed evaluation is appropriate. Assume $N = 10$ and $\alpha = 0.05_{1 \text{ tail}}$. By setting $\alpha = 0.05$ at the beginning of the experiment, the researcher desires to limit the probability of a Type I error to 5 in 100. Suppose the results of the experiment turn out to be 8 pluses and 2 minuses. Is this a result that allows rejection of H_0 consistent with the alpha level? Your first impulse is no doubt to answer "yes" because $p(8 \text{ pluses}) = 0.0439$. However, if we reject H_0 with 8 pluses, we must also reject it if the results are 9 or 10 pluses. Why? Because these outcomes are even more favorable to H_1 than 8 pluses and 2 minuses. Certainly, if marijuana really does increase appetite, obtaining 10 pluses and 0 minuses is better evidence than 8 pluses and 2 minuses, and similarly for 9 pluses and 1 minus. Thus, if we reject with 8 pluses, we must also reject with 9 and 10 pluses. But what is the probability of getting 8, 9, or 10 pluses if chance alone is operating?

$$p(8, 9, \text{ or } 10 \text{ pluses}) = p(8) + p(9) + p(10)$$
$$= 0.0439 + 0.0098 + 0.0010$$
$$= 0.0547$$

The probability is greater than alpha. Therefore, we can't allow 8 pluses to be a result for which we could reject H_0; the probability of falsely rejecting H_0 would be greater than the alpha level. Note that this is true even though the probability of 8 pluses itself is less than alpha. Therefore, we don't evaluate the exact outcome, but rather we evaluate the tail so as to limit the probability of a Type I error to the alpha level set at the beginning of the experiment. The reason we use a two-tailed evaluation with a nondirectional alternative hypothesis is that results at both ends of the distribution are legitimate candidates for rejecting the null hypothesis.

ONE- AND TWO-TAILED PROBABILITY EVALUATIONS

When setting the alpha level, we must decide whether the probability evaluation should be one- or two-tailed. When making this decision, use the following rule:

The evaluation should always be two-tailed unless the experimenter will retain H_0 when results are extreme in the direction opposite to the predicted direction.

In following this rule, there are two situations commonly encountered that warrant directional hypotheses. First, when it makes no practical difference if the results turn out to be in the opposite direction, it is legitimate to use a directional hypothesis and a one-tailed evaluation. For example, if a manufacturer of automobile tires is testing a new type of tire that is supposed to last longer, a one-tailed evaluation is legitimate because it doesn't make any practical difference if the experimental results turn out in the opposite direction. The conclusion will be to retain H_0, and the manufacturer will continue to use the old tires. Another situation in which it seems permissible to use a one-tailed evaluation is when there are good theoretical reasons, as well as strong supporting data, to justify the predicted direction. In this case, if the experimental results turn out to be in the opposite direction, the experimenter again will conclude by retaining H_0 (at least until the experiment is replicated) because the results fly in the face of previous data and theory.

In situations in which the experimenter will reject H_0 if the results of the experiment are extreme in the direction opposite to the prediction direction, a two-tailed evaluation should be used. To understand why, let's assume the researcher goes ahead and uses a directional prediction, setting $\alpha = 0.05_{1\,\text{tail}}$, and the results turn out to be extreme in the opposite direction. If he is unwilling to conclude by retaining H_0, what he will probably do is shift, after seeing the data, to using a nondirectional hypothesis employing $\alpha = 0.05_{2\,\text{tail}}$ (0.025 under each tail) to be able to reject H_0. In the long run, following this procedure will result in a Type I error probability of 0.075 (0.05 under the tail in the predicted direction and 0.025 under the other tail). Thus, switching alternative hypotheses after seeing the data produces an inflated Type I error probability. It is of course even worse if, after seeing that the data are in the direction opposite to that predicted, the experimenter switches to $\alpha = 0.05_{1\,\text{tail}}$ in the direction of the outcome so as to reject H_0. In this case, the probability of a Type I error, in the long run, would be 0.10 (0.05 under each tail). For example, an experimenter following this procedure for 100 experiments, assuming all involved true null hypotheses, would be expected to falsely reject the null hypothesis 10 times. Since each of these rejections would be a Type I error, following this procedure leads to the probability of a Type I error of 0.10 (10/100 = 0.10). Therefore, to maintain the Type I error probability at the desired level, it is important to decide at the beginning of the experiment whether H_1 should be directional or nondirectional and to set the alpha level accordingly. If a directional H_1 is used, the predicted direction must be adhered to, even if the results of the experiment turn out to be extreme in the opposite direction. Consequently, H_0 must be retained in such cases.

For solving the problems and examples contained in this textbook, we shall indicate whether a one- or two-tailed evaluation is appropriate; we would like you to practice both. Be careful when solving these problems. When a scientist conducts an experiment, he or she is often following a hunch that predicts a directional effect. The problems in this textbook are often stated in terms of the scientist's directional hunch. Nonetheless, *unless the scientist will conclude by retaining H_0 if the results turn out to be extreme in the opposite direction*, he or she should use a nondirectional H_1 and a two-tailed evaluation, even though his or her hunch is directional. Each textbook problem will tell you whether you should use a nondirectional or directional H_1 when it asks for the alternative hypothesis. If you are asked for a nondirectional H_1, you should assume that the appropriate criterion for a directional alternative hypothesis has not been met, regardless of whether the scientist's hunch in the problem is directional. If you are asked for a directional H_1, assume that the appropriate criterion has been met and it is proper to use a directional H_1.

We are now ready to do a complete problem in exactly the same way any scientist would if he or she were using the sign test to evaluate the data.

MENTORING TIP

Caution: when answering any of the end-of-chapter problems, use the direction specified by the H_1 or alpha level given in the problem to determine if the evaluation is to be one-tailed or two-tailed.

Practice Problem 10.1

Assume we have conducted an experiment to test the hypothesis that marijuana affects the appetites of AIDS patients. The procedure and population are the same as we described previously, except this time we have sampled 12 AIDS patients. The results are shown here (the scores are in calories):

Condition	Patient											
	1	**2**	**3**	**4**	**5**	**6**	**7**	**8**	**9**	**10**	**11**	**12**
THC	1051	1066	963	1179	1144	912	1093	1113	985	1271	978	951
Placebo	872	943	912	1213	1034	854	1125	1042	922	1136	886	902

a. What is the nondirectional alternative hypothesis?
b. What is the null hypothesis?
c. Using $\alpha = 0.05_{2\,tail}$, what do you conclude?
d. What error might you be making by your conclusion in part **c**?
e. To what population does your conclusion apply?

The solution follows.

SOLUTION

a. Nondirectional alternative hypothesis: Marijuana affects appetites of AIDS patients who are being treated at your hospital.
b. Null hypothesis: Marijuana has no effect on appetites of AIDS patients who are being treated at your hospital.
c. Conclusion, using $\alpha = 0.05_{2\,tail}$:

STEP 1: **Calculate the number of pluses and minuses.** The first step is to calculate the number of pluses and minuses in the sample. We have subtracted the "placebo" scores from the corresponding "THC" scores. The reverse could also have been done. There are 10 pluses and 2 minuses.

STEP 2: **Evaluate the number of pluses and minuses.** Once we have calculated the obtained number of pluses and minuses, we must determine the probability of getting this outcome or any even more extreme in both directions because this is a two-tailed evaluation. The binomial distribution is appropriate for this determination. N = the number of difference scores (pluses and minuses) = 12. We can let P = the probability of a plus with any subject. If marijuana has no effect on appetite, chance alone accounts for whether any subject scores a plus or a minus. Therefore, $P = 0.50$. The obtained result was 10 pluses and 2 minuses, so the number of P events = 10. The probability of getting an outcome as extreme as or more

(continued)

extreme than 10 pluses (two-tailed) equals the probability of 0, 1, 2, 10, 11, or 12 pluses. Since the distribution is symmetrical, p(0, 1, 2, 10, 11, or 12 pluses) equals p(10, 11, or 12 pluses) $\times$ 2. Thus, from Table B:

Table B entry

N	No. of P Events	P 0.50
12	10	0.0161
	11	0.0029
	12	0.0002

p(0, 1, 2, 10, 11 or 12 pluses)

$$= p(10, 11, \text{ or } 12 \text{ pluses}) \times 2$$
$$= [p(10) + p(11) + p(12)] \times 2$$
$$= (0.0161 + 0.0029 + 0.0002) \times 2$$
$$= 0.0384$$

The same value would have been obtained if we had added the six probabilities together rather than finding the one-tailed probability and multiplying by 2. Since 0.0384 < 0.05, we reject the null hypothesis. It is not a reasonable explanation of the results. Therefore, we conclude that marijuana affects appetite. It appears to increase it.

d. Possible error: By rejecting the null hypothesis, you might be making a Type I error. In reality, the null hypothesis may be true and you have rejected it.

e. Population: These results apply to the population of AIDS patients from which the sample was taken.

Practice Problem 10.2

You have good reason to believe a particular TV program is causing increased violence in teenagers. To test this hypothesis, you conduct an experiment in which 15 individuals are randomly sampled from the teenagers attending your neighborhood high school. Each subject is run in an experimental and a control condition. In the experimental condition, the teenagers watch the TV program for 3 months, during which you record the number of violent acts committed. The control condition also lasts for 3 months, but the teenagers are not allowed to watch the program during this period. At the end of each 3-month period, you total the number of violent acts committed. The results are given here:

Condition	Subject 1	2	3	4	5	6	7	8	9	10	11	12	13	14	15
Viewing the program	25	35	10	8	24	40	44	18	16	25	32	27	33	28	26
Not viewing the program	18	22	7	11	13	35	28	12	20	18	38	24	27	21	22

a. What is the directional alternative hypothesis?
b. What is the null hypothesis?
c. Using $\alpha = 0.01_{1\ tail}$, what do you conclude?
d. What error might you be making by your conclusion in part **c**?
e. To what population does your conclusion apply?

The solution follows.

SOLUTION

a. Directional alternative hypothesis: Watching the TV program causes increased violence in teenagers.
b. Null hypothesis: Watching the TV program does not cause increased violence in teenagers.
c. Conclusion, using $\alpha = 0.01_{1\ tail}$:

STEP 1: **Calculate the number of pluses and minuses.** The first step is to calculate the number of pluses and minuses in the sample from the data. We have subtracted the scores in the "not viewing" condition from the scores in the "viewing" condition. The obtained result is 12 pluses and 3 minuses.

STEP 2: **Evaluate the number of pluses and minuses.** Next, we must determine the probability of getting this outcome or any even more extreme in the direction of the alternative hypothesis. This is a one-tailed evaluation because the alternative hypothesis is directional. The binomial distribution is appropriate. N = the number of difference scores = 15. Let P = the probability of a plus with any subject. We can evaluate the null hypothesis by assuming chance alone accounts for whether any subject scores a plus or minus. Therefore, $P = 0.50$. The obtained result was 12 pluses and 3 minuses, so the number of P events = 12. The probability of 12 pluses or more equals the probability of 12, 13, 14, or 15 pluses. This can be found from Table B. Thus,

Table B entry

		No. of P	P
$p(12, 13, 14$ or 15 pluses$)$	N	Events	**0.50**
$= p(12) + p(13) + p(14) + p(15)$	15	12	0.0139
$= 0.0139 + 0.0032 + 0.0005 + 0.0000$		13	0.0032
$= 0.0176$		14	0.0005
		15	0.0000

Since $0.0176 > 0.01$, we fail to reject the null hypothesis. Therefore, we retain H_0 and cannot conclude that the TV program causes increased violence in teenagers.

d. Possible error: By retaining the null hypothesis, you might be making a Type II error. The TV program may actually cause increased violence in teenagers.
e. Population: These results apply to the population of teenagers attending your neighborhood school.

Practice Problem 10.3

A corporation psychologist believes that exercise affects self-image. To investigate this possibility, 14 employees of the corporation are randomly selected to participate in a jogging program. Before beginning the program, they are given a questionnaire that measures self-image. Then they begin the jogging program. The program consists of jogging at a moderately taxing rate for 20 minutes a day, 4 days a week. Each employee's self-image is measured again after 2 months on the program. The results are shown here (the higher the score, the higher the self-image); a score of 20 is the highest score possible.

Subject	Before Jogging	After Jogging	Subject	Before Jogging	After Jogging
1	14	20	8	16	13
2	13	16	9	10	16
3	8	15	10	14	18
4	14	12	11	6	14
5	12	15	12	15	17
6	7	13	13	12	18
7	10	12	14	9	15

a. What is the alternative hypothesis? Use a nondirectional hypothesis.
b. What is the null hypothesis?
c. Using $\alpha = 0.05_{2\,tail}$, what do you conclude?
d. What error might you be making by your conclusion in part **c**?
e. To what population does your conclusion apply?

The solution follows.

SOLUTION

a. Nondirectional alternative hypothesis: Jogging affects self-image.
b. Null hypothesis: Jogging has no effect on self-image.
c. Conclusion, using $\alpha = 0.05_{2\,tail}$:

STEP 1: **Calculate the number of pluses and minuses.** We have subtracted the "Before Jogging" from the "After Jogging" scores. There are 12 pluses and 2 minuses.

STEP 2: **Evaluate the number of pluses and minuses.** Because H_1 is nondirectional, we must determine the probability of getting a result as extreme as or more extreme than 12 pluses (two-tailed), assuming chance alone accounts for the differences. The binomial distribution is appropriate. $N = 14$, $P = 0.50$, and number of P events $= 0, 1, 2,$ 12, 13, or 14. Thus, from Table B:

$$p(0, 1, 2, 12, 13, \text{ or } 14 \text{ pluses}) = p(0) + p(1) + p(2) + p(12)$$
$$+ p(13) + p(14)$$
$$= 0.0001 + 0.0009 + 0.0056 + 0.0056$$
$$+ 0.0009 + 0.0001$$
$$= 0.0132$$

Table B entry

N	No. of P Events	P 0.50
14	0	0.0001
	1	0.0009
	2	0.0056
	12	0.0056
	13	0.0009
	14	0.0001

The same value would have been obtained if we had found the one-tailed probability and multiplied by 2. Since $0.0132 < 0.05$, we reject the null hypothesis. It appears that jogging improves self-image.

  d. Possible error: By rejecting the null hypothesis, you might be making a Type I error. The null hypothesis may be true, and it was rejected.
  e. Population: These results apply to all the employees of the corporation who were employed at the time of the experiment.

SIZE OF EFFECT: SIGNIFICANT VERSUS IMPORTANT

The procedure we have been following in assessing the results of an experiment is first to evaluate directly the null hypothesis and then to conclude indirectly with regard to the alternative hypothesis. If we are able to reject the null hypothesis, we say the results are *significant*. What we really mean by "significant" is "statistically significant." That is, the results are probably not due to chance, the independent variable has had a real effect, and if we repeat the experiment, we would again get results that would allow us to reject the null hypothesis. It might have been better to use the term *reliable* to convey this meaning rather than significant. However, the usage of *significant* is well established, so we will have to live with it. The point is that we must not confuse statistically significant with practically or theoretically "important." A statistically significant effect says little about whether the effect is an important one. For example, suppose the real effect of marijuana is to increase appetite by only 10 calories. Using careful experimental design and a large enough sample, it is possible that we would be able to detect even this small an effect.

MENTORING TIP

The importance of an effect generally depends on the size of the effect.

If so, we would conclude that the result is significant (reliable), but then we still need to ask, "How important is this real effect?" For most purposes, except possibly theoretical ones, the importance of an effect increases directly with the size of the effect. For further discussion of this point, see "What Is the Truth? Much Ado About Almost Nothing," in Chapter 15.

WHAT IS THE TRUTH?   Chance or Real Effect?—1

 An article appeared in *Time* magazine concerning the "Pepsi Challenge Taste Test." A Pepsi ad, shown on the facing page, appeared in the article. Taste Test participants were Coke drinkers from Michigan who were asked to drink from a glass of Pepsi and another glass of Coke and say which they preferred. To avoid obvious bias, the glasses were not labeled "Coke" or "Pepsi." Instead, to facilitate a "blind" administration of the drinks, the Coke glass was marked with a "Q" and the Pepsi glass with an "M." The results as stated in the ad are,

"More than half the Coca-Cola drinkers tested in Michigan preferred Pepsi." Aside from a possible real preference for Pepsi in the population of Michigan Coke drinkers, can you think of any other possible explanation of these sample results?

Answer The most obvious alternative explanation of these results is that they are due to chance alone; that in the population, the preference for Pepsi and Coke is equal ($P = 0.50$). You, of course, recognize this as the null hypothesis explanation. This explanation could and, in our opinion, should have been ruled out (within the limits of Type I error) by analyzing the sample data with the appropriate inference test. If the results really are significant, it doesn't take much space in an ad to say so. This ad is like many that state sample results favoring their product without evaluating chance as a reasonable explanation.

As an aside, Coke did not cry "chance alone," but instead claimed the study was invalid because people like the letter "M" better than "Q." Coke conducted a study to test its contention by putting Coke in both the "M" and "Q" glasses. Sure enough, more people preferred the drink in the "M" glass, even though it was Coke in both glasses. Pepsi responded by doing another Pepsi Challenge round, only this time revising the letters to "S" and "L," with Pepsi always in the "L" glass. The sample results again favored Pepsi. Predictably, Coke executives again cried foul, claiming an "L" preference. A noted motivational authority was then consulted and he reported that he knew of no studies showing a bias in favor of the letter "L." As a budding statistician, how might you design an experiment to determine whether there is a preference for Pepsi or Coke in the population and at the same time eliminate glass-preference as a possible explanation? ∎

Take the Pepsi Challenge.
Let your taste decide.

Pepsi-Cola's blind taste test.
Maybe you've seen "The Pepsi Challenge" on TV.
It's a simple, straightforward taste test where Coca-Cola
drinkers taste Coca-Cola and Pepsi
without knowing which is which.
Then we ask them which one they prefer.

More than half the Coca-Cola drinkers tested in Michigan preferred Pepsi.
Hundreds of Coca-Cola
drinkers from Michigan were
tested and we found that
more than half the people
tested preferred the taste
of Pepsi.

DAN SIMMONS

Let your taste decide.
We're not asking you to take our word for it. Or anyone
else's. Just try it yourself. Take The Pepsi Challenge
and let your taste decide.

WHAT IS THE TRUTH? | Chance or Real Effect?—2

The research reported in the previous "What Is the Truth?" section did not report any significance levels. Sometimes experiments are reported that do contain significance levels but that nonetheless raise suspicions of Type I error. Consider the following newspaper article on research done at a leading U.S. university.

STUDY: MILDLY DEPRESSED WOMEN LIVE LONGER

If you want to live a long, happy life, think again.

Researchers at Duke University followed more than 4,000 older people for a decade and found that women who lived longest were mildly depressed. The study is in the May-June issue of American Journal of Geriatric Psychiatry.

"We don't quite understand the finding," said Dr. Dan Blazer, a professor of psychiatry and behavioral science at Duke and co-author of the study. This jump in longevity was seen only in women. Women with mild depression, determined by a diagnostic questionnaire, were 40 percent less likely to die during the study period compared with women with no depression or those with more serious forms of depression. For men, mild depression was found to have no effect on mortality.

Blazer suspects mild forms of depression could protect against death by slowing people down, giving them more time to pay attention to their minds and bodies.

Author Unknown. Article from *The Coloradoan*, May 4, 2002, p. A3.

Although the newspaper article didn't state any significance levels, let's assume that in the *American Journal of Geriatric Psychiatry* article, it was stated that using $\alpha = 0.05$, the women with mild depression were *significantly* more likely to die during the study period compared with women with no depression or those with more serious forms of depression. Yes, this result reached significance at the 0.05 level, but it still seems questionable. First, this seems like a surprising, unplanned result that the authors admit they don't understand. Second, neither women with no depression nor women with more serious depression showed it. Finally, mildly depressed men didn't show the result either. What do you think, chance or real effect? The point being made is that just because a result is statistically significant, it does not automatically mean that the result is a real effect. It could have been a Type I error. Of course, replication would help resolve the issue. ■

WHAT IS THE TRUTH?   "No Product Is Better Than Our Product"

Often we see advertisements that present no data and make the assertion, "No product is better in doing X than our product." An ad regarding Excedrin, which was published in a national magazine, is an example of this kind of advertisement. The ad showed a large picture of a bottle of Excedrin tablets along with the statements,

> "Nothing you can buy is stronger."
> "Nothing you can buy works harder."
> "Nothing gives you bigger relief."

The question is, "How do we interpret these claims?" Do we rush out and buy Excedrin because it is stronger, works harder, and gives bigger relief than any other headache remedy available? If there are experimental data that form the basis of this ad's claims, we wonder what the results really are. What is your guess?

Answer Of course, we really don't know in every case, and therefore we don't intend our remarks to be directed at any specific ad. We have just chosen the Excedrin ad as an illustration of many such ads. However, we can't help but be suspicious that in most, if not all, cases where sample data exist, the actual data show that there is no significant difference in doing X between the

advertiser's product and the other products tested.

For the sake of discussion, let's call the advertiser's product "A." If the data had shown that "A" was better than the competing products, it seems reasonable that the advertiser would directly claim superiority for its product, rather than implying this indirectly through the weaker statement that no other product is better.

Why, then, would the advertiser make this weaker statement? Probably because the actual data do not show product "A" to be superior at all. Most likely, the sample data show product "A" to be either equal to or inferior to the others and the inference test shows no significant difference between the products. Given such data, rather than saying

that the research shows our product to be inferior or, at best, equal to the other products at doing X (which clearly would not sell a whole bunch of product "A"), the results are stated in this more positive, albeit, in our opinion, misleading way. Saying "No other product is better than ours in doing X" will obviously sell more products than "All products tested were equal in doing X." And after all, if you read the weaker statement closely, it does not really say that product "A" is superior to the others.

Thus, in the absence of reported data to the contrary, we believe the most accurate interpretation of the claim "No other competitor's product is superior to ours at doing X" is that *the products are equal at doing X.* ∎

WHAT IS THE TRUTH?   Anecdotal Reports Versus Systematic Research

The following article appeared in an issue of *The New York Times* on the Web. It is representative of a fairly common occurrence: individual testimony about a phenomenon conflicts with systematic research concerning the phenomenon. In this case, we are looking at the effects of the hormone secretin on autism. After reading the article, if you were asked if secretin has a beneficial effect on autism, how would you answer? Would you be satisfied with anecdotal reports? Would negative results from systematic research satisfy you? How could the issue be resolved?

A DRUG USED FOR AUTISM IS UNDER FIRE IN NEW STUDY

By SANDRA BLAKESLEE

A hormone trumpeted in the news media and on the Internet as a potential cure for autism worked no better than saltwater in its first controlled clinical trial, scientists are reporting.

The study, described in the is-sue of The New England Journal of Medicine *that is being published on Thursday, is one of a dozen efforts sponsored by the National Institutes of Health to test the hormone, se-cretin, which gained wide attention after a 3-year-old New Hampshire boy showed rapid improvement after taking the drug in 1996.*

"Secretin is in widespread use across the country," said Dr. Marie Bristol-Power, special assistant for autism programs at the Institute of Child Health and Human Develop-ment at the N.I.H. "Parents need to

know if it is promising and should be pursued or if they are being taken advantage of."

The study was not definitive, Dr. Bristol-Power said, and re-searchers will know more by the middle of next year, when other studies are completed.

Defenders of secretin were quick to denounce the new study, saying it did not give the drug a fair test.

"No one has ever claimed that secretin is a miracle cure for autism," said Victoria Beck, the mother of the boy who improved after taking it.

Ms. Beck, who has written a book on secretin, called the hor-mone "a clue that deserves careful investigation and not dismissal from autism gurus, who have a vested interest in the way things have been done in the past."

Autism is a serious brain disorder that begins in infancy and prevents children from developing normally. Symptoms, which can be mild to severe, include an inability to communicate, a refusal to make eye contact, repetitive behavior like head banging or hand flapping, and

a preoccupation with unusual activi-ties or interests.

Half a million Americans have the disorder, which is typically treat-ed with educational and behavioral therapy; while intensive therapy has proved effective in some patients, two-thirds of adults with autism cannot live independently.

Secretin is a hormone made in the intestine that stimulates the re-lease of digestive fluids. Ms. Beck's son, Parker, was given an injection to explore the cause of his chronic diarrhea. Soon afterward, his behav-ior improved; he could hold still and interact better with adults.

When his story was reported on television, in newspapers and on the Internet last year, parents began clamoring to get the drug.

Dr. Bristol-Power said that the supply of secretin was soon exhausted and that a black market developed involving sham secretin, price gouging and other forms of profiteering; Ms. Beck said that these stories were exaggerated and that no one today had any problem obtaining secretin at a fair price, which is about $200 an injection.

Anecdotes about secretin are thriving on the Internet. Some parents report dramatic effects, others see no effect and some report that their child's symptoms have worsened. Both sides in the secretin controversy estimate that 2,000 to 3,000 children have taken or are taking the drug.

The controlled study of secretin was led by Dr. James Bodfish of the Western Carolina Center and the University of North Carolina, and by Dr. Adrian Sandler of the Center for Child Development at Thoms Rehabilitation Hospital in Asheville, N.C.

In all, 56 autistic children 3 to 14 years of age took part in the study. Half received a single injection of synthetic human secretin, and the rest were injected with

saltwater. Researchers administered a battery of behavioral tests at intervals of one day, one week and four weeks after the treatment. Statistically, the two groups showed no differences, Dr. Sandler said.

A third of the children in both the placebo and treatment groups improved on some of the same measures. The only explanation, Dr. Sandler said, is a "significant placebo effect."

"Parents are extremely invested in the possibility of new treatments and have high expectancies," he said. "They are looking for subtle improvements. Kids with autism show variability in their day-to-day behavior and it would be easy to attribute normal variations to the secretin."

Ms. Beck disputes this interpretation, contending that by concentrating on standard behavioral tests the researchers have failed to look [at] a wide range of physiological responses seen in children given secretin, and that those should be studied.

"It really bothers me that parents are portrayed as this desperate group of people with vacant minds," she said. "They claim we can't tell the difference between an improvement and a mirage. But when a child stops vomiting, sleeps through the night and makes eye contact, we know something is happening."

∎ SUMMARY

In this chapter, I have discussed the topic of hypothesis testing, using the sign test as our vehicle. The sign test is used in conjunction with the repeated measures design. The essential features of the repeated measures design are that there are paired scores between conditions, and difference scores are analyzed.

In any hypothesis-testing experiment, there are always two hypotheses that compete to explain the results: the alternative hypothesis and the null hypothesis. The alternative hypothesis specifies that the independent variable is responsible for the differences in score values between the conditions. The alternative hypothesis may be directional or nondirectional. It is legitimate to use a directional hypothesis when there is a good theoretical basis and good supporting evidence in the literature. If the experiment is a basic fact-finding experiment, ordinarily a nondirectional hypothesis should be used. A directional alternative hypothesis is evaluated with a one-tailed probability value and a nondirectional hypothesis with a two-tailed probability value.

The null hypothesis is the logical counterpart to the alternative hypothesis such that if the null hypothesis is false, the alternative hypothesis must be true. If the alternative hypothesis is nondirectional, the null hypothesis specifies that the independent variable has no effect on the dependent variable. If the alternative hypothesis is directional, the null hypothesis states that the independent variable has no effect in the direction specified.

In evaluating the data from an experiment, we never directly evaluate the alternative hypothesis. We always first evaluate the null hypothesis. The null hypothesis is evaluated by assuming chance alone is responsible for the differences in scores between conditions. In doing this evaluation, we calculate the probability of getting the obtained result or a result even more extreme if chance alone is responsible. If this obtained probability is equal to or lower than the alpha level, we consider the null hypothesis explanation unreasonable and reject the null hypothesis. We conclude by accepting the alternative hypothesis because it is the only other explanation. If the obtained probability is greater than the alpha level, we retain the null hypothesis. It is still considered a reasonable explanation of the data. Of course, if the null hypothesis is not rejected, the alternative hypothesis cannot be accepted. The conclusion applies legitimately only to the population from which the sample was randomly drawn. We must be careful to distinguish "statistically significant" from practically or theoretically "important."

The alpha level is usually set at 0.05 or 0.01 to minimize the probability of making a Type I error. A Type I error occurs when the null hypothesis is rejected and it is actually true. The alpha level limits the probability of making a Type I error. It is also possible to make a Type II error. This occurs when we retain the null hypothesis and it is false. Beta is defined as the probability of making a Type II error. When alpha is made more stringent, beta increases. By minimizing alpha and beta, it is possible to have a high probability of correctly concluding from an experiment regardless of whether H_0 or H_1 is true. A significant result really says that it is a reliable result but gives little information about the size of the effect. The larger the effect, the more likely it is to be an important effect.

In analyzing the data of an experiment with the sign test, we ignore the magnitude of difference scores and just consider their direction. There are only two possible scores for each subject: a plus or a minus. We sum the pluses and minuses for all subjects, and the obtained result is the total number of pluses and minuses. To test the null hypothesis, we calculate the probability of getting the total number of pluses or a number of pluses even more extreme if chance alone is responsible. The binomial distribution, with P (the probability of a plus) $= 0.50$ and $N =$ the number of difference scores, is appropriate for making this determination. An illustrative problem and several practice problems were given to show how to evaluate the null hypothesis using the binomial distribution.

■ IMPORTANT NEW TERMS

Alpha (α) level (p. 252, 255)
Alternative hypothesis (H_1) (p. 252)
Beta (β) (p. 255)
Correct decision (p. 255)
Correlated groups design (p. 251)
Directional hypothesis (p. 252)
Fail to reject null hypothesis
  (p. 253)

Importance of an effect (p. 265)
Nondirectional hypothesis (p. 252)
Null hypothesis (H_0) (p. 252)
One-tailed probability (p. 259)
Reject null hypothesis (p. 254)
Repeated measures design (p. 251)
Replicated measures design (p. 251)
Retain null hypothesis (p. 252)

Sign test (p. 250)
Significant (p. 253, 265)
Size of effect (p. 265)
State of reality (p. 255)
Two-tailed probability (p. 258)
Type I error (p. 254)
Type II error (p. 254)

■ QUESTIONS AND PROBLEMS

Caution: Remember, when answering any of the end-of-chapter problems, in this and the remaining chapters, use the direction specified by the H_1 or α-level given in the problem to determine if the evaluation is to be 1-tailed or 2-tailed.

1. Briefly define or explain each of the terms in the Important New Terms section.

2. Briefly describe the process involved in hypothesis testing. Be sure to include the alternative hypothesis, the null hypothesis, the decision rule, the possible type of error, and the population to which the results can be generalized.

3. Explain in your own words why it is important to know the possible errors we might make when rejecting or failing to reject the null hypothesis.

4. Does the null hypothesis for a nondirectional H_1 differ from the null hypothesis for a directional H_1? Explain.

5. Under what conditions is it legitimate to use a directional H_1? Why is it not legitimate to use a directional H_1 just because the experimenter has a "hunch" about the direction?

6. If the obtained probability in an experiment equals 0.0200, does this mean that the probability that H_0 is true equals 0.0200? Explain.

7. Discuss the difference between "significant" and "important." Include "effect size" in your discussion.

8. What considerations go into determining the best alpha level to use? Discuss.

9. A primatologist believes that rhesus monkeys possess curiosity. She reasons that, if this is true, then they should prefer novel stimulation to repetitive stimulation. An experiment is conducted in which 12 rhesus monkeys are randomly selected from the university colony and taught to press two bars. Pressing bar 1 always produces the same sound, whereas bar 2 produces a

novel sound each time it is pressed. After learning to press the bars, the monkeys are tested for 15 minutes, during which they have free access to both bars. The number of presses on each bar during the 15 minutes is recorded. The resulting data are as follows:

Subject	Bar 1	Bar 2
1	20	40
2	18	25
3	24	38
4	14	27
5	5	31
6	26	21
7	15	32
8	29	38
9	15	25
10	9	18
11	25	32
12	31	28

a. What is the alternative hypothesis? In this case, assume a nondirectional hypothesis is appropriate because there is insufficient empirical basis to warrant a directional hypothesis.
b. What is the null hypothesis?
c. Using $\alpha = 0.05_{2 \text{ tail}}$, what is your conclusion?
d. What error might you be making by your conclusion in part **c**?
e. To what population does your conclusion apply? cognitive, biological

10. A school principal is interested in a new method for teaching eighth-grade social studies, which he believes will increase the amount of material learned. To test this method, the principal conducts the following experiment. The eighth-grade students in the school district are grouped into pairs based on matching their IQs and past grades. Twenty matched pairs are randomly selected for the experiment. One member of each pair is randomly assigned to a group that receives the new method, and the other member of each pair to a group that receives the standard instruction. At the end of the course, all students take a common final exam. The following are the results:

Pair No.	New Method	Standard Instruction
1	95	83
2	75	68
3	73	80
4	85	82
5	78	84
6	86	78
7	93	85
8	88	82
9	75	84
10	84	68
11	72	81
12	84	91
13	75	72
14	87	81
15	94	83
16	82	87
17	70	65
18	84	76
19	72	63
20	83	80

a. What is the alternative hypothesis? Use a directional hypothesis.
b. What is the null hypothesis?
c. Using $\alpha = 0.05_{1 \text{ tail}}$, what is your conclusion?
d. What error might you be making by your conclusion in part **c**?
e. To what population does your conclusion apply? education

11. A physiologist believes that the hormone angiotensin II is important in regulating thirst. To investigate this belief, she randomly samples 16 rats from the vivarium of the drug company where she works and places them in individual cages with free access to food and water. After they have grown acclimated to their new "homes," the experimenter measures the amount of water each rat drinks in a 20-minute period. Then she injects each animal intravenously with a known concentration (100 micrograms per kilogram) of angiotensin II. The rats are then put back into their home cages, and the amount each drinks for another

20-minute period is measured. The results are shown in the following table. Scores are in milliliters drunk per 20 minutes.

Subject	Before Injection	After Injection
1	1.2	11.3
2	0.8	10.7
3	0.5	10.3
4	1.3	11.5
5	0.6	9.6
6	3.5	3.3
7	0.7	10.5
8	0.4	11.4
9	1.1	12.0
10	0.3	12.8
11	0.6	11.4
12	0.3	9.8
13	0.5	10.6
14	4.1	3.2
15	0.4	12.1
16	1.0	11.2

a. What is the nondirectional alternative hypothesis?
b. What is the null hypothesis?
c. Using $\alpha = 0.05_{2 \text{ tail}}$, what is your conclusion? Assume the injection itself had no effect on drinking behavior.
d. What error might you be making by your conclusion in part **c**?
e. To what population does your conclusion apply? biological

12. A leading toothpaste manufacturer advertises that, in a recent medical study, 70% of the people tested had brighter teeth after using its toothpaste (called Very Bright) as compared to using the leading competitor's brand (called Brand X). The advertisement continues, "Therefore, use Very Bright and get brighter teeth." In point of fact, the data upon which these statements were based were collected from a random sample of 10 employees from the manufacturer's Pasadena plant. In the experiment, each employee used both toothpastes. Half of the employees used Brand X for 3 weeks, followed by Very Bright for the same time period. The other half used Very Bright first, followed by Brand X. A brightness test was given at the end of each 3-week period. Thus, there were two scores for each employee, one from the brightness test following the use of Brand X and one following the use of Very Bright. The following table shows the scores (the higher, the brighter):

Subject	Very Bright	Brand X
1	5	4
2	4	3
3	4	2
4	2	3
5	3	1
6	4	1
7	1	3
8	3	4
9	6	5
10	6	4

a. What is the alternative hypothesis? Use a directional hypothesis.
b. What is the null hypothesis?
c. Using $\alpha = 0.05_{1 \text{ tail}}$, what do you conclude?
d. What error might you be making by your conclusion in part **c**?
e. To what population does your conclusion apply?
f. Does the advertising seem misleading? I/O

13. A researcher is interested in determining whether acupuncture affects pain tolerance. An experiment is performed in which 15 students are randomly chosen from a large pool of university undergraduate volunteers. Each subject serves in two conditions. In both conditions, each subject receives a short-duration electric shock to the pulp of a tooth. The shock intensity is set to produce a moderate level of pain to the unanesthetized subject. After the shock is terminated, each subject rates the perceived level of pain on a scale of 0–10, with 10 being the highest level. In the experimental condition, each subject receives the appropriate acupuncture treatment prior to receiving the shock. The control condition is made as similar to the experimental condition as possible, except a placebo treatment is given instead of acupuncture. The two conditions are run on separate days at the same time of day. The pain ratings in the accompanying table are obtained.
a. What is the alternative hypothesis? Assume a nondirectional hypothesis is appropriate.

b. What is the null hypothesis?
c. Using $\alpha = 0.05_{2\,tail}$, what is your conclusion?
d. What error might you be making by your conclusion in part **c**?
e. To what population does your conclusion apply?

Subject	Acupuncture	Placebo
1	4	6
2	2	5
3	1	5
4	5	3
5	3	6
6	2	4
7	3	7
8	2	6
9	1	8
10	4	3
11	3	7
12	4	8
13	5	3
14	2	5
15	1	4

cognitive, health

What Is the Truth? Questions

1. *Chance or Real Effect? − 1*
   The main question to be answered is, "In the population, is there a preference for Coke or for Pepsi?"
   As a budding statistician, design an experiment that will answer this question and at the same time eliminate glass-preference as a possible explanation. Use the repeated measures design, and an N of 20 with each subject tasting both Coke and Pepsi once in separate tastings. For this question, make up the sample data such that the results are significant favoring Pepsi. Evaluate the data using the sign test with $\alpha = 0.05_{2\,tail}$. To make the analysis interesting, use sample scores such that the obtained probability exceeds alpha by the smallest amount possible with an N of 20.

2. *Chance or Real Effect? − 2*
   When we reject the null hypothesis, it is possible we made a Type I error. How can we reduce the probability of making a Type I error by manipulating the alpha level? Explain.

3. *"No Product Is Better Than Our Product"*
   If you read an advertisement that states, "No other competitor's product is superior to ours" and the advertisement does not show comparative data, what is most likely the true state of affairs regarding the competitor's and the advertiser's products? Discuss.

4. *Anecdotal Reports Versus Systematic Research*
   After reading this *What is the Truth?* section, if you were asked if secretin has a beneficial effect on autism, how would you answer? Would you be satisfied with anecdotal reports? Would negative results from systematic research satisfy you? How can the issue be resolved?

■ NOTES

10.1 If the null hypothesis is false, then chance does not account for the results. Strictly speaking, this means that something systematic differs between the two groups. Ideally, the only systematic difference is due to the independent variable. Thus, we say that if the null hypothesis is false, the alternative hypothesis must be true. Practically speaking, however, the reader should be aware that it is hard to do the perfect experiment. Consequently, in addition to the alternative hypothesis, there are often additional possible explanations of the systematic difference. Therefore, when we say "we accept H_1," you should be aware that there may be additional explanations of the systematic difference.

10.2 If the alternative hypothesis is directional, the null hypothesis asserts that the independent variable does not have an effect in the direction specified by the alternative hypothesis. This is true in the overwhelming number of experiments conducted. Occasionally, an experiment is conducted in which the alternative hypothesis specifies not only the direction but also the magnitude of the effect. For example, in connection with the marijuana experiment, an alternative hypothesis of this type might be "Marijuana increases appetite so as to increase average daily eating by more than 200 calories." The null hypothesis for this alternative hypothesis is "Marijuana increases appetite so as to increase daily eating by 200 or fewer calories."

■ ONLINE STUDY RESOURCES

CENGAGE**brain**.com

Login to CengageBrain.com to access the resources your instructor has assigned. For this book, you can access the book's companion website for chapter-specific learning tools including Know and Be Able to Do, practice quizzes, flash cards, and glossaries, and a link to Statistics and Research Methods Workshops.

aplia

If your professor has assigned Aplia homework:

1. Sign in to your account.
2. Complete the corresponding homework exercises as required by your professor.
3. When finished, click "Grade It Now" to see which areas you have mastered and which need more work, and for detailed explanations of every answer.

Visit **www.cengagebrain.com** to access your account and to purchase materials.

Power

LEARNING OBJECTIVES

After completing this chapter, you should be able to:

- Define power, in terms of both H_1 and H_0.
- Define P_{null} and P_{real}, and specify what P_{real} measures.
- Specify the effect that N, size of real effect, and alpha level have on power.
- Explain the relationship between power and beta.
- Explain why we never "accept" H_0, but instead "fail to reject," or "retain" it.
- Calculate power using the sign test.
- Understand the illustrative examples, do the practice problems, and understand the solutions.

INTRODUCTION

MENTORING TIP

Caution: many students find this is a difficult chapter. You may need to give it some extra time.

We have seen in Chapter 10 that there are two errors we might make when testing hypotheses. We have called them Type I and Type II errors. We have further pointed out that the alpha level limits the probability of making a Type I error. By setting alpha to 0.05 or 0.01, experimenters can limit the probability that they will falsely reject the null hypothesis to these low levels. But what about Type II errors? We defined beta (β) as the probability of making a Type II error. We shall see later in this chapter that $\beta = 1 - $ power. By maximizing power, we minimize beta, which means we minimize the probability of making a Type II error. Thus, power is a very important topic.

WHAT IS POWER?

Conceptually, the power of an experiment is a measure of the sensitivity of the experiment to detect a *real effect* of the independent variable. *By "a real effect of the independent variable," we mean an effect that produces a change in the dependent variable.* If the independent variable does not produce a change in the dependent variable, it has no effect and we say that the independent variable does not have a real effect.

In analyzing the data from an experiment, we "detect" a real effect of the independent variable by rejecting the null hypothesis. Thus, power is defined in terms of rejecting H_0.

definition ■ *Mathematically,* **the power of an experiment** *is defined as the probability that the results of an experiment will allow rejection of the null hypothesis if the independent variable has a real effect.*

Another way of stating the definition is that the power of an experiment is the probability that the results of an experiment will allow rejection of the null hypothesis if the null hypothesis is false.

Since power is a probability, its value can vary from 0.00 to 1.00. The higher the power, the more sensitive the experiment is to detect a real effect of the independent variable. Experiments with power as high as 0.80 or higher are very desirable but rarely seen in the behavioral sciences. Values of 0.40 to 0.60 are much more common. It is especially useful to determine the power of an experiment when (1) initially designing the experiment and (2) interpreting the results of experiments that fail to detect any real effects of the independent variable (i.e., experiments that retain H_0).

P_{null} AND P_{real}

When computing the power of an experiment using the sign test, it is useful to distinguish between P_{null} and P_{real}.

P_{null} always equals 0.50. For experiments where H_1 is nondirectional, P_{real} equals any one of the other possible values of P (i.e., any value of P that does not equal 0.50).

P_{real}: A Measure of the Real Effect

The actual value of P_{real} will depend on the size and direction of the real effect. To illustrate, let's use the marijuana experiment of Chapter 10. (Refer to Figure 11.1 for the rest of this discussion.) Let us for the moment assume that the marijuana experiment was conducted on the entire population of 10,000 AIDS patients being treated at your hospital, not just on the sample of 10. If the effect of marijuana is to increase appetite and the size of the real effect is large enough to overcome all the

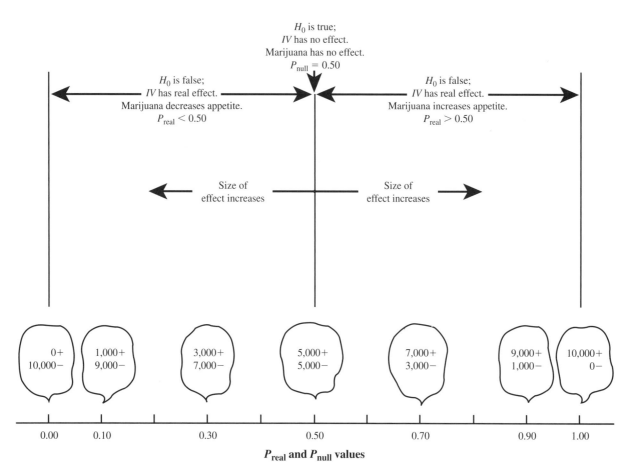

figure 11.1   Relationship among null hypothesis, size of marijuana effect, and P values for a nondirectional H_1.

variables that might be acting to decrease appetite, we would get pluses from all 10,000 patients. Accordingly, there would be 10,000 pluses and 0 minuses in the population. Thus, for this size and direction of marijuana effect, $P_{real} = 1.00$. This is because there are all pluses in the population and the scores of the 10 subjects in the actual experiment would have to be a random sample from this population of scores. We can now elaborate further on the definition of P_{real}.

definition   ■ *As we defined it earlier, $\mathbf{P_{real}}$ is the probability of a plus with any subject in the sample of the experiment if the independent variable has a real effect. However, it is also the proportion of pluses in the population if the experiment were done on the entire population and the independent variable has a real effect.*

Of course, the value P_{real} is the same whether defined in terms of the population proportion of pluses or the probability of a plus with any subject in the sample. Let us return now to our discussion of P_{real} and the size of the effect of the independent variable.

If marijuana increases appetite less strongly than to produce all pluses—say, to produce 9 pluses for every 1 minus—in the population, there would be 9000 pluses and 1000 minuses and $P_{real} = 0.90$.* If the increasing effect of marijuana were of even smaller size—say, 7 pluses for every 3 minuses—the population would have 7000 pluses and 3000 minuses. In this case, $P_{real} = 0.70$. Finally, if marijuana had no effect on appetite, then there would be 5000 pluses and 5000 minuses, and $P_{null} = 0.50$. Of course, this is the chance-alone prediction.

On the other hand, if marijuana decreases appetite, we would expect fewer pluses than minuses. Here, $P_{real} < 0.50$. To illustrate, if the decreasing effect on appetite is large enough, there would be all minuses (10,000 minuses and 0 pluses) in the population and $P_{real} = 0.00$. A decreasing effect of smaller size, such that there were 1000 pluses and 9000 minuses, would yield $P_{real} = 0.10$. A still weaker decreasing effect on appetite—say, 3 pluses for every 7 minuses—would yield $P_{real} = 0.30$. As the decreasing effect on appetite weakens still further, we finally return to the null hypothesis specification of $P_{null} = 0.50$ (marijuana has no effect).

MENTORING TIP

The further P_{real} is from 0.50, the greater is the size of the effect.

From the previous discussion, we can see that P_{real} is a measure of the size and direction of the independent variable's real effect. The further P_{real} is from 0.50, the greater is the size of the real effect. It turns out that the power of the experiment varies with the size of the real effect. Thus, when doing a power analysis with the sign test, we must consider all P_{real} values of possible interest.

POWER ANALYSIS OF THE AIDS EXPERIMENT

Suppose you are planning an experiment to test the hypothesis that "marijuana affects appetite in AIDS patients." You plan to randomly select five AIDS patients from your hospital AIDS population and conduct the experiment as previously

*The reader should note that, even though there are minuses in the population, we assume that the effect of marijuana is the same on all subjects (namely, it increases appetite in all subjects). The minuses are assumed to have occurred due to randomly occurring variables that decrease appetite.

described. Since you want to limit the probability of falsely rejecting the null hypothesis to a low level, you set $\alpha = 0.05_{2\text{ tail}}$. Given this stringent alpha level, if you reject H_0, you can be reasonably confident your results are due to marijuana and not to chance. But what is the probability that you will reject the null hypothesis as a result of doing this experiment? To answer this question, we must first determine what sample results, if any, will allow H_0 to be rejected. The results most favorable for rejecting the null hypothesis are all pluses or all minuses. Suppose you got the strongest possible result—all pluses in the sample. Could you reject H_0? Since H_1 is nondirectional, a two-tailed evaluation is appropriate. With $N = 5$ and $P_{\text{null}} = 0.50$, from Table B in Appendix D,

p(5 pluses or 5 minuses) $= p$(5 pluses or 0 pluses)

$\qquad\qquad\qquad\qquad\quad = 0.0312 + 0.0312$

$\qquad\qquad\qquad\qquad\quad = 0.0624$

Table B entry

N	No. of P Events	P 0.50
5	0	0.0312
	5	0.0312

Since 0.0624 is greater than alpha, if we obtained these results in the experiment, we must conclude by retaining H_0. Thus, even if the results were the *most favorable possible for rejecting H_0*, we still can't reject it!

Let's look at the situation a little more closely. Suppose, in fact, that marijuana has a very large effect on appetite and that it increases appetite so much that, if the experiment were conducted on the entire population, there would be all pluses. For example, if the population were 10,000 patients, there would be 10,000 pluses. The five scores in the sample would be a random sample from this population of scores, and the sample would have all pluses. But we've just determined that, even with five pluses in the sample, we would be unable to reject the null hypothesis. Thus, no matter how large the marijuana effect really is, we would not be able to reject H_0. With $N = 5$ and $\alpha = 0.05_{2\text{ tail}}$, there is no sample result that would allow H_0 to be rejected. This is the most insensitive experiment possible. Power has been defined as the probability of rejecting the null hypothesis if the independent variable has a real effect. In this experiment, the probability of rejecting the null hypothesis is zero, no matter how large the independent variable effect really is. *Thus, the power of this experiment is zero for all P_{real} values.* We can place very little value on results from such an insensitive experiment.

Effect of *N* and Size of Real Effect

Next, suppose N is increased to 10. Are there now any sample results that will allow us to reject the null hypothesis? The solution is shown in Table 11.1. If the sample outcome is 0 pluses, from Table B, with $N = 10$ and using $P_{\text{null}} = 0.50$, p(0 or 10 pluses) $= 0.0020$. Note we included 10 pluses because the alternative hypothesis is nondirectional, requiring a two-tailed evaluation. Since 0.0020 is less than alpha, we would reject H_0 if we got this sample outcome. Since the two-tailed probability for 10 pluses is also 0.0020, we would also reject H_0 with this outcome. From Table B, we can see that, if the sample outcome were 1 plus or 9 pluses,

table 11.1 Determining the sample outcomes that will allow rejection of the null hypothesis with $N = 10$, $P_{null} = 0.50$, and $\alpha = 0.05_{2\ tail}$

Sample Outcome	Probability	Decision
0 pluses	$p(0 \text{ or } 10 \text{ pluses}) = 2(0.0010)$ $= 0.0020$	Reject H_0
10 pluses	$p(0 \text{ or } 10 \text{ pluses}) = 2(0.0010)$ $= 0.0020$	Reject H_0
1 plus	$p(0, 1, 9, \text{ or } 10 \text{ pluses}) = 2(0.0010 + 0.0098)$ $= 0.0216$	Reject H_0
9 pluses	$p(0, 1, 9, \text{ or } 10 \text{ pluses}) = 2(0.0010 + 0.0098)$ $= 0.0216$	Reject H_0
2 pluses	$p(0, 1, 2, 8, 9, \text{ or } 10 \text{ pluses}) = 2(0.0010 + 0.0098 + 0.0439)$ $= 0.1094$	Retain H_0
8 pluses	$p(0, 1, 2, 8, 9, \text{ or } 10 \text{ pluses}) = 2(0.0010 + 0.0098 + 0.0439)$ $= 0.1094$	Retain H_0

we would also reject H_0 ($p = 0.0216$). However, if the sample outcome were 2 pluses or 8 pluses, the two-tailed probability value (0.1094) would be greater than alpha. Hence, we would retain H_0 with 2 or 8 pluses. If we can't reject H_0 with 2 or 8 pluses, we certainly can't reject H_0 if we get an outcome less extreme, such as 3, 4, 5, 6, or 7 pluses. Thus, the only outcomes that will allow us to reject H_0 are 0, 1, 9, or 10 pluses. Note that, in making this determination, since we were evaluating the null hypothesis, we used $P_{null} = 0.50$ (which assumes no effect) and began at the extremes, working in toward the center of the distribution until we reached the first outcome for which the two-tailed probability exceeded alpha. The outcomes allowing rejection of H_0 are the ones more extreme than this first outcome for which we retain H_0.

How can we use these outcomes to determine power? Power equals the probability of rejecting H_0 if the independent variable has a real effect. We've just determined that the only way we shall reject H_0 is if we obtain a sample outcome of 0, 1, 9, or 10 pluses. Therefore, power equals the probability of getting 0, 1, 9, or 10 pluses in our sample if the independent variable has a real effect. Thus,

Power = probability of rejecting H_0 if the independent variable (IV) has a real effect
= $p(0, 1, 9, \text{ or } 10 \text{ pluses})$ if IV has a real effect

But the probability of getting 0, 1, 9, or 10 pluses depends on the size of marijuana's real effect on appetite. Therefore, power differs for different sizes of effect. To illustrate this point, we shall calculate power for several possible sizes of real effect. Using P_{real} as our measure of the magnitude and direction of the marijuana effect, we will calculate power for $P_{real} = 1.00$, 0.90, 0.70, 0.30, 0.10, and 0.00. These values have been chosen to span the full range of possible real effects.

First, let's assume marijuana has such a large increasing effect on appetite that, if it were given to the entire population, it would produce all pluses. In this case, $P_{real} = 1.00$. Determining power for $P_{real} = 1.00$ is as follows:

Power = probability of rejecting H_0 if IV has a real effect
= $p(0, 1, 9,$ or 10 pluses) as the sample outcome if $P_{real} = 1.00$
= $p(0) + p(1) + p(9) + p(10)$ if $P_{real} = 1.00$
= $0.0000 + 0.0000 + 0.0000 + 1.0000$
= 1.0000

If $P_{real} = 1.00$, the only possible scores are pluses. Therefore, the sample of 10 scores must be all pluses. Thus, $p(0$ pluses$) = p(1$ plus$) = p(9$ pluses$) = 0.0000$, and $p(10$ pluses$) = 1.0000$. Thus, by the addition rule, power = 1.0000. The probability of rejecting the null hypothesis when it is false, such that $P_{real} = 1.00$, is equal to 1.0000. It is certain that if the effect of marijuana is as large as described, the experiment with 10 subjects will detect its effect. H_0 will be rejected with certainty.

Suppose, however, that the effect of marijuana on appetite is not quite as large as has been described—that is, if it were given to the population, there would still be many more pluses than minuses, but this time there would be 9 pluses on the average for every 1 minus. In this case, $P_{real} = 0.90$. The power for this somewhat lower magnitude of real effect is found from Table B, using $P = 0.90$ ($Q = 0.10$). Thus,

Power = probability of rejecting H_0 if IV has a real effect
= $p(0, 1, 9,$ or 10 pluses) as the sample outcome if $P_{real} = 0.90$
= $p(0) + p(1) + p(9) + p(10)$ if $P_{real} = 0.90$
= $0.000 + 0.0000 + 0.3874 + 0.3487$
= 0.7361

Table B entry

N	No. of Q Events	Q 0.10
10	0	0.3487
	1	0.3874
	9	0.0000
	10	0.0000

The power of this experiment to detect an effect represented by $P_{real} = 0.90$ is 0.7361. Thus, the power of the experiment has decreased. Note that in determining the power for $P_{real} = 0.90$, the sample outcomes for rejecting H_0 haven't changed. As before, they are 0, 1, 9, or 10 pluses. Since these are the outcomes that will allow rejection of H_0, they are dependent on only N and α. Remember that we find these outcomes for the given N and α level by assuming chance alone is at work ($P_{null} = 0.50$) and determining the sample outcomes for which the obtained probability is equal to or less than α using P_{null}.

What happens to the power of the experiment if the marijuana has only a medium effect such that $P_{real} = 0.70$?

Power = probability of rejecting H_0 if IV has a real effect

= $p(0, 1, 9,$ or 10 pluses) as the sample outcome if $P_{real} = 0.70$

= $p(0) + p(1) + p(9) + p(10)$ if $P_{real} = 0.70$

= $0.0000 + 0.0001 + 0.1211 + 0.0282$

= 0.1494

Table B entry

N	No. of Q Events	Q 0.30
10	0	0.0282
	1	0.1211
	9	0.0001
	10	0.0000

Power has decreased to 0.1494. Power calculations have also been made for effect sizes represented by $P_{real} = 0.30$, $P_{real} = 0.10$, and $P_{real} = 0.00$. The results are summarized in Table 11.2.

At this point, several generalizations are possible. First, as N increases, power goes up. Second, for a particular N, say $N = 10$, power varies directly with the size of the real effect. As the size decreases, the power of the experiment decreases. When the size of the effect approaches that predicted by the null hypothesis, power gets very low. This relationship is shown in Figure 11.2.

MENTORING TIP

Power varies directly with N and directly with size of real effect.

table **11.2** Calculation of power and beta

N	H_0	α	Sample Outcomes*	Size of Marijuana Effect	Power	β
5	$P_{null} = 0.50$	$0.05_{2\ tail}$	None	For all P_{real} values	0	1.0000
10	$P_{null} = 0.50$	$0.05_{2\ tail}$	0, 1, 9, or 10 pluses	$P_{real} = 1.00$	1.0000	0.0000
				$P_{real} = 0.90$	0.7361	0.2639
				$P_{real} = 0.70$	0.1494	0.8506
				$P_{null} = 0.50$	†	
				$P_{real} = 0.30$	0.1494	0.8506
				$P_{real} = 0.10$	0.7361	0.2639
				$P_{real} = 0.00$	1.0000	0.0000
20	$P_{null} = 0.50$	$0.05_{2\ tail}$	0–5 or 15–20 pluses	$P_{real} = 0.30$	0.4163	0.5837
20	$P_{null} = 0.50$	$0.01_{2\ tail}$	0–3 or 17–20 pluses	$P_{real} = 0.30$	0.1070	0.8930

*Sample outcomes that would result in rejecting H_0.
†See Note 11.1.

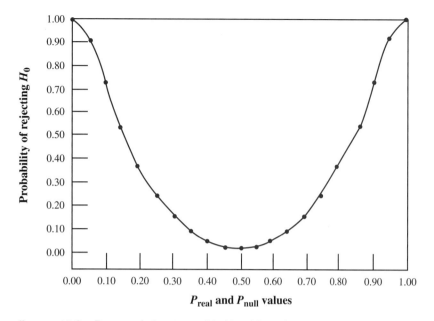

figure 11.2   Power of sign test with $N = 10$ and $\alpha = 0.05_{2\,tail}$.

Power and Beta (β)

As the power of an experiment increases, the probability of making a Type II error decreases. This can be shown as follows.

When we draw a conclusion from an experiment, there are only two possibilities: We either reject H_0 or retain H_0. These possibilities are also mutually exclusive. Therefore, the sum of their probabilities must equal 1. Assuming H_0 is false,

$$p(\text{rejecting } H_0 \text{ if it is false}) + p(\text{retaining } H_0 \text{ if it is false}) = 1$$

but

$$\text{Power} = p(\text{rejecting } H_0 \text{ if it is false})$$
$$\text{Beta} = p(\text{retaining } H_0 \text{ if it is false})$$

Thus,

$$\text{Power} + \text{Beta} = 1$$

or

$$\text{Beta} = 1 - \text{Power}$$

Thus, as power increases, beta decreases. The appropriate beta values are shown in the last column of Table 11.2.

You will note that Table 11.2 has some additional entries. When $N = 20$, the power for this experiment to detect an effect of $P_{real} = 0.30$ is equal to 0.4163. When $N = 10$, the power is only 0.1494. This is another demonstration that as N increases, power increases.

Power and Alpha (α)

The last row of Table 11.2 demonstrates the fact that, by making alpha more stringent, power goes down and beta is increased. With $N = 20$, $P_{real} = 0.30$, and $\alpha = 0.01_{2\,tail}$:

Power = p(0–3 or 17–20 pluses)

= 0.1070

$\beta = 1 -$ Power

= 1 − 0.1070

= 0.8930

Table B entry

N	No. of P Events	P 0.30
20	0	0.0008
	1	0.0068
	2	0.0278
	3	0.0716
	17	0.0000
	18	0.0000
	19	0.0000
	20	0.0000

By making alpha more stringent, the possible sample outcomes for rejecting H_0 are decreased. Thus, for $\alpha = 0.01_{2\,tail}$, only 0–3 or 17–20 pluses will allow rejection of H_0, whereas for $\alpha = 0.05_{2\,tail}$, 0–5 or 15–20 pluses will result in rejection of H_0. This naturally reduces the probability of rejecting the null hypothesis. The decrease in power results in an increase in beta.

Let's summarize a little.

1. The power of an experiment is the probability that the experiment will result in rejecting the null hypothesis if the independent variable has a real effect.
2. Power = 1 – Beta. Therefore, the higher the power is, the lower beta is.
3. Power varies directly with N. Increasing N increases power.
4. Power varies directly with the size of the real effect of the independent variable.
5. Power varies directly with alpha level. Power decreases with more stringent alpha levels.

MENTORING TIP

Summary: Power varies directly with *N*, size of real effect, and alpha level.

The reader should be aware that the experimenter never knows how large the effect of the independent variable actually is before doing the experiment. Otherwise, why do the experiment? In practice, we estimate its size from pilot work or other research and then design an experiment that has high power to detect that size of effect. For example, if a medium effect ($P_{real} = 0.70$) is expected, by selecting the appropriate N we can arrive at a decent sensitivity (e.g., power = 0.8000 or higher). How high should power be? What size of effect should be expected? These are questions that must be answered by the researcher based on experience and available resources. It should be pointed out that by designing the experiment to have a power = 0.8000 for $P_{real} = 0.70$, the power of the experiment will be even higher if the effect of the independent variable is larger than expected. *Thus, the strategy is to design the experiment for the maximum power that resources will allow for the minimum size of real effect expected.*

ALPHA–BETA AND REALITY

When one does an experiment, there are only two possibilities: Either H_0 is really true or it is false. By minimizing alpha and beta, we maximize the likelihood that our conclusions will be correct. For example, if H_0 is really true, the probability of correctly concluding from the experiment is

$$p(\text{correctly concluding}) = p(\text{retaining } H_0) = 1 - \alpha$$

If alpha is at a stringent level (say, 0.05), then $p(\text{correctly concluding})$ is

$$p(\text{correctly concluding}) = 1 - \alpha = 1 - 0.05 = 0.95$$

On the other hand, if H_0 is really false, the probability of correctly concluding is

$$p(\text{correctly concluding}) = p(\text{rejecting } H_0) = \text{power} = 1 - \beta$$

If beta is low (say, equal to 0.10 for the minimum real effect of interest), then

$$p(\text{correctly concluding}) = 1 - \beta = 1 - 0.10 = 0.90$$

Thus, whichever is the true state of affairs (H_0 is true or H_0 is false), there is a high probability of correctly concluding when α is set at a stringent level and β is low. One way of achieving a low beta level when α is set at a stringent level is to have a large N. Another way is to use the statistical inference test that is the most powerful for the data. A third way is to control the external conditions of the experiment such that the variability of the data is reduced. We shall discuss the latter two methods when we cover Student's t test in Chapters 13 and 14.

INTERPRETING NONSIGNIFICANT RESULTS

MENTORING TIP

Caution: it is not valid to conclude by accepting H_0 when the results fail to reach significance.

Although power aids in designing an experiment, it is much more often used when interpreting the results of an experiment that has already been conducted and that has yielded nonsignificant results. Failure to reject H_0 may occur because (1) H_0 is in fact true or (2) H_0 is false, but the experiment was of low power. It is due to the second possible reason that we can *never accept H_0* as being correct when an experiment fails to yield significance. Instead, we say the experiment has failed to allow the null hypothesis to be rejected. It is possible that H_0 is indeed false, but the experiment was insensitive; that is, it didn't give H_0 much of a chance to be rejected. A case in point is the example we presented before with $N = 5$. In that experiment, whatever results we obtained, they would not reach significance. We could not reject H_0 no matter how large the real effect actually was. It would be a gross error to accept H_0 as a result of doing that experiment. The experiment did not give H_0 any chance of being rejected. *From this viewpoint, we can see that every experiment exists to give H_0 a chance to be rejected. The higher the power, the more the experiment allows H_0 to be rejected if it is false.*

Perhaps an analogy will help in understanding this point. We can liken the power of an experiment to the use of a microscope. Physiological psychologists have long been interested in what happens in the brain to allow the memory of an event to be recorded. One hypothesis states that a group of neurons fires together as a result of the stimulus presentation. With repeated firings (trials), there is growth across the synapses of the cells, so after a while, they become activated together whenever the stimulus is presented. This "cell assembly" then becomes the physiological engram of the stimuli (i.e., it is the memory trace).

To test this hypothesis, an experiment is done involving visual recognition. After some animals have practiced a task, the appropriate brain cells from each are prepared on slides so as to look for growth across the synapses. H_0 predicts no growth; H_1 predicts growth. First, the slides are examined with the naked eye; no growth is seen. Can we therefore accept H_0? No, because the eye is not powerful enough to see growth even if it were there. The same holds true for a low-power experiment. If the results are not significant, we cannot conclude by accepting H_0 because even if H_0 is false, the low power makes it unlikely that we would reject the null hypothesis. So next, a light microscope is used, and still there is no growth seen between synapses. Even though this is a more powerful experiment, can we conclude that H_0 is true? No, because a light microscope doesn't have enough power to see the synapses clearly. So finally, an electron microscope is used, producing a very powerful experiment in which all but the most minute structures at the synapse can be seen clearly. If H_0 is false (that is, if there is growth across the synapse), this powerful experiment has a higher probability of detecting it. Thus, the higher the power of an experiment is, the more the experiment allows H_0 to be rejected if it is false.

In light of the foregoing discussion, whenever an experiment fails to yield significant results, we must be careful in our interpretation. Certainly, we can't assert that the null hypothesis is correct. However, if the power of the experiment is high, we can say a little more than just that the experiment has failed to allow rejection of H_0. For example, if power is 1.0000 for an effect represented by $P_{real} = 1.00$ and we fail to reject H_0, we can at least conclude that the independent variable does not have that large an effect. If the power is, say, 0.8000 for a medium effect ($P_{real} = 0.70$), we can be reasonably confident the independent variable is not that effective. On the other hand, if power is low, nonsignificant results tell us little about the true state of reality. Thus, a power analysis tells us how much confidence to place in experiments that fail to reject the null hypothesis. When we fail to reject the null hypothesis, the higher the power is to detect a given real effect, the more confident we are that the effect of the independent variable is not that large. However, note that as the real effect of the independent variable gets very small, the power of the experiment to detect it gets very low (see Figure 11.2). *Thus, it is impossible to ever prove that the null hypothesis is true because the power to detect very small but real effects of the independent variable is always low.*

CALCULATION OF POWER

Calculation of power involves a two-step process for each level of P_{real}:

STEP 1: Assume the null hypothesis is true. Using $P_{null} = 0.50$, determine the possible sample outcomes in the experiment that allow H_0 to be rejected.

STEP 2: For the level of P_{real} under consideration (e.g., $P_{real} = 0.30$), determine the probability of getting any of the sample outcomes arrived at in Step 1. This probability is the power of the experiment to detect this level of real effect.

Practice Problem 11.1

You are interested in determining whether word recall is better when (1) the words are just directly memorized or (2) a story that includes all the words is made up by the subjects. In the second method, the story, from which the words could be recaptured, would be recalled. You plan to run 14 subjects in a repeated measures experiment and analyze the data with the sign test. Each subject will use both methods with equivalent sets of words. The number of words remembered in each condition will be the dependent variable; $\alpha = 0.05_{2\,tail}$.

a. What is the power of the experiment to detect this large* effect of $P_{real} = 0.80$ or 0.20?

b. What is the probability of a Type II error?

The solution follows. From the solution, we see that the power to detect a large difference ($P_{real} = 0.80$ or 0.20) in the effect on word recall between memorizing the words and making up a story including the words is 0.4480. This means that we have about a 45% chance of rejecting H_0 if the effect is as large as $P_{real} = 0.80$ or 0.20 and a 55% chance of making a Type II error. If the effect is smaller than $P_{real} = 0.80$ or 0.20, then the probability of making a Type II error is even higher. Of course, increasing N in the experiment will increase the probability of rejecting H_0 and decrease the probability of making a Type II error.

SOLUTION

a. Calculation of power: The calculation of power involves a two-step process:

MENTORING TIP

Remember: for Step 1, $P = 0.50$. For Step 2, P is a value other than 0.50. For this example, in Step 2, $P = 0.80$ or $P = 0.20$.

STEP 1: **Assume the null hypothesis is true ($P_{null} = 0.50$) and determine the possible sample outcomes in the experiment that will allow H_0 to be rejected.** $\alpha = 0.05_{2\,tail}$. With $N = 14$ and $P = 0.50$, from Table B,

p(0 pluses)	$= 0.0001$	p(0 pluses)	$= 0.0001$
p(1 plus)	$= 0.0009$	p(1 plus)	$= 0.0009$
p(2 pluses)	$= 0.0056$	p(2 pluses)	$= 0.0056$
p(12 pluses)	$= 0.0056$	p(3 pluses)	$= 0.0222$
p(13 pluses)	$= 0.0009$	p(11 pluses)	$= 0.0222$
p(14 pluses)	$= 0.0001$	p(12 pluses)	$= 0.0056$
p(0, 1, 2, 12, 13, or 14)	$= 0.0132$	p(13 pluses)	$= 0.0009$
		p(14 pluses)	$= 0.0001$
		p(0, 1, 2, 3, 11, 12, 13, or 14)	$= 0.0576$

*Following Cohen (1988), we have divided the size of effect range into the following three intervals: for a large effect $P_{real} = 0.00$–0.25 or 0.75–1.00; a medium effect, $P_{real} = 0.26$–0.35 or 0.65–0.74; and a small effect, $P_{real} = 0.36$–0.49 or 0.51–0.64. For reference, see footnote in Chapter 13, p. 339.

(continued)

Beginning at the extremes and moving toward the middle of the distribution, we find that we can reject H_0 if we obtain 2 or 12 pluses ($p = 0.0132$), but we fail to reject H_0 if we obtain 3 or 11 pluses ($p = 0.0576$) in the sample. Therefore, the outcomes that will allow rejection of H_0 are 0, 1, 2, 12, 13, or 14 pluses.

STEP 2: **For $P_{real} = 0.20$, determine the probability of getting any of the aforementioned sample outcomes.** This probability is the power of the experiment to detect this hypothesized real effect. With $N = 14$ and $P_{real} = 0.20$, from Table B,

Power = probability of rejecting H_0 if IV has a real effect

$= p(0, 1, 2, 12, 13,$ or 14 pluses) as sample outcomes if $P_{real} = 0.20$

$= 0.0440 + 0.1539 + 0.2501 + 0.0000 + 0.0000 + 0.0000$

$= 0.4480$

Table B entry

N	No. of P Events	P 0.20
14	0	0.0440
	1	0.1539
	2	0.2501
	12	0.0000
	13	0.0000
	14	0.0000

Note that the same answer would result for $P_{real} = 0.80$.

b. Calculation of beta:

$$\beta = 1 - \text{Power}$$
$$= 1 - 0.4480$$
$$= 0.5520$$

Practice Problem 11.2

Assume you are planning an experiment to evaluate a drug. The alternative hypothesis is directional, in the direction to produce mostly pluses. You will use the sign test to analyze the data; $\alpha = 0.05_{1 \text{ tail}}$. You want to be able to detect a small effect of $P_{real} = 0.60$ in the same direction as the alternative hypothesis. There will be 16 subjects in the experiment.

a. What is the power of the experiment to detect this small effect?
b. What is the probability of making a Type II error?

SOLUTION

a. Calculation of power: There are two steps involved in calculating power:

STEP 1: **Assume the null hypothesis is true ($P_{null} = 0.50$) and determine the possible sample outcomes in the experiment that will allow H_0 to be rejected.** $\alpha = 0.05_{1\,tail}$. With $N = 16$ and $P = 0.50$, from Table B,

$p(12 \text{ pluses})$	$= 0.0278$	$p(11 \text{ pluses})$	$= 0.0667$
$p(13 \text{ pluses})$	$= 0.0085$	$p(12 \text{ pluses})$	$= 0.0278$
$p(14 \text{ pluses})$	$= 0.0018$	$p(13 \text{ pluses})$	$= 0.0085$
$p(15 \text{ pluses})$	$= 0.0002$	$p(14 \text{ pluses})$	$= 0.0018$
$p(16 \text{ pluses})$	$= 0.0000$	$p(15 \text{ pluses})$	$= 0.0002$
$p(12, 13, 14, 15, \text{ or } 16)$	$= 0.0383$	$p(16 \text{ pluses})$	$= 0.0000$
		$p(11, 12, 13, 14, 15, \text{ or } 16)$	$= 0.1050$

Since the alternative hypothesis is in the direction of mostly pluses, outcomes for rejecting H_0 are found under the tail with the higher numbers of pluses. Beginning with 16 pluses and moving toward the middle of the distribution, we find that we shall reject H_0 if we obtain 12 pluses ($p = 0.0383$), but we shall fail to reject H_0 if we obtain 11 pluses ($p = 0.1050$) in the sample. Therefore, the outcomes that will allow rejection of H_0 are 12, 13, 14, 15, or 16 pluses.

STEP 2: **For $P_{real} = 0.60$, determine the probability of getting any of the aforementioned sample outcomes.** This probability is the power of the experiment to detect this hypothesized real effect. With $N = 16$ and $P_{real} = 0.60$ ($Q = 0.40$), from Table B,

$$\begin{aligned}
\text{Power} &= \text{probability of rejecting } H_0 \text{ if } IV \text{ has a real effect} \\
&= p(12, 13, 14, 15, \text{ or } 16 \text{ pluses}) \text{ as sample outcomes if } P_{real} = 0.60 \\
&= 0.1014 + 0.0468 + 0.0150 + 0.0030 + 0.0003 \\
&= 0.1665
\end{aligned}$$

Table B entry

N	No. of Q Events	Q 0.40
16	0	0.0003
	1	0.0030
	2	0.0150
	3	0.0468
	4	0.1014

b. Calculation of beta:

$$\begin{aligned}
\beta &= 1 - \text{Power} \\
&= 1 - 0.1665 \\
&= 0.8335
\end{aligned}$$

(continued)

This experiment is very insensitive to a small drug effect of $P_{real} = 0.60$. The probability of a Type II error is too high. The N should be made larger to increase the power of the experiment to detect the small drug effect.

Practice Problem 11.3

In Practice Problem 10.2 (p. 262), you conducted an experiment testing the directional alternative hypothesis that watching a particular TV program caused increased violence in teenagers. The experiment included 15 subjects, and $\alpha = 0.01_{1\ tail}$. The data were analyzed with the sign test, and we retained H_0.

a. In that experiment, what was the power to detect a medium effect of $P_{real} = 0.70$ in the direction of the alternative hypothesis?

b. What was the probability of a Type II error?

SOLUTION

a. Calculation of power: There are two steps involved in calculating power:

> **STEP 1:** **Assume the null hypothesis is true ($P_{null} = 0.50$) and determine the possible sample outcomes in the experiment that will allow H_0 to be rejected.** $\alpha = 0.01_{1\ tail}$. With $N = 15$ and $P = 0.50$, from Table B,

p(13 pluses)	= 0.0032	p(12 pluses)	= 0.0139
p(14 pluses)	= 0.0005	p(13 pluses)	= 0.0032
p(15 pluses)	= 0.0000	p(14 pluses)	= 0.0005
p(13, 14, or 15)	= 0.0037	p(15 pulses)	= 0.0000
		p(12, 13, 14, or 15)	= 0.0176

> Since the alternative hypothesis is in the direction of mostly pluses, outcomes for rejecting H_0 are found under the tail with the higher numbers of pluses. Beginning with 15 pluses and moving toward the middle of the distribution, we find that we shall reject H_0 if we obtain 13 pluses ($p = 0.0037$), but we shall retain H_0 if we obtain 12 pluses ($p = 0.0176$) in the sample. Therefore, the outcomes that will allow rejection of H_0 are 13, 14, or 15 pluses.

> **STEP 2:** **For $P_{real} = 0.70$, determine the probability of getting any of the aforementioned sample outcomes.** This probability is the power of the experiment to detect this hypothesized real effect. With $N = 15$ and $P_{real} = 0.70$, from Table B,

$$
\begin{aligned}
\text{Power} &= \text{probability of rejecting } H_0 \text{ if } IV \text{ has a real effect} \\
&= p(13, 14, \text{ or } 15 \text{ pluses}) \text{ as sample outcomes if } P_{real} = 0.70 \\
&= 0.0916 + 0.0305 + 0.0047 \\
&= 0.1268
\end{aligned}
$$

Table B entry		
N	No. of Q Events	Q 0.30
15	0	0.0047
	1	0.0305
	2	0.0916

b. Calculation of beta:

$$\beta = 1 - \text{Power}$$
$$= 1 - 0.1268$$
$$= 0.8732$$

Note that since the power to detect a medium effect of $P_{real} = 0.70$ is very low, even though we retained H_0 in the experiment, we can't conclude that the program does not affect violence. The experiment should be redone with increased power to allow a better evaluation of the program's effect on violence.

WHAT IS THE TRUTH? Astrology and Science

 A newspaper article appeared in a recent issue of the Pittsburgh *Post-Gazette* with the headline, "When Clinical Studies Mislead." Excerpts from the article are reproduced here:

Shock waves rolled through the medical community two weeks ago when researchers announced that a frequently prescribed triad of drugs previously shown to be helpful after a heart attack had proved useless in new studies....

"People are constantly dazzled by numbers, but they don't know what lies behind the numbers," said Alvan R. Feinstein, a professor of medicine and epidemiology at the Yale University School of Medicine. "Even scientists and physicians have been brainwashed into thinking that the magic phrase 'statistical significance' is the answer to everything."

The recent heart-drug studies belie that myth. Clinical trials involving thousands of patients over a period of several years had shown previously that nitrate-containing drugs such as nitroglycerine, the enzyme inhibitor captopril and magnesium all helped save lives when administered after a heart attack.

Comparing the life spans of those who took the medicines with those who didn't, researchers found the difference to be statistically significant, and the drugs became part of the standard medical practice. In the United States, more than 80 percent of heart attack patients are given nitrate drugs.

But in a new study involving more than 50,000 patients, researchers found no benefit from nitrates or magnesium and captopril's usefulness was marginal. Oxford epidemiologist Richard Peto, who oversaw the latest study, said the positive results from the previous trial must have been due to "the play of chance." ... Faulty number crunching, Peto said, can be a matter of life and death.

He and his colleagues drove that point home in 1988 when they submitted a paper to the British medical journal The Lancet. *Their landmark report showed that heart attack victims had a better chance of surviving if they were given aspirin within a few hours after their attacks. As Peto tells the story, the journal's editors wanted the researchers to break down the data into various subsets, to see whether certain kinds of patients who differed from each*

WHAT IS THE TRUTH? *(continued)*

other by age or other characteristics were more or less likely to benefit from aspirin.

Peto objected, arguing that a study's validity could be compromised by breaking it into too many pieces. If you compare enough subgroups, he said, you're bound to get some kind of correlation by chance alone. When the editors insisted, Peto capitulated, but among other things he divided his patients by zodiac birth signs and demanded that his findings be included in the published paper. Today, like a warning sign to the statistically uninitiated, the wacky numbers are there for all

to see: Aspirin is useless for Gemini and Libra heart attack victims but is a lifesaver for people born under any other sign....

Studies like these exemplify two of the more common statistical offenses committed by scientists— making too many comparisons and paying too little attention to whether something makes sense— said James L. Mills, chief of the pediatric epidemiology section of the National Institute of Child Health and Human Development.

"People search through their results for the most exciting and positive things," he said. "But you also have to look at the biological plausibility. A lot of findings that don't withstand the test of time

didn't really make any sense in the first place...."

In the past few years, many scientists have embraced larger and larger clinical trials to minimize the chances of being deceived by a fluke.

What do you think? If you were a physician, would you continue to prescribe nitrates to heart attack patients? Is it really true that the early clinical trials are an example of Type I error, as suggested by Dr. Peto? Will larger and larger clinical trials minimize the chances of being deceived by a fluke? Finally, is aspirin really useless for Gemini and Libra heart attack victims but a lifesaver for people born under any other sign? ∎

■ SUMMARY

In this chapter, I discussed the topic of power. Power is defined as the probability of rejecting the null hypothesis when the independent variable has a real effect. Since power varies with the size of the real effect, it should be calculated for the smallest real effect of interest. The power will be even higher for larger effects. Calculation of power involves two steps. In the first step, the null hypothesis is assumed true ($P_{null} = 0.50$), and all the possible sample outcomes in the experiment that would allow the null hypothesis to be rejected are determined. Next, for the real effect under consideration (e.g., the effect represented by $P_{real} = 0.30$), the probability of getting any of these sample outcomes is calculated. This probability is the power of the experiment to detect this effect ($P_{real} = 0.30$).

With other factors held constant, power increases with increases in N and increases in the size of the real effect of the independent variable. Power decreases as the alpha level is made more stringent. Power equals 1 – beta. Therefore, maximizing power minimizes the probability of a Type II error. Thus, by minimizing alpha and beta, we maximize the probability of correctly determining the true effect of the independent variable, no matter what the state of reality.

A power analysis is useful when (1) initially designing an experiment and (2) interpreting the results of experiments that retain the null hypothesis. When an experiment is conducted and the results are not significant, it may be because the null hypothesis is true or because the experiment has low power. It is for this reason that, when the results are not significant, we do not conclude by *accepting* the null hypothesis but rather by *failing to reject* it. The null hypothesis actually may be false, but the experiment did not have high enough power to detect it. Every experiment exists to give the null hypothesis a chance to be rejected. The more powerful the experiment, the higher the probability the null hypothesis will be rejected if it is false. Since power gets low as the real effect of the independent variable decreases, it is impossible to prove that H_0 is true.

■ IMPORTANT NEW TERMS

P_{null} (p. 279)
P_{real} (p. 279)

Power (p. 278)
Real effect (p. 278)

■ QUESTIONS AND PROBLEMS

1. What is power? How is it defined?
2. In what two situations is a power analysis especially useful? Explain.
3. In hypothesis testing experiments, why is the conclusion "We retain H_0" preferable to "We accept H_0 as true"?
4. In hypothesis-testing experiments, is it ever correct to conclude that the independent variable has had *no* effect? Explain.
5. In computing power, why do we always compute the sample outcomes that will allow rejection of H_0?
6. Using α and β, explain how we can maximize the probability of correctly concluding from an experiment, regardless of whether H_0 is true or false. As part of your explanation, choose values for α and β and determine the probability of correctly concluding when H_0 is true and when H_0 is false.
7. You are considering testing a new drug that is supposed to facilitate learning in mentally retarded children. Because there is relatively little known about the drug, you plan to use a nondirectional alternative hypothesis. Your resources are limited, so you can test only 15 subjects. The subjects will be run in a repeated measures design and the data analyzed with the sign test using $\alpha = 0.05_{2\,tail}$. If the drug has a medium effect on learning such that $P_{real} = 0.70$, what is the probability you will detect it when doing your experiment? What is the probability of a Type II error? cognitive
8. In Chapter 10, Problem 10 (p. 273), a new teaching method was evaluated. Twenty pairs of subjects were run in a repeated measures design. The results were in favor of the new method but did not reach significance (H_0 was not rejected) using the sign test

with $\alpha = 0.05_{1 \text{ tail}}$. In trying to interpret why the results were not significant, you reason that there are two possibilities: either (1) the two teaching methods are really equal in effectiveness (H_0 is true) or (2) the new method is better, but the experiment was insensitive. To evaluate the latter possibility, you conduct an analysis to determine the power of the experiment to detect a large difference favoring the new method such that $P_{\text{real}} = 0.80$. What is the power of the experiment to detect this effect? What is beta? education

9. A researcher is going to conduct an experiment to determine whether one night's sleep loss affects performance. Assume the requirements are met for a directional alternative hypothesis. Fourteen subjects will be run in a repeated measures design. The data will be analyzed with the sign test, using $\alpha = 0.05_{1 \text{ tail}}$. Each subject will receive two conditions: condition 1, where the performance of the subject is measured after a good night's sleep, and condition 2, where performance is measured after one night's sleep deprivation. The better the performance, the higher the score. When the data are analyzed, the scores of condition 2 will be subtracted from those of condition 1. If one night's loss of sleep has a large detrimental effect on performance such that $P_{\text{real}} = 0.90$, what is the power of the experiment to detect this effect? What is the probability of a Type II error? cognitive

10. In Chapter 10, Problem 12 (p. 274), what is the power of the experiment to detect a medium effect such that $P_{\text{real}} = 0.70$? I/O

11. A psychiatrist is planning an experiment to determine whether stimulus isolation affects depression. Eighteen subjects will be run in a repeated measures design. The data will be analyzed with the sign test, using $\alpha = 0.05_{2 \text{ tail}}$. Each subject will receive two conditions: condition 1, one week of living in an environment with a normal amount of external stimulation, and condition 2, one week in an environment where external stimulation has been radically curtailed. A questionnaire measuring

depression will be administered after each condition. The higher the score on the questionnaire, the greater the subject's depression. In analyzing the data, the scores of condition 1 will be subtracted from the scores of condition 2. If one week of stimulus isolation has an effect on depression such that $P_{\text{real}} = 0.60$, what is the power of the experiment to detect this small effect? What is beta? If the results of the experiment are not significant, is it legitimate for the psychiatrist to conclude that stimulus isolation has no effect on depression? Why? cognitive, clinical, health

12. In Chapter 10, Practice Problem 10.2 (p. 262), an experiment was conducted to determine whether watching a particular TV program resulted in increased violence in teenagers. In that experiment, 15 subjects were run with each subject serving in an experimental and control condition. The sign test was used to analyze the data, with $\alpha = 0.01_{1 \text{ tail}}$. Suppose the TV program does increase violence and that the effect size is medium ($P_{\text{real}} = 0.70$). Before running the experiment, what is the probability that the experiment will detect at least this level of real effect? What is the probability of a Type II error? The data collected in this experiment failed to allow rejection of H_0. Are we therefore justified in concluding that the TV program has no effect on violence in teenagers? Explain. social

What Is the Truth? Questions

1. *Astrology and Science*
   a. If you were a physician, would you continue to prescribe nitrates to heart attack patients? Why or why not?
   b. Is it really true that the early clinical trials are an example of Type I error, as suggested by Dr. Peto? Discuss.
   c. Will larger and larger clinical trials minimize the chances of being deceived by a fluke? Explain
   d. Is aspirin really useless for Gemini and Libra heart attack victims but a lifesaver for people born under another sign? Explain.

■ NOTES

11.1 This probability is not equal to power because when $P = 0.50$, H_0 is true. Power is calculated when H_0 is false. The probability of rejecting H_0 when H_0 is true is defined as the probability of making a Type I error. For this example:

$$p(\text{reject } H_0 \text{ when } P = 0.50) = p(\text{Type I error})$$
$$= p(0) + p(1) + p(9) + p(10)$$
$$= 0.0010 + 0.0098 + 0.0098 + 0.0010$$
$$= 0.0216$$

Note that the probability of making a Type I error (0.0216) is not equal to the α level (0.05) because the number of pluses is a discrete variable rather than a continuous variable. To have p(Type I error) equal alpha, we would need an outcome between 8 and 9 pluses. Of course, this is impossible because the number of pluses can only be 8 or 9 (discrete values). The probability of making a Type I error is equal to alpha when the variable is continuous.

■ ONLINE STUDY RESOURCES

CENGAGE**brain**.com

Login to CengageBrain.com to access the resources your instructor has assigned. For this book, you can access the book's companion website for chapter-specific learning tools including Know and Be Able to Do, practice quizzes, flash cards, and glossaries, and a link to Statistics and Research Methods Workshops.

aplia™

If your professor has assigned Aplia homework:

1. Sign in to your account.
2. Complete the corresponding homework exercises as required by your professor.
3. When finished, click "Grade It Now" to see which areas you have mastered and which need more work, and for detailed explanations of every answer.

Visit **www.cengagebrain.com** to access your account and to purchase materials.

12

Sampling Distributions, Sampling Distribution of the Mean, the Normal Deviate (z) Test

© Strmko / Dreamstime.com

CHAPTER OUTLINE

LEARNING OBJECTIVES

After completing this chapter, you should be able to:

- Specify the two basic steps involved in analyzing data.
- Define *null-hypothesis population*, and explain how to generate sampling distributions empirically.
- Define the sampling distribution of a statistic, define the sampling distribution of the mean and specify its characteristics, and state the Central Limit Theorem.
- Define critical region, critical value(s) of a statistic, critical value(s) of $\overline{X}$, and critical value(s) of z.
- Solve inference problems using the z test and specify the conditions under which the z test is appropriate.
- Define μ_{null} and μ_{real}.
- Compute power using the z test.
- Specify the relationship between power and the following: N, size of real effect, and alpha level.
- Understand the illustrative examples, do the practice problems, and understand the solutions.

INTRODUCTION

In Chapters 10 and 11, we have seen how to use the scientific method to investigate hypotheses. We have introduced the replicated measures and the independent groups designs and discussed how to analyze the resulting data. At the heart of the analysis is the ability to answer the question, *what is the probability of getting the obtained result or results even more extreme if chance alone is responsible for the differences between the experimental and control scores?*

Although it hasn't been emphasized, the answer to this question involves two steps: (1) calculating the appropriate statistic and (2) evaluating the statistic based on its sampling distribution. In this chapter, we shall more formally discuss the topic of a statistic and its sampling distribution. Then we shall begin our analysis of single sample experiments, using the mean of the sample as a statistic. This involves the sampling distribution of the mean and the normal deviate (z) test.

SAMPLING DISTRIBUTIONS

What is a sampling distribution?

definition ■ *The* **sampling distribution of a statistic** *gives* (1) *all the values that the statistic can take and* (2) *the probability of getting each value under the assumption that it resulted from chance alone.*

In the replicated measures design, we used the sign test to analyze the data. The statistic calculated was the *number of pluses* in the sample of N difference scores. In one version of the "marijuana and appetite" experiment, we obtained nine pluses and one minus. This result was evaluated by using the binomial distribution. The binomial distribution with $P = 0.50$ lists all the possible values of the statistic, the number of pluses, along with the probability of getting each value under the assumption that chance alone produced it. *The binomial distribution with $P = 0.50$ is the sampling distribution of the statistic used in the sign test.* Note that there is a different sampling distribution for each sample size (N).

MENTORING TIP

This is the essential process underlying all of hypothesis testing, no matter what inference test is used. I suggest you spend a little extra time here to be sure you understand it.

Generalizing from this example, it can be seen that data analysis basically involves two steps:

1. Calculating the appropriate statistic—for example, number of pluses and minuses for the sign test
2. Evaluating the statistic based on its sampling distribution

If the probability of getting the obtained value of the statistic or any value more extreme is equal to or less than the alpha level, we reject H_0 and accept H_1. If not, we retain H_0. If we reject H_0 and it is true, we've made a Type I error. If we retain H_0 and it's false, we've made a Type II error. This process applies to all experiments involving hypothesis testing. *What changes from experiment to experiment is the statistic used and its accompanying sampling distribution.* Once you understand this concept, you can appreciate that a large part of teaching inferential statistics is devoted to presenting the most often used statistics, their sampling distributions, and the conditions under which each statistic is appropriately used.

Generating Sampling Distributions

We have defined a sampling distribution as a probability distribution of all the possible values of a statistic under the assumption that chance alone is operating. One way of deriving sampling distributions is from basic probability considerations. We used this approach in generating the binomial distribution. Sampling distributions can also be derived from an empirical sampling approach. In this approach, we have an actual or theoretical set of population scores that exists if the independent variable has no effect. We derive the sampling distribution of the statistic by

1. Determining all the possible different samples of size N that can be formed from the population of scores
2. Calculating the statistic for each of the samples
3. Calculating the probability of getting each value of the statistic if chance alone is operating

To illustrate the sampling approach, let's suppose we are conducting an experiment with a sample size $N = 2$, using the sign test for analysis. We can imagine a theoretical set of scores that would result if the experiment were done on the entire population and the independent variable had no effect. This population set of scores is called the *null-hypothesis population*.

definition   ■   *The **null-hypothesis population** is an actual or theoretical set of population scores that would result if the experiment were done on the entire population and the independent variable had no effect. It is called the null-hypothesis population because it is used to test the validity of the null hypothesis.*

In the case of the sign test, if the independent variable had no effect, the null-hypothesis population would have an equal number of pluses and minuses ($P = Q = 0.50$).

For computational ease in generating the sampling distribution, let's assume there are only six scores in the population: three pluses and three minuses. To derive the sampling distribution of "the number of pluses" with $N = 2$, we must first determine all the different samples of size N that can be formed from the population. Sampling is one at a time, with replacement. Figure 12.1 shows the population and, schematically, the different samples of size 2 that can be drawn from it. It turns out that there are 36 different samples of size 2 possible. These are listed in the table of Figure 12.1, column 2. Next, we must calculate the value of the statistic for each sample. This information is presented in the table of Figure 12.1, columns 3 and 4. Note that of the 36 different samples possible, 9 have two pluses, 18 have one plus, and 9 have no pluses. The last step is to calculate the probability of getting each value of the statistic. If chance alone is operating, each sample is equally likely. Thus,

$$p(2 \text{ pluses}) = \frac{9}{36} = 0.2500$$

$$p(1 \text{ plus}) = \frac{18}{36} = 0.5000$$

$$p(0 \text{ pluses}) = \frac{9}{36} = 0.2500$$

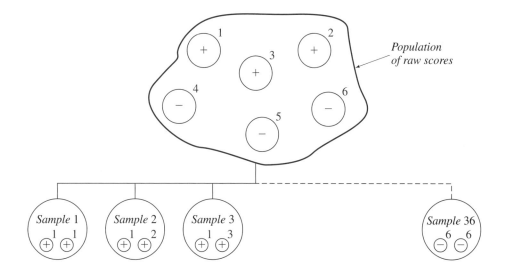

Sample Composition				Sample Composition			
Sample number (1)	Element numbers (2)	Actual scores (3)	Statistic, no. of pluses (4)	Sample number (1)	Element numbers (2)	Actual scores (3)	Statistic, no. of pluses (4)
1	1, 1	+ +	2+	19	4, 1	− +	1+
2	1, 2	+ +	2+	20	4, 2	− +	1+
3	1, 3	+ +	2+	21	4, 3	− +	1+
4	1, 4	+ −	1+	22	4, 4	− −	0+
5	1, 5	+ −	1+	23	4, 5	− −	0+
6	1, 6	+ −	1+	24	4, 6	− −	0+
7	2, 1	+ +	2+	25	5, 1	− +	1+
8	2, 2	+ +	2+	26	5, 2	− +	1+
9	2, 3	+ +	2+	27	5, 3	− +	1+
10	2, 4	+ −	1+	28	5, 4	− −	0+
11	2, 5	+ −	1+	29	5, 5	− −	0+
12	2, 6	+ −	1+	30	5, 6	− −	0+
13	3, 1	+ +	2+	31	6, 1	− +	1+
14	3, 2	+ +	2+	32	6, 2	− +	1+
15	3, 3	+ +	2+	33	6, 3	− +	1+
16	3, 4	+ −	1+	34	6, 4	− −	0+
17	3, 5	+ −	1+	35	6, 5	− −	0+
18	3, 6	+ −	1+	36	6, 6	− −	0+

figure 12.1 All of the possible samples of size 2 that can be drawn from a population of three pluses and three minuses. Sampling is one at a time, with replacement.

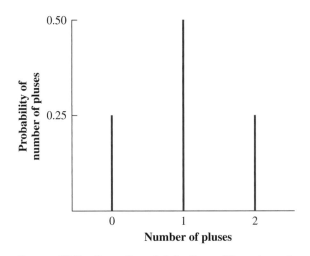

figure 12.2 Sampling distribution of "number of pluses" with $N = 2$ and $P = 0.50$.

We have now derived the sampling distribution for $N = 2$ of the statistic "number of pluses." The distribution is plotted in Figure 12.2. In this example, we used a population in which there were only six scores. The identical sampling distribution would have resulted (even though there would be many more "different" samples) had we used a larger population as long as the number of pluses equaled the number of minuses and the sample size equaled 2. Note that this is the same sampling distribution we arrived at through basic probability considerations when we were discussing the binomial distribution with $N = 2$ (see Figure 12.3 for a comparison). This time, however, we generated it by sampling from the null-hypothesis population. The sampling distribution of a statistic is often defined in terms of this process. Viewed in this manner, we obtain the following definition.

Empirical Sampling Approach			**A Priori Approach**		
Draw 1	*Draw 2*	*Number of ways*	*Coin 1*	*Coin 2*	*Number of ways*
+	+	9	H	H	1
+ −	− + $\}$	18	H T	T H $\}$	2
−	−	9	T	T	1
		$\overline{36}$			$\overline{4}$

$$p(2+) = \frac{9}{36} = 0.2500 \qquad\qquad p(2\text{H}) = \frac{1}{4} = 0.2500$$

$$p(1+) = \frac{18}{36} = 0.5000 \qquad\qquad p(1\text{H}) = \frac{2}{4} = 0.5000$$

$$p(0+) = \frac{9}{36} = 0.2500 \qquad\qquad p(0\text{H}) = \frac{1}{4} = 0.2500$$

figure 12.3 Comparison of empirical sampling approach and *a priori* approach for generating sampling distributions.

THE NORMAL DEVIATE (z) TEST

Although much of the foregoing has been abstract and seemingly impractical, it is necessary to understand the sampling distributions underlying many of the statistical tests that follow. One such test, the normal deviate (z) test, is used when we know the parameters of the null-hypothesis population. The z test employs the mean of the sample as a basic statistic. Let's consider an experiment where the z test is appropriate.

experiment

Evaluating a School Reading Program

Assume you are superintendent of public schools for the city in which you live. Recently, local citizens have been concerned that the reading program in the public schools may be an inferior one. Since this is a serious issue, you decide to conduct an experiment to investigate the matter. You set $\alpha = 0.05_{1\,\text{tail}}$ for making your decision. You begin by comparing the reading level of current high school seniors with established norms. The norms are based on scores from a reading proficiency test administered nationally to a large number of high school seniors. The scores of this population are normally distributed with $\mu = 75$ and $\sigma = 16$. For your experiment, you administer the reading test to 100 randomly selected high school seniors in your city. The obtained mean of the sample $(\overline{X}_{\text{obt}}) = 72$. What is your conclusion?

There is no doubt that the sample mean of 72 is lower than the national population mean of 75. Is it significantly lower, however? If chance alone is at work, then we can consider the 100 sample scores to be a random sample from a population with $\mu = 75$ and $\sigma = 16$. What is the probability of getting a mean score as low as or even lower than 72 if the 100 scores are a random sample from a normally distributed population having a mean of 75 and standard deviation of 16? If the probability is equal to or lower than alpha, we reject H_0 and accept H_1. If not, we retain H_0. It is clear that the statistic we are using is the *mean* of the sample. Therefore, to determine the appropriate probability, we must know the *sampling distribution of the mean.*

MENTORING TIP

Remember: to evaluate a statistic, we must know its sampling distribution.

In the following section, we shall discuss the sampling distribution of the mean. For the time being, set aside the "Super" and his problem. We shall return to him soon enough. For now, it is sufficient to realize that we are going to use the mean of a sample to evaluate H_0, and to do that, we must know the sampling distribution of the mean.

Sampling Distribution of the Mean

Applying the definition of the sampling distribution of a statistic to the mean, we obtain the following:

The sampling distribution of the mean can be determined empirically and theoretically, the latter through use of the Central Limit Theorem. The theoretical derivation is complex and beyond the level of this textbook. Therefore, for pedagogical reasons, we prefer to present the empirical approach. When we follow this approach, we can determine the sampling distribution of the mean by actually taking a specific population of raw scores having a mean μ and standard deviation and (1) drawing all possible different samples of a fixed size N, (2) calculating the mean of each sample, and (3) calculating the probability of getting each mean value if chance alone were operating. This process is shown in Figure 12.4. After performing these three steps, we would have derived the sampling distribution of the mean for samples of size N taken from a specific population with mean μ and standard deviation σ. This sampling distribution of the mean would give all the values that the mean could take for samples of size N, along with the probability of getting each value if sampling is random from the specified population. By repeating the three-step process for populations of different score values and by systematically varying N, we can determine that the sampling distribution of the mean has the

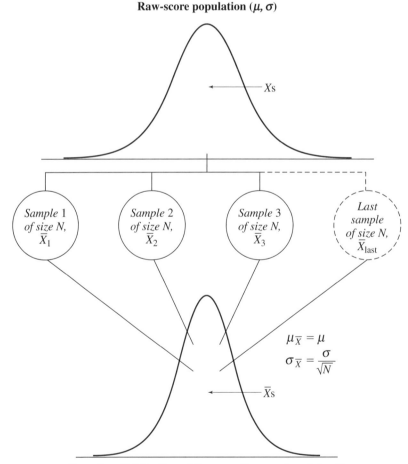

Raw-score population (μ, σ)

$\mu_{\bar{X}} = \mu$

$\sigma_{\bar{X}} = \dfrac{\sigma}{\sqrt{N}}$

Sampling distribution of the mean for samples of size N

figure 12.4   Generating the sampling distribution of the mean for samples of size N taken from a population of raw scores.

following general characteristics. For samples of any size N, the sampling distribution of the mean

(1) is a distribution of scores, each score of which is a sample mean of N scores. This distribution has a mean and a standard deviation. The distribution is shown in the bottom part of Figure 12.4. You should note that this is a *population* set of scores even though the scores are based on samples, because the distribution contains the *complete* set of sample means. We shall symbolize the mean of the distribution as $\mu_{\overline{X}}$ and the standard deviation as $\sigma_{\overline{X}}$. Thus,

$$\mu_{\overline{X}} = \text{mean of the sampling distribution of the mean}$$
$$\sigma_{\overline{X}} = \text{standard deviation of the sampling distribution of the mean}$$
$$= \text{standard error of the mean}$$

$\sigma_{\overline{X}}$ is also called the *standard error of the mean* because each sample mean can be considered an estimate of the mean of the raw-score population. Variability between sample means then occurs due to errors in estimation—hence the phrase standard *error* of the mean for $\sigma_{\overline{X}}$.

(2) has a mean equal to the mean of the raw-score population. In equation form,

$$\mu_{\overline{X}} = \mu$$

(3) has a standard deviation equal to the standard deviation of the raw-score population divided by $\sqrt{N}$. In equation form,

$$\sigma_{\overline{X}} = \frac{\sigma}{\sqrt{N}}$$

(4) is normally shaped, depending on the shape of the raw-score population and on the sample size, N.

The first characteristic is rather obvious. It merely states that the sampling distribution of the mean is made up of sample mean scores. As such, it, too, must have a mean and a standard deviation. The second characteristic says that the *mean of the sampling distribution of the mean* is equal to the mean of the raw scores ($\mu_{\overline{X}} = \mu$). We can gain some insight into this relationship by recognizing that each sample mean is an estimate of the mean of the raw-score population. Each will differ from the mean of the raw-score population due to chance. Sometimes the sample mean will be greater than the population mean, and sometimes it will be smaller because of chance factors. As we take more sample means, the average of these sample means will get closer to the mean of the raw-score population because the chance factors will cancel. Finally, when we have all of the possible different sample means, their average will equal the mean of the raw-score population ($\mu_{\overline{X}} = \mu$).

The third characteristic says that the standard deviation of the sampling distribution of the mean is equal to the standard deviation of the raw-score population divided by $\sqrt{N}$ ($\sigma_{\overline{X}} = \sigma/\sqrt{N}$). This says that the standard deviation of the sampling distribution of the mean varies directly with the standard deviation of the raw-score population and inversely with $\sqrt{N}$. It is fairly obvious why $\sigma_{\overline{X}}$ should vary directly with σ. If the scores in the population are more variable, σ goes up and so does the variability between the means based on these scores. Understanding why $\sigma_{\overline{X}}$ varies inversely with $\sqrt{N}$ is a little more difficult. Recognizing that each sample mean is an estimate of the mean of the raw-score population is the key. As N (the number of scores in each sample) goes up, each sample mean becomes a more accurate estimate of μ. Since the sample means are more accurate, they will vary less from sample to sample, causing the variance ($\sigma_{\overline{X}}^2$) of the sample means to decrease. Thus, $\sigma_{\overline{X}}^2$ varies inversely with N. Since

$\sigma_{\overline{X}} = \sqrt{\sigma_{\overline{X}}^2}$, then $\sigma_{\overline{X}}$ varies inversely with $\sqrt{N}$. We would like to further point out that, since the standard deviation of the sampling distribution of the mean ($\sigma_{\overline{X}}$) changes with sample size, there is a different sampling distribution of the mean for each different sample size. This seems reasonable, because if the sample size changes, then the scores in each sample change and, consequently, so do the sample means. Thus, the sampling distribution of the mean for samples of size 10 should be different from the sampling distribution of the mean for samples of size 20 and so forth.

Regarding the fourth point, there are two factors that determine the shape of the sampling distribution of the mean: (1) the shape of the population raw scores and (2) the sample size (*N*). Concerning the first factor, if the population of raw scores is normally distributed, the sampling distribution of the mean will also be normally distributed, regardless of sample size. However, if the population of raw scores is not normally distributed, the shape of the sampling distribution depends on the sample size. *The Central Limit Theorem tells us that, regardless of the shape of the population of raw scores, the sampling distribution of the mean approaches a normal distribution as sample size N increases.* If *N* is sufficiently large, the sampling distribution of the mean is approximately normal. How large must *N* be for the sampling distribution of the mean to be considered normal? This depends on the shape of the raw-score population. The further the raw scores deviate from normality, the larger the sample size must be for the sampling distribution of the mean to be normally shaped. If $N \geq 300$, the shape of the population of raw scores is no longer important. With this size *N*, regardless of the shape of the raw-score population, the sampling distribution of the mean will deviate so little from normality that, for statistical calculations, we can consider it normally distributed. Since most populations encountered in the behavioral sciences do not differ greatly from normality, if $N \geq 30$, it is usually assumed that the sampling distribution of the mean will be normally shaped.*

Although it is beyond the scope of this text to prove these characteristics, we can demonstrate them, as well as gain more understanding about the sampling distribution of the mean, by considering a population and deriving the sampling distribution of the mean for samples taken from it. To simplify computation, let's use a population with a small number of scores. For the purposes of this illustration, assume the population raw scores are 2, 3, 4, 5, and 6. The mean of the population (μ) equals 4.00, and the standard deviation (σ) equals 1.41. We want to derive the sampling distribution of the mean for samples of size 2 taken from this population. Again, assume sampling is one score at a time, with replacement. The first step is to draw all possible different samples of size 2 from the population. Figure 12.5 shows the population raw scores and, schematically, the different samples of size 2 that can be drawn from it. There are 25 different samples of size 2 possible. These are listed in the table of Figure 12.5, column 2. Next, we must calculate the mean of each sample. The results are shown in column 3 of this table. It is now a simple matter to calculate the probability of getting each mean value. Thus,

$$p(\overline{X} = 2.0) = \frac{\text{Number of possible } \overline{X}\text{s of 2.0}}{\text{Total number of } \overline{X}\text{s}} = \frac{1}{25} = 0.04$$

$$p(\overline{X} = 2.5) = \frac{2}{25} = 0.8$$

$$p(\overline{X} = 3.0) = \frac{3}{25} = 0.12$$

$$p(\overline{X} = 3.5) = \frac{4}{25} = 0.16$$

$$p(\overline{X} = 4.0) = \frac{5}{25} = 0.20$$

---

*There are some notable exceptions to this rule, such as reaction-time scores.

$$p(\overline{X} = 4.5) = \frac{4}{25} = 0.16$$

$$p(\overline{X} = 5.0) = \frac{3}{25} = 0.12$$

$$p(\overline{X} = 5.5) = \frac{2}{25} = 0.08$$

$$p(\overline{X} = 6.0) = \frac{1}{25} = 0.04$$

We have now derived the sampling distribution of the mean for samples of $N = 2$ taken from a population comprising the raw scores 2, 3, 4, 5, and 6. We have determined

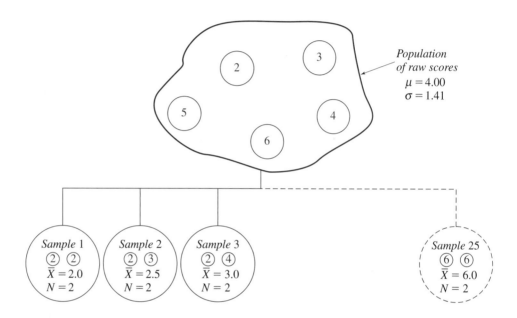

Sample Number (1)	Sample Scores (2)	$\overline{X}$ (3)	Sample Number (1)	Sample Scores (2)	$\overline{X}$ (3)
1	2, 2	2.0	14	4, 5	4.5
2	2, 3	2.5	15	4, 6	5.0
3	2, 4	3.0	16	5, 2	3.5
4	2, 5	3.5	17	5, 3	4.0
5	2, 6	4.0	18	5, 4	4.5
6	3, 2	2.5	19	5, 5	5.0
7	3, 3	3.0	20	5, 6	5.5
8	3, 4	3.5	21	6, 2	4.0
9	3, 5	4.0	22	6, 3	4.5
10	3, 6	4.5	23	6, 4	5.0
11	4, 2	3.0	24	6, 5	5.5
12	4, 3	3.5	25	6, 6	6.0
13	4, 4	4.0			

figure 12.5 All of the possible samples of size 2 that can be drawn from a population comprising the raw scores 2, 3, 4, 5, and 6. Sampling is one at a time, with replacement.

$\overline{X}$	$p(\overline{X})$
2.0	0.04
2.5	0.08
3.0	0.12
3.5	0.16
4.0	0.20
4.5	0.16
5.0	0.12
5.5	0.08
6.0	0.04

all the mean values possible from sampling two scores from the given population, along with the probability of obtaining each mean value if sampling is random from the population. The complete sampling distribution is shown in Table 12.1.

Suppose, for some reason, we wanted to know the probability of obtaining an $\overline{X} \geq 5.5$ due to randomly sampling two scores, one at a time, with replacement, from the raw-score population. We can determine the answer by consulting the sampling distribution of the mean for $N = 2$. Why? Because this distribution contains all of the possible sample mean values and their probability under the assumption of random sampling. Thus,

$$p(\overline{X} \geq 5.5) = 0.08 + 0.04 = 0.12$$

Now, let's consider the characteristics of this distribution: first, its shape. The original population of raw scores and the sampling distribution have been plotted in Figure 12.6(a) and (b). In part (c), we have plotted the sampling distribution of the mean with $N = 3$. Note that the shape of the two sampling distributions differs greatly from the population of raw scores. *Even with an N as small as 3 and a very nonnormal population of raw scores, the sampling distribution of the mean has a shape that approaches normality.* This is an illustration of what the Central Limit Theorem is telling us—namely, that as N increases, the shape of the sampling distribution of the mean approaches that of a normal distribution. Of course, if the shape of the raw-score population were normal, the shape of the sampling distribution of the mean would be too.

Next, let's demonstrate that $\mu_{\overline{X}} = \mu$:

$$\mu = \frac{\Sigma X}{\text{Number of raw scores}} \qquad\qquad \mu_{\overline{X}} = \frac{\Sigma \overline{X}}{\text{Number of mean scores}}$$

$$= \frac{20}{5} = 4.00 \qquad\qquad\qquad = \frac{100}{25} = 4.00$$

Thus,

$$\mu_{\overline{X}} = \mu$$

The mean of the raw scores is found by dividing the sum of the raw scores by the number of raw scores: $\mu = 4.00$. The mean of the sampling distribution of the mean is found by dividing the sum of the sample mean scores by the number of mean scores: $\mu_{\overline{X}} = 4.00$. Thus, $\mu_{\overline{X}} = \mu$.

Finally, we need to show that $\sigma_{\overline{X}} = \sigma/\sqrt{N}$. $\sigma_{\overline{X}}$ can be calculated in two ways: (1) from the equation $\sigma_{\overline{X}} = \sigma/\sqrt{N}$ and (2) directly from the sample mean scores themselves. Our demonstration will involve calculating $\sigma_{\overline{X}}$ in both ways, showing that they lead to the same value. The calculations are shown in Table 12.2. Since both methods yield the same value ($\sigma_{\overline{X}} = 1.00$), we have demonstrated that

$$\sigma_{\overline{X}} = \frac{\sigma}{\sqrt{N}}$$

Note that N in the previous equation is the number of scores in each sample. Thus, we have demonstrated that

1. $\mu_{\overline{X}} = \mu$
2. $\sigma_{\overline{X}} = \sigma/\sqrt{N}$
3. The sampling distribution of the mean takes on a shape similar to normal even if the raw scores are nonnormal.

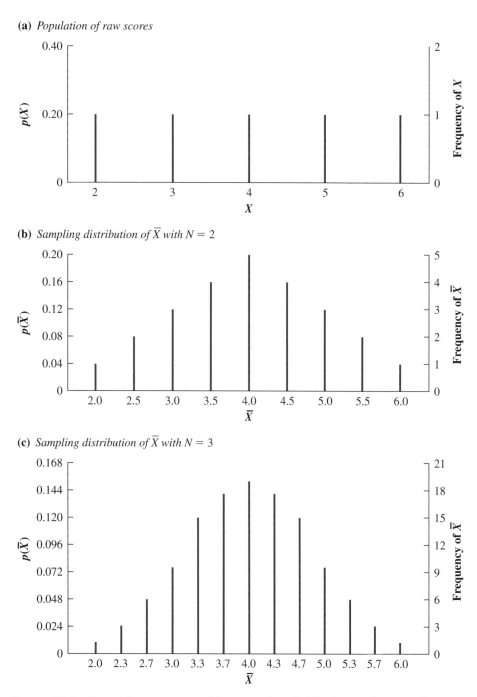

(a) *Population of raw scores*

(b) *Sampling distribution of $\overline{X}$ with N = 2*

(c) *Sampling distribution of $\overline{X}$ with N = 3*

figure 12.6   Population scores and the sampling distribution of the mean for samples of size $N = 2$ and $N = 3$.

The Reading Proficiency Experiment Revisited

We are now in a position to return to the "Super" and evaluate the data from the experiment evaluating reading proficiency. Let's restate the experiment.

You are superintendent of public schools and have conducted an experiment to investigate whether the reading proficiency of high school seniors living in your city is deficient.

table 12.2 Demonstration that $\sigma_{\overline{X}} = \dfrac{\sigma}{\sqrt{N}}$

Using $\sigma_{\overline{X}} = \sigma/\sqrt{N}$	Using the Sample Mean Scores
$\sigma_{\overline{X}} = \dfrac{\sigma}{\sqrt{N}}$	$\sigma_{\overline{X}} = \sqrt{\dfrac{\Sigma(\overline{X} - \mu_{\overline{X}})^2}{\text{Number of mean scores}}}$
$= \dfrac{1.41}{\sqrt{2}}$	$= \sqrt{\dfrac{(2.0 - 4.0)^2 + (2.5 - 4.0)^2 + \cdots + (6.0 - 4.0)^2}{25}}$
$= 1.00$	$= \sqrt{\dfrac{25}{25}} = 1.00$

A random sample of 100 high school seniors from this population had a mean reading score of 72 ($\overline{X}_{\text{obt}} = 72$). National norms of reading proficiency for high school seniors show a normal distribution of scores with a mean of 75 ($\mu = 75$) and a standard deviation of 16 ($\sigma = 16$). Is it reasonable to consider the 100 scores a random sample from a normally distributed population of reading scores where $\mu = 75$ and $\sigma = 16$? Use $\alpha = 0.05_{1\text{ tail}}$.

If we take all possible samples of size 100 from the population of normally distributed reading scores, we can determine the sampling distribution of the mean samples with $N = 100$. From what has been said before, this distribution (1) is normally shaped, (2) has a mean $\mu_{\overline{X}} = \mu = 75$, and (3) has a standard deviation $\sigma_{\overline{X}} = \sigma/\sqrt{N} = 16/\sqrt{100} = 1.6$. The two distributions are shown in Figure 12.7. Note that the sampling distribution of the mean contains all the possible mean scores from samples of size 100 drawn from the null-hypothesis population ($\mu = 75$, $\sigma = 16$). For the sake of clarity in the following exposition, we have redrawn the sampling distribution of the mean alone in Figure 12.8.

The shaded area of Figure 12.8 contains all the mean values of samples of $N = 100$ that are as low as or lower than $\overline{X}_{\text{obt}} = 72$. The proportion of shaded area to total area will tell us the probability of obtaining a sample mean equal to or less than 72 if chance alone is at work (another way of saying this is, "if the sample is a random sample from the null-hypothesis population"). Since the sampling distribution of the mean is normally shaped, we can find the proportion of the shaded area by (1) calculating the z transform (z_{obt}) for $\overline{X}_{\text{obt}} = 72$ and (2) determining the appropriate area from Table A, Appendix D, using z_{obt}.

The equation for z_{obt} is very similar to the z equation in Chapter 5, but instead of dealing with raw scores, we are dealing with mean values. The two equations are shown in Table 12.3.

table 12.3 z equations

Raw Scores	Mean Scores
$z = \dfrac{X - \mu}{\sigma}$	$z_{\text{obt}} = \dfrac{\overline{X}_{\text{obt}} - \mu_{\overline{X}}}{\sigma_{\overline{X}}}$

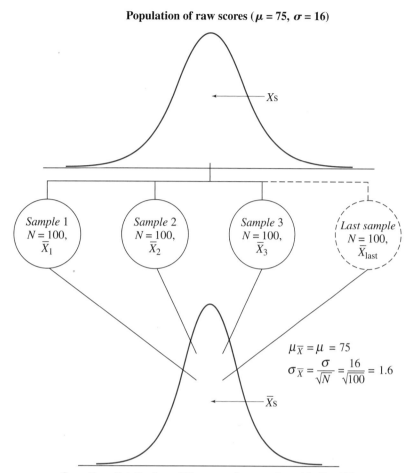

Population of raw scores ($\mu = 75$, $\sigma = 16$)

Xs

Sample 1
$N = 100$,
$\overline{X}_1$

Sample 2
$N = 100$,
$\overline{X}_2$

Sample 3
$N = 100$,
$\overline{X}_3$

Last sample
$N = 100$,
$\overline{X}_{last}$

$\mu_{\overline{X}} = \mu = 75$

$\sigma_{\overline{X}} = \dfrac{\sigma}{\sqrt{N}} = \dfrac{16}{\sqrt{100}} = 1.6$

$\overline{X}$s

Sampling distribution of the mean for samples with $N = 100$

figure 12.7 Sampling distribution of the mean for samples of size $N = 100$ drawn from a population of raw scores with $\mu = 75$ and $\sigma = 16$.

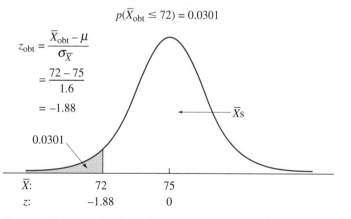

$p(\overline{X}_{obt} \leq 72) = 0.0301$

$$z_{obt} = \frac{\overline{X}_{obt} - \mu}{\sigma_{\overline{X}}}$$

$$= \frac{72 - 75}{1.6}$$

$$= -1.88$$

0.0301

$\overline{X}$s

$\overline{X}$:	72	75
z:	-1.88	0

figure 12.8 Evaluation of reading proficiency data comparing the obtained probability with the alpha level.

Since $\mu_{\overline{X}} = \mu$, the z_{obt} equation for sample means simplifies to

$$z_{obt} = \frac{\overline{X}_{obt} - \mu}{\sigma_{\overline{X}}} \qquad z \text{ transformation for } \overline{X}_{obt}$$

Calculating z_{obt} for the present experiment, we obtain

$$z_{obt} = \frac{\overline{X}_{obt} - \mu}{\sigma_{\overline{X}}} = \frac{72 - 75}{1.6} = -1.88 \quad for \ \overline{X}_{obt} = 72$$

From Table A, column C, in Appendix D,

$$p(\overline{X}_{obt} \leq 72) = 0.0301$$

Since $0.0301 < 0.05$, we reject H_0 and conclude that it is unreasonable to assume that the 100 scores are a random sample from a population where $\mu = 75$. The reading proficiency of high school seniors in your city appears deficient.

Alternative Solution Using z_{obt} and z_{crit}

MENTORING TIP

This is the preferred method.

The results of this experiment can be analyzed in another way. This method is actually the preferred method because it is simpler and it sets the pattern for the inference tests to follow. However, it builds upon the previous method and therefore couldn't be presented until now. To use this method, we must first define some terms.

definitions

■ *The* **critical region for rejection of the null hypothesis** *is the area under the curve that contains all the values of the statistic that allow rejection of the null hypothesis.*

■ *The* **critical value of a statistic** *is the value of the statistic that bounds the critical region.*

To analyze the data using the alternative method, all we need do is calculate z_{obt}, determine the critical value of z (z_{crit}), and assess whether z_{obt} falls within the critical region for rejection of H_0. We already know how to calculate z_{obt}.

The critical region for rejection of H_0 is determined by the alpha level. For example, if $\alpha = 0.05_{1 \ tail}$ in the direction predicting a negative z_{obt} value, as in the previous example, then the critical region for rejection of H_0 is the area under the left tail of the curve that equals 0.0500. We find z_{crit} for this area by using Table A in a reverse manner. Referring to Table A and skimming column C until we locate 0.0500, we can determine the z value that corresponds to 0.0500. It turns out that 0.0500 falls midway between the z scores of 1.64 and 1.65. Therefore, the z value corresponding to 0.0500 is 1.645. Since we are dealing with the left tail of the distribution,

$$z_{crit} = -1.645$$

This score defines the critical region for rejection of H_0 and, hence, is called z_{crit}. If z_{obt} falls in the critical region for rejection, we will reject H_0. These relationships are shown in Figure 12.9(a). If $\alpha = 0.05_{1 \ tail}$ in the direction predicting a positive z_{obt} value, then

$$z_{crit} = 1.645$$

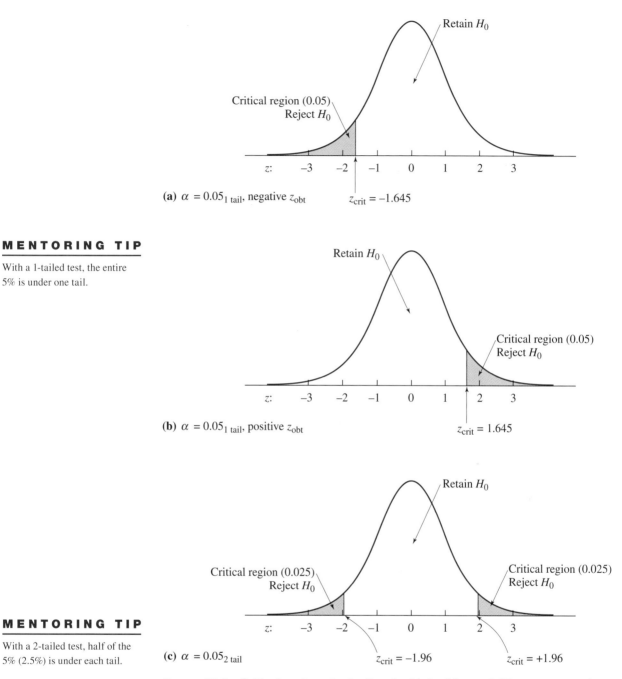

figure 12.9   Critical region of rejection for H_0 for (a) $\alpha = 0.05_{1 \text{ tail}}$, z_{obt} negative; (b) $\alpha = 0.05_{1 \text{ tail}}$, z_{obt} positive; and (c) $\alpha = 0.05_{2 \text{ tail}}$.

Adapted from *Fundamental Statistics for Behavioral Sciences* by Robert B. McCall, © 1998 by Brooks/Cole.

This is shown in Figure 12.9(b). If $\alpha = 0.05_{2 \text{ tail}}$, then the combined area under the two tails of the curve must equal 0.0500. Thus, the area under each tail must equal 0.0250, as in Figure 12.9(c). For this area,

$$z_{crit} = \pm 1.96$$

To reject H_0, the obtained sample mean ($\overline{X}_{obt}$) must have a *z*-transformed value (z_{obt}) that falls within the critical region for rejection.

Let's now use these concepts to analyze the reading data. First, we calculate z_{obt}:

$$z_{obt} = \frac{\overline{X}_{obt} - \mu}{\sigma_{\overline{X}}} = \frac{72 - 75}{1.6} = \frac{-3}{1.6} = -1.88$$

The next step is to determine z_{crit}. Since $\alpha = 0.05_{1\,tail}$, the area under the left tail equals 0.0500. For this area, from Table A we obtain

$$z_{crit} = -1.645$$

Finally, we must determine whether z_{obt} falls within the critical region. If it does, we reject the null hypothesis. If it doesn't, we retain the null hypothesis. The decision rule states the following:

If $|z_{obt}| \geq |z_{crit}|$, *reject the null hypothesis. If not, retain the null hypothesis.*

Note that this equation is just a shorthand way of specifying that, if z_{obt} is positive, it must be equal to or greater than $+z_{crit}$ to fall within the critical region. If z_{obt} is negative, it must be equal to or less than $-z_{crit}$ to fall within the critical region.

In the present example, since $|z_{obt}| > 1.645$, we reject the null hypothesis. The complete solution using this method is shown in Figure 12.10. We would like to point out that, in using this method, we are following the two-step procedure outlined previously in this chapter for analyzing data: (1) calculating the appropriate statistic and (2) evaluating the statistic based on its sampling distribution. Actually, the experimenter calculates two statistics: $\overline{X}_{obt}$ and z_{obt}. The final one evaluated is z_{obt}. If the sampling

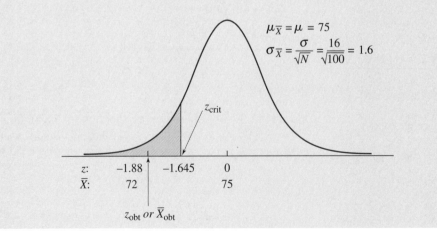

STEP 1: Calculate the appropriate statistic: $z_{obt} = \dfrac{\overline{X}_{obt} - \mu}{\sigma_{\overline{X}}} = \dfrac{72 - 75}{1.6} = -1.88$

STEP 2: Evaluate the statistic based on its sampling distribution. The decision rule is as follows: If $|z_{obt}| \geq |z_{crit}|$, reject H_0. Since $\alpha = 0.05_{1\,tail}$, from Table A,

$$z_{crit} = -1.645$$

Since $|z_{obt}| < 1.645$, it fall within the critical region for rejection of H_0. Therefore, we reject H_0.

$\mu_{\overline{X}} = \mu = 75$

$\sigma_{\overline{X}} = \dfrac{\sigma}{\sqrt{N}} = \dfrac{16}{\sqrt{100}} = 1.6$

z_{crit}

| *z*: | −1.88 | −1.645 | 0 |
| *X̄*: | 72 | | 75 |

z_{obt} *or* $\overline{X}_{obt}$

figure 12.10 Solution to reading proficiency experiment using z_{obt} and the critical region.

distribution of $\overline{X}$ is normally shaped, then the *z* distribution will also be normal and the appropriate probabilities will be given by Table A. Of course, the *z* distribution has a mean of 0 and a standard deviation of 1, as discussed in Chapter 5.

Let's try another problem using this approach.

Practice Problem 12.1

A university president believes that, over the past few years, the average age of students attending his university has changed. To test this hypothesis, an experiment is conducted in which the age of 150 students who have been randomly sampled from the student body is measured. The mean age is 23.5 years. A complete census taken at the university a few years before the experiment showed a mean age of 22.4 years, with a standard deviation of 7.6.

a. What is the nondirectional alternative hypothesis?
b. What is the null hypothesis?
c. Using $\alpha = 0.05_{2\,\text{tail}}$, what is the conclusion?

SOLUTION

a. Nondirectional alternative hypothesis: Over the past few years, the average age of students at the university has changed. Therefore, the sample with $\overline{X}_{\text{obt}} = 23.5$ is a random sample from a population where $\mu \neq 22.4$.
b. Null hypothesis: The null hypothesis asserts that it is reasonable to consider the sample with $\overline{X}_{\text{obt}} = 23.5$ a random sample from a population with $\mu = 22.4$.
c. Conclusion, using: $\alpha = 0.05_{2\,\text{tail}}$:

> **STEP 1:** **Calculate the appropriate statistic.** The data are given in the problem.
>
> $$z_{\text{obt}} = \frac{\overline{X}_{\text{obt}} - \mu}{\sigma_{\overline{X}}}$$
>
> $$= \frac{\overline{X}_{\text{obt}} - \mu}{\sigma/\sqrt{N}} = \frac{23.5 - 22.4}{7.6/\sqrt{150}}$$
>
> $$= \frac{1.1}{0.6205} = 1.77$$

> **STEP 2:** **Evaluate the statistic based on its sampling distribution.** The decision rule is as follows: If $|z_{\text{obt}}| \geq |z_{\text{crit}}|$, reject H_0. If not, retain H_0. Since $\alpha = 0.05_{2\,\text{tail}}$, from Table A,
>
> $$z_{\text{crit}} = \pm 1.96$$

Since $|z_{\text{obt}}| < 1.96$, it does not fall within the critical region for rejection of H_0. Therefore, we retain H_0. We cannot conclude that the average age of students attending the university has changed.

Practice Problem 12.2

A gasoline manufacturer believes a new additive will result in more miles per gallon. A large number of mileage measurements on the gasoline without the additive have been made by the company under rigorously controlled conditions. The results show a mean of 24.7 miles per gallon and a standard deviation of 4.8. Tests are conducted on a sample of 75 cars using the gasoline plus the additive. The sample mean equals 26.5 miles per gallon.

a. Let's assume there is adequate basis for a one-tailed test. What is the directional alternative hypothesis?

b. What is the null hypothesis?

c. What is the conclusion? Use $\alpha = 0.05_{1\,\text{tail}}$.

SOLUTION

a. Directional alternative hypothesis: The new additive increases the number of miles per gallon. Therefore, the sample with $\overline{X}_{\text{obt}} = 26.5$ is a random sample from a population where $\mu > 24.7$.

b. Null hypothesis H_0: The sample with $\overline{X}_{\text{obt}} = 26.5$ is a random sample from a population with $\mu \leq 24.7$.

c. Conclusion, using $\alpha = 0.05_{1\,\text{tail}}$:

STEP 1: **Calculate the appropriate statistic.** The data are given in the problem.

$$z_{\text{obt}} = \frac{\overline{X}_{\text{obt}} - \mu}{\sigma/\sqrt{N}}$$

$$= \frac{26.5 - 24.7}{4.8/\sqrt{75}} = \frac{1.8}{0.5543}$$

$$= 3.25$$

STEP 2: **Evaluate the statistic based on its sampling distribution.** The decision rule is as follows: If $|z_{\text{obt}}| \geq |z_{\text{crit}}|$, reject H_0. If not, retain H_0. Since $\alpha = 0.05_{1\,\text{tail}}$, from Table A,

$$z_{\text{crit}} = 1.645$$

Since $|z_{\text{obt}}| > 1.645$, it falls within the critical region for rejection of H_0. Therefore, we reject the null hypothesis and conclude that the gasoline additive does increase miles per gallon.

Conditions Under Which the z Test Is Appropriate

The z test is appropriate when the experiment involves a single sample mean ($\overline{X}_{\text{obt}}$) and the parameters of the null-hypothesis population are known (i.e., when μ and σ are known). In addition, to use this test, the sampling distribution of the mean should be normally distributed. This, of course, requires that $N \geq 30$ or that the null-hypothesis population itself be normally distributed.* This normality requirement is spoken of as "the mathematical assumption underlying the *z* test."

Power and the *z* Test

MENTORING TIP

This is a difficult section. Please be prepared to spend more time on it.

In Chapter 11, we discussed power in conjunction with the sign test. Let's review some of the main points made in that chapter.

1. Conceptually, power is the sensitivity of the experiment to detect a real effect of the independent variable, if there is one.
2. Power is defined mathematically as the probability that the experiment will result in rejecting the null hypothesis if the independent variable has a real effect.
3. Power + Beta = 1. Thus, power varies inversely with beta.
4. Power varies directly with *N*. Increasing *N* increases power.
5. Power varies directly with the size of the real effect of the independent variable. The power of an experiment is greater for large effects than for small effects.
6. Power varies directly with alpha level. If alpha is made more stringent, power decreases.

These points about power are true regardless of the inference test. In this section, we will again illustrate these conclusions, only this time in conjunction with the normal deviate test. We will begin with a discussion of power and sample size.

example

Power and Sample Size (*N*)

Let's return to the illustrative experiment at the beginning of this chapter. We'll assume you are again wearing the hat of superintendent of public schools. This time, however, you are just designing the experiment. It has not yet been conducted. You want to determine whether the reading program for high school seniors in your city is deficient. As described previously, the national norms of reading proficiency of high school seniors is a normal distribution of population scores with $\mu = 75$ and $\sigma = 16$. You plan to test a random sample of high school seniors from your city, and you are trying to determine how large the sample size should be. You will use $\alpha = 0.05_{1 \text{ tail}}$ in evaluating the data when collected. You want to be able to detect proficiency deficiencies in your program of 3 or more mean points from the national norms. That is, if the mean reading proficiency of the population of high school seniors in your city is lower than the national norms by 3 or more points, you want your experiment to have a high probability to detect it.

a. If you decide to use a sample size of 25 ($N = 25$), what is the power of your experiment to detect a population deficiency in reading proficiency of 3 mean points from the national norms?

b. If you increase the sample size to $N = 100$, what is the power now to detect a population deficiency in reading proficiency of 3 mean points?

c. What size *N* should you use for the power to be approximately 0.9000 to detect a population deficiency in reading proficiency of 3 mean points?

SOLUTION

a. Power with $N = 25$.

As discussed in Chapter 11, power is the probability of rejecting H_0 if the independent variable has a real effect. In computing the power to detect a hypothesized real effect, we

---

*Many authors would limit the use of the *z* test to data that are of interval or ratio scaling. Please see the footnote in Chapter 2, p. 35, for references discussing this point.

must first determine the sample outcomes that will allow rejection of H_0. Then, we must determine the probability of getting any of these sample outcomes if the independent variable has the hypothesized real effect. The resulting probability is the power to detect the hypothesized real effect. Thus, there are two steps in computing power:

STEP 1: Determine the possible sample mean outcomes in the experiment that would allow H_0 to be rejected. With the z test, this means determining the critical region for rejection of H_0, using $\overline{X}$ as the statistic.

STEP 2: Assuming the hypothesized real effect of the independent variable is the true state of affairs, determine the probability of getting a sample mean in the critical region for rejection of H_0.

Let's now compute the power to detect a population deficiency in reading proficiency of 3 mean points from the national norms, using $N = 25$.

STEP 1: Determine the possible sample mean outcomes in the experiment that would allow H_0 to be rejected. With the z test, this means determining the critical region for rejection of H_0, using $\overline{X}$ as the statistic.

When evaluating H_0 with the z test, we assume the sample is a random sample from the null-hypothesis population. We will symbolize the mean of the null-hypothesis population as μ_{null}. In the present example, the null-hypothesis population is the set of scores established by national testing, that is, a normal population of scores with $\mu_{null} = 75$. With $\alpha = 0.05_{1\ tail}$, $z_{crit} = -1.645$. To determine the *critical value of* $\overline{X}$, we can use the z equation solved for $\overline{X}_{crit}$:

$$z_{crit} = \frac{\overline{X}_{crit} - \mu_{null}}{\sigma_{\overline{X}}}$$

$$\overline{X}_{crit} = \mu_{null} + \sigma_{\overline{X}}(z_{crit})$$

Substituting the data with $N = 25$,

$$\overline{X}_{crit} = 75 + 3.2(-1.645) \qquad \sigma_{\overline{X}} = \frac{\sigma}{\sqrt{N}} = \frac{16}{\sqrt{25}} = 3.2$$

$$= 75 - 5.264$$

$$= 69.74$$

Thus, with $N = 25$, we will reject H_0 if, when we conduct the experiment, the mean of the sample ($\overline{X}_{obt}$) ≤ 69.74. See Figure 12.11 for a pictorial representation of these relationships.

STEP 2: Assuming the hypothesized real effect of the independent variable is the true state of affairs, determine the probability of getting a sample mean in the critical region for rejection of H_0.

If the independent variable has the hypothesized real effect, then the sample scores in your experiment are not a random sample from the null-hypothesis population. Instead, they are a random sample from a population having a mean as specified by the hypothesized real effect. We shall symbolize this mean as μ_{real}. Thus, if the reading proficiency of the population of seniors in your city is 3 mean points lower than the national norms, then the sample in your experiment is a random sample from a population where $\mu_{real} = 72$. The probability of your sample mean falling in the critical region if the sample is actually a random sample from a population where $\mu_{real} = 72$ is found by obtaining the z transform for $\overline{X}_{obt} \leq 69.74$ and looking up its corresponding area in Table A. Thus,

$$z_{obt} = \frac{\overline{X}_{obt} - \mu_{real}}{\sigma_{\overline{X}}} \qquad \sigma_{\overline{X}} = \frac{\sigma}{\sqrt{N}} = \frac{16}{\sqrt{25}} = 3.2$$

$$= \frac{69.74 - 72}{3.2}$$

$$= -0.71$$

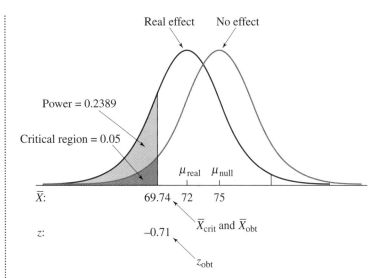

figure 12.11   Power for $N = 25$.

From Table A,

$$p(\overline{X}_{\text{obt}} \leq 69.74) = p(z_{\text{obt}} \leq -0.71) = 0.2389$$

Thus,

$$\text{Power} = 0.2389$$

$$\text{Beta} = 1 - \text{Power} = 1 - 0.2389 = 0.7611$$

Thus, the power to detect a deficiency of 3 mean points with $N = 25$ is 0.2389 and beta = 0.7611.

   Since the probability of a Type II error is too high, you decide not to go ahead and run the experiment with $N = 25$. Let's now see what happens to power and beta if N is increased to 100.

b. If $N = 100$, what is the power to detect a population difference in reading proficiency of 3 mean points?

STEP 1: Determine the possible sample mean outcomes in the experiment that would allow H_0 to be rejected. With the z test, this means determining the critical region for rejection of H_0, using $\overline{X}$ as the statistic:

$$\overline{X}_{\text{crit}} = \mu_{\text{null}} + \sigma_{\overline{X}}(z_{\text{crit}}) \qquad \sigma_{\overline{X}} = \frac{\sigma}{\sqrt{N}} = \frac{16}{\sqrt{100}} = 1.6$$

$$= 75 + 1.6(-1.645)$$

$$= 75 - 2.632$$

$$= 72.37$$

STEP 2: Assuming the hypothesized real effect of the independent variable is the true state of affairs, determine the probability of getting a sample mean in the critical region for rejection of H_0.

$$z_{\text{obt}} = \frac{\overline{X}_{\text{obt}} - \mu_{\text{real}}}{\sigma_{\overline{X}}} \qquad \sigma_{\overline{X}} = \frac{\sigma}{\sqrt{N}} = \frac{16}{\sqrt{100}} = 1.6$$

$$= \frac{72.37 - 72}{1.6}$$

$$= 0.23$$

From Table A,

$$p(\overline{X}_{obt} \leq 72.37) = 0.5000 + 0.0910 = 0.5910$$

Thus,

$$Power = 0.5910$$

$$Beta = 1 - Power = 1 - 0.5910 = 0.4090$$

Thus, by increasing N from 25 to 100, the power to detect a deficiency of 3 mean points has increased from 0.2389 to 0.5910. Beta has decreased from 0.7611 to 0.4090. This is a demonstration that power varies directly with N and beta varies inversely with N. Thus, increasing N causes an increase in power and a decrease in beta. Figure 12.12 summarizes the relationships for this problem.

c. What size N should you use for the power to be approximately 0.9000?

For the power to be 0.9000 to detect a population deficiency of 3 mean points, the probability that $\overline{X}_{obt}$ will fall in the critical region must be equal to 0.9000. As shown in Figure 12.13, this dictates that the area between z_{obt} and $\mu_{real} = 0.4000$. From Table A, $z_{obt} = 1.28$. (Note that we have taken the closest table reading rather than interpolating. This will result in a power close to 0.9000, but not exactly equal to 0.9000.) By solving the z_{obt} equation for $\overline{X}_{obt}$ and setting $\overline{X}_{obt}$ equal to $\overline{X}_{crit}$, we can determine N. Thus,

$$\overline{X}_{obt} = \mu_{real} + \sigma_{\overline{X}}(z_{obt})$$

$$\overline{X}_{crit} = \mu_{null} + \sigma_{\overline{X}}(z_{crit})$$

Setting $\overline{X}_{obt} = \overline{X}_{crit}$, we have

$$\mu_{real} + \sigma_{\overline{X}}(z_{obt}) = \mu_{null} + \sigma_{\overline{X}}(z_{crit})$$

Solving for N,

$$\mu_{real} - \mu_{null} = \sigma_{\overline{X}}(z_{crit}) - \sigma_{\overline{X}}(z_{obt})$$

$$\mu_{real} - \mu_{null} = \sigma_{\overline{X}}(z_{crit} - z_{obt})$$

$$\mu_{real} - \mu_{null} = \frac{\sigma}{\sqrt{N}}(z_{crit} - z_{obt})$$

$$\sqrt{N}(\mu_{real} - \mu_{null}) = \sigma(z_{crit} - z_{obt})$$

$$N = \left[\frac{\sigma(z_{crit} - z_{obt})}{\mu_{real} - \mu_{null}}\right]^2$$

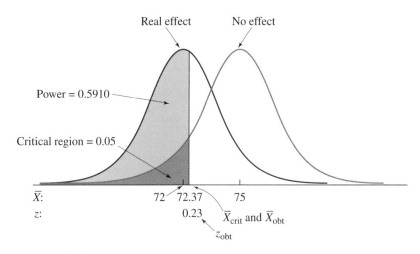

figure 12.12 Power for $N = 100$.

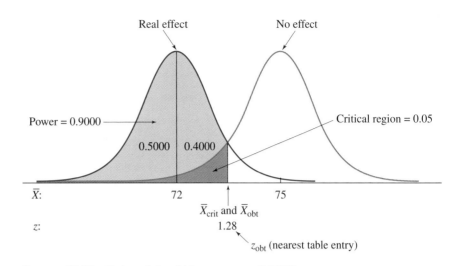

figure 12.13 Determining *N* for power = 0.9000.

Thus, to determine *N*, the equation we use is

$$N = \left[\frac{\sigma(z_{\text{crit}} - z_{\text{obt}})}{\mu_{\text{real}} - \mu_{\text{null}}} \right]^2 \qquad \textit{equation for determining N}$$

Applying this equation to the problem we have been considering, we get

$$N = \left[\frac{\sigma(z_{\text{crit}} - z_{\text{obt}})}{\mu_{\text{real}} - \mu_{\text{null}}} \right]^2$$

$$= \left[\frac{16(-1.645 - 1.28)}{72 - 75} \right]^2$$

$$= 243$$

Thus, if you increase *N* to 243 subjects, the power will be approximately 0.9000 (power = 0.8997) to detect a population deficiency in reading proficiency of 3 mean points. I suggest you confirm this power calculation yourself, using *N* = 243 as a practice exercise.

Power and alpha level   Next, let's take a look at the relationship between power and alpha. Suppose you had set $\alpha = 0.01_{1\text{ tail}}$ instead of $0.05_{1\text{ tail}}$. What happens to the resulting power? (We'll assume *N* = 100 in this question.)

SOLUTION

STEP 1: Determine the possible sample mean outcomes in the experiment that would allow H_0 to be rejected. With the *z* test, this means determining the critical region for rejection of H_0, using $\overline{X}$ as the statistic:

$$\overline{X}_{\text{crit}} = \mu_{\text{null}} + \sigma_{\overline{X}}(z_{\text{crit}}) \qquad \sigma_{\overline{X}} = \frac{\sigma}{\sqrt{N}} = \frac{16}{100} = 1.6$$

$$= 75 + 1.6(-2.33) \qquad z_{\text{crit}} = -2.33$$

$$= 71.27$$

STEP 2: Assuming the hypothesized real effect of the independent variable is the true state of affairs, determine the probability of getting a sample mean in the critical region for rejection of H_0.

$$z_{obt} = \frac{\overline{X}_{obt} - \mu_{real}}{\sigma_{\overline{X}}} \qquad \sigma_{\overline{X}} = \frac{\sigma}{\sqrt{N}} = \frac{16}{100} = 1.6$$

$$= \frac{71.27 - 72}{1.6}$$

$$= -0.46$$

From Table A,

$$p(\overline{X}_{obt} \leq 71.27) = 0.3228$$

Thus,

$$Power = 0.3228$$

$$Beta = 1 - Power = 0.6772$$

Thus, by making alpha more stringent (changing it from $0.05_{1\text{ tail}}$ to $0.01_{1\text{ tail}}$), power has decreased from 0.5910 to 0.3228. Beta has increased from 0.4090 to 0.6772. This demonstrates that there is a direct relationship between alpha and power and an inverse relationship between alpha and beta. Figure 12.14 shows the relationships for this problem.

Relationship between size of real effect and power Next, let's investigate the relationship between the size of the real effect and power. To do this, let's calculate the power to detect a population deficiency in reading proficiency of 5 mean points from the national norms. We'll assume $N = 100$ and $\alpha = 0.05_{1\text{ tail}}$. Figure 12.15 shows the relationships for this problem.

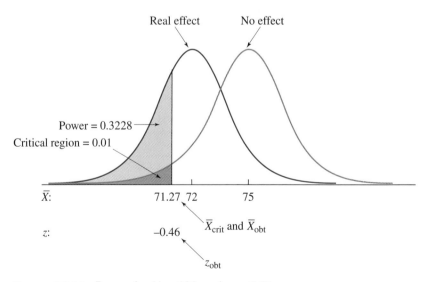

figure 12.14 Power for $N = 100$ and $\alpha = 0.01_{1\text{ tail}}$.

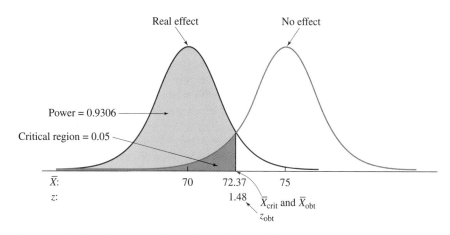

figure 12.15   Power for $N = 100$ and $\mu_{real} = 70$.

SOLUTION

STEP 1: Determine the possible sample mean outcomes in the experiment that would allow H_0 to be rejected. With the z test, this means determining the critical region for rejection of H_0, using $\overline{X}$ as the statistic:

$$\overline{X}_{crit} = \mu_{null} + \sigma_{\overline{X}}(z_{crit}) \qquad \sigma_{\overline{X}} = \frac{\sigma}{\sqrt{N}} = \frac{16}{100} = 1.6$$

$$= 75 + 1.6(-1.645)$$

$$= 72.37$$

STEP 2: Assuming the hypothesized real effect of the independent variable is the true state of affairs, determine the probability of getting a sample mean in the critical region for rejection of H_0:

$$z_{obt} = \frac{\overline{X}_{obt} - \mu_{real}}{\sigma_{\overline{X}}} \qquad \sigma_{\overline{X}} = \frac{\sigma}{\sqrt{N}} = \frac{16}{100} = 1.6$$

$$= \frac{72.37 - 70}{1.6}$$

$$= 1.48$$

From Table A,

$$p(\overline{X}_{obt} \le 72.37) = 0.5000 + 0.4306 = 0.9306$$

Thus,

$$Power = 0.9306$$

$$Beta = 1 - Power = 1 - 0.9306 = 0.0694$$

Thus, by increasing the size of the real effect from 3 to 5 mean points, power has increased from 0.5910 to 0.9306. Beta has decreased from 0.4090 to 0.0694. This demonstrates that there is a direct relationship between the size of the real effect and the power to detect it.

■ SUMMARY

In this chapter, I discussed the topics of the sampling distribution of a statistic, how to generate sampling distributions from an empirical sampling approach, the sampling distribution of the mean, and how to analyze single sample experiments with the *z* test. I pointed out that the procedure for analyzing data in most hypothesis-testing experiments is to calculate the appropriate statistic and then evaluate the statistic based on its sampling distribution. The sampling distribution of a statistic gives all the values that the statistic can take, along with the probability of getting each value if sampling is random from the null-hypothesis population. The sampling distribution can be generated theoretically with the Central Limit Theorem or empirically by (1) determining all the possible different samples of size N that can be formed from the raw-score population, (2) calculating the statistic for each of the samples, and (3) calculating the probability of getting each value of the statistic if sampling is random from the null-hypothesis population.

The sampling distribution of the mean is a distribution of sample mean values having a mean ($\mu_{\overline{X}}$) equal to μ and a standard deviation ($\sigma_{\overline{X}}$) equal to $\sigma/\sqrt{N}$. It is normally distributed if the raw-score population is normally distributed or if $N \geq 30$, assuming the raw-score population is not radically different from normality. The *z* test is appropriate for analyzing single-sample experiments, where μ and σ are known and the sample mean is used as the basic statistic. When this test is used, z_{obt} is calculated and then evaluated to determine whether it falls in the critical region for rejecting the null hypothesis. To use the *z* test, the sampling distribution of the mean must be normally distributed. This in turn requires that the null-hypothesis population be normally distributed or that $N \geq 30$.

Finally, I discussed power in conjunction with the *z* test. Power is the probability of rejecting H_0 if the independent variable has a real effect. To calculate power, we followed a two-step procedure: determining the possible sample means that allowed rejection of H_0 and finding the probability of getting any of these sample means, assuming the hypothesized real effect of the independent variable is true. Power varies directly with N, alpha, and the size of the real effect of the independent variable. Power varies inversely with beta.

■ IMPORTANT NEW TERMS

Critical region (p. 312)
Critical value of a statistic (p. 312)
Critical value of $\overline{X}$ (p. 318)
Critical value of *z* (p. 312)
Mean of the sampling distribution
  of the mean (p. 305)

Normal deviate (*z*) test (p. 303)
Null-hypothesis population (p. 300)
Sampling distribution of a statistic
  (p. 299)
Sampling distribution of the mean
  (p. 303)

Standard error of the mean (p. 305)
μ_{null} (p. 318)
μ_{real} (p. 318)

■ QUESTIONS AND PROBLEMS

1. Define each of the terms in the Important New Terms section.
2. Why is the sampling distribution of a statistic important to be able to use the statistic in hypothesis testing? Explain in a short paragraph.
3. How are sampling distributions generated using the empirical sampling approach?
4. What are the two basic steps used when analyzing data?
5. What are the assumptions underlying the use of the *z* test?
6. What are the characteristics of the sampling distribution of the mean?

7. Explain why the standard deviation of the sampling distribution of the mean is sometimes referred to as the "standard error of the mean."
8. How do each of the following differ?
   a. s and $s_{\overline{X}}$
   b. s^2 and σ^2
   c. μ and $\mu_{\overline{X}}$
   d. σ and $\sigma_{\overline{X}}$
9. Explain why $\sigma_{\overline{X}}$ should vary directly with σ and inversely with N.
10. Why should $\mu_{\overline{X}} = \mu$?

11. Is the shape of the sampling distribution of the mean always the same as the shape of the null-hypothesis population? Explain.

12. When using the z test, why is it important that the sampling distribution of the mean be normally distributed?

13. If the assumptions underlying the z test are met, what are the characteristics of the sampling distribution of z?

14. Define power, both conceptually and mathematically.

15. Explain what happens to the power of the z test when each of the following variables increases.
    a. N
    b. Alpha level
    c. Size of real effect of the independent variable
    d. σ

16. How does increasing the N of an experiment affect the following?
    a. Power
    b. Beta
    c. Alpha
    d. Size of real effect

17. Given the population set of scores 3, 4, 5, 6, 7,
    a. Determine the sampling distribution of the mean for sample sizes of 2. Assume sampling is one at a time, with replacement.
    b. Demonstrate that $\mu_{\bar{X}} = \mu$.
    c. Demonstrate that $\sigma_{\bar{X}} = \sigma/\sqrt{N}$.

18. If a population of raw scores is normally distributed and has a mean $\mu = 80$ and a standard deviation $\sigma = 8$, determine the parameters ($\mu_{\bar{X}}$ and $\sigma_{\bar{X}}$) of the sampling distribution of the mean for the following sample sizes.
    a. $N = 16$
    b. $N = 35$
    c. $N = 50$
    d. Explain what happens as N gets larger. other

19. Is it reasonable to consider a sample of 40 scores with $\bar{X}_{obt} = 65$ to be a random sample from a population of scores that is normally distributed, with $\mu = 60$ and $\sigma = 10$? Use $\alpha = 0.05_{2\,tail}$ in making your decision. other

20. A set of sample scores from an experiment has an $N = 30$ and an $\bar{X}_{obt} = 19$.
    a. Can we reject the null hypothesis that the sample is a random sample from a normal population with $\mu = 22$ and $\sigma = 8$? Use $\alpha = 0.01_{1\,tail}$. Assume the sample mean is in the correct direction.
    b. What is the power of the experiment to detect a real effect such that $\mu_{real} = 20$?
    c. What is the power to detect a $\mu_{real} = 20$ if N is increased to 100?

d. What value does N have to equal to achieve a power of 0.8000 to detect a $\mu_{real} = 20$? Use the nearest table value for z_{obt}. other

21. On the basis of her newly developed technique, a student believes she can reduce the amount of time schizophrenics spend in an institution. As director of training at a nearby institution, you agree to let her try her method on 20 schizophrenics, randomly sampled from your institution. The mean duration that schizophrenics stay at your institution is 85 weeks, with a standard deviation of 15 weeks. The scores are normally distributed. The results of the experiment show that the patients treated by the student stay a mean duration of 78 weeks, with a standard deviation of 20 weeks.
    a. What is the alternative hypothesis? In this case, assume a nondirectional hypothesis is appropriate because there are insufficient theoretical and empirical bases to warrant a directional hypothesis.
    b. What is the null hypothesis?
    c. What do you conclude about the student's technique? Use $\alpha = 0.05_{2\,tail}$. clinical, health

22. A professor has been teaching statistics for many years. His records show that the overall mean for final exam scores is 82, with a standard deviation of 10. The professor believes that this year's class is superior to his previous ones. The mean for final exam scores for this year's class of 65 students is 87. What do you conclude? Use $\alpha = 0.05_{1\,tail}$. education

23. An automotive engineer believes that her newly designed engine will be a great gas saver. A large number of tests on engines of the old design yielded a mean gasoline consumption of 27.5 miles per gallon, with a standard deviation of 5.2. Fifteen new engines are tested. The mean gasoline consumption is 29.6 miles per gallon. What is your conclusion? Use $\alpha = 0.05_{1\,tail}$. other

24. In Practice Problem 12.2 (p. 316), we presented data testing a new gasoline additive. A large number of mileage measurements on the gasoline without the additive showed a mean of 24.7 miles per gallon and a standard deviation of 4.8. An experiment was performed in which 75 cars were tested using the gasoline plus the additive. The results showed a sample mean of 26.5 miles per gallon. To evaluate these data, a directional test with $\alpha = 0.05_{1\,tail}$ was used. Suppose that before doing the experiment, the manufacturer wants to determine the probability that he will be able to detect a real mean increase of 2.0 miles per gallon with the additive if the additive is at least that effective.

a. If he tests 20 cars, what is the power to detect a mean increase of 2.0 miles per gallon?

b. If he increases the *N* to 75 cars, what is the power to detect a mean increase of 2.0 miles per gallon?

c. How many cars should he use if he wants to have a 99% chance of detecting a mean increase of 2.0 miles per gallon? I/O

25. A physical education professor believes that exercise can slow the aging process. For the past 10 years, he has been conducting an exercise class for 14 individuals who are currently 50 years old. Normally, as one ages, maximum oxygen consumption decreases. The national norm for maximum oxygen consumption in 50-year-old individuals is 30 milliliters per kilogram per minute, with a standard deviation of 8.6. The mean of the 14 individuals is 40 milliliters per kilogram per minute. What do you conclude? Use $\alpha = 0.05_{1\ \text{tail}}$. biological, health

■ ONLINE STUDY RESOURCES

CENGAGE**brain**.com

Login to CengageBrain.com to access the resources your instructor has assigned. For this book, you can access the book's companion website for chapter-specific learning tools including Know and Be Able to Do, practice quizzes, flash cards, and glossaries, and a link to Statistics and Research Methods Workshops.

aplia

If your professor has assigned Aplia homework:

1. Sign in to your account.
2. Complete the corresponding homework exercises as required by your professor.
3. When finished, click "Grade It Now" to see which areas you have mastered and which need more work, and for detailed explanations of every answer.

Visit **www.cengagebrain.com** to access your account and to purchase materials.

13

Student's *t* Test for Single Samples

LEARNING OBJECTIVES

After completing this chapter, you should be able to:

- Contrast the *t* test and the *z* test for single samples.
- Define degrees of freedom.
- Define the sampling distribution of *t*, and state its characteristics.
- Compare the *t* and *z* distributions.
- Solve problems using the *t* test for single samples and specify the conditions under which the *t* test for single samples is appropriate.
- Compute size of effect using Cohen's *d*.
- Contrast point and interval estimation.
- Define confidence interval and confidence limits.
- Define and construct the 95% and 99% confidence limits for the population mean.
- Determine for the significance of Pearson *r* using two methods.
- Understand the illustrative examples, do the practice problems, and understand the solutions.

INTRODUCTION

In Chapter 12, we discussed the *z* test and determined that it was appropriate in situations in which both the mean and the standard deviation of the null-hypothesis population were known. However, these situations are relatively rare. It is more common to encounter situations in which the mean of the null-hypothesis population can be specified and the standard deviation is *unknown*. In these cases, the *z* test cannot be used. Instead, another test, called *Student's t test*, is employed. The *t* test is very similar to the *z* test. It was developed by W. S. Gosset, writing under the pen name of "Student." Student's *t* test is a practical, quite powerful test widely used in the behavioral sciences. In this chapter, we shall discuss the *t* test in conjunction with experiments involving a single sample. In Chapter 14, we shall discuss the *t* test as it applies to experiments using two samples or conditions.

COMPARISON OF THE *z* AND *t* TESTS

The *z* and *t* tests for single sample experiments are quite alike. The equations for each are shown in Table 13.1.

In comparing these equations, we can see that the only difference is that the *z* test uses the standard deviation of the null-hypothesis population (σ), whereas the *t* test uses the standard deviation of the sample (*s*). When σ is unknown, we estimate it using the estimate given by *s*, and the resulting statistic is called *t*. Thus, the denominator of the *t* test is $s/\sqrt{N}$ rather than $\sigma/\sqrt{N}$. The symbol $s_{\overline{X}}$ replaces $\sigma_{\overline{X}}$ where

MENTORING TIP

The *t* test is like the *z* test, except it uses *s* instead of σ.

$$s_{\overline{X}} = \frac{s}{\sqrt{N}} \qquad \textit{estimated standard error of the mean}$$

We are ready now to consider an experiment using the *t* test to analyze the data.

table 13.1 Comparison of equations for the *z* and *t* tests

	z Test	*t* Test
	$z_{\text{obt}} = \dfrac{\overline{X}_{\text{obt}} - \mu}{\sigma/\sqrt{N}}$	$t_{\text{obt}} = \dfrac{\overline{X}_{\text{obt}} - \mu}{s/\sqrt{N}}$
	$= \dfrac{\overline{X}_{\text{obt}} - \mu}{\sigma_{\overline{X}}}$	$= \dfrac{\overline{X}_{\text{obt}} - \mu}{s_{\overline{X}}}$

where s = estimate of σ
$s_{\overline{X}}$ = estimate of $\sigma_{\overline{X}}$

experiment

Increasing Early Speaking in Children

Suppose you have a technique that you believe will affect the age at which children begin speaking. In your locale, the average age of first word utterances is 13.0 months. The standard deviation is unknown. You apply your technique to a random sample of 15 children. The results show that the sample mean age of first word utterances is 11.0 months, with a standard deviation of 3.34.

1. What is the nondirectional alternative hypothesis?
2. What is the null hypothesis?
3. Did the technique work? Use $\alpha = 0.05_{2\,\text{tail}}$.

SOLUTION

1. Alternative hypothesis: The technique affects the age at which children begin speaking. Therefore, the sample with $\overline{X}_{\text{obt}} = 11.0$ is a random sample from a population where $\mu \neq 13.0$.
2. Null hypothesis: H_0: The sample with $\overline{X}_{\text{obt}} = 11.0$ is a random sample from a population with $\mu = 13.0$.
3. Conclusion using $\alpha = 0.05_{2\,\text{tail}}$:

STEP 1: Calculate the appropriate statistic. Since σ is unknown, it is impossible to determine z_{obt}. However, s is known, so we can calculate t_{obt}. Thus,

$$t_{\text{obt}} = \frac{\overline{X}_{\text{obt}} - \mu}{s/\sqrt{N}}$$

$$= \frac{11.0 - 13.0}{3.34/\sqrt{15}}$$

$$= \frac{-2}{0.862}$$

$$= -2.32$$

The next step ordinarily would be to evaluate t_{obt} using the sampling distribution of *t*. However, because this distribution is not yet familiar, we need to discuss it before we can proceed with the evaluation.

THE SAMPLING DISTRIBUTION OF *t*

Using the definition of sampling distribution developed in Chapter 12, we note the following.

definition

■ *The **sampling distribution of** **t** is a probability distribution of the t values that would occur if all possible different samples of a fixed size N were drawn from the null-hypothesis population. It gives (1) all the possible different t values for samples of size N and (2) the probability of getting each value if sampling is random from the null-hypothesis population.*

As with the sampling distribution of the mean, the sampling distribution of *t* can be determined theoretically or empirically. Again, for pedagogical reasons, we prefer the empirical approach. The sampling distribution of *t* can be derived empirically by taking a specific population of raw scores, drawing all possible different samples of a fixed size *N*, and then calculating the *t* value for each sample. Once all the possible *t* values are obtained, it is a simple matter to calculate the probability of getting each different *t* value under the assumption of random sampling from the population. By varying *N* and the population scores, one can derive sampling distributions for various populations and sample sizes. Empirically or theoretically, it turns out that, if the null-hypothesis population is normally shaped, or if $N \geq 30$, the *t* distribution looks very much like the *z* distribution except that there is a *family* of *t* curves that vary with sample size. You will recall that the *z* distribution has only one curve for all sample sizes (the values represented in Table A in Appendix D). On the other hand, the *t* distribution, like the sampling distribution of the mean, has many curves depending on sample size. Since we are estimating by using *s* in the *t* equation and the size of the sample influences the accuracy of the estimate, it makes sense that there should be a different sampling distribution of *t* for different sample sizes.

Degrees of Freedom

Although the *t* distribution varies with sample size, Gosset found that it varies *uniquely* with the *degrees of freedom* associated with *t*, rather than simply with sample size. Why this is so will not be apparent until Chapter 14. For now, let's just pursue the concept of degrees of freedom.

definition   ■   *The* **degrees of freedom (df)** *for any statistic is the number of scores that are free to vary in calculating that statistic.*

For example, there are *N* degrees of freedom associated with the mean. How do we know this? For any set of scores, *N* is given. If there are three scores and we know the first two scores, the last score can take on any value. It has no restrictions. There is no way to tell what it must be by knowing the other two scores. The same is true for the first two scores. Thus, all three scores are free to vary when calculating the mean. Thus, there are *N* degrees of freedom.

Contrast this with calculating the standard deviation:

$$s = \sqrt{\frac{\Sigma (X - \overline{X})^2}{N - 1}}$$

Since the sum of deviations about the mean must equal zero, only $N - 1$ of the deviation scores are free to take on any value. Thus, there are $N - 1$ degrees of freedom associated with *s*. Why is this so? Consider the raw scores 4, 8, and 12. The mean is 8. Table 13.2 shows what happens when calculating *s*.

table 13.2 Number of deviation scores free to vary

X	$\overline{X}$	$X - \overline{X}$
4	8	−4
8	8	0
12	8	?

Since the mean is 8, the deviation score for the raw score of 4 is –4 and for the raw score of 8 is 0. Since $\Sigma(X - \overline{X}) = 0$, the last deviation is fixed by the other deviations. It must be +4 (see the "?" in Table 13.2). It cannot take on any value; instead it is fixed at +4 by the other two deviation scores. Therefore, only two of the three deviation scores are free to vary. Whatever value these take, the third is fixed. In calculating *s*, only $N - 1$ deviation scores are free to vary. Thus, there are $N - 1$ degrees of freedom associated with the standard deviation.

In calculating *t* for single samples, we must first calculate *s*. We lose 1 degree of freedom in calculating *s*, so there are $N - 1$ degrees of freedom associated with *t*. Thus, for the *t* test,

$$\text{df} = N - 1 \quad \textit{degrees of freedom for t test (single sample)}$$

t AND *z* DISTRIBUTIONS COMPARED

Figure 13.1 shows the *t* distribution for various degrees of freedom. The *t* distribution is symmetrical about zero and becomes closer to the normally distributed *z* distribution with increasing df. Notice how quickly it approaches the normal curve. Even with df as small as 20, the *t* distribution rather closely approximates the normal

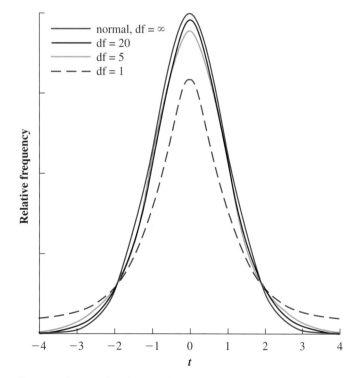

figure 13.1   *t* distribution for various degrees of freedom.

MENTORING TIP

The *t* test is less powerful than
the *z* test.

curve. Theoretically, when df $= \infty$,* the *t* distribution is identical to the *z* distribution. This makes sense because as the df increases, sample size increases and the estimate *s* gets closer to σ. At any df other than ∞, the *t* distribution has more extreme *t* values than the *z* distribution, since there is more variability in *t* because we used *s* to estimate σ. Another way of saying this is that the tails of the *t* distribution are elevated relative to the *z* distribution. Thus, for a given alpha level, the critical value of *t* is higher than for *z*, making the *t* test less powerful than the *z* test. That is, for any alpha level, t_{obt} must be higher than z_{obt} to reject the null hypothesis. Table 13.3 shows the critical values of *z* and *t* at the 0.05 and 0.01 alpha levels. As the df increases, the critical value of *t* approaches that of *z*. The critical *z* value, of course, doesn't change with sample size.

Critical values of t for various alpha levels and df are contained in Table D of Appendix D. These values have been obtained from the sampling distribution of *t* for each df and alpha level. In Table D, the degrees of freedom range from 1 to ∞. One-tailed alpha levels vary from .0005 to .10; two-tailed alpha levels vary from .001 to .20. A portion of Table D is shown below in Table 13.4. As can be seen from Table 13.4, df values are given in the leftmost column and alpha levels are given as column headings. t_{crit} values are given in the cell at the intersection of a given df and alpha level. For example, with df $= 13$ and $\alpha = 0.05_{2\text{-tail}}$, $t_{crit} = \pm 2.160$ (gray shading). Note that the table doesn't supply the "$\pm$" sign; you must do so yourself for all two-tailed alpha levels. If df $= 13$ and $\alpha = 0.05_{1\text{-tail}}$, then $t_{crit} = 1.771$ (blue shading)

table 13.3 Critical values of *z* and *t* at the 0.05
and 0.01 alpha levels, one-tailed

df	$z_{0.05}$	$t_{0.05}$	$z_{0.01}$	$t_{0.01}$
5	1.645	2.015	2.326	3.365
30	1.645	1.697	2.326	2.457
60	1.645	1.671	2.326	2.390
∞	1.645	1.645	2.326	2.326

MENTORING TIP

When using Table D, if alpha
is two-tailed, you need to
supply the "$\pm$" sign for t_{crit}.
If alpha is one-tailed, and the
predicted direction of the real
effect is a decrease in the
DV, t_{crit} is negative.

table 13.4 t_{crit} values, a portion of Table D

	Level of Significance for One-Tailed Test, $\alpha_{1\text{-tail}}$			
	.05	**.025**	**.01**	**.005**
df	Level of Significance for Two-Tailed Test, $\alpha_{2\text{-tail}}$			
	.10	**.05**	**.02**	**.01**
12	1.782	2.179	2.681	3.055
13	1.771	2.160	2.650	3.012
14	1.761	2.145	2.624	2.977
15	1.753	2.131	2.602	2.947

---

*As df approaches infinity, the *t* distribution approaches the normal curve.

or $t_{crit} = -1.771$, depending on the direction of the predicted effect. Again, you must supply the appropriate sign. Table D is used for evaluating t_{obt} for any experiment. We are now ready to return to the illustrative example.

EARLY SPEAKING EXPERIMENT REVISITED

You are investigating a technique purported to affect the age at which children begin speaking: $\mu = 13.0$ months; σ is unknown; the sample of 15 children using your technique has a mean for first word utterances of 11.0 months and a standard deviation of 3.34.

1. What is the nondirectional alternative hypothesis?
2. What is the null hypothesis?
3. Did the technique work? Use $\alpha = 0.05_{2\,tail}$.

SOLUTION

1. Alternative hypothesis: The technique affects the age at which children begin speaking. Therefore, the sample with $\overline{X}_{obt} = 11.0$ is a random sample from a population where $\mu \neq 13.0$.
2. Null hypothesis: H_0: The sample with $\overline{X}_{obt} = 11.0$ is a random sample from a population with $\mu = 13.0$.
3. Conclusion using $\alpha = 0.05_{2\,tail}$:

STEP 1: **Calculate the appropriate statistic.** Since this is a single sample experiment with unknown σ, t_{obt} is appropriate:

$$t_{obt} = \frac{\overline{X}_{obt} - \mu}{s/\sqrt{N}}$$

$$= \frac{11.0 - 13.0}{3.34/\sqrt{15}}$$

$$= -2.32$$

STEP 2: **Evaluate the statistic based on its sampling distribution.** Just as with the z test, if

$$|t_{obt}| \geq |t_{crit}|$$

then it falls within the critical region for rejection of the null hypothesis. t_{crit} is found in Table D under the appropriate alpha level and df. For this example, with $\alpha = 0.05_{2\,tail}$ and df $= N - 1 = 15 - 1 = 14$, from Table D,

$$t_{crit} = \pm 2.145$$

Since $|t_{obt}| > 2.145$, we reject H_0 and conclude that the technique does affect the age at which children in your locale first begin speaking. It appears to increase early speaking. The solution is shown in Figure 13.2.

STEP 1: **Calculate the appropriate statistic.** Since σ is unknown, t_{obt} is appropriate.

$$t_{obt} = \frac{\overline{X}_{obt} - \mu}{s/\sqrt{N}} = \frac{11.0 - 13.0}{3.34/\sqrt{15}} = -2.32$$

STEP 2: **Evaluate the statistic.** If $|t_{obt}| \geq |t_{crit}|$, reject H_0. Since $\alpha = 0.05_{2\,tail}$ and df $= N - 1 = 15 - 1 = 14$, from Table D,

$$t_{crit} = \pm 2.145$$

Since $|t_{obt}| > 2.145$, it falls within the critical region. Therefore, we reject H_0.

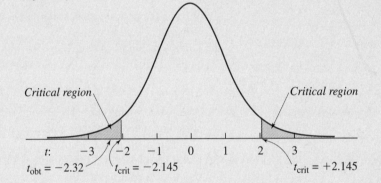

figure 13.2   Solution to the first word utterance experiment using Student's *t* test.

CALCULATING t_{obt} FROM ORIGINAL SCORES

If in a given situation the original scores are available, *t* can be calculated directly without first having to calculate *s*. The appropriate equation is given here:*

$$t_{obt} = \frac{\overline{X}_{obt} - \mu}{\sqrt{\dfrac{SS}{N(N-1)}}} \quad equation\ for\ computing\ t_{obt}\ from\ raw\ scores$$

where $$SS = \Sigma(X - \overline{X})^2 = \Sigma X^2 - \frac{(\Sigma X)^2}{N}$$

example

Suppose the original data in the previous problem were as shown in Table 13.5. Let's calculate t_{obt} directly from these raw scores.

SOLUTION

$$t_{obt} = \frac{\overline{X}_{obt} - \mu}{\sqrt{\dfrac{SS}{N(N-1)}}} = \frac{11.0 - 13.0}{\sqrt{\dfrac{156}{15(14)}}} = \frac{-2}{0.862} \qquad SS = \Sigma X^2 - \frac{(\Sigma X)^2}{N} = 1971 - \frac{(165)^2}{15}$$

$$= -2.32 \qquad\qquad\qquad\qquad\qquad\qquad\qquad\qquad = 156$$

*The derivation is presented in Note 13.1.

table 13.5 Raw scores for first word utterances example

Age (months)	
X	X^2
8	64
9	81
10	100
15	225
18	324
17	289
12	144
11	121
7	49
8	64
10	100
11	121
8	64
9	81
12	144
165	1971
N = 15	$\overline{X}_{obt} = \frac{165}{15} = 11.0$

This is the same value arrived at previously. Note that it is all right to first calculate s and then use the original t_{obt} equation. However, the answer is more subject to rounding error.

Let's try another problem.

Practice Problem 13.1

A researcher believes that in recent years women have been getting taller. She knows that 10 years ago the average height of young adult women living in her city was 63 inches. The standard deviation is unknown. She randomly samples eight young adult women currently residing in her city and measures their heights. The following data are obtained:

Height (in.)	
X	X^2
64	4,096
66	4,356
68	4,624
60	3,600

(continued)

Height (in.)	
X	**X²**
62	3,844
65	4,225
66	4,356
63	3,969
514	33,070

$$N = 8 \qquad \overline{X}_{obt} = \frac{514}{8} = 64.25$$

a. What is the alternative hypothesis? In evaluating this experiment, assume a nondirectional hypothesis is appropriate because there are insufficient theoretical and empirical bases to warrant a directional hypothesis.
b. What is the null hypothesis?
c. What is your conclusion? Use $\alpha = 0.01_{2\,tail}$.

SOLUTION

a. Nondirectional alternative hypothesis: In recent years, the height of women has been changing. Therefore, the sample with $\overline{X}_{obt} = 64.25$ is a random sample from a population where $\mu \neq 63$.
b. Null hypothesis: The null hypothesis asserts that it is reasonable to consider the sample with $\overline{X}_{obt} = 64.25$ a random sample from a population with $\mu = 63$.
c. Conclusion, using $\alpha = 0.01_{2\,tail}$:

> **STEP 1: Calculate the appropriate statistic.** The data were given previously. Since σ is unknown, t_{obt} is appropriate. There are two ways to find t_{obt}: (1) by calculating s first and then t_{obt} and (2) by calculating t_{obt} directly from the raw scores. Both methods are shown here:

s first and then t_{obt}:

$$s = \sqrt{\frac{SS}{N-1}} = \sqrt{\frac{45.5}{7}} \qquad\qquad SS = \Sigma X^2 - \frac{(\Sigma X)^2}{N} = 33,070 - \frac{(514)^2}{8}$$

$$= \sqrt{6.5} = 2.550 \qquad\qquad\qquad = 33,070 - 33,024.5 = 45.5$$

$$t_{obt} = \frac{\overline{X}_{obt} - \mu}{s/\sqrt{N}} = \frac{64.25 - 63}{2.550/\sqrt{8}}$$

$$= \frac{1.25}{0.902} = 1.39$$

directly from the raw scores:

$$t_{obt} = \frac{\overline{X}_{obt} - \mu}{\sqrt{\dfrac{SS}{N(N-1)}}} = \frac{64.25 - 63}{\sqrt{\dfrac{45.5}{8(7)}}} \qquad\qquad SS = \Sigma X^2 - \frac{(\Sigma X)^2}{N}$$

$$= \frac{1.25}{\sqrt{0.812}} = 1.39 \qquad\qquad = 33{,}070 - \frac{(514)^2}{8}$$

$$= 33{,}070 - 33{,}024.5 = 45.5$$

STEP 2:   **Evaluate the statistic.** If $|t_{obt}| \geq |t_{crit}|$, reject H_0. If not, retain H_0. With $\alpha = 0.01_{2\,tail}$ and df $= N - 1 = 8 - 1 = 7$, from Table D,

$$t_{crit} = \pm 3.499$$

Since $|t_{obt}| < 3.499$, it doesn't fall in the critical region. Therefore, we retain H_0. We cannot conclude that young adult women in the researcher's city have been changing in height in recent years.

Practice Problem 13.2

A friend of yours has been "playing" the stock market. He claims he has spent years doing research in this area and has devised an empirically successful method for investing. Since you are not averse to becoming a little richer, you are considering giving him some money to invest for you. However, before you do, you decide to evaluate his method. He agrees to a "dry run" during which he will use his method, but instead of actually buying and selling, you will just monitor the stocks he recommends to see whether his method really works. During the trial time period, the recommended stocks showed the following price changes (a plus score means an increase in price, and a minus indicates a decrease):

Stock	Price Change ($) X	X^2
A	+4.52	20.430
B	+5.15	26.522
C	+3.28	10.758
D	+4.75	22.562
E	+6.03	36.361
F	+4.09	16.728
G	+3.82	14.592
	31.64	147.953

$$N = 7 \qquad \overline{X}_{obt} = 4.52$$

During the same time period, the average price change of the stock market as a whole was +\$3.25. Since you want to know whether the method does better or worse than chance, you decide to use a two-tailed evaluation.

(continued)

a. What is the nondirectional alternative hypothesis?
b. What is the null hypothesis?
c. What is your conclusion? Use $\alpha = 0.05_{2 \text{ tail}}$.

SOLUTION

a. Nondirectional alternative hypothesis: Your friend's method results in a choice of stocks whose change in price differs from that expected due to random sampling from the stock market in general. Thus, the sample with $\overline{X}_{\text{obt}} = \$4.52$ cannot be considered a random sample from a population where $\mu = \$3.25$.

b. Null hypothesis: Your friend's method results in a choice of stocks whose change in price doesn't differ from that expected due to random sampling from the stock market in general. Therefore, the sample with $\overline{X}_{\text{obt}} = \$4.52$ can be considered a random sample from a population where $\mu = \$3.25$.

c. Conclusion, using $\alpha = 0.05_{2 \text{ tail}}$:

STEP 1: Calculate the appropriate statistic. The data are given in the previous table. Since σ is unknown, t_{obt} is appropriate.

$$t_{\text{obt}} = \frac{\overline{X}_{\text{obt}} - \mu}{\sqrt{\dfrac{SS}{N(N-1)}}}$$

$$SS = \Sigma X^2 - \frac{(\Sigma X)^2}{N}$$

$$= \frac{4.52 - 3.25}{\sqrt{\dfrac{4.940}{7(6)}}}$$

$$= 147.953 - \frac{(31.64)^2}{7}$$

$$= 4.940$$

$$= \frac{1.27}{0.343} = 3.70$$

STEP 2: Evaluate the statistic. With $\alpha = 0.05_{2 \text{ tail}}$ and df $= N - 1 = 7 - 1 = 6$, from Table D,

$$t_{\text{crit}} = \pm 2.447$$

Since $|t_{\text{obt}}| > 2.447$, we reject H_0. Your friend appears to be a winner. His method does seem to work! However, before investing heavily, we suggest you run the experiment at least one more time to guard against Type I error. Remember that replication is essential before accepting a result as factual. Better to be safe than poor.

CONDITIONS UNDER WHICH THE *t* TEST IS APPROPRIATE

The *t* test (single sample) is appropriate when the experiment has only one sample, μ is specified, σ is unknown, and the mean of the sample is used as the basic statistic. Like the *z* test, the *t* test requires that the sampling distribution of $\overline{X}$ be normal. For the sampling distribution of $\overline{X}$ to be normal, N must be ≥ 30 or the population of raw scores must be normal.*

---

*Many authors would limit the use of the *t* test to data that are of interval or ratio scaling. Please see the footnote in Chapter 2, p. 35, for references discussing this point.

SIZE OF EFFECT USING COHEN'S *d*

Thus far, we have discussed the *t* test for single samples and shown how to use it to determine whether the independent variable has a real effect on the dependent variable being measured. To determine whether the data show a real effect, we calculate t_{obt}; if t_{obt} is significant, we conclude there is a real effect. Of course, this gives us very important information. It allows us to support and further delineate theories involving the independent and dependent variables as well as provide information that may have important practical consequences.

In addition to determining whether there is a real effect, it is often desirable to determine the size of the effect. For example, in the experiment dealing with early speaking (p. 333), t_{obt} was significant and we were able to conclude that the technique had a real effect. Although we might be content with finding that there is a real effect, we might also be interested in determining the magnitude of the effect. Is it so small as to have negligible practical consequences, or is it a large and important discovery?

Cohen (1988)* has provided a simple method for determining the magnitude of real effect. Used with the *t* test, the method relies on the fact that there is a direct relationship between the size of real effect and the size of the mean difference. With the *t* test for single samples, the mean difference of interest is $\overline{X}_{obt} - \mu$. As the size of the real effect gets greater, so does the difference between $\overline{X}_{obt}$ and μ. Since *size* of real effect is the variable of interest, not direction of real effect, the statistic measuring size of real effect is given a positive value by taking the absolute value of the mean difference. We have symbolized this by "|mean difference|." The statistic used is labeled *d* and is a standardized measure of |mean difference|. Standardization is achieved by dividing by the population standard deviation, similar to what was done with the *z* score in Chapter 5. Thus, *d* has a positive value that indicates the size (magnitude) of the mean difference in standard deviation units. For example, for the *t* test for single samples, a value of $d = 0.42$ tells us that the sample mean differs from the population mean by 0.42 standard deviation units.

In its generalized form, the equation for *d* is given by

$$d = \frac{|\text{mean difference}|}{\text{population standard deviation}} \quad \textit{general equation for size of effect}$$

The general equation for *d* is the same whether we are considering the *t* test for single samples, the *t* test for correlated groups or the *t* test for independent groups (Chapter 14). What differs from test to test is the mean difference and population standard deviation used in each test. For the *t* test for single samples, *d* is given by the following conceptual equation:

$$d = \frac{|\overline{X}_{obt} - \mu|}{\sigma} \quad \textit{conceptual equation for size of effect, single sample t test}$$

Taking the absolute value of $\overline{X}_{obt} - \mu$ in the previous equation keeps *d* positive regardless of whether $\overline{X}_{obt} > \mu$ or $\overline{X}_{obt} < \mu$. Of course in situations in which we use

---

*J. Cohen, *Statistical Power Analysis for the Behavioral Sciences,* 2nd ed., Lawrence Erlbaum Associates, Hillsdale, NJ, 1988.

the *t* test for single samples, we don't know σ, so we estimate it with *s*. The resulting equation yields an estimate of *d* that is given by

$$\hat{d} = \frac{|\overline{X}_{obt} - \mu|}{s} \qquad \textit{computational equation for size of effect, single sample t test}$$

where $\hat{d}$ = estimated *d*
$\overline{X}_{obt}$ = the sample mean
μ = the population mean
s = the sample standard deviation

This is the computational equation for computing size of effect for the single samples *t* test. Please note that when applying this equation, if H_1 is directional, $\overline{X}_{obt}$ must be in the direction predicted by H_1. If it is not in the predicted direction, when analyzing the data of the experiment, the conclusion would be to retain H_0 and, ordinarily, it would make no sense to inquire about the size of the real effect. The larger $\hat{d}$, the greater is the size of effect. How large should $\hat{d}$ be for a small, medium, or large effect? Cohen has provided criteria for answering this question. These criteria are presented in Table 13.6 below.

example

Early Speaking Experiment

Let's now apply this theoretical discussion to some data. We will use the experiment evaluating the technique for affecting early speaking (p. 333). You will recall when we evaluated the data, we obtained a significant *t* value; we rejected H_0 and concluded that the technique had a real effect. Now the question is, "What is the size of the effect?"

To answer this question, we compute $\hat{d}$. $\overline{X}_{obt} = 11.0$, $\mu = 13.0$, and $s = 3.34$. Substituting these values in the equation for $\hat{d}$, we obtain

$$\hat{d} = \frac{|\overline{X}_{obt} - \mu|}{s} = \frac{|11.0 - 13.0|}{3.34} = \frac{2}{3.34} = 0.60$$

The obtained value of $\hat{d}$ is 0.60. This falls in the range of 0.21–0.79 of Table 13.6 and therefore indicates a medium effect. Although there is a fair amount of theory to get through to understand $\hat{d}$, computation and interpretation of $\hat{d}$ are quite easy!

table 13.6 Cohen's criteria for interpreting the value of $\hat{d}$*

Value of $\hat{d}$	Interpretation of $\hat{d}$
0.00–0.20	Small effect
0.21–0.79	Medium effect
≥ 0.80	Large effect

*Cohen's criteria have some limitations, and should not be interpreted too rigidly. For a discussion of this point, see D.C. Howell, *Statistical Methods for Psychology*, 6th ed., Thompson/Wadsworth, Belmont, CA, 2007.

CONFIDENCE INTERVALS FOR THE POPULATION MEAN

Sometimes it is desirable to know the value of a population mean. Since it is very uneconomical to measure everyone in the population, a random sample is taken and the sample mean is used as an estimate of the population mean. To illustrate, suppose a university administrator is interested in the average IQ of professors at her university. A random sample is taken, and $\overline{X} = 135$. The estimate, then, would be 135. The value 135 is called a *point estimate* because it uses only one value for the estimate. However, if we asked the administrator whether she thought the population mean was exactly 135, her answer would almost certainly be "no." Well, then, how close is 135 to the population mean?

The usual way to answer this question is to give a range of values for which one is reasonably confident that the range includes the population mean. This is called *interval* estimation. For example, the administrator might have some confidence that the population mean lies within the range 130–140. Certainly, she would have more confidence in the range of 130–140 than in the single value of 135. How about the range 110–160? Clearly, there would be more confidence in this range than in the range 130–140. Thus, the wider the range, the greater is the confidence that it contains the population mean.

definitions ■　*A* **confidence interval** *is a range of values that probably contains the population value.*

■　**Confidence limits** *are the values that bound the confidence interval.*

It is possible to be more quantitative about the degree of confidence we have that the interval contains the population mean. In fact, we can construct confidence intervals about which there are specified degrees of confidence. For example, we could construct the 95% confidence interval:

The 95% confidence interval is an interval such that the probability is 0.95 that the interval contains the population value.

Although there are many different intervals we could construct, in practice the 95% and 99% confidence intervals are most often used. Let's consider how to construct these intervals.

Construction of the 95% Confidence Interval

Suppose we have randomly sampled a set of N scores from a population of raw scores having a mean $= \mu$ and have calculated t_{obt}. Assuming the assumptions of t are met, we see that the probability is 0.95 that the following inequality is true:

$$-t_{0.025} \le t_{\text{obt}} \le t_{0.025}$$

$t_{0.025}$ is the critical value of t for $\alpha = 0.025_{1 \text{ tail}}$ and df $= N - 1$. All this inequality says is that if we randomly sample N scores from a population of raw scores having a mean of μ and calculate t_{obt}, the probability is 0.95 that t_{obt} will lie between $-t_{0.025}$ and $+t_{0.025}$. The truth of this statement can be understood best by referring to Figure 13.3. This figure shows the t distribution for $N - 1$ degrees of

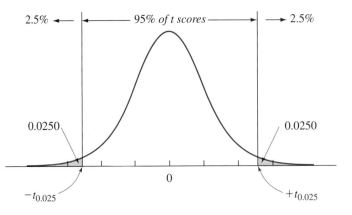

figure 13.3 Percentage of *t* scores between $\pm\ t_{\text{crit}}$ for $\alpha = 0.05_{2\ \text{tail}}$ and df = $N - 1$.

freedom. We've located $+t_{0.025}$ and $-t_{0.025}$ on the distribution. Remember that these values are the critical values of *t* for $\alpha = 0.025_{1\ \text{tail}}$. By definition, 2.5% of the *t* values must lie under each tail, and 95% of the values must lie between $-t_{0.025}$ and $+t_{0.025}$. It follows, then, that the probability is 0.95 that t_{obt} will lie between $-t_{0.025}$ and $+t_{0.025}$.

We can use the previously given inequality to derive an equation for estimating the value of an unknown μ. Thus,

$$-t_{0.025} \le t_{\text{obt}} \le t_{0.025}$$

but

$$t_{\text{obt}} = \frac{\overline{X}_{\text{obt}} - \mu}{s_{\overline{X}}}$$

Therefore,

$$-t_{0.025} \le \frac{\overline{X}_{\text{obt}} - \mu}{s_{\overline{X}}} \le t_{0.025}$$

Solving this inequality for μ, we obtain*

$$\overline{X}_{\text{obt}} - s_{\overline{X}} t_{0.025} \le \mu \le \overline{X}_{\text{obt}} + s_{\overline{X}} t_{0.025}$$

This states that the chances are 95 in 100 that the interval $\overline{X}_{\text{obt}} \pm s_{\overline{X}}\ t_{0.025}$ contains the population mean. Thus, the interval $\overline{X}_{\text{obt}} \pm s_{\overline{X}}\ t_{0.025}$ is the 95% confidence interval. The lower and upper confidence limits are given by

$$\mu_{\text{lower}} = \overline{X}_{\text{obt}} - s_{\overline{X}}\ t_{0.025} \quad \textit{lower limit for 95\% confidence interval}$$

$$\mu_{\text{upper}} = \overline{X}_{\text{obt}} + s_{\overline{X}}\ t_{0.025} \quad \textit{upper limit for 95\% confidence interval}$$

We are now ready to do an example. Let's return to the university administrator.

---

*See Note 13.2 for the intermediate steps in this derivation.

Estimating the Mean IQ of Professors

Suppose a university administrator is interested in determining the average IQ of professors at her university. It is too costly to test all of the professors, so a random sample of 20 is drawn from the population. Each professor is given an IQ test, and the results show a sample mean of 135 and a sample standard deviation of 8. Construct the 95% confidence interval for the population mean.

SOLUTION

The 95% confidence interval for the population mean can be found by solving the equations for the upper and lower confidence limits. Thus,

$$\mu_{\text{lower}} = \overline{X}_{\text{obt}} - s_{\overline{X}}\, t_{0.025} \quad \text{and} \quad \mu_{\text{upper}} = \overline{X}_{\text{obt}} + s_{\overline{X}}\, t_{0.025}$$

Solving for $s_{\overline{X}}$,

$$s_{\overline{X}} = \frac{s}{\sqrt{N}} = \frac{8}{\sqrt{20}} = 1.789$$

From Table D, with $\alpha = 0.025_{1\,\text{tail}}$ and $df = N - 1 = 20 - 1 = 19$,

$$t_{0.025} = 2.093$$

Substituting the values for $s_{\overline{X}}$ and $t_{0.025}$ in the confidence limit equations, we obtain

$$\mu_{\text{lower}} = \overline{X}_{\text{obt}} - s_{\overline{X}}\, t_{0.025} \qquad \text{and} \qquad \mu_{\text{upper}} = \overline{X}_{\text{obt}} + s_{\overline{X}}\, t_{0.025}$$

$$= 135 - 1.789(2.093) \qquad\qquad\qquad = 135 + 1.789(2.093)$$

$$= 135 - 3.744 \qquad\qquad\qquad\qquad = 135 + 3.744$$

$$= 131.26 \quad \textit{lower limit} \qquad\qquad = 138.74 \quad \textit{upper limit}$$

Thus, the 95% confidence interval = 131.26–138.74.

What precisely does it mean to say that the 95% confidence interval equals a certain range? In the case of the previous sample, the range is 131.26–138.74. A second sample would yield a different $\overline{X}_{\text{obt}}$ and a different range, perhaps $\overline{X}_{\text{obt}} = 138$ and a range of 133.80–142.20. If we took all of the different possible samples of $N = 20$ from the population, we would have derived the sampling distribution of the 95% confidence interval for samples of size 20. *The important point here is that 95% of these intervals will contain the population mean; 5% of the intervals will not.* Thus, when we say "the 95% confidence interval is 131.26–138.74," we mean the probability is 0.95 that the interval contains the population mean. Note that the probability value applies to the interval and not to the population mean. The population mean is constant. What varies from sample to sample is the interval. Thus, it is not technically proper to state "the probability is 0.95 that the population mean lies within the interval." Rather, the proper statement is "the probability is 0.95 that the interval contains the population mean."

General Equations for Any Confidence Interval

The equations we have presented thus far deal only with the 95% confidence interval. However, they are easily extended to form general equations for any confidence interval. Thus,

$$\mu_{\text{lower}} = \overline{X}_{\text{obt}} - s_{\overline{X}}\, t_{\text{crit}} \quad \textit{general equation for lower confidence limit}$$

$$\mu_{\text{upper}} = \overline{X}_{\text{obt}} + s_{\overline{X}}\, t_{\text{crit}} \quad \textit{general equation for upper confidence limit}$$

where t_{crit} = the critical one-tailed value of t corresponding to the desired confidence interval.

Thus, if we were interested in the 99% confidence interval, $t_{crit} = t_{0.005}$ = the critical value of t for $\alpha = 0.005_{1\ tail}$. To illustrate, let's solve the previous problem for the 99% confidence interval.

SOLUTION

From Table D, with df = 19 and $\alpha = 0.005_{1\ tail}$,

$$t_{0.005} = 2.861$$

From the previous solution, $s_{\overline{X}} = 1.789$. Substituting these values into the equations for confidence limits, we have

$$\mu_{lower} = \overline{X}_{obt} - s_{\overline{X}}\,t_{crit} \qquad \text{and} \qquad \mu_{upper} = \overline{X}_{obt} + s_{\overline{X}}\,t_{crit}$$

$$= \overline{X}_{obt} - s_{\overline{X}}\,t_{0.005} \qquad\qquad\qquad = \overline{X}_{obt} + s_{\overline{X}}\,t_{0.005}$$

$$= 135 - 1.789(2.861) \qquad\qquad\qquad = 135 + 1.789(2.861)$$

$$= 135 - 5.118 \qquad\qquad\qquad\qquad = 135 + 5.118$$

$$= 129.88 \quad lower\ limit \qquad\qquad = 140.12 \quad upper\ limit$$

Thus, the 99% confidence interval = 129.88–140.12.

Note that this interval is larger than the 95% confidence interval (131.26–138.74). As discussed previously, the larger the interval, the more confidence we have that it contains the population mean.

Let's try a practice problem.

MENTORING TIP

Remember: the larger the interval, the more confidence we have that the interval contains the population mean.

Practice Problem 13.3

An ethologist is interested in determining the average weight of adult Olympic marmots (found only on the Olympic Peninsula in Washington). It would be expensive and impractical to trap and measure the whole population, so a random sample of 15 adults is trapped and weighed. The sample has a mean of 7.2 kilograms and a standard deviation of 0.48. Construct the 95% confidence interval for the population mean.

SOLUTION

The data are given in the problem. The 95% confidence interval for the population mean is found by determining the upper and lower confidence limits. Thus,

$$\mu_{lower} = \overline{X}_{obt} - s_{\overline{X}}\,t_{0.025} \qquad \text{and} \qquad \mu_{upper} = \overline{X}_{obt} + s_{\overline{X}}\,t_{0.025}$$

Solving for $s_{\overline{X}}$,

$$s_{\overline{X}} = \frac{s}{\sqrt{N}} = \frac{0.48}{\sqrt{15}} = 0.124$$

From Table D, with $\alpha = 0.025_{1\ \text{tail}}$ and df $= N - 1 = 15 - 1 = 14$,

$$t_{0.025} = 2.145$$

Substituting the values for $s_{\overline{X}}$ and $t_{0.025}$ in the confidence limit equations, we obtain

$$\mu_{\text{lower}} = \overline{X}_{\text{obt}} - s_{\overline{X}} t_{0.025} \quad \text{and} \quad \mu_{\text{upper}} = \overline{X}_{\text{obt}} + s_{\overline{X}} t_{0.025}$$

$$= 7.2 - 0.124(2.145) \qquad\qquad = 7.2 + 0.124(2.145)$$

$$= 7.2 - 0.266 \qquad\qquad\qquad = 7.2 + 0.266$$

$$= 6.93 \quad \textit{lower limit} \qquad\qquad = 7.47 \quad \textit{upper limit}$$

Thus, the 95% confidence interval $= 6.93–7.47$ kilograms.

Practice Problem 13.4

To estimate the average life of its 100-watt light bulbs, the manufacturer randomly samples 200 light bulbs and keeps them lit until they burn out. The sample has a mean life of 215 hours and a standard deviation of 8 hours. Construct the 99% confidence limits for the population mean. In solving this problem, use the closest table value for degrees of freedom.

SOLUTION

The data are given in the problem.

$$\mu_{\text{lower}} = \overline{X}_{\text{obt}} - s_{\overline{X}} t_{0.005} \quad \text{and} \quad \mu_{\text{upper}} = \overline{X}_{\text{obt}} + s_{\overline{X}} t_{0.005}$$

$$s_{\overline{X}} = \frac{s}{\sqrt{N}} = \frac{8}{\sqrt{200}} = 0.567$$

From Table D, with $\alpha = 0.005_{1\ \text{tail}}$ and df $= N - 1 = 200 - 1 = 199$,

$$t_{0.005} = 2.617$$

Note that this is the closest table value available from Table D. Substituting the values for $s_{\overline{X}}$ and $t_{0.005}$ in the confidence limit equations, we obtain

$$\mu_{\text{lower}} = \overline{X}_{\text{obt}} - s_{\overline{X}} t_{0.005} \quad \text{and} \quad \mu_{\text{upper}} = \overline{X}_{\text{obt}} + s_{\overline{X}} t_{0.005}$$

$$= 215 - 0.567(2.617) \qquad\qquad = 215 + 0.567(2.617)$$

$$= 213.52 \quad \textit{lower limit} \qquad\qquad = 216.48 \quad \textit{upper limit}$$

Thus, the 99% confidence interval $= 213.52–216.48$ hours.

TESTING THE SIGNIFICANCE OF PEARSON *r*

When a correlational study is conducted, it is rare for the whole population to be involved. Rather, the usual procedure is to randomly sample from the population and calculate the correlation coefficient on the sample data. To determine whether a correlation exists in the population, we must test the significance of the obtained $r(r_{obt})$. Of course, this is the same procedure we have used all along for testing hypotheses. The population correlation coefficient is symbolized by the Greek letter ρ (rho). A nondirectional alternative hypothesis asserts that $\rho \neq 0$. A directional alternative hypothesis asserts that ρ is positive or negative depending on the predicted direction of the relationship. The null hypothesis is tested by assuming that the sample set of X and Y scores having a correlation equal to r_{obt} is a random sample from a population where $\rho = 0$. The sampling distribution of r can be generated empirically by taking all samples of size N from a population in which $\rho = 0$ and calculating r for each sample. By systematically varying the population scores and N, the sampling distribution of r is generated.

The significance of r can be evaluated using the t test. Thus,

$$t_{obt} = \frac{r_{obt} - \rho}{s_r} \qquad \text{t test for testing the significance of r}$$

where r_{obt} = correlation obtained on a sample of N subjects
        ρ = population correlation coefficient
        s_r = estimate of the standard deviation of the sampling distribution of r

Note that this is very similar to the t equation used when dealing with the mean of a single sample. The only difference is that the statistic we are dealing with is r rather than $\overline{X}$.

$$t_{obt} = \frac{r_{obt} - \rho}{s_r} = \frac{r_{obt}}{\sqrt{\dfrac{1 - r_{obt}^2}{N - 2}}}$$

where $\rho = 0$
$$s_r = \sqrt{(1 - r_{obt}^2)/(N - 2)}$$
$$\text{df} = N - 2$$

Let's use this equation to test the significance of the correlation obtained in the "IQ and grade point average" problem presented in Chapter 6, p. 135. Assume that the 12 students were a random sample from a population of university undergraduates and that we want to determine whether there is a correlation in the population. We'll use $\alpha = 0.05_{2\text{ tail}}$ in making our decision.

Ordinarily, the first step in a problem of this sort is to calculate r_{obt}. However, we have already done this and found that $r_{obt} = 0.856$. Substituting this value into the t equation, we obtain

$$t_{obt} = \frac{r_{obt}}{\sqrt{\dfrac{1 - r_{obt}^2}{N - 2}}} = \frac{0.856}{\sqrt{\dfrac{1 - (0.856)^2}{10}}} = \frac{0.856}{0.163} = 5.252 = 5.25$$

From Table D, with df $= N - 2 = 10$ and $\alpha = 0.05_{2\text{ tail}}$,

$$t_{crit} = \pm 2.228$$

Since $|t_{obt}| > 2.228$, we reject H_0 and conclude that there is a significant positive correlation in the population.

Although the foregoing method works, there is an even easier way to solve this problem. By substituting t_{crit} into the t equation, r_{crit} can be determined for any df and any α level. Once r_{crit} is known, all we need do is compare r_{obt} with r_{crit}. The decision rule is

$$\text{If } |r_{obt}| \geq |r_{crit}|, \text{ reject } H_0.$$

Statisticians have already calculated r_{crit} for various df and α levels. These are shown in Table E in Appendix D. This table is used in the same way as the t table (Table D) except the entries list r_{crit} rather than t_{crit}.

Applying the r_{crit} method to the present problem, we would first calculate r_{obt} and then determine r_{crit} from Table E. Finally, we would compare r_{obt} with r_{crit} using the decision rule. In the present example, we have already determined that $r_{obt} = 0.856$. From Table E, with df $= 10$ and $\alpha = 0.05_{2\,tail}$,

$$r_{crit} = \pm 0.5760$$

Since $|r_{obt}| > 0.5760$, we reject H_0, as before. This solution is preferred because it is shorter and easier than the solution that involves comparing t_{obt} with t_{crit}.

Let's try some problems for practice.

Practice Problem 13.5

Folklore has it that there is an inverse correlation between mathematical and artistic ability. A psychologist decides to determine whether there is anything to this notion. She randomly samples 15 undergraduates and gives them tests measuring these two abilities. The resulting data are shown here. Is there a correlation in the population between mathematical ability and artistic ability? Use $\alpha = 0.01_{2\,tail}$.

Subject No.	Math Ability X	Artistic Ability Y	X^2	Y^2	XY
1	15	19	225	361	285
2	30	22	900	484	660
3	35	17	1,225	289	595
4	10	25	100	625	250
5	28	23	784	529	644
6	40	21	1,600	441	840
7	45	14	2,025	196	630
8	24	10	576	100	240
9	21	18	441	324	378
10	25	19	625	361	475
11	18	30	324	900	540
12	13	32	169	1,024	416
13	9	16	81	256	144
14	30	28	900	784	840
15	23	24	529	576	552
Total	366	318	10,504	7,250	7,489

(continued)

SOLUTION

STEP 1: **Calculate the appropriate statistic:**

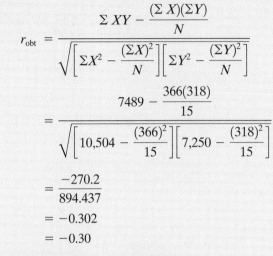

$$r_{obt} = \frac{\Sigma XY - \dfrac{(\Sigma X)(\Sigma Y)}{N}}{\sqrt{\left[\Sigma X^2 - \dfrac{(\Sigma X)^2}{N}\right]\left[\Sigma Y^2 - \dfrac{(\Sigma Y)^2}{N}\right]}}$$

$$= \frac{7489 - \dfrac{366(318)}{15}}{\sqrt{\left[10{,}504 - \dfrac{(366)^2}{15}\right]\left[7{,}250 - \dfrac{(318)^2}{15}\right]}}$$

$$= \frac{-270.2}{894.437}$$

$$= -0.302$$

$$= -0.30$$

STEP 2: **Evaluate the statistic.** From Table E, with df = $N - 2 = 15 - 2 = 13$ and $\alpha = 0.01_{2\,tail}$,

$$r_{crit} = \pm 0.6411$$

Since $|r_{obt}| < 0.6411$, we conclude by retaining H_0.

Practice Problem 13.6

In Chapter 6, Practice Problem 6.2, we calculated the Pearson *r* for the relationship between similarity of attitudes and attraction in a sample of 15 college students. In that example, $r_{obt} = 0.94$. Using $\alpha = 0.05_{2\,tail}$, let's now determine whether this is a significant value for r_{obt}.

SOLUTION

From Table E, with df = $N - 2 = 13$ and $\alpha = 0.05_{2\,tail}$,

$$r_{crit} = \pm 0.5139$$

Since $|r_{obt}| > 0.5139$, we reject H_0 and conclude there is a significant correlation in the population.

■ SUMMARY

In this chapter, I discussed the use of Student's t test for (1) testing hypotheses involving single sample experiments, (2) estimating the population mean by constructing confidence intervals, and (3) testing the significance of Pearson r.

In testing hypotheses involving single sample experiments, the t test is appropriate when the mean of the null-hypothesis population is known and the standard deviation is unknown. In this situation, we estimate σ by using the sample standard deviation. The equation for calculating t_{obt} is very similar to z_{obt}, but we use s instead of σ. The sampling distribution of t is a family of curves that varies with the degrees of freedom associated with calculating t. There are $N-1$ degrees of freedom associated with the t test for single samples. The sampling distribution curves are symmetrical, bell-shaped curves having a mean equal to 0. However, these are elevated at the tails relative to the normal distribution. In using the t test, t_{obt} is computed and then evaluated to determine whether it falls within the critical region. The t test is appropriate when the sampling distribution of $\overline{X}$ is normal. For the sampling distribution of $\overline{X}$ to be normal, the population of raw scores must be normally distributed, or $N \geq 30$.

After discussing how to evaluate t_{obt} to determine if there is a real effect, I discussed how to compute the size of the effect, using Cohen's d statistic. Cohen's d, for the single samples t test, is a standardized measure of the absolute difference between $\overline{X}$ and μ, with standardization being achieved by dividing this difference by σ. Since we don't know σ when using the t test, we estimate it using s and, hence, compute $\hat{d}$, instead of d. The larger $\hat{d}$ is, the greater the real effect. Criteria were also given for determining if the obtained value of $\hat{d}$ represents a small, medium, or large effect.

Next I discussed constructing confidence intervals for the population mean. A confidence interval was defined as a range of values that probably contains the population value. Confidence limits are the values that bound the confidence interval. In discussing this topic, we showed how to construct confidence intervals about which we have a specified degree of confidence that the interval contains the population mean. Illustrative and practice problems were given for constructing the 95% and 99% confidence intervals.

The last topic involved testing the significance of Pearson r. I pointed out that, because most correlative data are collected on samples, we must evaluate the sample r value (r_{obt}) to see whether there is a correlation in the population. The evaluation involves the t test. However, by substituting t_{crit} into the t equation, we can determine r_{crit} for any df and any alpha level. The value of r_{obt} is evaluated by comparing it with r_{crit} for the given df and alpha level. Several problems were given for practice in evaluating r_{obt}.

■ IMPORTANT NEW TERMS

Cohen's d (p. 339)
Confidence interval (p. 341)
Confidence limits (p. 341)

Critical value of r (p. 347)
Critical value of t (p. 332)
Degrees of freedom (p. 330)

Sampling distribution of t (p. 329)
Student's t test for single samples (p. 328)

■ QUESTIONS AND PROBLEMS

1. Define each of the terms in the Important New Terms section.
2. Assuming the assumptions underlying the t test are met, what are the characteristics of the sampling distribution of t?
3. Elaborate on what is meant by *degrees of freedom*. Use an example.
4. What are the assumptions underlying the proper use of the t test?
5. Discuss the similarities and differences between the z and t tests.
6. Explain in a short paragraph why the z test is more powerful than the t test.
7. Which of the following two statements is technically more correct? (1) We are 95% confident that the population mean lies in the interval 80–90, or (2) We are 95% confident that the interval 80–90 contains the population mean. Explain.

8. Explain why df $= N - 1$ when the *t* test is used with single samples.

9. If the sample correlation coefficient has a value different from zero (e.g., $r = 0.45$), this automatically means that the correlation in the population is also different from zero. Is this statement correct? Explain.

10. For the same set of sample scores, is the 99% confidence interval for the population mean greater or smaller than the 95% confidence interval? Does this make sense? Explain.

11. A sample set of 30 scores has a mean equal to 82 and a standard deviation of 12. Can we reject the hypothesis that this sample is a random sample from a normal population with $\mu = 85$? Use $\alpha = 0.01_{2 \text{ tail}}$ in making your decision. other

12. A sample set of 29 scores has a mean of 76 and a standard deviation of 7. Can we accept the hypothesis that the sample is a random sample from a population with a mean greater than 72? Use $\alpha = 0.01_{1 \text{ tail}}$ in making your decision. other

13. Is it reasonable to consider a sample with $N = 22$, $\overline{X}_{\text{obt}} = 42$, and $s = 9$ to be a random sample from a normal population with $\mu = 38$? Use $\alpha = 0.05_{1 \text{ tail}}$ in making your decision. Assume $\overline{X}_{\text{obt}}$ is in the right direction. other

14. Using each of the following random samples, determine the 95% and 99% confidence intervals for the population mean:
    a. $\overline{X}_{\text{obt}} = 25$, $s = 6$, $N = 15$
    b. $\overline{X}_{\text{obt}} = 120$, $s = 8$, $N = 30$
    c. $\overline{X}_{\text{obt}} = 30.6$, $s = 5.5$, $N = 24$
    d. Redo part **a** with $N = 30$. What happens to the confidence interval as N increases? other

15. In Problem 21 of Chapter 12, a student conducted an experiment on 25 schizophrenic patients to test the effect of a new technique on the amount of time schizophrenics need to stay institutionalized. The results showed that under the new treatment, the 25 schizophrenic patients stayed a mean duration of 78 weeks, with a standard deviation of 20 weeks. Previously collected data on a large number of schizophrenic patients showed a normal distribution of scores, with a mean of 85 weeks and a standard deviation of 15 weeks. These data were evaluated using $\alpha = 0.05_{2 \text{ tail}}$. The results showed a significant effect. For the present problem, assume that the standard deviation of the population is unknown. Again, using $\alpha = 0.05_{2 \text{ tail}}$, what do you conclude about the new technique? Explain the difference in conclusion between Problem 21 and this one. clinical, health

16. As the principal of a private high school, you are interested in finding out how the training in mathematics at your school compares with that of the public schools in your area. For the last 5 years, the public schools have given all graduating seniors a mathematics proficiency test. The distribution has a mean of 78. You give all the graduating seniors in your school the same mathematics proficiency test. The results show a distribution of 41 scores, with a mean of 83 and a standard deviation of 12.2.
    a. What is the alternative hypothesis? Use a nondirectional hypothesis.
    b. What is the null hypothesis?
    c. Using $\alpha = 0.05_{2 \text{ tail}}$, what do you conclude? education

17. A college counselor wants to determine the average amount of time first-year students spend studying. He randomly samples 61 students from the freshman class and asks them how many hours a week they study. The mean of the resulting scores is 20 hours, and the standard deviation is 6.5 hours.
    a. Construct the 95% confidence interval for the population mean.
    b. Construct the 99% confidence interval for the population mean. education

18. A professor in the women's studies program believes that the amount of smoking by women has increased in recent years. A complete census taken 2 years ago of women living in a neighboring city showed that the mean number of cigarettes smoked daily by the women was 5.4 with a standard deviation of 2.5. To assess her belief, the professor determined the daily smoking rate of a random sample of 200 women currently living in that city. The data show that the number of cigarettes smoked daily by the 200 women has a mean of 6.1 and a standard deviation of 2.7.
    a. Is the professor's belief correct? Assume a directional H_1 is appropriate and use $\alpha = 0.05_{1 \text{ tail}}$ in making your decision. Be sure that the most sensitive test is used to analyze the data.
    b. Assume the population mean is unknown and reanalyze the data using the same alpha level. What is your conclusion this time?
    c. Explain any differences between part **a** and part **b**.
    d. Determine the size of the effect found in part **b**. social

19. A cognitive psychologist believes that a particular drug improves short-term memory. The drug is safe, with no side effects. An experiment is conducted in which 8 randomly selected subjects are given the drug and then given a short time to memorize a list of 10 words. The subjects are then tested for retention

15 minutes after the memorization period. The number of words correctly recalled by each subject is as follows: 8, 9, 10, 6, 8, 7, 9, 7. Over the past few years, the psychologist has collected a lot of data using this task with similar subjects. Although he has lost the original data, he remembers that the mean was 6 words correctly recalled and that the data were normally distributed.

a. On the basis of these data, what can we conclude about the effect of the drug on short-term memory? Use $\alpha = 0.05_{2\,tail}$ in making your decision.

b. Determine the size of the effect. cognitive

20. A physician employed by a large corporation believes that due to an increase in sedentary life in the past decade, middle-age men have become fatter. In 1995, the corporation measured the percentage of fat in their employees. For the middle-age men, the scores were normally distributed, with a mean of 22%. To test her hypothesis, the physician measures the fat percentage in a random sample of 12 middle-age men currently employed by the corporation. The fat percentages found were as follows: 24, 40, 29, 32, 33, 25, 15, 22, 18, 25, 16, 27. On the basis of these data, can we conclude that middle-age men employed by the corporation have become fatter? Assume a directional H_1 is legitimate and use $\alpha = 0.05_{1\,tail}$ in making your decision. health

21. A local business school claims that its graduating seniors get higher-paying jobs than the national average for business school graduates. Last year's figures for salaries paid to all business school graduates on their first job showed a mean of $10.20 per hour. A random sample of 10 graduates from last year's class of the local business school showed the following hourly salaries for their first job: $9.40, $10.30, $11.20, $10.80, $10.40, $9.70, $9.80, $10.60, $10.70, $10.90. You are skeptical of the business school claim and decide to evaluate the salary of the business school graduates, using $\alpha = 0.05_{2\,tail}$. What do you conclude? education

22. You wanted to estimate the mean number of vehicles crossing a busy bridge in your neighborhood each morning during rush hour for the past year. To accomplish this, you stationed yourself and a few assistants at one end of the bridge on 18 randomly selected mornings during the year and counted the number of vehicles crossing the bridge in a 10-minute period during rush hour. You found the mean to be 125 vehicles per minute, with a standard deviation of 32.

a. Construct the 95% confidence limits for the population mean (vehicles per minute).

b. Construct the 99% confidence limits for the population mean (vehicles per minute). other

23. In Chapter 6, Problem 17, data were presented from a study conducted to investigate the relationship between cigarette smoking and illness. The number of cigarettes smoked daily and the number of days absent from work in the last year due to illness was determined for 12 individuals employed at the company where the researcher worked. The data are shown again here.

Subject	Cigarettes Smoked	Days Absent
1	0	1
2	0	3
3	0	8
4	10	10
5	13	4
6	20	14
7	27	5
8	35	6
9	35	12
10	44	16
11	53	10
12	60	16

a. Construct a scatter plot for these data.

b. Calculate the value of Pearson r.

c. Is the correlation between cigarettes smoked and days absent significant? Use $\alpha = 0.05_{2\,tail}$. health

24. In Chapter 6, Problem 18, an educator evaluated the reliability of a test for mechanical aptitude that she had constructed. Two administrations of the test, spaced 1 month apart, were given to 10 students. The data are again shown here.

Student	Administration 1	Administration 2
1	10	10
2	12	15
3	20	17
4	25	25
5	27	32
6	35	37
7	43	40
8	40	38
9	32	30
10	47	49

a. Calculate the value of Pearson *r* for the two administrations of the mechanical aptitude test.

b. Is the correlation significant? Use $\alpha = 0.05_{2\,\text{tail}}$.
I/O

25. In Chapter 6, Problem 15, a sociology professor gave two exams to 8 students. The results are again shown here.

Student	Exam 1	Exam 2
1	60	60
2	75	100
3	70	80
4	72	68
5	54	73
6	83	97
7	80	85
8	65	90

a. Calculate the value of Pearson *r* for the two exams.

b. Using $\alpha = 0.05_{2\,\text{tail}}$, determine whether the correlation is significant. If not, does this mean that $\rho = 0$? Explain.

c. Assume you increased the number of students to 20, and now *r* = 0.653. Using the same alpha level as in part **b**, what do you conclude this time? Explain. education

26. A developmental psychologist is interested in whether tense parents tend to have tense children. A study is done involving one parent for each of 15 families and the oldest child in each family, measuring tension in each pair. Pearson *r* = 0.582. Using $\alpha = 0.05_{2\,\text{tail}}$, is the relationship significant? developmental, clinical

■ SPSS ILLUSTRATIVE EXAMPLE 13.1

The general operation of SPSS and data entry are described in Appendix E, *Introduction to SPSS*. This chapter of the textbook discusses the *t* test for single samples. SPSS calls Student's *t* test for Single Samples the *One-Sample T Test*. When analyzing the data for *t* tests, SPSS computes the obtained *t* value (we call it t_{obt}; SPSS just calls it **t**) and the two-tailed probability of getting **t or a value more extreme** if chance alone is at work. SPSS calls this probability **Sig. (2-tailed).** We call this probability *p*(2-tailed).

SPSS does not provide a value for t_{crit}. However, since SPSS gives the two-tailed probability of getting **t or a value more extreme** assuming chance alone is at work, it is not necessary to compare **t** with t_{crit} when evaluating H_0, as is done in the textbook. Instead, we compare **Sig. (2-tailed)** with the alpha level and conclude following the decision rules given below.

For non-directional H_1's:

if **Sig. (2-tailed)** $\le \alpha$, reject H_0
if **Sig. (2-tailed)** $> \alpha$, retain H_0

For directional H_1's where the sample mean is in the predicted direction:

if **Sig. (2-tailed)/2** $\le \alpha$, reject H_0
if **Sig. (2-tailed)/2** $> \alpha$, retain H_0

As you are no doubt aware, for directional H_1's, if the sample mean is in the direction opposite to that predicted, we don't even need to do a probability analysis. The conclusion must be to retain H_1.

example

Use SPSS to do the illustrative problem in the textbook Chapter 13, p. 333, involving a technique for increasing early speaking in children. For your convenience, the example is repeated here:

Suppose you have a technique that you believe will affect the age at which children begin speaking. In your locale, the average age of first word utterances is 13.0 months. The standard deviation is unknown. You apply your technique to a random sample of 15 children. The results are shown in the table below.

Child No.	1	2	3	4	5	6	7	8	9	10	11	12	13	14	15
Age (months)	8	9	10	15	18	17	12	11	7	8	10	11	8	9	12

Did the technique work? Use $\alpha = 0.05_{2\,\text{tail}}$. Compare your conclusion based on the SPSS analysis and the value of t_{obt} computed by SPSS with that given in the textbook. In solving the example, name the variable *Age*.

SOLUTION

STEP 1: **Enter the Data.** Enter the *Age* scores in the first column (**VAR00001**) of the SPSS Data Editor, beginning with the first score listed above in the first cell of the first column of the Data Editor.

STEP 2: **Name the Variables.** In this example, we will give the default variable **VAR00001** the name of **Age**. Here's how it is done.

1. **Click** the **Variable View** tab in the lower left corner of the Data Editor.	This displays the **Variable View** on screen with **VAR00001** displayed in the first cell of the **Name** column.
2. **Click VAR00001**; then **type Age** in the highlighted cell and then **press Enter**.	**Age** is entered as the variable name, replacing **VAR00001**.

STEP 3: **Analyze the Data.** The appropriate test for this example is Student's *t* test for Single Samples. To have SPSS do the analysis using this test,

1. **Click** on **Analyze**; then **select Compare Means**; then **click** on **One-Sample T Test…**.	This produces the **One-Sample T Test** dialog box with **Age** highlighted in the large box on the left.
2. **Click** the **arrow** in the middle of the dialog box.	This moves **Age** into the **Test Variable(s):** box on the right.
3. In the **Test Value: box, replace 0** with **13.0**.	This puts the value **13.0** in the **Test Value: box. 13.0** is the value of the null hypothesis population mean.
4. **Click OK**.	SPSS analyzes the data using the **One-Sample T Test** and outputs the results in the two tables shown below.

Analysis Results

One-Sample Statistics

	N	Mean	Std. Deviation	Std. Error Mean
Age	15	11.000	3.33809	.86189

One-Sample Test

	Test Value = 13.0					
					95% Confidence Interval of the Difference	
	t	df	Sig. (2-tailed)	Mean Difference	Lower	Upper
Age	-2.320	14	.036	-2.00000	-3.8486	-.1514

The **One-Sample Statistics** table gives values for **N**, **Mean**, **Std. Deviation** and the **Std. Error mean** for the sample. The **One-Sample Test** table gives the results of the inferential analysis. Our conclusion is based on this table. It shows that **t = −2.320**, and that **Sig. (2-tailed) = .036$_{2\text{-tailed}}$**. Since **.036 < 0.05**, we reject H_0 and affirm H_1. Note that the SPSS and textbook *t* values are the same. The conclusion reached in both cases is also the same. The **One-Sample Test** table contains additional information that is not germane to our analysis, so we will ignore it.

■ SPSS ADDITIONAL PROBLEMS

1. Use SPSS to do Practice Problem 13.1, p. 335 of the textbook. Compare your answer with that given in the textbook. Name the variable *Height*.
2. A physician employed by a large corporation believes that due to an increase in sedentary life in the past decade, middle-age men have become fatter. In 1995, the corporation measures the percentage of fat in its employees. For the middle-age men, the scores were normally distributed, with a mean of 22%. To test her hypothesis, the physician measures the fat percentage in a random sample of 12 middle-age men currently employed by the corporation. The fat percentages found were as follows: 24, 40, 29, 32, 33, 25, 15, 22, 18, 25, 16, 27.
   a. On the basis of these data, can we conclude that middle-age men employed by the corporation have become fatter? Assume a directional H_1 is legitimate and use $\alpha = 0.05_{1\text{ tail}}$ in making your decision.
   b. Although H_0 was retained in part **a**, the physician notes that the sample mean was in the right direction. She suspects that the failure to reject H_0 was due to running too few subjects (low power). Therefore she conducts the experiment again, this time using a random sample of 36 men from her corporation. The following percentages of fat were obtained: 26, 24, 31, 40, 31, 29, 30, 32, 35, 33, 23, 25, 17, 15, 20, 22, 20, 18, 23, 25, 18, 16, 25, 27, 22, 42, 27, 34, 31, 27, 13, 24, 16, 27, 14, 29. Using $\alpha = 0.05_{1\text{ tail}}$, what is your conclusion this time?

■ NOTES

13.1 $t_{obt} = \dfrac{\overline{X}_{obt} - \mu}{s/\sqrt{N}}$ *Given*

$= \dfrac{\overline{X}_{obt} - \mu}{\sqrt{\dfrac{SS}{N-1}} \Big/ \sqrt{N}}$ *Substituting* $\sqrt{\dfrac{SS}{N-1}}$ *for s*

$(t_{obt})^2 = \dfrac{(\overline{X}_{obt} - \mu)^2}{\left(\dfrac{SS}{N-1}\right)\left(\dfrac{1}{N}\right)} = \dfrac{(\overline{X}_{obt} - \mu)^2}{\dfrac{SS}{N(N-1)}}$ *Squaring both sides of the equation and rearranging terms*

$t_{obt} = \dfrac{\overline{X}_{obt} - \mu}{\sqrt{\dfrac{SS}{N(N-1)}}}$ *Taking the square root of both sides of the equation*

13.2 $-t_{0.025} \leq \dfrac{\overline{X}_{obt} - \mu}{s_{\overline{X}}} \leq t_{0.025}$ *Given*

$-s_{\overline{X}}t_{0.025} \leq \overline{X}_{obt} - \mu \leq s_{\overline{X}}t_{0.025}$ *Multiplying by* $s_{\overline{X}}$

$-\overline{X}_{obt} - s_{\overline{X}}t_{0.025} \leq -\mu \leq -\overline{X}_{obt} + s_{\overline{X}}t_{0.025}$ *Subtracting* $\overline{X}_{obt}$

$\overline{X}_{obt} + s_{\overline{X}}t_{0.025} \geq \mu \geq \overline{X}_{obt} - s_{\overline{X}}t_{0.025}$ *Multiplying by* -1

$\overline{X}_{obt} - s_{\overline{X}}t_{0.025} \leq \mu \leq \overline{X}_{obt} + s_{\overline{X}}t_{0.025}$ *Rearranging terms*

■ ONLINE STUDY RESOURCES

CENGAGE**brain**.com

Login to CengageBrain.com to access the resources your instructor has assigned. For this book, you can access the book's companion website for chapter-specific learning tools including Know and Be Able to Do, practice quizzes, flash cards, and glossaries, and a link to Statistics and Research Methods Workshops.

aplia

If your professor has assigned Aplia homework:

1. Sign in to your account.
2. Complete the corresponding homework exercises as required by your professor.
3. When finished, click "Grade It Now" to see which areas you have mastered and which need more work, and for detailed explanations of every answer.

Visit **www.cengagebrain.com** to access your account and to purchase materials.

14

Student's *t* Test for Correlated and Independent Groups

© Strmko / Dreamstime.com

CHAPTER OUTLINE

LEARNING OBJECTIVES

After completing this chapter, you should be able to:

- Contrast the single sample and correlated groups *t* tests.
- Understand that the correlated groups *t* test is an extension of the single sample t test, only using difference scores rather than raw scores.
- Solve problems involving the *t* test for correlated groups.
- Compute size of effect using Cohen's *d*, with the *t* test for correlated groups.
- Specify which test is generally more powerful, the *t* test for correlated groups or the sign test, and justify your answer.
- Compare the repeated measures and the independent groups designs.
- Specify H_0 and H_1 in terms of μ_1 and μ_2 for the independent groups design.
- Define and specify the characteristics of the sampling distribution of the difference between sample means.
- Understand the derivation of s_w^2, and explain why df $= N - 2$ for the independent groups *t* test.
- Solve problems using the *t* test for independent groups, state the assumptions underlying this test, and state the effect on the test of violations of its assumptions.
- Compute size of effect using Cohen's *d* with the independent groups *t* test.
- Determine the relationship between power and *N*, size of real effect, and sample variability, using *t* equations.
- Compare the correlated groups and independent groups *t* tests regarding their relative power.
- Explain the difference between the null hypothesis approach and the confidence interval approach, and specify an advantage of the confidence interval approach.

- Construct the 95% and 99% confidence interval for $\mu_1 - \mu_2$ for data from the two-group, independent groups design, and interpret these results.
- Understand the illustrative examples, do the practice problems, and understand the solutions.

INTRODUCTION

In Chapters 12 and 13, we have seen that hypothesis testing basically involves two steps: (1) calculating the appropriate statistic and (2) evaluating the statistic using its sampling distribution. We further discussed how to use the *z* and *t* tests to evaluate hypotheses that have been investigated with single sample experiments. In this chapter, we shall present the *t* test in conjunction with experiments involving two conditions or two samples.

We have already encountered the two-condition experiment when using the sign test. The two-condition experiment, whether of the correlated groups or independent groups design, has great advantages over the single sample experiment previously discussed. A major limitation of the single sample experiment is the requirement that at least one population parameter (μ) must be specified. In the great majority of cases, this information is not available. As will be shown later in this chapter, the two-treatment experiment completely eliminates the need to measure population parameters when testing hypotheses. This has obvious widespread practical utility.

A second major advantage of the two-condition experiment has to do with interpreting the results of the study. Correct scientific methodology does not often allow an investigator to use previously acquired population data when conducting an experiment. For example, in the illustrative problem involving early speaking in children (p. 329), we used a population mean value of 13.0 months. How do we really know the mean is 13.0 months? Suppose the figures were collected 3 to 5 years before performing the experiment. How do we know that infants haven't changed over those years? And what about the conditions under which the population data were collected? Were they the same as in the experiment? Isn't it possible that the people collecting the population data were not as motivated as the experimenter and, hence, were not as careful in collecting the data? Just how were the data collected? By being on hand at the moment that the child spoke the first word? Quite unlikely. The data probably were collected by asking parents when their children first spoke. How accurate, then, is the population mean?

Even if the foregoing problems didn't exist, there are others having to do with the experimental method itself. For example, assuming 13.0 months is accurate and applies properly to the sample of 15 infants, how can we be sure it was the experimenter's technique that produced the early utterances? Couldn't they have been due to the extra attention or handling or stimulation given the children in conjunction with the method rather than the method itself?

Many of these problems can be overcome by the use of the two-condition experiment. By using two groups of infants (arrived at by matched pairs [correlated groups design] or random assignment [independent groups design]), giving each group the same treatment except for the experimenter's particular technique (same in attention, handling, etc.), running both groups concurrently, using the same people to collect the data from both groups, and so forth, most alternative explanations of results can be ruled out. In the discussion that follows, we shall first consider the *t* test for the correlated groups design and then for the independent groups design.

STUDENT'S *t* TEST FOR CORRELATED GROUPS*

You will recall that, in the repeated measures or correlated groups design, each subject gets two or more treatments: a difference score is calculated for each subject, and the resulting difference scores are analyzed. The simplest experiment of this type uses two conditions, often called *control* and *experimental*, or *before* and *after*. In a variant of this design, instead of the same subject being used in both conditions, pairs of subjects that are matched on one or more characteristics serve in the two conditions. Thus, pairs might be matched on IQ, age, gender, and so forth. The difference scores between the matched pairs are then analyzed in the same manner as when the same subject serves in both conditions. This design is also referred to as a *correlated groups design* because the subjects in the groups are not independently assigned; that is, the pairs share specifically matched common characteristics. In the independent groups design, which is discussed later in this chapter, there is no pairing.

We first encountered the correlated groups design when using the sign test. However, the sign test had low power because it ignored the magnitude of the difference scores. We used the sign test because of its simplicity. In the analysis of actual experiments, another test, such as the *t* test, would probably be used. The *t test for correlated groups* allows utilization of *both* the magnitude and direction of the difference scores. Essentially, it treats the difference scores as though they were raw scores and tests the assumption that the difference scores are a random sample from a population of difference scores having a mean of zero. This can best be seen through an example.

MENTORING TIP

Remember: analysis is done on the difference scores.

experiment

Brain Stimulation and Eating

To illustrate, suppose a neuroscientist believes that a brain region called the lateral hypothalamus is involved in eating behavior. One way to test this belief is to use a group of animals (e.g., rats) and electrically stimulate the lateral hypothalamus through a chronically

---

*See Note 14.1.

indwelling electrode. If the lateral hypothalamus is involved in eating behavior, electrical stimulation of the lateral hypothalamus might alter the amount of food eaten. To control for the effect of brain stimulation per se, another electrode would be implanted in each animal in a neutral brain area. Each area would be stimulated for a fixed period of time, and the amount of food eaten would be recorded. A difference score for each animal would then be calculated.

Let's assume there is insufficient supporting evidence to warrant a directional alternative hypothesis. Therefore, a two-tailed evaluation is planned. The alternative hypothesis states that electrical stimulation of the lateral hypothalamus affects the amount of food eaten. The null hypothesis specifies that electrical stimulation of the lateral hypothalamus does not affect the amount of food eaten. If H_0 is true, the difference score for each rat would be due to chance factors. Sometimes it would be positive, and other times it would be negative; sometimes it would be large in magnitude and other times small. If the experiment were done on a large number of rats, say, the entire population, the mean of the difference scores would equal zero.* Figure 14.1 shows such a distribution. Note, carefully, that the mean of this population is known ($\mu_D = 0$) and that the standard deviation is unknown ($\sigma_D = ?$). The chance explanation assumes that the difference scores of the sample in the experiment are a random sample from this population of difference scores. Thus, we have a situation in which there is one set of scores (e.g., the sample difference scores), and we are interested in determining whether it is reasonable to consider these scores a random sample from a population of difference scores having a known mean ($\mu_D = 0$) and unknown standard deviation.

Comparison Between Single Sample and Correlated Groups *t* Tests

The situation just described is almost identical to those we have previously considered regarding the *t* test with single samples. The only change is that in the correlated groups experiment we are analyzing *difference* scores rather than *raw* scores. It follows, then, that the equations for each should be quite similar. These equations are presented in Table 14.1.

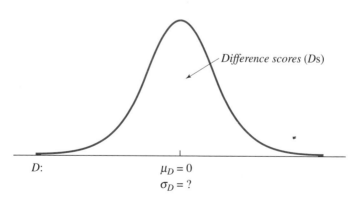

D:

$\mu_D = 0$
$\sigma_D = ?$

figure 14.1 Null-hypothesis population of difference scores.

*See Note 14.2.

table 14.1   *t* Test for single samples and correlated groups

t Test for Single Samples	*t* Test for Correlated Groups
$t_{obt} = \dfrac{\overline{X}_{obt} - \mu}{s/\sqrt{N}}$	$t_{obt} = \dfrac{\overline{D}_{obt} - \mu_D}{s_D/\sqrt{N}}$
$t_{obt} = \dfrac{\overline{X}_{obt} - \mu}{\sqrt{\dfrac{SS}{N(N-1)}}}$	$t_{obt} = \dfrac{\overline{D}_{obt} - \mu_D}{\sqrt{\dfrac{SS_D}{N(N-1)}}}$
$SS = \Sigma X^2 - \dfrac{(\Sigma X)^2}{N}$	$SS_D = \Sigma D^2 - \dfrac{(\Sigma D)^2}{N}$

where

D = *difference score*

$\overline{D}_{obt}$ = mean of the sample difference scores

μ_D = mean of the population of difference scores

s_D = standard deviation of the sample difference scores

N = number of difference scores

$SS_D = \Sigma(D - \overline{D})^2$ = sum of squares of sample difference scores

It is obvious that the two sets of equations are identical except that, in the single sample case, we are dealing with raw scores, whereas in the correlated groups experiment, we are analyzing difference scores. Let's now add some numbers to the brain stimulation experiment and see how to use the *t* test for correlated groups.

Brain Stimulation Experiment Revisited and Analyzed

A neuroscientist believes that the lateral hypothalamus is involved in eating behavior. If so, then electrical stimulation of that area might affect the amount eaten. To test this possibility, chronic indwelling electrodes are implanted in 10 rats. Each rat has two electrodes: one implanted in the lateral hypothalamus and the other in an area where electrical stimulation is known to have no effect. After the animals have recovered from surgery, they each receive 30 minutes of electrical stimulation to each brain area, and the amount of food eaten during the stimulation is measured. The amount of food in grams that was eaten during stimulation is shown in Table 14.2.

1. What is the alternative hypothesis? Assume a nondirectional hypothesis is appropriate.
2. What is the null hypothesis?
3. What is the conclusion? Use $\alpha = 0.05_{2\,tail}$.

SOLUTION

1. Alternative hypothesis: The alternative hypothesis specifies that electrical stimulation of the lateral hypothalamus affects the amount of food eaten. The sample difference scores having a mean $\overline{D}_{obt} = 5.3$ are a random sample from a population of difference scores having a mean $\mu_D \neq 0$.

table 14.2 Data from brain stimulation experiment

Subject	Lateral hypothalamus (g)	Neutral area (g)	Difference D	Difference D^2
	Food Eaten		**Difference**	
1	10	6	4	16
2	18	8	10	100
3	16	11	5	25
4	22	14	8	64
5	14	10	4	16
6	25	20	5	25
7	17	10	7	49
8	22	18	4	16
9	12	14	−2	4
10	21	13	8	64
			53	379

$$N = 10 \qquad \overline{D}_{\text{obt}} = \frac{\Sigma D}{N} = \frac{53}{10} = 5.3$$

2. Null hypothesis: The null hypothesis states that electrical stimulation of the lateral hypothalamus has no effect on the amount of food eaten. The sample difference scores having a mean $\overline{D}_{\text{obt}} = 5.3$ are a random sample from a population of difference scores having a mean $\mu_D = 0$.

3. Conclusion, using $\alpha = 0.05_{2\,\text{tail}}$:

   STEP 1: Calculate the appropriate statistic. Since this is a correlated groups design, we are interested in the difference between the paired scores rather than the scores in each condition per se. The difference scores are shown in Table 14.2. Of the tests covered so far, both the sign test and the *t* test are possible choices. We want to use the test that is most powerful, so the *t* test has been chosen. From the data table, $N = 10$ and $\overline{D}_{\text{obt}} = 5.3$. The calculation of t_{obt} is as follows:

$$t_{\text{obt}} = \frac{\overline{D}_{\text{obt}} - \mu_D}{\sqrt{\dfrac{SS_D}{N(N-1)}}}$$

$$= \frac{5.3 - 0}{\sqrt{\dfrac{98.1}{10(9)}}}$$

$$= \frac{5.3}{\sqrt{1.04}}$$

$$= 5.08$$

$$SS_D = \Sigma D^2 - \frac{(\Sigma D)^2}{N}$$

$$= 379 - \frac{(53)^2}{10}$$

$$= 98.1$$

STEP 2: **Evaluate the statistic.** As with the *t* test for single samples, if t_{obt} falls within the critical region for rejection of H_0, the conclusion is to reject H_0. Thus, the same decision rule applies, namely,

$$\text{if } |t_{obt}| \geq |t_{crit}|, \text{ reject } H_0.$$

The degrees of freedom are equal to the number of difference scores minus 1. Thus, df $= N - 1 = 10 - 1 = 9$. From Table D in Appendix D, with $\alpha = 0.05_{2\,tail}$ and df $= 9$,

$$t_{crit} = \pm 2.262$$

Since $|t_{obt}| > 2.262$, we reject H_0 and conclude that electrical stimulation of the lateral hypothalamus affects eating behavior. It appears to increase the amount eaten.

Practice Problem 14.1

To motivate citizens to conserve gasoline, the government is considering mounting a nationwide conservation campaign. However, before doing so on a national level, it decides to conduct an experiment to evaluate the effectiveness of the campaign. For the experiment, the conservation campaign is conducted in a small but representative geographical area. Twelve families are randomly selected from the area, and the amount of gasoline they use is monitored for 1 month before the advertising campaign and for 1 month after the campaign. The following data are collected:

Family	Before the Campaign (gal/mo.)	After the Campaign (gal/mo.)	Difference D	D^2
A	55	48	7	49
B	43	38	5	25
C	51	53	−2	4
D	62	58	4	16
E	35	36	−1	1
F	48	42	6	36
G	58	55	3	9
H	45	40	5	25
I	48	49	−1	1
J	54	50	4	16
K	56	58	−2	4
L	32	25	7	49
			35	235

$$N = 12 \qquad \overline{D}_{obt} = \frac{\Sigma D}{N} = \frac{35}{12} = 2.917$$

a. What is the alternative hypothesis? Use a nondirectional hypothesis.
b. What is the null hypothesis?
c. What is the conclusion? Use $\alpha = 0.05_{2\,\text{tail}}$.

SOLUTION

a. Alternative hypothesis: The conservation campaign affects the amount of gasoline used. The sample with $\overline{D}_{\text{obt}} = 2.917$ is a random sample from a population of difference scores where $\mu_D \neq 0$.
b. Null hypothesis: The conservation campaign has no effect on the amount of gasoline used. The sample with $\overline{D}_{\text{obt}} = 2.917$ is a random sample from a population of difference scores where $\mu_D = 0$.
c. Conclusion, using $\alpha = 0.05_{2\,\text{tail}}$:

> **STEP 1:** **Calculate the appropriate statistic.** The difference scores are included in the previous table. We have subtracted the "after" scores from the "before" scores. Assuming the assumptions of *t* are met, the appropriate statistic is t_{obt}. From the data table, $N = 12$ and $\overline{D}_{\text{obt}} = 2.917$.

$$t_{\text{obt}} = \frac{\overline{D}_{\text{obt}} - \mu_D}{\sqrt{\dfrac{SS_D}{N(N-1)}}} \qquad SS_D = \Sigma D^2 - \frac{(\Sigma D)^2}{N}$$

$$= \frac{2.917 - 0}{\sqrt{\dfrac{132.917}{12(11)}}} \qquad = 235 - \frac{(35)^2}{12}$$

$$\qquad\qquad\qquad\qquad = 132.917$$

$$= 2.91$$

> **STEP 2:** **Evaluate the statistic.** Degrees of freedom $= N - 1 = 12 - 1 = 11$. From Table D, with $\alpha = 0.05_{2\,\text{tail}}$ and 11 df,

$$t_{\text{crit}} = \pm 2.201$$

> Since $|t_{\text{obt}}| > 2.201$, we reject H_0. The conservation campaign affects the amount of gasoline used. It appears to decrease gasoline consumption.

Size of Effect Using Cohen's *d*

As we pointed out in the discussion of size of effect in conjunction with the *t* test for single samples, in addition to determining whether there is a real effect, it is often desirable to determine the size of the effect. For example, in the experiment investigating the involvement of the lateral hypothalamus in eating behavior (p. 360), t_{obt} was significant and we were able to conclude that electrical stimulation of the lateral hypothalamus had a real effect on eating behavior. It seems reasonable that we would also like to know the size of the effect.

To evaluate the size of effect we will again use Cohen's method involving the statistic *d*.* For convenience, we have repeated below the general equation for *d*, given in Chapter 13, p. 339.

$$d = \frac{|\text{mean difference}|}{\text{population standard deviation}} \quad \textit{General equation for size of effect}$$

In the correlated groups design, it is the magnitude of the mean of the difference scores $(\overline{D})$ that varies directly with the size of effect, and the standard deviation of the population difference scores (σ_D) that are of interest. Thus, for this design,

$$d = \frac{|\overline{D}_{\text{obt}}|}{\sigma_D} \quad \begin{array}{l} \textit{Conceptual equation for size of effect,} \\ \textit{correlated groups t test} \end{array}$$

Taking the absolute value of $\overline{D}_{\text{obt}}$ in the previous equation keeps *d* positive regardless of whether the convention used in subtracting the two scores for each subject produces a positive or negative $\overline{D}_{\text{obt}}$. Please note that when applying this equation, if H_1 is directional, $\overline{D}_{\text{obt}}$ must be in the direction predicted by H_1. If it is not in the predicted direction, then when analyzing the data of the experiment, the conclusion would be to retain H_0 and ordinarily, as with the single sample *t* test, it would make no sense to inquire about the size of the real effect.

Since we don't know σ_D, as usual, we estimate it with s_D, the standard deviation of the sample difference scores. The resulting equation is given by

$$\hat{d} = \frac{|\overline{D}_{\text{obt}}|}{s_D} \quad \begin{array}{l} \textit{Computational equation for size} \\ \textit{of effect, correlated groups t test} \end{array}$$

where $\hat{d}$ = estimated *d*

$|\overline{D}_{\text{obt}}|$ = the absolute value of the mean of the sample difference scores

s_D = the standard deviation of the sample difference scores

experiment

Lateral Hypothalamus and Eating Behavior

Let's now apply this theory to some data. For the experiment investigating the effect of electrical stimulation of the lateral hypothalamus on eating behavior (p. 360), we concluded that the electrical stimulation had a real effect. Now, let's determine the size of the effect. In that experiment,

$$\overline{D}_{\text{obt}} = 5.3 \text{ and } s_D = \sqrt{\frac{SS_D}{N-1}} = \sqrt{\frac{98.1}{10-1}} = 3.30$$

Substituting these values in the equation for $\hat{d}$, we obtain

$$\hat{d} = \frac{\overline{D}_{\text{obt}}}{s_D} = \frac{5.3}{3.30} = 1.61$$

To interpret the $\hat{d}$ value, we use the same criterion of Cohen that was presented in Table 13.6 on p. 340. For convenience we have reproduced the table again here.

---

*For reference, see footnote in Chapter 13, p. 339.

table 14.3   Cohen's criteria for interpreting the value of $\hat{d}$

Value of $\hat{d}$	Interpretation of $\hat{d}$
0.00–0.20	Small effect
0.21–0.79	Medium effect
≥0.80	Large effect

Since the $\hat{d}$ value of 1.61 is higher than 0.80, we conclude that the electrical stimulation of the lateral hypothesis had a large effect on eating behavior.*

t Test for Correlated Groups and Sign Test Compared

It would have been possible to solve either of the previous two problems using the sign test. We chose the *t* test because it is more powerful. To illustrate this point, let's use the sign test to solve the problem dealing with gasoline conservation.

SOLUTION USING SIGN TEST

STEP 1: Calculate the statistic. There are 8 pluses in the sample.

STEP 2: Evaluate the statistic. With $N = 12$, $P = 0.50$, and $\alpha = 0.05_{2\,\text{tail}}$,

$$p(8 \text{ or more pluses}) = p(8) + p(9) + p(10) + p(11) + p(12)$$

From Table B in Appendix D,

$$p(8 \text{ or more pluses}) = 0.1208 + 0.0537 + 0.0161 + 0.0029 + 0.0002$$
$$= 0.1937$$

Since the alpha level is two-tailed,

$$p(\text{outcome at least as extreme as 8 pluses}) = 2[(p8 \text{ or more pluses})]$$
$$= 2(0.1937)$$
$$= 0.3874$$

Since $0.3874 > 0.05$, we conclude by retaining H_0.

We are unable to reject H_0 with the sign test, but we were able to reject H_0 with the *t* test. Does this mean the campaign is effective if we analyze the data with the *t* test and ineffective if we use the sign test? Obviously not. With the low power of the sign test, there is a high chance of making a Type II error (i.e., retaining H_0 when it is false). The *t* test is usually more powerful than the sign test. The additional power gives H_0 a better chance to be rejected if it is false. In this case, the additional power resulted in rejection of H_0. When several tests are appropriate for analyzing the data, *it is a general rule of statistical analysis to use the most powerful one*, because this gives the highest probability of rejecting H_0 when it is false.

MENTORING TIP

When analyzing real data, always use the most powerful test that the data and assumptions of the test allow.

---

*See Chapter 13 footnote on p. 340 for a reference discussing some cautions in using Cohen's criteria.

Assumptions Underlying the *t* Test for Correlated Groups

The assumptions are very similar to those underlying the *t* test for single samples. The *t* test for correlated groups requires that the sampling distribution of $\overline{D}$ be normally distributed. This means that *N* should be $\geq$ 30, assuming the population shape doesn't differ greatly from normality, or the population scores themselves should be normally distributed.*

z AND *t* TESTS FOR INDEPENDENT GROUPS

Independent Groups Design

Two basic experimental designs are used most frequently in studying behavior. The first was introduced when discussing the sign test and the *t* test for correlated groups. This design is called the repeated or replicated measures design. The simplest form of the design uses two conditions: an experimental and a control condition. The essential feature of the design is that there are paired scores between conditions, and *difference* scores from each score pair are analyzed to determine whether chance alone can reasonably explain them.

The other type of design is called the *independent groups design*. Like the correlated groups design, the independent groups design involves experiments using two or more conditions. Each condition uses a different level of the independent variable. The most basic experiment has only two conditions: an experimental and a control condition. In this chapter, we shall consider this basic experiment involving only two conditions. More complicated experiments will be considered in Chapter 15.

In the independent groups design, subjects are randomly selected from the subject population and then randomly assigned to either the experimental or the control condition. There is no basis for pairing of subjects, and each subject is tested only once. All of the subjects in the experimental condition receive the level of the independent variable appropriate for the experimental condition, and the subjects themselves are referred to as the "experimental group." All of the subjects in the control condition receive the level of the independent variable appropriate for the control condition and are referred to as the "control group."

MENTORING TIP

Remember: for the independent groups design, the samples (groups) are separate; there is no basis for pairing of scores, and the raw scores within each group are analyzed separately.

When analyzing the data, since subjects are randomly assigned to conditions, there is no basis for pairing scores between the conditions. Rather, a statistic is computed for each group *separately*, and the two group statistics are compared to determine whether chance alone is a reasonable explanation of the differences between the group statistics. The statistic that is computed on each group depends on the inference test being used. The *t* test for independent groups computes the mean of each group and then analyzes the difference between these two group means to determine whether chance alone is a reasonable explanation of the difference between the two means.

---

*Many authors limit the use of the *t* test to data that are of interval or ratio scaling. Please see the footnote in Chapter 2, p. 35, for references discussing this point.

H$_1$ **and** ***H***$_0$ The sample scores in one of the conditions (say, condition 1) can be considered a random sample from a normally distributed population of scores that would result if all the individuals in the population received that condition (condition 1). Let's call the mean of this hypothetical population μ_1 and the standard deviation σ_1. Similarly, the sample scores in condition 2 can be considered a random sample from a normally distributed population of scores that would result if all the individuals in the population were given condition 2. We can call the mean of this second population μ_2 and the standard deviation σ_2. Thus,

μ_1 = mean of the population that receives condition 1

σ_1 = standard deviation of the population that receives condition 1

μ_2 = mean of the population that receives condition 2

σ_2 = standard deviation of the population that receives condition 2

Changing the level of the independent variable is assumed to affect the mean of the distribution (μ_2) but not the standard deviation (σ_2) or variance (σ_2^2). Thus, under this assumption, if the independent variable has a real effect, the means of the populations will differ but their variances will stay the same. Hence, σ_1^2 is assumed equal to σ_2^2. One way in which this assumption would be met is if the independent variable has an equal effect on each individual. A directional alternative hypothesis would predict that the samples are random samples from populations where $\mu_1 > \mu_2$ or $\mu_1 < \mu_2$, depending on the direction of the effect. A nondirectional alternative hypothesis would predict $\mu_1 \neq \mu_2$. If the independent variable has no effect, the samples would be random samples from populations where $\mu_1 = \mu_2$* and chance alone would account for the differences between the sample means.

MENTORING TIP

Remember: for a directional H_1: $\mu_1 > \mu_2$ or $\mu_1 < \mu_2$; for a nondirectional H_1: $\mu_1 \neq \mu_2$.

z TEST FOR INDEPENDENT GROUPS

Before discussing the *t* test for independent groups, we shall present the *z* test. In the two-group situation, the *z* test is almost never used because it requires that σ_1^2 or σ_2^2 be known. However, it provides an important conceptual foundation for understanding the *t* test. After presenting the *z* test, we shall move to the *t* test.

Let's begin with an experiment.

experiment

Hormone X and Sexual Behavior

A physiologist has the hypothesis that hormone X is important in producing sexual behavior. To investigate this hypothesis, 20 male rats were randomly sampled and then randomly assigned to two groups. The animals in group 1 were injected with hormone X and then were placed in individual housing with a sexually receptive female. The animals in group 2 were given similar treatment except they were injected with a placebo solution. The number of matings was counted over a 20-minute period. The results are shown in Table 14.4.

As shown in Table 14.4, the mean of group 1 is higher than the mean of group 2. The difference between the means of the two samples is 2.8 and is in the direction that

---

*In this case, there would be two null-hypothesis populations: one with a mean μ_1 and a standard deviation of σ_1 and the other with a mean μ_2 and a standard deviation σ_2. However, since $\mu_1 = \mu_2$ and $\sigma_1 = \sigma_2$, the populations would be identical.

table 14.4 Data from hormone X and sexual behavior experiment

Hormone X Group 1	Placebo Group 2
8	5
10	6
12	3
6	4
6	7
7	8
9	6
8	5
7	4
11	8
84	56

$n_1 = 10$ $n_2 = 10$

$\overline{X}_1 = 8.4$ $\overline{X}_2 = 5.6$

$\overline{X}_1 - \overline{X}_2 = 2.8$

indicates a positive effect. Is it legitimate to conclude that hormone X was responsible for the difference in means? The answer, of course, is no. Before drawing this conclusion, we must evaluate the null-hypothesis explanation. The statistic we are using for this evaluation is the difference between the means of the two samples. As with all other statistics, we must know its sampling distribution before we can evaluate the null hypothesis.

The Sampling Distribution of the Difference Between Sample Means ($\overline{X}_1 - \overline{X}_2$)

Like the sampling distribution of the mean, this sampling distribution can be determined theoretically or empirically. Again, for pedagogical purposes, we prefer the empirical approach. To empirically derive the sampling distribution of $\overline{X}_1 - \overline{X}_2$, all possible different samples of size n_1 would be drawn from a population with a mean of μ_1 and variance of σ_1^2. Likewise, all possible samples of size n_2 would be drawn from another population with a mean of μ_2 and variance of σ_2^2. The values of $\overline{X}_1$ and $\overline{X}_2$ would then be calculated for each sample. Next, $\overline{X}_1 - \overline{X}_2$ would be calculated for all possible pairings of samples of size n_1 and n_2. The resulting distribution would contain all the possible different $\overline{X}_1 - \overline{X}_2$ scores that could be obtained from the populations when sample sizes are n_1 and n_2. Once this distribution has been obtained, it is a simple matter to calculate the probability of obtaining each mean difference score $\overline{X}_1 - \overline{X}_2$, assuming sampling is random of n_1 and n_2 scores from their respective populations. This then would be the *sampling distribution of the difference between sample means* for samples of n_1 and n_2 taken from the specified populations. This process would be repeated for different sample sizes and population scores. Whether determined theoretically or empirically, the sampling distribution of the difference between sample means has the following characteristics:

1. If the populations from which the samples are taken are normal, then the sampling distribution of the difference between sample means is also normal.

2. $\mu_{\overline{X}_1 - \overline{X}_2} = \mu_1 - \mu_2$

   where $\mu_{\overline{X}_1 - \overline{X}_2} =$ *the mean of the sampling distribution of the difference between sample means*

3. $\sigma_{\overline{X}_1 - \overline{X}_2} = \sqrt{\sigma_{\overline{X}_1}^2 + \sigma_{\overline{X}_2}^2}$

   where $\sigma_{\overline{X}_1 - \overline{X}_2} =$ *standard deviation of the sampling distribution of the difference between sample means*; alternatively, *standard error of the difference between sample means*

   $\sigma_{\overline{X}_1}^2 =$ *variance of the sampling distribution of the mean for samples of size n_1 taken from the first population*

   $\sigma_{\overline{X}_2}^2 =$ *variance of the sampling distribution of the mean for samples of size n_2 taken from the second population*

If, as mentioned previously, we assume that the variances of the two populations are equal ($\sigma_1^2 = \sigma_2^2$), then the equation for $\sigma_{\overline{X}_1 - \overline{X}_2}$ can be simplified as follows:

$$\sigma_{\overline{X}_1 - \overline{X}_2} = \sqrt{\sigma_{\overline{X}_1}^2 + \sigma_{\overline{X}_2}^2}$$

$$= \sqrt{\frac{\sigma_1^2}{n_1} + \frac{\sigma_2^2}{n_2}}$$

$$= \sqrt{\sigma^2\left(\frac{1}{n_1} + \frac{1}{n_2}\right)}$$

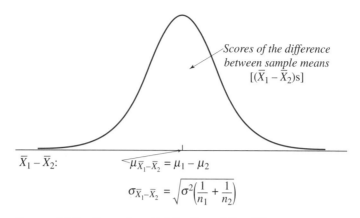

figure 14.2 Sampling distribution of the difference between sample mean scores.

where $\quad \sigma^2 = \sigma_1{}^2 = \sigma_2{}^2 =$ the variance of each population

The distribution is shown in Figure 14.2. Now let's return to the illustrative example.

experiment

Hormone X Experiment Revisited

The results of the experiment showed that 10 rats injected with hormone X had a mean of 8.4 matings, whereas the mean of the 10 rats injected with a placebo was 5.6. Is the mean difference ($\overline{X}_1 - \overline{X}_2 = 2.8$) significant? Use $\alpha = 0.05_{2\text{ tail}}$.

SOLUTION

The sampling distribution of $\overline{X}_1 - \overline{X}_2$ is shown in Figure 14.3. The shaded area contains all the mean difference scores of 2.8 or more. Assuming the sampling distribution of $\overline{X}_1 - \overline{X}_2$ is normal, if the sample mean difference (2.8) can be converted to its *z* value, we can use the *z* test to solve the problem. The equation for z_{obt} is similar to the other *z* equations we have already considered. However, here the value we are converting is $\overline{X}_1 - \overline{X}_2$. Thus,

$$z_{\text{obt}} = \frac{(\overline{X}_1 - \overline{X}_2) - \mu_{\overline{X}_1 - \overline{X}_2}}{\sigma_{\overline{X}_1 - \overline{X}_2}} \qquad equation\ for\ z_{\text{obt}},\ independent\ groups\ design$$

If hormone X had no effect on mating behavior, then both samples are random samples from populations where $\mu_1 = \mu_2$ and $\mu_{\overline{X}_1 - \overline{X}_2} = \mu_1 - \mu_2 = 0$. Thus,

$$z_{\text{obt}} = \frac{(\overline{X}_1 - \overline{X}_2) - 0}{\sqrt{\sigma^2\left(\dfrac{1}{n_1} + \dfrac{1}{n_2}\right)}} = \frac{2.8}{\sqrt{\sigma^2\left(\dfrac{1}{10} + \dfrac{1}{10}\right)}}$$

Note that the variance of the populations (σ^2) must be known before z_{obt} can be calculated. Since σ^2 is almost never known, this limitation severely restricts the practical use of the *z* test in this design. However, as you might guess, σ^2 can be estimated from the sample data. When this is done, we have the *t* test for independent groups.

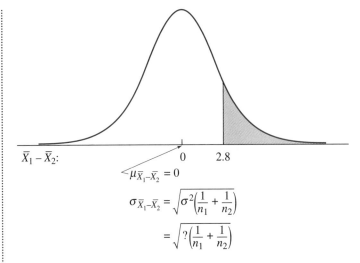

$$\overline{X}_1 - \overline{X}_2:$$

$$\mu_{\overline{X}_1-\overline{X}_2} = 0$$

$$\sigma_{\overline{X}_1-\overline{X}_2} = \sqrt{\sigma^2\left(\frac{1}{n_1} + \frac{1}{n_2}\right)}$$

$$= \sqrt{?\left(\frac{1}{n_1} + \frac{1}{n_2}\right)}$$

figure 14.3 Sampling distribution of the difference between sample mean scores for the hormone problem.

STUDENT'S *t* TEST FOR INDEPENDENT GROUPS

Comparing the Equations for z_{obt} and t_{obt}

The equations for the *z* and *t* test are shown in Table 14.5. The *z* and *t* equations are identical except that the *t* equation uses s_W^2 to estimate the population variance (σ^2). This situation is analogous to the *t* test for single samples. You will recall in that situation we used the sample standard deviation (*s*) to estimate σ. However, in the *t* test for independent groups, there are two samples, and we wish to estimate σ^2. Since *s* is an accurate estimate of σ, s^2 is an accurate estimate of σ^2. There are two samples, and either could be used to estimate σ^2, but we can get a more precise estimate by using both. It turns out the most precise estimate of σ^2 is obtained by using a weighted

MENTORING TIP

The *t* test is used instead of the *z* test because the value of σ^2 is almost never known.

table 14.5 *z* and *t* equations compared

z Test	*t* Test
$z_{obt} = \dfrac{(\overline{X}_1 - \overline{X}_2) - \mu_{\overline{X}_1-\overline{X}_2}}{\sigma_{\overline{X}_1-\overline{X}_2}}$	$t_{obt} = \dfrac{(\overline{X}_1 - \overline{X}_2) - \mu_{\overline{X}_1-\overline{X}_2}}{s_{\overline{X}_1-\overline{X}_2}}$
$= \dfrac{(\overline{X}_1 - \overline{X}_2) - \mu_{\overline{X}_1-\overline{X}_2}}{\sqrt{\sigma^2\left(\dfrac{1}{n_1} + \dfrac{1}{n_2}\right)}}$	$= \dfrac{(\overline{X}_1 - \overline{X}_2) - \mu_{\overline{X}_1-\overline{X}_2}}{\sqrt{s_W^2\left(\dfrac{1}{n_1} + \dfrac{1}{n_2}\right)}}$

Where s_W^2 = *weighted estimate of* σ^2

$s_{\overline{X}_1-\overline{X}_2}$ = *estimate of* $\sigma_{\overline{X}_1-\overline{X}_2}$ = *estimated standard error of the difference between sample means*

average of the sample variances s_1^2 and s_2^2. Weighting is done using *degrees of freedom* as the weights. Thus,

$$s_W^2 = \frac{\mathrm{df}_1 s_1^2 + \mathrm{df}_2 s_2^2}{\mathrm{df}_1 + \mathrm{df}_2} = \frac{(n_1 - 1)\left(\dfrac{SS_1}{n_1 - 1}\right) + (n_2 - 1)\left(\dfrac{SS_2}{n_2 - 1}\right)}{(n_1 - 1) + (n_2 - 1)} = \frac{SS_1 + SS_2}{n_1 + n_2 - 2}$$

where s_W^2 = weighted estimate of σ^2

s_1^2 = variance of the first sample

s_2^2 = variance of the second sample

SS_1 = sum of squares of the first sample

SS_2 = sum of squares of the second sample

Substituting for s_W^2 in the *t* equation, we arrive at the computational equation for t_{obt}. Thus,

$$t_{obt} = \frac{(\overline{X}_1 - \overline{X}_2) - \mu_{\overline{X}_1 - \overline{X}_2}}{\sqrt{s_W^2\left(\dfrac{1}{n_1} + \dfrac{1}{n_2}\right)}} = \frac{(\overline{X}_1 - \overline{X}_2) - \mu_{\overline{X}_1 - \overline{X}_2}}{\sqrt{\left(\dfrac{SS_1 + SS_2}{n_1 + n_2 - 2}\right)\left(\dfrac{1}{n_1} + \dfrac{1}{n_2}\right)}}$$

computational equation for t_{obt}
independent groups design

To evaluate the null hypothesis, we assume both samples are random samples from populations having the same mean value ($\mu_1 = \mu_2$). Therefore, $\mu_{\overline{X}_1 - \overline{X}_2} = 0$.* The previous equation reduces to

$$t_{obt} = \frac{\overline{X}_1 - \overline{X}_2}{\sqrt{\left(\dfrac{SS_1 + SS_2}{n_1 + n_2 - 2}\right)\left(\dfrac{1}{n_1} + \dfrac{1}{n_2}\right)}}$$

computational equation for t_{obt},
assuming $\mu_{\overline{X}_1 - \overline{X}_2} = 0$

We could go ahead and calculate t_{obt} for the hormone X example, but to evaluate t_{obt}, we must know the sampling distribution of *t*. It turns out that, when one derives the sampling distribution of *t* for independent groups, the same family of curves is obtained as with the sampling distribution of *t* for single samples, except that there is a different number of degrees of freedom. You will recall that 1 degree of freedom is lost each time a standard deviation is calculated. Since we calculate s_1^2 and s_2^2 for the two-sample case, we lose 2 degrees of freedom, one from each sample. Thus,

$$\mathrm{df} = (n_1 - 1) + (n_2 - 1) = n_1 + n_2 - 2 = N - 2$$

where $N = n_1 + n_2$

MENTORING TIP

Remember: the *t* distribution varies uniquely with df, not with *N*.

Table D, then, can be used in the same manner as with the *t* test for single samples, except in the two-sample case, we enter the table with $N - 2$ df. Thus, the *t* distribution varies both with *N* and degrees of freedom, but it varies uniquely only with degrees of freedom. That is, the *t* distribution corresponding to 13 df is the same whether it is derived from the single sample situation with $N = 14$ or the two-sample situation with $N = 15$.

---

*See Note 14.3.

Analyzing the Hormone X Experiment

At long last, we are ready to evaluate the hormone data. The problem and data are restated for convenience.

A physiologist has conducted an experiment to evaluate the effect of hormone X on sexual behavior. Ten rats were injected with hormone X, and 10 other rats received a placebo injection. The number of matings was counted over a 20-minute period. The results are shown in Table 14.6.

1. What is the alternative hypothesis? Use a nondirectional hypothesis.
2. What is the null hypothesis?
3. What do you conclude? Use $\alpha = 0.05_{2\,\text{tail}}$.

SOLUTION

1. Alternative hypothesis: The alternative hypothesis specifies that hormone X affects sexual behavior. The sample mean difference of 2.8 is due to random sampling from populations where $\mu_1 \neq \mu_2$.
2. Null hypothesis: The null hypothesis states that hormone X is not related to sexual behavior. The sample mean difference of 2.8 is due to random sampling from populations where $\mu_1 = \mu_2$.
3. Conclusion, using $\alpha = 0.05_{2\,\text{tail}}$:

   STEP 1: Calculate the appropriate statistic. For now, assume *t* is appropriate. We shall discuss the assumptions of *t* in a later section. From

table 14.6 Data from hormone X experiment

Hormone X Group 1		Placebo Group 2	
X_1	X_1^2	X_2	X_2^2
8	64	5	25
10	100	6	36
12	144	3	9
6	36	4	16
6	36	7	49
7	49	8	64
9	81	6	36
8	64	5	25
7	49	4	16
11	121	8	64
84	744	56	340

$$n_1 = 10 \qquad n_2 = 10$$
$$\overline{X}_1 = 8.4 \qquad \overline{X}_2 = 5.6$$
$$\overline{X}_1 - \overline{X}_2 = 2.8$$

Table 14.6, $n_1 = 10$, $n_2 = 10$, $\overline{X}_1 = 8.4$, and $\overline{X}_2 = 5.6$. Solving for SS_1 and SS_2,

$$SS_1 = \Sigma X_1^2 - \frac{(\Sigma X_1)^2}{n_1} \qquad SS_2 = \Sigma X_2^2 - \frac{(\Sigma X_2)^2}{n_2}$$

$$= 744 - \frac{(84)^2}{10} \qquad\qquad = 340 - \frac{(56)^2}{10}$$

$$= 38.4 \qquad\qquad\qquad = 26.4$$

Substituting these values in the equation for t_{obt}, we have

$$t_{obt} = \frac{\overline{X}_1 - \overline{X}_2}{\sqrt{\left(\dfrac{SS_1 + SS_2}{n_1 + n_2 - 2}\right)\left(\dfrac{1}{n_1} + \dfrac{1}{n_2}\right)}} = \frac{8.4 - 5.6}{\sqrt{\left(\dfrac{38.4 + 26.4}{10 + 10 - 2}\right)\left(\dfrac{1}{10} + \dfrac{1}{10}\right)}} = 3.30$$

STEP 2: Evaluate the statistic. As with the previous *t* tests, if t_{obt} falls in the critical region for rejection of H_0, we reject H_0. Thus,

If $|t_{obt}| \geq |t_{crit}|$, reject H_0.

If not, retain H_0.

The number of degrees of freedom is df = $N - 2 = 20 - 2 = 18$. From Table D, with $\alpha = 0.05_{2\,tail}$ and df = 18,

$$t_{crit} = \pm 2.101$$

Since $|t_{obt}| > 2.101$, we conclude by rejecting H_0.

Calculating t_{obt} When $n_1 = n_2$

When the sample sizes are equal, the equation for t_{obt} can be simplified. Thus,

$$t_{obt} = \frac{\overline{X}_1 - \overline{X}_2}{\sqrt{\left(\dfrac{SS_1 + SS_2}{n_1 + n_2 - 2}\right)\left(\dfrac{1}{n_1} + \dfrac{1}{n_2}\right)}}$$

but $n_1 = n_2 = n$. Substituting n for n_1 and n_2 in the equation for t_{obt}

$$t_{obt} = \frac{\overline{X}_1 - \overline{X}_2}{\sqrt{\left(\dfrac{SS_1 + SS_2}{n + n - 2}\right)\left(\dfrac{1}{n} + \dfrac{1}{n}\right)}} = \frac{\overline{X}_1 - \overline{X}_2}{\sqrt{\left(\dfrac{SS_1 + SS_2}{2(n - 1)}\right)\left(\dfrac{2}{n}\right)}} = \frac{\overline{X}_1 - \overline{X}_2}{\sqrt{\left(\dfrac{SS_1 + SS_2}{n(n - 1)}\right)}}$$

Thus,

$$t_{obt} = \frac{\overline{X}_1 - \overline{X}_2}{\sqrt{\dfrac{SS_1 + SS_2}{n(n - 1)}}} \qquad \textit{equation for calculating } t_{obt} \textit{ when } n_1 = n_2$$

Since $n_1 = n_2$ in the previous problem, we can use the simplified equation to calculate t_{obt}. Thus,

$$t_{obt} = \frac{\overline{X}_1 - \overline{X}_2}{\sqrt{\dfrac{SS_1 + SS_2}{n(n-1)}}} = \frac{8.4 - 5.6}{\sqrt{\dfrac{38.4 + 26.4}{10(9)}}} = 3.30$$

This is the same value for t_{obt} that we obtained when using the more complicated equation. Whenever $n_1 = n_2$, it's easier to use the simplified equation. When $n_1 \neq n_2$, the more complicated equation *must* be used.

Let's do one more problem for practice.

Practice Problem 14.2

A neurosurgeon believes that lesions in a particular area of the brain, called the thalamus, will decrease pain perception. If so, this could be important in the treatment of terminal illness that is accompanied by intense pain. As a first attempt to test this hypothesis, he conducts an experiment in which 16 rats are randomly divided into two groups of eight each. Animals in the experimental group receive a small lesion in the part of the thalamus thought to be involved with pain perception. Animals in the control group receive a comparable lesion in a brain area believed to be unrelated to pain. Two weeks after surgery, each animal is given a brief electrical shock to the paws. The shock is administered in an ascending series, beginning with a very low intensity level and increasing until the animal first flinches. In this manner, the pain threshold to electric shock is determined for each rat. The following data are obtained. Each score represents the current level (milliamperes) at which flinching is first observed. The higher the current level is, the higher is the pain threshold. Note that one animal died during surgery and was not replaced.

Neutral Area Lesions Control Group Group 1		Thalamic Lesions Experimental Group Group 2	
X_1	X_1^2	X_2	X_2^2
0.8	0.64	1.9	3.61
0.7	0.49	1.8	3.24
1.2	1.44	1.6	2.56
0.5	0.25	1.2	1.44
0.4	0.16	1.0	1.00
0.9	0.81	0.9	0.81
1.4	1.96	1.7	2.89
1.1	1.21	10.1	15.55
7.0	6.96		

$$n_1 = 8 \qquad\qquad n_2 = 7$$
$$\overline{X}_1 = 0.875 \qquad\qquad \overline{X}_2 = 1.443$$
$$\overline{X}_1 - \overline{X}_2 = -0.568$$

a. What is the alternative hypothesis? In this problem, assume there is sufficient theoretical and experimental basis to use a directional hypothesis.
b. What is the null hypothesis?
c. What do you conclude? Use $\alpha = 0.05_{1 \text{ tail}}$.

SOLUTION

a. Alternative hypothesis: The alternative hypothesis states that lesions of the thalamus decrease pain perception. The difference between sample means of -0.568 is due to random sampling from populations where $\mu_1 < \mu_2$.
b. Null hypothesis: The null hypothesis states that lesions of the thalamus either have no effect or they increase pain perception. The difference between sample means of -0.568 is due to random sampling from populations where $\mu_1 \geq \mu_2$.
c. Conclusion, using $\alpha = 0.05_{1 \text{ tail}}$:

> **STEP 1: Calculate the appropriate statistic.** Assuming the assumptions of t are met, t_{obt} is the appropriate statistic. From the data table, $n_1 = 8$, $n_2 = 7$, $\overline{X}_1 = 0.875$, and $\overline{X}_2 = 1.443$. Solving for SS_1 and SS_2, we obtain
>
> $$SS_1 = \Sigma X_1^2 - \frac{(\Sigma X_1)^2}{n_1} \qquad SS_2 = \Sigma X_2^2 - \frac{(\Sigma X_2)^2}{n_2}$$
>
> $$= 6.960 - \frac{(7)^2}{8} \qquad\qquad = 15.550 - \frac{(10.1)^2}{7}$$
>
> $$= 0.835 \qquad\qquad\qquad = 0.977$$
>
> Substituting these values into the general equation for t_{obt}, we have
>
> $$t_{obt} = \frac{\overline{X}_1 - \overline{X}_2}{\sqrt{\left(\dfrac{SS_1 + SS_2}{n_1 + n_2 - 2}\right)\left(\dfrac{1}{n_1} + \dfrac{1}{n_2}\right)}}$$
>
> $$= \frac{0.875 - 1.443}{\sqrt{\left(\dfrac{0.835 + 0.977}{8 + 7 - 2}\right)\left(\dfrac{1}{8} + \dfrac{1}{7}\right)}} = -2.94$$

> **STEP 2: Evaluate the statistic.** Degrees of freedom $= N - 2 = 15 - 2 = 13$. From Table D, with $\alpha = 0.05_{1 \text{ tail}}$ and df $= 13$,
>
> $$t_{crit} = -1.771$$
>
> Since $|t_{obt}| > 1.771$, we reject H_0 and conclude that lesions of the thalamus decrease pain perception.

Assumptions Underlying the *t* Test

The assumptions underlying the *t* test for independent groups are as follows:
  1. The sampling distribution of $\overline{X}_1 - \overline{X}_2$ is normally distributed. This means the populations from which the samples were taken should be normally distributed.
  2. There is *homogeneity of variance*. You will recall that, at the beginning of our discussion concerning the *t* test for independent groups, we pointed out that the

t test assumes that the independent variable affects the means of the populations but not their standard deviations ($\sigma_1 = \sigma_2$). Since the variance is just the square of the standard deviation, the *t* test for independent groups also assumes that the variances of the two populations are equal; that is, $\sigma_1^2 = \sigma_2^2$. This is spoken of as the *homogeneity of variance assumption*. If the variances of the samples in the experiment (s_1^2 and s_2^2) are very different (e.g., if one variance is more than 4 times larger than the other), the two samples probably are not random samples from populations where $\sigma_1^2 = \sigma_2^2$. If this is true, the homogeneity of variance assumption is violated ($\sigma_1^2 \neq \sigma_2^2$).*

Violation of the Assumptions of the *t* Test

Experiments have been conducted to determine the effect on the *t* test for independent groups of violating the assumptions of normality of the raw-score populations and homogeneity of variance. Fortunately, it turns out that the *t* test is a *robust* test. *A test is said to be robust if it is relatively insensitive to violations of its underlying mathematical assumptions.* The *t* test is relatively insensitive to violations of normality and homogeneity of variance, depending on sample size and the type and magnitude of the violation.[†] If $n_1 = n_2$ and the size of each sample is equal to or greater than 30, the *t* test for independent groups may be used without appreciable error despite moderate violation of the normality and/or the homogeneity of variance assumptions. If there are extreme violations of these assumptions, then an alternative test such as the Mann–Whitney *U* test should be used. This test is discussed in Chapter 17.

Before leaving this topic, it is worth noting that, when the two samples show large differences in their variances, it may indicate that the independent variable is not having an equal effect on all the subjects within a condition. This can be an important finding in its own right, leading to further experimentation into how the independent variable varies in its effects on different types of subjects.

Size of Effect Using Cohen's *d*

As has been previously discussed, in addition to determining whether there is a real effect, it is often desirable to determine the *size of the effect*. For example, in the experiment investigating the role of the thalamus in pain perception (p. 374), t_{obt} was significant and we were able to conclude that lesions of the thalamus decrease pain perception. But surely, it would also be desirable to know how large a role the thalamus plays. Does the thalamus totally control pain perception such that if the relevant thalamic nuclei were completely destroyed, the subject would no longer feel pain? On the other hand, is the effect so small that for any practical purposes, it can be ignored? After all, even small effects are likely to be significant if *N* is large enough. Determining the size of the thalamic effect would be particularly important for the neurosurgeon doing this

---

*There are many inference tests to determine whether the data meet homogeneity of variance assumptions. However, this topic is beyond the scope of this textbook. See R. E. Kirk, *Experimental Design*, 3rd ed., Brooks/Cole, Pacific Grove, CA, 1995, pp. 100–103. Some statisticians also require that the data be of interval or ratio scaling to use the *z* test, Student's *t* test, and the analysis of variance (covered in Chapter 15). For a discussion of this point, see the references contained in the Chapter 2 footnote on p. 35.
[†]For a review of this topic, see C. A. Boneau, "The Effects of Violations of Assumptions Underlying the *t* Test," *Psychological Bulletin, 57* (1960), 49–64.

research in hope of developing a treatment for reducing the intense pain felt by some terminal patients.

To evaluate the size of effect we will again use Cohen's method involving the statistic *d*.* With the *t* test for independent groups, it is the magnitude of the difference between the two sample means $(\overline{X}_1 - \overline{X}_2)$ that varies directly with the size of effect. Thus, for this design,

$$d = \frac{|\overline{X}_1 - \overline{X}_2|}{\sigma} \qquad \textit{Conceptual equation for size of effect, independent groups t test}$$

Taking the absolute value of $\overline{X}_1 - \overline{X}_2$ in the previous equation keeps *d* positive regardless of whether the convention used in assigning treatments to condition 1 and condition 2 results in a positive or negative value for $\overline{X}_1 - \overline{X}_2$. Again, please note that when applying this equation, if H_1 is directional, $\overline{X}_1 - \overline{X}_2$ must be in the direction predicted by H_1. If it is not in the predicted direction, then when analyzing the data of the experiment, the conclusion would be to retain H_0 and, as with the other *t* tests, it would make no sense to inquire about the size of the real effect.

Since we don't know σ, we estimate it with $\sqrt{s_W{}^2}$. The resulting equation is given by

$$\hat{d} = \frac{|\overline{X}_1 - \overline{X}_2|}{\sqrt{s_W{}^2}} \qquad \textit{Computational equation for size of effect, independent groups t test}$$

where

$$\hat{d} = \text{estimated } d$$

$$|\overline{X}_1 - \overline{X}_2| = \text{the absolute value of the difference between the two sample means}$$

$$\sqrt{s_W{}^2} = \text{weighted estimate of } \sigma$$

experiment

Thalamus and Pain Perception

Let's now apply this theory to some data. For the experiment investigating whether thalamic lesions decrease pain perception (Practice Problem 14.2, p. 374),

$$|\overline{X}_1 - \overline{X}_2| = |0.875 - 1.443| = 0.568$$

and

$$s_W{}^2 = \frac{SS_1 + SS_2}{n_1 + n_2 - 2} = \frac{0.835 + 0.977}{8 + 7 - 2} = 0.139$$

Substituting these values into the equation for $\hat{d}$, we obtain

$$\hat{d} = \frac{|\overline{X}_1 - \overline{X}_2|}{\sqrt{s_W{}^2}} = \frac{0.568}{\sqrt{0.139}} = 1.52$$

To interpret the $\hat{d}$ value, we again use the same criterion of Cohen that was presented in Table 13.6 on p. 340. For convenience we have reproduced the table here.

---

*See Chapter 13 footnote, p. 339 for reference.

table 14.7 Cohen's criteria for interpreting the value of $\hat{d}$*

Value of $\hat{d}$	Interpretation of $\hat{d}$
0.00–0.20	Small effect
0.21–0.79	Medium effect
≥0.80	Large effect

Since the $\hat{d}$ value of 1.52 is higher than 0.80, we conclude that the thalamic lesions had a large effect on pain perception.

POWER OF THE *t* TEST

The three equations for t_{obt} are as follows:

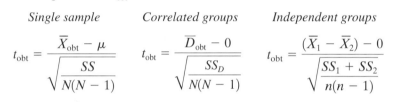

$$\textit{Single sample} \qquad \textit{Correlated groups} \qquad \textit{Independent groups}$$

$$t_{obt} = \frac{\overline{X}_{obt} - \mu}{\sqrt{\dfrac{SS}{N(N-1)}}} \qquad t_{obt} = \frac{\overline{D}_{obt} - 0}{\sqrt{\dfrac{SS_D}{N(N-1)}}} \qquad t_{obt} = \frac{(\overline{X}_1 - \overline{X}_2) - 0}{\sqrt{\dfrac{SS_1 + SS_2}{n(n-1)}}}$$

It seems fairly obvious that the larger t_{obt} is, the more likely H_0 will be rejected. Hence, anything that increases the likelihood of obtaining high values of t_{obt} will result in a more powerful *t* test. This can occur in several ways. First, the larger the real effect of the independent variable is, the more likely $\overline{X}_{obt} - \mu$, $\overline{D}_{obt} - 0$, or $(\overline{X}_1 - \overline{X}_2) - 0$ will be large. Since these difference scores are in the numerator of the *t* equation, it follows that *the greater the effect of the independent variable, the higher the power of the t test* (other factors held constant). Of course, we don't know before doing the experiment what the actual effect of the independent variable is. If we did, then why do the experiment? Nevertheless, this analysis is useful because it suggests that, when designing an experiment, it is desirable to use the level of independent variable that the experimenter believes is the most effective to maximize the chances of detecting its effect. This analysis further suggests that, given meager resources for conducting an experiment, the experiment may still be powerful enough to detect the effect if the independent variable has a large effect.

The denominator of the *t* equation varies as a function of sample size and sample variability. As sample size increases, the denominator decreases. Therefore,

$$\sqrt{\frac{SS}{N(N-1)}}, \quad \sqrt{\frac{SS_D}{N(N-1)}}, \quad \text{and} \quad \sqrt{\frac{SS_1 + SS_2}{n(n-1)}}$$

decrease, causing t_{obt} to increase. *Thus, increasing sample size increases the power of the t test.*

---

*See Chapter 13 footnote on p. 339 for a reference discussing some cautions in using Cohen's criteria.

The denominator also varies as a function of sample variability. In the single sample case, SS is the measure of variability. SS_D in the correlated groups experiments and $SS_1 + SS_2$ in the independent groups experiments reflect the variability. As the variability increases, the denominator in each case also increases, causing t_{obt} to decrease. *Thus, high sample variability decreases power.* Therefore, it is desirable to *decrease* variability as much as possible. One way to decrease variability is to carefully control the experimental conditions. For example, in a reaction-time experiment, the experimenter might use a warning signal that directly precedes the stimulus to which the subject must respond. In this way, variability due to attention lapses could be eliminated. Another way is to use the appropriate experimental design. For example, in certain situations, using a correlated groups design rather than an independent groups design will decrease variability.

MENTORING TIP

Remember: power varies directly with N and size of effect, and inversely with sample variability.

CORRELATED GROUPS AND INDEPENDENT GROUPS DESIGNS COMPARED

You are probably aware that many of the hypotheses presented in illustrative examples could have been investigated with either the correlated groups design or the independent groups design. For instance, in Practice Problem 14.1, we presented an experiment that was conducted to evaluate the effect of a conservation campaign on gasoline consumption. The experiment used a correlated groups design, and the data were analyzed with the t test for correlated groups. For convenience, the data and analysis are provided again in Table 14.8.

The conservation campaign could also have been evaluated using the independent groups design. Instead of using the same subjects in each condition, there would be two groups of subjects. One group would be monitored before the campaign and the other group monitored after the campaign. To evaluate the null hypothesis, each sample would be treated as an independent sample randomly selected from populations where $\mu_1 = \mu_2$. The basic statistic calculated would be $\overline{X}_1 - \overline{X}_2$. For the sake of comparison, let's analyze the conservation campaign data as though they were collected by using an independent groups design.* Assume there are two different groups. The families in group 1 (before) are monitored for 1 month with respect to the amount of gasoline used before the conservation campaign is conducted, whereas the families in group 2 are monitored for 1 month after the campaign has been conducted.

Since $n_1 = n_2$, we can use the t_{obt} equation for equal n. In this experiment, $n_1 = n_2 = 12$. Solving for $\overline{X}_1$, $\overline{X}_2$, SS_1, and SS_2, we obtain

$$\overline{X}_1 = \frac{\Sigma X_1}{n_1} = \frac{587}{12} = 48.917 \qquad \overline{X}_2 = \frac{\Sigma X_2}{n_2} = \frac{552}{12} = 46.000$$

$$SS_1 = \Sigma X_1{}^2 - \frac{(\Sigma X_1)^2}{n_1} \qquad SS_2 = \Sigma X_2{}^2 - \frac{(\Sigma X_2)^2}{n_2}$$

$$= 29{,}617 - \frac{(587)^2}{12} \qquad = 26{,}496 - \frac{(552)^2}{12}$$

$$= 902.917 \qquad\qquad\qquad = 1104$$

---

*Of course you can't do this with actual data. Once the data have been collected according to a particular experimental design, you must use inference tests appropriate to that design.

table 14.8 Data and analysis from conservation campaign experiment

Family	Before the Campaign (gal) (1)	After the Campaign (gal) (2)	Difference D	D^2
A	55	48	7	49
B	43	38	5	25
C	51	53	−2	4
D	62	58	4	16
E	35	36	−1	1
F	48	42	6	36
G	58	55	3	9
H	45	40	5	25
I	48	49	−1	1
J	54	50	4	16
K	56	58	−2	4
L	32	25	7	49
			35	235

$$N = 12 \qquad \overline{D}_{obt} = \frac{\Sigma D}{N} = \frac{35}{12} = 2.917$$

$$t_{obt} = \frac{\overline{D}_{obt} - \mu_D}{\sqrt{\dfrac{SS_D}{N(N-1)}}} \qquad SS_D = \Sigma D^2 - \frac{(\Sigma D)^2}{N}$$

$$= \frac{2.917 - 0}{\sqrt{\dfrac{132.917}{12(11)}}} \qquad = 235 - \frac{(35)^2}{12}$$

$$= 2.907 \qquad\qquad\qquad = 132.917$$

$$= 2.91$$

From Table D, with df $= 11$ and $\alpha = 0.05_{2\,\text{tail}}$,

$$t_{crit} = \pm 2.201$$

Since $|t_{obt}| > 2.201$, we rejected H_0 and concluded that the conservation campaign does indeed affect gasoline consumption. It significantly lowered the amount of gasoline used.

Substituting these values into the equation for t_{obt} with equal *n*, we obtain

$$t_{obt} = \frac{\overline{X}_1 - \overline{X}_2}{\sqrt{\dfrac{SS_1 + SS_2}{n(n-1)}}} = \frac{48.917 - 46.000}{\sqrt{\dfrac{902.917 + 1104}{12(11)}}} = 0.748 = 0.75$$

From Table D, with df $= N - 2 = 24 - 2 = 22$ and $\alpha = 0.05_{2\,\text{tail}}$,

$$t_{crit} = \pm 2.074$$

Since $|t_{obt}| < 2.074$, we retain H_0.

Something seems strange here. When the data were collected with a correlated groups design, we were able to reject H_0. However, with an independent groups design, we were unable to reject H_0, even though the data were identical. Why? The correlated groups design allows us to use the subjects as their own control. This maximizes the possibility that there will be a high correlation between the scores in the two conditions. In the present illustration, Pearson r for the correlation between the paired before and after scores equals 0.938. When the correlation is high,* the difference scores will be much less variable than the original scores. For example, consider the scores of families A and L. Family A uses quite a lot of gasoline (55 gallons), whereas family L uses much less (32 gallons). As a result of the conservation campaign, the scores of both families decrease by 7 gallons. Their difference scores are identical (7). There is no variability between the difference scores for these families, whereas there is great variability between their raw scores. It is the potential for a high correlation and, hence, decreased variability that causes the correlated groups design to be potentially more powerful than the independent groups design. The decreased variability in the present illustration can be seen most clearly by viewing the two solutions side by side. This is shown in Table 14.9.

The two equations yield the same values except for SS_D in the correlated groups design and $SS_1 + SS_2$ in the independent groups design. SS_D is a measure of the variability of the difference scores. $SS_1 + SS_2$ are measures of the variability of the raw scores. $SS_D = 132.917$, whereas $SS_1 + SS_2 = 2006.917$. SS_D is much smaller than $SS_1 + SS_2$. It is this decreased variability that causes t_{obt} to be greater in the correlated groups analysis.

If the correlated groups design is potentially more powerful, why not always use this design? First, the independent groups design is much more efficient from a df per measurement analysis. The degrees of freedom are important because the higher the df, the lower t_{crit}. In the present illustration, for the correlated groups design, there were 24 measurements taken, but only 11 df resulted. For the independent groups design, there were 24 measurements and 22 df. Thus, the independent groups design results in twice the df for the same number of measurements.

Second, many experiments preclude using the same subject in both conditions. For example, suppose we are interested in investigating whether men and women differ in aggressiveness. Obviously, the same subject could not be used in both

table 14.9 Solutions for correlated and independent groups designs

Correlated Groups	Independent Groups
$t_{obt} = \dfrac{\overline{D}}{\sqrt{\dfrac{SS_D}{N(N-1)}}}$	$t_{obt} = \dfrac{\overline{X}_1 - \overline{X}_2}{\sqrt{\dfrac{SS_1 + SS_2}{n(n-1)}}}$
$= \dfrac{2.917}{\sqrt{\dfrac{132.917}{12(11)}}}$	$= \dfrac{2.917}{\sqrt{\dfrac{902.917 + 1104}{12(11)}}}$
$= 2.91$	$= 0.75$

---

*See Note 14.4 for a direct comparison between the two t equations that involves Pearson r.

conditions. Sometimes the effect of the first condition persists too long over time. If the experiment calls for the two conditions to be administered closely in time, it may not be possible to run the same subject in both conditions without the first condition affecting performance in the second condition. Often, when the subject is run in the first condition, he or she is "used up" or can't be run in the second condition. This is particularly true in *learning experiments*. For example, if we are interested in the effects of exercise on learning how to ski, we know that once the subjects have learned to ski, they can't be used in the second condition because they already know how to ski. When the same subject can't be used in the two conditions, then it is still possible to match subjects. However, matching is time-consuming and costly. Furthermore, it is often true that the experimenter doesn't know which variables are important for matching so as to produce a higher correlation. For all these reasons, the independent groups design is used more often than the correlated groups design.

ALTERNATIVE ANALYSIS USING CONFIDENCE INTERVALS

Thus far in the inferential statistics part of the textbook, we have been evaluating the effect of the independent variable by determining if it is reasonable to reject the null hypothesis, given the data of the experiment. If it is reasonable to reject H_0, then we can conclude by affirming that the independent variable has a real effect. We will call this the *null-hypothesis approach*. A limitation of the null-hypothesis approach is that by itself it does not tell us anything about the size or the effect.

An alternative approach also allows us to determine if it is reasonable to affirm that the independent variable has a real effect and at the same time gives us an estimate of the size of the real effect. This method uses confidence intervals. Not surprisingly, we will call this method the *confidence-interval approach*. We will illustrate this confidence-interval approach using the two-group, independent groups design.

You will recall that in Chapter 13, when we were discussing the *t* test for single samples, we showed how to construct confidence intervals for the population mean μ. Typically, we constructed the 95% or the 99% confidence interval for μ. Of course in that chapter, we were discussing single sample experiments. In the two-group, independent groups design, we have not one but two samples, and each sample is considered to be a random sample from its own population. We have designated the population mean of sample 1 as μ_1 and the population mean of sample 2 as μ_2. The difference $\mu_1 - \mu_2$ is a measure of the real effect of the independent variable. If there is no real effect, then $\mu_1 = \mu_2$ and $\mu_1 - \mu_2 = 0$. By constructing the 95% or 99% confidence interval for the difference $\mu_1 - \mu_2$, we can determine if it is reasonable to affirm that there is a real effect, and if so, we can estimate its size.

Constructing the 95% Confidence Interval for $\mu_1 - \mu_2$

Constructing the 95% or 99% confidence interval for $\mu_1 - \mu_2$ is very much like constructing these intervals for μ. We will illustrate by comparing the equations for both used to construct the 95% confidence interval. These equations are shown in Table 14.10.

As you can see from the table, the equations for constructing the 95% confidence interval for μ and for $\mu_1 - \mu_2$ are identical, except that in the two-sample experiment, $(\overline{X}_1 - \overline{X}_2)$ is used instead of $\overline{X}_{\text{obt}}$ and $s_{\overline{X}_1} - s_{\overline{X}_2}$ is used instead of $s_{\overline{X}}$.

table 14.10 Comparison of equations for constructing the 95% confidence interval for μ and $\mu_1 - \mu_2$

Single Sample Experiment	Two Sample Experiment
95% Confidence Interval for μ	**95% Confidence Interval for $\mu_1 - \mu_2$**
$\mu_{\text{lower}} = \overline{X}_{\text{obt}} - s_{\overline{X}}\, t_{0.025}$	$\mu_{\text{lower}} = (\overline{X}_1 - \overline{X}_2) - s_{\overline{X}_1 - \overline{X}_2}\, t_{0.025}$
$\mu_{\text{upper}} = \overline{X}_{\text{obt}} + s_{\overline{X}}\, t_{0.025}$	$\mu_{\text{upper}} = (\overline{X}_1 - \overline{X}_2) + s_{\overline{X}_1 - \overline{X}_2}\, t_{0.025}$
where $s_{\overline{X}} = \dfrac{s}{\sqrt{n}}$	where $s_{\overline{X}_1 - \overline{X}_2} = \sqrt{\left(\dfrac{SS_1 + SS_2}{n_1 + n_2 - 2}\right)\left(\dfrac{1}{n_1} + \dfrac{1}{n_2}\right)}$

So far, we have been rather theoretical. Now let's try an example. Let's assume we are interested in analyzing the data from the hormone X experiment, using the confidence-interval approach. For your convenience, we have repeated the experiment below.

A physiologist has conducted an experiment to evaluate the effect of hormone X on sexual behavior. Ten male rats were injected with hormone X, and 10 other male rats received a placebo injection. The animals were then placed in individual housing with a sexually receptive female. The number of matings was counted over a 20-minute period. The results are shown in Table 14.11.

Evaluate the data of this experiment by constructing the 95% confidence interval for $\mu_1 - \mu_2$. What is your conclusion?

table 14.11 Results from hormone X experiment

Hormone X Group 1	Placebo Group 2
X_1	X_2
8	5
10	6
12	3
6	4
6	7
7	8
9	6
8	5
7	4
11	8
84	56
$n_1 = 10$	$n_2 = 10$
$\Sigma X_1 = 84$	$\Sigma X_2 = 56$
$\overline{X}_1 = 8.4$	$\overline{X}_2 = 5.6$
$\Sigma X_1^2 = 744$	$\Sigma X_2^2 = 340$

The equations used to construct the 95% confidence interval are

$$\mu_{\text{lower}} = (\overline{X}_1 - \overline{X}_2) - s_{\overline{X}_1 - \overline{X}_2} t_{0.025} \text{ and } \mu_{\text{upper}} = (\overline{X}_1 - \overline{X}_2) + s_{\overline{X}_1 - \overline{X}_2} t_{0.025}$$

Solving first for SS_1 and SS_2,

$$SS_1 = \Sigma X_1^2 - \frac{(\Sigma X_1)^2}{n_1} \qquad SS_2 = \Sigma X_2^2 + \frac{(\Sigma X_2)^2}{n_2}$$

$$= 744 - \frac{(84)^2}{10} \qquad = 340 + \frac{(56)^2}{10}$$

$$= 38.4 \qquad = 26.4$$

Solving next for $s_{\overline{X}_1 - \overline{X}_2}$,

$$s_{\overline{X}_1 - \overline{X}_2} = \sqrt{\left(\frac{SS_1 + SS_2}{n_1 + n_2 - 2}\right)\left(\frac{1}{n_1} + \frac{1}{n_2}\right)}$$

$$= \sqrt{\left(\frac{38.4 + 26.4}{10 + 10 - 2}\right)\left(\frac{1}{10} + \frac{1}{10}\right)} = 0.849$$

The last value we need to compute μ_{lower} and μ_{upper} is the value of $t_{0.025}$. From Table D, with $\alpha = 0.025_{1 \text{ tail}}$ and df $= N - 2 = 20 - 2 = 18$,

$$t_{0.025} = 2.101$$

We now have all the values we need to compute μ_{lower} and μ_{upper}. For convenience, we've listed them again here. $\overline{X}_1 = 8.4$, $\overline{X}_2 = 5.6$, $s_{\overline{X}_1 - \overline{X}_2} = 0.849$, and $t_{0.025} = 2.101$. Substituting these values in the equations for μ_{lower} and μ_{upper}, we obtain

$$\mu_{\text{lower}} = (\overline{X}_1 - \overline{X}_2) - s_{\overline{X}_1 - \overline{X}_2} t_{0.025} \qquad \mu_{\text{upper}} = (\overline{X}_1 - \overline{X}_2) + s_{\overline{X}_1 - \overline{X}_2} t_{0.025}$$

$$= (8.4 - 5.6) - 0.849(2.101) \qquad = (8.4 - 5.6) + 0.849(2.101)$$

$$= 1.02 \qquad = 4.58$$

Thus, the 95% confidence interval for $\mu_1 - \mu_2 = 1.02$–4.58.

Conclusion Based on the Obtained Confidence Interval

Having computed the 95% confidence interval for $\mu_1 - \mu_2$, we can both come to a conclusion with regard to the null hypothesis and also give an estimate of the size of the real effect of hormone X. The 95% confidence interval corresponds to $\alpha = 0.05_{2 \text{ tail}}$ (0.025 under each tail; see Figure 13.3, p. 342). The nondirectional null hypothesis predicts that $\mu_1 - \mu_2 = 0$. Since the obtained 95% confidence interval does not include a value of 0, we can reject the null hypothesis and affirm that hormone X appears to have a real effect. This is the conclusion we reached when we analyzed the data using the null-hypothesis approach with $\alpha = 0.05_{2 \text{ tail}}$.

In addition, we have an estimate of the size of the real effect. We are 95% confident that the range of 1.02–4.58 contains the real effect of hormone X. If so, then the real effect of hormone X is to cause 1.0–4.58 more matings than the placebo. Note that if the interval contained the value 0, we would not be able to reject H_0, in which case we couldn't affirm that hormone X has a real effect.

Constructing the 99% Confidence Interval for $\mu_1 - \mu_2$

Constructing the 99% confidence interval for $\mu_1 - \mu_2$ is very much like constructing the 95% confidence interval. The one difference is that for the 99% confidence interval we use $t_{0.005}$ in the equations for the μ_{lower} and μ_{upper} instead of $t_{0.025}$. This corresponds to $\alpha = 0.01_{2\,\text{tail}}$. The equations used to compute the 99% confidence interval are shown here.

$$\mu_{\text{lower}} = (\overline{X}_1 - \overline{X}_2) - s_{\overline{X}_1 - \overline{X}_2}t_{0.005} \text{ and } \mu_{\text{upper}} = (\overline{X}_1 - \overline{X}_2) + s_{\overline{X}_1 - \overline{X}_2}t_{0.005}$$

For the hormone X experiment, $\overline{X}_1 = 8.4$, $\overline{X}_2 = 5.6$, and $s_{\overline{X}_1 - \overline{X}_2} = 0.849$. From Table D, with a $\alpha = 0.005_{1\,\text{tail}}$ and df $= N - 2 = 20 - 2 = 18$,

$$t_{0.005} = 2.878$$

Using these equations with the data of the hormone X experiment, we obtain

$$\mu_{\text{lower}} = (\overline{X}_1 - \overline{X}_2) - s_{\overline{X}_1 - \overline{X}_2}t_{0.005} \qquad \mu_{\text{upper}} = (\overline{X}_1 - \overline{X}_2) + s_{\overline{X}_1 - \overline{X}_2}t_{0.005}$$

$$= (8.4 - 5.6) - 0.849(2.878) \qquad\qquad = (8.4 - 5.6) + 0.849(2.878)$$

$$= 0.36 \qquad\qquad\qquad\qquad\qquad = 5.24$$

Thus, the 99% confidence interval for the data of the hormone X experiment is 0.36–5.24. Since the obtained 99% confidence interval does not contain the value 0, we can reject H_0 at $\alpha = 0.01_{2\,\text{tail}}$. In addition, we are 99% confident that the size of the real effect of hormone X falls in the interval of 0.36–5.24. If this interval does contain the real effect, then the real effect is somewhere between 0.36 and 5.24 more matings than the placebo. As was true for the 95% confidence interval, if the 99% confidence interval did contain the value 0, we would not be able to reject H_0, and therefore we couldn't affirm H_1. Notice also that the 99% confidence interval (0.36–5.24) is larger than the 95% confidence interval (1.02–4.58). This is what we would expect from our discussion in Chapter 13, because the larger the interval, the more confidence we have that it contains the population value being estimated.

■ SUMMARY

In this chapter, I have discussed the t test for correlated and independent groups. I pointed out that the t test for correlated groups was really just a special case of the t test for single samples. In the correlated groups design, the differences between paired scores are analyzed. If the independent variable has no effect and chance alone is responsible for the difference scores, then they can be considered a random sample from a population of difference scores where $\mu_D = 0$ and σ_D is unknown. But these are the exact conditions in which the t test for single samples applies. The only change is that, in the correlated groups design, we analyze difference scores, whereas in the single sample design, we analyze raw scores. After presenting some illustrative and practice problems, I discussed computing size of effect. Using Cohen's method with the t test for correlated groups, we again estimate d using $\hat{d}$. With the correlated groups

t test, the magnitude of real effect varies directly with the size of $\overline{D}_{\text{obt}}$. The statistic $\hat{d}$ gives a standardized value, achieved by dividing $\overline{D}_{\text{obt}}$ by s_D; the greater $\hat{d}$, the greater is the real effect. In addition to explaining Cohen's method for determining size of real effect, criteria were given for assessing whether the obtained value of $\hat{d}$ represents a small, medium, or large effect. After discussing size of effect, I concluded our discussion of the t test for correlated groups by comparing it with the sign test. I showed that although both are appropriate for the correlated groups design, as long as its assumptions are met, the t test should be used because it is more powerful.

The t test for independent groups is used when there are two independent groups in the experiment. The statistic that is analyzed is the difference between the means of the two samples ($\overline{X}_1 - \overline{X}_2$). The scores of sample 1 can be considered a random sample from a population having a mean μ_1 and a standard deviation σ_1. The scores of sample 2 are a random sample from a population having a mean μ_2 and a standard deviation σ_2.

If the independent variable has a real effect, then the difference between sample means is due to random sampling from populations where $\mu_1 \neq \mu_2$. Changing the level of the independent variable is assumed to affect the means of the populations but not their standard deviations or variances. If the independent variable has no effect and chance alone is responsible for the differences between the two samples, then the difference between sample means is due to random sampling from populations where $\mu_1 = \mu_2$. Under these conditions, the sampling distribution of $\overline{X}_1 - \overline{X}_2$ has a mean of zero and a standard deviation whose value depends on knowing the variance of the populations from which the samples were taken. Since this value is never known, the *z* test cannot be used. However, we can estimate the variance using a weighted estimate taken from both samples. When this is done, the resulting statistic is t_{obt}.

The *t* statistic, then, is also used for analyzing the data from the two-sample, independent groups experiment. The sampling distribution of *t* for this design is the same as for the single sample design, except the degrees of freedom are different. In the independent groups design, df = $N - 2$. After presenting some illustrative and practice problems, I discussed the assumptions underlying the *t* test for independent groups. I pointed out that this test requires that (1) the raw-score populations be normally distributed and (2) there be homogeneity of variance. I also pointed out that the *t* test is robust with regard to violations of the population normality and homogeneity of variance assumptions. In addition to determining whether there is a significant effect, it is also important to determine the size of the effect. In an independent groups experiment, the size of effect of the independent variable may be found by estimating Cohen's *d* with $\hat{d}$. The statistic $\hat{d}$ gives a standardized value, achieved by dividing $|\overline{X}_1 - \overline{X}_2|$ by the weighted estimate of σ, $\sqrt{s_W{}^2}$. The greater $\hat{d}$, the greater is the real effect. Again, criteria were given for assessing whether the obtained value of $\hat{d}$ represents a small, medium, or large effect.

Next, I discussed the power of the *t* test. I showed that its power varies directly with the size of the real effect of the independent variable and the *N* of the experiment but varies inversely with the variability of the sample scores.

Then, I compared the correlated groups and independent groups designs. When the correlation between paired scores is high, the correlated groups design is more powerful than the independent groups design. However, it is easier and more efficient regarding degrees of freedom to conduct an independent groups experiment. In addition, there are many situations in which the correlated groups design is inappropriate.

Finally, I showed how to evaluate the effect of the independent variable using a confidence-interval approach in experiments employing the two-group, independent groups design. This approach is more complicated than the basic hypothesis testing approach used throughout the inference part of this textbook but has the advantage that it allows both the evaluation of H_0 and an estimation of the size of effect of the independent variable.

■ IMPORTANT NEW TERMS

Confidence-interval approach (p. 382)

Degrees of freedom (p. 371)

Estimated standard error of the difference between sample means (p. 368)

Homogeneity of variance (p. 375)

Independent groups design (p. 366)

Mean of the population of difference scores (p. 360)

Mean of the sampling distribution of the difference between sample means (p. 368)

Null-hypothesis approach (p. 382)

Sampling distribution of the difference between sample means (p. 368)

Size of effect (p. 376)

Standard deviation of the sampling distribution of the difference between sample means (p. 368)

Standard error of the difference between sample means (p. 368)

t test for correlated groups (p. 358)

t test for independent groups (p. 366, 370)

■ QUESTIONS AND PROBLEMS

1. Identify or define the terms in the Important New Terms section.
2. Discuss the advantages of the two-condition experiment compared with the advantages of the single sample experiment.
3. The *t* test for correlated groups can be thought of as a special case of the *t* test for single samples, discussed in the previous chapter. Explain.
4. What is the main advantage of using the *t* test for correlated groups over using the sign test to analyze data from a correlated groups experiment?
5. What are the characteristics of the sampling distribution of the difference between sample means?
6. Why is the *z* test for independent groups never used?
7. What is estimated in the *t* test for independent groups? How is the estimate obtained?
8. It is said that the variance of the sample data has an important bearing on the power of the *t* test. Is this statement true? Explain.
9. What are the advantages and disadvantages of using a correlated groups design as compared with using an independent groups design?
10. What are the assumptions underlying the *t* test for independent groups?
11. Having just made what you believe to be a Type II error, using an independent groups design and a *t* test analysis, name all the things you might do in the next experiment to reduce the probability of a Type II error.
12. Is the size of effect of the independent variable important? Explain.
13. If the effect of the independent variable is significant, does that necessarily mean the effect is a large one? Explain.

For each of the following problems, unless otherwise told, assume normality in the population.

14. You are interested in determining whether an experimental birth control pill has the side effect of changing blood pressure. You randomly sample ten women from the city in which you live. You give five of them a placebo for a month and then measure their diastolic blood pressure. Then you switch them to the birth control pill for a month and again measure their blood pressure. The other five women receive the same treatment except they are given the birth control pill first for a month, followed by the placebo for a month. The blood pressure readings are shown here. Note that to safeguard the women from unwanted pregnancy, another means of birth control that does not interact with the pill was used for the duration of the experiment.

	Diastolic Blood Pressure	
Subject No.	**Birth control pill**	**Placebo**
1	108	102
2	76	68
3	69	66
4	78	71
5	74	76
6	85	80
7	79	82
8	78	79
9	80	78
10	81	85

a. What is the alternative hypothesis? Assume a nondirectional hypothesis is appropriate.
b. What is the null hypothesis?
c. What do you conclude? Use $\alpha = 0.01_{2\,tail}$. social, biological, health

15. On the basis of previous research and sound theoretical considerations, a cognitive psychologist believes that memory for pictures is superior to memory for words. To test this hypothesis, the psychologist performs an experiment in which students from an introductory psychology class are used as subjects. Eight randomly selected students view 30 slides with nouns printed on them, and another group of eight randomly selected students views 30 slides with pictures of the same nouns. Each slide contains either one noun or one picture and is viewed for 4 seconds. After viewing the slides, subjects are given a recall test, and the number of

correctly remembered items is measured. The data follow:

No. of Pictures Recalled	No. of Nouns Recalled
18	12
21	9
14	21
25	17
23	16
19	10
26	19
15	22

a. What is the alternative hypothesis? Assume that a directional hypothesis is warranted.
b. What is the null hypothesis?
c. Using $\alpha = 0.05_{1\,tail}$, what is your conclusion?
d. Estimate the size of the real effect. cognitive

16. A nurse was hired by a governmental ecology agency to investigate the impact of a lead smelter on the level of lead in the blood of children living near the smelter. Ten children were chosen at random from those living near the smelter. A comparison group of seven children was randomly selected from those living in an area relatively free from possible lead pollution. Blood samples were taken from the children and lead levels determined. The following are the results (scores are in micrograms of lead per 100 milliliters of blood):

Lead Levels	
Children living near smelter	*Children living in unpolluted area*
18	9
16	13
21	8
14	15
17	17
19	12
22	11
24	
15	
18	

a. Using $\alpha = 0.01_{1\,tail}$, what do you conclude?
b. Estimate the size of the real effect. health

17. The manager of the cosmetics section of a large department store wants to determine whether newspaper advertising really does affect sales. For her experiment, she randomly selects 15 items currently in stock and proceeds to establish a baseline. The 15 items are priced at their usual competitive values, and the quantity of each item sold for a 1-week period is recorded. Then, without changing their price, she places a large ad in the newspaper, advertising the 15 items. Again, she records the quantity sold for a 1-week period. The results follow.

Item	No. Sold Before Ad	No. Sold After Ad
1	25	32
2	18	24
3	3	7
4	42	40
5	16	19
6	20	25
7	23	23
8	32	35
9	60	65
10	40	43
11	27	28
12	7	11
13	13	12
14	23	32
15	16	28

a. Using $\alpha = 0.05_{2\,tail}$, what do you conclude?
b. What is the size of the effect? I/O

18. Since muscle tension in the head region has been associated with tension headaches, you reason that if the muscle tension could be reduced, perhaps the headaches would decrease or go away altogether. You design an experiment in which nine subjects with tension headaches participate. The subjects keep daily logs of the number of headaches they experience during a 2-week baseline period. Then you train them to lower their muscle tension in the head region, using a biofeedback device. For this experiment, the biofeedback device is connected to the frontalis muscle, a muscle in the forehead region. The device tells the subject the amount of tension in the muscle

to which it is attached (in this case, frontalis) and helps them achieve low tension levels. After 6 weeks of training, during which the subjects have become successful at maintaining low frontalis muscle tension, they again keep a 2-week log of the number of headaches experienced. The following are the numbers of headaches recorded during each 2-week period.

| Subject No. | No. of Headaches | |
	Baseline	*After training*
1	17	3
2	13	7
3	6	2
4	5	3
5	5	6
6	10	2
7	8	1
8	6	0
9	7	2

a. Using $\alpha = 0.05_{2\,tail}$, what do you conclude? Assume the sampling distribution of the mean of the difference scores $(\overline{D})$ is normally distributed. Assume a nondirectional hypothesis is appropriate, because there is insufficient empirical basis to warrant a directional hypothesis.
b. If the sampling distribution of $\overline{D}$ is not normally distributed, what other test could you use to analyze the data? What would your conclusion be? clinical, health

19. There is an interpretation difficulty with Problem 18. It is clear that the headaches decreased significantly. However, it is possible that the decrease was not due to the biofeedback training but rather to some other aspect of the situation, such as the attention shown to the subjects. What is really needed is a group to control for this possibility. Assume another group of nine headache patients was run at the same time as the group in Problem 18. This group was treated in the same way except the subjects did not receive any training involving biofeedback. They just talked with you about their headaches each week for 6 weeks, and you showed them lots of warmth, loving care, and attention. The numbers of headaches for the baseline and 2-week

follow-up period for the control group were as follows:

| Subject No. | No. of Headaches | |
	Baseline	*Follow-up*
1	5	4
2	8	9
3	14	12
4	16	15
5	6	4
6	5	3
7	8	7
8	10	6
9	9	7

Evaluate the effect of these other factors, such as attention, on the incidence of headaches. Use $\alpha = 0.05_{2\,tail}$. clinical, health

20. Since the control group in Problem 19 also showed significant reductions in headaches, the interpretation of the results in Problem 18 is in doubt. Did relaxation training contribute to the headache decrease, or was the decrease due solely to other factors, such as attention? To answer this question, we can compare the change scores between the two groups. These scores are shown here:

| Headache Change Scores | |
Relaxation training group	*Control group*
14	1
6	−1
4	2
2	1
−1	2
8	2
7	1
6	4
5	2

What is your conclusion? Use $\alpha = 0.05_{2\,tail}$. clinical, health

21. The director of human resources at a large company is considering hiring part-time employees to fill jobs

previously staffed with full-time workers. However, he wonders if doing so will affect productivity. Therefore, he conducts an experiment to evaluate the idea before implementing it factory-wide. Six full-time job openings, from the parts manufacturing division of the company, are each filled with two employees hired to work half-time. The output of these six half-time pairs is compared with the output of a randomly selected sample of six full-time employees from the same division. Note that all employees in the experiment are engaged in manufacturing the same parts. The average number of parts produced per day by the half-time pairs and full-time workers is shown here:

Parts Produced per Day

Half-time pairs	Full-time workers
24	20
26	28
46	40
32	36
30	24
36	30

Does the hiring of part-time workers affect productivity? Use $\alpha = 0.05_{2\,tail}$ in making your decision. I/O

22. On the basis of her experience with clients, a clinical psychologist thinks that depression may affect sleep. She decides to test this idea. The sleep of nine depressed patients and eight normal controls is monitored for three successive nights. The average number of hours slept by each subject during the last two nights is shown in the following table:

Hours of Sleep

Depressed patients	Normal controls
7.1	8.2
6.8	7.5
6.7	7.7
7.3	7.8
7.5	8.0
6.2	7.4
6.9	7.3
6.5	6.5
7.2	

a. Is the clinician correct? Use $\alpha = 0.05_{2\,tail}$ in making your decision.
b. If the effect is significant, estimate the size of the effect. Using Cohen's criterion, is the effect a large one? clinical, health

23. An educator wants to determine whether early exposure to school will affect IQ. He enlists the aid of the parents of 12 pairs of preschool-age identical twins who agree to let their twins participate in this experiment. One member of each twin pair is enrolled in preschool for 2 years while the other member of each pair remains at home. At the end of the 2 years, the IQs of all the children are measured. The results follow.

| | IQ | |
| | Twin at | Twin at |
Pair	preschool	home
1	110	114
2	121	118
3	107	103
4	117	112
5	115	117
6	112	106
7	130	125
8	116	113
9	111	109
10	120	122
11	117	116
12	106	104

Does early exposure to school affect IQ? Use $\alpha = 0.05_{2\,tail}$.
cognitive, developmental, education

24. Researchers at a leading university were interested in the effect of sleep on memory consolidation. Twenty-four student volunteers from an introductory psychology course were randomly assigned to either a "Sleep" or "No-Sleep" group, such that there were 12 students in each group. On the first day, all students were flashed pictures of 15 different objects, for 200 milliseconds each, on a computer screen and asked to remember as many of the objects as possible. That night, the "Sleep" group got an ordinary night's sleep. The "No-Sleep" group was kept awake until the second night. All

subjects got an ordinary night's sleep on the second and third nights. On the fourth day, all subjects were tested to see how many of the original 15 objects they remembered. The following are the number of objects remembered by each subject on the test:

Sleep Group	No-Sleep Group
14	8
13	9
8	6
9	13
11	7
10	9
9	10
13	12
12	8
11	11
14	9
13	12

a. Using $\alpha = 0.05_{2\,tail}$, what do you conclude?
b. Using the confidence-interval approach, construct the 95% confidence interval for $\mu_1 - \mu_2$. What do you conclude regarding H_0? What is your estimate of the size of the effect?
c. Using the confidence-interval approach, construct the 99% confidence interval for $\mu_1 - \mu_2$. What do you conclude regarding H_0? What is your estimate of the size of the effect? cognitive

25. Developmental psychologists at a prominent California university conducted a longitudinal study investigating the effect of high levels of curiosity in early childhood on intelligence. The local population of 3-year-olds was screened via a test battery assessing curiosity. Twelve of the 3-year-olds scoring in the upper 90% of this variable were given an IQ test at age 3 and again at age 11. The following IQ scores were obtained.

Subject Number	IQ (Age 3)	IQ (Age 11)
1	100	114
2	105	116
3	125	139
4	140	151
5	108	106
6	122	119
7	117	131
8	112	136
9	135	148
10	128	139
11	104	122
12	98	113

a. Using $\alpha = 0.01_{2\,tail}$, what do you conclude? In drawing your conclusion, assume that it is well established that IQ stays relatively constant over these years for individuals with average or below-average levels of curiosity.
b. What is the size of the effect? cognitive, developmental

26. Noting that women seem more interested in emotions than men, a researcher in the field of women's studies wondered if women recall emotional events better than men. She decides to gather some data on the matter. An experiment is conducted in which eight randomly selected men and women are shown 20 highly emotional photographs and then asked to recall them 1 week after the showing. The following recall data are obtained. Scores are percent correct; one man failed to show up for the recall test.

Men	Women
75	85
85	92
67	78
77	80
83	88
88	94
86	90
	89

Using $\alpha = 0.05_{2\,tail}$, what do you conclude? cognitive, social

27. Since the results of the experiment in Problem 26 were very close to being significant, the researcher decides to replicate that experiment, only this time increasing the power by increasing *N*. This study included 10 men and 10 women. The following results were obtained.

Men	Women
74	87
87	90
64	80
76	77
85	91
86	95
84	89
78	92
77	90
80	94

Using $\alpha = 0.05_{2\,tail}$, what do you conclude this time? cognitive, social

28. A physics instructor believes that natural lighting in classrooms improves student learning. He conducts an experiment in which he teaches the same physics unit to two groups of seven randomly assigned students in each group. Everything is similar for the groups, except that one of the groups receives the instruction in a classroom that admits a lot of natural light in addition to the fluorescent lighting, while the other uses a classroom with only fluorescent lighting. At the end of the unit, both groups are given the same end-of-unit exam. There are 20 possible points on the exam; the higher the score, the better the performance. The following scores are obtained.

Natural Plus Fluorescent Lighting	Fluorescent Lighting Only
16	17
18	13
14	12
17	14
16	13
19	15
17	14

Using $\alpha = 0.05_{2\,tail}$, what do you conclude? education

■ SPSS ILLUSTRATIVE EXAMPLE 14.1

The general operation of SPSS and data entry are described in Appendix E, *Introduction to SPSS*. Chapter 14 of the textbook discusses the *t* test for correlated and independent groups. As discussed in the SPSS material for Chapter 13, when analyzing the data for *t* tests, SPSS computes **t** and the two-tailed probability of getting **t or a value more extreme** if chance alone is at work. SPSS calls this probability **Sig. (2-tailed).** Remember:

Sig. (2-tailed) = *p*(2-tailed)

The decision rules we will follow to evaluate H_0 and H_1 are the same as in Chapter 13.

For non-directional H_1's:

if **Sig. (2-tailed)** $\leq \alpha$, reject H_0 and affirm H_1
if **Sig. (2-tailed)** $> \alpha$, retain H_0; cannot affirm H_1

For directional H_1's where the sample mean difference is in the predicted direction:

if **Sig. (2-tailed)/2** $\leq \alpha$, reject H_0 and affirm H_1
if **Sig. (2-tailed)/2** $> \alpha$, retain H_0; cannot affirm H_1

For directional H_1's where sample mean difference is not in the predicted direction, we always retain H_0.

example

Use SPSS to solve Practice Problem 14.1, p. 362, in the text. Compare your answer with that given in the practice problem. For convenience, the problem is repeated here.

To motivate citizens to conserve gasoline, the government is considering mounting a nationwide conservation campaign. However, before doing so on a national level, it decides to conduct an experiment to evaluate the effectiveness of the campaign. For the experiment, the conservation campaign is conducted in a small but representative geographical area. Twelve families are randomly selected from the area, and the amount of gasoline they use is monitored for 1 month before the advertising campaign and for 1 month after the campaign. The following data are collected:

Family	Before the Campaign (gal/month) Group 1	After the Campaign (gal/month) Group 2
A	55	48
B	43	38
C	51	53
D	62	58
E	35	36
F	48	42
G	58	55
H	45	40
I	48	49
J	54	50
K	56	58

What is your conclusion? Use $\alpha = 0.05_{2\,tail}$.

SOLUTION

STEP 1: Enter the Data.

1. Enter the scores of Group 1 in the first column (**VAR00001**) of the Data Editor, beginning with the first Group 1 score in the top cell of the first column.
2. Enter the scores of Group 2 in the second column (**VAR00002**) of the Data Editor, beginning with the first Group 2 score in the top cell of the second column.

STEP 2: Name the Variables. In this example, we will give the default variables **VAR00001** and **VAR00002** the new names of **Before_C** and **After_C**, respectively.

1. **Click** the **Variable View** tab in the lower left corner of the Data Editor.	This displays the **Variable View** on screen with **VAR00001** and **VAR00002** displayed in the first and second cells of the **Name** column, respectively.

2. **Click VAR00001**; then **type Before_C** in the highlighted cell and then **press Enter**.

3. **Replace VAR00002** with **After_C** and then **press Enter**.

Before_C is entered as the variable name, replacing **VAR00001**. The cursor then moves to the next cell, highlighting **VAR00002**.

After_C is entered as the variable name, replacing **VAR00002**.

STEP 3: **Analyze the Data.** The appropriate test for this example is Student's *t* test for Correlated Groups. SPSS calls this test the "Paired-Samples T test." To have SPSS do the analysis using the Paired-Samples T test,

---

1. **Click Analyze**; then **select Compare Means**; then **click Paired-Samples T Test....**

This produces the **Paired-Samples T Test** dialog box with **Before_C** and **After_C** displayed in the large box on the left. **Before_C** is highlighted.

2. **Click** the **arrow** in the middle of the dialog box.

This moves **Before_C** to the **Paired Variables:** box under the column heading of **Variable1**.

3. **Click After_C;** then **click** the **arrow** in the middle of the dialog box.

This moves **After_C** to the **Paired Variables:** box under the column heading of **Variable2**. The SPSS default is to subtract **Variable2** scores from **Variable1** scores. Since we entered **Before_C** under **Variable1** and **After_C** under **Variable2**, SPSS will subtract the **After_C** scores from the **Before_C** scores. This will result in a positive *t* value, since in our example, the **Before_C** scores are higher than the **After_C** scores.

4. **Click OK**.

SPSS analyzes the data using the **Paired-Samples T Test** and outputs the results.

Analysis Results

The results are displayed in three tables: the **Paired-Samples Statistics** table, **the Paired Samples Correlations** table, and the **Paired-Samples Test** table. Since for this analysis we are interested in just the **Paired-Samples Test** table, we have displayed it here alone.

Paired Samples Test

| | | Paired Differences | | | | | | | |
| | | | | | 95% Confidence Interval of the Difference | | | | |
		Mean	Std. Deviation	Std. Error Mean	Lower	Upper	t	df	Sig. (2-tailed)
Pair 1	Before_C - After_C	2.91667	3.47611	1.00347	.70805	5.12528	2.907	11	.014

The **Paired-Samples Test** table shows that *t* = **2.907**, and that **Sig. (2-tailed)** = **.014**$_{2tailed}$. Since **.014** < **0.05**, we reject H_0 and affirm H_1. The campaign appears to reduce gasoline consumption. Please note that the SPSS **t** value when rounded to two decimal places is in agreement with the t_{obt} value arrived at in the textbook. The conclusions are the same as well.

■ SPSS ADDITIONAL PROBLEMS

1. Use SPSS to solve Problem 23 in Chapter 14, p. 390 of the textbook. When solving this problem, name the variables *Preschool* and *Home*. Compare your answer with the answer to Problem 23, given in Appendix C of the textbook.

2. Use SPSS to do Chapter 14, Problem 25, p. 391 of textbook. When solving this problem, name the variables *IQ_3* and *IQ_11*.

■ SPSS ILLUSTRATIVE EXAMPLE 14.2

example

Use SPSS to analyze the Illustrative Problem, *Hormone X* and *Sexual Behavior*, given in Chapter 14, p. 395 of the textbook. For convenience, the problem is repeated here.

A physiologist has the hypothesis that *hormone X* is important in producing sexual behavior. To investigate this hypothesis, 20 male rats were randomly sampled and then randomly assigned to two groups. The animals in group 1 were injected with *hormone X* and then were placed in individual housing with a sexually receptive female. The animals in group 2 were given similar treatment except they were injected with a placebo solution. The number of matings was counted over a 20-minute period. The results are shown in the table below.

Hormone X Group 1	Placebo Group 2
8	5
10	6
12	3
6	4
6	7
7	8
9	6
8	5
7	4
11	8

What do you conclude? Use $\alpha = 0.05_{2\,tail}$.

SOLUTION

STEP 1:   Enter the Data.

1. For an independent groups analysis, the scores of each group are stacked vertically, one after the other, in a single column of the Data Editor. To enter all the scores in the first column (**VAR00001**) of the Data Editor, do as follows:

    a. First, enter the scores of Group 1 in the first column (**VAR00001**) of the Data Editor, beginning with the first Group 1 score in the first cell of the first column.

b. Next**,** enter the scores of Group 2 in the first column, directly beneath the last Group 1 score. There should be no spaces or empty cells after the last Group 1 score. The first column should now contain the scores of both groups, beginning with the Group 1 scores and ending with the Group 2 scores, and there should be no spaces or empty cells between any of the scores.

2. In the second column (**VAR00002**) enter a coding number to designate to which group each score belongs. Do this as follows. In the second column (**VAR00002**) enter the number **1** next to each Group 1 score, and the number **2** next to each Group 2 score. These coding numbers identify the group to which each score belongs.

The resulting Data Editor is shown here. For the moment, ignore the variable name row; we will name the variables in the next step.

	Matings	Group	var
1	8.00	1.00	
2	10.00	1.00	
3	12.00	1.00	
4	6.00	1.00	
5	6.00	1.00	
6	7.00	1.00	
7	9.00	1.00	
8	8.00	1.00	
9	7.00	1.00	
10	11.00	1.00	
11	5.00	2.00	
12	6.00	2.00	
13	3.00	2.00	
14	4.00	2.00	
15	7.00	2.00	
16	8.00	2.00	
17	6.00	2.00	
18	5.00	2.00	
19	4.00	2.00	
20	8.00	2.00	

STEP 2: **Name the Variables.** In this example, we will give the default variables **VAR00001** and **VAR00002** the new names of **Matings** and **Group,** respectively. To do so,

1. Click the **Variable View** tab in the lower left corner of the Data Editor.	This displays the **Variable View** on screen, with **VAR00001** and **VAR00002** displayed in the first and second cells of the **Name** column, respectively.
2. Click VAR00001; then **type Matings** in the highlighted cell and then **press Enter**.	**Matings** is entered as the variable name, replacing **VAR00001**. The cursor then moves to the next cell, highlighting **VAR00002**.
3. Replace VAR00002 with **Group** and then **press Enter**.	**Group** is entered as the variable name, replacing **VAR00002**.

STEP 3:   **Analyze the Data.** The appropriate test for this example is Student's *t* test for Independent Groups. SPSS calls this test the **Independent-Samples T Test**. To have SPSS do the analysis using the **Independent-Samples T Test**,

1. Click **A**nalyze; then **select Co**mpare **Means**; then **click on Independent-Samples T Test....**	This produces the **Independent-Samples T Test** dialog box with **Matings** and **Group** displayed in the large box on the left. **Matings** is highlighted.
2. Click the **top arrow** for the **T**est **Variable(s):** box.	This moves **Matings** to the **T**est **Variable(s):** dialog box. SPSS calls the dependent variable, the **Test Variable**.
3. Click **Group**; then **click** the **bottom arrow** for the **G**rouping Variable box.	This results in **Group(??)** being displayed in the **G**rouping **Variable** box. This is SPSS's way of identifying **Group** as the grouping variable and telling you that you need to define the groups.
4. Click **D**efine Groups....	This produces the **Define Groups...** dialog box. This is where you tell SPSS the number that you assigned **Group 1** and **Group 2** scores. SPSS needs to know to which group each score belongs to do the analysis.
5. Type **1** in the **Group 1:** box.	This tells SPSS that **Group 1** scores are coded with a **1**.
6. Type **2** in the **Group 2:** box.	This tells SPSS that **Group 2** scores are coded with a **2**. SPSS now has all the information it needs to carry out the analysis.
7. Click on **Continue**.	This returns you to the **Independent-Samples T Test** dialog box so that you can give the **OK** command.
8. Click **OK**.	SPSS analyzes the **Matings** data and displays the **Group Statistics** and the **Independent Samples Test** tables. Since for this analysis we are interested in just the **Independent Samples Test** table, we have displayed it here alone.

Analysis Results

Independent Samples Test

		Levene's Test for Equality of Variances		t-test for Equality of Means						
									95% Confidence Interval of the Difference	
		F	Sig.	t	df	Sig. (2-tailed)	Mean Difference	Std. Error Difference	Lower	Upper
Matings	Equal variances assumed	.416	.527	3.300	10	.004	2.80000	.84853	1.01731	4.58269
	Equal variances not assumed			3.300	17.403	.004	2.80000	.84853	1.01292	4.58708

To conclude regarding H_0, we are interested in **t** and **Sig. (2-tailed)**. This information is contained in the **Equal variances assumed** row of the table (remember, one of the assumptions of the independent groups *t* test is homogeneity of variance). This row shows that *t* = **3.300**, and that **Sig. (2-tailed)** = **.004$_{2\text{ tailed}}$**. Since **.004** < **0.05**, you conclude by rejecting H_0 and affirming H_1. *Hormone X* appears to increase the number of matings. Note, the value obtained for t_{obt} in the textbook and the value of **t** given by SPSS are the same; so are the conclusions reached by using each.

■ SPSS ADDITIONAL PROBLEMS

1. Use SPSS to do Chapter 14, problem 24, part a, p. 390. Name the grouping variable *Group* and name the scores *Memory*.
2. A health psychologist wants to evaluate the effects of a particular diet on weight. Thirty-five obese male volunteers are randomly selected and put on the diet for 3 months. Baseline and end-of-program weights are recorded for each subject. The weights (in pounds) shown here are obtained:

Subject #	1	2	3	4	5	6	7	8	9	10	11	12
Baseline	210	213	190	220	235	265	258	198	271	195	241	212
End of Program	200	201	185	210	230	245	263	193	261	184	237	200

Subject #	13	14	15	16	17	18	19	20	21	22	23	24
Baseline	261	272	185	242	266	247	232	170	243	180	275	248
End of Program	246	252	167	231	251	241	227	155	240	185	262	223

Subject #	25	26	27	28	29	30	31	32	33	34	35
Baseline	261	269	278	210	252	258	249	214	182	246	271
End of Program	240	245	250	196	237	238	241	201	167	232	256

   a. Use SPSS and $\alpha = 0.05_{2\,tail}$ to evaluate the nondirectional H_1.
   b. Next, analyze these data with the *t* test for independent groups, calling the *Baseline* scores *Group 1*, and calling the *End of Program* scores *Group 2*. How do you explain the difference in outcome between part **a** and part **b**?
3. The random numbers in the table below were obtained from the table J, the table of random numbers in Appendix G.

Group 1	8	2	7	5	2	6		
Group 2	7	5	6	3	1	7	4	1

   a. Use SPSS to do a one-way, independent groups *t* test on the data. Use $\alpha = 0.05_{2\,tail}$.
   b. Add a constant of 5 to the scores of Group 2. This process is analogous to what concerning the population scores? Reanalyze the data. What do you conclude this time? Explain the difference between conclusions in part **a** and part **b**.

■ NOTES

14.1 Most textbooks present two methods for finding t_{obt} for the correlated groups design: (1) the direct-difference method and (2) a method that requires calculations of the degree of relationship (the correlation coefficient) existing between the two sets of raw scores. We have omitted the latter method because it is rarely used in practice and, in our opinion, confuses many students. The direct-difference method flows naturally and logically from the discussion of the *t* test for single samples. It is much easier to use and much more frequently employed in practice.

14.2 Occasionally, in a repeated measures experiment in which the alternative hypothesis is directional, the researcher may want to test whether the independent variable has an effect greater than some specified value other than 0. For example, assuming a directional hypothesis was justified in the present experiment, the researcher might want to test whether the average reward value of area A was greater than five bar presses per minute more than area B. In this case, the null hypothesis would be that the reward value of area A is not greater than five bar presses per minute more than area B. In this case, $\mu_D = 5$ rather than 0.

14.3 Occasionally, in an experiment involving independent groups, the alternative hypothesis is directional and specifies that the independent variable has an effect greater than some specified value other than 0. For example, in the "hormone X" experiment, assuming a directional hypothesis was legitimate, the physiologist might want to test whether hormone X has an average effect of over three matings more than the placebo. In this case, the null hypothesis would be that the average effect of the hormone X is ≤ 3 matings more than the placebo. We would test this hypothesis by assuming that the sample which received hormone X was a random sample from a population having a mean 3 units more than the population from which the placebo sample was taken. In this case, $\mu_{\overline{X}_1 - \overline{X}_2} = 3$ and

$$t_{obt} = \frac{(\overline{X}_1 - \overline{X}_2) - 3}{\sqrt{\left(\dfrac{SS_1 + SS_2}{n_1 + n_2 - 2}\right)\left(\dfrac{1}{n_1} + \dfrac{1}{n_2}\right)}}$$

14.4 Although we haven't previously presented the t equations in this form, the following can be shown:

t Test for Independent Groups	t Test for Correlated Groups
$t_{obt} = \dfrac{\overline{X}_1 - \overline{X}_2}{s_{\overline{X}_1 - \overline{X}_2}}$	$t_{obt} = \dfrac{\overline{D}}{s_D}$
$= \dfrac{\overline{X}_1 - \overline{X}_2}{\sqrt{s_{\overline{X}_1}^2 + s_{\overline{X}_2}^2}}$	$= \dfrac{\overline{D}}{\sqrt{s_{\overline{X}_1}^2 + s_{\overline{X}_2}^2 - 2rs_{\overline{X}_1}s_{\overline{X}_2}}}$

Since $\overline{X}_1 - \overline{X}_2$ is equal to $\overline{D}$, the t_{obt} equations for independent groups and correlated groups are identical except for the term $-2rs_{\overline{X}_1 - s_{\overline{X}_2}}$ in the denominator of the correlated groups equation. Thus, the higher power of the correlated groups design depends on the magnitude of r. The higher the value of r is, the more powerful the correlated groups design will be relative to the independent groups design. Using the data of the conservation film experiment to illustrate the use of these equations, we obtain the following:

Independent Groups

$$t_{obt} = \frac{\overline{X}_1 - \overline{X}_2}{\sqrt{s_{\overline{X}_1}^2 + s_{\overline{X}_2}^2}}$$

where $s_{\overline{X}_1}^2 = \dfrac{s_1^2}{n_1} = \dfrac{82.083}{12} = 6.840$

and $s_{\overline{X}_2}^2 = \dfrac{s_2^2}{n_2} = \dfrac{100.364}{12} = 8.364$

Correlated Groups

$$t_{obt} = \frac{\overline{D}}{\sqrt{s_{\overline{X}_1}^2 + s_{\overline{X}_2}^2 - 2rs_{\overline{X}_1}s_{\overline{X}_2}}}$$

$$= \frac{2.917}{\sqrt{6.840 + 8.364 - 2(0.938)(2.615)(2.892)}}$$

$$= \frac{2.917}{\sqrt{1.017}}$$

$$= 2.91$$

Thus,

$$t_{obt} = \frac{48.917 - 46.000}{\sqrt{6.840 + 8.364}} = \frac{2.917}{\sqrt{15.204}} = 0.748$$

$$= 0.75$$

Note that these are the same values obtained previously.

■ ONLINE STUDY RESOURCES

CENGAGE**brain**.com

Login to CengageBrain.com to access the resources your instructor has assigned. For this book, you can access the book's companion website for chapter-specific learning tools including Know and Be Able to Do, practice quizzes, flash cards, and glossaries, and a link to Statistics and Research Methods Workshops.

aplia™

If your professor has assigned Aplia homework:

1. Sign in to your account.
2. Complete the corresponding homework exercises as required by your professor.
3. When finished, click "Grade It Now" to see which areas you have mastered and which need more work, and for detailed explanations of every answer.

Visit **www.cengagebrain.com** to access your account and to purchase materials.

© Strmko / Dreamstime.com

Introduction to the Analysis of Variance

CHAPTER OUTLINE

LEARNING OBJECTIVES

After completing this chapter, you should be able to:

- Define the sampling distribution of F and specify its characteristics.
- Specify the H_0 and H_1 for one-way, independent groups ANOVA.
- Solve problems using one-way ANOVA; understand the derivation of MS_{within} and $MS_{between}$, and explain why $MS_{between}$ is always put in the numerator; explain why $MS_{between}$ is sensitive to the real effects of the IV and MS_{within} is not; and specify the assumptions underlying one-way ANOVA.
- Explain why H_1 in one-way ANOVA is always nondirectional and why we evaluate it with a one-tailed evaluation.
- Calculate the size of effect for a one-way ANOVA using $\hat{\omega}^2$ and η^2, and explain the difference between the values obtained by each.
- Specify how power using one-way ANOVA varies with changes in N, size of the real effect, and sample variability.
- Specify the difference between *planned* and *post hoc* comparisons; specify which is more powerful, and explain why.
- Do multiple comparisons using *planned* comparisons and explain why MS_{within} from the ANOVA is used.
- Do *post hoc* multiple comparisons using the Tukey HSD and the Scheffé tests; understand how each test accomplishes its goal of controlling Type I error.
- Understand the differences in power between *planned* comparisons, the Tukey HSD, and the Scheffé test.

- Rank order planned comparisons, the Tukey HSD and the Scheffé test, with regard to power.
- Understand the illustrative examples, do the practice problems, and understand the solutions.

INTRODUCTION: THE *F* DISTRIBUTION

In Chapters 12, 13, and 14, we have been using the *mean* as the basic statistic for evaluating the null hypothesis. It's also possible to use the *variance* of the data for hypothesis testing. One of the most important tests that does this is called the *F* test, after R. A. Fisher, the statistician who developed it. In using this test, we calculate the statistic F_{obt}, which fundamentally is the ratio of two independent variance estimates of the same population variance σ^2. In equation form,

$$F_{obt} = \frac{\text{Variance estimate 1 of } \sigma^2}{\text{Variance estimate 2 of } \sigma^2}$$

The sampling distribution of *F* can be generated empirically by (1) taking all possible samples of size n_1 and n_2 from the same population, (2) estimating the population variance σ^2 from each of the samples using s_1^2 and s_2^2, (3) calculating F_{obt} for all possible combinations of s_1^2 and s_2^2, and then (4) calculating $p(F)$ for each different value of F_{obt}. The resulting distribution is the *sampling distribution of F*. Thus, as with all sampling distributions,

definition ■ *The **sampling distribution of F** gives all the possible F values along with the $p(F)$ for each value, assuming sampling is random from the population.*

Like the *t* distribution, the *F* distribution varies with *degrees of freedom*. However, the *F* distribution has two values for degrees of freedom, one for the numerator and one for the denominator. As you might guess, we lose 1 degree of freedom for each calculation of variance. Thus,

$$\text{df for the numerator} = \text{df}_1 = n_1 - 1$$
$$\text{df for the denominator} = \text{df}_2 = n_2 - 1$$

Figure 15.1 shows an *F* distribution with 3 df in the numerator and 16 df in the denominator. Several features are apparent. First, since *F* is a ratio of variance

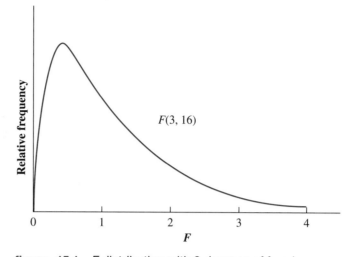

figure 15.1 *F* distribution with 3 degrees of freedom in the numerator and 16 degrees of freedom in the denominator.

estimates, it never has a negative value (s_1^2 and s_2^2 will always be positive). Second, the *F* distribution is positively skewed. Finally, the median *F* value is approximately equal to 1.

Like the *t* test, there is a family of *F* curves. With the *F test*, however, there is a different curve for each combination of df_1 and df_2. Table F in Appendix D gives the critical values of *F* for various combinations of df_1 and df_2. There are two entries for every cell. The light entry gives the critical *F* value for the 0.05 level. The dark entry gives the critical *F* value for the 0.01 level. Note that these are one-tailed values for the right-hand tail of the *F* distribution. To illustrate, Figure 15.2 shows the *F* distribution for 4 df in the numerator and 20 df in the denominator. From Table F, F_{crit} at the 0.05 level equals 2.87. This means that 5% of the *F* values are equal to or greater than 2.87. The area containing these values is shown shaded in Figure 15.2.

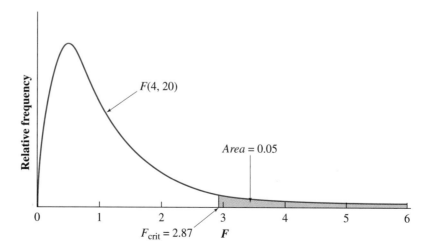

figure 15.2 Illustration showing that F_{crit} in Table F is one-tailed for the right-hand tail.

F TEST AND THE ANALYSIS OF VARIANCE (ANOVA)

The *F* test is appropriate in any experiment in which the scores can be used to form two independent estimates of the population variance. One quite frequent situation in the behavioral sciences for which the *F* test is appropriate occurs when analyzing the data from experiments that use more than two groups or conditions.

Thus far in the text, we have discussed the most fundamental experiment: the two-group study involving a control group and an experimental group. Although this design is still used frequently, it is more common to encounter experiments that involve three or more groups. A major limitation of the two-group study is that often two groups are not sufficient to allow a clear interpretation of the findings. For example, the "thalamus and pain perception" experiment (p. 374) included two groups. One received lesions in the thalamus and the other in an area "believed" to be unrelated to pain. The results showed a significantly higher pain threshold for the rats with thalamic lesions. Our conclusion was that lesions of the thalamus increased pain threshold. However, the difference between the two groups could just as well have been due to a lowering of pain threshold as a result of the other lesion rather than a raising of threshold because of the thalamic damage. This ambiguity could have been dispelled if three groups had been run rather than two. The third group would be a non-lesion control group. Comparing the pain threshold of the two lesion groups with the non-lesion group would help resolve the issue.

Another class of experiments requiring more than two groups involves experiments in which the independent variable is varied as a factor; that is, a predetermined range of the independent variable is selected, and several values spanning the range are used in the experiment. For example, in the "hormone X and sexual behavior" experiment, rather than arbitrarily picking one value of the hormone, the experimenter would probably pick several levels across the range of possible effective values. Each level would be administered to a different group of subjects, randomly sampled from the population. There would be as many groups in the experiment as there are levels of the hormone. This type of experiment has the advantage of allowing the experimenter to determine how the dependent variable changes with several different levels of the independent variable. In this example, the experimenter would find out how mating behavior varies in frequency with different levels of hormone X. Not only does using several levels allow a lawful relationship to emerge if one exists, but when the experimenter is unsure of what single level might be effective, using several levels increases the possibility of a positive result occurring from the experiment.

Given that it is frequently desirable to do experiments with more than two groups, you may wonder why these experiments aren't analyzed in the usual way. For example, if the experiment used four independent groups, why not simply compare the group means two at a time using the *t* test for independent groups? That is, why not just calculate *t* values comparing group 1 with 2, 3, and 4; 2 with 3 and 4; and 3 with 4?

The answer involves considerations of Type I error. You will recall that, when we set alpha at the 0.05 level, we are in effect saying that we are willing to risk being wrong 5% of the time when we reject H_0. In an experiment with two groups, there would be just one *t* calculation, and we would compare t_{obt} with t_{crit} to see whether t_{obt} fell in the critical region for rejecting H_0. Let's assume alpha = 0.05. The critical value of *t* at the 0.05 level was originally determined by taking the sampling distribution of

t for the appropriate df and locating the t value such that the proportion of the total number of t values that were equal to or more extreme than it equaled 0.05. That is, if we were randomly sampling one t score from the t distribution, the probability it would be $\geq t_{\text{crit}}$ is 0.05.

Now what happens when we do an experiment involving many t comparisons, say, 20 of them? We are no longer sampling just one t value from the t distribution but 20. The probability of getting t values equal to or greater than t_{crit} obviously goes up. It is no longer equal to 0.05. The probability of making a Type I error has increased as a result of doing an experiment with many groups and analyzing the data with more than one comparison.

OVERVIEW OF ONE-WAY ANOVA

The *analysis of variance* is a statistical technique used to analyze multigroup experiments. Using the F test allows us to make *one* overall comparison that tells whether there is a significant difference between the *means* of the groups. Thus, it avoids the problem of an increased probability of Type I error that occurs when assessing many t values. The analysis of variance, or ANOVA, as it is frequently called, is used in both independent groups and repeated measures designs. It is also used when one or more factors (variables) are investigated in the same experiment. In this section, we shall consider the simplest of these designs: the *simple randomized-group design*. This design is also often referred to as the *one-way analysis of variance*, independent groups design. A third designation often used is the *single factor experiment, independent groups design*.* According to this design, subjects are randomly sampled from the population and then randomly assigned to the conditions, preferably such that there are an equal number of subjects in each condition. There are as many independent groups as there are conditions. If the study is investigating the effect of an independent variable as a factor, then the conditions would be the different levels of the independent variable used. Each group would receive a different level of the independent variable (e.g., a different concentration of hormone X). Thus, in this design, scores from several independent groups are analyzed.

The alternative hypothesis used in the analysis of variance is nondirectional. It states that one or more of the conditions have different effects from at least one of the others on the dependent variable. The null hypothesis states that the different conditions are all equally effective, in which case the scores in each group are random samples from populations having the same mean value. If there are k groups, then the null hypothesis specifies that

$$\mu_1 = \mu_2 = \mu_3 = \cdots = \mu_k$$

where μ_1 = mean of the population from which group 1 is taken
μ_2 = mean of the population from which group 2 is taken
μ_3 = mean of the population from which group 3 is taken
μ_k = mean of the population from which group k is taken

---

*The use of ANOVA with repeated measures designs is covered in D. C. Howell, *Statistical Methods for Psychology*, Wadsworth/Cengage Learning, 2010, p. 461–513.

Like the t test, the analysis of variance assumes that only the mean of the scores is affected by the independent variable, not the variance. Therefore, the analysis of variance assumes that

$$\sigma_1^2 = \sigma_2^2 = \sigma_3^2 = \cdots = \sigma_k^2$$

where σ_1^2 = variance of the population from which group 1 is taken
σ_2^2 = variance of the population from which group 2 is taken
σ_3^2 = variance of the population from which group 3 is taken
σ_k^2 = variance of the population from which group k is taken

Essentially, the analysis of variance partitions the *total variability* of the data (SS_{total}) into two sources: the variability that exists within each group, called the *within-groups sum of squares* (SS_{within}), and the variability that exists between the groups, called the *between-groups sum of squares* ($SS_{between}$) (See Figure 15.3). Each sum of squares is used to form an independent estimate of the H_0 population variance. The estimate based on the within-groups variability is called the *within-groups variance estimate* (MS_{within}), and the estimate based on the between-groups variability is called the *between-groups variance estimate* ($MS_{between}$). Finally, an F ratio is calculated where

$$F_{obt} = \frac{\text{Between-groups variance estimate}}{\text{Within-groups variance estimate}} = \frac{MS_{between}}{MS_{within}}$$

This process is shown in Figure 15.3. The between-groups variance estimate increases with the magnitude of the independent variable's effect, whereas the within-groups variance estimate is unaffected. Thus, the larger the F ratio is, the more unreasonable the null hypothesis becomes. As with the other statistics, we evaluate F_{obt} by comparing it with F_{crit}. If F_{obt} is equal to or exceeds F_{crit}, we reject H_0. Thus, the decision rule states the following:

If $F_{obt} \geq F_{crit}$, reject H_0.
If $F_{obt} < F_{crit}$, retain H_0.

Within-Groups Variance Estimate, MS_{within}

Remember from Chapter 4, variance is a concept that quantifies how large the differences are between the scores and the mean of a set of scores. The within-groups variance estimate of the H_0 population variance σ^2 is based on the variability *within* each

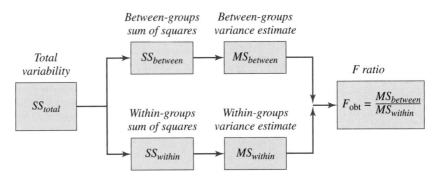

figure 15.3 Overview of the analysis of variance technique, simple randomized-groups design.

group. It tells us how large the differences are between the scores within each group and the group mean. It is symbolized by MS_{within}, which stands for *Mean Square within*. This variance is the same variance estimate used in the t test for independent groups that was symbolized by s_w^2. Statistical tradition is quite strong that when dealing with the analysis of variance, the symbol MS_{within} is employed to denote the within-groups variance estimate, instead of s_w^2. For various reasons I have decided to follow tradition and use the symbol MS_{within}.

Since MS_{within} and s_w^2 symbolize the same variance estimate, you would expect the equations for each to be the same. This is indeed the case when there are only two groups in the experiment or study. The equations for MS_{within} and s_w^2 are given below.

$$s_W^2 = \frac{SS_1 + SS_2}{(n_1 - 1) + (n_2 - 1)} \qquad \textit{t test, only 2 groups}$$

$$MS_{within} = \frac{SS_1 + SS_2 + SS_3 + \cdots + SS_k}{(n_1 - 1) + (n_2 - 1) + (n_3 - 1) + \cdots + (n_k - 1)}$$

$$\textit{analysis of variance, usually more than 2 groups}$$

where k = the number of groups in the experiment

From the above two equations, it can be seen that for a two-group experiment,

$$MS_{within} = s_W^2$$

The conceptual equation for MS_{within} is given below.

$$MS_{within} = \frac{SS_1 + SS_2 + SS_3 + \cdots + SS_k}{(n_1 - 1) + (n_2 - 1) + (n_3 - 1) + \cdots + (n_k - 1)}$$

$$\textit{conceptual equation for within-groups variance estimate}$$

This equation can be simplified to

$$MS_{within} = \frac{SS_1 + SS_2 + SS_3 + \cdots + SS_k}{N - k}$$

where $N = n_1 + n_2 + n_3 + \cdots + n_k$

The numerator of this equation is called the *within-groups sum of squares*. It is symbolized by SS_{within}. The denominator equals the degrees of freedom for the within-groups variance estimate. Since we lose 1 degree of freedom for each sample variance calculated and there are k variances, there are $N - k$ degrees of freedom. Thus,

$$MS_{within} = \frac{SS_{within}}{df_{within}} \qquad \textit{within-groups variance estimate}$$

where $SS_{within} = SS_1 + SS_2 + SS_3 + \ldots + SS_k$ *within-groups sum of squares*

$df_{within} = N - k$ *within-groups degrees of freedom*

This equation for SS_{within} is fine conceptually, but when actually computing SS_{within}, it is better to use another equation. This equation is the algebraic equivalent of the conceptual equation, but it is easier to use and leads to fewer rounding errors. The computational equation is given here and will be used subsequently when we analyze the data from an experiment:

$$SS_{within} = \sum^{\substack{all \\ scores}} X^2 - \left[\frac{(\Sigma X_1)^2}{n_1} + \frac{(\Sigma X_2)^2}{n_2} + \frac{(\Sigma X_3)^2}{n_3} + \cdots + \frac{(\Sigma X_k)^2}{n_k} \right]$$

computational equation for within-groups sum of squares

Between-Groups Variance Estimate, $MS_{between}$

The second estimate of the variance of the null-hypothesis populations, σ^2 is based on the variability *between* the groups. It is symbolized by $MS_{between}$ and is called the *between-groups variance estimate*.

The null hypothesis states that each group is a random sample from populations where $\mu_1 = \mu_2 = \mu_3 = \ldots = \mu_k$. If the null hypothesis is correct, then we can use the variability between the means of the samples to estimate the variance of these populations, σ^2.

We know from Chapter 12 that, if we take all possible samples of size n from a population and calculate their mean values, the resulting sampling distribution of means has a variance of

$$\sigma_{\overline{X}}^2 = \sigma^2/n.$$

Solving for σ^2, we arrive at

$$\sigma^2 = n\sigma_{\overline{X}}^2$$

Estimating $\sigma_{\overline{X}}^2$, the previous equation becomes

$$\text{estimate of } \sigma^2 = n(\text{estimate of } \sigma_{\overline{X}}^2)$$

If $\sigma_{\overline{X}}^2$ can be estimated, we can substitute the estimate in the previous equation to arrive at an independent estimate of σ^2. It turns out that $\sigma_{\overline{X}}^2$ can be estimated because in the actual experiment there are several sample mean scores. We can use the variance of these mean scores to estimate the variance of the full set of sample mean scores, $\sigma_{\overline{X}}^2$. This estimate of $\sigma_{\overline{X}}^2$ is symbolized by $s_{\overline{X}}^2$. Since there are k sample means, to use the variability of the sample means as an estimate, we divide by $k - 1$, just as when we have N raw scores in a sample we divide by $N - 1$ to estimate the variance of the raw score population [with sample raw scores, the estimate of σ^2 is $\frac{\Sigma(X - \overline{X})^2}{N - 1}$]. Thus,

$$s_{\overline{X}}^2 = \frac{\Sigma(\overline{X} - \overline{X}_G)^2}{k - 1} \qquad \textit{estimate of } \sigma_{\overline{X}}^2 \textbf{ and}$$

$$MS_{between} = \text{estimate of } \sigma^2 = n\,(\text{estimate of } \sigma_{\overline{X}}^2) = ns_{\overline{X}}^2 = n\left[\frac{\Sigma(\overline{X} - \overline{X}_G)^2}{k - 1} \right]$$

where $\overline{X}_G$ = *grand mean* (overall mean of all the scores combined)

k = number of sample means = number of groups

Expanding the summation, we arrive at the conceptual equation for $MS_{between}$.

$$MS_{between} = \frac{n\left[(\overline{X}_1 - \overline{X}_G)^2 + (\overline{X}_2 - \overline{X}_G)^2 + (\overline{X}_3 - \overline{X}_G)^2 + \cdots + (\overline{X}_k - \overline{X}_G)^2\right]}{k - 1}$$

conceptual equation for the between-groups variance estimate

The numerator of this equation is called the *between-groups sum of squares*. It is symbolized by $SS_{between}$. The denominator is the degrees of freedom for the between-groups variance estimate. It is symbolized by $df_{between}$. Thus,

$$MS_{between} = \frac{SS_{between}}{df_{between}} \qquad \textit{between-groups variance estimate}$$

where
$$SS_{between} = n[(\overline{X}_1 - \overline{X}_G)^2 + (\overline{X}_2 - \overline{X}_G)^2 + (\overline{X}_3 - \overline{X}_G)^2 + \cdots$$
$$+ (\overline{X}_k - \overline{X}_G)^2] \qquad \textit{between-groups sum of squares}$$

$$df_{between} = k - 1 \qquad \textit{between-groups degrees of freedom}$$

It should be clear that, as the effect of the independent variable increases, the differences between the sample means increase. This causes $(\overline{X}_1 - \overline{X}_G)^2$, $(\overline{X}_2 - \overline{X}_G)^2$, ..., $(\overline{X}_k - \overline{X}_G)^2$ to increase, which in turn produces an increase in $SS_{between}$. Since $SS_{between}$ is in the numerator, increases in it produce increases in $MS_{between}$. Thus, the between-groups variance estimate ($MS_{between}$) increases with the effect of the independent variable.

This equation for $SS_{between}$ is fine conceptually, but when actually computing $SS_{between}$, it is better to use another equation. As with SS_{within}, there is a computational equation for $SS_{between}$ that is the algebraic equivalent of the conceptual equation but is easier to use and leads to fewer rounding errors. The computational equation is given here and will be discussed shortly when we analyze the data from an experiment:

$$SS_{between} = \left[\frac{(\Sigma X_1)^2}{n_1} + \frac{(\Sigma X_2)^2}{n_2} + \frac{(\Sigma X_3)^2}{n_3} + \cdots + \frac{(\Sigma X_k)^2}{n_k}\right] - \frac{\left(\overset{\text{all scores}}{\underset{}{\Sigma}} X\right)^2}{N}$$

computational equation for between-groups sum of squares

The *F* Ratio

We noted earlier that $MS_{between}$ increases with the effect of the independent variable. However, since an assumption of the analysis of variance is that the independent variable affects only the mean and not the variance of each group, the within-groups variance estimate does not change with the effect of the independent variable. Since $F_{obt} = MS_{between}/MS_{within}$, F increases with the effect of the independent variable. Thus, the larger the F ratio is, the more reasonable it is that the independent variable has had a real effect. Another way of saying this is that $MS_{between}$ is really an estimate of σ^2 plus the effects of the independent variable, whereas MS_{within} is just an estimate of σ^2. Thus,

$$F_{obt} = \frac{MS_{between}}{MS_{within}} = \frac{\sigma^2 + \text{independent variable effects}}{\sigma^2}$$

The larger F_{obt} becomes, the more reasonable it is that the independent variable has had a real effect. Of course, F_{obt} must be equal to or exceed F_{crit} before H_0 can be rejected. If F_{obt} is less than 1, we don't even need to compare it with F_{crit}. It is obvious the treatment has not had a significant effect, and we can immediately conclude by retaining H_0.

ANALYZING DATA WITH THE ANOVA TECHNIQUE

So far, we have been quite theoretical. Now let's do a problem to illustrate the analysis of variance technique.

experiment

Different Situations and Stress

Suppose you are interested in determining whether certain situations produce differing amounts of stress. You know the amount of the hormone corticosterone circulating in the blood is a good measure of how stressed a person is. You randomly assign 15 students into three groups of 5 each. The students in group 1 have their corticosterone levels measured immediately after returning from vacations (low stress). The students in group 2 have their corticosterone levels measured after they have been in class for a week (moderate stress). The students in group 3 are measured immediately before final exam week (high stress). All measurements are taken at the same time of day. You record the data shown in Table 15.1. Scores are in milligrams of corticosterone per 100 milliliters of blood.

1. What is the alternative hypothesis?
2. What is the null hypothesis?
3. What is the conclusion? Use $\alpha = 0.05$.

table 15.1   Stress experiment data

Vacation Group 1		Class Group 2		Final Exam Group 3	
X_1	X_1^2	X_2	X_2^2	X_3	X_3^2
2	4	10	100	10	100
3	9	8	64	13	169
7	49	7	49	14	196
2	4	5	25	13	169
6	36	10	100	15	225
20	102	40	338	65	859

$n_1 = 5$ $n_2 = 5$ $n_3 = 5$

$\overline{X}_1 = 4.00$ $\overline{X}_2 = 8.00$ $\overline{X}_3 = 13.00$

$$\sum_{\substack{\text{all} \\ \text{scores}}} X = 125 \qquad \sum_{\substack{\text{all} \\ \text{scores}}} X^2 = 1299 \qquad \overline{X}_G = \frac{\sum_{\substack{\text{all} \\ \text{scores}}} X}{N} = 8.333$$

$N = 15$

SOLUTION

1. Alternative hypothesis: The alternative hypothesis states that at least one of the situations affects stress differently than at least one of the remaining situations. Therefore, at least one of the means (μ_1, μ_2, or μ_3) differs from at least one of the others.

2. Null hypothesis: The null hypothesis states that the different situations affect stress equally. Therefore, the three sample sets of scores are random samples from populations where $\mu_1 = \mu_2 = \mu_3$.

3. Conclusion, using $\alpha = 0.05$: The conclusion is reached in the same general way as with the other inference tests. First, we calculate the appropriate statistic, in this case, F_{obt}, and then we evaluate F_{obt} based on its sampling distribution.

A. Calculate F_{obt}.

STEP 1: Calculate the between-groups sum of squares, $SS_{between}$. To calculate $SS_{between}$, we shall use the following computational equation:

$$SS_{between} = \left[\frac{(\Sigma X_1)^2}{n_1} + \frac{(\Sigma X_2)^2}{n_2} + \frac{(\Sigma X_3)^2}{n_3} + \cdots + \frac{(\Sigma X_k)^2}{n_k} \right] - \frac{\left(\overset{\text{all scores}}{\underset{}{\Sigma}} X \right)^2}{N}$$

computational equation for $SS_{between}$

In this problem, since $k = 3$, this equation reduces to

$$SS_{between} = \left[\frac{(\Sigma X_1)^2}{n_1} + \frac{(\Sigma X_2)^2}{n_2} + \frac{(\Sigma X_3)^2}{n_3} \right] - \frac{\left(\overset{\text{all scores}}{\underset{}{\Sigma}} X \right)^2}{N}$$

where $\overset{\text{all scores}}{\underset{}{\Sigma}} X = $ sum of all the scores

Substituting the appropriate values from Table 15.1 into this equation, we obtain

$$SS_{between} = \left[\frac{(20)^2}{5} + \frac{(40)^2}{5} + \frac{(65)^2}{5} \right] - \frac{(125)^2}{15} = 203.333$$

STEP 2: Calculate the within-groups sum of squares, SS_{within}. The computational equation for SS_{within} is as follows:

$$SS_{within} = \overset{\text{all scores}}{\underset{}{\Sigma}} X^2 - \left[\frac{(\Sigma X_1)^2}{n_1} + \frac{(\Sigma X_2)^2}{n_2} + \frac{(\Sigma X_3)^2}{n_3} + \cdots + \frac{(\Sigma X_k)^2}{n_k} \right]$$

computational equation for SS_{within}

where $\overset{\text{all scores}}{\underset{}{\Sigma}} X^2 = $ sum of all the squared scores

Since $k = 3$, for this problem the equation reduces to

$$SS_{within} = \overset{\text{all scores}}{\underset{}{\Sigma}} X^2 - \left[\frac{(\Sigma X_1)^2}{n_1} + \frac{(\Sigma X_2)^2}{n_2} + \frac{(\Sigma X_3)^2}{n_3} \right]$$

Substituting the appropriate values into this equation, we obtain

$$SS_{within} = 1299 - \left[\frac{(20)^2}{5} + \frac{(40)^2}{5} + \frac{(65)^2}{5} \right] = 54.000$$

MENTORING TIP

Step 3 is just a check on calculations in steps 1 and 2; it does not have to be done, but probably is a good idea before going on to Step 4.

STEP 3: **Calculate the total sum of squares, SS_{total}.** This step is just a check to be sure the calculations in Steps 1 and 2 are correct. You will recall that at the beginning of the analysis of variance section, p. 406, we said this technique partitions the total variability into two parts: the within variability and the between variability. The measure of total variability is SS_{total}, the measure of within variability is SS_{within}, and the measure of between variability is $SS_{between}$. Thus,

$$SS_{total} = SS_{within} + SS_{between}$$

By independently calculating SS_{total}, we can check to see whether this relationship holds true for the calculations in steps 1 and 2. The equation for independent computation of SS_{total} is

$$SS_{total} = \sum_{}^{\substack{all \\ scores}} X^2 - \frac{\left(\sum_{}^{\substack{all \\ scores}} X \right)^2}{N}$$

You will recognize that this equation is quite similar to the sum of squares with each sample, except here we are using the scores of all the samples as a single group. Calculating SS_{total}, we obtain

$$SS_{total} = 1299 - \frac{(125)^2}{15} = 257.333$$

Substituting the values of SS_{total}, SS_{within}, and $SS_{between}$ into the equation, we obtain

$$SS_{total} = SS_{within} + SS_{between}$$
$$257.333 = 54.000 + 203.333$$
$$257.333 = 257.333$$

Note that, if the within sum of squares plus the between sum of squares does not equal the total sum of squares, you've made a calculation error. Go back and check steps 1, 2, and 3 until the equation balances (within rounding error).

STEP 4: **Calculate the degrees of freedom for each estimate.**

$$df_{between} = k - 1 = 3 - 1 = 2$$
$$df_{within} = N - k = 15 - 3 = 12$$
$$df_{total} = N - 1 = 15 - 1 = 14$$

STEP 5: **Calculate the between-groups variance estimate, $MS_{between}$.** The variance estimates are just the sums of squares divided by their degrees of freedom. Thus,

$$MS_{between} = \frac{SS_{between}}{df_{between}} = \frac{203.333}{2} = 101.667$$

STEP 6: **Calculate the within-groups variance estimate, MS_{within}.**

$$MS_{within} = \frac{SS_{within}}{df_{within}} = \frac{54.000}{12} = 4.500$$

STEP 7: **Calculate F_{obt}.** We have calculated two independent estimates of σ^2, $MS_{between}$ and MS_{within}. The F value is the ratio of $MS_{between}$ to MS_{within}. Thus,

$$F_{obt} = \frac{MS_{between}}{MS_{within}} = \frac{101.667}{4.500} = 22.59$$

Note that $MS_{between}$ is always put in the numerator and MS_{within} in the denominator.

B. Evaluate F_{obt}. Since $MS_{between}$ is a measure of the effect of the independent variable as well as an estimate of σ^2, it should be larger than MS_{within}, unless chance alone is at work. If $F_{obt} \leq 1$, it is clear that the independent variable has not had a significant effect and we conclude by retaining H_0 without even bothering to compare F_{obt} with F_{crit}. If $F_{obt} > 1$, we must compare it with F_{crit}. If $F_{obt} \geq F_{crit}$, we reject H_0.

Table F in Appendix D is used to determine F_{crit}. Table F presents the critical values of F for *Degrees of Freedom Numerator* and *Degrees of Freedom Denominator*. These values have been obtained from the sampling distribution of F for each combination of *Degrees of Freedom Numerator* and *Degrees of Freedom Denominator*.

$$\text{Degrees of Freedom Numerator} = \text{df}_{numerator} = \text{df}_{between}$$
$$\text{Degrees of Freedom Denominator} = \text{df}_{denominator} = \text{df}_{within}$$

In Table F, the degrees of freedom for both the numerator and denominator range from 1 to ∞. A portion of Table F is shown below in Table 15.2. As can be seen from Table 15.2, each cell in the table contains two values; both are critical values of F. The roman type entry is F_{crit} for $\alpha = 0.05$, and the bolded entry is F_{crit} for $\alpha = 0.01$. To use Table F, you must know *Degrees of Freedom Numerator*, *Degrees of Freedom Denominator*, and the alpha level.

The table is entered using the degrees of freedom for the numerator and the denominator for a particular experiment or study. The critical values of F for these degrees of freedom are found in the cell located at the intersection of row and column appropriate for the degrees of freedom. For example, referring to Table 15.2, if $\text{df}_{numerator} = $ *Degrees of Freedom Numerator* $= 3$ and $\text{df}_{denominator} = $ *Degrees of Freedom Denominator* $= 11$, the appropriate F_{crit} values are found in the cell at the intersection of the row with heading "11" and the column with heading "3." For these df values, the two numbers in the cell are 3.59 and **6.22**. In this example, if $\text{df}_{numerator} = 3$, $\text{df}_{denominator} = 11$, and $\alpha = 0.05$, then $F_{crit} = 3.59$; if $\alpha = 0.01$, then $F_{crit} = $ **6.22**.

We are now ready to determine F_{crit} for the stress experiment. From Table F in Appendix D, with $\text{df}_{numerator} = 2$, $\text{df}_{denominator} = 12$, and $\alpha = 0.05$, we find that

$$F_{crit} = 3.88$$

table 15.2 F_{crit} values, a portion of table F

Degrees of Freedom Denominator	Degrees of Freedom Numerator			
	1	2	3	4
10	4.96	4.10	3.71	3.48
	10.04	**7.56**	**6.55**	**5.99**
11	4.84	3.98	3.59	3.36
	9.65	**7.20**	**6.22**	**5.67**
12	4.75	3.88	3.49	3.26
	9.33	**6.93**	**5.95**	**5.41**
13	4.67	3.80	3.41	3.18
	9.07	**6.70**	**5.74**	**5.20**
14	4.60	3.74	3.34	3.11
	8.86	**6.51**	**5.56**	**5.03**

Note that, in looking up F_{crit} in Table F, it is important to keep the df for the numerator and denominator straight. If by mistake you had entered the table with 2 df for the denominator and 12 df for the numerator, F_{crit} would equal 19.41, which is quite different from 3.88.

Now that we have determined F_{crit}, we can evaluate F_{obt}. In the stress experiment, $F_{obt} = 22.59$. Since $F_{obt} > 3.88$, we reject H_0. The three situations are not all the same in the stress levels they produce. A summary of the solution is shown in Table 15.3.

table 15.3 Summary table for ANOVA problem involving stress

Source	SS	df	MS	F_{obt}
Between groups	203.333	2	101.667	22.59*
Within groups	54.000	12	4.500	
Total	257.333	14		

*With $\alpha = 0.05$, $F_{crit} = 3.88$. Therefore, H_0 is rejected.

LOGIC UNDERLYING THE ONE-WAY ANOVA

Now that we have worked through the calculations of an illustrative example, I would like to discuss in more detail the logic underlying the one-way ANOVA. Earlier, I pointed out that the one-way ANOVA partitions the total variability (SS_{total}) into two parts: the within-groups sum of squares (SS_{within}) and the between-groups sum of squares ($SS_{between}$). We can gain some insight into this partitioning by recognizing that it is based on the simple idea that the deviation of each score from the grand mean is made up of two parts: the deviation of the score from its own group mean and the deviation of that group mean from the grand mean. Applying this idea to the first score in group 1, we obtain

Deviation of each score from the grand mean	=	Deviation of the score from its own group mean	+	Deviation of that group mean from the grand mean
$2 - 8.33$	=	$2 - 4.00$	+	$4.00 - 8.33$
$X - \overline{X}_G$	=	$X - \overline{X}_1$	+	$\overline{X}_1 - \overline{X}_G$
$\downarrow$	=	$\downarrow$	+	$\downarrow$
SS_{total}	=	SS_{within}	+	$SS_{between}$

Note that the term on the left $(\overline{X}_1 - \overline{X}_G)$ when squared and summed over all the scores becomes SS_{total}. Thus,

$$SS_{total} = \sum_{}^{\substack{all \\ scores}} (X - \overline{X}_G)^2$$

The term in the middle $(X - \overline{X}_1)$ when squared and summed for all the scores (of course, we must subtract the appropriate group mean from each score) becomes SS_{within}. Thus,

$$SS_{within} = SS_1 + SS_2 + SS_3$$
$$= \Sigma(X - \overline{X}_1)^2 + \Sigma(X - \overline{X}_2)^2 + \Sigma(X - \overline{X}_3)^2$$

It is important to note that, since the subjects *within* each group receive the same treatment, variability among the scores within each group cannot be due to differences in the effect of the independent variable. Thus, the within-groups sum of squares (SS_{within}) is not a measure of the effect of the independent variable. Since $SS_{within}/df_{within} = MS_{within}$, this means that the within-groups variance estimate MS_{within} also is not a measure of the real effect of the independent variable. Rather, it provides us with an estimate of the inherent variability of the scores themselves. Thus, MS_{within} is an estimate of σ^2 that is unaffected by treatment differences.

The last term in the equation partitioning the variability of score 2 from the grand mean is $\overline{X}_1 - \overline{X}_G$. When this term is squared and summed for all the scores, it becomes $SS_{between}$. Thus,

$$SS_{between} = n(\overline{X}_1 - \overline{X}_G)^2 + n(\overline{X}_2 - \overline{X}_G)^2 + n(\overline{X}_3 - \overline{X}_G)^2$$

As discussed previously, $SS_{between}$ is sensitive to the effect of the independent variable, because the greater the effect of the independent variable, the more the means of each group will differ from each other and hence, will differ from $\overline{X}_G$. Since $SS_{between}/df_{between} = MS_{between}$, this means that the between-groups variance estimate $MS_{between}$ is also sensitive to the real effect of the independent variable. Thus, $MS_{between}$ gives us an estimate of σ^2 plus the effects of the independent variable. Since

$$F_{obt} = \frac{MS_{between}}{MS_{within}} = \frac{\sigma^2 + \text{effects of the independent variable}}{\sigma^2}$$

MENTORING TIP

$MS_{between}$ is sensitive to the effects of the independent variable; MS_{within} is not.

the larger F_{obt} is, the less reasonable the null-hypothesis explanation is. If the independent variable has no effect, then both $MS_{between}$ and MS_{within} are independent estimates of σ^2 and their ratio is distributed as F with df = $df_{between}$ (numerator) and df_{within} (denominator). We evaluate the null hypothesis by comparing F_{obt} with F_{crit}. If $F_{obt} \geq F_{crit}$, we reject H_0.

Let's try one more problem for practice.

Practice Problem 15.1

A college professor wants to determine the best way to present an important topic to his class. He has the following three choices: (1) he can lecture, (2) he can lecture and assign supplementary reading, or (3) he can show a film and assign supplementary reading. He decides to do an experiment to evaluate the three options. He solicits 27 volunteers from his class and randomly assigns 9 to each of three conditions. In condition 1, he lectures to the students. In condition 2,

(continued)

he lectures plus assigns supplementary reading. In condition 3, the students see a film on the topic plus receive the same supplementary reading as the students in condition 2. The students are subsequently tested on the material. The following scores (percentage correct) were obtained:

Lecture Condition 1/Group 1		Lecture + Reading Condition 2/Group 2		Film + Reading Condition 3/Group 3	
X_1	X_1^2	X_2	X_2^2	X_3	X_3^2
92	8,464	86	7,396	81	6,561
86	7,396	93	8,649	80	6,400
87	7,569	97	9,409	72	5,184
76	5,776	81	6,561	82	6,724
79	6,241	94	8,836	83	6,889
86	7,396	89	7,921	89	7,921
91	8,281	98	9,604	76	5,776
81	6,561	90	8,100	88	7,744
83	6,889	91	8,281	83	6,889
761	64,573	819	74,757	734	60,088

$$n_1 = 9 \qquad n_2 = 9 \qquad n_3 = 9$$

$$\overline{X}_1 = 84.556 \qquad \overline{X}_2 = 91.000 \qquad \overline{X}_3 = 81.556$$

$$\overset{\text{all}}{\underset{\text{scores}}{\sum}} X = 2314 \qquad \overset{\text{all}}{\underset{\text{scores}}{\sum}} X^2 = 199{,}418 \qquad \overline{X}_G = \frac{\overset{\text{all}}{\underset{\text{scores}}{\sum}} X}{N} = 85.704$$

$$N = 27$$

a. What is the overall null hypothesis?
b. What is the conclusion? Use $\alpha = 0.05$.

SOLUTION

a. Null hypothesis: The null hypothesis states that the different methods of presenting the material are equally effective. Therefore, $\mu_1 = \mu_2 = \mu_3$.
b. Conclusion, using $\alpha = 0.05$: To assess H_0, we must calculate F_{obt} and then evaluate it based on its sampling distribution.
   A. Calculate F_{obt}.

     STEP 1: Calculate $SS_{between}$.

$$SS_{between} = \left[\frac{(\sum X_1)^2}{n_1} + \frac{(\sum X_2)^2}{n_2} + \frac{(\sum X_3)^2}{n_3} \right] - \frac{\left(\overset{\text{all}}{\underset{\text{scores}}{\sum}} X \right)^2}{N}$$

$$= \left[\frac{(761)^2}{9} + \frac{(819)^2}{9} + \frac{(734)^2}{9} \right] - \frac{(2314)^2}{27}$$

$$= 419.185$$

STEP 2: Calculate SS_{within}.

$$SS_{within} = \overset{\overset{\text{all}}{\text{scores}}}{\sum} X^2 - \left[\frac{(\sum X_1)^2}{n_1} + \frac{(\sum X_2)^2}{n_2} + \frac{(\sum X_3)^2}{n_3} \right]$$

$$= 199{,}418 - \left[\frac{(761)^2}{9} + \frac{(819)^2}{9} + \frac{(734)^2}{9} \right]$$

$$= 680.444$$

STEP 3: Calculate SS_{total}.

$$SS_{total} = \overset{\overset{\text{all}}{\text{scores}}}{\sum} X^2 - \frac{\left(\overset{\overset{\text{all}}{\text{scores}}}{\sum} X \right)^2}{N}$$

$$= 199{,}418 - \frac{(2314)^2}{27}$$

$$= 1099.630$$

This step is a check to see whether $SS_{between}$ and SS_{within} were correctly calculated. If so, then $SS_{total} = SS_{between} + SS_{within}$. This check is shown here:

$$SS_{total} = SS_{between} + SS_{within}$$
$$1099.630 \cong 419.185 + 680.444$$
$$1099.630 \cong 1099.629 \text{ (checks within rounding error)}$$

STEP 4: Calculate df.

$$\text{df}_{between} = k - 1 = 3 - 1 = 2$$
$$\text{df}_{within} = N - k = 27 - 3 = 24$$
$$\text{df}_{total} = N - 1 = 27 - 1 = 26$$

STEP 5: Calculate $MS_{between}$.

$$MS_{between} = \frac{SS_{between}}{\text{df}_{between}} = \frac{419.185}{2} = 209.593$$

STEP 6: Calculate MS_{within}.

$$MS_{within} = \frac{SS_{within}}{\text{df}_{within}} = \frac{680.444}{24} = 28.352$$

STEP 7: Calculate F_{obt}.

$$F_{obt} = \frac{MS_{between}}{MS_{within}} = \frac{209.593}{28.352} = 7.39$$

B. Evaluate F_{obt}. With $\alpha = 0.05$, $\text{df}_{numerator} = 2$, and $\text{df}_{denominator} = 24$, from Table F,

$$F_{crit} = 3.40$$

(continued)

Since $F_{obt} > 3.40$, we reject H_0. The methods of presentation are not equally effective. The solution is summarized in Table 15.4.

table **15.4** Summary ANOVA table for methods of presentation experiment

Source	SS	df	MS	F_{obt}
Between groups	419.185	2	209.593	7.39*
Within groups	680.444	24	28.352	
Total	1099.630	26		

*With $\alpha = 0.05$, $F_{crit} = 3.40$. Therefore, H_0 is rejected.

RELATIONSHIP BETWEEN ANOVA AND THE *t* TEST

When a study involves just two independent groups and we are testing the null hypothesis that $\mu_1 = \mu_2$, we can use either the *t* test for independent groups or the analysis of variance. In such situations, it can be shown algebraically that $t^2 = F$. For a demonstration of this point, go to the Online Study Resources website.

ASSUMPTIONS UNDERLYING THE ANALYSIS OF VARIANCE

The assumptions underlying the analysis of variance are similar to those of the *t* test for independent groups:

1. The populations from which the samples were taken are normally distributed.
2. The samples are drawn from populations of equal variances. As pointed out in Chapter 14 in connection with the *t* test for independent groups, this is called the homogeneity of variance assumption. The analysis of variance also assumes homogeneity of variance.*

Like the *t* test, the analysis of variance is a robust test. It is minimally affected by violations of population normality. It is also relatively insensitive to violations of homogeneity of variance, provided the samples are of equal size.[†]

---

*See Chapter 14 footnote (*) on p. 376. Some statisticians would also limit the use of ANOVA to data that are interval or ratio in scaling. For a discussion of this point, see the references in the Chapter 2 footnote on p. 35.

[†]For an extended discussion of these points, see G. V. Glass, P. D. Peckham, and J. R. Sanders, "Consequences of Failure to Meet the Assumptions Underlying the Use of Analysis of Variance and Covariance," *Review of Educational Research, 42* (1972), 237–288.

SIZE OF EFFECT USING $\hat{\omega}^2$ OR η^2

Omega Squared, $\hat{\omega}^2$

We have already discussed the size of the effect of the X variable on the Y variable in conjunction with correlational research when we discussed the coefficient of determination (r^2) in Chapter 6, p. 139. You will recall that r^2 is a measure of the proportion of the total variability of Y accounted for by X and hence is a measure of the strength of the relationship between X and Y. If the X variable is causal with regard to the Y variable, the coefficient of determination is also a measure of the size of the effect of X on Y.

The situation is very similar when we are dealing with the one-way, independent groups ANOVA. In this situation, the independent variable is the X variable and the dependent variable is the Y variable. One of the statistics computed to measure size of effect in the one-way, independent groups ANOVA is omega squared ($\hat{\omega}^2$). The other is eta squared (η^2), which we discuss in the next section. Conceptually, $\hat{\omega}^2$ and η^2 are like r^2 in that each provides an estimate of the proportion of the total variability of Y that is accounted for by X. $\hat{\omega}^2$ is a relatively unbiased estimate of this proportion in the population, whereas the estimate provided by η^2 is more biased. The conceptual equation for $\hat{\omega}^2$ is given by

$$\hat{\omega}^2 = \frac{\sigma^2_{between}}{\sigma^2_{between} + \sigma^2_{within}} \qquad \textit{Conceptual equation}$$

Since we do not know the values of these population variances, we estimate them from the sample data. The resulting equation is

$$\hat{\omega}^2 = \frac{SS_{between} - (k-1)MS_{within}}{SS_{total} + MS_{within}} \qquad \textit{Computational equation}$$

Cohen (1988) suggests the criteria shown in Table 15.5 for interpreting $\hat{\omega}^2$ or η^2.

table 15.5 Cohen's criteria for interpreting the value of $\hat{\omega}^2$ or η^2*

$\hat{\omega}^2$ or η^2 (Proportion of Variance Accounted for)	Interpretation
0.01–0.05	Small effect
0.06–0.13	Medium effect
≥ 0.14	Large effect

* See Chapter 13 footnote on p. 340 for a reference discussing some cautions in using Cohen's criteria.

example

Stress Experiment

Let's compute the size of effect using $\hat{\omega}^2$ for the stress experiment, p. 410. For this experiment, $SS_{between} = 203.333$, $SS_{total} = 257.333$, $MS_{within} = 4.500$, and $k = 3$. The size of effect for these data, using $\hat{\omega}^2$, is

MENTORING TIP

Caution: compute $\hat{\omega}^2$ to 3-decimal-place accuracy, since this proportion is often converted to a percentage.

$$\hat{\omega}^2 = \frac{203.333 - (3 - 1)4.500}{257.333 + 4.500} = 0.742$$

Thus, the estimate provided by $\hat{\omega}^2$ tells us that the stress situations account for 0.742 or 74.2% of the variance in corticosterone levels. Referring to Table 15.5, since the value of $\hat{\omega}^2$ is greater than 0.14, this is considered a large effect.

Eta Squared, η^2

Eta squared is an alternative measure for determining size of effect in one-way, independent groups ANOVA experiments. It also provides an estimate of the proportion of the total variability of Y that is accounted for by X, and is very similar to $\hat{\omega}^2$. However, it gives a more biased estimate than $\hat{\omega}^2$, and the biased estimate is usually larger than the true size of the effect. Nevertheless, it is quite easy to calculate, has been around longer than $\hat{\omega}^2$, and is still commonly used. Hence, we have included a discussion of it here. The equation for computing η^2 is given by

$$\eta^2 = \frac{SS_{between}}{SS_{total}} \qquad \textit{Conceptual and computational equation}$$

example

Stress Experiment

This time, let's compute η^2 for the data of the stress experiment. As previously mentioned, $SS_{between} = 203.333$, and $SS_{total} = 257.333$. Computing the value of η^2 for these data, we obtain

MENTORING TIP

Caution: compute η^2 to 3-decimal-place accuracy, since this proportion is often converted to a percentage.

$$\eta^2 = \frac{SS_{between}}{SS_{total}} = \frac{203.333}{257.333} = 0.790$$

Based on η^2, the stress situations account for 0.790 or 79.0% of the variance in corticosterone levels. According to Cohen's criteria (see Table 15.5), this value of η^2 also indicates a large effect. Note, however, that the value of η^2 is larger than the value obtained for $\hat{\omega}^2$, even though both were calculated on the same data. Because $\hat{\omega}^2$ provides a more accurate estimate of the size of effect, we recommend its use over η^2.

POWER OF THE ANALYSIS OF VARIANCE

The power of the analysis of variance is affected by the same variables and in the same manner as was the case with the t test for independent groups. You will recall that for the t test for independence groups, power is affected as follows:

1. Power varies directly with N. Increasing N increases power.
2. Power varies directly with the size of the real effect of the independent variable. The power of the t test to detect a real effect is greater for large effects than for smaller ones.
3. Power varies inversely with sample variability. The greater the sample variability is, the lower the power to detect a real effect is.

Let's now look at each of these variables and how they affect the analysis of variance. This discussion is most easily understood by referring to the following equation for F_{obt}, for an experiment involving three groups.

$$F_{obt} = \frac{MS_{between}}{MS_{within}} = \frac{n[(\overline{X}_1 - \overline{X}_G)^2 + (\overline{X}_2 - \overline{X}_G)^2 + (\overline{X}_3 - \overline{X}_G)^2]/2}{(SS_1 + SS_2 + SS_3)/(N - 3)}$$

Power and *N*

Obviously, anything that increases F_{obt} also increases power. As N, the total number of subjects in the experiment, increases, so must n, the number of subjects in each group. Increases in each of these variables results in an increase in F_{obt}. This can be seen as follows. Referring to the F_{obt} equation, as N increases, since it is in the denominator of the equation for MS_{within}, MS_{within} decreases. Since MS_{within} is in the denominator of the F_{obt} equation, F_{obt} increases. Regarding n, since n is in the numerator of the F_{obt} equation and is a multiplier of positive values, increases in n result in an increase in $MS_{between}$. Since $MS_{between}$ is in the numerator of the F_{obt} equation, increases in $MS_{between}$ cause an increase in F_{obt}. As stated earlier, anything that increases F_{obt} also increases power. *Thus, increases in N and n result in increased power.*

Power and the Real Effect of the Independent Variable

The larger the real effect of the independent variable is, the larger will be the values of $(\overline{X}_1 - \overline{X}_G)^2$, $(\overline{X}_2 - \overline{X}_G)^2$, and $(\overline{X}_3 - \overline{X}_G)^2$. Increases in these values produce an increase in $MS_{between}$. Since $MS_{between}$ is in the numerator of the F_{obt} equation, increases in $MS_{between}$ result in an increase in F_{obt}. *Thus, the larger the real effect of the independent variable is, the higher is the power.*

Power and Sample Variability

MENTORING TIP

Summary: power varies directly with *N* and real effect of independent variable, and inversely with *within*-group variability.

SS_1 (the sum of squares of group 1), SS_2 (the sum of squares of group 2), and SS_3 (the sum of squares of group 3) are measures of the variability within each group. Increases in SS_1, SS_2, and SS_3 result in an increase in the within-variance estimate MS_{within}. Since MS_{within} is in the denominator of the F_{obt} equation, increases in MS_{within} result in a decrease in F_{obt}. *Thus, increases in within-group variability result in decreases in power.*

MULTIPLE COMPARISONS

In one-way ANOVA, a significant F value indicates that all the conditions do not have the same effect on the dependent variable. For example, in the stress experiment presented earlier in the chapter, a significant F value was obtained and we concluded that the three situations were not the same in the stress levels they produced. For pedagogical reasons, we stopped the analysis at this conclusion. However, in actual practice, the analysis does not ordinarily end at this point. Usually, we are also interested in determining which of the conditions differ from each other. A significant F value tells us that at least one condition differs from at least one of the others. It is also possible that they are all different or any combination in between may be true.

To determine which conditions differ, multiple comparisons between pairs of group means are usually made. In the remainder of this chapter, we shall discuss two types of comparisons that may be made: *a priori* or *planned* comparisons and *a posteriori* or *post hoc* comparisons.

A Priori, or Planned, Comparisons

A priori, or, as they are often called, *planned* comparisons are specific comparisons that are planned in advance of the experiment and often arise from predictions based on theory and prior research. These comparisons may be directional or nondirectional, depending on the prediction. *A posteriori* or *post hoc* comparisons are not planned before conducting the experiment, and are not based on theory and prior research. Instead, they are done because the investigator looks at the data and then decides which groups to compare, or because the investigator wants to do all possible comparisons of interest to see what the data might reveal. With *planned* comparisons, we do not correct for the higher probability of Type I error that arises due to multiple comparisons, as is done with the *post hoc* methods. This correction, which we shall cover in the next section, in effect makes it harder for the null hypothesis to be rejected. Because *planned* comparisons do not involve correcting for the higher probability of Type I error, *planned* comparisons have higher power than *post hoc* comparisons.

In doing *planned* comparisons, the t test for independent groups is used. We could calculate t_{obt} in the usual way. For example, in comparing conditions 1 and 2, we could use the equation

$$t_{obt} = \frac{\overline{X}_1 - \overline{X}_2}{\sqrt{\left(\dfrac{SS_1 + SS_2}{n_1 + n_2 - 2}\right)\left(\dfrac{1}{n_1} + \dfrac{1}{n_2}\right)}}$$

However, remembering that $(SS_1 + SS_2)/(n_1 + n_2 - 2)$ is an estimate of σ^2 based on the within variance of the two groups, we can use a better estimate since we have three or more groups in the ANOVA experiment. Instead of $(SS_1 + SS_2)/(n_1 + n_2 - 2)$ we can use the within-variance estimate MS_{within}, which is based on all of the groups. Substituting MS_{within} for $(SS_1 + SS_2)/(n_1 + n_2 - 2)$, we arrive at the general t equation for *planned* comparisons.

MENTORING TIP

This equation is just like the t equation for independent groups, except MS_{within} from the ANOVA analysis replaces $\left(\dfrac{SS_1 + SS_2}{n_1 + n_2 - 2}\right)$.

$$t_{obt} = \frac{\overline{X}_1 - \overline{X}_2}{\sqrt{MS_{within}\left(\dfrac{1}{n_1} + \dfrac{1}{n_2}\right)}}$$ *General* t *equation for planned comparisons*

The t equation for independent groups and the general t equation for *planned* comparisons are shown below for comparison purposes.

$$t_{obt} = \frac{\overline{X}_1 - \overline{X}_2}{\sqrt{\left(\dfrac{SS_1 + SS_2}{n_1 + n_2 - 2}\right)\left(\dfrac{1}{n_1} + \dfrac{1}{n_2}\right)}}$$ t *Equation for independent groups*

$$t_{obt} = \frac{\overline{X}_1 - \overline{X}_2}{\sqrt{MS_{within}\left(\dfrac{1}{n_1} + \dfrac{1}{n_2}\right)}}$$ *General* t *equation for* **planned** *comparisons,*

With $n_1 = n_2 = n$, the general t equation for *planned* comparisons becomes

$$t_{obt} = \frac{\overline{X}_1 - \overline{X}_2}{\sqrt{2MS_{within}/n}}$$ **t** *equation for* **planned** *comparisons with equal* **n** *in the two groups*

Let's apply this to the stress experiment presented on p. 410. For convenience, the relevant statistics from that experiment are repeated here: $\overline{X}_1 = 4.00$, $\overline{X}_2 = 8.00$, $\overline{X}_3 = 13.00$, $n = 5$, $MS_{within} = 4.50$.

Suppose we have the *a priori* hypothesis based on theoretical grounds that the effect of Condition 3, *Final Exam*, will result in more stress than either Condition 1 or Condition 2. Therefore, prior to collecting any data, we have planned to compare the scores of Group 3 with those of Group 1 and Group 2, using a directional H_1. Accordingly, we will use $\alpha = 0.05_{1\,tail}$ for the evaluation.

To perform the *planned* comparisons, we first calculate the appropriate t_{obt} values and then compare them with t_{crit}. The calculations are as follows:

Group 1 and Group 3:

$$t_{obt} = \frac{\overline{X}_1 - \overline{X}_3}{\sqrt{2MS_{within}/n}} = \frac{4.00 - 13.00}{\sqrt{2(4.50)/5}} = -6.71$$

Group 2 and Group 3:

$$t_{obt} = \frac{\overline{X}_2 - \overline{X}_3}{\sqrt{2MS_{within}/n}} = \frac{8.00 - 13.00}{\sqrt{2(4.50)/5}} = -3.73$$

Are any of these t values significant? The value of t_{crit} is found from Table D in Appendix D, using the degrees of freedom for MS_{within}. Thus, with df = $df_{within} = N - k$ = 12 and $\alpha = 0.05_{1\,tail}$,

$$t_{crit} = -1.78$$

Both of the obtained t scores have absolute values greater than 1.78, and the direction of mean differences are in the predicted direction. Therefore, we conclude that condition 3 is significantly more stressful than Conditions 1 and 2.

A Posteriori, or Post Hoc, Comparisons

When the comparisons are not planned in advance, we must use an *a posteriori* or *post hoc* test. These comparisons usually arise after the experimenter sees the data and picks groups with mean scores that are far apart, or else they arise from doing all the mean comparisons possible with no theoretical *a priori* basis. Since these comparisons were not planned before the experiment, we must correct for the inflated probability values that occur when doing multiple comparisons, as mentioned in the previous section.

Many methods are available for achieving this correction.* This topic is complex, and it is beyond the scope of this textbook to present all of the methods. Instead, we shall present two commonly accepted methods: a method devised by Tukey called the

---

*For a detailed discussion of these methods, see R. E. Kirk, *Experimental Design*, 3rd ed., Brooks/Cole, Pacific Grove, CA, 1995, p. 144–159.

Tukey HSD (Honestly Significant Difference) test and the Scheffé test. Both of these tests are *post hoc* multiple comparison tests. They both maintain the Type I error rate at α. However, the Tukey HSD test maintains the Type I error rate at α when controlling for *all possible comparisons between pairs of means*, while the Scheffé test maintains the Type I error rate at α when controlling for *all possible comparisons, not just pairwise mean comparisons*. Since the Scheffé test controls for more comparisons, it is less powerful than the Tukey HSD test.

The Tukey Honestly Significant Difference (HSD) Test

The Tukey Honestly Significant Difference test is designed to compare all possible pairs of means while maintaining the Type I error for making the complete set of pair-wise comparisons at α. This test avoids the inflated probability of making a Type I error that would result from making these comparisons if t and the sampling distribution of t were used, by using a new statistic Q and the sampling distribution of Q.

The sampling distribution of Q is also called the Studentized range distribution. It was developed by randomly taking k samples (rather than just two as with the t test) of equal n from the same population and determining the difference between the highest and lowest sample means. The differences were then divided by $\sqrt{MS_{within}/n}$, producing Q distributions that were like the t distributions except that these provide the basis for making multiple comparisons between sample means, not just one comparison as in the t test. The 95th and 99th percentile points for the Q distribution are given in Table G in Appendix D. These values are the critical values of Q for the 0.05 and 0.01 alpha levels. As you might guess, the critical values depend on the number of sample means and the degrees of freedom associated with MS_{within}.

The statistic calculated for the HSD test is Q_{obt}. It is defined by the following equation:

$$Q_{obt} = \frac{\overline{X}_i - \overline{X}_j}{\sqrt{MS_{within}/n}}$$

where $\overline{X}_i$ = larger of the two means being compared
$\overline{X}_j$ = smaller of the two means being compared
MS_{within} = within-groups variance estimate
n = number of subjects in each group

Note that in calculating Q_{obt}, the smaller mean is always subtracted from the larger mean. This always makes Q_{obt} positive. Otherwise, the Q statistic is very much like the t statistic. In fact, it can be shown that $Q = \sqrt{2}t$. In performing the HSD test, we first calculate Q_{obt} for the desired comparisons and then compare Q_{obt} with Q_{crit}, determined from Table G. The decision rule states that

if $Q_{obt} \geq Q_{crit}$, reject H_0. If not, then retain H_0.

To illustrate the use of the HSD test, we shall apply it to the data of the stress experiment. There are two steps in using the HSD test. First, we must calculate Q_{obt} for each comparison and then compare each Q_{obt} value with Q_{crit}. For convenience, the relevant statistics from the stress experiment are repeated here: $\overline{X}_1 = 4.00$, $\overline{X}_2 = 8.00$, $\overline{X}_3 = 13.00$, $n = 5$, $MS_{within} = 4.50$.

MENTORING TIP

The HSD test uses Q sampling distributions instead of t sampling distributions.

STEP 1: Calculating the value of Q_{obt}.

Groups 1 and 2.

$$Q_{obt} = \frac{\overline{X}_2 - \overline{X}_1}{\sqrt{MS_{within}/n}} = \frac{8.00 - 4.00}{\sqrt{4.50/5}} = 4.21$$

Groups 1 and 3.

$$Q_{obt} = \frac{\overline{X}_3 - \overline{X}_1}{\sqrt{MS_{within}/n}} = \frac{13.00 - 4.00}{\sqrt{4.50/5}} = 9.48$$

Groups 2 and 3.

$$Q_{obt} = \frac{\overline{X}_3 - \overline{X}_2}{\sqrt{MS_{within}/n}} = \frac{13.00 - 8.00}{\sqrt{4.50/5}} = 5.27$$

STEP 2: Evaluating Q_{obt}. The next step is to compare the Q_{obt} values with Q_{crit}. The value of Q_{crit} is determined from Table G. To locate the appropriate value, we must know the df, k, and the alpha level. The df are the degrees of freedom associated with MS_{within}. In this experiment, df = df_{within} = 12. As mentioned earlier, k stands for the number of groups in the experiment. In the present experiment, $k = 3$. For this experiment, alpha was set at 0.05. From Table G, with df = 12, $k = 3$, and $\alpha = 0.05$, we obtain

$$Q_{crit} = 3.77$$

Since $Q_{obt} > 3.77$ for each comparison, we reject H_0 in each case and conclude that $\mu_1 \neq \mu_2 \neq \mu_3$. All three conditions differ in stress-inducing value.

The Scheffé Test

The Scheffé test is the most conservative of all the possible *post hoc* tests. It controls Type I error for *doing all possible post hoc comparisons*, not just pair-wise mean comparisons. There are many different *post hoc* comparisons that could be performed. For example, in the stress experiment, it is conceivable that we might want to compare Group 1 with Group 2 & Group 3 combined, or Group 2 with Groups 1 & 3 combined. We might also be interested to test whether the means of the three groups form a linear trend, and so forth. In theory, there are a great many *post hoc* comparisons that could be tested. The Scheffé test limits the probability of making a Type I error to the alpha level for all possible *post hoc* comparisons. Since it controls for all possible comparisons, the Scheffé test is the safest *post hoc* test one can use in protecting against making a Type I error.

Even though the Scheffé test controls for doing all possible *post hoc* comparisons, it is very often used to perform *post hoc* analysis on only pair-wise mean comparisons. Since this allows a good comparison with the Tukey HSD test, we will illustrate the Scheffé test by performing a *post hoc* analysis of all possible pair-wise mean comparisons, as we did with the HSD test. As we shall see, the Scheffé test provides its extra protection against making Type I error by using $df_{between}$, MS_{within} and F_{crit} from the entire ANOVA, rather than from the two groups being compared. Using $df_{between}$, MS_{within}, and F_{crit} from the entire ANOVA makes it harder to reject H_0 for any of the *post hoc* comparisons.

To perform the analysis, the Scheffé test computes F_{obt} for each pair-wise comparison and then evaluates against F_{crit}. The numerator of each F_{obt} is a between-groups variance estimate $MS_{between\ (groups\ i\ and\ j)}$ derived *from the two groups that are being compared*, and the denominator is MS_{within}, the within-groups variance estimate that we calculated in doing the entire ANOVA. To determine $MS_{between\ (groups\ i\ and\ j)}$ for each comparison, $SS_{between\ (groups\ i\ and\ j)}$ for the two groups being compared is divided by the $df_{between}$ we calculated in doing the entire ANOVA. To highlight the fact that the F value the Scheffé test computes is based on the two groups being compared, and the $df_{between}$ and MS_{within} are from the entire ANOVA rather than from two groups, we shall call the F value, "$F_{Scheffé}$" instead of F_{obt}.

In equation form, for groups i and j

$$SS_{between\ (groups\ i\ and\ j)} = \left[\frac{(\Sigma X_i)^2}{n_i} + \frac{(\Sigma X_j)^2}{n_j} \right] - \frac{\left(\overset{\substack{groups \\ i\ and\ j}}{\Sigma} X \right)^2}{n_i + n_j}$$

computational equation
for $SS_{between\ (groups\ i\ and\ j)}$

where ΣX_i = sum of the scores of one of the groups being compared

ΣX_j = sum of the scores of the other group being compared

$\overset{\substack{groups \\ i\ and\ j}}{\Sigma} X$ = sum of the scores of the two groups being compared

$$df_{between\ (entire\ ANOVA)} = k - 1$$

where k = the number of groups in the entire ANOVA

$$MS_{between\ (groups\ i\ and\ j)} = \frac{SS_{between\ (groups\ i\ and\ j)}}{df_{between\ (entire\ ANOVA)}}$$

$$F_{Scheffé} = \frac{MS_{between\ (groups\ i\ and\ j)}}{MS_{within\ (entire\ ANOVA)}}$$

$$F_{crit} = F_{crit\ (entire\ ANOVA)}$$

Next, let's use the stress experiment to see how to do a *post hoc* analysis using the Scheffé test. For convenience, the relevant statistics from that experiment are repeated here: $\Sigma X_1 = 20$, $\Sigma X_2 = 40$, $\Sigma X_3 = 65$, $n_1 = n_2 = n_3 = 5$, $df_{between} = 2$, $MS_{within} = 4.50$.

A. Calculate $F_{Scheffé}$ for each paired comparison.

>   **STEP 1:   Calculate the between-groups sum of squares, $SS_{between\ (groups\ i\ and\ j)}$ for each paired comparison.** In computing $SS_{between\ (groups\ i\ and\ j)}$ remember that the data come *from the two groups being compared*,

MENTORING TIP

The values for $df_{between}$, MS_{within} and F_{crit} are the values taken from the entire ANOVA.

MENTORING TIP

Remember the data used in determining $SS_{between\ (group\ i\ and\ j)}$ come from just the two groups being compared.

and not from the entire ANOVA. For example, in determining $SS_{between\ (groups\ i\ and\ j)}$ for Groups 1 and 2, all the data come from Group 1 and Group 2. Let's now do the calculations.

Groups 1 and 2

$$SS_{between\ (groups\ 1\ and\ 2)} = \left[\frac{(\Sigma X_1)^2}{n_1} + \frac{(\Sigma X_2)^2}{n_2}\right] - \frac{\left(\overset{\overset{\text{groups}}{\text{1 and 2}}}{\sum X}\right)^2}{n_1 + n_2}$$

$$= \left[\frac{(20)^2}{5} + \frac{(40)^2}{5}\right] - \frac{(60)^2}{10}$$

$$= 80 + 320 - 360 = 40.00$$

Groups 1 and 3

$$SS_{between\ (groups\ 1\ and\ 3)} = \left[\frac{(\Sigma X_1)^2}{n_1} + \frac{(\Sigma X_3)^2}{n_3}\right] - \frac{\left(\overset{\overset{\text{groups}}{\text{1 and 3}}}{\sum X}\right)^2}{n_1 + n_3}$$

$$= \left[\frac{(20)^2}{5} + \frac{(65)^2}{5}\right] - \frac{(85)^2}{10}$$

$$= 80 + 845 - 722.50 = 202.50$$

Groups 2 and 3

$$SS_{between\ (groups\ 2\ and\ 3)} = \left[\frac{(\Sigma X_2)^2}{n_2} + \frac{(\Sigma X_3)^2}{n_3}\right] - \frac{\left(\overset{\overset{\text{groups}}{\text{2 and 3}}}{\sum X}\right)^2}{n_2 + n_3}$$

$$= \left[\frac{(40)^2}{5} + \frac{(65)^2}{5}\right] - \frac{(105)^2}{10}$$

$$= 320 + 845 - 1102.50 = 62.50$$

Step 2: Calculate $MS_{between\ (groups\ i\ and\ j)}$ for each paired comparison.
Groups 1 and 2

$$MS_{between\ (groups\ 1\ and\ 2)} = \frac{SS_{between\ (groups\ 1\ and\ 2)}}{df_{between\ (entire\ ANOVA)}} = \frac{40}{2} = 20.00$$

Groups 1 and 3

$$MS_{between\ (groups\ 1\ and\ 3)} = \frac{SS_{between\ (groups\ 1\ and\ 3)}}{df_{between\ (entire\ ANOVA)}} = \frac{202.50}{2} = 101.25$$

Groups 2 and 3

$$MS_{between\ (groups\ 2\ and\ 3)} = \frac{SS_{between\ (groups\ 2\ and\ 3)}}{df_{between\ (entire\ ANOVA)}} = \frac{62.50}{2} = 31.25$$

Step 3. Calculate $F_{\text{Scheffé}}$ for each paired comparison. $F_{\text{Scheffé}}$ for each comparison is formed by dividing each $MS_{between\ (groups\ i\ and\ j)}$ by MS_{within} obtained from the entire ANOVA.

Groups 1 and 2

$$F_{\text{Scheffé}} = \frac{MS_{between\ (groups\ 1\ and\ 2)}}{MS_{within\ (entire\ ANOVA)}} = \frac{20.00}{4.50} = 4.44$$

Groups 1 and 3

$$F_{\text{Scheffé}} = \frac{MS_{between\ (groups\ 1\ and\ 3)}}{MS_{within\ (entire\ ANOVA)}} = \frac{101.25}{4.50} = 22.50$$

Groups 2 and 3

$$F_{\text{Scheffé}} = \frac{MS_{between\ (groups\ 2\ and\ 3)}}{MS_{within\ (entire\ ANOVA)}} = \frac{31.25}{4.50} = 6.94$$

B. Evaluate each $F_{\text{Scheffé}}$ value.

To evaluate each $F_{\text{Scheffé}}$ value, we compare each with the value of F_{crit} from the entire ANOVA analysis. The decision rule is the same, namely,

If $F_{\text{Scheffé}} \geq F_{crit}$, reject H_0; if not, retain H_0.

In the stress experiment,

$$F_{crit} = 3.88$$

Since $F_{\text{Scheffé}} \geq 3.88$ for all three comparisons, we can reject H_0 in each case and conclude that $\mu_1 \neq \mu_2 \neq \mu_3$. The three stress conditions are significantly different from each other. This is the same result that we obtained with the Tukey HSD test. Even though the Scheffé test has lower power than the Tukey HSD test, in this example, it had enough power to result in significance for all three comparisons.

The analysis is summarized in table 15.6.

Now let's do a practice problem.

table 15.6 Summary table for Scheffé test *post hoc* analysis of stress experiment

Groups	$SS_{between}$ (groups i and j)	$df_{between}$ from ANOVA	$MS_{between}$ (groups i and j)	MS_{within} from ANOVA	$F_{\text{Scheffé}}$
1 and 2	40.00	2	20.00	4.50	4.44*
1 and 3	202.50	2	101.25	4.50	22.50*
2 and 3	62.50	2	31.25	4.50	6.94*

*F_{crit} from ANOVA $= 3.88$. Therefore, reject H_0.

Practice Problem 15.2

Using the data of Practice Problem 15.1 (p. 415),

a. Test the planned comparisons that (1) *Lecture* + *Reading* and *Lecture* have different effects and (2) that *Lecture* + *Reading* and *Film* + *Reading* have different effects. Assume there is a good theoretical basis for making these comparisons, but the predicted direction of the effect is unclear. Consequently, use $\alpha = 0.05_{2\ \text{tail}}$ in concluding.

b. Make all possible *post hoc* mean comparisons using the HSD test. Use $\alpha = 0.05$.

c. Make all possible *post hoc* mean comparisons using the Scheffé test. Use $\alpha = 0.05$.

SOLUTION

a. *Planned* **comparisons.** For convenience, the relevant statistics from Practice Problem 15.1 are shown here: $\overline{X}_1 = 84.556$, $\overline{X}_2 = 91.000$, $\overline{X}_3 = 81.556$, $n = 9$, $MS_{within} = 28.352$, $df_{within} = 24$, $k = 3$. The comparisons are as follows:

Lecture **(1)** and *Lecture* + *Reading* **(2)**:

$$t_{obt} = \frac{\overline{X}_2 - \overline{X}_1}{\sqrt{2MS_{within}/n}} = \frac{91.000 - 84.556}{\sqrt{2(28.352)/9}} = 2.57$$

Lecture + *Reading* **(2)** and *Film* + *Reading* **(3)**:

$$t_{obt} = \frac{\overline{X}_2 - \overline{X}_3}{\sqrt{2MS_{within}/n}} = \frac{91.000 - 81.556}{\sqrt{2(28.352)/9}} = 3.76$$

To evaluate these values of t_{obt}, we must determine t_{crit}. From Table D, with $\alpha = 0.05_{2\ \text{tail}}$ and $df = df_{within} = 24$,

$$t_{crit} = \pm 2.064$$

Since $|t_{obt}| > 2.064$ in both comparisons, we reject H_0 in each case and conclude that $\mu_1 \neq \mu_2$ and $\mu_2 \neq \mu_3$. By using *a priori* tests, *Lecture* + *Reading* appears to be the most effective method.

b. *Post hoc* **comparisons using the HSD test.** With the HSD test, Q_{obt} is determined for each comparison and then evaluated against Q_{crit}. The value of Q_{crit} is the same for each comparison and is such that the Type I error probability is maintained at α. For convenience, the relevant statistics from Practice Problem 15.1 are shown here: $\overline{X}_1 = 84.556$, $\overline{X}_2 = 91.000$, $\overline{X}_3 = 81.556$, $n = 9$, $MS_{within} = 28.352$. $df_{within} = 24$, $k = 3$. The calculations for Q_{obt} are as follows:

Lecture **(1)** and *Lecture* + *Reading* **(2)**:

$$Q_{obt} = \frac{\overline{X}_2 - \overline{X}_1}{\sqrt{MS_{within}/n}} = \frac{91.000 - 84.556}{\sqrt{(28.352)/9}} = 3.63$$

(continued)

Lecture (1) and *Film + Reading* (3):

$$Q_{obt} = \frac{\overline{X}_1 - \overline{X}_3}{\sqrt{MS_{within}/n}} = \frac{84.556 - 81.556}{\sqrt{(28.352)/9}} = 1.69$$

Lecture + Reading (2) and *Film + Reading* (3):

$$Q_{obt} = \frac{\overline{X}_2 - \overline{X}_3}{\sqrt{MS_{within}/n}} = \frac{91.000 - 81.566}{\sqrt{(28.352)/9}} = 5.32$$

Next, we must determine Q_{crit}. From Table G, with df = df_{within} = 24, $k = 3$, and $\alpha = 0.05$, we obtain

$$Q_{crit} = 3.53$$

Comparing the three values of Q_{obt} with Q_{crit}, we find that the comparisons between conditions 1 and 2, and between 2 and 3 are significant. For these comparisons, $Q_{obt} > 3.53$. However, the remaining comparison between conditions 1 and 3 is not significant, because $Q_{obt} < 3.53$. Thus, on the basis of the HSD test, we may reject H_0 for the *Lecture* and *Lecture + Reading* comparison and for the *Lecture + Reading* and *Film + Reading* comparison. *Lecture + Reading* appears to be the most effective condition. However, we cannot reject H_0 with regard to the *Lecture* and *Film + Reading* comparison.

c. *Post hoc* **comparisons using the Scheffé test:** To do the Scheffé test we first compute $F_{Scheffé}$ for each paired comparison and then evaluate each $F_{Scheffé}$ against F_{crit}. The value of F_{crit} is the value used in the one-way ANOVA. For convenience, the relevant statistics from Practice Problem 15.1 are presented here: $\Sigma X_1 = 761$, $\Sigma X_2 = 819$, $\Sigma X_3 = 734$, $n_1 = n_2 = n_3 = 9$, $k = 3$, $df_{between} = 2$, $MS_{within} = 28.352$.

1. **Calculate $F_{Scheffé}$ for each paired comparison.**

   Step 1. Calculate the between-groups sum of squares, $SS_{between\ (groups\ i\ and\ j)}$, for each paired comparison.

   Groups 1 and 2:

$$SS_{between\ (groups\ 1\ and\ 2)} = \left[\frac{(\Sigma X_1)^2}{n_1} + \frac{(\Sigma X_2)^2}{n_2}\right] - \frac{\left(\overset{\substack{groups\\1\ and\ 2}}{\sum} X\right)^2}{n_1 + n_2}$$

$$= \left[\frac{(761)^2}{9} + \frac{(819)^2}{9}\right] - \frac{(1580)^2}{18} = 186.89$$

   Groups 1 and 3:

$$SS_{between\ (groups\ 1\ and\ 3)} = \left[\frac{(\Sigma X_1)^2}{n_1} + \frac{(\Sigma X_3)^2}{n_3}\right] - \frac{\left(\overset{\substack{groups\\1\ and\ 3}}{\sum} X\right)^2}{n_1 + n_3}$$

$$= \left[\frac{(761)^2}{9} + \frac{(734)^2}{9}\right] - \frac{(1495)^2}{18} = 40.50$$

Groups 2 and 3:

$$SS_{between\ (groups\ 2\ and\ 3)} = \left[\frac{(\Sigma X_2)^2}{n_2} + \frac{(\Sigma X_3)^2}{n_3} \right] - \frac{\left(\overset{\overset{\text{groups}}{\text{2 and 3}}}{\sum} X \right)^2}{n_2 + n_3}$$

$$= \left[\frac{(819)^2}{9} + \frac{(734)^2}{9} \right] - \frac{(1553)^2}{18} = 401.39$$

Step 2. Calculate $MS_{between\ (groups\ i\ and\ j)}$ for each paired comparison.

Groups 1 and 2:

$$MS_{between\ (groups\ 1\ and\ 2)} = \frac{SS_{between\ (groups\ 1\ and\ 2)}}{df_{between\ (entire\ ANOVA)}} = \frac{186.89}{2} = 93.44$$

Groups 1 and 3:

$$MS_{between\ (groups\ 1\ and\ 3)} = \frac{SS_{between\ (groups\ 1\ and\ 3)}}{df_{between\ (entire\ ANOVA)}} = \frac{40.50}{2} = 20.25$$

Groups 2 and 3:

$$MS_{between\ (groups\ 2\ and\ 3)} = \frac{SS_{between\ (groups\ 2\ and\ 3)}}{df_{between\ (entire\ ANOVA)}} = \frac{410.39}{2} = 200.69$$

Step 3. Calculate $F_{\text{Scheffé}}$ for each paired comparison. $F_{\text{Scheffé}}$ for each comparison is formed by dividing each $MS_{between\ (groups\ i\ and\ j)}$ by MS_{within} obtained from the entire ANOVA.

Groups 1 and 2:

$$F_{\text{Scheffé}} = \frac{MS_{between\ (groups\ 1\ and\ 2)}}{MS_{within\ (entire\ ANOVA)}} = \frac{93.44}{28.35} = 3.30$$

Groups 1 and 3:

$$F_{\text{Scheffé}} = \frac{MS_{between\ (groups\ 1\ and\ 3)}}{MS_{within\ (entire\ ANOVA)}} = \frac{20.25}{28.35} = 0.71$$

Groups 2 and 3:

$$F_{\text{Scheffé}} = \frac{MS_{between\ (groups\ 2\ and\ 3)}}{MS_{within\ (entire\ ANOVA)}} = \frac{200.69}{28.35} = 7.08$$

2. Evaluate each $F_{\text{Scheffé}}$ value.

To evaluate each $F_{\text{Scheffé}}$ value, we compare each with the value of F_{crit} from the entire one-way ANOVA analysis. The decision rule is

If $F_{\text{Scheffé}} \geq F_{\text{crit}}$, reject H_0; if not, retain H_0.

(continued)

In this experiment,

$$F_{crit} = 3.40$$

Comparing each $F_{Scheffé}$ value with F_{crit}, we find that $F_{Scheffé} \geq 3.40$ only for the Groups 2 and 3 comparison. For this comparison, we reject H_0 and conclude that there is a real difference between the scores in the *Lecture* and those in the *Lecture + Reading* conditions. Since $F_{Scheffé} < 3.40$ for the other two comparisons, we must retain H_0 for these comparisons. Please note that this is not the same result that we obtained with the Tukey HSD test. Because the Tukey HSD test is more powerful than the Scheffé test, we were able to reject H_0 for both the *Lecture and Lecture + Reading* comparison and for the *Lecture + Reading* and *Film + Reading* comparison. Due to loss of power, the Scheffé test was not able to reject H_0 for the *Lecture* and *Lecture + Reading* comparison. The analysis is summarized in table 15.7.

table **15.7** Summary table for Scheffé test *post hoc* analysis of Practice Problem 15.2

Groups	$SS_{between}$ (groups i and j)	$df_{between}$ from ANOVA	$MS_{between}$ (groups i and j)	MS_{within} from ANOVA	$F_{Scheffé}$
1 and 2	186.89	2	93.44	28.35	3.30
1 and 3	40.50	2	20.25	28.35	0.71
2 and 3	401.39	2	200.69	28.35	7.08*

*F_{crit} from ANOVA = 3.40. Therefore, reject H_0 for the groups 2 and 3 comparison; retain H_0 for the other two comparisons.

MENTORING TIP

Planned comparisons are the most powerful of the multiple comparison tests.

Comparison Between *Planned* Comparisons, the Tukey HSD Test, and the Scheffé Test

Table 15.8, shown here, summarizes the multiple comparisons that were done on the data of the stress experiment. This table presents the value of the statistic computed for each comparison and the one- or two-tailed probability of getting that value of the statistic or one even more extreme, assuming chance alone is at work. You will recall that we called this probability the *obtained probability* in Chapter 10, p. 252, when discussing the sign test. If the obtained probability $\leq \alpha$, then we reject H_0; if not, we retain H_0.

table **15.8** Summary of Multiple Comparisons for the Stress Experiment

Group Comparisons	*Planned* Comparisons — *t* test		*Post Hoc* Comparisons — Tukey HSD test		Scheffé test	
	t	Obtained Probability	*Q*	Obtained Probability	*F*	Obtained Probability
1 and 2			4.21	0.029	4.44	0.036
1 and 3	−6.71	0.00001	9.48	0.00006	22.50	0.00009
2 and 3	−3.73	0.001	5.27	0.008	6.94	0.010

Obviously, the higher the obtained probability, the lower is the chance of rejecting H_0, and hence the lower the power. From the obtained probabilities shown in Table 15.8 we can see that the obtained probabilities for the Tukey HSD test range from 6 to 8 times higher than those of the *planned* comparisons, for the same comparisons. The table also shows that the obtained probabilities for the Scheffé test range from 1.2 to 1.5 times higher than those of the Tukey HSD test, for the same comparisons. This is an illustration of the general principle that *planned* comparisons are more powerful than *post-hoc* comparisons, and within the category of *post-hoc* comparisons, the Tukey HSD test is more powerful than the Scheffé test.

Since *planned* comparisons do not correct for an increased probability of making a Type I error, they are more powerful than either of the *post hoc* tests we have discussed. Moreover, *planned* comparisons can be directional, additionally increasing their power over *post hoc* tests, which by their very nature must be nondirectional. Because of its greater power, *planned* comparisons is the method of choice when applicable. It is important to note, however, that the *planned* comparisons should be relatively few and should flow meaningfully and logically from the experimental design.

If one is doing only *post hoc*, pair-wise comparisons between group means, deciding between the Tukey HSD and the Scheffé tests really depends on one's research philosophy. If interest centers on getting the most power while at the same time reasonably controlling Type I error, then the Tukey HSD test is preferable to the Scheffé test. Because it controls Type I error only for pair-wise mean comparisons, the HSD test is more powerful than the Scheffé test. Controlling Type I error for all possible comparisons, rather than just for all pair-wise mean comparisons, reduces the power of the Scheffé test. For this reason, if interest is limited to doing only pair-wise mean comparisons, I recommend using the Tukey HSD test over the Scheffé test. Of course, if one is interested in doing other than *post hoc* pair-wise mean comparisons, the Scheffé test is the only appropriate test considered here.

WHAT IS THE TRUTH?   Much Ado About Almost Nothing

In a magazine advertisement placed by Rawlings Golf Company, Rawlings claims to have developed a new golf ball that travels a greater distance. The ball is called Tony Penna DB (DB stands for distance ball). To its credit, Rawlings not only offered terms such as *high rebound core, Surlyn cover, centrifugal action,* and so forth to explain why it is reasonable to believe that its ball would travel farther but also

hired a consumer testing institute to conduct an experiment to determine whether, in fact, the Tony Penna DB ball does travel farther. In this experiment, six different brands of balls were evaluated. Fifty-one golfers each hit 18 new balls (3 of each brand) off a driving tee with a driver. The mean distance traveled for each ball was reported as follows:

1. Tony Penna DB — 254.57 yd
2. Titleist Pro Trajectory — 252.50 yd
3. Wilson Pro Staff — 249.24 yd
4. Titleist DT — 249.16 yd
5. Spalding Top-Flite — 247.12 yd
6. Dunlop Blue Max — 244.22 yd

Although no inference testing was reported, the ad concludes, "as you can see, while we can't promise you 250 yards off the tee, we can offer you a competitive edge, if only a yard or two. But an edge is an edge." Since you are by now thoroughly grounded in inferential statistics, how do you respond to this ad?

(continued)

WHAT IS THE TRUTH? *(continued)*

Answer First, I think you should commend the company on conducting evaluative research that compares its product with competitors' on a very important dependent variable. It is to be further commended in engaging an impartial organization to conduct the research. Finally, it is to be commended for reporting the results and calling readers' attention to the fact that the differences between balls are quite small (although, admittedly, the wording of the ad tries to achieve a somewhat different result).

A major criticism of this ad (and an old friend by now) is that we have not been told whether these results are statistically significant. Without establishing this point, the most reasonable explanation of the differences may be "chance." Of course, if chance is the correct explanation, then using the Tony Penna DB ball

won't even give you a yard or two advantage! Before we can take the superiority claim seriously, the manufacturer must report that the differences were statistically significant. Without this statement, as a general rule, I believe we should assume the differences were tested and were not significant, in which case chance alone remains a reasonable explanation of the data. (By the way, what inference test would you have used? Did you answer ANOVA? Nice going!)

For the sake of my second point, let's say the appropriate inference testing has been done, and the data are statistically significant. We still need to ask: "*So what? Even if the results are statistically significant, is the size of the effect worth bothering about?*" Regarding the difference in yardage between the first two brands, I think the answer is "no." Even if I were an avid golfer, I fail to see how a yard or two would make

any practical difference in my golf game. In all likelihood, my 18-hole score would not change by even one stroke, regardless of which ball I used. On the other hand, if I had been using a Dunlop Blue Max ball, these results would cause me to try one of the top two brands. Regarding the third-, fourth-, and fifth-place brands, a reasonable person could go either way. If there were no difference in cost, I think I would switch to one of the first two brands, on a trial basis.

In summary, there are two points I have tried to make. The first is that *product claims of superiority based on sample data should report whether the results are statistically significant.* The second is that "*statistical significance*" and "*importance*" *are different issues.* Once statistical significance has been established, we must look at the size of the effect to see if it is large enough to warrant changing our behavior. ■

■ SUMMARY

In this chapter, I discussed the F test and the analysis of variance. The F test is fundamentally the ratio of two independent variance estimates of the same population variance, σ^2. The F distribution is a family of curves that varies with degrees of freedom. Since F_{obt} is a ratio, there are two values for degrees of freedom: one for the numerator and one for the denominator. The F distribution (1) is positively skewed, (2) has no negative values, and (3) has a median approximately equal to 1, depending on the ns of the estimates.

The analysis of variance technique is used in conjunction with experiments involving more than two independent groups. Basically, it allows the means of the various groups to be compared in one overall evaluation, thus avoiding the inflated probability of making a Type I error when doing many t tests. In the one-way analysis of variance, the total variability of the data (SS_{total}) is partitioned into two parts: the variability that exists within each group, called the within-groups sum of squares (SS_{within}), and the variability that exists between the groups, called the between-groups sum of squares ($SS_{between}$). Each sum of squares is used to form an independent estimate of the variance of the null-hypothesis populations. Finally, an F ratio is calculated, where the between-groups variance estimate ($MS_{between}$) is in the numerator and the within-groups variance estimate (MS_{within}) is in the denominator. Since the between-groups variance estimate increases with the effect of the independent variable and the within-groups variance estimate remains constant, the larger the F ratio, the more unreasonable the null hypothesis becomes. We evaluate F_{obt} by comparing it with F_{crit}. If $F_{obt} \geq F_{crit}$, we reject the null hypothesis and conclude that at least one of the conditions differs significantly from at least one of the other conditions.

Next, I discussed the assumptions underlying the analysis of variance. There are two assumptions: (1) The populations from which the samples are drawn should be normal, and (2) there should be homogeneity of variance. The F test is robust with regard to violations of normality and homogeneity of variance.

After discussing assumptions, I presented two methods for estimating the size of effect of the independent variable. One of the statistics computed to measure size of effect in the one-way, independent groups ANOVA is omega squared ($\hat{\omega}^2$). The other is eta squared (η^2). Conceptually, $\hat{\omega}^2$ and η^2 are like r^2 in that each provides an estimate of the proportion of the total variability of

Y that is accounted for by X. The larger the proportion, the larger is the size of the effect. $\hat{\omega}^2$ gives a relatively unbiased estimate of this proportion in the population, whereas the estimate provided by η^2 is more biased. In addition to explaining how to compute $\hat{\omega}^2$ and η^2, criteria were given to determine if the computed size of effect was small, medium, or large.

Next, I presented a section on the power of the analysis of variance. As with the t test, power of the ANOVA varies directly with N and the size of the real effect and varies inversely with the sample variability.

Finally, I presented a section on multiple comparisons. In experiments using the ANOVA technique, a significant F value indicates that the conditions are not all equal in their effects. To determine which conditions differ from each other, multiple comparisons between pairs of group means are usually performed. There are two approaches to doing multiple comparisons: *a priori*, or *planned*, comparisons and *a posteriori*, or *post hoc*, comparisons.

In the *a priori* approach, there are between-groups comparisons that have been planned in advance of collecting the data. These may be done in the usual way, regardless of whether the obtained F value is significant, by calculating t_{obt} for the two groups and evaluating t_{obt} by comparing it with t_{crit}. In conducting the analysis, we use the within-groups variance estimate calculated in doing the analysis of variance. Since this estimate is based on more groups than the two-group estimate used in the ordinary t test, it is more accurate. There is no correction necessary for *a priori* multiple comparisons.

A posteriori, or *post hoc*, comparisons were not planned before conducting the experiment and arise after looking at the data. As a result, we must be very careful about Type I error considerations. *Post hoc* comparisons must be made with a method that corrects for the inflated Type I error probability. Many methods do this.

For *post hoc* comparisons, I described Tukey's HSD test and the Scheffé test. The HDS test maintains the Type I error rate at α for making all possible comparisons between pairs of sample means. The Scheffé test maintains the Type I error rate at α for making all possible comparisons. The HSD test controls for Type I error by using the Q or Studentized range statistic. As with the t test, Q_{obt} is calculated for each comparison and evaluated against Q_{crit} determined from the sampling distribution of Q. If $Q_{obt} \geq Q_{crit}$, the null hypothesis is rejected. The Scheffé test uses the F statistic and the sampling distribution of

F. For pair-wise mean comparisons, it computes $MS_{between}$ for the two groups being compared using $df_{between}$ from the overall ANOVA and $F_{Scheffé}$ for each pair-wise comparison. $F_{Scheffé}$ is computed by dividing $MS_{between}$ for the two groups being compared with MS_{within} obtained from the one-way ANOVA. Each $F_{Scheffé}$ is evaluated by comparing it with F_{crit} from the overall ANOVA. The Scheffé test controls for Type I error by using $df_{between}$, MS_{within}, and F_{crit} from the one-way ANOVA, rather than from the two groups being compared.

■ IMPORTANT NEW TERMS

A posteriori comparisons (p. 423)
A priori comparisons (p. 422)
Analysis of variance (p. 405)
Between-groups sum of squares
   ($SS_{between}$) (p. 406, 409)
Between-groups variance estimate
   ($MS_{between}$) (p. 406, 408)
Eta squared (η^2) (p. 420)
F test (p. 402)
F_{crit} (p. 403, 413)
$F_{Scheffé}$ (p. 426)

Grand mean ($\overline{X}_G$) (p. 408)
Omega squared ($\hat{\omega}^2$) (p. 419)
One-way analysis of variance,
   independent groups design
   (p. 405)
Planned comparisons (p. 422)
Post hoc comparisons (p. 423)
Q_{crit} (p. 424)
Q_{obt} (p. 424)
Sampling distribution of *F* (p. 402)
Scheffé test (p. 425)

Simple randomized-group design
   (p. 406)
Single factor experiment, independent groups design (p. 405)
Total variability (SS_{total}) (p. 406,
   412)
Tukey HSD test (p. 424)
Within-groups sum of squares
   (SS_{within}) (p. 406, 407)
Within-groups variance estimate
   (MS_{within}) (p. 406)

■ QUESTIONS AND PROBLEMS

1. Identify or define the terms in the Important New Terms section.
2. What are the characteristics of the *F* distribution?
3. What advantages are there in doing experiments with more than two groups or conditions?
4. When doing an experiment with many groups, what is the problem with doing *t* tests between all possible groups without any correction? Why does use of the analysis of variance avoid that problem?
5. The analysis of variance technique analyzes the variability of the data. Yet a significant *F* value indicates that there is at least one significant mean difference between the conditions. How does analyzing the variability of the data allow conclusions about the means of the conditions?
6. What are the steps in forming an *F* ratio in using the one-way analysis of variance technique?
7. In the analysis of variance, if F_{obt} is less than 1, we don't even need to compare it with F_{crit}. It is obvious that the independent variable has not had a significant effect. Why is this so?

8. What are the assumptions underlying the analysis of variance?
9. The analysis of variance is a nondirectional technique, yet it uses a one-tailed evaluation. Is this statement correct? Explain.
10. Find F_{crit} for the following situations:
    a. df(numerator) = 2, df(denominator) = 16, $\alpha = 0.05$
    b. df(numerator) = 3, df(denominator) = 36, $\alpha = 0.05$
    c. df(numerator) = 3, df(denominator) = 36, $\alpha = 0.01$
    What happens to F_{crit} as the degrees of freedom increase and alpha is held constant? What happens to F_{crit} when the degrees of freedom are held constant and alpha is made more stringent?
11. In Chapter 14, Practice Problem 14.2, an independent groups experiment was conducted to investigate whether lesions of the thalamus decrease pain perception. $\alpha = 0.05_{1\ tail}$ was used in the analysis. The data are again presented here. Scores are pain threshold (milliamps) to electric shock. Higher scores indicate decreased pain perception.

Neutral Area Lesions	Thalamic Lesions
0.8	1.9
0.7	1.8
1.2	1.6
0.5	1.2
0.4	1.0
0.9	0.9
1.4	1.7
1.1	

Using these data, verify that $F = t^2$ when there are just two groups in the independent groups experiment.

12. What are the variables that affect the power of the one-way analysis of variance technique?

13. For each of the variables identified in Question 12, state how power is affected if the variable is increased. Use the equation for F_{obt} on p. 421 to justify your answer.

14. Explain why we must correct for doing multiple comparisons when doing *post hoc* comparisons.

15. How do planned comparisons, *post hoc* comparisons using the Tukey HSD test, and *post hoc* comparisons using the Scheffé test differ with regard to
    a. Power? Explain.
    b. The probability of making a Type I error? Explain.

16. What are the Q or Studentized range distributions? How do they avoid the problem of inflated Type I errors that result from doing multiple comparisons with the t distribution?

17. In doing planned comparisons, it is better to use MS_{within} from the ANOVA rather than the weighted variance estimate from the two groups being compared. Is this statement correct? Why?

18. What three factors does the Scheffé test use to make it more difficult to reject H_0?

19. The accompanying table is a one-way, independent groups ANOVA summary table with part of the material missing.

Source	SS	df	MS	F_{obt}
Between groups	1253.68	3		
Within groups				
Total	5016.40	39		

a. Fill in the missing values.
b. How many groups are there in the experiment?

c. Assuming an equal number of subjects in each group, how many subjects are there in each group?
d. What is the value of F_{crit}, using $\alpha = 0.05$?
e. Is there a significant effect?

20. Assume you are a nutritionist who has been asked to determine whether there is a difference in sugar content among the three leading brands of breakfast cereal (brands A, B, and C). To assess the amount of sugar in the cereals, you randomly sample six packages of each brand and chemically determine their sugar content. The following grams of sugar were found:

Breakfast Cereal		
A	B	C
1	7	6
2	5	4
3	3	4
3	7	5
2	4	7
6	7	8

a. Using the conceptual equations of the one-way ANOVA, determine whether any of the brands differ in sugar content. Use $\alpha = 0.05$.
b. Same as part **a**, except use the computational equations. Which do you prefer? Why?
c. Do a *post hoc* analysis on each pair of means using the Tukey HSD test with $\alpha = 0.05$ to determine which cereals are different in sugar content.
d. Same as part **c**, but use the Scheffé test.
e. Explain any differences between the results of part **c** and part **d**. health

21. A sleep researcher conducts an experiment to determine whether sleep loss affects the ability to maintain sustained attention. Fifteen individuals are randomly divided into the following three groups of five subjects each: group 1, which gets the normal amount of sleep (7–8 hours); group 2, which is sleep-deprived for 24 hours; and group 3, which is sleep-deprived for 48 hours. All three groups are tested on the same auditory vigilance task. Subjects are presented with half-second tones spaced at irregular intervals over a 1-hour duration. Occasionally, one of the tones is slightly shorter than the rest. The subject's task is to detect the shorter tones.

The following percentages of correct detections were observed:

Normal Sleep	Sleep-Deprived for 24 Hours	Sleep-Deprived for 48 Hours
85	60	60
83	58	48
76	76	38
64	52	47
75	63	50

a. Determine whether there is an overall effect for sleep deprivation, using the conceptual equations of the one-way ANOVA. Use $\alpha = 0.05$.
b. Same as part **a**, except use the computational equations.
c. Which do you prefer? Why?
d. Determine the size of effect, using $\hat{\omega}^2$.
e. Determine the size of effect, using η^2.
f. Explain the difference in answers between part **d** and part **e**.
g. Do a planned comparison between the means of the 48-hour sleep-deprived group and the normal sleep group to see whether these conditions differ in their effect on the ability to maintain sustained attention. Use $\alpha = 0.05_{2\ tail}$. What do you conclude?
h. Do *post hoc* comparisons, comparing each pair of means using the Tukey HSD test and $\alpha = 0.05$. What do you conclude?
i. Same as part **h**, but use the Scheffé test. Compare your answers to parts **h** and **i**. Explain any difference. cognitive

22. To test whether memory changes with age, a researcher conducts an experiment in which there are four groups of six subjects each. The groups differ according to the age of subjects. In group 1, the subjects are each 30 years old; group 2, 40 years old; group 3, 50 years old; and group 4, 60 years old. Assume that the subjects are all in good health and that the groups are matched on other important variables such as years of education, IQ, gender, motivation, and so on. Each subject is shown a series of nonsense syllables (a meaningless combination of three letters such as DAF or FUM) at a rate of one syllable every 4 seconds. The series is shown twice, after which the subjects are asked to write down as many of the syllables as they can remember. The number of syllables remembered by each subject is shown here:

30 Years Old	40 Years Old	50 Years Old	60 Years Old
14	12	17	13
13	15	14	10
15	16	14	7
17	11	9	8
12	12	13	6
10	18	15	9

a. Use the analysis of variance with $\alpha = 0.05$ to determine whether age has an effect on memory.
b. If there is a significant effect in part **a**, determine the size of effect, using $\hat{\omega}^2$.
c. Determine the size of effect, using η^2.
d. Explain the difference in answers between part **b** and part **c**.
e. Using planned comparisons with $\alpha = 0.05_{2\ tail}$, compare the means of the 60-year-old and the 30-year-old groups. What do you conclude?
f. Use the Scheffé test with $\alpha = 0.05$ to compare all possible pairs of means. What do you conclude? cognitive

23. Assume you are employed by a consumer-products rating service and your assignment is to assess car batteries. For this part of your investigation, you want to determine whether there is a difference in useful life among the top-of-the-line car batteries produced by three manufacturers (A, B, and C). To provide the database for your assessment, you randomly sample four batteries from each manufacturer and run them through laboratory tests that allow you to determine the useful life of each battery. The following are the results given in months of useful battery life:

Battery Manufacturer		
A	B	C
56	46	44
57	52	53
55	51	50
59	50	51

a. Use the analysis of variance with $\alpha = 0.05$ to determine whether there is a difference among these three brands of batteries.

b. Suppose you are asked to make a recommendation regarding the batteries based on useful life. Use the Tukey HSD test with $\alpha = 0.05$ to help you with your decision. I/O

24. In Chapter 14, an illustrative experiment involved investigating the effect of hormone X on sexual behavior. Although we presented only two concentrations in that problem, let's assume the experiment actually involved four different concentrations of the hormone. The full data are shown here, where the concentrations are arranged in ascending order; that is, 0 concentration is where there is zero amount of hormone X (this is the placebo group), and concentration 3 represents the highest amount of the hormone:

Concentration of Hormone X			
0	*1*	*2*	*3*
5	4	8	13
6	5	10	10
3	6	12	9
4	4	6	12
7	5	6	12
8	7	7	14
6	7	9	9
5	8	8	13
4	4	7	10
8	8	11	12

a. Using the analysis of variance with $\alpha = 0.05$, determine whether hormone X affects sexual behavior.

b. If there is a real effect, estimate the size of the effect using $\hat{\omega}^2$.

c. Using planned comparisons with $\alpha = 0.05_{2 \text{ tail}}$, compare the mean of concentration 3 with that of concentration 0. What do you conclude?

d. Using the Tukey HSD test with $\alpha = 0.05$, compare all possible pairs of means. What do you conclude? biological

25. A clinical psychologist is interested in evaluating the effectiveness of the following three techniques for treating mild depression: cognitive restructuring, assertiveness training, and an exercise/nutrition

program. Forty undergraduate students suffering from mild depression are randomly sampled from the university counseling center's waiting list and randomly assigned ten each to the three techniques previously mentioned, and the remaining ten to a placebo control group. Treatment is conducted for 10 weeks, after which depression is measured using the Beck Depression Inventory. The post-treatment depression scores are given here. Higher scores indicate greater depression.

Treatment			
Placebo	*Cognitive Restructuring*	*Assertiveness Training*	*Exercise/ nutrition*
27	10	16	26
16	8	18	24
18	14	12	17
26	16	15	23
18	18	9	25
28	8	13	22
25	12	17	16
20	14	20	15
24	9	21	18
26	7	19	23

a. What is the overall null hypothesis?

b. Using $\alpha = 0.05$, what do you conclude?

c. Do *post hoc* comparisons, using the Tukey HSD test, with $\alpha = 0.05$. What do you conclude? clinical, health

26. A university researcher knowledgeable in Chinese medicine conducted a study to determine whether acupuncture can help reduce cocaine addiction. In this experiment, 18 cocaine addicts were randomly assigned to one of three groups of 6 addicts per group. One group received 10 weeks of acupuncture treatment in which the acupuncture needles were inserted into points on the outer ear where stimulation is believed to be effective. Another group, a placebo group, had acupuncture needles inserted into points on the ear believed not to be effective. The third group received no acupuncture treatment; instead, addicts in this group received relaxation therapy. All groups also received counseling over the 10-week treatment period. The dependent variable was craving for cocaine as measured by the number

of cocaine urges experienced by each addict in the last week of treatment. The following are the results.

Acupuncture + Counseling	Placebo + Counseling	Relaxation Therapy + Counseling
4	8	12
7	12	7
6	11	9
5	8	6
2	10	11
3	7	6

a. Using $\alpha = 0.05$, what do you conclude?
b. If there is a significant effect, estimate the size of effect, using $\hat{\omega}^2$.
c. This time estimate the size of the effect, using η^2.
d. Explain the difference in answers between part **b** and part **c**. clinical, health

27. An instructor is teaching three sections of Introductory Psychology, each section covering the same material. She has made up a different final exam for each section, but she suspects that one of the versions is more difficult than the other two. She decides to conduct an experiment to evaluate the difficulty of the exams. During the review period, just before finals, she randomly selects five volunteers from each class. Class 1 volunteers are given version 1 of the exam; class 2 volunteers get version 2, and class 3 volunteers receive version 3. Of course, all volunteers are sworn not to reveal any of the exam questions, and also, of course, all of the volunteers will receive a different final exam from the one they took in the experiment. The following are the results.

Exam Version 1	Exam Version 2	Exam Version 3
70	95	88
92	75	76
85	81	84
83	83	93
78	72	77

Using $\alpha = 0.05$, what do you conclude? education

What Is the Truth? Questions
Much Ado About Almost Nothing

a. The golf company reports mean data, but no significance analysis. Do you think it important to do a significance test on these data? Explain *why* or *why not*. In general, do you believe advertisements should include sample data and significance testing? Explain.
b. The *What Is the Truth?* section says that "statistical inference" and "importance" are different issues. Are they really different issues? Discuss.

■ SPSS ILLUSTRATIVE EXAMPLE 15.1

The general operation of SPSS and data entry are discussed in Appendix E, *Introduction to SPSS*. As it did with the *t* test, in doing a one-way ANOVA, SPSS computes **F** (the same as our F_{obt}) and the probability of getting **F or a value more extreme** if chance alone is at work. It calls this probability **Sig.** Again, using **Sig.** to denote a probability value is a bit confusing. However, that is the term SPSS chose, so we are stuck with it. Remember:

Sig. = the probability of getting F_{obt} or a value more extreme, assuming chance.

The decision rule we will use to evaluate H_0 and H_1 is as follows.

If **Sig.** $\leq \alpha$, reject H_0 and affirm H_1.
If **Sig.** $> \alpha$, retain H_0; cannot affirm H_1.

example

For this illustrative example, we'll use the stress experiment described in Chapter 15 of the textbook, p. 410. For convenience, this experiment is repeated here.

Suppose you are interested in determining whether certain situations produce differing amounts of stress. You know the amount of hormone corticosterone circulating in the blood is a good measure of how stressed a person is. You randomly assign 15 students into three groups of 5 each. The students in group 1 have their corticosterone levels measured immediately after returning from vacations (low stress). The students in group 2 have their corticosterone levels measured after they have been in class for a week (moderate stress). The students in group 3 are measured immediately before final exam week (high stress). All measurements are taken at the same time of day. You record the data shown in the table below. Scores are in milligrams of corticosterone per 100 milliliters of blood.

Vacation Group 1	Class Group 2	Final Exam Group 3
2	10	10
3	8	13
7	7	14
2	5	13
6	10	15

What is your conclusion, using $\alpha = 0.05$?

SOLUTION

STEP 1: Enter the Data. The data are entered in the same manner that we did for the *t* test for Independent Groups. The only difference is that in the present experiment there are three groups instead of two.

1. For an independent groups analysis, the scores of each group are stacked vertically, one after the other, in a single column of the Data Editor (like we did for the *t* test for Independent Groups). To enter all the scores in the first column (**VAR00001**) of the Data Editor, do as follows:
   a. First, enter the scores of Group 1 in the first column (**VAR00001**) of the Data Editor, beginning with the first Group 1 score in the first cell of the first column.
   b. Next, enter the scores of Group 2 in the first column, directly beneath the last Group 1 score. There should be no spaces or empty cells after the last Group 1 score.
   c. Finally, enter the scores of Group 3 in the first column directly beneath the last Group 2 score. There should be no spaces or empty cells after the last Group 2 score. The first column should now contain the scores of all three groups, beginning with the Group 1 scores and ending with the Group 3 scores, with no spaces or empty cells in between any of the scores.
2. In the second column (**VAR00002**), enter a coding number to designate to which group each score belongs. Do this as follows. In the second column (**VAR00002**) enter the number **1** next to each Group 1 score, the number **2** next to each Group 2 score, and the number **3** next to each Group 3 score. These coding numbers identify the group to which each score belongs.

The resulting Data Editor is shown here. For the moment, ignore the names of the variables. We will name the variables in the next step. **Stress** replaces **VAR00001** and **Group** replaces **VAR00002**.

	Stress	Group	v
1	2.00	1.00	
2	3.00	1.00	
3	7.00	1.00	
4	2.00	1.00	
5	6.00	1.00	
6	10.00	2.00	
7	8.00	2.00	
8	7.00	2.00	
9	5.00	2.00	
10	10.00	2.00	
11	10.00	3.00	
12	13.00	3.00	
13	14.00	3.00	
14	13.00	3.00	
15	15.00	3.00	
16			

STEP 2: **Name the Variables.** In this example, we will give the default variables **VAR00001** and **VAR00002** the new names of **Stress** and **Group**, respectively. To do so,

1. **Click** the **Variable View** tab in the lower left corner of the Data Editor.	This displays the **Variable View** on screen, with **VAR00001** and **VAR00002** displayed in the first and second cells of the **Name** column, respectively.
2. **Click VAR00001**; then **type Stress** in the highlighted cell and then **press Enter**.	**Stress** is entered as the variable name, replacing **VAR00001**. The cursor then moves to the next cell, highlighting **VAR00002**.
3. **Replace VAR00002** with **Group** and then **press Enter**.	**Group** is entered as the variable name, replacing **VAR00002**.

STEP 3: **Analyze the Data.** To do a one-way ANOVA on the stress scores,

1. **Click Analyze**; then **select Compare Means**; then **click One-Way ANOVA....**	This produces the **One-Way ANOVA** dialog box with **Stress** highlighted.
2. **Click** the **top arrow** for the **Dependent List:** box.	This moves **Stress** into the **Dependent List:** box. This identifies **Stress** as the dependent variable.
3. **Click Group** in the large box on the left; then **click** the **bottom arrow** for the **Factor:** box.	This moves **Group** into the **Factor:** box. This identifies **Group** as the independent variable, or **Factor,** as SPSS calls it.

4. Click the Options button on the right.

This produces the **One-Way ANOVA: Options** dialog box. This and the next two steps are not really necessary for the overall analysis. They instruct SPSS to compute some descriptive statistics. While not necessary, the descriptive statistics are often useful, so I recommend you include these two steps.

5. Click Descriptive.

This puts a **check** in the **Descriptive** box, telling SPSS to compute default descriptive statistics and include them in the output.

6. Click Continue.

This returns you to the **One-Way ANOVA** dialog box. SPSS is now ready to do the analysis once it gets the **OK** command.

7. Click OK.

SPSS analyzes the **Stress** data and displays the results.

Analysis Results

The results are displayed in two tables, the **Descriptives** and **ANOVA** tables. These are shown here.

Descriptives

Stress

	N	Mean	Std. Deviation	Std. Error	95% Confidence Interval for Mean		Minimum	Maximum
					Lower Bound	Upper Bound		
1.00	5	4.0000	2.34521	1.04881	1.0880	6.9120	2.00	7.00
2.00	5	8.0000	2.12132	.94868	5.3660	10.6340	5.00	10.00
3.00	5	13.0000	1.87083	.83666	10.6771	15.3229	10.00	15.00
Total	15	8.3333	4.28730	1.10698	5.9591	10.7076	2.00	15.00

ANOVA

Stress

	Sum of Squares	df	Mean Square	F	Sig.
Between Groups	203.333	2	101.667	22.593	.000
Within Groups	54.000	12	4.500		
Total	257.333	14			

The **Descriptives** table gives descriptive statistics on the groups. The **ANOVA** table should be familiar to you. It is the ANOVA summary table given in the textbook, with the exception that SPSS adds a column at the end labeled "Sig." (for significance). This is the table we use to conclude about the overall effect of the independent variable. It shows that **F = 22.593** and **Sig. = .000** (p = .000). Since **.000 < 0.05**, our conclusion is to reject H_0. The three situations are not all the same in the stress levels they produce. Please note that the value SPSS gives for **F** is the same as F_{obt} computed in the textbook. The conclusions are also the same.

■ SPSS ADDITIONAL PROBLEMS

1. Use SPSS to analyze the data of Chapter 15, Problem 21, p. 437, of the textbook.
   Name the scores *Pct_Corr,* and name the grouping variable *Group.*
2. A computer monitor manufacturer is interested in determining the effects of various display foreground and background colors on visual clarity. An experiment is conducted in which 60 subjects are randomly divided into four groups of 15 subjects each. Each group is exposed to a different combination of foreground and background colors. The combinations used are the ones the manufacturer believes will promote the greatest clarity. All subjects are given an acuity test via the computer screen using the combination they received. The following visual acuity results were obtained. The higher the score, the higher is the visual acuity.

Group 1	75	85	71	77	87	67	89	63	73	76	92	86	68	65	72
Group 2	73	74	81	70	95	85	68	67	85	69	78	73	87	75	71
Group 3	94	78	93	92	82	90	69	86	75	94	98	69	76	84	76
Group 4	81	83	77	91	84	67	82	90	65	67	78	88	66	77	69

   What is your conclusion, using $\alpha = 0.05$? Name the scores *Acuity,* and name the grouping variable *Group.*
3. The sets of scores shown in the table here are random numbers obtained from Table J, the table of random numbers, in Appendix G.

Group 1	8	2	7	5	2	6
Group 2	7	5	6	3	1	7
Group 3	0	7	0	2	6	2

   a. Use SPSS to do a one-way, independent groups ANOVA on the data, with $\alpha = 0.05$. What do you conclude?
   b. Add a constant of 4 to each score in Group 1. This is analogous to what concerning the population scores? Reanalyze the data, again using $\alpha = 0.05$. What do you conclude this time? Explain any difference in conclusions between part **a** and part **b**.

■ ONLINE STUDY RESOURCES

CENGAGE**brain**.com

Login to CengageBrain.com to access the resources your instructor has assigned. For this book, you can access the book's companion website for chapter-specific learning tools including Know and Be Able to Do, practice quizzes, flash cards, and glossaries, and a link to Statistics and Research Methods Workshops.

aplia™

If your professor has assigned Aplia homework:

1. Sign in to your account.
2. Complete the corresponding homework exercises as required by your professor.
3. When finished, click "Grade It Now" to see which areas you have mastered and which need more work, and for detailed explanations of every answer.

Visit **www.cengagebrain.com** to access your account and to purchase materials.

16

© Strmko / Dreamstime.com

Introduction to Two-Way Analysis of Variance

CHAPTER OUTLINE

LEARNING OBJECTIVES

After completing this chapter, you should be able to:

- Define factorial experiment, main effect, and interaction effect.
- Correctly label graphs showing no effect and various combinations of main and interaction effects.
- Understand the partitioning of SS_{total} into its four components, the formation of variance estimates, and the formation of the three F ratios.
- Understand the derivation of the row, column, interaction, and the within-cells variance estimates.
- Solve problems involving two-way ANOVA and specify the assumptions underlying this technique.
- Understand the illustrative example, do the practice problems, and understand the solutions.

INTRODUCTION TO TWO-WAY ANOVA–QUALITATIVE PRESENTATION

In Chapter 15, we discussed the most elementary analysis of variance design. We called it the simple randomized-groups design, the one-way analysis of variance, independent groups design, or the single factor experiment, independent groups design. The characteristics of this design are that there is only one independent variable (one factor) that is being investigated, there are two or more levels or conditions of the independent variable, and subjects are randomly assigned to each condition.

Actually, the analysis of variance design is not limited to single factor experiments. In fact, the effect of many different factors may be investigated at the same time in one experiment. Such experiments are called factorial experiments.

definition ■ *A* **factorial experiment** *is one in which the effects of two or more factors or independent variables are assessed in one experiment.*

In a factorial experiment, the conditions or treatments used are combinations of the levels of the factors. For example, in an experiment investigating the effects on sleep on two levels of exercise (light and heavy), carried out at two times of the day (morning and evening), there would be four treatments or conditions. One of the treatments would be the combination of light exercise done in the morning; another would be light exercise done in the evening; a third treatment would be heavy exercise done in the morning, and the remaining treatment would be heavy exercise done in the evening.

The two-way analysis of variance is more complicated than the one-way design. However, we get a lot more information from the two-way design. *The* **two-way analysis of variance** *allows us in one experiment to evaluate the effect of two independent variables and the interaction between them.*

Let's examine this design in more detail, using the example given previously. Suppose a professor in physical education conducts an experiment to compare the effects on nighttime sleep of different intensities of exercise and the time of day when the exercise is done. As mentioned previously, let's assume that there are two levels of exercise (light and heavy) and two times of day (morning and evening). The experiment is depicted diagrammatically in Figure 16.1. From this figure, we can see that there are two factors or independent variables: factor A, which is *time of day*, and factor B, which is *exercise intensity*. Each factor has two levels. Thus, this design is referred to as a 2×2 (read "two by two") design. The first number refers to variable A and tells us there are two levels of variable A. The second number refers to variable B and again tells us there are two levels of variable B. If factor A had three levels, the experiment would be called a 3×2, two-way ANOVA. In a $2 \times 4 \times 3$ design, there would be three factors having two, four, and three levels, respectively. This would be called a three-way ANOVA. Don't worry about how we would assign letters to the three variables. This is quite a complicated design and beyond the scope of this textbook.

In the present example, there are two factors, each having two levels. This results in four cells (conditions or treatments): a_1b_1 (*morning–light exercise*), a_1b_2 (*morning–heavy exercise*), a_2b_1 (*evening–light exercise*), and a_2b_2 (*evening–heavy exercise*). Since this is an independent groups design, subjects would be randomly assigned to each

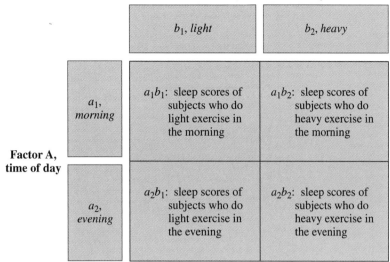

figure 16.1 Schematic diagram of two-way analysis of variance example involving exercise intensity and time of day.

of the cells so that a different group of subjects occupies each cell. Since the levels of each factor were systematically chosen by the experimenter rather than being randomly chosen, this is called a *fixed effects* design.

There are three analyses done in this design. First, we want to determine whether factor A has a significant effect, disregarding the effect of factor B. This is called the **main effect** of factor A. In this experiment, we are interested in determining whether *time of day* makes a difference in the effect of exercise on sleep, disregarding the effect of *exercise intensity*. Second, we want to determine whether factor B has a significant effect, without considering the effect of factor A. This is called the **main effect** of factor B. For this experiment, we are interested in determining whether the *exercise intensity* makes a difference in sleep activity, disregarding the effect of *time of day*. Finally, we want to determine whether there is an interaction between factors A and B. This is called the **interaction effect** of factors A and B. In the present experiment, we want to determine whether there is an interaction between *time of day* and *exercise intensity* in their effect on sleep.

definitions   ■   *The effect of factor A (averaged over the levels of factor B) and the effect of factor B (averaged over the levels of factor A) are called* **main effects**. *An* **interaction effect** *occurs when the effect of one factor is not the same at all levels of the other factor.*

Figure 16.2 shows some possible outcomes of this experiment. In part (a), there are no significant effects. In part (b), there is a significant main effect for *time of day* but no effect for *exercise intensity* and no interaction. Thus, the subjects get significantly more sleep if the exercise is done in the morning rather than in the evening. However, it doesn't seem to matter if the exercise is light or heavy. In part (c), there is a significant main effect for *exercise intensity* but no effect for *time of day* and no interaction.

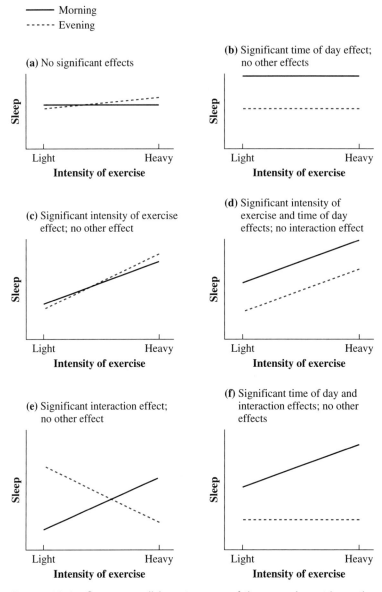

figure 16.2 Some possible outcomes of the experiment investigating the effects of intensity of exercise and time of day.

In this example, heavy exercise results in significantly more sleep than light exercise, and it doesn't matter whether the exercise is done in the morning or evening; the effect appears to be the same. Part (d) shows a significant main effect for *exercise intensity* and *time of day*, with no interaction effect.

Both parts (e) and (f) show significant interaction effects. As stated previously, the essence of an interaction is that the effect of one factor is not the same at all levels of the other factor. This means that, when an interaction occurs between factors A and B, the differences in the dependent variable due to changes in one factor are not the same for each level of the other factor. In part (e), there is a significant interaction effect between *exercise intensity* and time of day. The effect of different intensities of exercise is not the same for all levels of *time of day.* Thus, if the exercise is done in the evening, light exercise results in significantly more sleep than heavy exercise. On the other hand, if the exercise is done in the morning, light exercise results in significantly less sleep than heavy exercise. In part (f), there is a significant main effect for *time of day* and a significant interaction effect. Thus, when the exercise is done in the morning, it results in significantly more sleep than when done in the evening, regardless of whether it is light or heavy exercise. In addition to this main effect, there is an interaction between *exercise intensity* and *time of day.* Thus, there is no difference in the effect of the two intensities when the exercise is done in the evening, but when done in the morning, heavy exercise results in more sleep than light exercise.

In analyzing the data from a two-way analysis of variance design, we determine four variance estimates: $MS_{within\text{-}cells}$, MS_{rows}, $MS_{columns}$, and $MS_{interaction}$. The estimate $MS_{within\text{-}cells}$ is the *within-cells variance estimate* and corresponds to the within-groups variance estimate used in the one-way ANOVA. It becomes the standard against which the other estimates, MS_{rows}, $MS_{columns}$, and $MS_{interaction}$, are compared. The other estimates are sensitive to the effects of the independent variables. The estimate MS_{rows} is called the *row variance estimate.* It is based on the variability of the row means (see Figure 16.1) and, hence, is sensitive to the effects of variable A. The estimate $MS_{columns}$ is called the *column variance estimate.* It is based on the variability of the column means and, hence, is sensitive to the effects of variable B. The estimate $MS_{interaction}$ is the *interaction variance estimate.* It is based on the variability of the cell means and, hence, is sensitive to the interaction effects of variables A and B. If variable A has no effect, MS_{rows} is an independent estimate of σ^2. If variable B has no effect, then $MS_{columns}$ is an independent estimate of σ^2. Finally, if there is no interaction between variables A and B, $MS_{interaction}$ is also an independent estimate of σ^2. Thus, the estimates MS_{rows}, $MS_{columns}$, and $MS_{interaction}$ are analogous to the between-groups variance estimate of the one-way design. To test for significance, three F ratios are formed:

For variable A,
$$F_{\text{obt}} = \frac{MS_{rows}}{MS_{within\text{-}cells}}$$

For variable B,
$$F_{\text{obt}} = \frac{MS_{columns}}{MS_{within\text{-}cells}}$$

MENTORING TIP

This interaction would typically be called an "*A by B*" interaction.

For the interaction between A and B,
$$F_{\text{obt}} = \frac{MS_{interaction}}{MS_{within\text{-}cells}}$$

Each F_{obt} value is evaluated against F_{crit} as in the one-way analysis. For the rows comparison, if $F_{\text{obt}} \geq F_{\text{crit}}$, there is a significant main effect for factor A. If $F_{\text{obt}} \geq F_{\text{crit}}$ for the columns comparison, there is a significant main effect for factor B. Finally, if $F_{\text{obt}} \geq F_{\text{crit}}$ for the interaction comparison, there is a significant interaction effect. Thus, there are many similarities between the one-way and two-way designs. The biggest difference is that, with a two-way design, we can do essentially two one-way experiments, plus we are able to evaluate the interaction between the two independent variables.

Thus far, the two-way analysis of variance, independent groups, fixed effects design has been discussed in a qualitative way. In the remainder of this chapter, we shall present a more detailed, quantitative discussion of the data analysis for this design.

QUANTITATIVE PRESENTATION OF TWO-WAY ANOVA

In the one-way analysis of variance, the total sum of squares is partitioned into two components: the within-groups sum of squares and the between-groups sum of squares. These two components are divided by the appropriate degrees of freedom to form two variance estimates: the within-groups variance estimate, MS_{within} and the between-groups variance estimate $MS_{between}$. If the null hypothesis is correct, then both estimates are estimates of the null-hypothesis population variance (σ^2), and the ratio $MS_{between}/MS_{within}$ will be distributed as F. If the independent variable has a real effect, then $MS_{between}$ will tend to be larger than otherwise and so will the F ratio. Thus, the larger the F ratio is, the more unreasonable the null hypothesis becomes. When $F_{obt} \geq F_{crit}$, we reject H_0 as being too unreasonable to entertain as an explanation of the data.

The situation is quite similar in the two-way analysis of variance. However, in the two-way analysis of variance, we partition the total sum of squares, SS_{total}, into four components: the within-cells sum of squares ($SS_{within-cells}$), the row sum of squares (SS_{rows}), the column sum of squares ($SS_{columns}$), and the interaction sum of squares ($SS_{interaction}$). This partitioning is shown in Figure 16.3. When these sums of squares are divided by the appropriate degrees of freedom, they form four variance estimates. These estimates are the within-cells variance estimate ($MS_{within-cells}$), the row variance estimate (MS_{rows}), the column variance estimate ($MS_{columns}$), and the interaction variance estimate ($MS_{interaction}$). In discussing each of these variance estimates, it will be useful to refer to Figure 16.4, which shows the notation and general layout of data for a two-way analysis of variance, independent groups design. We have assumed in the following discussion that the number of subjects in each cell is the same.

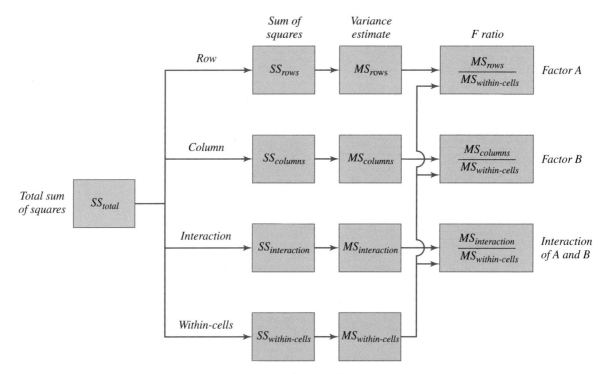

figure 16.3 Overview of two-way analysis of variance technique, independent groups design.

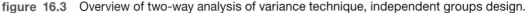

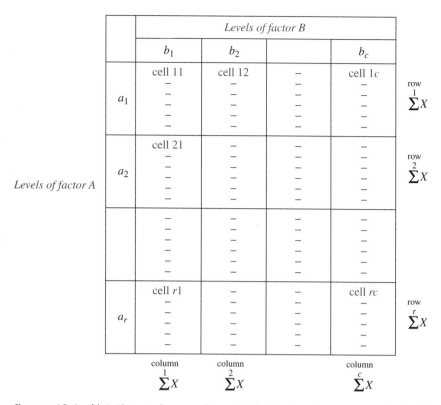

figure 16.4   Notation and general layout of data for a two-way analysis of variance design.

Within-Cells Variance Estimate, $MS_{within\text{-}cells}$

This estimate is derived from the variability of the scores within each cell. Since all the subjects within each cell receive the same level of variables A and B, the variability of their scores cannot be due to treatment differences. The within-cells variance estimate is analogous to the within-groups variance estimate used in the one-way analysis of variance. It is a measure of the inherent variability of the scores within each cell; hence, it is unaffected by the effects of factors A and B or their interaction. Therefore, it gives us an estimate of the null-hypothesis population variance (σ^2) alone. It is the yardstick against which we compare each of the other variance estimates. In equation form,

$$MS_{within\text{-}cells} = \frac{SS_{within\text{-}cells}}{\text{df}_{within\text{-}cells}} \qquad \textit{conceptual equation for within-cells} \\ \textit{variance estimate}$$

where $SS_{within\text{-}cells}$ = within-cells sum of squares
            $\text{df}_{within\text{-}cells}$ = within-cells degrees of freedom

The *within-cells sum of squares* ($SS_{within\text{-}cells}$) is just the sum of squares within each cell added together. Conceptually,

$$SS_{within\text{-}cells} = SS_{11} + SS_{12} + \cdots + SS_{rc} \qquad \textit{conceptual equation for the within-cells} \\ \textit{sum of squares}$$

where SS_{11} = sum of squares for the scores in the cell defined by the intersection of row 1 and column 1

SS_{12} = sum of squares for the scores in the cell defined by the intersection of row 1 and column 2

SS_{rc} = sum of squares for the scores in the cell defined by the intersection of row r and column c; this is the last cell in the matrix

As has been the case so often previously, the conceptual equation is not the best equation to use for computational purposes. The computational equation is given here:

$$SS_{within\text{-}cells} = \overset{\substack{\text{all} \\ \text{scores}}}{\sum} X^2 - \left[\frac{\left(\overset{\substack{\text{cell} \\ 11}}{\sum} X\right)^2 + \left(\overset{\substack{\text{cell} \\ 12}}{\sum} X\right)^2 + \cdots + \left(\overset{\substack{\text{cell} \\ rc}}{\sum} X\right)^2}{n_{\text{cell}}} \right]$$

computational equation for the within-cells sum of squares

Note the similarity of these equations to the comparable equations for the within-groups variance estimate used in the one-way ANOVA. The only difference is that in the two-way ANOVA, summation is with regard to the cells, whereas in the one-way ANOVA, summation is with regard to the groups.

In computing $SS_{within\text{-}cells}$, the number of deviation scores = n_{cell} = n. Therefore, there are $n - 1$ degrees of freedom contributed by each cell. Since we sum over all cells in calculating $SS_{within\text{-}cells}$, the *within-cells degrees of freedom* equal $n - 1$ multiplied by the number of cells. If we let r equal the number of rows and c equal the number of columns, then rc equals the number of cells. Therefore, the within-cells degrees of freedom equal $rc(n - 1)$. Thus,

$$\text{df}_{within\text{-}cells} = rc(n - 1) \qquad \textit{within-cells degrees of freedom}$$

where r = number of rows
c = number of columns

Row Variance Estimate, MS_{rows}

This estimate is based on the differences between the row means. It is analogous to the between-groups variance estimate ($MS_{between}$) in the one-way ANOVA. You will recall that $MS_{between}$ is an estimate of σ^2 plus the effect of the independent variable. Similarly, the row variance estimate (MS_{rows}) in the two-way ANOVA is an estimate of σ^2 plus the effect of factor A. If factor A has no effect, then the population row means are equal ($\mu_{a_1} = \mu_{a_2} = \cdots = \mu_{a_r}$), and the differences between sample row means will just be due to random sampling from identical populations. In this case, MS_{rows} becomes an estimate of just σ^2 alone. If factor A has an effect, then the differences among the row means, and hence MS_{rows}, will tend to be larger than otherwise. In equation form,

$$MS_{rows} = \frac{SS_{rows}}{\text{df}_{rows}} \qquad \textit{conceptual equation for the row variance estimate}$$

where SS_{rows} = row sum of squares
df_{rows} = row degrees of freedom

The *row sum of squares* is very similar to the between-groups sum of squares in the one-way ANOVA. The only difference is that with the row sum of squares we use the row means, whereas the between-groups sum of squares used the group means.

The conceptual equation for SS_{rows} follows. Note that in computing row means, all the scores in a given row are combined and averaged. This is referred to as computing the row means "averaged over the columns" (see Figure 16.4). Thus, the row means are arrived at by averaging over the columns:

MENTORING TIP

Caution: you can't use this equation unless n_{row} is the same for all rows.

$$SS_{rows} = n_{row}\left[(\overline{X}_{row\,1} - \overline{X}_G)^2 + (\overline{X}_{row\,2} - \overline{X}_G)^2 + \cdots + (\overline{X}_{row\,r} - \overline{X}_G)^2 \right]$$

*conceptual equation for
the row sum of squares*

where
$$\overline{X}_{row\,1} = \frac{\sum\limits_{}^{row\,1} X}{n_{row\,1}}$$

$$\overline{X}_{row\,r} = \frac{\sum\limits_{}^{row\,r} X}{n_{row\,r}}$$

$$\overline{X}_G = \text{grand mean}$$

From the conceptual equation, it is easy to see that SS_{rows} increases with the effect of variable A. As the effect of variable A increases, the row means become more widely separated, which in turn causes $(\overline{X}_{row\,1} - \overline{X}_G)^2$, $(\overline{X}_{row\,2} - \overline{X}_G)^2 \ldots (\overline{X}_{row\,r} - \overline{X}_G)^2$, to increase. Since these terms are in the numerator, SS_{rows} increases. Of course, if SS_{rows} increases, so does MS_{rows}.

In calculating SS_{rows}, there are r deviation scores. Thus, the *row degrees of freedom* equal $r - 1$. In equation form,

$$\text{df}_{rows} = r - 1 \qquad \textit{row degrees of freedom}$$

Recall that the between-groups degrees of freedom ($\text{df}_{between}$) $= k - 1$ for the one-way ANOVA. The row degrees of freedom are quite similar except we are using rows rather than groups.

Again, the conceptual equation turns out not to be the best equation to use for computing SS_{rows}. The computational equation is given here:

$$SS_{rows} = \left[\frac{\left(\sum\limits^{row\,1} X\right)^2 + \left(\sum\limits^{row\,2} X\right)^2 + \cdots + \left(\sum\limits^{row\,r} X\right)^2}{n_{row}} \right] - \frac{\left(\sum\limits^{all\,scores} X\right)^2}{N}$$

*computational equation for the
row sum of squares*

Column Variance Estimate, $MS_{columns}$

This estimate is based on the differences between the column means. It is exactly the same as MS_{rows}, except that it uses the column means rather than the row means. Since factor B affects the column means, the column variance estimate ($MS_{columns}$) is an estimate of σ^2 plus the effects of factor B. If the levels of factor B have no differential effect, then the population column means are equal ($\mu_{b_1} = \mu_{b_2} = \mu_{b_3} = \ldots = \mu_{b_c}$)

and the differences between the sample column means are due to random sampling from identical populations. In this case, $MS_{columns}$ will be an estimate of σ^2 alone. If factor B has an effect, then the differences among the column means, and hence $MS_{columns}$, will tend to be larger than otherwise.

The equation for $MS_{columns}$ is

$$MS_{columns} = \frac{SS_{columns}}{df_{columns}} \quad \textit{column variance estimate}$$

where $SS_{columns}$ = column sum of squares
              $df_{columns}$ = column degrees of freedom

The *column sum of squares* is also very similar to the row sum of squares. The only difference is that we use the column means in calculating the column sum of squares rather than the row means. The conceptual equation for $SS_{columns}$ is shown here. Note that, in computing the column means, all the scores in a given column are combined and averaged. Thus, the column means are arrived at by averaging over the rows.

MENTORING TIP

Caution: you can't use this equation unless n_{column} is the same for all columns.

$$SS_{columns} = n_{column}\left[(\overline{X}_{column\,1} - \overline{X}_G)^2 + (\overline{X}_{column\,2} - \overline{X}_G)^2 + \cdots + (\overline{X}_{column\,c} - \overline{X}_G)^2 \right]$$

                                                    conceptual equation for the
                                                    column sum of squares

where $\overline{X}_{column\,1} = \dfrac{\overset{column\,1}{\sum} X}{n_{column\,1}}$

                $\overline{X}_{column\,c} = \dfrac{\overset{column\,c}{\sum} X}{n_{column\,c}}$

Again, we can see from the conceptual equation that $SS_{columns}$ increases with the effect of variable B. As the effect of variable B increases, the column means become more widely spaced, which in turn causes $(\overline{X}_{column\,1} - \overline{X}_G)^2$, $(\overline{X}_{column\,2} - \overline{X}_G)^2 \ldots (\overline{X}_{column\,c} - \overline{X}_G)^2$ to increase. Since these terms are in the numerator of the equation for $SS_{columns}$, the result is an increase in $SS_{columns}$. Of course, an increase in $SS_{columns}$ results in an increase in $MS_{columns}$.

Since there are c deviation scores used in calculating $SS_{columns}$, the *column degrees of freedom* equal $c - 1$. Thus,

$$df_{columns} = c - 1 \quad \textit{column degrees of freedom}$$

The computational equation for $SS_{columns}$ is

$$SS_{columns} = \left[\frac{\left(\overset{column\,1}{\sum} X\right)^2 + \left(\overset{column\,2}{\sum} X\right)^2 + \cdots + \left(\overset{column\,c}{\sum} X\right)^2}{n_{column}} \right] - \frac{\left(\overset{all\,scores}{\sum} X\right)^2}{N}$$

                                                    computational equation for the
                                                    column sum of squares

Interaction Variance Estimate, $MS_{interaction}$

Earlier in this chapter, we pointed out that an interaction exists when the effect of one of the variables is not the same at all levels of the other variable. Another way of saying this is that an interaction exists when the effect of the combined action of the variables is different from that which would be predicted by the individual effects of the variables. To illustrate this point, consider Figure 16.2(f), p. 448, where there is an interaction between *time of day* and *exercise intensity*. An interaction exists because the sleep score for heavy exercise done in the morning is higher than would be predicted based on the individual effects of the *time of day* and *exercise intensity* variables. If there were no interaction, then we would expect the lines to be parallel. The *exercise intensity* variable would have the same effect when done in the evening as when done in the morning.

The interaction variance estimate ($MS_{interaction}$) is used to evaluate the interaction of variables A and B. As such, it is based on the differences between the cell means beyond that which is predicted by the individual effects of the two variables. The interaction variance estimate is an estimate of σ^2 plus the interaction of A and B. If there is no interaction and any main effects are removed, then the population cell means are equal ($\mu_{a_1b_1} = \mu_{a_1b_2} = \cdots = \mu_{a_rb_c}$) and differences among cell means must be due to random sampling from identical populations. In this case, $MS_{interaction}$ will be an estimate of σ^2 alone. If there is an interaction between factors A and B, then the differences among the cell means and, hence $MS_{interaction}$, will tend to be higher than otherwise.

The equation for $MS_{interaction}$ is

$$MS_{interaction} = \frac{SS_{interaction}}{df_{interaction}} \qquad \textit{interaction variance estimate}$$

where $SS_{interaction}$ = interaction sum of squares
$df_{interaction}$ = interaction degrees of freedom

The *interaction sum of squares* is equal to the variability of the cell means when the variability due to the individual effects of factors A and B has been removed. Both the conceptual and computational equations are given here:

MENTORING TIP

Caution: you can't use this equation unless n_{cell} is the same for all cells.

$$SS_{interaction} = n_{cell}[(\overline{X}_{cell\ 11} - \overline{X}_G)^2 + (\overline{X}_{cell\ 12} - \overline{X}_G)^2 + \cdots + (\overline{X}_{cell\ rc} - \overline{X}_G)^2]$$
$$- SS_{rows} - SS_{columns}$$

*conceptual equation for the
interaction sum of squares*

$$SS_{interaction} = \left[\frac{\left(\overset{cell\ 11}{\sum} X\right)^2 + \left(\overset{cell\ 12}{\sum} X\right)^2 + \cdots + \left(\overset{cell\ rc}{\sum} X\right)^2}{n_{cell}} \right] - \frac{\left(\overset{all\ scores}{\sum} X\right)^2}{N}$$
$$- SS_{rows} - SS_{columns}$$

*computational equation for the
interaction sum of squares*

The *degrees of freedom* for the interaction variance estimate equal $(r-1)(c-1)$. Thus,

$$df_{interaction} = (r-1)(c-1) \qquad \textit{interaction degrees of freedom}$$

Computing *F* Ratios

Once the variance estimates have been determined, they are used in conjunction with $MS_{within\text{-}cells}$ to form *F* ratios (see Figure 16.3, p 450) to test the main effects of the variables and their interaction. The following three *F* ratios are computed:

To test the main effect of variable A (row effect):

$$F_{\text{obt}} = \frac{MS_{rows}}{MS_{within\text{-}cells}} = \frac{\sigma^2 + \text{effects of variable A}}{\sigma^2}$$

To test the main effect of variable B (column effect):

$$F_{\text{obt}} = \frac{MS_{columns}}{MS_{within\text{-}cells}} = \frac{\sigma^2 + \text{effects of variable B}}{\sigma^2}$$

To test the interaction of variables A and B (interaction effect):

$$F_{\text{obt}} = \frac{MS_{interaction}}{MS_{within\text{-}cells}} = \frac{\sigma^2 + \text{interaction effects of variable A and B}}{\sigma^2}$$

The ratio $MS_{rows}/MS_{within\text{-}cells}$ is used to test the main effect of variable A. If variable A has no main effect, MS_{rows} is an independent estimate of σ^2 and $MS_{rows}/MS_{within\text{-}cells}$ is distributed as *F* with degrees of freedom equal to df_{rows} and $df_{within\text{-}cells}$. If variable A has a main effect, MS_{rows} will be larger than otherwise and the F_{obt} value for rows will increase.

The ratio $MS_{columns}/MS_{within\text{-}cells}$ is used to test the main effect of variable B. If this variable has no main effect, then is an independent estimate of σ^2 and $MS_{columns}/MS_{within\text{-}cells}$ is distributed as *F* with degrees of freedom equal to $df_{columns}$ and $df_{within\text{-}cells}$. If variable B has a main effect, $MS_{columns}$ will be larger than otherwise, causing an increase in the F_{obt} value for columns.

The interaction between A and B is tested using the ratio $MS_{interaction}/MS_{within\text{-}cells}$. If there is no interaction, $MS_{interaction}$ is an independent estimate of σ^2 and $MS_{interaction}/MS_{within\text{-}cells}$ is distributed as *F* with degrees of freedom equal to $df_{interaction}$ and $df_{within\text{-}cells}$. If there is an interaction, $MS_{interaction}$ will be larger than otherwise, causing an increase in the F_{obt} value for interaction.

The main effect of each variable and their interaction are tested by comparing the appropriate F_{obt} value with F_{crit}. F_{crit} is found in Table F in Appendix D, using α and the degrees of freedom of the *F* value being evaluated. The decision rule is the same as with the one-way ANOVA, namely,

If $F_{\text{obt}} \geq F_{\text{crit}}$, reject H_0.   *decision rule for evaluating H_0 in two-way ANOVA*

ANALYZING AN EXPERIMENT WITH TWO-WAY ANOVA

We are now ready to analyze the data from an illustrative example.

experiment

Effect of Exercise on Sleep

Let's assume a professor in physical education conducts an experiment to compare the effects on nighttime sleep of *exercise intensity* and of the *time of day* when the exercise is done. The experiment uses a fixed effects, 3×2 factorial design with independent groups. There are three levels of *exercise intensity* (light, moderate, and heavy) and two levels of *time of day* (morning and evening). Thirty-six college students in good physical condition are randomly assigned to the six cells such that there are six subjects per cell. The subjects who do heavy exercise jog for 3 miles, the subjects who do moderate exercise jog for 1 mile, and the subjects in the light exercise condition jog for only $\frac{1}{4}$ mile. Morning exercise is done at 7:30 A.M., whereas evening exercise is done at 7:00 P.M. Each subject exercises once, and the number of hours slept that night is recorded. The data are shown in Table 16.1.

1. What are the null hypotheses for this experiment?
2. Using $\alpha = 0.05$, what do you conclude?

SOLUTION

1. Null hypotheses:
   a. *For the A variable (main effect)*: The *time of day* when exercise is done does not affect nighttime sleep. The population row means for morning and evening exercise averaged over the different levels of *exercise intensity* are equal ($\mu_{a_1} = \mu_{a_2}$).
   b. *For the B variable (main effect)*: The different levels of *exercise intensity* have the same effect on nighttime sleep. The population column means for light, medium, and heavy *exercise intensity* averaged over *time of day* conditions are equal ($\mu_{b_1} = \mu_{b_2} = \mu_{b_3}$).
   c. *For the interaction between A and B*: There is no interaction between *time of day* and *exercise intensity*. With any main effects removed, the population cell means are equal ($\mu_{a_1b_1} = \mu_{a_1b_2} = \mu_{a_1b_3} = \mu_{a_2b_1} = \mu_{a_2b_2} = \mu_{a_2b_3}$).
2. Conclusion, using $\alpha = 0.05$:
   a. Calculate F_{obt} for each hypothesis.

STEP 1: Calculate the row sum of squares, SS_{rows}. Note that *time of day* is the row variable.

$$
SS_{rows} = \left[\frac{\left(\sum_{}^{\text{row}_1} X\right)^2 + \left(\sum_{}^{\text{row}_2} X\right)^2}{n_{row}} \right] - \frac{\left(\sum_{}^{\substack{\text{all}\\\text{scores}}} X\right)^2}{N}
$$

$$
= \left[\frac{(129.2)^2 + (147.2)^2}{18} \right] - \frac{(276.4)^2}{36} = 9.000
$$

table 16.1   Data from exercise experiment

Time of Day Factor A (row variable)	Exercise Intensity Factor B (column variable)		
	Light (1)	**Moderate (2)**	**Heavy (3)**
Morning (1)	6.5 7.4	7.4 7.3	8.0 7.6	$\Sigma X = 129.20$
	7.3 7.2	6.8 7.6	7.7 6.6	$\Sigma X^2 = 930.50$
	6.6 6.8	6.7 7.4	7.1 7.2	$n = 18$
	$\overline{X} = 6.97$	$\overline{X} = 7.20$	$\overline{X} = 7.37$	$\overline{X} = 7.18$
Evening (2)	7.1 7.7	7.4 8.0	8.2 8.7	$\Sigma X = 147.20$
	7.9 7.5	8.1 7.6	8.5 9.6	$\Sigma X^2 = 1212.68$
	8.2 7.6	8.2 8.0	9.5 9.4	$n = 18$
	$\overline{X} = 7.67$	$\overline{X} = 7.88$	$\overline{X} = 8.98$	$\overline{X} = 8.18$
	$\Sigma X = 87.80$	$\Sigma X = 90.50$	$\Sigma X = 98.10$	$\Sigma X = 276.40$
	$\Sigma X^2 = 645.30$	$\Sigma X^2 = 685.07$	$\Sigma X^2 = 812.81$	$\Sigma X^2 = 2143.18$
	$n = 12$	$n = 12$	$n = 12$	$N = 36$
	$\overline{X} = 7.32$	$\overline{X} = 7.54$	$\overline{X} = 8.18$	

STEP 2: Calculate the column sum of squares, $SS_{columns}$. Note that *exercise intensity* is the column variable.

$$SS_{column} = \left[\frac{\left(\overset{\text{column}}{\underset{1}{\sum}}X\right)^2 + \left(\overset{\text{column}}{\underset{2}{\sum}}X\right)^2 + \left(\overset{\text{column}}{\underset{3}{\sum}}X\right)^2}{n_{\text{column}}}\right] - \frac{\left(\overset{\text{all}}{\underset{\text{scores}}{\sum}}X\right)^2}{N}$$

$$= \left[\frac{(87.8)^2 + (90.5)^2 + (98.1)^2}{12}\right] - \frac{(276.4)^2}{36} = 4.754$$

STEP 3: Calculate the interaction sum of squares, $SS_{interaction}$.

$$SS_{interaction} = \left[\frac{\left(\overset{\text{cell}}{\underset{11}{\sum}}X\right)^2 + \left(\overset{\text{cell}}{\underset{12}{\sum}}X\right)^2 + \left(\overset{\text{cell}}{\underset{13}{\sum}}X\right)^2 + \left(\overset{\text{cell}}{\underset{21}{\sum}}X\right)^2 + \left(\overset{\text{cell}}{\underset{22}{\sum}}X\right)^2 + \left(\overset{\text{cell}}{\underset{23}{\sum}}X\right)^2}{n_{\text{cell}}}\right]$$

$$- \frac{\left(\overset{\text{all}}{\underset{\text{scores}}{\sum}}X\right)^2}{N} - SS_{rows} - SS_{columns}$$

$$= \left[\frac{(41.8)^2 + (43.2)^2 + (44.2)^2 + (46.0)^2 + (47.3)^2 + (53.9)^2}{6}\right] - \frac{(276.4)^2}{36}$$

$$- 9.000 - 4.754$$

$$= 1.712$$

STEP 4: Calculate the within-cells sum of squares, $SS_{within\text{-}cells}$.

$$SS_{within\text{-}cells} = \overset{\text{all}}{\underset{\text{scores}}{\sum}}X^2$$

$$- \left[\frac{\left(\overset{\text{cell}}{\underset{11}{\sum}}X\right)^2 + \left(\overset{\text{cell}}{\underset{12}{\sum}}X\right)^2 + \left(\overset{\text{cell}}{\underset{13}{\sum}}X\right)^2 + \left(\overset{\text{cell}}{\underset{21}{\sum}}X\right)^2 + \left(\overset{\text{cell}}{\underset{22}{\sum}}X\right)^2 + \left(\overset{\text{cell}}{\underset{23}{\sum}}X\right)^2}{n_{\text{cell}}}\right]$$

$$= 2143.18 - \left[\frac{(41.8)^2 + (43.2)^2 + (44.2)^2 + (46.0)^2 + (47.3)^2 + (53.9)^2}{6}\right]$$

$$= 5.577$$

MENTORING TIP

Again, this step is a check on the previous calculations. It is not necessary to do this step for the analysis.

STEP 5: Calculate the total sum of squares, SS_{total}. This step is a check to be sure the previous calculations are correct. Once we calculate SS_{total}, we can use the following equation to check the other calculations:

$$SS_{total} = SS_{rows} + SS_{columns} + SS_{interaction} + SS_{within\text{-}cells}$$

First, we must independently calculate SS_{total}.

$$SS_{total} = \overset{\text{all}}{\underset{\text{scores}}{\sum}}X^2 - \frac{\left(\overset{\text{all}}{\underset{\text{scores}}{\sum}}X\right)^2}{N}$$

$$= 2143.18 - \frac{(276.4)^2}{36} = 21.042$$

Substituting the obtained values of SS_{total}, SS_{rows}, $SS_{columns}$, $SS_{interaction}$, and $SS_{within\text{-}cells}$ into the partitioning equation for SS_{total}, we obtain

$$SS_{total} = SS_{rows} + SS_{columns} + SS_{interaction} + SS_{within\text{-}cells}$$

$$21.042 \cong 9.000 + 4.754 + 1.712 + 5.577$$

$$21.042 \cong 21.043$$

The equation checks within rounding accuracy. Therefore, we can assume our calculations up to this point are correct.

STEP 6: Calculate the degrees of freedom for each variance estimate.

$$\mathrm{df}_{rows} = r - 1 = 2 - 1 = 1$$

$$\mathrm{df}_{columns} = c - 1 = 3 - 1 = 2$$

$$\mathrm{df}_{interaction} = (r-1)(c-1) = (1)2 = 2$$

$$\mathrm{df}_{within\text{-}cells} = rc(n_{cell} - 1) = 2(3)(5) = 30$$

$$\mathrm{df}_{total} = N - 1 = 35$$

Note that

$$\mathrm{df}_{total} = \mathrm{df}_{rows} + \mathrm{df}_{columns} + \mathrm{df}_{interaction} + \mathrm{df}_{within\text{-}cells}$$

$$35 = 1 + 2 + 2 + 30$$

$$35 = 35$$

STEP 7: Calculate the variance estimates MS_{rows}, $MS_{columns}$, $MS_{interaction}$, and $MS_{within\text{-}cells}$. Each variance estimate is equal to the sum of squares divided by the appropriate degrees of freedom. Thus,

$$\text{Row variance estimate} = MS_{rows} = \frac{SS_{rows}}{\mathrm{df}_{rows}} = \frac{9.000}{1} = 9.000$$

$$\text{Column variance estimate} = MS_{columns} = \frac{SS_{columns}}{\mathrm{df}_{columns}} = \frac{4.754}{2} = 2.377$$

$$\text{Interaction variance estimate} = MS_{interaction} = \frac{SS_{interaction}}{\mathrm{df}_{interaction}} = \frac{1.712}{2} = 0.856$$

$$\text{Within-cells variance estimate} = MS_{within\text{-}cells} = \frac{SS_{within\text{-}cells}}{\mathrm{df}_{within\text{-}cells}} = \frac{5.577}{30} = 0.186$$

STEP 8: a. Calculate the F ratios: For the row effect,

$$F_{\mathrm{obt}} = \frac{MS_{rows}}{MS_{within\text{-}cells}} = \frac{9.000}{0.186} = 48.42$$

For the column effect,

$$F_{\mathrm{obt}} = \frac{MS_{columns}}{MS_{within\text{-}cells}} = \frac{2.377}{0.186} = 12.78$$

For the interaction effect,

$$F_{\mathrm{obt}} = \frac{MS_{interaction}}{MS_{within\text{-}cells}} = \frac{0.856}{0.186} = 4.60$$

b. Evaluate the F_{obt} values.

For the row effect: From Table F, with $\alpha = 0.05$, $df_{numerator} = df_{rows} = 1$, and $df_{denominator} = df_{within\text{-}cells} = 30$, $F_{crit} = 4.17$. Since F_{obt} (48.42) > 4.17, we reject H_0 with respect to the A variable, which in this experiment is *time of day*. There is a significant main effect for *time of day*.

For the column effect: From Table F, with $\alpha = 0.05$, $df_{numerator} = df_{columns} = 2$, and $df_{denominator} = df_{within\text{-}cells} = 30$, $F_{crit} = 3.32$. Since F_{obt} (12.78) > 3.32, we reject H_0 with respect to the B variable, which in this experiment is *exercise intensity*. There is a significant main effect for *exercise intensity*.

For the interaction effect: From Table F, with $\alpha = 0.05$, $df_{numerator} = df_{interaction} = 2$, and $df_{denominator} = df_{within\text{-}cells} = 30$, $F_{crit} = 3.32$. Since F_{obt} (4.60) > 3.32, we reject H_0 regarding the interaction of variables A and B. There is a significant interaction between *time of day* and *exercise intensity*.

The analysis is summarized in Table 16.2.

table **16.2** Summary ANOVA table for exercise and time of day experiment

Source	SS	df	MS	F_{obt}	F_{crit}
Rows (*time of day*)	9.000	1	9.000	48.42*	4.17
Columns (*exercise intensity*)	4.754	2	2.377	12.78*	3.32
Interaction	1.712	2	0.856	4.60*	3.32
Within-cells	5.577	30	0.186		
Total	21.042	35			

*Since $F_{obt} > F_{crit}$, H_0 is rejected.

Interpreting the Results

In the preceding analysis, we have rejected the null hypothesis for both the row and column effects. A significant effect for rows indicates that variable A has had a significant main effect. The differences between the row means, averaged over the columns, were too great to attribute to random sampling from populations where $\mu_{a_1} = \mu_{a_2}$. In the present experiment, the significant row effect indicates that there was a significant main effect for the *time of day* factor. The differences between the means for the *time of day* levels averaged over the *exercise intensity* levels were too great to attribute to chance. We have plotted the mean of each cell in Figure 16.5. From this figure, it can be seen that evening exercise resulted in greater sleep than morning exercise.

A significant effect for columns indicates that variable B has had a significant main effect—that the differences among the column means, computed by averaging over the rows, were too great to attribute to random sampling from the null-hypothesis population. In the present experiment, the significant effect for columns tells us that the differences between the means of the three *exercise intensity* levels computed by averaging over the *time of day* levels were too great to attribute to random sampling fluctuations. From Figure 16.5, it can be seen that the effect of

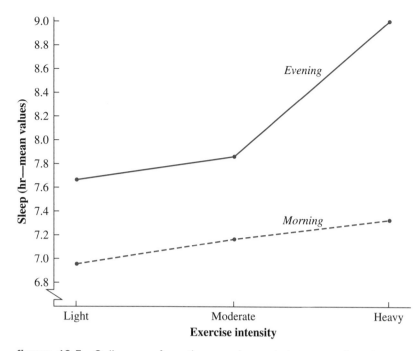

figure 16.5 Cell means from the exercise and sleep experiment.

MENTORING TIP

This is spoken of as an *exercise intensity by time of day* interaction.

increasing the intensity of exercise averaged over the *time of day* levels was to increase the amount of sleep.

The results of this experiment also showed a significant interaction effect. As discussed previously, a significant interaction effect indicates that the effects of one of the variables are not the same at all the levels of the other factor. Plotting the mean for each cell is particularly helpful for interpreting an interaction effect. From Figure 16.5, we can see that the increase in the amount of sleep is about the same in going from light to moderate exercise, whether the exercise is done in the morning or evening. However, the difference in the amount of sleep in going from moderate to heavy exercise varies depending on whether the exercise is done in the morning or evening. Heavy exercise results in a much greater increase in the amount of sleep when the exercise is done in the evening than when it is done in the morning.

MENTORING TIP

If there is a significant interaction effect, caution must be exercised in interpreting main effects. If there is a significant interaction effect, a main effect might be due solely to the independent variable having an effect at only one level of the other variable, and no effect at the other levels.

When there is a significant interaction effect, care needs to be taken when interpreting main effects. Without a significant interaction effect, it is usually assumed that a main effect indicates that the independent variable has a significant effect at each level of the other variable, and that the effect is uniform over the levels. However, if there is a significant interaction effect, it is possible that the main effect is not uniform over all the values of the other independent variable. It is even possible that the entire main effect is due to the interaction effect.

Graphing the data helps interpret what a significant main effect means when there is a significant interaction effect. For example, to interpret the significant main effect for *time of day*, it is useful to plot the sleep scores separately for morning and evening exercise, plotting *exercise intensity* on the X axis as is done in Figure 16.5. Referring to this figure, it doesn't appear that there is an interaction at light and moderate *exercise intensity*. At these levels the morning and evening lines are parallel. However, at the heavy *exercise intensity* level, the difference in mean values between the evening and morning groups is much greater than you would expect based on the mean value differences at the other two levels. It is this "higher than expected" mean value for the heavy exercise–evening group that caused the greater difference, which in turn caused the lines to diverge. It is this "higher than expected" mean value that is responsible for the significant interaction effect. It is also possible that the greatly elevated mean value for the heavy exercise–evening group is responsible for producing the significant main effect for *time of day*, that the *time of day* effect is significant only at the heavy *exercise intensity* level, and that mean differences at the other two levels of *exercise intensity* are not significant at all. If so, then the significant main effect for *time of day* would be due solely to its effect at the heavy level of *exercise intensity*. This is hardly an outcome that would be consistent with the interpretation ordinarily given to a significant main effect, namely, of a uniform effect that is significant at all levels of the other variable.

To resolve these issues, it is necessary to statistically compare mean differences at each level of the other variable. Analyzing these mean differences falls within the topic of *multiple comparisons*. Conceptually this topic is similar to that discussed in conjunction with one-way ANOVA, but it is too complicated to be discussed here (see the section in this chapter titled *Multiple Comparisons*, p. 471). However, given the significant interaction effect in the present experiment, without graphing the data and doing these comparisons, it would be premature to conclude that the significant main effect for *time of day* indicates that the effect is uniform or that the effect is significant at all levels of *exercise intensity*.

Let's do another problem for practice.

Practice Problem 16.1

A statistics professor conducts an experiment to compare the effectiveness of two methods of teaching his course. Method I is the usual way he teaches the course: lectures, homework assignments, and a final exam. Method II is the same as Method I, except that students receiving Method II get 1 additional hour per week in which they solve illustrative problems under the guidance of the professor. The professor is also interested in how the methods affect students of differing mathematical abilities, so 30 volunteers for the experiment are subdivided according to mathematical ability into superior, average, and poor groups of 10 each. Five students from each mathematics ability group are randomly assigned to Method I and the remaining 5 students from each group to Method II. This random assignment results in 15 students receiving Teaching Method 1 and another 15 students receiving Method II. At the end of the course, all 30 students take the same final exam. The following final exam scores resulted:

Mathematical Ability Factor A (row variable)	Teaching Method Factor B (column variable)			
	Method I (1)		Method II (2)	
Superior (1)	39*	41	49	47
	48	42	47	48
	44		43	
Average (2)	43	36	38	46
	40	35	45	44
	42		42	
Poor (3)	30	33	37	41
	29	36	34	33
	37		40	

*Scores are the number of points received out of a total of 50 possible points.

a. What are the null hypotheses for this experiment?
b. Using $\alpha = 0.05$, what do you conclude?

SOLUTION

a. Null hypotheses:
1. *For the A variable (main effect)*: The three levels of *mathematical ability* do not differentially affect final exam scores in this course. The population row means for the three levels of *mathematical ability* averaged over *teaching method* are equal ($\mu_{a_1} = \mu_{a_2} = \mu_{a_3}$).
2. *For the B variable (main effect)*: The two levels of *teaching method* are equal in their effects on final exam scores in this course. The population column means for teaching methods I and II averaged over the three levels of mathematical ability are equal ($\mu_{b_1} = \mu_{b_2}$).
3. *For the interaction between variables A and B:* There is no interaction effect between *mathematical ability* and teaching method. With any main effects removed, the population cell means are equal ($\mu_{a_1b_1} = \mu_{a_1b_2} = \mu_{a_2b_1} = \mu_{a_2b_2} = \mu_{a_3b_1} = \mu_{a_3b_2}$).

b. Conclusion, using $\alpha = 0.05$:
1. Calculating F_{obt}:

STEP 1: Calculate SS_{rows}.

$$SS_{rows} = \left[\frac{\left(\overset{\text{row}_1}{\sum X} \right)^2 + \left(\overset{\text{row}_2}{\sum X} \right)^2 + \left(\overset{\text{row}_3}{\sum X} \right)^2}{n_{\text{row}}} \right] - \frac{\left(\overset{\text{all scores}}{\sum X} \right)^2}{N}$$

$$= \left[\frac{(448)^2 + (411)^2 + (350)^2}{10} \right] - \frac{(1209)^2}{30} = 489.800$$

(continued)

STEP 2: Calculate $SS_{columns}$.

$$SS_{columns} = \left[\frac{\left(\overset{column}{\underset{1}{\sum}}X\right)^2 + \left(\overset{column}{\underset{2}{\sum}}X\right)^2}{n_{column}}\right] - \frac{\left(\overset{all\ scores}{\sum}X\right)^2}{N}$$

$$= \left[\frac{(575)^2 + (634)^2}{15}\right] - \frac{(1209)^2}{30} = 116.033$$

STEP 3: Calculate $SS_{interaction}$.

$$SS_{interaction} = \left[\frac{\left(\overset{cell}{\underset{11}{\sum}}X\right)^2 + \left(\overset{cell}{\underset{12}{\sum}}X\right)^2 + \left(\overset{cell}{\underset{21}{\sum}}X\right)^2 + \left(\overset{cell}{\underset{22}{\sum}}X\right)^2 + \left(\overset{cell}{\underset{31}{\sum}}X\right)^2 + \left(\overset{cell}{\underset{32}{\sum}}X\right)^2}{n_{cell}}\right]$$

$$- \frac{\left(\overset{all\ scores}{\sum}X\right)^2}{N} - SS_{rows} - SS_{columns}$$

$$= \left[\frac{(214)^2 + (234)^2 + (196)^2 + (215)^2 + (165)^2 + (185)^2}{5}\right] - \frac{(1209)^2}{30}$$

$$- 498.8 - 116.083 = 49,328.6 - 48,722.7 - 489.8 - 116.033 = 0.067$$

STEP 4: Calculate $SS_{within\text{-}cells}$.

$$SS_{within\text{-}cells} = \overset{all\ scores}{\sum}X^2$$

$$- \left[\frac{\left(\overset{cell}{\underset{11}{\sum}}X\right)^2 + \left(\overset{cell}{\underset{12}{\sum}}X\right)^2 + \left(\overset{cell}{\underset{21}{\sum}}X\right)^2 + \left(\overset{cell}{\underset{22}{\sum}}X\right)^2 + \left(\overset{cell}{\underset{31}{\sum}}X\right)^2 + \left(\overset{cell}{\underset{32}{\sum}}X\right)^2}{n_{cell}}\right]$$

$$= 49,587 - \left[\frac{(214)^2 + (234)^2 + (196)^2 + (215)^2 + (165)^2 + (185)^2}{5}\right]$$

$$= 49,587 - 49,328.6 = 258.4$$

STEP 5: Calculate SS_{total}. This step is to check the previous calculations.

$$SS_{total} = \overset{all\ scores}{\sum}X^2 - \frac{\left(\overset{all\ scores}{\sum}X\right)^2}{N}$$

$$= 49,587 - \frac{(1209)^2}{30} = 864.3$$

$$SS_{total} = SS_{rows} + SS_{columns} + SS_{interaction} + SS_{within\text{-}cells}$$
$$864.3 = 489.8 + 116.033 + 0.067 + 258.4$$
$$864.3 = 864.3$$

Since the partitioning equation checks, we can assume our calculations thus far are correct.

STEP 6: Calculate df.

$$df_{rows} = r - 1 = 3 - 1 = 2$$
$$df_{columns} = c - 1 = 2 - 1 = 1$$
$$df_{interaction} = (r - 1)(c - 1) = (3 - 1)(2 - 1) = 2$$
$$df_{within\text{-}cells} = rc(n_{cell} - 1) = 6(4) = 24$$
$$df_{total} = N - 1 = 29$$

STEP 7: Calculate MS_{rows}, $MS_{columns}$, $MS_{interaction}$, and $MS_{within\text{-}cells}$.

$$MS_{rows} = \frac{SS_{rows}}{df_{rows}} = \frac{489.8}{2} = 244.9$$

$$MS_{columns} = \frac{SS_{columns}}{df_{columns}} = \frac{116.033}{1} = 116.033$$

$$MS_{interaction} = \frac{SS_{interaction}}{df_{interaction}} = \frac{0.067}{2} = 0.034$$

$$MS_{within\text{-}cells} = \frac{SS_{within\text{-}cells}}{df_{within\text{-}cells}} = \frac{258.4}{24} = 10.767$$

STEP 8: Calculate F_{obt}. For the row effect,

$$F_{obt} = \frac{MS_{rows}}{MS_{within\text{-}cells}} = \frac{244.9}{10.767} = 22.75$$

For the column effect,

$$F_{obt} = \frac{MS_{columns}}{MS_{within\text{-}cells}} = \frac{116.033}{10.767} = 10.78$$

For the interaction effect,

$$F_{obt} = \frac{MS_{interaction}}{MS_{within\text{-}cells}} = \frac{0.034}{10.767} = 0.003$$

2. Evaluate the F_{obt} values.

For the row effect: From Table F, with $\alpha = 0.05$, $df_{numerator} = df_{rows} = 2$, and $df_{denominator} = df_{within\text{-}cells} = 24$, $F_{crit} = 3.40$. Since F_{obt} (22.75) > 3.40, we reject H_0 with respect to the A variable. There is a significant effect for *mathematical ability*.

For the column effect: From Table F, with $\alpha = 0.05$, $df_{numerator} = df_{columns} = 1$, and $df_{denominator} = df_{within\text{-}cells} = 24$, $F_{crit} = 4.26$. Since F_{obt} (10.78) > 4.26, we reject H_0 with respect to the B variable. There is a significant main effect for *teaching method*.

For the interaction effect: Since F_{obt} (0.003) < 1, we retain H_0 and conclude that the data do not support the hypothesis that there is an interaction between *mathematical ability* and *teaching method*.

(continued)

The solution to this problem is summarized in Table 16.3.

table 16.3 Summary ANOVA table for *mathematical ability* and *teaching method* experiment

Source	SS	df	MS	F_{obt}	F_{crit}
Rows (*mathematical ability*)	489.800	2	244.900	22.75*	3.40
Columns (*teaching method*)	116.033	1	116.033	10.78*	4.26
Interaction	0.067	2	0.034	0.003	3.40
Within-cells	258.400	24	10.767		
Total	864.300	29			

*Since $F_{obt} > F_{crit}$, H_0 is rejected.

Interpreting the results of Practice Problem 16.1 In the preceding analysis, we rejected the null hypothesis for both the row and column effects. Rejecting H_0 for rows means that there was a significant main effect for variable A, *mathematical ability*. The differences between the means for the different levels of *mathematical ability* averaged over *teaching method* were too great to attribute to chance. The mean of each cell has been plotted in Figure 16.6. From this figure, it can be seen that increasing the level of *mathematical ability* results in increased final exam scores.

Rejecting H_0 for columns indicates that there was a significant main effect for the B variable, *teaching method*. The difference between the means for Teaching Method I and Teaching Method II averaged over *mathematical ability* was too great to attribute to random sampling fluctuations. From Figure 16.6, we can see that method II was superior to method I.

In this experiment, there was no significant interaction effect. This means that, within the limits of sensitivity of this experiment, the effect of each variable was the

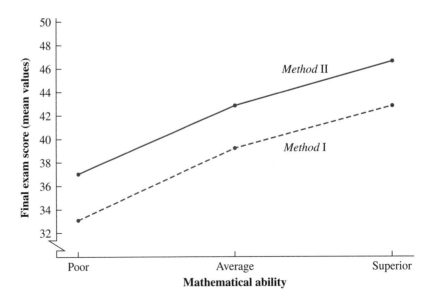

figure 16.6 Cell means from the teaching method and mathematical ability experiment.

same over all levels of the other variable. This can be most clearly seen by viewing Figure 16.6 with regard to variable A. The lack of a significant interaction effect indicates that the effect of different levels of mathematical ability on final exam scores was the same for Teaching Methods I and II. This results in parallel lines when the means of the cells are plotted (see Figure 16.6). *In fact, it is a general rule that, when the lines are parallel in a graph of the individual cell means, there probably is no interaction effect.* For there to be an interaction effect, the lines must diverge significantly from parallel. In this regard, it will be useful to review Figure 16.2 to see whether you can determine which graphs show interaction effects.*

Practice Problem 16.2

A clinical psychologist is interested in the effect that *anxiety level* has on the ability of individuals to learn new material. She is also interested in whether the effect of *anxiety level* depends on the difficulty of the new material. An experiment is conducted in which there are three levels of anxiety (low, medium, and high) and three levels of *material difficulty* (low, medium, and high). Out of a pool of volunteers, 15 low-anxious, 15 medium-anxious, and 15 high-anxious subjects are selected and randomly assigned 5 each to the three *material difficulty* levels. Each subject is given 30 minutes to learn the new material, after which the subjects are tested to determine the amount learned.

The following data are collected:

Difficulty of Material Factor A (row variable)	Anxiety Level* Factor B (column variable)					
	Low (1)		Medium (2)		High (3)	
Low (1)	18	17	18	18	18	17
	20	16	19	15	16	18
	17		17		19	
Medium (2)	18	14	18	17	14	15
	17	16	18	15	17	12
	14		14		16	
High (3)	11	6	15	12	9	8
	10	10	13	11	7	8
	8		12		5	

*Each score is the total points obtained out of a possible 20 points.

a. What are the null hypotheses?
b. Using $\alpha = 0.05$, what do you conclude?

(continued)

---

*Of course, you can't really be sure if the interaction is significant without doing a statistical analysis.

SOLUTION

a. Null hypotheses:

1. *For variable A (main effect):* The null hypothesis states that *material difficulty* has no effect on the amount learned. The population row means for *material difficulty* averaged over *anxiety level* are equal ($\mu_{a_1} = \mu_{a_2} = \mu_{a_3}$).
2. *For variable B (main effect):* The null hypothesis states that *anxiety level* has no effect on the amount learned. The population column means for low, medium, and high anxiety levels averaged over *material difficulty* are equal ($\mu_{b_1} = \mu_{b_2} = \mu_{b_3}$).
3. *For the interaction between variables A and B:* The null hypothesis states that there is no interaction between *material difficulty* and *anxiety level*. With any main effects removed, the population cell means are equal ($\mu_{a_1b_1} = \mu_{a_1b_2} = \ldots = \mu_{a_3b_3}$).

b. Conclusion, using $\alpha = 0.05$:

1. Calculate F_{obt}:

STEP 1: Calculate SS_{rows}.

$$SS_{rows} = \left[\frac{\left(\overset{row\,1}{\sum X} \right)^2 + \left(\overset{row\,2}{\sum X} \right)^2 + \left(\overset{row\,3}{\sum X} \right)^2}{n_{row}} \right] - \frac{\left(\overset{all\,scores}{\sum X} \right)^2}{N}$$

$$= \left[\frac{(263)^2 + (235)^2 + (145)^2}{15} \right] - \frac{(643)^2}{45} = 506.844$$

STEP 2: Calculate $SS_{columns}$.

$$SS_{columns} = \left[\frac{\left(\overset{column\,1}{\sum X} \right)^2 + \left(\overset{column\,2}{\sum X} \right)^2 + \left(\overset{column\,3}{\sum X} \right)^2}{n_{column}} \right] - \frac{\left(\overset{all\,scores}{\sum X} \right)^2}{N}$$

$$= \left[\frac{(212)^2 + (232)^2 + (199)^2}{15} \right] - \frac{(643)^2}{45} = 36.844$$

STEP 3: Calculate $SS_{interaction}$.

$$SS_{interaction} = \left[\frac{\left(\overset{cell\,11}{\sum X} \right)^2 + \left(\overset{cell\,12}{\sum X} \right)^2 + \cdots + \left(\overset{cell\,33}{\sum X} \right)^2}{n_{cell}} \right]$$

$$- \frac{\left(\overset{all\,scores}{\sum X} \right)^2}{N} - SS_{rows} - SS_{columns}$$

$$= \left[\frac{(88)^2 + (87)^2 + (88)^2 + (79)^2 + (82)^2 + (74)^2 + (45)^2 + (63)^2 + (37)^2}{5} \right]$$

$$- \frac{(643)^2}{45} - 506.844 - 36.844 = 40.756$$

STEP 4: Calculate $SS_{within\text{-}cells}$.

$$SS_{within\text{-}cells} = \overset{\overset{\text{all}}{\text{scores}}}{\sum} X^2 - \left[\frac{\left(\overset{\text{cell}}{\overset{11}{\sum}} X \right)^2 + \left(\overset{\text{cell}}{\overset{12}{\sum}} X \right)^2 + \cdots + \left(\overset{\text{cell}}{\overset{33}{\sum}} X \right)^2}{n_{cell}} \right] = 9871 -$$

$$\left[\frac{(88)^2 + (87)^2 + (88)^2 + (79)^2 + (82)^2 + (74)^2 + (45)^2 + (63)^2 + (37)^2}{5} \right]$$

$$= 98.800$$

STEP 5: Calculate SS_{total}. This step is a check on the previous calculations:

$$SS_{total} = \overset{\overset{\text{all}}{\text{scores}}}{\sum} X^2 - \frac{\left(\overset{\overset{\text{all}}{\text{scores}}}{\sum} X \right)^2}{N} = 9871 - \frac{(643)^2}{45} = 683.244$$

$$SS_{total} = SS_{rows} + SS_{columns} + SS_{interaction} + SS_{within\text{-}cells}$$

$$683.244 = 506.844 + 36.844 + 40.756 + 98.800$$

$$683.244 = 683.244$$

Since the partitioning equation checks, we can assume our calculations thus far are correct.

STEP 6: Calculate df.

$$df_{rows} = r - 1 = 3 - 1 = 2$$
$$df_{columns} = c - 1 = 3 - 1 = 2$$
$$df_{interaction} = (r - 1)(c - 1) = 2(2) = 4$$
$$df_{within\text{-}cells} = rc(n_{cell} - 1) = 3(3)(5 - 1) = 36$$
$$df_{total} = N - 1 = 45 - 1 = 44$$

STEP 7: Calculate MS_{rows}, $MS_{columns}$, $MS_{interaction}$, and $MS_{within\text{-}cells}$.

$$MS_{rows} = \frac{SS_{rows}}{df_{rows}} = \frac{506.844}{2} = 253.422$$

$$MS_{columns} = \frac{SS_{columns}}{df_{columns}} = \frac{36.844}{2} = 18.442$$

$$MS_{interaction} = \frac{SS_{interaction}}{df_{interaction}} = \frac{40.756}{4} = 10.189$$

$$MS_{within\text{-}cells} = \frac{SS_{within\text{-}cells}}{df_{within\text{-}cells}} = \frac{98.800}{36} = 2.744$$

STEP 8: Calculate F_{obt}. For the row effect,

$$F_{obt} = \frac{MS_{rows}}{MS_{within\text{-}cells}} = \frac{253.422}{2.744} = 92.34$$

(continued)

For the column effect,

$$F_{\text{obt}} = \frac{MS_{columns}}{MS_{within\text{-}cells}} = \frac{18.442}{2.744} = 6.71$$

For the interaction effect,

$$F_{\text{obt}} = \frac{MS_{interaction}}{MS_{within\text{-}cells}} = \frac{10.189}{2.744} = 3.71$$

Evaluate F_{obt}:

For the row effect: With $\alpha = 0.05$, $df_{numerator} = df_{rows} = 2$, and $df_{denominator} = df_{within\text{-}cells} = 36$, from Table F, $F_{crit} = 3.26$. Since F_{obt} (92.34) > 3.26, we reject H_0 for the A variable. There is a significant main effect for *material difficulty*.

For the column effect: With $\alpha = 0.05$, $df_{numerator} = df_{columns} = 2$, and $df_{denominator} = df_{within\text{-}cells} = 36$, from Table F, $F_{crit} = 3.26$. Since F_{obt} (6.71) > 3.26, H_0 is rejected for the B variable. There is a significant main effect for *anxiety level*.

For the interaction effect: With $\alpha = 0.05$, $df_{numerator} = df_{interaction} = 4$, and $df_{denominator} = df_{within\text{-}cells} = 36$, from Table F, $F_{crit} = 2.63$. Since F_{obt} (3.71) > 2.63, H_0 is rejected. There is a significant interaction between *material difficulty* and *anxiety level*.

The solution is summarized in Table 16.4.

table 16.4   Summary ANOVA table for *anxiety level* and *material difficulty* experiment

Source	SS	df	MS	F_{obt}	F_{crit}
Rows (*material difficulty*)	506.844	2	253.244	92.35*	3.26
Columns (*anxiety level*)	36.844	2	18.442	6.71*	3.26
Interaction	40.756	4	10.189	3.71*	2.63
Within-cells	98.800	36	2.744		
Total	683.244	44			

*Since $F_{\text{obt}} > F_{\text{crit}}$, H_0 is rejected.

Interpreting the results of Practice Problem 16.2   In the preceding analysis, there was a significant main effect for both *material difficulty* and *anxiety level*. The significant main effect for *material difficulty* indicates that the differences among the means for the three difficulty levels averaged over anxiety levels were too great to attribute to chance. The cell means have been plotted in Figure 16.7. From this figure, it can be seen that increasing the difficulty of the material results in lower mean values when the scores are averaged over anxiety levels.

The significant main effect for *anxiety level* is more difficult to interpret. Of course, at the operational level, this main effect tells us that the differences among the means for the three levels of anxiety when averaged over difficulty levels were

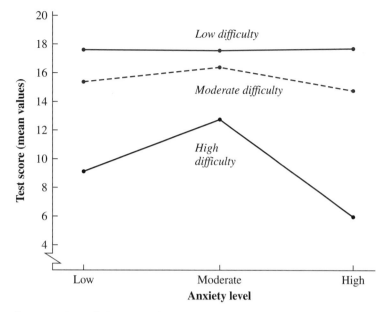

figure 16.7 Cell means from the difficulty of material and anxiety level experiment.

too great to attribute to chance. However, beyond this, the interpretation is not clear because of the interaction between the two variables. From Figure 16.7, we can see that the effect of different anxiety levels depends on the difficulty of the material. At the low level of difficulty, differences in anxiety level seem to have no effect on the test scores. However, for the other two difficulty levels, differences in anxiety levels do affect performance. The interaction is a complicated one such that both low and high levels of anxiety seem to interfere with performance when compared with moderate anxiety. This is an example of the inverted U-shaped curve that occurs frequently in psychology when relating performance and arousal levels.

MULTIPLE COMPARISONS

In the three examples we have just analyzed, we have ended the analyses by evaluating the F_{obt} values. In actual practice, the analysis is usually carried further by doing multiple comparisons on the appropriate pairs of means. For example, in Practice Problem 16.2, there was a significant F_{obt} value for *material difficulty*. The next step ordinarily is to determine which difficulty levels are significantly different from each other. Conceptually, this topic is very similar to that which we presented in Chapter 15 when discussing multiple comparisons in conjunction with the one-way ANOVA. One main difference is that in the two-way ANOVA we are often evaluating pairs of *row* means or *column* means rather than pairs of group means. Further exposition of this topic is beyond the scope of this textbook.*

—————————

*The interested reader should consult B. J. Winer et al., *Statistical Principles in Experimental Design*, 3rd ed., McGraw-Hill, New York, 1991.

ASSUMPTIONS UNDERLYING TWO-WAY ANOVA

The assumptions underlying the two-way ANOVA are the same as those for the one-way ANOVA:

1. The populations from which the samples were taken are normally distributed.
2. The population variances for each of the cells are equal. This is the *homogeneity of variance* assumption.

As with the one-way ANOVA, the two-way ANOVA is robust with regard to violations of these assumptions, provided the samples are of equal size.*

■ SUMMARY

First, I presented a qualitative discussion of the two-way analysis of variance, independent groups design. Like the one-way design, in the two-way design, subjects are randomly assigned to the conditions. However, the two-way design allows us to investigate two independent variables and the interaction between them in one experiment. The effect of either independent variable (averaged over the levels of the other variable) is called a main effect. An interaction occurs when the effect of one of the variables is not the same at each level of the other variable.

The two-way analysis of variance is very similar to the one-way ANOVA. However, in the two-way ANOVA, the total sum of squares (SS_{total}) is partitioned into four components: the within-cells sum of squares ($SS_{within\text{-}cells}$), the row sum of squares (SS_{rows}), the column sum of squares ($SS_{columns}$), and the interaction sum of squares ($SS_{interaction}$). When these sums of squares are divided by the appropriate degrees of freedom, they form four variance estimates: the within-cells variance estimate ($MS_{within\text{-}cells}$), the row variance estimate (MS_{rows}), the column variance estimate ($MS_{columns}$), and the interaction variance estimate ($MS_{interaction}$).

The within-cells variance estimate ($MS_{within\text{-}cells}$) is the yardstick against which the other variance estimates are compared. Since all the subjects within each cell receive the same level of variables A and B, the within-cells variability cannot be due to treatment differences. Rather, it is a measure of the inherent variability of the scores within each cell and, hence, gives us an estimate of the null-hypothesis population variance (σ^2) alone. The row variance estimate (MS_{rows}) is based on the differences between the row means. It is an estimate of σ^2 plus the effect of factor A and is used to evaluate the main effect of variable A. The column variance estimate ($MS_{columns}$) is based on the differences between the column means. It is an estimate of σ^2 plus the effect of factor B and is used to evaluate the main effect of variable B. The interaction variance estimate ($MS_{interaction}$) is based on the differences between the cell means beyond that which is predicted by the individual effects of the two variables. It is an estimate of σ^2 plus the interaction of A and B. As such, it is used to evaluate the interaction of variables A and B.

In addition to presenting the conceptual basis for the two-way ANOVA, equations for computing each of the four variance estimates were developed, and several illustrative examples were given for practice in using the two-way ANOVA technique. It was further pointed out that multiple comparison techniques similar to those used with the one-way ANOVA are used with the two-way ANOVA. Finally, the assumptions underlying the two-way ANOVA were presented.

---

*Some statisticians also require the data to be interval or ratio in scaling. For a discussion of this point, see the footnoted references in Chapter 2, p. 35.

■ IMPORTANT NEW TERMS

Column degrees of freedom
 (df$_{columns}$) (p. 454)
Column sum of squares ($SS_{columns}$)
 (p. 454)
Column variance estimate
 ($MS_{columns}$) (p. 449, 454)
Factorial experiment (p. 446)
Interaction degrees of freedom
 (df$_{interaction}$) (p. 455)
Interaction effect (p. 447)

Interaction sum of squares
 ($SS_{interaction}$) (p. 455)
Interaction variance estimate
 ($MS_{interaction}$) (p. 449, 455)
Main effect (p. 447)
Row degrees of freedom (df$_{rows}$)
 (p. 452)
Row sum of squares (SS_{rows}) (p. 452)
Row variance estimate (MS_{rows})
 (p. 449, 452)

Two-way analysis of variance
 (p. 446, 450)
Within-cells degrees of freedom
 (df$_{within-cells}$) (p. 451)
Within-cells sum of squares
 ($SS_{within-cells}$) (p. 451)
Within-cells variance estimate
 ($MS_{within-cells}$) (p. 449, 451)

■ QUESTIONS AND PROBLEMS

1. Define or identify each of the terms in the Important New Terms section.
2. What are the advantages of the two-way ANOVA compared with the one-way ANOVA?
3. What is a factorial experiment?
4. In the two-way ANOVA, what is a *main* effect? What is an *interaction*? Is it possible to have a main effect without an interaction? An interaction without a main effect? Explain.
5. In the two-way ANOVA, the total sum of squares is partitioned into four components. What are the four components?
6. Why is the within-cells variance estimate used as the yardstick against which the other variance estimates are compared?
7. The four variance estimates (MS_{rows}, $MS_{columns}$, $MS_{interaction}$, and $MS_{within-cells}$) are also referred to as *mean squares*. Can you explain why?
8. If the A variable's effect increased, what do you expect would happen to the differences among the row means? What would happen to MS_{rows}? Explain. Assuming there is no interaction, what would happen to the differences among the column means?
9. If the B variable's effect increased, what would happen to the differences among the column means? What would happen to $MS_{columns}$? Explain. Assuming there is no interaction, what would happen to the differences among the row means?
10. What are the assumptions underlying the two-way ANOVA, independent groups design?
11. It is theorized that repetition aids recall and that the learning of new material can interfere with the recall of previously learned material. A professor interested in human learning and memory conducts a 2 × 3 factorial experiment to investigate the effects of these

two variables on recall. The material to be recalled consists of a list of 16 nonsense syllable pairs. The pairs are presented one at a time, for 4 seconds, cycling through the entire list, before the first pair is shown again. There are three levels of repetition: level 1, in which each pair is shown 4 times; level 2, in which each pair is shown 8 times; and level 3, in which each pair is shown 12 times. After being presented the list the requisite number of times and prior to testing for recall, each subject is required to learn some intervening material. The intervening material is of two types: type 1, which consists of number pairs, and type 2, which consists of nonsense syllable pairs. After the intervening material has been presented, the subjects are tested for recall of the original list of 16 nonsense syllable pairs. Thirty-six college freshmen serve as subjects. They are randomly assigned so that there are six per cell. The following scores are recorded; each is the number of syllable pairs from the original list correctly recalled.

Intervening Material (row variable)	Number of Repetitions (column variable)					
	4 times		*8 times*		*12 times*	
Number pairs	10	11	16	12	16	14
	12	15	11	15	16	13
	14	10	13	14	15	16
Nonsense	8	7	11	13	14	12
syllable pairs	4	5	9	10	16	15
	5	6	8	9	12	13

a. What are the null hypotheses for this experiment?

b. Using $\alpha = 0.05$, what do you conclude? Plot a graph of the cell means to help you interpret the results. cognitive

12. Assume you have just accepted a position as chief scientist for a leading agricultural company. Your first assignment is to make a recommendation concerning the best type of grass to grow in the Pacific Northwest and the best fertilizer for it. To provide the database for your recommendation, having just graduated *summa cum laude* in statistics, you decide to conduct an experiment involving a factorial independent groups design. Since there are three types of grass and two fertilizers under active consideration, the experiment you conduct is 2×3 factorial, where the A variable is the type of fertilizer and the B variable is the type of grass. In your field station, you duplicate the soil and the climate of the Pacific Northwest. Then you divide the soil into 30 equal areas and randomly set aside 5 for each combination of treatments. Next, you fertilize the areas with the appropriate fertilizer and plant in each area the appropriate grass seed. Thereafter, all areas are treated alike. When the grass has grown sufficiently, you determine the number of grass blades per square inch in each area. Your recommendation is based on this dependent variable. The "denser" the grass is, the better. The following scores are obtained:

	Number of Grass Blades Per Square Inch					
Fertilizer	Red Fescue		Kentucky Blue		Green Velvet	
Type 1	14	15	15	17	20	19
	16	17	12	18	15	22
	10		11		25	
Type 2	11	7	10	6	15	11
	11	8	8	13	18	10
	14		12		19	

a. What are the null hypotheses for this experiment?

b. Using $\alpha = 0.05$, what are your conclusions? Draw a graph of the cell means to help you interpret the results. I/O

13. A sleep researcher conducts an experiment to determine whether a hypnotic drug called Drowson, which is advertised as a remedy for insomnia, actually does promote sleep. In addition, the researcher is interested in whether a tolerance to the drug develops with chronic use. The design of the experiment is a 2×2 factorial independent groups design. One of the variables is the concentration of Drowson. There are two levels: (1) zero concentration (placebo) and (2) the manufacturer's minimum recommended dosage. The other variable concerns the previous use of Drowson. Again there are two levels: (1) subjects with no previous use and (2) chronic users. Sixteen individuals with sleep-onset insomnia (difficulty in falling asleep) who have had no previous use of Drowson are randomly assigned to the two concentration conditions, such that there are eight subjects in each condition. Sixteen chronic users of Drowson are also assigned randomly to the two conditions, eight subjects per condition. All subjects take their prescribed "medication" for 3 consecutive nights, and the time to fall asleep is recorded. The scores shown in the following table are the mean times in minutes to fall asleep for each subject, averaged over the 3 days:

	Concentration of Drowson			
Previous Use	Placebo		Minimum Recommended Dosage	
No previous use	45	53	30	47
	48	58	33	35
	62	55	40	31
	70	64	50	39
Chronic users	47	68	52	46
	52	64	60	49
	55	58	58	50
	62	59	68	55

a. What are the null hypotheses for this experiment?

b. Using $\alpha = 0.05$, what do you conclude? Plot a graph of the cell means to help you interpret the results. clinical, health

■ SPSS ILLUSTRATIVE EXAMPLE 16.1 _____

The general operation of SPSS and data entry are presented in Appendix E, *Introduction to SPSS*. As it did in doing a one-way ANOVA, SPSS computes **F** (the same as our F_{obt}) and the probability of getting **F** or a value even greater, if chance alone is at work. It calls this probability, **Sig.** (meaning "significance"; this is another way to represent what we have been calling, "the *p* value"). The decision rule we will follow to evaluate H_0 and H_1 is as follows.

If **Sig.** $\leq \alpha$, reject H_0 and affirm H_1.

If **Sig.** $> \alpha$, retain H_0; cannot affirm H_1.

example

Use SPSS to analyze the data given in the illustrative experiment described in Chapter 16 of the textbook, p. 456. For convenience the experiment is repeated here.

Let's assume a professor in physical education conducts an experiment to compare the effects on sleep of different amounts of exercise and the time of day when the exercise is done. The experiment uses a fixed effects, 3 × 2 factorial design with independent groups. There are three levels of exercise (light, moderate, and heavy) and two times of day (morning and evening). Thirty-six college students in good physical condition are randomly assigned to the six cells such that there are six subjects per cell. The subjects who do heavy exercise jog for 3 miles; the subjects who do moderate exercise jog for 1 mile; and the subjects in the light exercise condition jog for only $\frac{1}{4}$ mile. Morning exercise is done at 7:30 A.M., whereas evening exercise is done at 7:00 P.M. Each subject exercises once and the number of hours slept that night is recorded. The data are shown in the table below.

| | Exercise | | |
Time of Day	Light (1)	Moderate (2)	Heavy (3)
Morning (1)	6.5 7.4	7.4 7.3	8.0 7.6
	7.3 7.2	6.8 7.6	7.7 6.6
	6.6 6.8	6.7 7.4	7.1 7.2
Evening (2)	7.1 7.7	7.4 8.0	8.2 8.7
	7.9 7.5	8.1 7.6	8.5 9.6
	8.2 7.6	8.2 8.0	9.5 9.4

Using SPSS, what do you conclude regarding the main effect for the column variable *Exercise*, the main effect for the row variable *Time of Day*, and the interaction effect between the row and column variables *Exercise* and *Time of Day*? Use $\alpha = 0.05$.

SOLUTION

STEP 1:   **Enter the Data.** In the one-way ANOVA independent groups design there was only one independent variable. For SPSS to analyze the data, we had to tell SPSS which treatment each score received. We did this by specifying to which group each score belonged. In the two-way ANOVA, since there are two independent variables, we need to tell SPSS which combination of treatments each score received. We do this by specifying the level of Factor A (the row variable) and the level of Factor B (the column variable) to which each score belongs. Thus, for each dependent variable score, we need to enter two grouping values, the level of Factor A and the level of Factor B associated with that dependent variable score. In all, there are three variables, the dependent variable score, the level of the Factor A variable, and the level of the Factor B variable.

The dependent variable scores are entered in a stacked manner in the first column (**VAR00001**); coding scores that identify the level of Factor A are stacked in the second column (**VAR00002**); and coding scores that identify the level of Factor B (**VAR00003**), are stacked in the third column. In the present experiment, the dependent variable is *Sleep*, Factor A is *Time of Day*, and Factor B is *Exercise*.

Entering The *Sleep* Scores.

1. In the first column (**VAR00001**) of the Data Editor, enter the *Sleep* scores of all subjects who received the *Time of Day* treatment in the morning (*Time of Day-Morning*); there are 18 of these scores, the first score entered is "6.5," and the last score is "7.2." Then, directly under the last score of "7.2" (no empty cells), enter the sleep scores for all subjects who received the *Time of Day-Evening* treatment; these are the remaining 18 scores, beginning with "7.1" and ending with "9.4." When you have completed this step, all of the sleep scores will have been entered in the first column of the Data Editor in a stacked manner, with all the scores of *Time of Day-Morning* preceding the scores of *Time of Day-Evening*. There should be no gaps or empty cells between the scores of *Time of Day-Morning* and the scores of *Time of Day-Evening*.

Entering the Coding Numbers for *Time of Day*

1. The code we will use is as follows: **1** = *Morning* and **2** = *Evening*. The numbers **1** and **2** are entered into the second column (**VAR00002**) of the Data Editor in a stacked manner in the order of **1**, followed by **2**. To do the coding, enter the code number **1** in the second column (**VAR00002**) next to each sleep score that received the *Time of Day-Morning* treatment. Then, directly under the last **1** just entered, enter the code number **2** next to each sleep score that received the *Time of Day-Evening* treatment. There should be no gaps or empty cells between the last **1** and the first **2**. When this step is completed, all of the sleep scores will have been identified with respect to the level of the *Time of Day* treatment they received.

Entering the Coding Numbers for *Exercise*

1. We will enter the coding for *Exercise* in the third column (**VAR00003**). The code we will use is as follows: **1** = *Light*, **2**= *Moderate,* and **3** = *Heavy*. The numbers **1**, **2**, and **3** for the *Time of Day-Morning* scores are entered first into the third column (**VAR00003**) in a stacked manner, followed by the numbers **1**, **2**, and **3** for the *Time of Day-Evening* scores.

To do the coding, we will first enter the codes for the sleep scores that received the *Time of Day-Morning* treatment. For these scores, enter the code number **1** in the third column (**VAR00003**) on the same row of each sleep score that received the *Exercise-Light* treatment. Then, directly under the last **1** just entered, enter the code number **2** on the same row of each *Time of Day-Morning* sleep score that received the *Exercise-Moderate* treatment. Finally, directly under the last **2** just entered, enter the code number **3** on the same row of each *Time of Day-Morning* sleep score that received the *Exercise-Heavy* treatment.

Next, we will code the sleep scores that received the *Time of Day-Evening* treatment. For these scores, directly under the last **3** previously entered, enter a **1** for each sleep score that received the *Exercise-Light* treatment. Then, directly under the last **1** just entered, enter the number **2** on the same row of each the *Time of Day-Evening* sleep score that received the *Exercise-Moderate* treatment. Finally, directly under the last **2** just entered, enter the number **3** on the same row of each *Time of Day-Evening* sleep score that received the *Exercise-Heavy* treatment. There should be no gaps or blank cells between the coding scores. When this step is completed, all of the *Sleep* scores will have been identified with respect to the level of the *Exercise* treatment they received. SPSS is now able to identify each sleep score with regard to the level of *Time of Day* and *Exercise* it received.

The resulting Data Editor is shown here. For now, ignore the variable name row; we name the variables in the next step.

	Sleep	T_Day	Exercise	var	va
1	6.50	1.00	1.00		
2	7.30	1.00	1.00		
3	6.60	1.00	1.00		
4	7.40	1.00	1.00		
5	7.20	1.00	1.00		
6	6.80	1.00	1.00		
7	7.40	1.00	2.00		
8	6.80	1.00	2.00		
9	6.70	1.00	2.00		
10	7.30	1.00	2.00		
11	7.60	1.00	2.00		
12	7.40	1.00	2.00		
13	8.00	1.00	3.00		
14	7.70	1.00	3.00		
15	7.10	1.00	3.00		
16	7.60	1.00	3.00		
17	6.60	1.00	3.00		
18	7.20	1.00	3.00		
19	7.10	2.00	1.00		
20	7.90	2.00	1.00		
21	8.20	2.00	1.00		
22	7.70	2.00	1.00		
23	7.50	2.00	1.00		
24	7.60	2.00	1.00		
25	7.40	2.00	2.00		
26	8.10	2.00	2.00		
27	8.20	2.00	2.00		
28	8.00	2.00	2.00		
29	7.60	2.00	2.00		
30	8.00	2.00	2.00		
31	8.20	2.00	3.00		
32	8.50	2.00	3.00		
33	9.50	2.00	3.00		
34	8.70	2.00	3.00		

STEP 2: **Name the Variables.** In this example, we will give the default variables **VAR00001**, **VAR00002**, and **VAR00003**, the new names of *Sleep*, *T_Day*, and *Exercise*, respectively. To do so,

1. **Click** the **Variable View** tab in the lower left corner of the Data Editor.	This displays the **Variable View** on screen, with **VAR00001, VAR00002, and VAR00003** displayed in the first, second, and third cells of the **Name** column, respectively.
2. **Click VAR00001**; then **type Sleep** in the highlighted cell and then **press Enter**.	**Sleep** is entered as the variable name, replacing **VAR00001**. The cursor then moves to the next cell, highlighting **VAR00002**.
3. **Replace VAR00002** with **T_Day** and then **press Enter**.	**T_Day** is entered as the variable name, replacing **VAR00002**. The cursor then moves to the next cell, highlighting **VAR00003**.
4. **Replace VAR00003** with **Exercise** and then **press Enter**.	**Exercise** is entered as the variable name, replacing **VAR00003**.

STEP 3: **Analyze the Data.** The appropriate test is the two-way, independent groups analysis of variance. To have SPSS do the analysis using this test,

1. **Click Analyze**; then **select General Linear Model**; then **click on Univariate…**.	This produces the **Univariate** dialog box with **Sleep** highlighted in the large box on the left.
2. **Click** the **arrow** for the **Dependent Variable:** box.	This moves **Sleep** into the **Dependent Variable:** box. This identifies **Sleep** as the dependent variable.
3. **Click T_Day** in the large box on the left; then **click** the **arrow** for the **Fixed Factor(s):** box.	This moves **T_Day** into the **Fixed Factor(s):** box. This identifies **T_Day** as a fixed factor. SPSS differentiates between random and fixed factors. All the independent variables discussed in Chapters 15 and 16 are fixed factors. If you take an advanced course, you will learn the difference between fixed and random factors.
4. **Click Exercise** in the large box on the left; then **click** the **arrow** for the **Fixed Factor(s):** box.	This moves **Exercise** into the **Fixed Factor(s):** box. This identifies **Exercise** as a fixed factor.
5. **Click Options…**.	This produces the **Univariate: Options** dialog box. This and the next two steps are not really necessary for the overall analysis. They instruct SPSS to compute some descriptive statistics. While not necessary, the descriptive statistics are often useful, so I recommend you include these two steps.
6. **Click Descriptive statistics**.	This puts a **check** in the **Descriptive statistics** box, telling SPSS to compute some descriptive statistics and include them in the output.

7. Click Continue.

This returns you to the **Univariate** dialog box.

8. Click OK.

SPSS analyzes the **Sleep** data and outputs the results.

Analysis Results

The results are displayed in three tables, the **Between-Subjects Factors** table, the **Descriptive Statistics** table, and the **Tests of Between-Subjects Effects** table. Since for this analysis we are interested in just the **Tests of Between-Subjects Effects** table, we have displayed it here alone.

Tests of Between-Subjects Effects

Dependent Variable:Sleep

Source	Type III Sum of Squares	df	Mean Square	F	Sig.
Corrected Model	15.466[a]	5	3.093	16.640	.000
Intercept	2122.138	1	2122.138	11416.163	.000
T_Day	9.000	1	9.000	48.416	.000
Exercise	4.754	2	2.377	12.787	.000
T_Day * Exercise	1.712	2	.856	4.604	.018
Error	5.577	30	.186		
Total	2143.180	36			
Corrected Total	21.042	35			

a. R Squared = .735 (Adjusted R Squared = .691)

The **Tests of Between-Subjects Effects** table gives us information about the main and interaction effects. This is a summary table, very much like the summary table shown in the textbook. This table shows that for the **T_Day** main effect, **F = 48.416** and **Sig. = .000**. Since **.000 < .05**, we reject H_0. There is a significant **T_Day** main effect. This table also shows that for the **Exercise** main effect, **F = 12.787,** and **Sig. = .000**. Since .000 < .05, we reject H_0. There is a significant main effect for **Exercise**. Finally, the table shows that for the **T_Day*Exercise*** interaction, **F = 4.604,** and **Sig. = .018**. Since .018 < .05, we reject H_0. There is a significant interaction between **Time of Day and Exercise**.

----

*T_Day*Exercise means T_Day by Exercise

■ SPSS ADDITIONAL PROBLEMS

1. Use SPSS to analyze the data of the experiment presented in Chapter 16, Problem 13, p. 474, in the textbook. Using SPSS, what do you conclude regarding the main effect for *Previous Use*, the main effect for *Drowson Concentration*, and the interaction effect between *Previous Use* and *Drowson Concentration*? Use $\alpha = 0.05$. In solving this problem, name the dependent variable *Sleep*, the row variable *P_Use*, and the column variable *D_Conc*.

2. A researcher is interested in whether the effects of marijuana vary with prior usage of the drug. An experiment is conducted in which 12 moderate users, 12 high users, and 12 nonusers (no prior use) are randomly sampled from the college population. Within each usage level, half of the subjects are randomly assigned to a placebo condition and the other half to an experimental condition. In the placebo condition, each subject smokes two regular cigarettes that taste and smell like marijuana cigarettes. In the experimental condition, each subject smokes two marijuana cigarettes. Immediately after finishing their cigarettes, each subject is given a reaction time test. The following scores in milliseconds are obtained.

Marijuana Prior Usage	Placebo		Experimental (Marijuana)	
None	795	605	695	878
	700	752	865	916
	648	710	811	840
Moderate	800	610	843	665
	705	757	765	810
	645	712	713	776
Heavy	790	600	815	635
	695	752	735	782
	634	705	683	744

Perform a two-way, independent groups ANOVA on the data, using $\alpha = 0.05$. Name the dependent variable scores *RT*, the row variable *P_Usage*, and the column variable *M_amt*.

3. The sets of numbers given below in each table cell were obtained from the table of random numbers, Table J, in Appendix D. There are 6 table cells, one at the intersection of each combination of the levels of Factor A and Factor B, e.g., the cell located at the intersection of *Factor A—Level 1* and *Factor B—Level 1* contains the numbers 7, 5, 6, 4, 6.

Factor A	Factor B		
	Level 1	Level 2	Level 3
Level 1	7	5	2
	5	7	2
	6	3	7
	4	4	4
	6	6	5
Level 2	2	4	7
	6	9	8
	2	6	4
	8	0	5
	1	8	2

a. Use SPSS and do a two-way independent groups ANOVA with $\alpha = 0.05$ to determine if there are any significant main or interaction effects.

b. Add a constant of 3 to each score in row 1(Level 1 row of Factor A in the above data table). This process is analogous to what concerning a main or interaction effect? Compute the two-way analysis of variance again, and explain any differences between this result and that of part **a**.

c. Using the original scores, add a constant of 3 to each score in column 2 (Level 2) and a constant of 6 to each score in column 3 (Level 3). This process is analogous to what concerning a main or interaction effect? Compute the two-way analysis of variance again, and explain any differences between this result and that of part **a**.

d. Using the original scores, add a constant of 3 to all the scores of cell 11 (the scores in the cell at the intersection of *Factor A—Level 1* and *Factor B—Level 1*) of the above data table. The original scores in this cell are 7, 5, 6, 4, 6. This process is analogous to what concerning a main or interaction effect? Compute the two-way analysis of variance again, and explain any differences between this result and that of part **a**.

■ ONLINE STUDY RESOURCES

CENGAGE**brain**.com

Login to CengageBrain.com to access the resources your instructor has assigned. For this book, you can access the book's companion website for chapter-specific learning tools including Know and Be Able to Do, practice quizzes, flash cards, and glossaries, and a link to Statistics and Research Methods Workshops.

aplia™

If your professor has assigned Aplia homework:

1. Sign in to your account.
2. Complete the corresponding homework exercises as required by your professor.
3. When finished, click "Grade It Now" to see which areas you have mastered and which need more work, and for detailed explanations of every answer.

Visit **www.cengagebrain.com** to access your account and to purchase materials.

17

Chi-Square and Other Nonparametric Tests

© Strmko / Dreamstime.com

LEARNING OBJECTIVES

After completing this chapter, you should be able to:

- Specify the distinction between parametric and nonparametric tests, when to use each, and give an example of each.
- Specify the level of variable scaling that chi-square requires for its use; understand that chi-square uses sample frequencies and predicts to population proportions.
- Define a contingency table; specify the H_1 and H_0 for chi-square analyses.
- Understand that chi-square basically computes the difference between f_e and f_o, and the larger this difference, the more likely we can reject H_0.
- Solve problems using chi-square, and specify the assumptions underlying this test.

The following objective applies to the Wilcoxon matched-pairs signed ranks test, the Mann–Whitney U test, and the Kruskal–Wallis test.

- Specify the parametric test that each substitutes for, solve problems using each test, and specify the assumptions underlying each test.
- Rank-order the sign test, the Wilcoxon match-pairs signed ranks test, and the t test for correlated groups with regard to power.
- Understand the illustrative examples, do the practice problems, and understand the solutions.

INTRODUCTION: DISTINCTION BETWEEN PARAMETRIC AND NONPARAMETRIC TESTS

Statistical inference tests are often classified as to whether they are parametric or non-parametric. You will recall from our discussion in Chapter 1 that a parameter is a characteristic of a population. A parametric inference test is one that depends considerably on population characteristics, or parameters, for its use. The *z* test, *t* test, and *F* test are examples of parametric tests. The *z* test, for instance, requires that we specify the mean and standard deviation of the null-hypothesis population, as well as requiring that the population scores must be normally distributed for small *N*s. The *t* test for single samples has the same requirements, except that we don't specify *σ*. The *t* tests for two samples or conditions (correlated *t* or independent *t*) both require that the population scores be normally distributed when the samples are small. The independent *t* test further requires that the population variances be equal. The analysis of variance has requirements quite similar to those for the independent *t* test.

Although all inference tests depend on population characteristics to some extent, the requirements of nonparametric tests are minimal. For example, the sign test is a nonparametric test. To use the sign test, it is not necessary to know the mean, variance, or shape of the population scores. Because nonparametric tests depend little on knowing population distributions, they are often referred to as *distribution-free tests*.

Since nonparametric inference tests have fewer requirements or assumptions about population characteristics, the question arises as to why we don't use them all the time and forget about parametric tests. The answer is twofold. First, many of the parametric inference tests are robust with regard to violations of underlying assumptions. You will recall that a test is robust if violations in the assumptions do not greatly disturb the sampling distribution of its statistic. Thus, the *t* test is robust regarding the violation of normality in the population. Even though, theoretically, normality in the population is required with small samples, it turns out empirically that unless the departures from normality are substantial, the sampling distribution of *t* remains essentially the same. Thus, the *t* test can be used with data even though the data violate the assumptions of normality.

MENTORING TIP

Parametric tests are more powerful than nonparametric tests. When analyzing real data, always use a parametric over a nonparametric test if the data meet the assumptions of the parametric test.

The main reasons for preferring parametric to nonparametric tests are that, in general, they are more powerful and more versatile than nonparametric tests. We saw an example of the higher power of parametric tests when we compared the *t* test with the sign test for correlated groups. The factorial design discussed in Chapter 16 provides a good example of the versatility of parametric statistics. With this design, we can test two, three, four, or more variables and their interactions. No comparable technique exists with nonparametric statistics.

As a general rule, investigators use parametric tests whenever possible. However, when there is an extreme violation of an assumption of the parametric test or if the investigator believes the scaling of the data makes the parametric test inappropriate, a nonparametric inference test will be employed. We have already presented one nonparametric test: the sign test. In the remaining sections of this chapter, we shall present four more: chi-square, the Wilcoxon matched-pairs signed ranks test, the Mann–Whitney U test, and the Kruskal–Wallis test.*

CHI-SQUARE (χ^2)

Single-Variable Experiments

Thus far, we have presented inference tests used primarily in conjunction with ordinal, interval, or ratio data. But what about nominal data? Experiments involving nominal data occur fairly often, particularly in social psychology. You will recall that with this type of data, observations are grouped into several discrete, mutually exclusive categories, and one counts the frequency of occurrence in each category. The inference test most often used with nominal data is a nonparametric test called *chi-square* (χ^2). As has been our procedure throughout the text, we shall begin our discussion of chi-square with an experiment.

experiment

Preference for Different Brands of Light Beer

Suppose you are interested in determining whether there is a difference among beer drinkers living in the Puget Sound area in their preference for different brands of light beer. You decide to conduct an experiment in which you randomly sample 150 beer drinkers and let them taste the three leading brands. Assume all the precautions of good experimental design are followed, such as not disclosing the names of the brands to the subjects and so forth. The resulting data are presented in Table 17.1.

table 17.1 Preference for brands of light beer

Brand A	Brand B	Brand C	Total
[1] 45	[2] 40	[3] 65	150

SOLUTION

The entries in each cell are the number or frequency of subjects appropriate to that cell. Thus, 45 subjects preferred brand A (cell 1); 40, brand B (cell 2); and 65, brand C (cell 3). Can we conclude from these data that there is a difference in preference in the population? The null hypothesis for this experiment states that there is no difference in preference among the brands in the population. More specifically, in the population, the proportion of individuals favoring brand A is equal to the proportion favoring brand B, which is equal to the proportion favoring brand C. Referring to the table, it is clear that in the sample the number of individuals preferring each brand is different. However, it doesn't necessarily follow that there is a difference in the population. Isn't it possible that these scores could be due to random sampling from a population of beer drinkers in which the proportion of individuals favoring each brand is equal? Of course, the answer is "yes." Chi-square allows us to evaluate this possibility.

---

*Although we cover several nonparametric tests, there are many more. The interested reader should consult S. Siegel and N. Castellan, Jr., *Nonparametric Statistics for the Behavioral Sciences*, McGraw-Hill, New York, 1988, or W. Daniel, *Applied Nonparametric Statistics*, 2nd ed., PWS-Kent, Boston, 1990.

Computation of χ^2_{obt}   To calculate χ^2_{obt}, we must first determine the frequency we would expect to get in each cell if sampling is random from the null-hypothesis population. These frequencies are called *expected frequencies* and are symbolized by f_e. The frequencies actually obtained in the experiment are called *observed frequencies* and are symbolized by f_o. Thus,

f_e = expected frequency under the assumption sampling
is random from the null-hypothesis population
f_o = observed frequency in the sample

It should be clear that the closer the observed frequency of each cell is to the expected frequency for that cell, the more reasonable is H_0. On the other hand, the greater the difference between f_o and f_e is, the more reasonable H_1 becomes.

After determining f_e for each cell, we obtain the difference between f_o and f_e, square the difference, and divide by f_e. In symbolic form, $(f_o - f_e)^2 / f_e$ is computed for each cell. Finally, we sum the resultant values from each of the cells. In equation form,

$$\chi^2_{obt} = \Sigma \; \frac{(f_o - f_e)^2}{f_e} \qquad \textit{equation for calculating } \chi^2$$

where f_o = observed frequency in the cell
f_e = expected frequency in the cell, and Σ is over all the cells

From this equation, you can see that χ^2 is basically a measure of how different the observed frequencies are from the expected frequencies.

To calculate the value of χ^2_{obt} for the present experiment, we must determine f_e for each cell. The values of f_o are given in the table. If the null hypothesis is true, then the proportion of beer drinkers in the population that prefers brand A is equal to the proportion that prefers brand B, which in turn is equal to the proportion that prefers brand C. This means that one-third of the population must prefer brand A; one-third, brand B; and one-third, brand C. Therefore, if the null hypothesis is true, we would expect one-third of the individuals in the population and, hence, in the sample to prefer brand A, one-third to prefer brand B, and one-third to prefer brand C. Since there are 150 subjects in the sample, f_e for each cell $= \frac{1}{3}(150) = 50$. We have redrawn the data table and entered the f_e values in parentheses:

Brand A	Brand B	Brand C	Total
45	40	65	**150**
(50)	(50)	(50)	**(150)**

Now that we have determined the value of f_e for each cell, we can calculate χ^2_{obt}. All we need do is sum the values of $(f_o - f_e)^2 / f_e$ for each cell. Thus,

MENTORING TIP

Because $(f_o - f_e)$ is squared, χ^2_{obt} is always positive.

$$\chi^2_{obt} = \Sigma \; \frac{(f_o - f_e)^2}{f_e}$$

$$= \frac{(45 - 50)^2}{50} + \frac{(40 - 50)^2}{50} + \frac{(65 - 50)^2}{50}$$

$$= 0.50 + 2.00 + 4.50 = 7.00$$

Evaluation of χ^2_{obt} The theoretical sampling distribution of χ^2 is shown in Figure 17.1. The χ^2 distribution consists of a family of curves that, like the t distribution, varies with degrees of freedom. For the lower degrees of freedom, the curves are positively skewed. The degrees of freedom are determined by the number of f_o scores that are free to vary. In the present experiment, two of the f_o scores are free to vary. Once two of the f_o scores are known, the third f_o score is fixed, since the sum of the three f_o scores must equal N. Therefore, df = 2. In general, with experiments involving just one variable, there are $k-1$ degrees of freedom, where k equals the number of groups or categories. When we take up the use of χ^2 with contingency tables, there will be another equation for determining degrees of freedom. We shall discuss it when the topic arises.

Table H in Appendix D gives the critical values of χ^2 for different alpha levels. Since χ^2 is basically a measure of the overall discrepancy between f_o and f_e, it follows that the larger the discrepancy between the observed and expected frequencies is, the larger the value of χ^2_{obt} will be. Therefore, the larger the value of χ^2_{obt} is, the more unreasonable the null hypothesis is. As with the t and F tests, if χ^2_{obt} falls within the critical region for rejection, then we reject the null hypothesis. The decision rule states the following:

$$\text{If } \chi^2_{obt} \geq \chi^2_{crit}, \text{ reject } H_0.$$

It should be noted that in calculating χ^2_{obt} it doesn't matter whether f_o is greater or less than f_e. The difference is *squared*, divided by f_e, and added to the other cells to obtain χ^2_{obt}. Since the direction of the difference is immaterial, the χ^2 test is a nondirectional test (Please see Note 17.1 for further discussion of this point). Furthermore, since each difference adds to the value of χ^2, the critical region for rejection always lies under the right-hand tail of the χ^2 distribution.

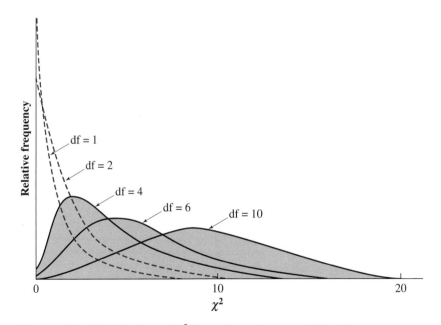

figure 17.1 Distribution of χ^2 for various degrees of freedom.
From *Design and Analysis of Experiments in Psychology and Education* by E.F. Lindquist. Copyright © 1953 Houghton Mifflin Company. Reproduced by permission.

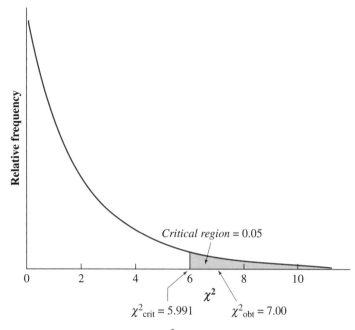

figure 17.2 Evaluation of χ^2_{obt} for the light beer drinking problem, df = 2 and α = 0.05.

In the present experiment, we determined that χ^2_{obt} = 7.00. To evaluate it, we must determine χ^2_{crit} to see if χ^2_{obt} falls into the critical region for rejection of H_0. From Table H, with df = 2 and α = 0.05,

$$\chi^2_{crit} = 5.991$$

Figure 17.2 shows the χ^2 distribution with df = 2 and the critical region for α = 0.05. Since $\chi^2_{obt} > 5.991$, it falls within the critical region and we reject H_0. There is a difference in the population regarding preference for the three brands of light beer tested. It appears as though brand C is the favored brand.

Let's try one more problem for practice.

Practice Problem 17.1

A political scientist believes that, in recent years, the ethnic composition of the city in which he lives has changed. The most current figures (collected a few years ago) show that the inhabitants were 53% Norwegian, 32% Swedish, 8% Irish, 5% Hispanic, and 2% Italian. (Note that nationalities with percentages under 2% have not been included.) To test his belief, a random sample of 750 inhabitants is taken; the results are shown in the following table:

Norwegian	Swedish	Irish	Hispanic	Italian	Total
[1] 399	[2] 193	[3] 63	[4] 82	[5] 13	750

(continued)

a. What is the null hypothesis?
b. What do you conclude? Use $\alpha = 0.05$.

SOLUTION

a. Null hypothesis: The ethnic composition of the city has not changed. Therefore, the sample of 750 individuals is a random sample from a population in which 53% are Norwegian, 32% Swedish, 8% Irish, 5% Hispanic, and 2% Italian.
b. Conclusion, using $\alpha = 0.05$:

> **STEP 1:** **Calculate the appropriate statistic.** The appropriate statistic is χ^2_{obt}. The calculations are shown here:

Cell No.	f_o	f_e	$\dfrac{(f_o - f_e)^2}{f_e}$
1	399	$0.53(750) = 397.5$	$\dfrac{(399 - 397.5)^2}{397.5} = 0.006$
2	193	$0.32(750) = 240.0$	$\dfrac{(193 - 240)^2}{240} = 9.204$
3	63	$0.08(750) = 60.0$	$\dfrac{(63 - 60)^2}{60} = 0.150$
4	82	$0.05(750) = 37.5$	$\dfrac{(82 - 37.5)^2}{37.5} = 52.807$
5	13	$0.02(750) = 15.0$	$\dfrac{(13 - 15)^2}{15} = 0.267$

$$\chi^2_{obt} = \sum \frac{(f_o - f_e)^2}{f_e} = 62.434$$

$$= 62.43$$

> **STEP 2:** **Evaluate the statistic.** Degrees of freedom $= 5 - 1 = 4$. With df $= 4$ and $\alpha = 0.05$, from Table H,
>
> $$\chi^2_{crit} = 9.488$$
>
> Since $\chi^2_{obt} > 9.488$, we reject H_0. The ethnic composition of the city appears to have changed. There has been an increase in the proportion of Hispanics and a decrease in the Swedish.

Test of Independence Between Two Variables

One of the main uses of χ^2 is in determining whether two categorical variables are independent or are related. To illustrate, let's consider the following example.

experiment

Political Affiliation and Attitude

Suppose a bill that proposes to lower the legal age for drinking to eighteen is pending before the state legislature. A political scientist living in the state is interested in determining whether there is a relationship between political affiliation and attitude toward the bill. A random sample of 200 registered Republicans and 200 registered Democrats is sent letters explaining the scientist's interest and asking the recipients whether they are in favor of the bill, are undecided, or are against the bill. Strict confidentiality is assured. A self-addressed envelope is included to facilitate responding. Answers are received from all 400 Republicans and Democrats. The results are shown in Table 17.2.

table 17.2   Political affiliation and attitude data

	Attitude			Row Marginal
	For	*Undecided*	*Against*	
Republican	68	22	110	200
Democrat	92	18	90	200
Column Marginal	160	40	200	400

The entries in each cell are the frequency of subjects appropriate to the cell. For example, with the Republicans, 68 are for the bill, 22 are undecided, and 110 are against. With the Democrats, 92 are for the bill, 18 are undecided, and 90 are against. This type of table is called a *contingency table*.

definition

■   *A* **contingency table** *is a two-way table showing the contingency between two variables where the variables have been classified into mutually exclusive categories and the cell entries are frequencies.*

Note that in constructing a contingency table, it is essential that the categories be mutually exclusive. Thus, if an entry is appropriate for one of the cells, the categories must be such that it cannot appropriately be entered in any other cell.

This contingency table contains the data bearing on the contingency between political affiliation and attitude toward the bill. The null hypothesis states that there is no contingency between the variables in the population. For this example, H_0 states that, in the population, attitude toward the bill and political affiliation are independent. If this is true, then both the Republicans and Democrats in the population should have the same proportion of individuals "for," "undecided," and "against" the bill. It is clear that in the contingency table, the frequencies in these three columns are different for Republicans and Democrats. The null hypothesis states that these frequencies are due to random sampling from a population in which the proportion of Republicans is equal to the proportion of Democrats in each of the categories. The alternative hypothesis is that Republicans and Democrats do differ in their attitudes toward the bill. If so, then in the population, the proportions would be different.

Computation of χ^2_{obt}   To test the null hypothesis, we must calculate χ^2_{obt} and compare it with χ^2_{crit}. With experiments involving two variables, the most difficult part of the process is in determining f_e for each cell. As discussed, the null hypothesis states that, in the population, the proportion of Republicans for each category is the same as the proportion of Democrats. If we knew these proportions, we could just multiply them by the number of Republicans or Democrats in the sample to find f_e for each cell. For example, suppose that, if H_0 is true, the proportion of Republicans in the population against the bill equals 0.50. To find f_e for that cell, all we would have to do is multiply 0.50 by the number of Republicans in the sample. Thus, for the "Republican-against" cell, f_e would equal $0.50(200) = 100$.

Since we do not know the population proportions, we estimate them from the sample. In the present experiment, 160 Republicans and Democrats out of 400 were for the bill, 40 out of 400 were undecided, and 200 out of 400 were against the bill. Since the null hypothesis assumes independence between political party and attitude, we can use these sample proportions as our estimates of the null-hypothesis population proportions. Then, we can use these estimates to calculate the expected frequencies. Our estimates for the null-hypothesis population proportions are as follows:

$$\text{Estimated } H_0 \text{ population proportion for the bill} = \frac{\text{Number of subjects for the bill}}{\text{Total number of subjects}} = \frac{160}{400}$$

$$\text{Estimated } H_0 \text{ population proportion undecided} = \frac{\text{Number of subjects undecided}}{\text{Total number of subjects}} = \frac{40}{400}$$

$$\text{Estimated } H_0 \text{ population proportion against the bill} = \frac{\text{Number of subjects against the bill}}{\text{Total number of subjects}} = \frac{200}{400}$$

Using these estimates to calculate the expected frequencies, we obtain the following values for f_e:

For the Republican-for cell (cell 1 in the table on p. 491):

$$f_e = \left(\begin{array}{c}\text{Estimated } H_0 \text{ population} \\ \text{proportion for the bill}\end{array}\right)\left(\begin{array}{c}\text{Total number of} \\ \text{Republicans}\end{array}\right) = \frac{160}{400}(200) = 80$$

For the Republican-undecided cell (cell 2):

$$f_e = \left(\begin{array}{c}\text{Estimated } H_0 \text{ population} \\ \text{proportion undecided}\end{array}\right)\left(\begin{array}{c}\text{Total number of} \\ \text{Republicans}\end{array}\right) = \frac{40}{400}(200) = 20$$

For the Republican-against cell (cell 3):

$$f_e = \left(\begin{array}{c}\text{Estimated } H_0 \text{ population} \\ \text{proportion against} \\ \text{the bill}\end{array}\right)\left(\begin{array}{c}\text{Total number of} \\ \text{Republicans}\end{array}\right) = \frac{200}{400}(200) = 100$$

For the Democrat-for cell (cell 4):

$$f_e = \left(\begin{array}{c}\text{Estimated } H_0 \text{ population} \\ \text{proportion for the bill}\end{array}\right)\left(\begin{array}{c}\text{Total number of} \\ \text{Democrats}\end{array}\right) = \frac{160}{400}(200) = 80$$

For the Democrat-undecided cell (cell 5):

$$f_e = \left(\begin{matrix} \text{Estimated } H_0 \text{ population} \\ \text{proportion undecided} \end{matrix} \right) \left(\begin{matrix} \text{Total number of} \\ \text{Democrats} \end{matrix} \right) = \frac{40}{400}(200) = 20$$

For the Democrat-against cell (cell 6):

$$f_e = \left(\begin{matrix} \text{Estimated } H_0 \text{ population} \\ \text{proportion against} \\ \text{the bill} \end{matrix} \right) \left(\begin{matrix} \text{Total number of} \\ \text{Democrats} \end{matrix} \right) = \frac{200}{400}(200) = 100$$

For convenience, the 2×3 contingency table has been redrawn here, and the f_e values entered within parentheses in the appropriate cells:

	Attitude			Row Marginal
	For	*Undecided*	*Against*	
Republican	¹ 68 (80)	² 22 (20)	³ 110 (100)	200
Democrat	⁴ 92 (80)	⁵ 18 (20)	⁶ 90 (100)	200
Column Marginal	160	40	200	400

The same values for f_e can also be found directly by multiplying the *marginals* for that cell and dividing by N. The marginals are the row and column totals lying outside the table. For example, the marginals for cell 1 are 160 (column total) and 200 (row total). Let's use this method to find f_e for each cell. Multiplying the marginals and dividing by N, we obtain

$$f_e(\text{cell 1}) = \frac{160(200)}{400} = 80$$

$$f_e(\text{cell 2}) = \frac{40(200)}{400} = 20$$

$$f_e(\text{cell 3}) = \frac{200(200)}{400} = 100$$

$$f_e(\text{cell 4}) = \frac{160(200)}{400} = 80$$

$$f_e(\text{cell 5}) = \frac{40(200)}{400} = 20$$

$$f_e(\text{cell 6}) = \frac{200(200)}{400} = 100$$

These values are, of course, the same ones we arrived at previously. Although using the marginals doesn't give much insight into why f_e should be that value, from a practical standpoint it is the best way to calculate f_e for the various cells. You should note that a good check to make sure your calculations of f_e are correct is to see whether the row and column totals of f_e equal the row and column marginals.

Once f_e for each cell has been determined, the next step is to calculate χ^2_{obt}. As before, this is done by summing $(f_o - f_e)^2 / f_e$ for each cell. Thus, for the present experiment,

$$\chi^2_{obt} = \Sigma \frac{(f_o - f_e)^2}{f_e}$$

$$= \frac{(68 - 80)^2}{80} + \frac{(22 - 20)^2}{20} + \frac{(110 - 100)^2}{100} + \frac{(92 - 80)^2}{80}$$

$$+ \frac{(18 - 20)^2}{20} + \frac{(90 - 100)^2}{100}$$

$$= 1.80 + 0.20 + 1.00 + 1.80 + 0.20 + 1.00 = 6.00$$

Evaluation of χ^2_{obt}   To evaluate χ^2_{obt}, we must compare it with χ^2_{crit} for the appropriate df. As discussed previously, the degrees of freedom are equal to the number of f_o scores that are free to vary while keeping the totals constant. In the two-variable experiment, we must keep both the column and row marginals at the same values. Thus, the degrees of freedom for experiments involving a contingency between two variables are equal to the number of f_o scores that are free to vary while at the same time keeping the column and row marginals the same. In the case of a 2 × 3 table, there are only 2 degrees of freedom. Only two f_o scores are free to vary, and all the remaining f_o and f_e scores are fixed.

To illustrate, consider the 2 × 3 table shown here:

¹ 68	² 22	³	200
⁴	⁵	⁶	200
160	40	200	400

If we fill in any two f_o scores, all the remaining f_o scores are fully determined, provided the marginals are kept at the same values. For example, in the table, we have filled in the f_o scores for cells 1 and 2. Note that all the other scores are fixed in value once two f_o scores are given; for example, the f_o score for cell 3 must be 110 [200 − (68 + 22)].

There is also an equation to calculate the df for contingency tables. It states

$$df = (r - 1)(c - 1)$$

where r = number of rows
                c = number of columns

Applying the equation to the present experiment, we obtain

$$df = (r - 1)(c - 1) = (2 - 1)(3 - 1) = 2$$

The χ^2 test is not limited to 2 × 3 tables. It can be used with contingency tables containing any number of rows and columns. This equation is perfectly general and applies to all contingency tables. Thus, if we did an experiment involving two variables and had four rows and six columns in the table, df $= (r - 1)(c - 1) = (4 - 1)(6 - 1) = 15$.

Returning to the evaluation of the present experiment, let's assume $\alpha = 0.05$. With df $= 2$ and $\alpha = 0.05$, from Table H,

$$\chi^2_{crit} = 5.991$$

Since $\chi^2_{obt} > 5.991$, we reject H_0. Political affiliation is related to attitude toward the bill. The Democrats appear to be more favorably disposed toward the bill than the Republicans. The complete solution is shown in Table 17.3.

table 17.3   Solution to political affiliation and attitude problem

a. Null hypothesis: Political affiliation and attitude toward the bill are independent. The frequency obtained in each cell is due to random sampling from a population where the proportions of Republicans and Democrats that are for, undecided about, and against the bill are equal.

b. Conclusion, using $\alpha = 0.05$:

> **STEP 1:** **Calculate the appropriate statistic.** The appropriate statistic is χ^2_{obt}. The data are shown on p. 489. The calculations are shown here.
>
> **STEP 2:** **Evaluate the statistic.** Degrees of freedom $= (r - 1)(c - 1) = (2 - 1)(3 - 1) = 2$. With df $= 2$ and $\alpha = 0.05$, from Table H,
>
> $$\chi^2_{crit} = 5.991$$
>
> Since $\chi^2_{obt} > 5.991$, we reject H_0. Political affiliation and attitude toward the bill are related. Democrats appear to favor the bill more than Republicans.

Cell No.	f_o	f_e	$\dfrac{(f_o - f_e)^2}{f_e}$
1	68	$\dfrac{160(200)}{400} = 80$	$\dfrac{(68 - 80)^2}{80} = 1.80$
2	22	$\dfrac{40(200)}{400} = 20$	$\dfrac{(22 - 20)^2}{20} = 0.20$
3	110	$\dfrac{200(200)}{400} = 100$	$\dfrac{(110 - 100)^2}{100} = 1.00$
4	92	$\dfrac{160(200)}{400} = 80$	$\dfrac{(92 - 80)^2}{80} = 1.80$
5	18	$\dfrac{40(200)}{400} = 20$	$\dfrac{(18 - 20)^2}{20} = 0.20$
6	90	$\dfrac{200(200)}{400} = 100$	$\dfrac{(90 - 100)^2}{100} = 1.00$

$$\chi^2_{obt} = \sum \frac{(f_o - f_e)^2}{f_e} = 6.00$$

Let's try a problem for practice.

Practice Problem 17.2

A university is considering implementing one of the following three grading systems: (1) All grades are pass–fail, (2) all grades are on the 4.0 system, and (3) 90% of the grades are on the 4.0 system and 10% are pass–fail. A survey is taken to determine whether there is a relationship between undergraduate major and grading system preference. A random sample of 200 students with engineering majors, 200 students with arts and sciences majors, and 100 students with fine arts majors is selected. Each student is asked which of the three grading systems he or she prefers. The results are shown in the following 3×3 contingency table:

| | Grading System | | | |
	Pass–fail	4.0 and Pass–fail	4.0	Row Marginal
Fine arts	¹ 26	² 55	³ 19	100
Arts and sciences	⁴ 24	⁵ 118	⁶ 58	200
Engineering	⁷ 20	⁸ 112	⁹ 68	200
Column Marginal	70	285	145	500

a. What is the null hypothesis?
b. What do you conclude? Use $\alpha = 0.05$.

SOLUTION

a. Null hypothesis: Undergraduate major and grading system preference are independent. The frequency obtained in each cell is due to random sampling from a population where the proportions of fine arts, arts and sciences, and engineering majors who prefer each grading system are the same.

b. Conclusion, using $\alpha = 0.05$:

STEP 1: **Calculate the appropriate statistic.** The data are shown in the following table. The appropriate statistic is χ^2_{obt}. Before calculating χ^2_{obt}, we must first calculate f_e for each cell. The values of f_e were found using the marginals.

Cell No.	f_o	f_e	$\dfrac{(f_o - f_e)^2}{f_e}$
1	26	$\dfrac{70(100)}{500} = 14$	$\dfrac{(26 - 14)^2}{14} = 10.286$
2	55	$\dfrac{285(100)}{500} = 57$	$\dfrac{(55 - 57)^2}{57} = 0.070$

Cell No.	f_o	f_e	$\dfrac{(f_o - f_e)^2}{f_e}$
3	19	$\dfrac{145(100)}{500} = 29$	$\dfrac{(19 - 29)^2}{29} = 3.448$
4	24	$\dfrac{70(200)}{500} = 28$	$\dfrac{(24 - 28)^2}{28} = 0.571$
5	118	$\dfrac{285(200)}{500} = 114$	$\dfrac{(118 - 114)^2}{114} = 0.140$
6	58	$\dfrac{145(200)}{500} = 58$	$\dfrac{(58 - 58)^2}{58} = 0.000$
7	20	$\dfrac{70(200)}{500} = 28$	$\dfrac{(20 - 28)^2}{28} = 2.286$
8	112	$\dfrac{285(200)}{500} = 114$	$\dfrac{(112 - 114)^2}{114} = 0.035$
9	68	$\dfrac{145(200)}{500} = 58$	$\dfrac{(68 - 58)^2}{58} = 1.724$

$$\chi^2_{\text{obt}} = \sum \frac{(f_o - f_e)^2}{f_e} = 18.561$$

$$= 18.56$$

STEP 2: **Evaluate the statistic.** Degrees of freedom $= (r - 1)(c - 1) = (3 - 1)(3 - 1) = 4$. With df $= 4$ and $\alpha = 0.05$, from Table H,

$$\chi^2_{\text{crit}} = 9.488$$

Since $\chi^2_{\text{obt}} > 9.488$, we reject H_0. Undergraduate major and grading system preference are related.

In trying to determine what the differences in preference were between the groups, since the number of subjects differs considerably for the fine arts majors, it is necessary to convert the frequency entries into proportions. These proportions are shown in Table 17.4.

table 17.4 Preferences for grading systems expressed as proportions

	Pass–fail	4.0 and Pass–fail	4.0
Fine arts	0.26	0.55	0.19
Arts and sciences	0.12	0.59	0.29
Engineering	0.10	0.56	0.34

From this table, it appears that the differences between groups are in their pref-erences for the all pass–fail or all-4.0 grading systems. The fine arts students show a higher proportion favoring the pass–fail system rather than the all-4.0 system, whereas the arts and sciences and engineering students show the reverse pattern. All groups show about the same proportions favoring the system advocating a combina-tion of 4.0 and pass–fail grades.

Let's try one more problem for practice.

Practice Problem 17.3

A social psychologist is interested in determining whether there is a relationship between the education level of parents and the number of children they have. Accordingly, a survey is taken, and the following results are obtained:

	No. of Children		
	Two or less	More than two	Row Marginal
College education	1 53	2 22	75
High school education only	3 37	4 38	75
Column Marginal	90	60	150

a. What is the null hypothesis?
b. What is the conclusion? Use $\alpha = 0.05$.

SOLUTION

a. Null hypothesis: The educational level of parents and the number of children they have are independent. The frequency obtained in each cell is due to random sampling from a population where the proportions of college-educated and only high-school-educated parents that have (1) two or fewer and (2) more than two children are equal.
b. Conclusion, using $\alpha = 0.05$:

> **STEP 1:   Calculate the appropriate statistic.** The data are shown in the following table. The appropriate statistic is χ^2_{obt}. The calculations follow.

Cell No.	f_o	f_e	$\dfrac{(f_o - f_e)^2}{f_e}$
1	53	$\dfrac{90(75)}{150} = 45$	$\dfrac{(53 - 45)^2}{45} = 1.422$
2	22	$\dfrac{60(75)}{150} = 30$	$\dfrac{(22 - 30)^2}{30} = 2.133$
3	37	$\dfrac{90(75)}{150} = 45$	$\dfrac{(37 - 45)^2}{45} = 1.422$
4	38	$\dfrac{60(75)}{150} = 30$	$\dfrac{(38 - 30)^2}{30} = 2.133$

$$\chi^2_{\text{obt}} = \sum \frac{(f_o - f_e)^2}{f_e} = 7.110$$

STEP 2: **Evaluate the statistic.** Degrees of freedom $= (r - 1)(c - 1) = (2 - 1)(2 - 1) = 1$. With df $= 1$ and $\alpha = 0.05$, from Table H,

$$\chi^2_{\text{crit}} = 3.841$$

Since $\chi^2_{\text{obt}} > 3.841$, we reject H_0. The educational level of parents and the number of children they have are related.

Assumptions Underlying χ^2

A basic assumption in using χ^2 is that there is independence between each observation recorded in the contingency table. This means that each subject can have only one entry in the table. It is not permissible to take several measurements on the same subject and enter them as separate frequencies in the same or different cells. This error would produce a larger N than there are independent observations.

A second assumption is that the sample size must be large enough that the *expected* frequency in each cell is at least 5 for tables where r or c is greater than 2. If the table is a 1×2 or 2×2 table, then each expected frequency should be at least 10. If the sample size is small enough to result in expected frequencies that violate these requirements, then the actual sampling distribution of χ^2 deviates considerably from the theoretical one and the probability values given in Table H do not apply. If the experiment involves a 2×2 contingency table and the data violate this assumption, Fisher's exact probability test should be used.*

Although χ^2 is used frequently when the data are only of nominal scaling, it is not limited to nominal data. Chi-square can be used with ordinal, interval, and ratio data. However, regardless of the actual scaling, the data must be reduced to mutually exclusive categories and appropriate frequencies before χ^2 can be employed.

---

*This test is discussed in S. Siegel and N. Castellan, Jr., *Nonparametric Statistics for the Behavioral Sciences,* 2nd ed., McGraw-Hill, New York, 1988, pp. 103–111. It is also discussed in W. Daniel, *Applied Nonparametric Statistics,* 2nd ed., PWS-Kent, Boston, 1990, pp. 150–162.

THE WILCOXON MATCHED-PAIRS SIGNED RANKS TEST

The *Wilcoxon matched-pairs signed ranks test* is used in conjunction with the correlated groups design with data that are at least ordinal in scaling. It is a relatively powerful test sometimes used in place of the *t* test for correlated groups when there is an extreme violation of the normality assumption or when the data are not of appropriate scaling. The Wilcoxon signed ranks test considers both the magnitude of the difference scores and their direction, which makes it more powerful than the sign test. It is, however, less powerful than the *t* test for correlated groups. To illustrate this test, let's consider the following experiment.

experiment

Changing Attitudes Toward Wildlife Conservation

A prominent ecological group is planning to mount an active campaign to increase wildlife conservation in their country. As part of the campaign, they plan to show a film designed to promote more favorable attitudes toward wildlife conservation. Before showing the film to the public at large, they want to evaluate its effects. A group of 10 subjects are randomly sampled and given a questionnaire that measures an individual's attitude toward wildlife conservation. Next, they are shown the film, after which they are again given the attitude questionnaire. The questionnaire has 50 possible points, and the higher the score is, the more favorable is the attitude toward wildlife conservation. The results are shown in Table 17.5.

1. What is the alternative hypothesis? Use a nondirectional hypothesis.
2. What is the null hypothesis?
3. What do you conclude? Use $\alpha = 0.05_{2 \text{ tail}}$.

SOLUTION

1. The alternative hypothesis is usually stated without specifying any population parameters. For this example, it states that the film affects attitudes toward wildlife conservation.
2. The null hypothesis is also usually stated without specifying any population parameters. For this example, it states that the film has no effect on attitudes toward wildlife conservation.
3. Conclusion, using $\alpha = 0.05_{2 \text{ tail}}$: As with all the other inference tests, the first step is to calculate the appropriate statistic. The data have been obtained from questionnaires, so they are at least of ordinal scaling. To illustrate use of the Wilcoxon signed ranks test, we shall assume that the data meet the assumptions of this test (these will be discussed shortly). The statistic calculated by the Wilcoxon signed ranks test is T_{obt}. Determining T_{obt} involves four steps:
   a. Calculate the difference between each pair of scores.
   b. Rank the absolute values of the difference scores from the smallest to the largest.
   c. Assign to the resulting ranks the sign of the difference score whose absolute value yielded that rank.
   d. Compute the sum of the ranks separately for the positive and negative signed ranks. The lower sum is T_{obt}.

These four steps have been done with the data from the attitude questionnaire, and the resultant values have been entered in Table 17.5. Thus, the difference scores have been calculated and are shown in the fourth column of Table 17.5. The ranks of the absolute values of the difference scores are shown in the fifth column. Note that, as a check on whether the ranking has been done correctly, the sum of the unsigned ranks should equal $n(n + 1)/2$.

table 17.5   Data and solution for wildlife conservation problem

Subject	Attitude Before	Attitude After	Difference	Rank of [Difference]	Signed Rank of Difference	Sum of Positive Ranks	Sum of Negative Ranks
1	40	44	4	4	4	4	
2	33	40	7	6	6	6	
3	36	49	13	10	10	10	
4	34	36	2	2	2	2	
5	40	39	−1	1	−1		1
6	31	40	9	8	8	8	
7	30	27	−3	3	−3		3
8	36	42	6	5	5	5	
9	24	35	11	9	9	9	
10	20	28	8	7	7	7	
				55		51	4

$$\frac{n(n+1)}{2} = \frac{10(11)}{2} = 55 \qquad T_{\text{obt}} = 4$$

From Table I, with $N = 10$ and $\alpha = 0.05_{2\text{ tail}}$,

$$T_{\text{crit}} = 8$$

Since $T_{\text{obt}} < 8$, H_0 is rejected. The film appears to promote more favorable attitudes toward wildlife conservation.

In the present example, this sum should equal 55 [10(11)/2 = 55], which it does. Step **c** asks us to give each rank the sign of the difference score whose absolute value yielded that rank. This has been done in the sixth column. Thus, the ranks of 1 and 3 are assigned minus signs, and the rest are positive. The ranks of 1 and 3 received minus signs because their associated difference scores are negative. T_{obt} is determined by computing the sum of the positive ranks and the sum of the negative ranks. T_{obt} is the lower of the two sums. In this example, the sum of the positive ranks equals 51, and the sum of the negative ranks equals 4. Thus,

$$T_{\text{obt}} = 4$$

Note that often it is not necessary to compute both sums. Usually it is apparent by inspection which sum will be lower. The final step is to evaluate T_{obt}. Table I in Appendix D contains the critical values of T for various values of N. With $N = 10$ and $\alpha = 0.05_{2\text{ tail}}$, from Table I,

$$T_{\text{crit}} = 8$$

With the Wilcoxon signed ranks test, the decision rule is

$$\text{If } T_{\text{obt}} \leq T_{\text{crit}}, \text{ reject } H_0.$$

Note that this is opposite to the rule we have been using for most of the other tests. Since, $T_{\text{obt}} < 8$, we reject H_0 and conclude that the film does affect attitudes toward wildlife conservation. It appears to promote more favorable attitudes.

It is easy to see why the Wilcoxon signed ranks test is more powerful than the sign test but not as powerful as the *t* test for correlated groups. The Wilcoxon signed ranks test takes into account the magnitude of the difference scores, which makes it more powerful than the sign test. However, it considers only the rank order of the difference scores, not their actual magnitude, as does the *t* test. Therefore, the Wilcoxon signed ranks test is not as powerful as the *t* test.

Let's try another problem for practice.

Practice Problem 17.4

An investigator is interested in determining whether the difficulty of the material to be learned affects the anxiety level of college students. A random sample of 12 students is each given hard and easy learning tasks. Before doing each task, they are shown a few sample examples of the material to be learned. Then their anxiety level is assessed using an anxiety questionnaire. Thus, anxiety level is assessed before each learning task. The data are shown in the following table. The higher the score is, the greater is the anxiety level. What is the conclusion, using the Wilcoxon signed ranks test and $\alpha = 0.05_{2\,tail}$?

SOLUTION

The solution is shown in the following table. Note that there are ties in some of the difference scores. Generally, two kinds of ties are possible. First, the raw scores may be tied, yielding a difference score of 0. If this occurs, these scores are disregarded and the overall N is reduced by 1 for each 0 difference score. Ties can also occur in the difference scores, as in the present example. When this happens, the ranks of these scores are given a value equal to the mean of the tied ranks. This is the same procedure we followed for the Spearman rho correlation coefficient. Thus, in this example, the two tied difference scores of 3 are assigned ranks of 2.5 [(2 + 3)/2 = 2.5], and the tied difference scores of 10 receive the rank of 9.5. Otherwise, the solution is quite similar to that of the previous example.

Student No.	Anxiety Hard tasks	Anxiety Easy tasks	Difference	Rank of |Difference|	Signed Rank of Difference	Sum of Positive Ranks	Sum of Negative Ranks
1	48	40	8	7	7	7	
2	33	27	6	5	5	5	
3	46	34	12	11	11	11	
4	42	28	14	12	12	12	
5	40	30	10	9.5	9.5	9.5	
6	27	24	3	2.5	2.5	2.5	

Student No.	Anxiety Hard tasks	Anxiety Easy tasks	Difference	Rank of |Difference|	Signed Rank of Difference	Sum of Positive Ranks	Sum of Negative Ranks
7	31	33	−2	1	−1		1
8	42	39	3	2.5	2.5	2.5	
9	38	31	7	6	6	6	
10	34	39	−5	4	−4		4
11	38	29	9	8	8	8	
12	44	34	10	9.5	9.5	9.5	
				78.0		73.0	5

$$\frac{n(n+1)}{2} = \frac{12(13)}{2} = 78 \qquad\qquad T_{obt} = 5$$

From Table I, with $N = 12$ and $\alpha = 0.05_{2\,tail}$,

$$T_{crit} = 13$$

Since $T_{obt} < 13$, we reject H_0 and conclude that the difficulty of material does affect anxiety. It appears that more difficult material produces increased anxiety.

Assumptions of the Wilcoxon Signed Ranks Test

There are two assumptions underlying the Wilcoxon signed ranks test. First, the scores within each pair must be at least of ordinal measurement. Second, the difference scores must also have at least ordinal scaling. The second requirement arises because in computing T_{obt} we rank-order the difference scores. Thus, the magnitude of the difference scores must be at least ordinal so that they can be rank-ordered.

THE MANN–WHITNEY *U* TEST

The *Mann–Whitney U test* is used in conjunction with the independent groups design with data that are at least ordinal in scaling. It is a powerful nonparametric test used in place of the *t* test for independent groups when there is an extreme violation of the normality assumption or when the data are not of appropriate scaling for the *t* test.

To illustrate this inference test, let's consider the following experiment.

experiment

The Effect of a High-Protein Diet on Intellectual Development

A developmental psychologist, with special competence in nutrition, believes that a high-protein diet eaten during early childhood is important for intellectual development. The diet in the geographic area where the psychologist lives is low in protein. The psychologist believes the low-protein diet eaten during the first few years of childhood is detrimental to intellectual development. If she is correct, a high-protein diet should result in higher intelligence. An experiment is conducted in which 18 children are randomly chosen from the 1-year-old children living in a nearby city. The 18 children are then randomly divided into two groups of 9 children each. The control group is fed the usual low-protein diet for 3 years, whereas the experimental group receives a diet high in protein for the same duration. At the end of the 3 years, each child is given an IQ test. The resulting data are shown in Table 17.6. One child in the experimental group moved to a different city and was not replaced.

table 17.6 Data from the protein and IQ experiment

IQ Test Scores	
Control group low protein 1	*Experimental group high protein* 2
102	110
104	115
105	117
107	122
108	125
111	130
113	135
118	140
120	

a. What is the directional alternative hypothesis?
b. What is the null hypothesis?
c. What do you conclude? Use $\alpha = 0.05_{1\,\text{tail}}$.

SOLUTION

1. Alternative hypothesis: As with the t test for independent groups, the alternative hypothesis states that a high-protein diet eaten during infancy will increase intellectual functioning relative to a low-protein diet. In the same manner as with the t test for independent groups, each sample is considered a random sample from its own population set of scores, with parameters μ_1, σ_1^2, and μ_2, σ_2^2, respectively. However, since this is a rank-order test, the Mann–Whitney U test does not evaluate sample mean differences and, hence, makes no prediction about the relationship of μ_1 and μ_2. Thus, there are no population parameters included in the statement of the alternative hypothesis.

2. Null hypothesis: The null hypothesis is also stated without any population parameters. It states that the high-protein diet, eaten during infancy, either will have no effect on intellectual functioning or will decrease intellectual functioning.

3. Conclusion using $\alpha = 0.05_{1\,\text{tail}}$: As with the other inference tests, the conclusion involves a two-step process: Compute the appropriate statistic and then evaluate the statistic using its sampling distribution.

   STEP 1: Compute the appropriate statistic. The statistic calculated by the Mann–U test is U_{obt} or U'_{obt}. These statistics measure the *degree of separation* between the two sample sets of scores. As the real effect of the independent variable increases, the samples become more separated (the scores of the two samples overlap less). As the degree of sample separation increases, U_{obt} decreases and

Remember: $U_{obt} = 0$ indicates the greatest degree of separation possible for any data.

U'_{obt} increases. When there is complete separation between samples (no overlap), $U_{obt} = 0$. For any experiment, $U_{obt} + U'_{obt} = n_1 n_2$. Both U_{obt} and U'_{obt} measure the same degree of separation. Hence, in analyzing the data from any experiment, it is necessary to compute and evaluate only U_{obt} or U'_{obt}. U_{obt} and U'_{obt} are computed as follows:

a. Combine the scores from both groups, rank-order them, and assign each a rank score, using 1 for the lowest score:

Original Score	102	104	105	107	108	110	111	113	115
Rank	1	2	3	4	5	6	7	8	9

Original Score	117	118	120	122	125	130	135	140	
Rank	10	11	12	13	14	15	16	17	

b. Sum the ranks for each group; that is, determine R_1 and R_2, where R_1 = sum of the ranks for group 1 and R_2 = sum of the ranks for group 2.

Control Group 1		Experimental Group 2	
Original score	Rank	*Original score*	Rank
102	1	110	6
104	2	115	9
105	3	117	10
107	4	122	13
108	5	125	14
111	7	130	15
113	8	135	16
118	11	140	17
120	12		$R_2 = 100$
	$R_1 = 53$		$n_2 = 8$
	$n_1 = 9$		

c. Solve the equations for U_{obt} and U'_{obt}. U_{obt} and U'_{obt} are computed by solving the following equations:

$$U_{obt} = n_1 n_2 + \frac{n_1(n_1 + 1)}{2} - R_1 \quad \text{general equation for finding } U_{obt} \text{ or } U'_{obt}$$

$$U_{obt} = n_1 n_2 + \frac{n_2(n_2 + 1)}{2} - R_2 \quad \text{general equation for finding } U_{obt} \text{ or } U'_{obt}$$

where n_1 = number of scores in group 1
n_2 = number of scores in group 2
R_1 = sum of ranks for scores in group 1
R_2 = sum of ranks for scores in group 2

In solving these equations, we identify one of the samples as group 1 and the other as group 2. Then, we just go ahead and solve the equations. One of the equations will yield a number lower than the number from the other equation. Arbitrarily, the lower of the two numbers is assigned as U_{obt} and the higher of the two numbers as U'_{obt}. It doesn't matter which sample is labeled group 1 and which is labeled group 2. If we reversed the labels, we would still obtain the same numbers from the equations. What does change with labeling is which equation yields the higher number and which yields the lower number. Since this depends on which group is labeled group 1 and which group 2, these equations are both written initially in terms of U_{obt}. In an actual analysis, the equation that yields the lower number is the U_{obt} equation; the one that yields the higher number is the U'_{obt} equation. For the data in the present example,

$$U_{obt} = n_1 n_2 + \frac{n_1(n_1 + 1)}{2} - R_1 \qquad U_{obt} = n_1 n_2 + \frac{n_2(n_2 + 1)}{2} - R_2$$

$$= 9(8) + \frac{9(10)}{2} - 53 \qquad = 9(8) + \frac{8(9)}{2} - 100$$

$$= 72 + 45 - 53 \qquad = 72 + 36 - 100$$

$$= 64 \qquad = 8$$

Therefore,

$$U_{obt} = 8$$
$$U'_{obt} = 64$$

STEP 2: Evaluate U_{obt} or U'_{obt}. Tables C.1–C.4 in Appendix D give the critical values of U and U'. For each cell, there are two entries. The upper entry is the highest value of U_{obt} for various n_1 and n_2 combinations that will allow rejection of H_0. The lower entry is the lowest value of U'_{obt} that will allow rejection of H_0. The decision rule is as follows:

$$\text{If } U_{obt} \leq U_{crit}, \text{ reject } H_0 \text{ and affirm } H_1.$$
$$\text{If } U_{obt} \geq U_{crit}, \text{ reject } H_0 \text{ and affirm } H_1.$$

Since both U_{obt} and U'_{obt} measure the same degree of separation, we shall evaluate only U_{obt}. Each of the Tables C.1–C.4 is for a different alpha level. For the data of the present experiment, Table C.4 is appropriate. With $n_1 = 9$ and $n_2 = 8$, $U_{crit} = 18$ and $U'_{crit} = 54$. Evaluating U_{obt}, since $U_{obt} < 18$, we reject H_0 and affirm H_1. A high-protein diet eaten during infancy appears to increase intellectual functioning relative to a low-protein diet.

Tied Ranks

We've already shown how to rank-order tied scores when we discussed the Spearman rho correlation coefficient (p. 141) and the Wilcoxon signed ranks test (p. 500). To review, tied scores are handled by assigning them the average of the tied ranks. For example, consider the two sets of scores presented in Table 17.7. To rank-order the combined scores, we proceed as follows. First, the scores are arranged in ascending order. Thus,

Raw Score	11	12	12	14	15	16	17	17	17	18	20
Rank	1	2.5	2.5	4	5	6	8	8	8	10	11

table 17.7 Data to illustrate ranking tied scores

Group 1	Group 2
12	11
14	12
15	16
17	17
18	17
	20

Next, we assign each raw score its rank, beginning with 1 for the lowest score. This has been shown previously. Note that the two raw scores of 12 are tied at the ranks of 2 and 3. They are assigned the average of these tied ranks. Thus, they each get a rank of 2.5 [(2 + 3)/2 = 2.5]. We have already used the ranks of 2 and 3, so the next score gets a rank of 4. The raw scores of 17 are tied at the ranks of 7, 8, and 9. Therefore, they receive the rank of 8, which is the average of 7, 8, and 9 [(7 + 8 + 9)/3 = 8]. Note that the next rank is 10 (not 9) because we've already used ranks 7, 8, and 9 in computing the average. If the ranking is done correctly, unless there are tied ranks at the end, the last raw score should have a rank equal to *N*. In this case, $N = 11$ and so does the rank of the last score. Once the ranks have been assigned, U_{obt} and U'_{obt} are calculated in the usual way.

Let's do the following problem for practice.

Practice Problem 17.5

Someone has told you that men are better in abstract reasoning than women. You are skeptical, so you decide to test this idea using a nondirectional hypothesis. You randomly select eight men and eight women from the freshman class at your university and administer an abstract reasoning test. A higher score reflects better abstract reasoning abilities. You obtain the following scores:

Men	Women
70	82
86	80
60	50
92	95
82	93
65	85
74	90
94	75

(continued)

a. What is the alternative hypothesis? Assume a nondirectional hypothesis is appropriate.
b. What is the null hypothesis?
c. Using $\alpha = 0.05_{2\,\text{tail}}$, what do you conclude?

SOLUTION

a. Nondirectional alternative hypothesis: Men and women differ in abstract reasoning ability.
b. Null hypothesis: Men and women are equal in abstract reasoning ability.
c. Conclusion, using $\alpha = 0.05_{2\,\text{tail}}$:

STEP 1:   **Calculate U_{obt} for the data:**

   a. Combine the scores, rank-order them, and assign each a rank, using 1 for the lowest score:

Original Score	50	60	65	70	74	75	80	82
Rank	1	2	3	4	5	6	7	8.5

Original Score	82	85	86	90	92	93	94	95
Rank	8.5	10	11	12	13	14	15	16

   b. Sum the ranks for each group; that is, determine R_1 and R_2.

Men 1		Women 2	
Original score	*Rank*	*Original score*	*Rank*
60	2	50	1
65	3	75	6
70	4	80	7
74	5	82	8.5
82	8.5	85	10
86	11	90	12
92	13	93	14
94	15	95	16
	$R_1 = 61.5$		$R_2 = 74.5$
	$n_1 = 8$		$n_2 = 8$

   c. Solve the equations for and U_{obt} and U'_{obt}:

$$U_{\text{obt}} = n_1 n_2 + \frac{n_1(n_1 + 1)}{2} - R_1 \qquad U_{\text{obt}} = n_1 n_2 + \frac{n_2(n_2 + 1)}{2} - R_2$$

$$= 8(8) + \frac{8(9)}{2} - 61.5 \qquad\qquad = 8(8) + \frac{8(9)}{2} - 74.5$$

$$= 64 + 36 - 61.5 = 38.5 \qquad = 64 + 36 - 74.5 = 25.5$$

Thus,

$$U_{\text{obt}} = 25.5$$
$$U'_{\text{obt}} = 38.5$$

STEP 2:   **Evaluate U_{obt}.** With $\alpha = 0.05_{2\,\text{tail}}$, Table C.3 is appropriate. With $n_1 = n_2 = 8$, $U_{\text{crit}} = 13$ and $U'_{\text{crit}} = 51$. Since $U_{\text{obt}} > 13$, we fail to reject H_0, and hence, we can't affirm H_1. These data do not support the hypothesis that men and women differ in abstract reasoning ability.

Assumptions Underlying the Mann–Whitney *U* Test

Since we must be able to rank-order the data to compute U_{obt} or U'_{obt}, the Mann–Whitney U test requires that the data be at least ordinal in scaling. It does not depend on the population scores being of any particular shape (e.g., normal distributions), as does the t test for independent groups. Thus, the Mann–Whitney U test can be used instead of the t test for independent groups when there is a serious violation of the normality assumption or when the data are not of interval or ratio scaling. The Mann–Whitney U test is a powerful test. However, since it uses only the ordinal property of the scores, it is not as powerful as the t test for independent groups, which uses the interval property of the scores.

THE KRUSKAL–WALLIS TEST

The *Kruskal–Wallis test* is a nonparametric test that is used with an independent groups design employing k samples. It is used as a substitute for the parametric one-way ANOVA discussed in Chapter 15, when the assumptions of that test are seriously violated. The Kruskal–Wallis test does not assume population normality or homogeneity of variance, as does parametric ANOVA, and requires only ordinal scaling of the dependent variable. It is used when violations of population normality and/or homogeneity of variance are extreme or when interval or ratio scaling is required and not met by the data. To understand this test, let's begin with an experiment.

experiment

Evaluating Two Weight Reduction Programs

A health psychologist, employed by a large corporation, is interested in evaluating two weight reduction programs she is considering using with employees of her corporation. She conducts an experiment in which 18 obese employees are randomly assigned to three conditions, with 6 subjects per condition. The subjects in condition 1 are placed on a diet that reduces their daily caloric intake by 500 calories. The subjects in condition 2 receive the same restricted diet, but in addition are required to walk 2 miles each day. Condition 3 is a control condition, in which the subjects are asked to maintain their usual eating and exercise habits. The data presented in Table 17.8 are the number of pounds lost by each subject over a 6-month period. A positive number indicates weight loss and a negative number is weight gain. Assume the data show that there is a strong violation of population

table 17.8   Data from weight reduction experiment

1 Diet		2 Diet + Exercise		3 Control	
Pounds lost	*Rank*	*Pounds lost*	*Rank*	*Pounds lost*	*Rank*
2	5	12	12	8	9
15	14	9	10	3	6
7	8	20	16	−1	4
6	7	17	15	−3	2
10	11	28	17	−2	3
14	13	30	18	−8	1
$n_1 = 6$	$R_1 = 58$	$n_2 = 6$	$R_2 = 88$	$n_3 = 6$	$R_3 = 25$

normality such that the psychologist decides to analyze the data with the Kruskal–Wallis test, rather than using parametric ANOVA.

a. What is the alternative hypothesis?
b. What is the null hypothesis?
c. What is the conclusion? Use $\alpha = 0.05$.

SOLUTION

a. Alternative hypothesis: As with parametric ANOVA, the alternative hypothesis states that at least one of the conditions affects weight loss differently than at least one of the other conditions. In the same manner as parametric ANOVA, each sample is considered a random sample from its own population set of scores. If there are k samples, there are k populations. In this example, $k = 3$. However, since this is a nonparametric test, Kruskal–Wallis makes no prediction about the population means μ_1, μ_2, or μ_3. It merely asserts that at least one of the population distributions is different from at least one of the other population distributions.
b. Null hypothesis: The samples are random samples from the same or identical population distributions. There is no prediction specifically regarding μ_1, μ_2, or μ_3.
c. Conclusion, using $\alpha = 0.05$: As usual, in evaluating H_0, we follow the two-step process: Compute the appropriate statistic and then evaluate the statistic using its sampling distribution.

> **STEP 1: Compute the appropriate statistic**. The statistic we compute for the Kruskal–Wallis test is H_{obt}. The procedure is very much like computing U_{obt} for the Mann–Whitney U test. All of the scores are grouped together and rank-ordered, assigning the rank of 1 to the lowest score, 2 to the next to lowest, and N to the highest. When this is done, the ranks for each condition or sample are summed. These procedures have been carried out for the data of the present example and entered in Table 17.8.
>
> The sums of ranks for each group have been symbolized as R_1, R_2, and R_3, respectively. For these data, $R_1 = 58$, $R_2 = 88$, and $R_3 = 25$. The Kruskal–Wallis test assesses whether these sums of ranks differ so much that it is unreasonable to consider that they come from samples that were randomly selected from the same population. The larger the differences between the sums of the ranks of each sample are, the less likely it is that the samples are from the same population.

The equation for computing H_{obt} is as follows:

$$H_{obt} = \left[\frac{12}{N(N+1)}\right]\left[\sum_{i=1}^{k}\frac{(R_i)^2}{n_i}\right] - 3(N+1)$$

$$= \left[\frac{12}{N(N+1)}\right]\left[\frac{R_1^2}{n_1} + \frac{R_2^2}{n_2} + \frac{R_3^2}{n_3} + \cdots + \frac{R_k^2}{n_k}\right] - 3(N+1)$$

where $\quad \sum_{i=1}^{k}\dfrac{(R_i)^2}{n_i} \quad$ tells us to square the sum of ranks for each sample, divide each squared value by the number of scores in the sample, and sum over samples

k = number of samples or groups

n_i = number of scores in the ith sample

n_1 = number of scores in sample 1

n_2 = number of scores in sample 2

n_3 = number of scores in sample 3

n_k = number of scores in sample k

N = number of scores in all samples combined

R_i = sum of the ranks for the ith sample

R_1 = sum of the ranks for sample 1

R_2 = sum of the ranks for sample 2

R_3 = sum of the ranks for sample 3

R_k = sum of the ranks for sample k

Substituting the appropriate values from the table into this equation, we obtain

$$H_{obt} = \left[\frac{12}{N(N+1)}\right]\left[\frac{(R_1)^2}{n_1} + \frac{(R_2)^2}{n_2} + \frac{(R_3)^2}{n_3}\right] - 3(N+1)$$

$$= \left[\frac{12}{18(18+1)}\right]\left[\frac{(58)^2}{6} + \frac{(88)^2}{6} + \frac{(25)^2}{6}\right] - 3(18+1)$$

$$= 68.61 - 57$$

$$= 11.61$$

STEP 2: Evaluate the statistic. It can be shown that, if the number of scores in each sample is 5 or more, the sampling distribution of the statistic H is approximately the same as chi-square with df = $k - 1$. In the present experiment, df = $k - 1 = 3 - 1 = 2$. From Table H, with $\alpha = 0.05$, and df = 2,

$$H_{crit} = 5.991$$

As with parametric ANOVA, the Kruskal–Wallis test is a nondirectional test. The decision rule states that

If $H_{obt} \geq H_{crit}$, reject H_0.

If $H_{obt} < H_{crit}$, retain H_0.

Since $H_{obt} > 5.991$, we reject H_0. It appears that the conditions are not equal with regard to weight loss.

Practice Problem 17.6

A business consultant is doing research in the area of management train-ing. There are two effective managerial styles: One is people-oriented and a second is task-oriented. Well-defined, static jobs are better served by the people-oriented managers, and changing, newly created jobs are better served by the task-oriented managers. The experiment being conducted investigates whether it is better to try to train managers to have both styles or whether it is better to match managers to jobs with no attempt to train in a second style. The managers for this experiment are 24 army officers, randomly selected from a large army base. The experiment involves three conditions. In condition 1, the subjects receive training in both managerial styles. After training is com-pleted, these subjects are randomly assigned to new jobs without matching style and job. In condition 2, the subjects receive no additional training but are assigned to jobs according to a match between their single managerial style and the job requirements. Condition 3 is a control condition in which subjects receive no additional training and are assigned to new jobs, like those in con-dition 1, without matching. After they are in their new job assignments for 6 months, a performance rating is obtained on each officer. The data follow. The higher the score is, the better the performance. At the beginning of the experiment, there were eight subjects in each condition. However, one of the subjects in condition 2 dropped out midway into the experiment and was not replaced. Assume the data do not meet the assumptions for the parametric one-way ANOVA.

Condition 1 Training		Condition 2 Matching		Condition 3 Control	
Score	*Rank*	*Score*	*Rank*	*Score*	*Rank*
65	8	90	21	55	3
84	16	83	15	82	14
87	19.5	76	12	71	10
53	2	87	19.5	60	6
70	9	92	22	52	1
85	17	86	18	81	13
56	4	93	23	73	11
63	7			57	5
$n_1 = 8$	$R_1 = 82.5$	$n_2 = 7$	$R_2 = 130.5$	$n_3 = 8$	$R_3 = 63$

a. What is the alternative hypothesis?
b. What is the null hypothesis?
c. What is the conclusion? Use $\alpha = 0.05$.

SOLUTION

a. Alternative hypothesis: At least one of the conditions has a different effect on job performance than at least one of the other conditions. Therefore, at least one of the population distributions is different from one of the others.

b. Null hypothesis: The conditions have the same effect on job performance. Therefore, the samples are random samples from the same or identical population distributions.

c. Conclusion, using $\alpha = 0.05$:

STEP 1:   Compute the appropriate statistic.

$$H_{obt} = \left[\frac{12}{N(N+1)} \right] \left[\frac{(R_1)^2}{n_1} + \frac{(R_2)^2}{n_2} + \frac{(R_3)^2}{n_3} \right] - 3(N+1)$$

$$= \left[\frac{12}{23(23+1)} \right] \left[\frac{(82.5)^2}{8} + \frac{(130.5)^2}{7} + \frac{(63)^2}{8} \right] - 3(23+1)$$

$$= 82.17 - 72$$

$$= 10.17$$

STEP 2:   Evaluate the statistic. In the present experiment, df $= k - 1 = 3 - 1 = 2$. From Table H, with $\alpha = 0.05$, and df $= 2$,

$$H_{crit} = 5.991$$

Since $H_{obt} > 5.991$, we reject H_0. It appears that the conditions are not equal with regard to their effect on job performance.

Assumptions Underlying the Kruskal–Wallis Test

To use the Kruskal–Wallis test, the data must be of at least ordinal scaling. In addition, there must be at least five scores in each sample to use the probabilities given in the table of chi-square.*

---

*To analyze data with fewer than five scores in a sample, see S. Siegel and N. Castellan, Jr., *Nonparametric Statistics for the Behavioral Sciences,* 2nd ed., McGraw-Hill, New York, 1988, pp. 206–212.

WHAT IS THE TRUTH? Statistics and Applied Social Research— Useful or "Abuseful"?

A front-page article in *The Wall Street Journal* discussed the possible misuses of social science research in connection with federally mandated changes in the U.S. welfare system. Excerpts from the article are reproduced here.

THINK TANKS BATTLE TO JUDGE THE IMPACT OF WELFARE OVERHAUL

Now that welfare overhaul is under way across the country, so is something else: an ideologically charged battle of the experts to label the various innovations as successes or failures.

Each side—liberal and conservative—fears the other will use early, and perhaps not totally reliable, results of surveys and studies of the impact of welfare reform to push a political agenda. And conservatives, along with other supporters of the law Congress recently passed to allow states to tinker with welfare, worry that they have the most to lose, because of the overwhelming presence of liberal scholars in the field of social science ….

Critics of the overhaul process make no bones that they intend to use research into the bill's effects to turn yesterday's angry taxpayers into tomorrow's friends of the downtrodden ….

The climate of ideological suspicions is exemplified by the California-based Kaiser Family Foundation's handling of a survey it financed of New Jersey welfare recipients. The study focused on the effect of New Jersey's controversial

rule barring additional benefits for women who have children while on welfare. Last spring, shortly before Kaiser was to announce the results, it canceled a planned news conference. Some critics suspect that Kaiser didn't like some of the answers it got, such as one showing that most recipients considered the "family cap" fair. The Rutgers University researcher who conducted the survey says he has been forbidden to discuss it. Kaiser says the suspicions are unfounded and that methodological errors destroyed the survey's usefulness ….

The hopes and fears of both sides are embodied in one of the biggest private social-policy research projects ever undertaken: a five-year, $30-million study of welfare overhaul and other elements of the "New Federalism." The kick-off of the study will be announced today by the Washington-based Urban Institute.

The institute, founded three decades ago to examine the woes of the nation's cities, has assembled a politically balanced project staff and promises to post "nonpartisan, reliable data" on the Internet for all sides to examine. But memories are still fresh of the institute's prediction last year that the law would toss one million children into poverty, so even some of its top officials fret about how the research will be received in a political culture increasingly riven between opposing ideological camps.

"Everyone wants to attach a political label to everything that comes out," says Isabel Sawhill, a former Clinton administration

budget official …. For people on both sides, "there's no such thing as unbiased information or apolitical studies anymore," she says.

Playing the Numbers

"Hopefully, the data we produce will be unassailable," says Anna Kondratas, a one-time Reagan administration aide who is the project's co-director. Yet she recalls a decade-old admonition from a Democratic congressman who didn't like her testimony about the food-stamp program: "Everybody's entitled to his own statistics …."

In recent years, evaluations of high-profile social programs have often found them less beneficial than many political liberals had hoped. "We are in desperate need to learn about what works," says Doug Nelson, president of the Annie E. Casey Foundation, the philanthropic organization that is the largest single backer of the Urban Institute study.

Assessing the crazy quilt of state welfare innovations poses an immense challenge for researchers of any ideological stripe. With so many policies changing at once, all involving potential effects on employment, childbearing and family life, figuring out which questions to ask may be as difficult as finding the answers. And the swirling cross-currents over race, economics and values—the possible conflict between attempts to reduce illegitimate births and attempts to prevent increases in abortion, for example—make determinations of "success" or "failure" highly subjective.

That is precisely why backers of the new law, which aims to reduce welfare spending by some $54 billion over seven years, are bracing for a flood of critical studies.

"Social scientists want to help children and families, and they believe the way to do that is to give them more benefits," says Ron Haskins, a former developmental psychologist at the University of North Carolina who now heads the staff of a House welfare subcommittee. As a result, adds the Heritage Foundation's Robert Rector, most studies of the law's effect "will emphasize a bogus measure of material poverty" while shortchanging shifts in attitudes, values and behavior that may be hard to gauge and slower to manifest themselves ....

That contentious atmosphere has embroiled the Kaiser Foundation's $90,000 survey of

New Jersey welfare recipients. Ted Goertzel, a Rutgers University sociologist not involved in the survey, notes rumors that ideology played a role in Kaiser's decision not to release the data. "The recipients said they agreed with the family cap," he says. "If you do research and don't like the results, are you obligated to release it? There's an ethical question here."

I've included this article at the end of this chapter because chi-square is one of the major inference tests used in social science research. I think you will agree that this article poses some very interesting and important questions regarding applied research in this area. For example, in applied social science research, is it really true that "there is no such thing as unbiased information"? Was the Democratic congressman correct in stating, "Everybody's entitled to his own statistics"? Is it really true that "most studies of the law's effect 'will emphasize a bogus measure of material poverty, while shortchanging shifts in attitudes, values and behavior that may be hard to gauge and slower to manifest themselves'"? If so, how will we ever find out whether this and other social programs really work? Or doesn't it matter whether we ever find out what social programs really work? Would it be better just to "pontificate" out of one's own private

social values (Method of Authority), or perhaps just to cite individual case examples? Or do you agree with Doug Nelson that, "We are in desperate need to learn about what works"? If so, how can appropriate data be collected?

What do you think about the notion that social scientists with strong political views will go out and do research, deliberately biasing their research instruments so that the data will confirm their own political views? If it is true, does that mean we should stop funding such research? How about the ordinary layperson: How will he or she be able to intelligently interpret the results of such studies? Does this mean students should stop studying statistics, or quite the contrary, that students need to learn even more about statistics so that they will not be taken in by poor, pseudo, or biased research? Finally, what do you think about the situation where an organization conducts socially relevant research and the findings turn out to be against the interest of the organization? Does the organization have an obligation to inform the public? Is it unethical if it doesn't?

■ SUMMARY

In this chapter, I discussed nonparametric statistics. Nonparametric inference tests depend considerably less on population characteristics than do parametric tests. The z, t, and F tests are examples of parametric tests; the sign test and the Mann–Whitney U test are examples of nonparametric tests. Parametric tests are used when possible because they are more powerful and versatile. However, when the assumptions of the parametric tests are violated, nonparametric tests are often used.

One of the most frequently used inference tests for analyzing nominal data is the nonparametric test called chi-square (χ^2). It is appropriate for analyzing frequency data dealing with one or two variables. Chi-square essentially measures the discrepancy between the observed frequency (f_o) and the expected frequency (f_e) for each of the cells in a one-way or two-way table. In equation form, $\chi^2_{obt} = \Sigma(f_o - f_e)^2/f_e$, where the summation is over all the cells. In single-variable situations, the data are presented in a one-way table and the various expected frequency values are determined on an *a priori* basis. In two-variable situations, the frequency data are presented in a contingency table, and we are interested in determining whether there is a relationship between the two variables. The null hypothesis states that there is no relationship—that the two variables are independent. The alternative hypothesis states that the two variables are related. The expected frequency for each cell is the frequency that would be expected if sampling is random from a population where the proportions for each category of one variable are equal for each category of the other variable. Since the population proportions are unknown, their expected values under the null hypothesis are estimated from the sample data, and the expected frequencies are calculated using these estimates.

The obtained value of χ^2 is evaluated by comparing it with χ^2_{crit}. If $\chi^2_{obt} \geq \chi^2_{crit}$, we reject the null hypothesis. The critical value of χ^2 is determined by the sampling distribution of χ^2 and the alpha level. The sampling distribution of χ^2 is a family of curves that varies with the degrees of freedom. In the one-variable experiment, df $= k - 1$. In the two-variable situation, df $= (r - 1)(c - 1)$. A basic assumption of χ^2 is that each subject can have only one entry in the table. A second assumption is that the expected frequency in each cell must be of a certain minimum size. The use of χ^2 is not limited to nominal data, but regardless of the scaling, the data must finally be divided into mutually exclusive categories and the cell entries must be frequencies.

The Wilcoxon matched-pairs signed ranks test is a nonparametric test that is used with a correlated groups design. The statistic calculated is T_{obt}. Determination of T_{obt} involves four steps: (1) finding the difference between each pair of scores, (2) ranking the absolute values of the difference scores, (3) assigning the appropriate sign to the ranks, and (4) separately summing the positive and negative ranks. T_{obt} is the lower sum. It is evaluated by comparison with T_{crit}. If $T_{obt} \leq T_{crit}$, we reject H_0. The Wilcoxon signed ranks test requires that (1) the within-pair scores be at least of ordinal scaling and (2) the difference scores also be at least of ordinal scaling. This test serves as an alternative to the t test for correlated groups when the assumptions of the t test have not been met. It is more powerful than the sign test, but not as powerful as the t test.

The Mann–Whitney U test analyzes the degree of separation between the samples in a two-group, independent groups experiment. The less the separation, the more reasonable chance is as the underlying explanation. For any analysis, two statistics are computed. Both indicate the same degree of separation. The lower value is arbitrarily called U_{obt}, and the higher value is called U'_{obt}. Tables C.1–C.4 give the critical values of U and U'. If $U_{obt} \leq U_{crit}$, reject H_0 and affirm H_1. If $U'_{obt} \geq U'_{crit}$, reject H_0 and affirm H_1. Otherwise, we retain H_0. The Mann–Whitney U test is appropriate for an independent groups design where the data are at least ordinal in scaling. It is a powerful test, often used in place of Student's t test when the data do not meet the assumptions of the t test.

The Kruskal–Wallis test is used as a substitute for one-way parametric ANOVA. It uses the independent groups design with k samples. The null hypothesis asserts that the k samples are random samples from the same or identical population distributions. No attempt is made to specifically test for population mean differences, as is the case with parametric ANOVA. The statistic computed is H_{obt}. If the number of scores in each sample is five or more, the sampling distribution of H_{obt} is close enough to that of chi-square to use the latter in determining H_{crit}. If $H_{obt} \geq H_{crit}$, H_0 is rejected. To compute H_{obt}, the scores of the k samples are combined and rank-ordered, assigning 1 to the lowest score. The ranks are then summed for each sample. Kruskal–Wallis tests whether it is reasonable to consider the summed ranks for each sample to be due to random sampling from a single population set of scores. The greater the differences between the sum of ranks for each sample are, the less tenable is the null hypothesis. This test assumes that the dependent variable is measured on a scale that is of at least ordinal scaling. There must also be five or more scores in each sample to validly use the chi-square sampling distribution.

■ IMPORTANT NEW TERMS

Chi-square (χ^2) (p. 484)
Contingency table (p. 489)
Degree of separation (p. 502)
Expected frequency (f_e) (p. 485)

Kruskal–Wallis test (H) (p. 507)
Mann–Whitney U test (U or U') (p. 501)
Marginals (p. 491)

Observed frequency (f_o) (p. 485)
Wilcoxon matched-pairs signed ranks test (T) (p. 498)

■ QUESTIONS AND PROBLEMS

1. Briefly identify or define the terms in the Important New Terms section.
2. What is the underlying rationale for the determination of f_e in the two-variable experiment?
3. What are the assumptions underlying chi-square?
4. In situations involving more than 1 degree of freedom, the χ^2 test is nondirectional. Is this statement correct? Explain.
5. What distinguishes parametric from nonparametric tests? Explain, giving some examples.
6. Are parametric tests preferable to nonparametric tests? Explain.
7. When might we use a nonparametric test? Give an example.
8. Under what conditions might one use the Wilcoxon signed ranks test?
9. Compare the Wilcoxon signed ranks test with the sign test and the t test for correlated groups with regard to power. Explain any differences.
10. What are the assumptions of the Wilcoxon signed ranks test?
11. In a two-condition, independent groups experiment, how is the degree of separation between samples affected by the size of real effect?
12. Under what conditions might one use the Mann–Whitney U test?
13. What are the assumptions underlying the Mann–Whitney U test?
14. Compare the power of Student's t test and the Mann–Whitney U test.
15. What are the assumptions underlying the Kruskal–Wallis test?
16. A researcher is interested in whether there really is a prevailing view that overweight people are more jolly. A random sample of 80 individuals was asked the question, "Do you believe fat people are more jolly?" The following results were obtained:

Yes	No	
44	36	80

Using $\alpha = 0.05$, what is your conclusion? social

17. A study was conducted to determine whether big-city and small-town dwellers differed in their helpfulness to strangers. In this study, the investigators rang the doorbells of strangers living in New York City or small towns in the vicinity. They explained they had misplaced the address of a friend living in the neighborhood and asked to use the phone. The following data show the number of individuals who admitted or did not admit the strangers (the investigators) into their homes:

	Helpfulness to Strangers		
	Admitted strangers into their home	Did not admit strangers into their home	
Big-city dweller	60	90	150
Small-town dweller	70	30	100
	130	120	250

Do big-city dwellers differ in their helpfulness to strangers? Use $\alpha = 0.05$ in making your decision. social

18. Because of rampant inflation, the government is considering imposing wage and price controls. A government economist, interested in determining whether there is a relationship between occupation and attitude toward wage and price controls, collects the following data. The data show for each occupation the number of individuals in the sample who were *for* or *against* the controls:

	Attitude Toward Wage and Price Controls		
	For	*Against*	
Labor	90	60	150
Business	100	150	250
Professions	110	90	200
	300	300	600

Do these occupations differ regarding attitudes toward wage and price controls? Use $\alpha = 0.01$ in making your decision. I/O

19. The head of the marketing division of a leading soap manufacturer must decide among four differently styled wrappings for the soap. To provide a database for the decision, he has the soap placed in the different wrapping styles and distributed to five supermarkets. At the end of 2 weeks, he finds that the following amounts of soap were sold:

Wrapping A	Wrapping B	Wrapping C	Wrapping D	
90	98	130	82	400

Is there sufficient basis for making a decision among wrappings? If so, which should he pick? Use $\alpha = 0.05$. I/O

20. A researcher believes that individuals in different occupations will show differences in their ability to be hypnotized. Six lawyers, six physicians, and six professional dancers are randomly selected for the experiment. A test of hypnotic susceptibility is administered to each. The results are shown in the next column. The higher the score, the higher the hypnotizability. Assume the data violate the assumptions required for use of the *F* test, but are at least of ordinal scaling. Using $\alpha = 0.05$, what is your conclusion?

Condition 1 Lawyers	Condition 2 Physicians	Condition 3 Dancers
26	14	30
17	19	21
27	28	35
32	22	29
20	25	37
25	15	34

cognitive, social

21. A professor of religious studies is interested in finding out whether there is a relationship between church attendance and educational level. Data are collected on a sample of individuals who completed only high school and on another sample who received a college education. The following are the resultant frequency data:

	Church Attendance		
	Attend regularly	*Do not attend regularly*	
High school	88	112	200
College	56	104	160
	144	216	360

What is your conclusion? Use $\alpha = 0.05$. social

22. A coffee manufacturer advertises that, in a recent experiment in which their brand (brand A) was compared with the other four leading brands of coffee, more people preferred their brand to the other four. The data from the experiment are given here:

	Coffee Brand				
A	*B*	*C*	*D*	*E*	
60	45	52	43	50	250

Do you believe the ad to be misleading? Use $\alpha = 0.05$ in making your decision. I/O

23. A study was conducted to determine whether there is a relationship between the amount of contact white housewives have with blacks and changes in their attitudes toward blacks. In this study, the changes in attitude toward blacks were measured for white housewives who had moved into segregated public housing projects where there was little daily contact with blacks and for white housewives who had

moved into fully integrated public housing projects where there was a great deal of contact. The following frequency data were recorded:

Attitude Toward Blacks

	Less favorable	No change	More favorable	
Segregated housing proj.	9	42	24	75
Integrated housing proj.	7	46	72	125
	16	88	96	200

Based on these data, what is your conclusion? Use $\alpha = 0.05$. social

24. A psychologist investigates the hypothesis that birth order affects assertiveness. Her subjects are 20 young adults between 20 and 25 years of age. There are seven first-born, six second-born, and seven third-born subjects. Each subject is given an assertiveness test, with the following results. High scores indicate greater assertiveness. Assume the data are so far from normally distributed that the F test can't be used, but the data are at least of ordinal scaling. Use $\alpha = 0.01$ to evaluate the data. What is your conclusion?

Condition 1 First-Born	Condition 2 Second-Born	Condition 3 Third-Born
18	18	7
8	12	19
4	3	2
21	24	30
28	22	18
32	1	5
10		14

social

25. An investigator believes that students who rank high in certain kinds of motives will behave differently in gambling situations. To investigate this hypothesis, the investigator randomly samples 50 students high in affiliation motivation, 50 students high in achievement motivation, and 50 students high in power motivation. The students are asked to play the game of roulette, and a record is kept of the bets they make. The data are then grouped into the number of subjects with each kind of motivation who make bets involving low, medium, and high risk.

Low risk means they make bets involving low odds (even money or less), medium risk involves bets of medium odds (from 2 to 1 to 5 to 1), and high risk involves playing long shots (from 17 to 1 to 35 to 1). The following data are obtained:

Kind of Motive

	Affiliation	Achievement	Power	
Low risk	26	13	9	48
Med. Risk	16	27	14	57
High risk	8	10	27	45
	50	50	50	150

Using $\alpha = 0.05$, is there a relationship between these different kinds of motives and gambling behavior? How do the groups differ? social

26. A major oil company conducts an experiment to assess whether a film designed to tell the truth about, and also promote more favorable attitudes toward, large oil companies really does result in more favorable attitudes. Twelve individuals are run in a replicated measures design. In the "Before" condition, each subject fills out a questionnaire designed to assess attitudes toward large oil companies. In the "After" condition, the subjects see the film, after which they fill out the questionnaire. The following scores were obtained. High scores indicate more favorable attitudes toward large oil companies.

Before	After
43	45
48	60
25	22
24	33
15	7
18	22
35	41
28	21
41	55
28	33
34	44
12	23

Analyze the data using the Wilcoxon signed ranks test with $\alpha = 0.05_{1 \text{ tail}}$. What do you conclude? I/O

27. In Chapter 14, Problem 18, p. 388, an experiment was conducted to evaluate the effect of decreases in frontalis muscle tension on headaches. The number of headaches experienced in a 2-week baseline period was recorded in nine subjects who had been experiencing tension headaches. Then the subjects were trained to lower frontalis muscle tension using biofeedback, after which the number of headaches in another 2-week period was again recorded. The data are again shown here.

| | No. of Headaches | |
Subject No.	Baseline	After training
1	17	3
2	13	7
3	6	2
4	5	3
5	5	6
6	10	2
7	8	1
8	6	0
9	7	2

In that problem, the sampling distribution of $\overline{D}$ was assumed to be normally distributed, and the analysis was conducted using the t test. For this problem assume the t test cannot be used because of an extreme violation of its normality assumption. Use the Wilcoxon signed ranks test to analyze the data. What do you conclude, using $\alpha = 0.05_{2 \text{ tail}}$? clinical, health

28. In Chapter 14, Problem 14, p. 387, an experiment was conducted to determine if an experimental birth control pill has the side effect of changing blood pressure. Ten women were randomly sampled from the city in which you live. Five of them were given a placebo for a month and then their diastolic blood pressure was measured. Then they were switched to the birth control pill for a month and again blood pressure was measured. The other five women were given the birth control pill first for a month, followed by the placebo for a month. The blood pressure readings are again shown here.

| | Diastolic Blood Pressure | |
Subject No.	Birth control pill	Placebo
1	108	102
2	76	68
3	69	66
4	78	71
5	74	76
6	85	80
7	79	82
8	78	79
9	80	78
10	81	85

In that problem, the sampling distribution of $\overline{D}$ was assumed to be normally distributed, and the analysis was conducted using the t test for correlated groups. For this problem, assume the data are so far from normally distributed as to invalidate use of the t test for correlated groups. Analyze the data with the Wilcoxon signed ranks test. What do you conclude, using $\alpha = 0.01_{2 \text{ tail}}$? biological, health, social

29. A social scientist believes that university theology professors are more conservative in political orientation than their colleagues in psychology. A random sample of 8 professors from the theology department and 12 professors from the psychology department at a local university are given a 50-point questionnaire that measures the degree of political conservatism. The following scores were obtained. Higher scores indicate greater conservatism.

a. What is the alternative hypothesis? In this case, assume a nondirectional hypothesis is appropriate because there are insufficient theoretical and empirical bases to warrant a directional hypothesis.

b. What is the null hypothesis?

c. What is your conclusion? Use the Mann–Whitney U test and $\alpha = 0.05_{2 \text{ tail}}$.

Theology Professors	Psychology Professors
36	13
42	25
22	40
48	29
31	10
35	26
47	43
38	17
	12
	32
	27
	32

social

30. An ornithologist thinks that injections of follicle-stimulating hormone (FSH) increase the singing rate of his captive male cotingas (birds). To test this hypothesis, he randomly selects 20 singing cotingas and divides them into two groups of 10 birds each. The first group receives injections of FSH and the second gets injections of saline solution, as a control for the trauma of receiving an injection. He then records the singing rate (in songs per hour) for both groups. The results are given in the following table. Note that two of the FSH birds escaped during injection and were not replaced.

Saline	FSH
17	10
31	29
14	37
12	41
29	16
23	45
7	34
19	57
28	
3	

a. What is the alternative hypothesis? Use a directional alternative hypothesis.
b. What is the null hypothesis?
c. Using the Mann–Whitney U test and $\alpha = 0.05_{1\ \text{tail}}$, what is your conclusion? biological

31. A psychologist is interested in determining whether left-handed and right-handed people differ in spatial ability. She randomly selects 10 left-handers and 10 right-handers from the students enrolled in the university where she works and administers a test that measures spatial ability. The following are the scores (a higher score indicates better spatial ability). Note that one of the subjects did not show up for the testing.

Left-Handers	Right-Handers
87	47
94	68
56	92
74	73
98	71
83	82
92	55
84	61
76	75
	85

a. What is the alternative hypothesis? Use a nondirectional hypothesis.
b. What is the null hypothesis?
c. Using the Mann–Whitney U test and $\alpha = 0.05_{2\ \text{tail}}$, what do you conclude? cognitive

32. A university counselor believes that hypnosis is more effective than the standard treatment given to students who have high test anxiety. To test his belief, he randomly divides 22 students with high test anxiety into two groups. One of the groups receives the hypnosis treatment, and the other group receives the standard treatment. When the treatments are concluded, each student is given a test anxiety questionnaire. High scores on the questionnaire indicate high anxiety. Following are the results:

Hypnosis Treatment	Standard Treatment
20	42
21	35
33	30
40	53
24	57
43	26
48	37
31	30
22	51
44	62
30	59

a. What is the alternative hypothesis? Assume there is sufficient basis for a directional hypothesis.
b. What is the null hypothesis?
c. Using the Mann–Whitney U test and $\alpha = 0.05_{1\ tail}$, what do you conclude? clinical, health

33. In Chapter 15, Problem 21, p. 437, an experiment was conducted to determine whether sleep loss affects the ability to maintain sustained attention. Fifteen individuals were randomly divided into the following three groups of five subjects each: group 1, which got the normal amount of sleep (7–8 hours); group 2, which was sleep-deprived for 24 hours; and group 3, which was sleep-deprived for 48 hours. All three groups were tested on the same auditory vigilance task. Half-second tones spaced at irregular intervals were presented over a 1-hour duration. Occasionally, one of the tones was slightly shorter than the rest. The subject's task was to detect the shorter tones. The following percentages of correct detections were observed:

Normal Sleep	Sleep-Deprived for 24 Hours	Sleep-Deprived for 48 Hours
85	60	60
83	58	48
76	76	38
64	52	47
75	63	50

In that problem, the normality assumption was assumed met, and the analysis was conducted using the F test. For this problem, assume the F test cannot be used because of an extreme violation of the normality assumption. Analyze the data with the Kruskal–Wallis test, using $\alpha = 0.05$. cognitive

34. A social psychologist is interested in whether there is a relationship between cohabitation before marriage and divorce. A random sample of 150 couples that were married in the past 10 years in a midwestern city were asked if they lived together before getting married and if their marriage was still intact. The following results were obtained.

	Divorced	Still married	
Cohabited before marriage	58	42	100
Did not cohabit before marriage	18	32	50
	76	74	150

Using $\alpha = 0.05$, what do you conclude? social

35. A political scientist conducts a study to determine whether there is a relationship between gender and attitude regarding government involvement in citizen affairs. A questionnaire is sent to a random sample of 1000 adult men and women, asking the question, "As a general policy, do you prefer the government to have a large, moderate, or small involvement in citizen affairs?" The following results were obtained.

	Attitude Regarding Federal Government Involvement			
	Large	*Moderate*	*Small*	
Women	240	30	230	500
Men	180	20	300	500
	420	50	530	1000

Using $\alpha = 0.05$, what do you conclude? I/O, social

36. Medical experts have long noticed that blacks do not receive the latest high-tech treatments. To determine whether physician bias contributed to this phenomenon, social psychologists analyzed Medicare records of 150 black and 150 white randomly selected heart attack patients who were treated either by a black or white physician. A different physician was required for each patient record used. The variable of interest

is whether the patients received an angiogram. The following data were collected.

Patients Receiving Angiograms

Physician	White	Black	
White	72	48	120
Black	52	28	80
	124	86	200

Using $\alpha = 0.05$, what do you conclude? health, social

37. A family therapist living in a large midwestern city is concerned that the proportion of single-father homes is increasing. The therapist finds that two relevant studies have been conducted. Both studies randomly surveyed 1000 families living in the city and gave information regarding single-father homes. The first was conducted in 1996; it reported there were 50 single-father homes. The second was conducted in 2002; it reported 76 such homes. If you were the therapist, what would you conclude? Use $\alpha = 0.05$ in making your decision. clinical, social

38. A public health researcher believes that smoking affects the gender of offspring. He records the gender of newborns that are delivered in local hospitals over a 1-year period. He also interviews the parents of the newborns to determine their degree of cigarette smoking. The following data are collected.

Offspring

Cigarette Smoking	Boys	Girls	
Neither parent smokes at least a pack a day	60	40	100
One parent smokes at least a pack a day	57	43	100
Both parents smoke at least a pack a day	18	32	50
	135	115	250

What is the conclusion? Use $\alpha = 0.05$. health, social

39. The director of the athletic department of a major state university is considering adding another women's varsity team. She is trying to decide among volleyball, soccer, and softball. A survey of 750 undergraduate women revealed the following first-choice preferences.

Volleyball	Soccer	Softball	
250	350	150	750

Does the survey reveal a reliable preference? Use $\alpha = 0.05$ in making your decision. I/O

40a. The Jones survey company conducted a national survey to see if religious sentiment in the United States changed after the terrorist attacks on the Twin Towers in New York City and the Pentagon in Washington, DC, on September 11, 2001. The survey of 1100 Americans was conducted 2 weeks after the attack; the question asked was, "Did you attend church in the past week?" Fortunately for comparison purposes, 6 months before the attack, the company had conducted a similar survey of 900 Americans, asking the same question. The data follow.

	Yes	No	
6 Months Preattack	360	540	900
2 Weeks Postattack	660	440	1100
	1020	980	2000

Using $\alpha = 0.05$, what do you conclude? I/O, social

40b. One year after the attacks, the Jones company conducted another national survey of 1100 Americans to determine whether the increase in religious sentiment following the attacks was still evident. To make this determination, the company used the data from their 6-month preattack and 1-year postattack surveys. The data follow.

	Yes	No	
6 Months Preattack	360	540	900
1 Year Postattack	420	680	1100

Using $\alpha = 0.05$, what do you conclude this time? I/O, social

What Is the Truth? **Questions**

Statistics and Applied Social Research — Useful or "Abuseful"?

1. a. In applied social science research, is it really true that "there is no such thing as unbiased information?" If so, why bother to do the research?

   b. Was the Democratic congressman correct in stating, "Everybody's entitled to his own statistics?" Discuss.

   c. What do you think about the notion that social scientists with strong political views will go out and do research, deliberately biasing their research instruments so that the data will confirm their own political views?

■ SPSS ILLUSTRATIVE EXAMPLE 17.1 _____

The general operation of SPSS and data entry are described in Appendix E, *Introduction to SPSS*. In analyzing the data of a Chi-square problem, SPSS computes the value of **Pearson Chi-Square** (the same thing as our χ^2_{obt}) and the probability of getting this value or greater if chance alone is at work. It calls this probability **Asymp. Sig. (2-sided)**. **Asymp. Sig. (2-sided)** is another way to represent p(2-sided) or p(2-tailed).

The decision rule we will follow to evaluate H_0 and H_1 is as follows.

> If **Asymp. Sig. (2-sided)** $\leq \alpha$, reject H_0 and affirm H_1.
> If **Asymp. Sig. (2-sided)** $> \alpha$, retain H_0; cannot affirm H_1.

example

Use SPSS to analyze the data given in the illustrative experiment described in Chapter 17 of the textbook, p. 489. For convenience the experiment is repeated here.

Suppose a bill that proposes to lower the legal age for drinking to eighteen is pending before the state legislature. A political scientist living in the state is interested in determining whether there is a relationship between political affiliation and attitude toward the bill. A random sample of 200 registered Republicans and 200 registered Democrats is sent letters explaining the scientist's interest and asking the recipients whether they are in favor of the bill, are undecided, or are against the bill. Strict confidentiality is assured. A self-addressed envelope is included to facilitate responding. Answers are received from all 400 Republicans and Democrats. The results are shown here.

	Attitude			Row Marginal
	For	Undecided	Against	
Republican	68	22	110	200
Democrat	92	18	90	200
Column Marginal	160	40	200	400

What do you conclude about the independence of *Attitude* and *Political Affiliation*? Use $\alpha = 0.05$.

SOLUTION

STEP 1: Enter the Data. To analyze the data, SPSS must be told the observed frequency scores (f_o), and the row and column number of each f_o score. Let's enter this information now.

1. Enter the observed frequency scores (f_o) in the first column (**VAR00001**) of the Data Editor. They may be entered in any order.
2. Enter the row number of each f_o score in the second column (**VAR00002**) of the Data Editor.
3. Enter the column number of each f_o score in the third column (**VAR00003**) of the Data Editor.

The resulting Data Editor is shown here. For now, ignore the variable name row; we name the variables in the next step.

	f_obs	row_num	col_umn	va
1	68.00	1.00	1.00	
2	22.00	1.00	2.00	
3	110.00	1.00	3.00	
4	92.00	2.00	1.00	
5	18.00	2.00	2.00	
6	90.00	2.00	3.00	
7				

STEP 2: Name the Variables. In this example, we will give the default variables **VAR00001**, **VAR00002**, and **VAR00003** the new names of *f_obs, row_num, and col_num*, respectively. To do so,

1. **Click** the **Variable View** tab in the lower left corner of the Data Editor; then **replace VAR00001** with **f_obs** and then **press Enter.**	**f_obs** is entered as the name of the dependent variable, replacing **VAR00001**.
2. **Replace VAR00002** with **row_num** and then **press Enter.**	**row_num** is entered as the variable name, replacing **VAR00002**.
3. **Replace VAR00003** with **col_num** and then **press Enter.**	**col_num** is entered as the variable name, replacing **VAR00003**.

STEP 3: Analyze the Data. The appropriate test is Chi-square. To have SPSS do the analysis using this test,

1. **Click Data** on the menu bar at the top, then **click Weight Cases….**	This produces the **Weight Cases** dialog box, with **f_obs** highlighted in the large box on the left.
2. **Click Weight cases by**; then **click** the **arrow** for the **Frequency Variable:** box.	This moves **f_obs** into the **Frequency Variable:** box, identifying **f_obs** as name of the observed frequency variable.
3. **Click OK.**	This moves you back to the Data Editor, from which you can carry out the rest of the analysis.
4. **Click Analyze** on the menu bar at the top; then **select Descriptive Statistics;** then **click Crosstabs….**	This produces the **Crosstabs** dialog box. This is where you tell SPSS the variable names in the Data Editor that contain the row and column coding values.

5. Click **row_num** in the large box on the left; then **click** the **arrow** for the **Row(s):** box on the right.

This moves **row_num** into the **Row(s):** box on the right, telling SPSS that **row_num** contains the row coding.

6. Click **col_num** in the large box on the left; then **click** the **arrow** for the **Column(s):** box on the right.

This moves **col_num** into the **Column(s):** box on the right, telling SPSS that **col_num** contains the column coding. This and the preceding step enable SPSS to identify the row and column location in the contingency table of each **f_obs** score.

7. Click the **Statistics…** button on the upper right; then **click Chi-square**; then **click Continue**.

This tells SPSS to do a chi-square analysis when it gets the **OK** command and then returns you to the **Crosstabs** dialog box where you can give the **OK** command.

8. Click **OK.**

SPSS computes and analyzes the data and outputs the results.

Analysis Results

The results are displayed in three tables, **the Case Processing Summary**, the **row_num*col_num Crosstabulation**, and **Chi-Square Tests** tables. We have only shown below the latter two tables because the **Case Processing Summary** table is not useful for our analysis.

row_num * col_num Crosstabulation

Count

		col_num			Total
		1.00	2.00	3.00	
row_num	1.00	68	22	110	200
	2.00	92	18	90	200
Total		160	40	200	400

Chi-Square Tests

	Value	df	Asymp. Sig. (2-sided)
Pearson Chi-Square	6.000[a]	2	.050
Likelihood Ratio	6.018	2	.049
Linear-by-Linear Association	5.425	1	.020
N of Valid Cases	400		

a. 0 cells (.0%) have expected count less than 5. The minimum expected count is 20.00.

The **row_num*col_num Crosstabulation** table displays the contingency table that SPSS used. You can use it to make sure your Data Editor entries were correct. You can see that it is the same as the contingency table shown in the statement of the illustrative example.

The **Chi-Square Tests** table contains the information we need for evaluating H_0. Instead of using the symbol χ^2_{obt} to designate the Chi-square statistic, SPSS uses *Pearson Chi-Square*. From the table, **Pearson Chi-Square = 6.000** and **Asymp. Sig. (2-sided) = .050**. Since $.050 \leq 0.05$, we reject H_0 and conclude that *Attitude* and *Political Affiliation* are not independent; there appears to be a relationship between them. Please note that the results and conclusions using SPSS are the same as given in the textbook.

■ SPSS ADDITIONAL PROBLEMS

1. A computer manufacturer believes that there is a relationship between the age of computer users and preference for desktop or laptop computers. An experiment is conducted in which a random sample is obtained of 50 customers for each age group for the following ages: 12–25, 40–45, and 60–65 years old. Subjects are asked their preference for desktop or laptop computers and the results are shown in the table below. Entries are the number of customers in each age group that preferred each computer type.

Computer Type	Age 20–25	40–45	60–65	
Laptops	42	30	22	94
Desktops	8	20	28	56
	50	50	50	150

What do you conclude? Use $\alpha = 0.05$.

2. A local microbrewery company is doing some marketing research. The company wants to determine if there is a relationship between *gender* and *preference for its light, dark, or regular beer*. A random sample of 100 women and 100 adult men is taken from the beer-drinking population living in the city where the company is located. Each individual is allowed to sample a small amount of each beer type and then is asked which one is preferred. The order in which the three types of beers are presented is counterbalanced over subjects. The following contingency table presents the resulting data. Entries are the number of men and women who preferred each type of beer.

Gender	Beer Types Light	Dark	Regular	
Men	25	35	40	100
Women	58	12	30	100
	83	47	70	200

Is there a relationship between *gender* and *preference for beer type*? Use $\alpha = 0.05$.

■ NOTES

17.1 When df = 1, directional alternative hypotheses can be tested with χ^2. With df = 1, $z_{obt} = \sqrt{\chi^2_{obt}}$. Therefore, we can convert χ^2_{obt} to z_{obt} and evaluate z_{obt} using z_{crit} for the appropriate one-tailed alpha level. Of course, the difference between f_o and f_e must be in the predicted direction to perform this test.

■ **ONLINE STUDY RESOURCES**

CENGAGE **brain** .com

Login to CengageBrain.com to access the resources your instructor has assigned. For this book, you can access the book's companion website for chapter-specific learning tools including Know and Be Able to Do, practice quizzes, flash cards, and glossaries, and a link to Statistics and Research Methods Workshops.

aplia™

If your professor has assigned Aplia homework:

1. Sign in to your account.
2. Complete the corresponding homework exercises as required by your professor.
3. When finished, click "Grade It Now" to see which areas you have mastered and which need more work, and for detailed explanations of every answer.

Visit **www.cengagebrain.com** to access your account and to purchase materials.

© Strmko / Dreamstime.com

18

Review of Inferential Statistics

CHAPTER OUTLINE

LEARNING OBJECTIVES

After completing this review chapter, you should be able to:

- Understand the big picture in regard to hypothesis testing and inferential statistics, utilizing the knowledge gained from this textbook.
- Select and use the appropriate inference test depending on scaling of data, experiment design, number of groups, and whether assumptions have been violated.
- Use this chapter to review important aspects of the inference tests covered in the textbook.

INTRODUCTION

We have covered a lot of material since we began our discussion of hypothesis testing with the sign test. I shall begin our review of this material with the most important terms and concepts pertaining to the general process of hypothesis testing. Then we shall discuss the general process itself. From there, we shall summarize the experimental designs and the inference tests used with each design. Since this material is very logical and interconnected, I hope this review will help bring closure and greater insight to the topic of inferential statistics.

TERMS AND CONCEPTS

Alternative hypothesis (H_1) The alternative hypothesis states that the differences in scores between conditions are due to the action of the independent variable. The alternative hypothesis may be nondirectional or directional. A nondirectional hypothesis states that the independent variable has an effect on the dependent variable but doesn't specify the direction of the effect. A directional hypothesis states the direction of the expected effect.

Null hypothesis (H_0) The null hypothesis is set up as the logical counterpart to the alternative hypothesis such that if the null hypothesis is false, the alternative hypothesis must be true. Conversely, if the null hypothesis is true, the alternative hypothesis must be false. The null hypothesis for a nondirectional alternative hypothesis is that the independent variable has no effect on the dependent variable. For a directional alternative hypothesis, the null hypothesis states that the independent variable does not have an effect in the direction specified.

Null-hypothesis population(s) The null-hypothesis population(s) is the set or sets of scores that would result if the experiment were done on the entire population and the independent variable had no effect. In a single sample design, it is the population with known μ. In a replicated measures design, it is the population of difference scores with $\mu_D = 0$ or $P = 0.50$. In an independent groups design, there are as many populations as there are groups and the samples are random samples from populations where $\mu_1 = \mu_2 = \mu_3 = \mu_k$.

Sampling distribution The sampling distribution of a statistic gives all the values the statistic can take, along with the probability of getting that value if chance alone is responsible or if sampling is random from the null-hypothesis population(s). This distribution can be derived theoretically from basic probability, as we did with the sign test, or empirically, as with the z, t, and F tests. Three steps are involved in constructing the sampling distribution of a statistic using the empirical approach. First, all possible different samples of size N that can be formed from the population are determined. Second, the statistic for each of the samples is calculated. Finally, the probability of getting each value of the statistic is calculated under the assumption that sampling is random from the null-hypothesis population(s).

Critical region for rejection of H_0   The critical region for rejection of H_0 is the area under the curve that contains all the values of the statistic that will allow rejection of the null hypothesis. The critical value of a statistic is that value of the statistic that bounds the critical region. It is determined by the alpha level.

Alpha level (α)   The alpha level is the threshold probability level against which the obtained probability is compared to determine the reasonableness of the null hypothesis. It also determines the critical region for rejection of the null hypothesis. Alpha is usually set at 0.05 or 0.01. The alpha level is set at the beginning of an experiment and limits the probability of making a Type I error.

Type I error   A Type I error occurs when the null hypothesis is rejected and it is true.

Type II error   A Type II error occurs when the null hypothesis is retained and it is false. Beta is equal to the probability of making a Type II error.

Power   The power of an experiment is equal to the probability of rejecting the null hypothesis if the independent variable has a real effect. It is useful to know the power of an experiment when designing the experiment and when interpreting non-significant results from an experiment that has already been conducted. Calculation of power involves two steps: (1) determining the sample outcomes that will allow rejection of the null hypothesis and (2) determining the probability of getting these outcomes under the assumed real effect of the independent variable. Power $= 1 - \beta$. Thus, as power increases, beta decreases. Power can be increased by increasing the number of subjects in the experiment, by increasing the size of real effect of the independent variable, by decreasing the variability of the data through careful experimental control and proper experimental design, and by using the most sensitive inference test possible for the design and data.

PROCESS OF HYPOTHESIS TESTING

We have seen that in every experiment involving hypothesis testing there are two hypotheses that attempt to explain the data. They are the alternative hypothesis and the null hypothesis. In analyzing the data, we always evaluate the null hypothesis and indirectly conclude with regard to the alternative hypothesis. If H_0 can be rejected, then H_1 is accepted. If H_0 is not rejected, then H_1 is not accepted.

Two steps are involved in assessing the null hypothesis. First, we calculate the appropriate statistic, and second, we evaluate the statistic. To evaluate the statistic, we assume that the independent variable has no effect and that chance alone is responsible for the score differences between conditions. Another way of saying this is that we assume sampling is random from the null-hypothesis population(s). Then we calculate the probability of getting the obtained result or any result more extreme under the previous assumption. This probability is one- or two-tailed depending on whether the alternative hypothesis is directional or nondirectional. To calculate the obtained probability, we must know the sampling distribution of the statistic. If the obtained probability is equal to or less than the alpha level, we reject H_0. Alternatively, we determine whether the obtained statistic falls in the critical region for rejecting H_0.

If it does, we reject the null hypothesis. Otherwise, H_0 remains a reasonable explanation, and we retain it.

If we reject H_0 and it is true, we have made a Type I error. The alpha level limits the probability of a Type I error. If we retain H_0 and it is false, we have made a Type II error. The power of the experiment determines the probability of making a Type II error. We have defined beta as the probability of making a Type II error. As power increases, beta decreases. By maintaining alpha sufficiently low and power sufficiently high, we achieve a high probability of making a correct decision when analyzing the data, no matter whether H_0 is true or false.

These statements apply to all experiments involving hypothesis testing. What varies from experiment to experiment is the inference test used, the statistic calculated, and its sampling distribution. The inference test used will depend on the experimental design and the data collected.

SINGLE SAMPLE DESIGNS

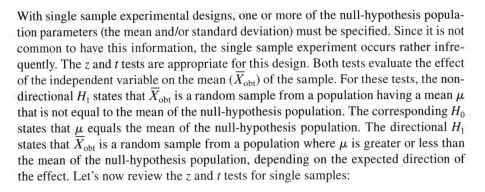

With single sample experimental designs, one or more of the null-hypothesis population parameters (the mean and/or standard deviation) must be specified. Since it is not common to have this information, the single sample experiment occurs rather infrequently. The z and t tests are appropriate for this design. Both tests evaluate the effect of the independent variable on the mean ($\overline{X}_{\text{obt}}$) of the sample. For these tests, the nondirectional H_1 states that $\overline{X}_{\text{obt}}$ is a random sample from a population having a mean μ that is not equal to the mean of the null-hypothesis population. The corresponding H_0 states that μ equals the mean of the null-hypothesis population. The directional H_1 states that $\overline{X}_{\text{obt}}$ is a random sample from a population where μ is greater or less than the mean of the null-hypothesis population, depending on the expected direction of the effect. Let's now review the z and t tests for single samples:

z Test for Single Samples

Test	Statistic Calculated	Decision Rule
z test for single samples	$z_{\text{obt}} = \dfrac{\overline{X}_{\text{obt}} - \mu}{\sigma/\sqrt{N}}$	If $|z_{\text{obt}}| \geq |z_{\text{crit}}|$, reject H_0.

General comments   The z test is used in situations in which both the mean and standard deviation of the null-hypothesis population can be specified. To evaluate H_0, we assume $\overline{X}_{\text{obt}}$ is a random sample from a population having a mean μ and standard deviation σ that are equal to the mean and standard deviation of the null-hypothesis population. The sampling distribution of $\overline{X}$ gives all the possible values of $\overline{X}$ for samples of size N and the probability of getting each value, if sampling is random from the population with a mean μ and a standard deviation σ. The sampling distribution of $\overline{X}$ has a mean $\mu_{\overline{X}} = \mu$, has a standard deviation $\sigma_{\overline{X}} = \sigma/\sqrt{N}$, and is normally shaped if the population from which the sample was drawn is normal or if $N \geq 30$, provided the population does not differ greatly from normality.

We can assess H_0 by (1) converting $\overline{X}_{\text{obt}}$ to its z-transformed value (z_{obt}) and determining the probability of getting a value as extreme as or more extreme than z_{obt} if chance alone is operating or (2) calculating z_{obt} and comparing it with z_{crit}. It is easier to do the latter. The equation for z_{obt} is given in the preceding table. The value of z_{obt} is evaluated by comparison with z_{crit}. The alpha level in conjunction with the sampling distribution of z determines the value of z_{crit}. The sampling distribution of z has a mean of 0 and a standard deviation of 1. If $\overline{X}_{obt}$ is normally distributed, then so is the corresponding z distribution and z_{crit} can be determined from Table A in Appendix D. Thus, the z test requires that $N \geq 30$ or that the population of raw scores be normally distributed.

t Test for Single Samples

Test	Statistic Calculated	Decision Rule
t test for single samples	$t_{obt} = \dfrac{\overline{X}_{obt} - \mu}{s/\sqrt{N}}$ $t_{obt} = \dfrac{\overline{X}_{obt} - \mu}{\sqrt{\dfrac{SS}{N(N-1)}}}$	If $|t_{obt}| \geq |t_{crit}|$, reject H_0.

General comments   The *t* test is used in situations in which the mean of the null-hypothesis population can be specified and standard deviation is unknown. In testing H_0, we assume $\overline{X}_{obt}$ is a random sample from a population having a mean μ equal to the mean of the null-hypothesis population and an unknown standard deviation. The *t* test is very much like the *z* test, except that since σ is unknown, we estimate it with s. When s is substituted for σ in the equation for t_{obt}, the first equation given in the table for t_{obt} results. The second equation in the table is a computational equation for t_{obt}, using the raw scores. To evaluate H_0, the value of t_{obt} is compared against t_{crit}, using the decision rule. The value of t_{crit} is determined by the alpha level and the sampling distribution of t. This distribution is a family of curves, shaped like the *z* distribution. The curves vary uniquely with degrees of freedom. The degrees of freedom for a statistic are equal to the number of scores that are free to vary in calculating the statistic. For the *t* test used with single samples, df $= N - 1$, because 1 degree of freedom is lost calculating s. The values of t_{crit} are found in Table D in Appendix D, using df and α. The *t* test has the same underlying assumptions as the *z* test. The population of raw scores should be normally distributed, or $N \geq 30$.

t Test for Testing the Significance of Pearson *r*

Test	Statistic Calculated	Decision Rule
t test for testing the significance of Pearson *r*	r_{obt}	If $|r_{obt}| \geq |r_{crit}|$, reject H_0.

General comments   To determine whether a correlation exists in the population, we must test the significance of r_{obt}. This can be done using the *t* test. The resulting equation is

$$t_{obt} = \frac{r_{obt} - \rho}{\sqrt{\dfrac{1 - r_{obt}^2}{N - 2}}}$$

By substituting t_{crit} for t_{obt} in this equation, r_{crit} can be determined for any df and α level. Once r_{crit} is known, all we need to do is compare r_{obt} with r_{crit}. The decision rule is given in the preceding table. The values of r_{crit} are found in Table E in Appendix D, using df and α. Degrees of freedom equal $N - 2$.

CORRELATED GROUPS DESIGN: TWO GROUPS

The essential feature of this design is that there are paired scores between the conditions, and the differences between the paired scores are analyzed. The paired scores can result from using the same subjects in each condition, from using identical twins, or from using subjects that have been matched in some other way. The most basic form of the design employs just two conditions: an experimental condition and a control condition. The two conditions are kept as alike as possible except for values of the independent variable, which are intentionally made different. We covered three tests for analyzing data from experiments of this design: the t test for correlated groups, the Wilcoxon matched-pairs signed ranks test, and the sign test.

t Test for Correlated Groups

Test	Statistic Calculated	Decision Rule
t test for correlated groups	$t_{obt} = \dfrac{\overline{D}_{obt} - \mu_D}{\sqrt{\dfrac{SS_D}{N(N-1)}}}$	If $|t_{obt}| \geq |t_{crit}|$, reject H_0.

General comments The t test for correlated groups analyzes the effect of the independent variable on the mean of the sample difference scores $(\overline{D}_{obt})$. If the independent variable has no effect, then $\overline{D}_{obt}$ is a random sample from a population of difference scores having a mean $\mu_D = 0$ and unknown σ_D. This situation is the same as what we encountered when using the t test for single samples (specifiable population mean but unknown standard deviation), except that we are dealing with difference scores rather than raw scores. Thus, the t test for correlated groups is identical to the t test for single samples, but it evaluates difference scores instead of raw scores.

The nondirectional H_1 states that the independent variable has an effect, in which case $\overline{D}_{obt}$ is due to random sampling from a population of difference scores where $\mu_D \neq 0$. The directional H_1 specifies that $\mu_D > 0$ (for which H_0 states that $\mu_D \leq 0$) or $\mu_D < 0$ (for which H_0 states that $\mu_D \geq 0$). H_0 is tested by assuming $\overline{D}_{obt}$ is a random sample from a population of difference scores where $\mu_D = 0$.

The statistic calculated is t_{obt} (see the preceding table), which is evaluated by comparing it with t_{crit}. The sampling distribution of t is the same as discussed in conjunction with the t test for single samples. The degrees of freedom are equal to $N - 1$, where $N =$ the number of difference scores. The values of t_{crit} are found in Table D, using df and α. The assumptions of this test are the same as those for the t test for single samples. This test is more sensitive than (1) the t test for independent groups when the correlation between the paired scores is high and (2) the Wilcoxon matched-pairs signed ranks test and the sign test, which are also appropriate for the correlated groups design.

Wilcoxon Matched-Pairs Signed Ranks Test

Test	Statistic Calculated	Decision Rule
Wilcoxon matched-pairs signed ranks test	T_{obt}	If $T_{obt} \leq T_{crit}$, reject H_0.

General comments   This is a nonparametric test that takes into account the magnitude and direction of the difference scores. It is therefore much more powerful than the sign test. Both the alternative and null hypotheses are usually stated without specifying population parameters. In analyzing the data, T_{obt} is calculated by (1) obtaining the difference score for each pair of scores, (2) rank-ordering the absolute values of the difference scores, (3) assigning the appropriate signs to the ranks, and (4) separately summing the positive and negative ranks. T_{obt} is the lower of the sums. T_{obt} is compared with T_{crit}. The values of T_{crit} are given in Table I in Appendix D, using N and α. The decision rule is shown in the preceding table. This test is recommended as an alternative to the t test for correlated groups when the assumptions of the t test are not met. The Wilcoxon signed ranks test requires that the within-pair scores be at least of ordinal scaling and that the difference scores also be at least of ordinal scaling.

Sign Test

Test	Statistic Calculated	Decision Rule
Sign test	Number of P events in a sample of size N	If the one- or two-tailed p(number of P events) $\leq \alpha$, reject H_0.

General comments   We used the sign test to introduce hypothesis testing because it is a simple test to understand. It is not commonly used in practice because it ignores the magnitude of the difference scores and considers only their direction.

   In analyzing data with the sign test, we determine the number of pluses in the sample and evaluate this statistic by using the binomial distribution. The binomial distribution is the appropriate sampling distribution when (1) there is a series of N trials, (2) there are only two possible outcomes on each trial, (3) there is independence between trials, (4) the outcomes on each trial are mutually exclusive, and (5) the probability of each possible outcome on any trial stays the same from trial to trial. The binomial distribution is given by $(P + Q)^N$, where P is the probability of a plus on any trial and Q is the probability of a minus. If the independent variable has no effect, then $P = Q = 0.50$. The nondirectional H_1 states that $P \neq Q \neq 0.50$. The directional H_1 specifies $P > 0.50$ or $P < 0.50$, depending on the expected direction of the effect.

   H_0 is tested by assuming that the number of P events in the sample is due to random sampling from a population where $P = Q = 0.50$. The one-or two-tailed p(number of P events) is compared with the alpha level to evaluate H_0. This probability is found in Table B in Appendix D, using N, number of P events, and $P = 0.50$. Alternatively, given alpha, N, and the binomial distribution, we could have also determined the critical region for rejecting H_0 (as we did with the other statistics), in which case we would compare the obtained number of P events with the critical number of P events. To use the sign test, the data must be at least ordinal in scaling and ties must be excluded from the analysis.

INDEPENDENT GROUPS DESIGN: TWO GROUPS

This design involves random sampling of subjects from the population and then random assignment of the subjects to each condition. There can be many conditions. The most basic form of the design uses two conditions, with each condition employing a different level of the independent variable. This design differs from the correlated groups design in that there is no basis for pairing scores between conditions. Analysis is performed separately on the raw scores of each sample, not on the difference scores. Both the t test and the Mann–Whitney U test are appropriate for this design.

t Test for Independent Groups

Test	Statistic Calculated	Decision Rule
t test for independent groups	$t_{obt} = \dfrac{(\overline{X}_1 - \overline{X}_2) - \mu_{\overline{X}_1 - \overline{X}_2}}{\sqrt{\left(\dfrac{SS_1 + SS_2}{n_1 + n_2 - 2}\right)\left(\dfrac{1}{n_1} + \dfrac{1}{n_2}\right)}}$	If $|t_{obt}| \geq |t_{crit}|$, reject H_0.

When $n_1 = n_2$,

$$t_{obt} = \frac{(\overline{X}_1 - \overline{X}_2) - \mu_{\overline{X}_1 - \overline{X}_2}}{\sqrt{\dfrac{SS_1 + SS_2}{n(n - 1)}}}$$

General comments   This test assumes that the independent variable affects the mean of the scores and not their variance. The mean of each sample is calculated, and then the difference between sample means $(\overline{X}_1 - \overline{X}_2)$ is determined. The t test for independent groups analyzes the effect of the independent variable on $(\overline{X}_1 - \overline{X}_2)$. The sample value $\overline{X}_1$ is due to random sampling from a population having a mean μ_1 and a variance σ_1^2. The sample value $\overline{X}_2$ is due to random sampling from a population having a mean μ_2 and a variance σ_2^2. The variance of both populations is assumed equal $(\sigma_1^2 = \sigma_2^2 = \sigma^2)$.

The sampling distribution of $\overline{X}_1 - \overline{X}_2$ has the following characteristics: (1) it has a mean $\mu_{\overline{X}_1 - \overline{X}_2} = \mu_1 - \mu_2$, (2) it has a standard deviation $\sigma_{\overline{X}_1 - \overline{X}_2} = \sqrt{\sigma^2[(1/n_1) + (1/n_2)]}$, and (3) it is normally shaped if the population from which the samples have been taken is normal. If the independent variable has no effect, then $\mu_1 = \mu_2$. The nondirectional H_1 states that $\mu_1 \neq \mu_2$. The directional H_1 states that $\mu_2 > \mu_1$ or $\mu_1 < \mu_2$, depending on the expected direction of the effect.

To assess H_0, we assume that the independent variable has no effect, in which case $\mu_1 = \mu_2$ and $\mu_{\overline{X}_1 - \overline{X}_2} = 0$. To test H_0, we could calculate z_{obt}, but we need to know σ^2 for this calculation. Since σ^2 is unknown, we estimate it using a weighted estimate from both samples. The resulting statistic is t_{obt}. Two equations for calculating t_{obt} are given in the table. The first is a general equation, and the second can be used when the ns in the two samples are equal. The degrees of freedom associated with calculating t_{obt} for the independent groups design is $N - 2$. We calculate two variances in determining t_{obt}, and we lose 1 degree of freedom for each calculation. The sampling distribution of t is as described earlier. The value of t_{obt} is evaluated by comparing it with t_{crit} according to the decision rule given in the preceding table. The values of t_{crit} are found in Table D, using df and α.

To use this test, the sampling distribution of $\mu_{\bar{X}_1 - \bar{X}_2}$ must be normally distributed. This means that the populations from which the samples were taken should be normally distributed.

In addition, to use the t test for independent groups, there should be homogeneity of variance. This test is considered robust with regard to violations of the normality and homogeneity of variance assumptions, provided $n_1 = n_2 \geq 30$. If there is a severe violation of an assumption, the Mann–Whitney U test serves as an alternative to the t test.

Mann–Whitney *U* Test

Test	Statistic Calculated	Decision Rule
Mann–Whitney U test	U_{obt} and U'_{obt}, where $$U_{\text{obt}} = n_1 n_2 + \frac{n_1(n_1 + 1)}{2} - R_1$$ $$U_{\text{obt}} = n_1 n_2 + \frac{n_2(n_2 + 1)}{2} - R_2$$	If $U_{\text{obt}} \leq U_{\text{crit}}$, reject H_0.

General comments   The Mann–Whitney U test is a nonparametric test that analyzes the degree of separation between the samples. The less the separation, the more reasonable chance is as the underlying explanation. For any analysis, there are two values that indicate the degree of separation. They both indicate the same degree of separation. The lower value is called U_{obt}, and the higher value is called U'_{obt}. The lower the U_{obt} value is, the greater the separation.

U_{obt} and U'_{obt} can be determined by using the equations given in the preceding table. For any analysis, one of the equations will yield U and the other U'. However, which yields U and which U' depends on which group is labeled group 1 and which is group 2. Since both U and U' are measures of the same degree of separation, it is necessary to evaluate only one of them.

To evaluate U_{obt}, it is compared with the critical values of U given in Tables C.1–C.4. Naturally, these values depend on the sampling distribution of U. The decision rule for rejecting H_0 is given in the preceding table.

The Mann–Whitney U test is appropriate for an independent groups design in which the data are at least ordinal in scaling. It is a powerful test, often used in place of Student's t test when the data do not meet the assumptions of the t test.

MULTIGROUP EXPERIMENTS

Although a two-group design is used fairly frequently in the behavioral sciences, it is more common to encounter experiments with three or more groups. Having more than two groups has two main advantages: (1) additional groups often clarify the interpretation of the results, and (2) additional groups allow many levels of the independent variable to be evaluated in one experiment. There is, however, one problem with doing multigroup experiments. Since many comparisons can be made, we run the risk of an inflated Type I error probability when analyzing the data. The analysis of variance technique allows us to analyze the data without incurring this risk.

One-Way Analysis of Variance, *F* Test

Test	Statistic Calculated	Decision Rule
Parametric one-way analysis of variance, *F* test	$F_{\text{obt}} = \dfrac{MS_{between}}{MS_{within}}$	If $F_{\text{obt}} \geq F_{\text{crit}}$, reject H_0.

General comments   The parametric analysis of variance uses the *F* test to evaluate the data. In using this test, we calculate F_{obt}, which is fundamentally the ratio of two independent variance estimates of a population variance σ^2. The sampling distribution of *F* is composed of a family of positively skewed curves that vary with degrees of freedom. There are two values for degrees of freedom: one for the numerator and one for the denominator. The *F* distribution (1) is positively skewed, (2) has no negative values, and (3) has a median approximately equal to 1.

The parametric analysis of variance technique can be used with both the independent groups and the correlated groups designs. We have considered only the one-way ANOVA independent groups design. The technique allows the means of all the groups to be compared in one overall evaluation, thus avoiding the inflated Type I error probability that occurs when doing many individual comparisons. Essentially, the analysis of variance partitions the total variability of the data into two parts: the variability that exists within each group (the within-groups sum of squares) and the variability that exists between the groups (the between-groups sum of squares). Each sum of squares is used to form an independent estimate of the variance of the null-hypothesis populations, σ^2. Finally, an *F* ratio is calculated where the between-groups variance estimate is in the numerator and the within-groups variance estimate is in the denominator.

The steps and equations for calculating F_{obt} are as follows:

STEP 1: Calculate the between-groups sum of squares, $SS_{between}$.

$$SS_{between} = \left[\frac{(\Sigma X_1)^2}{n_1} + \frac{(\Sigma X_2)^2}{n_2} + \frac{(\Sigma X_3)^2}{n_3} + \cdots + \frac{(\Sigma X_k)^2}{n_k} \right] - \frac{\left(\overset{\text{all scores}}{\Sigma X} \right)^2}{N}$$

STEP 2: Calculate the within-groups sum of squares, SS_{within}.

$$SS_{within} = \overset{\text{all scores}}{\Sigma} X^2 - \left[\frac{(\Sigma X_1)^2}{n_1} + \frac{(\Sigma X_2)^2}{n_2} + \frac{(\Sigma X_3)^2}{n_3} + \cdots + \frac{(\Sigma X_k)^2}{n_k} \right]$$

STEP 3: Calculate the total sum of squares, SS_{total}; check that $SS_{total} = SS_{within} + SS_{between}$.

$$SS_{total} = \overset{\text{all scores}}{\Sigma} X^2 - \frac{\left(\overset{\text{all scores}}{\Sigma X} \right)^2}{N}$$

$$SS_{total} = SS_{within} + SS_{between}$$

STEP 4: Calculate the degrees of freedom for each estimate.

$$\mathrm{df}_{between} = k - 1$$
$$\mathrm{df}_{within} = N - k$$
$$\mathrm{df}_{total} = N - 1$$

STEP 5: Calculate the between-groups variance estimate, $MS_{between}$.

$$S_B^2 \quad \underline{or} \quad MS_{between} = \frac{SS_{between}}{\mathrm{df}_{between}}$$

STEP 6: Calculate the within-groups variance estimate, MS_{within}.

$$S_W^2 \quad \underline{or} \quad MS_{within} = \frac{SS_{within}}{\mathrm{df}_{within}}$$

STEP 7: Calculate F_{obt}.

$$F_{obt} = \frac{MS_{between}}{MS_{within}}$$

The null hypothesis for the analysis of variance assumes that the independent variable has no effect and that the samples are random samples from populations where $\mu_1 = \mu_2 = \mu_3 = \mu_k$. Since the between-groups variance estimate increases with the effect of the independent variable and the within-groups variance estimate remains constant, the larger the F ratio is, the more unreasonable the null hypothesis becomes. We evaluate F_{obt} by comparing it with F_{crit}. If $F_{obt} \geq F_{crit}$, we reject H_0 and conclude that at least one of the conditions differs from at least one of the other conditions. Note that the analysis of variance technique is nondirectional.

Multiple comparisons   To determine which conditions differ from each other, *a priori* or *a posteriori comparisons* between pairs of groups are performed. *A priori comparisons* (also called *planned comparisons*) are appropriate when the comparisons have been planned in advance. No adjustment for multiple comparisons is made. *Planned* comparisons should be relatively few and should arise from the logic and meaning of the experiment. In doing the *planned* comparisons, we usually compare the means of the specified groups using the *t* test for independent groups. The value for t_{obt} is determined in the usual way, except we use MS_{within} from the analysis of variance in the denominator of the *t* equation. The t_{obt} value is compared with t_{crit}, using df_{within} and the alpha level and Table D to determine t_{crit}. The equations for calculating t_{obt} are given as follows:

$$t_{obt} = \frac{\overline{X}_1 - \overline{X}_2}{\sqrt{MS_{within}\left(\dfrac{1}{n_1} + \dfrac{1}{n_2}\right)}}$$

If $n_1 = n_2$,

$$t_{obt} = \frac{\overline{X}_1 - \overline{X}_2}{\sqrt{2\,MS_{within}/n}}$$

A posteriori, or *post hoc, comparisons* were not planned before conducting the experiment. They arise either after looking at the data or from assuming the "shotgun" approach of doing all possible mean comparisons in an attempt to gain as much information from the experiment as possible. For these reasons, comparisons made *post hoc* must correct for the increase in the probability of a Type I error that arises due to multiple comparisons. There are many techniques that do this. We have described Tukey's Honestly Significant Difference (HSD) test and the Scheffé test.

Tukey's HSD Test

Test	Statistic Calculated	Decision Rule
Tukey's HSD test	Q_{obt}, where $$Q_{obt} = \frac{\overline{X}_i - \overline{X}_j}{\sqrt{MS_{within}/n}}$$	If $Q_{obt} \geq Q_{crit}$, reject H_0.

General Comments   The HSD test is designed to compare all possible pairs of means while maintaining the Type I error rate for making the complete set of comparisons at α. The Q statistic is very much like the t statistic, but it is always positive and uses the Q distributions rather than the t distributions. The Q (Studentized range) distributions are derived by randomly taking k samples of equal n from the same population rather than just two samples as with the t distributions and determining the difference between the highest and lowest sample means. To use this test, we calculate Q_{obt} for the desired comparisons and compare Q_{obt} with Q_{crit}. The values of Q_{crit} are found in Table G in Appendix D, using k, df for MS_{within}, and α. The decision rule is given in the preceding table.

Scheffé Test

Test	Statistic Calculated	Decision Rule
Scheffé test	$$F_{Scheffé} = \frac{MS_{between \, (groups \, i \, and \, j)}}{MS_{within \, (entire \, ANOVA)}}$$	If $F_{Scheffé} \geq F_{crit}$, reject H_0.

General comments   The Scheffé test limits the probability of making a Type I error to the alpha level for all possible *post hoc* comparisons. In that regard, it is more versatile than the HSD test. However, in practice, it is often used to perform *post hoc* analysis on only pair-wise mean comparisons. It provides its extra protection against making Type I error by using $df_{between}$, MS_{within}, and F_{crit} from the entire ANOVA, rather than from the two groups being compared. Using $df_{between}$, MS_{within}, and F_{crit} from the entire ANOVA makes it harder to reject H_0 for any of the *post hoc* comparisons. Although more versatile, the Scheffé test is less powerful than the HSD test. Because of this, it is not recommended if only doing pair-wise, post hoc comparisons.

To perform the analysis on pair-wise mean comparisons, the Scheffé test computes $F_{Scheffé}$ for each pair-wise comparison and then evaluates against F_{crit}. The numerator of each $F_{Scheffé}$ is a between-groups variance estimate $MS_{between \, (groups \, i \, and \, j)}$ derived *from the two groups that are being compared*, and the denominator is MS_{within}, the within-groups variance estimate that was calculated in doing the entire ANOVA. To determine $MS_{between \, (groups \, i \, and \, j)}$ for each comparison, $SS_{between \, (groups \, i \, and \, j)}$ for the two groups being compared is divided by the $df_{between}$ calculated when doing the entire ANOVA.

The steps and equations for calculating $F_{\text{Scheffé}}$ are as follows:

STEP 1: Calculate the between-groups sum of squares, $SS_{between\,(groups\,i\,and\,j)}$, for each paired comparison.

$$SS_{between\,(groups\,i\,and\,j)} = \left[\frac{(\Sigma X_i)^2}{n_i} + \frac{(\Sigma X_j)^2}{n_j}\right] - \frac{\left(\overset{\overset{\text{groups}}{i\,\text{and}\,j}}{\sum} X\right)^2}{n_i + n_j}$$

STEP 2: Calculate $MS_{between\,(groups\,i\,and\,j)}$ for each paired comparison.

$$MS_{between\,(groups\,i\,and\,j)} = \frac{SS_{between\,(groups\,i\,and\,j)}}{\text{df}_{within\,(entire\,ANOVA)}}$$

STEP 3: Calculate $F_{\text{Scheffé}}$ for each paired comparison.

$$F_{\text{Scheffé}} = \frac{MS_{between\,(groups\,i\,and\,j)}}{MS_{within\,(entire\,ANOVA)}}$$

We evaluate $F_{\text{Scheffé}}$ by comparing it with F_{crit} from the entire ANOVA. If $F_{\text{Scheffé}} \geq F_{\text{crit}}$, we reject H_0 and conclude that the two groups being compared are significantly different. The procedure is repeated for each pair-wise comparison.

One-Way Analysis of Variance, Kruskal–Wallis Test

Test	Statistic Calculated	Decision Rule
Nonparametric one-way analysis of variance, Kruskal–Wallis test	$H_{\text{obt}} = \left[\dfrac{12}{N(N+1)}\right]\left[\displaystyle\sum_{i=1}^{k}\dfrac{(R_i)^2}{n_i}\right]$ $-3(N+1)$	If $H_{\text{obt}} > H_{\text{crit}}$, reject H_0.

General comments   The Kruskal–Wallis test is a nonparametric test, appropriate for a k group, independent groups design. It is used as an alternative test to one-way parametric ANOVA when the assumptions of that test are seriously violated. The Kruskal–Wallis test does not assume population normality and requires only ordinal scaling of the dependent variable. All the scores are grouped together and rank-ordered, assigning the rank of 1 to the lowest score, 2 to the next to lowest, and N to the highest. The ranks for each condition are then summed. The Kruskal–Wallis test assesses whether these sums of ranks differ so much that it is unreasonable to consider that they come from samples that were randomly selected from the same population.

Two-Way Analysis of Variance, *F* Test

Test	Statistic Calculated	Decision Rule
Parametric two-way analysis of variance, *F* test	$F_{\text{obt}} = \dfrac{MS_{rows}}{MS_{within\text{-}cells}}$	If $F_{\text{obt}} > F_{\text{crit}}$, reject H_0.
	$F_{\text{obt}} = \dfrac{MS_{columns}}{MS_{within\text{-}cells}}$	
	$F_{\text{obt}} = \dfrac{MS_{interaction}}{MS_{within\text{-}cells}}$	

The parametric two-way analysis of variance allows us to evaluate the effects of two variables and their interaction in one experiment. In the parametric two-way ANOVA, we partition the total sum of squares (SS_{total}) into four components: the within-cells sum of squares ($SS_{within\text{-}cells}$), the row sum of squares (SS_{rows}), the column sum of squares ($SS_{columns}$), and the interaction sum of squares ($SS_{interaction}$). When these sums of squares are divided by the appropriate degrees of freedom, they form four variance estimates: the within-cells variance estimate $MS_{within\text{-}cells}$, the row variance estimate MS_{rows}, the column variance estimate $MS_{columns}$, and the interaction variance estimate $MS_{interaction}$. The effect of each of the variables is determined by computing the appropriate F_{obt} value and comparing it with F_{crit}.

The steps and equations for calculating the various F_{obt} values are as follows:

STEP 1: Calculate the row sum of squares, SS_{rows}.

$$SS_{rows} = \left[\frac{\left(\sum_{}^{\overset{row}{1}} X \right)^2 + \left(\sum_{}^{\overset{row}{2}} X \right)^2 + \cdots + \left(\sum_{}^{\overset{row}{r}} X \right)^2}{n_{row}} \right] - \frac{\left(\sum_{}^{\overset{all}{scores}} X \right)^2}{N}$$

STEP 2: Calculate the column sum of squares, $SS_{columns}$.

$$SS_{columns} = \left[\frac{\left(\sum_{}^{\overset{column}{1}} X \right)^2 + \left(\sum_{}^{\overset{column}{2}} X \right)^2 + \cdots + \left(\sum_{}^{\overset{column}{c}} X \right)^2}{n_{column}} \right] - \frac{\left(\sum_{}^{\overset{all}{scores}} X \right)^2}{N}$$

STEP 3: Calculate the interaction sum of squares, $SS_{interaction}$.

$$SS_{interaction} = \left[\frac{\left(\sum_{}^{\overset{cell}{11}} X \right)^2 + \left(\sum_{}^{\overset{cell}{12}} X \right)^2 + \cdots + \left(\sum_{}^{\overset{cell}{rc}} X \right)^2}{n_{cell}} \right] - \frac{\left(\sum_{}^{\overset{all}{scores}} X \right)^2}{N}$$
$$- SS_{rows} - SS_{columns}$$

STEP 4: Calculate the within-cells sum of squares, $SS_{within\text{-}cells}$.

$$SS_{within\text{-}cells} = \sum_{}^{\overset{all}{scores}} X^2 - \left[\frac{\left(\sum_{}^{\overset{cell}{11}} X \right)^2 + \left(\sum_{}^{\overset{cell}{12}} X \right)^2 + \cdots + \left(\sum_{}^{\overset{cell}{rc}} X \right)^2}{n_{cell}} \right]$$

STEP 5: Calculate the total sum of squares, SS_{total}, and check that $SS_{total} = SS_{rows} + SS_{columns} + SS_{interaction} + SS_{within\text{-}cells}$.

$$SS_{total} = \sum_{}^{\overset{all}{scores}} X^2 - \frac{\left(\sum_{}^{\overset{all}{scores}} X \right)^2}{N}$$

$$SS_{total} = SS_{rows} + SS_{columns} + SS_{interaction} + SS_{within\text{-}cells}$$

STEP 6: Calculate the degrees of freedom for each variance estimate.

$$df_{rows} = r - 1$$

$$df_{columns} = c - 1$$

$$df_{interaction} = (r - 1)(c - 1)$$

$$df_{within\text{-}cells} = rc(n_{cell} - 1)$$

$$df_{total} = N - 1$$

STEP 7: Calculate the variance estimates.

$$\text{Row variance estimate} = MS_{rows} = \frac{SS_{rows}}{df_{rows}}$$

$$\text{Column variance estimate} = MS_{columns} = \frac{SS_{columns}}{df_{columns}}$$

$$\text{Interaction variance estimate} = MS_{interaction} = \frac{SS_{interaction}}{df_{interaction}}$$

$$\text{Within-cells variance estimate} = MS_{within\text{-}cells} = \frac{SS_{within\text{-}cells}}{df_{within\text{-}cells}}$$

STEP 8: Calculate the F ratios.

For the row effect,

$$F_{obt} = \frac{MS_{rows}}{MS_{within\text{-}cells}}$$

For the column effect,

$$F_{obt} = \frac{MS_{columns}}{MS_{within\text{-}cells}}$$

For the interaction effect,

$$F_{obt} = \frac{MS_{interaction}}{MS_{within\text{-}cells}}$$

After calculating the F_{obt} values, compare them with F_{crit} and conclude.

ANALYZING NOMINAL DATA

You will recall that with nominal data, observations are grouped into several discrete, mutually exclusive categories, and one counts the frequency of occurrence in each category. The inference test most often used with nominal data is *chi-square*.

Chi-Square Test

Test	Statistic Calculated	Decision Rule
Chi-square	$\chi^2_{\text{obt}} = \Sigma \dfrac{(f_o - f_e)^2}{f_e}$	If $\chi^2_{\text{obt}} \geq \chi^2_{\text{crit}}$, reject H_0.

General comments   This test is appropriate for analyzing frequency data involving one or two variables. In the two-variable situation, the frequency data are presented in a contingency table and we test to see whether there is a relationship between the two variables. The null hypothesis states that there is no relationship—that the variables are independent. The alternative hypothesis states that the two variables are related.

Chi-square measures the discrepancy between the observed frequency (f_o) and the expected frequency (f_e) for each cell in the table and then sums across cells. The equation for χ^2_{obt} is given in the table. When the data involve two variables, the expected frequency for each cell is the frequency that would be expected if sampling is random from a population where the two variables are equal in proportions for each category. Since the population proportions are unknown, their expected values under H_0 are estimated from the sample data, and the expected frequencies are calculated using these estimates. The simplest way to determine f_e for each cell is to multiply the marginals for that cell and divide by N. If the data involve only one variable, the population proportions are determined on some *a priori* basis (e.g., equal population proportions for each category).

The obtained value of χ^2 is evaluated by comparing it with χ^2_{crit} according to the decision rule given in the table. The critical value of χ^2 is determined by the sampling distribution of χ^2 and the alpha level. The sampling distribution of χ^2 is a family of curves that varies with the degrees of freedom. In the one-variable experiment, df = $k - 1$. In the two-variable situation, df = $(r - 1)(c - 1)$. The values of χ^2_{crit} are found in Table H in Appendix D, using df and α.

Proper use of this test assumes that (1) each subject has only one entry in the table (no repeated measures on the same subjects); (2) if r or c is greater than 2, f_e for each cell should be at least 5; and (3) if the table is a 1×2 or 2×2 table, each f_e should be at least 10.

Chi-square can also be used with ordinal, interval, and ratio data. However, regardless of the actual scaling, to use χ^2, the data must be reduced to mutually exclusive categories and appropriate frequencies.

CHOOSING THE APPROPRIATE TEST

One of the important aspects of statistical inference is choosing which test to use for any experiment or problem. Up to now, it has been easy. We just used the test that we were studying for the particular chapter. However, in this review chapter, the situation is more challenging. Since we have covered many inference tests, we now have the opportunity to choose among them in deciding which to use. This, of course, is much more like the situation we face when doing research.

In choosing an inference test, the fundamental rule that we should follow is:

Use the most powerful test possible.

To determine which tests are possible for a given experiment or problem, we must consider two factors: the measurement scale of the dependent variable and the design of the experiment. Referring to the flowchart of Figure 18.1, the first question we ask is, "What is the level of measurement used for the dependent variable?" If it is nominal, the only inference test we've covered that is appropriate for nominal data is χ^2. Thus, if the

MENTORING TIP

Refer to Figure 18.1 when deciding which tests are candidates for analyzing any given data set. If more than one test is possible, always choose the most powerful one whose assumptions are met by the data.

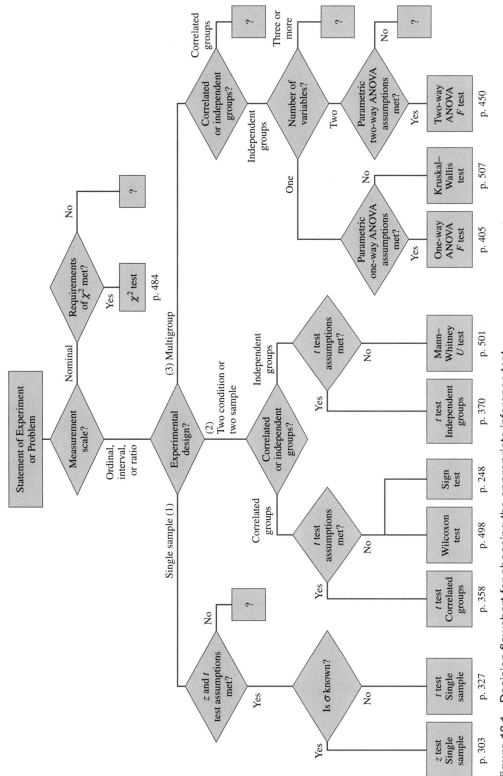

z test Single sample p. 303
t test Single sample p. 327
t test Correlated groups p. 358
Wilcoxon test p. 498
Sign test p. 248
t test Independent groups p. 370
Mann–Whitney U test p. 501
One-way ANOVA F test p. 405
Kruskal–Wallis test p. 507
Two-way ANOVA F test p. 450
χ² test p. 484

figure 18.1 Decision flowchart for choosing the appropriate inference test.

data are nominal in scaling and the requirements of χ^2 (frequency data, large enough N, mutually exclusive categories, and independent observations) are met, then we should choose the χ^2 test. If the assumptions are not met, then you probably don't know what test to use, because it hasn't been covered in this introductory text. In the flowchart, this regrettable state of affairs is indicated by a "?". I hasten to reassure you, however, that the inference tests we've covered are the most commonly encountered ones, with the possible exception of more complicated ANOVAs. Wherever you see a "?" you should assume appropriate tests do exist, but I believe they are too specialized or too complicated to be included in this introductory textbook.

If the data are not nominal, they must be ordinal, interval, or ratio in scaling. Having ruled out nominal data, we should next ask, "What is the experimental design?" The design used in the experiment limits the inference tests that we can use to analyze the data. We have covered three basic designs: single-sample, two-sample or two-condition, and multigroup experiments. If the design used is a single-sample design (path 1 in Figure 18.1), the two tests we have covered for this design are the z test and the t test for single samples. If the data meet the assumptions for these tests, to decide which to use we must ask the question, "Is σ known?" If the answer is "yes," then the appropriate test is the z test for single samples. If the answer is "no," then we must estimate σ and use the t test for single samples.

If the experimental design is a two-sample or two-condition design (path 2), we need to determine whether it is a correlated or independent groups design. If it is correlated groups and the assumptions of t are met, the appropriate test is the t test for correlated groups. Why? Because, if the assumptions are met, it is the most powerful test we can use for that design. If the assumptions are seriously violated, we should use an alternative test such as the Wilcoxon (if its assumptions are met) or the sign test. If it is an independent groups design and the assumptions of t are met, we should use the t test for independent groups. If the assumptions of t are seriously violated, we should use an alternative test such as the Mann–Whitney U test.

If the experimental design is a multigroup design (path 3), we need to determine whether it is an independent or correlated groups design. In this text, we have covered multigroup experiments that use the independent groups design. If the experiment is multigroup, uses an independent groups design, involves one variable, and the assumptions of parametric ANOVA are met, the appropriate test is parametric one-way ANOVA (F test). If the assumptions are seriously violated, we should use its alternative, the Kruskal–Wallis test. If the design is a multigroup, independent groups design, involving two variables, and the data meet the assumptions of parametric two-way ANOVA, we would use parametric two-way ANOVA (F test) to analyze the data. We have not considered the more complex designs involving three or more variables.

■ QUESTIONS AND PROBLEMS

Note to the student: In the previous chapters covering inferential statistics, when you were asked to solve an end-of-chapter problem, there was no question about which inference test you would use—you would use the test covered in the chapter. For example, if you were doing a problem in Chapter 13, you knew you should use the t test for single samples, because that was the test the chapter covered. Now you have reached the elevated position in which you know so much statistics that when solving a problem, there may be more than one inference test that could be used. Often, both a parametric and nonparametric test may be possible. This is a new challenge. The rule to follow is to use the most powerful test that the data will allow. For the problems in this chapter, always assume that the assumptions underlying the parametric test are met, unless the problem *explicitly* indicates otherwise.

1. Briefly define the following terms:
   Alternative hypothesis
   Null hypothesis
   Null-hypothesis population
   Sampling distribution
   Critical region for rejection of H_0
   Alpha level
   Type I error
   Type II error
   Power
2. Briefly describe the process of hypothesis testing. Be sure to include the terms listed in Question 1 in your discussion.
3. Why are sampling distributions important in hypothesis testing?
4. An educator conducts an experiment using an independent groups design to evaluate two methods of teaching third-grade spelling. The results are not significant, and the educator concludes that the two methods are equal. Is this conclusion sound? Assume that the study was properly designed and conducted; that is, proper controls were present, sample size was reasonably large, proper statistics were used, and so forth.
5. Why are parametric tests generally preferred over nonparametric tests?
6. List the factors that affect the power of an experiment and explain how they can be used to increase power.
7. What factors determine which inference test to use in analyzing the data of an experiment?
8. List the various experimental designs covered in this textbook. In addition, list the inference tests appropriate for each design in the order of their sensitivity.
9. What are the assumptions underlying each inference test?
10. What are the two steps followed in analyzing the data from any study involving hypothesis testing?
11. A new competitor in the scotch whiskey industry conducts a study to compare its scotch whiskey (called McPherson's Joy) to the other three leading brands. Two hundred scotch drinkers are randomly sampled from the scotch drinkers living in New York City. Each individual is asked to taste the four scotch whiskeys and pick the one they like the best. Of course, the whiskeys are unmarked, and the order in which they are tasted is balanced. The number of subjects that preferred each brand is shown in the following table:

McPherson's Joy	Brand X	Brand Y	Brand Z	
58	52	48	42	200

a. What is the alternative hypothesis? Use a nondirectional hypothesis.
b. What is the null hypothesis?
c. Using $\alpha = 0.05$, what do you conclude? I/O

12. A psychologist interested in animal learning conducts an experiment to determine the effect of adrenocorticotropic hormone (ACTH) on avoidance learning. Twenty 100-day-old male rats are randomly selected from the university vivarium for the experiment. Of the 20, 10 randomly chosen rats receive injections of ACTH 30 minutes before being placed in the avoidance situation. The other 10 receive placebo injections. The number of trials for each animal to learn the task is given here:

ACTH	Placebo
58	74
73	92
80	87
78	84
75	72
74	82
79	76
72	90
66	95
77	85

a. What is the nondirectional alternative hypothesis?
b. What is the null hypothesis?
c. Using $\alpha = 0.01_{2\,tail}$, what do you conclude?
d. What error may you have made by concluding as you did in part **c**?
e. To what population do these results apply?
f. What is the size of the effect? biological

13. A university nutritionist wonders whether the recent emphasis on eating a healthy diet has affected freshman students at her university. Consequently, she conducts a study to determine whether the diet of freshman students currently enrolled contains less fat than that of previous freshmen. To determine their percentage of daily fat intake, 15 students in this year's freshman class keep a record of everything they eat for 7 days. The results show that for the 15 students, the mean percentage of daily fat intake is 37%, with a standard deviation of 12%. Records kept on a large number of freshman students from previous years show a mean percentage of daily fat intake of 40%, a standard deviation of 10.5%, and a normal distribution of scores.

a. Based on these data, is the daily fat intake of currently enrolled freshmen less than that of previous years? Use $\alpha = 0.05_{1\,tail}$.

b. If the actual mean daily fat intake of currently enrolled freshmen is 35%, what is the power of the experiment to detect this level of real effect?

c. If N is increased to 30, what is the power to detect a real mean daily fat intake of 35%?

d. If the nutritionist wants a power of 0.9000 to detect a real effect of at least 5 mean points below the established population norms, what N should she run? health, I/O

14. A physiologist conducts an experiment designed to determine the effect of exogenous thyroxin (a hormone produced by the thyroid gland) on activity. Forty male rats are randomly assigned to four groups such that there are 10 rats per group. Each of the groups is injected with a different amount of thyroxin. Group 1 gets no thyroxin and receives saline solution instead. Group 2 receives a small amount, group 3 a moderate amount, and group 4 a high amount of thyroxin. After the injections, each animal is tested in an open-field apparatus to measure its activity level. The open-field apparatus is composed of a fairly large platform with sides around it to prevent the animal from leaving the platform. A grid configuration is painted on the surface of the platform such that the entire surface is covered with squares. To measure activity, the experimenter merely counts the number of squares that the animal has crossed during a fixed period of time. In the present experiment, each rat is tested in the open-field apparatus for 10 minutes. The results are shown in the table; the scores are the number of squares crossed per minute.

Amount of Thyroxin			
Zero	*Low*	*Moderate*	*High*
1	*2*	*3*	*4*
2	4	8	12
3	3	7	10
3	5	9	8
2	5	6	7
5	3	5	9
2	2	8	13
1	4	9	11
3	3	7	8
4	6	8	7
5	4	4	9

a. What is the overall null hypothesis?

b. Using $\alpha = 0.05$, what do you conclude?

c. What is the size of the effect, using $\hat{\omega}^2$?

d. Evaluate the *a priori* hypothesis that a high amount of exogenous thyroxin produces an effect on activity different from that of saline. Use $\alpha = 0.05_{2\,tail}$.

e. Use the Tukey HSD test with $\alpha = 0.05$ to compare all possible pairs of means. What do you conclude? biological

15. A study is conducted to determine whether dieting plus exercise is more effective in producing weight loss than dieting alone. Twelve pairs of matched subjects are run in the study. Subjects are matched on initial weight, initial level of exercise, age, and gender. One member of each pair is put on a diet for 3 months. The other member receives the same diet but, in addition, is put on a moderate exercise regimen. The following scores indicate the weight loss in pounds over the 3-month period for each subject:

Pair	Diet Plus Exercise	Diet Alone
1	24	16
2	20	18
3	22	19
4	15	16
5	23	18
6	21	18
7	16	17
8	17	19
9	19	13
10	25	18
11	24	19
12	13	14

In answering the following questions, assume the data are very nonnormal so as to preclude using the appropriate parametric test.

a. What is the alternative hypothesis? Use a directional hypothesis.

b. What is the null hypothesis?

c. Using $\alpha = 0.05_{1\,tail}$, what do you conclude? health

16 a. What other nonparametric test could you have used to analyze the data presented in Problem 15?

b. Use this test to analyze the data. What do you conclude with $\alpha = 0.05_{1\,tail}$?

c. Explain the difference between your conclusions for Problems 16b and 15c.

d. Let P equal the probability for each subject that diet plus exercise will yield greater weight loss. If $P_{real} = 0.75$, using the sign test with $\alpha = 0.05_{1\ tail}$, what is the power of the experiment to detect this level of effect? What is the probability of making a Type II error? health

17. A researcher in human sexuality is interested in determining whether there is a relationship between gender and time-of-day preference for having intercourse. A survey is conducted, and the results are shown in the following table; entries are the number of individuals who preferred morning or evening times:

Gender	Intercourse		
	Morning	*Evening*	
Male	36	24	60
Female	28	32	60
	64	56	120

a. What is the null hypothesis?
b. Using $\alpha = 0.05$, what do you conclude? social

18. A psychologist is interested in whether the internal states of individuals affect their perceptions. Specifically, the psychologist wants to determine whether hunger influences perception. To test this hypothesis, she randomly divides 24 subjects into three groups of 8 subjects per group. The subjects are asked to describe "pictures" that they are shown on a screen. Actually, there are no pictures, just ambiguous shapes or forms. Hunger is manipulated through food deprivation. One group is shown the pictures 1 hour after eating, another group 4 hours after eating, and the last group 12 hours after eating. The number of food-related objects reported by each subject is recorded. The following data are collected:

Food Deprivation		
1 hr	*4 hrs*	*12 hrs*
1	*2*	*3*
2	6	8
5	7	10
7	6	15
2	10	19
1	15	9
8	12	14
7	7	15
6	6	12

a. What is the overall null hypothesis?
b. What is your conclusion? Use $\alpha = 0.05$.
c. If there is a significant effect, estimate the size of the effect, using $\hat{\omega}^2$.
d. Estimate the size of the effect, using η^2.
e. Using the Scheffé test with $\alpha = 0.05$, do all possible *post hoc* comparisons between pairs of means. What is your conclusion? cognitive

19. An engineer working for a leading electronics firm claims to have invented a process for making longer-lasting LCD TVs. Tests run on 24 LCD TVs made with the new process show a mean life of 1725 hours and a standard deviation of 85 hours. Tests run over the last 3 years on a very large number of LCD TVs made with the old process show a mean life of 1538 hours.

a. Is the engineer correct in her claim? Use $\alpha = 0.01_{1\ tail}$ in making your decision.
b. If the engineer is correct, what is the size of the effect? I/O

20. In a study to determine the effect of alcohol on aggressiveness, 17 adult volunteers were randomly assigned to two groups: an experimental group and a control group. The subjects in the experimental group drank vodka disguised in orange juice, and the subjects in the control group drank only orange juice. After the drinks were finished, a test of aggressiveness was administered. The following scores were obtained. Higher scores indicate greater aggressiveness:

Orange Juice	Vodka Plus Orange Juice
11	14
9	13
14	19
15	16
7	15
10	17
8	11
10	18
8	

a. What is the alternative hypothesis? Use a nondirectional hypothesis.
b. What is the null hypothesis?
c. Using $\alpha = 0.05_{2\ tail}$, what is your conclusion? social, clinical

21. The dean of admissions at a large university wonders how strong the relationship is between high school

grades and college grades. During the 2 years that he has held this position, he has weighted high school grades heavily when deciding which students to admit to the university, yet he has never seen any data relating the two variables. Having a strong experimental background, he decides to conduct a study and find out for himself. He randomly samples 15 seniors from his university and obtains their high school and college grades. The following data are obtained:

	Grades	
Subject	**High school**	**College**
1	2.2	1.5
2	2.6	1.7
3	2.5	2.0
4	2.2	2.4
5	3.0	1.7
6	3.0	2.3
7	3.1	3.0
8	2.6	2.7
9	2.8	3.2
10	3.2	3.6
11	3.4	2.5
12	3.5	2.8
13	4.0	3.2
14	3.6	3.9
15	3.8	4.0

a. Compute r_{obt} for these data.
b. Is the correlation significant? Use $\alpha = 0.05_{2\ tail}$.
c. What proportion of the variability in college grades is accounted for by the high school grades?
d. Is the dean justified in weighting high school grades heavily when determining which students to admit to the university? education

22. An experiment is conducted to evaluate the effect of smoking on heart rate. Ten subjects who smoke cigarettes are randomly selected for the experiment. Each subject serves in two conditions. In condition 1, the subject rests for an hour, after which heart rate is measured. In condition 2, the subject rests for an hour and then smokes two cigarettes. In condition 2, heart rate is measured after the subject has finished smoking the cigarettes. The data follow.

	Heartbeats per Minute	
	No smoking	*Smoking*
Subject	*1*	*2*
1	72	76
2	80	84
3	68	75
4	74	73
5	80	86
6	85	88
7	86	84
8	78	80
9	68	72
10	67	70

a. What is the nondirectional alternative hypothesis?
b. What is the null hypothesis?
c. Using $\alpha = 0.05_{2\ tail}$, what do you conclude?
d. If your conclusion in part **c** is to affirm H_1, what is the size of the effect? biological, clinical

23. To meet the current oil crisis, the government must decide on a course of action to follow. There are two choices: (1) to allow the price of oil to rise or (2) to impose gasoline rationing. A survey is taken among individuals of various occupations to see whether there is a relationship between the occupations and the favored course of action. The results are shown in the following 3×2 table; cell entries are the number of individuals favoring the course of action that heads the cell:

	Course of Action		
	Oil price	*Gasoline*	
Occupation	*rise*	*rationing*	
Business	180	120	300
Homemaker	135	165	300
Labor	152	148	300
	467	433	900

a. What is the null hypothesis?
b. Using $\alpha = 0.05$, what do you conclude? I/O

24. You are interested in testing the hypothesis that adult men and women differ in logical reasoning ability. To do so, you randomly select 16 adults from the city in which you live and administer a logical reasoning test to them. A higher score indicates

better logical reasoning ability. The following scores are obtained:

Men	Women
70	80
60	50
82	81
65	75
83	95
92	85
85	93
	75
	90

In answering the following questions, assume that the data violate the assumptions underlying the use of the appropriate parametric test and that you must analyze the data with a nonparametric test.

a. What is the null hypothesis?

b. Using $\alpha = 0.05_{2 \text{ tail}}$, what is your conclusion?

c. What type error might have been made by the conclusion of part **b**? cognitive, social

25. For her doctoral thesis, a graduate student in women's studies investigated the effects of stress on the menstrual cycle. Forty-two women were randomly sampled and run in a two-condition replicated measures design. However, one of the women dropped out of the study. In the stress condition, the mean length of menstrual cycle for the remaining 41 women was 29 days, with a standard deviation of 14 days. Based on these data, determine the 95% confidence interval for the population mean length of menstrual cycle when under stress. health, social

26. A researcher interested in social justice believes that Hispanics are underrepresented in high school teachers in the part of the country in which she lives. A random sample of 150 high school teachers is taken from her geographical locale. The results show that there were 15 Hispanic teachers in the sample. The percentage of Hispanics living in the population of that locale equals 22%.

a. What is the null hypothesis?

b. Using $\alpha = 0.05$, what is your conclusion? social

27. A student believes that physical science professors are more authoritarian than social science professors. She conducts an experiment in which six physics, six psychology, and six sociology professors are randomly selected and given a questionnaire measuring authoritarianism. The results are shown here. The higher the score is, the more authoritarian is the individual. Assume the data seriously violate normality assumptions. What do you conclude, using $\alpha = 0.05$?

Professors		
Physics	*Psychology*	*Sociology*
75	73	71
82	80	80
80	85	90
97	92	78
94	70	94
76	69	68

social

28. A sleep researcher is interested in determining whether taking naps can improve performance and, if so, whether it matters if the naps are taken in the afternoon or evening. Thirty undergraduates are randomly sampled and assigned to one of six conditions: napping in the afternoon or evening, resting in the afternoon or evening, or engaging in a normal activity control condition, again in the afternoon or evening. There are five subjects in each condition. Each subject performs the activity appropriate for his or her assigned condition, after which a performance test is given. The higher the score is, the better is the performance. The following results were obtained. What is your conclusion? Use $\alpha = 0.05$ and assume the data are from normally distributed populations.

Time of Day	Activity					
	Napping		*Resting*		*Normal*	
Afternoon	8	9	7	8	3	4
	7	5	6	4	5	5
	6		5		6	
Evening	6	5	5	3	4	2
	7	4	4	5	3	4
	6		4		3	

cognitive

■ ONLINE STUDY RESOURCES

CENGAGE**brain**.com

Login to CengageBrain.com to access the resources your instructor has assigned. For this book, you can access the book's companion website for chapter-specific learning tools including Know and Be Able to Do, practice quizzes, flash cards, and glossaries, and a link to Statistics and Research Methods Workshops.

aplia™

If your professor has assigned Aplia homework:

1. Sign in to your account.
2. Complete the corresponding homework exercises as required by your professor.
3. When finished, click "Grade It Now" to see which areas you have mastered and which need more work, and for detailed explanations of every answer.

Visit **www.cengagebrain.com** to access your account and to purchase materials.

APPENDIXES

© Strmko / Dreamstime.com

Review of Prerequisite Mathematics

Introduction

Solving Equations with One Unknown

Linear Interpolation

INTRODUCTION

In this appendix, we shall present a review of some basic mathematical skills that we believe are important as background for an introductory course in statistics. This appendix is intended to be a review of material that you have already learned but that may be a little "rusty" from disuse. For students who have been away from mathematics for many years and who feel unsure of their mathematical background, we recommend the following books: H. M. Walker, *Mathematics Essential for Elementary Statistics* (rev. ed., Holt, New York, 1951) or A. J. Washington, *Arithmetic and Beginning Algebra* (Addison-Wesley, Reading, MA, 1984). In addition, there is a web site for online learning of basic and advanced mathematics that has received high praise. The web address is www.khanacedemy.org. For the prerequisite material required by this course, I recommend you view the modules that have "Algebra" or "Equations" in the title. Of course, it makes sense to study only those modules that deal with material that requires reviewing.

Algebraic Symbols

Symbol	Explanation
$>$	Is greater than
$5 > 4$	5 is greater than 4.
$X > 10$	X is greater than 10.
$a > b$	a is greater than b.
$<$	Is less than
$7 < 9$	7 is less than 9.
$X < 12$	X is less than 12.
$a < b$	a is less than b.
$2 < X < 20$	X is greater than 2 and less than 20, or the value of X lies between 2 and 20.
$X < 2$ or $X > 20$	X is less than 2 or greater than 20, or the value of X lies outside the interval of 2 to 20.
$\geq$	Is equal to or greater than
$X \geq 3$	X is equal to or greater than 3.
$a \geq b$	a is equal to or greater than b.
$\leq$	Is equal to or less than
$X \leq 5$	X is equal to or less than 5.
$a \leq b$	a is equal to or less than b.
$\neq$	Is not equal to
$3 \neq 5$	3 is not equal to 5.
$X \neq 8$	X is not equal to 8.
$a \neq b$	a is not equal to b.
$\lvert X \rvert$	The absolute value of X; the absolute value of X equals the magnitude of X irrespective of its sign.
$\lvert +7 \rvert$	The absolute value of 7; $\lvert +7 \rvert = 7$.
$\lvert -5 \rvert$	The absolute value of -5; $\lvert -5 \rvert = 5$.

Arithmetic Operations

Operation	*Example*
1. *Addition of two positive numbers:* To add two positive numbers, sum their absolute values and give the result a plus sign.	$2 + 8 = 10$
2. *Addition of two negative numbers:* To add two negative numbers, sum their absolute values and give the result a minus sign.	$-3 + (-4) = -7$
3. *Addition of two numbers with opposite signs:* To add two numbers with opposite signs, find the difference between their absolute values and give the number the sign of the larger absolute value.	$16 + (-10) = 6$ $3 + (-14) = -11$
4. *Subtraction of one number from another:* To subtract one number from another, change the sign of the number to be subtracted and proceed as in addition (operations 1, 2, or 3).	$16 - 4 = 16 + (-4) = 12$ $5 - 8 = 5 + (-8) = -3$ $9 - (-6) = 9 + (+6) = 15$ $-3 - 5 = -3 + (-5) = -8$
5. *Multiplying a series of numbers:* a. When multiplying a series of numbers, the result is positive if there are an even number of negative values in the series.	$2(-5)(-6)(3) = 180$ $-3(-7)(-2)(-1) = 42$ $-a(-b) = ab$
b. When multiplying a series of numbers, the result is negative if there are an odd number of negative values in the series.	$4(-5)(2) = -40$ $-8(-2)(-5)(3) = -240$ $-a(-b)(-c) = -abc$
6. *Dividing a series of numbers:* a. When dividing a series of numbers, the result is positive if there are an even number of negative values.	$\dfrac{-4}{-8} = \dfrac{1}{2}$ $\dfrac{-3(-4)(2)}{6} = 4$ $\dfrac{-a}{-b} = \dfrac{a}{b}$
b. When dividing a series of numbers, the result is negative if there are an odd number of negative values.	$\dfrac{-2}{5} = -0.40$ $\dfrac{-3(-2)}{-4} = -1.5$ $\dfrac{-a}{b} = -\dfrac{a}{b}$

Rules Governing the Order of Arithmetic Operations

Rule	Example
1. The order in which numbers are added does not change the result.	$6 + 4 + 11 = 4 + 6 + 11 = 11 + 6 + 4 = 21$ $6 + (-3) + 2 = -3 + 6 + 2 = 2 + 6 + (-3) = 5$
2. The order in which numbers are multiplied does not change the result.	$3 \times 5 \times 8 = 8 \times 5 \times 3 = 5 \times 8 \times 3 = 120$
3. If both multiplication and addition or subtraction are specified, the multiplication should be performed first unless parentheses or brackets indicate otherwise.	$4 \times 5 + 2 = 20 + 2 = 22$ $6 \times (14 - 12) \times 3 = 6 \times 2 \times 3 = 36$ $6 \times (4 + 3) \times 2 = 6 \times 7 \times 2 = 84$
4. If both division and addition or subtraction are specified, the division should be performed first unless parentheses or brackets indicate otherwise.	$12 \div 4 + 2 = 3 + 2 = 5$ $12 \div (4 + 2) = 12 \div 6 = 2$ $12 \div 4 - 2 = 3 - 2 = 1$ $12 \div (4 - 2) = 12 \div 2 = 6$

Rules Governing Parentheses and Brackets

Rule	Example

1. Parentheses and brackets indicate that whatever is shown within them is to be treated as a single number.

$(2 + 8)(6 - 3 + 2) = 10(5) = 50$

2. Where there are parentheses contained within brackets, perform the operations contained within the parentheses first.

$[(4)(6 - 3 + 2) + 6][2] = [(4)(5) + 6][2]$
$= [20 + 6][2] = [26][2] = 52$

3. When it is inconvenient to reduce whatever is contained within the parentheses to a single number, the parentheses may be removed as follows:

 a. If a positive sign precedes the parentheses, remove the parentheses without changing the sign of any number they contained.

$1 + (3 + 5 - 2) = 1 + 3 + 5 - 2 = 7$

 b. If a negative sign precedes the parentheses, remove the parentheses and change the signs of the numbers they contained.

$4 - (6 - 2 - 1) = 4 - 6 + 2 + 1 = 1$

 c. If a multiplier exists outside the parentheses, all the terms within the parentheses must be multiplied by the multiplier.

$3(2 + 3 - 4) = 6 + 9 - 12 = 3$
$a(b + c + d) = ab + ac + ad$

 d. The product of two sums is found by multiplying each element of one sum by the elements of the other sum.

$(a + b)(c + d) = ac + ad + bc + bd$

 e. If whatever is contained within the parentheses is operated on in any way, always do the operation first, before combining with other terms.

$5 + 4(3 + 1) + 6 = 5 + 4(4) + 6 = 5 + 16 + 6 = 27$
$4 + (3 + 1)/2 = 4 + 4/2 = 4 + 2 = 6$
$1 + (3 + 2)^2 = 1 + (5)^2 = 1 + 25 = 26$

Fractions

Operation	Example

1. *Addition of fractions:*
   To add two fractions, (1) find the least common denominator, (2) express each fraction in terms of the least common denominator, and (3) add the numerators and divide the sum by the common denominator.

   $\frac{1}{3} + \frac{1}{2} = \frac{2}{6} + \frac{3}{6} = \frac{5}{6}$

   $\frac{a}{b} + \frac{c}{d} = \frac{ad}{bd} + \frac{cb}{bd} = \frac{ad + cb}{bd}$

2. *Multiplication of fractions:*
   To multiply two fractions, multiply the numerators together and divide by the product of the denominators.

   $\frac{2}{5}\left(\frac{3}{7}\right) = \frac{6}{35}$

   $\frac{a}{b}\left(\frac{c}{d}\right) = \frac{ac}{bd}$

3. *Changing a fraction into its decimal equivalent:*
   To convert a fraction into a decimal, perform the indicated division, rounding to the required number of digits (two digits in the example shown).

   $\frac{3}{7} = 0.429 = 0.43$

4. *Changing a decimal into a percentage:*
   To convert a decimal fraction into a percentage, multiply the decimal fraction by 100.

   $0.43 \times 100 = 43\%$

5. *Multiplying an integer by a fraction:*
   To multiply an integer by a fraction, multiply the integer by the numerator of the fraction and divide the product by the denominator.

   $\frac{2}{5}(4) = \frac{8}{5} = 1.60$

6. *Cancellation:*
   When multiplying several fractions together, identical factors in the numerator and denominator may be canceled.

   $\frac{\overset{1}{\cancel{5}}}{\underset{3}{\cancel{12}}}\left(\frac{\overset{1}{\cancel{4}}}{\underset{3}{\cancel{9}}}\right)\left(\frac{\overset{1}{\cancel{3}}}{\underset{2}{\cancel{10}}}\right) = \frac{1}{18}$

Exponents

Operation	Example
1. *Multiplying a number by itself 2 times*	$(4)^2 = 4(4) = 16$ $a^2 = aa$
2. *Multiplying a number by itself 3 times*	$(4)^3 = 4(4)(4) = 64$ $a^3 = aaa$
3. *Multiplying a number by itself N times*	$4^N = \overbrace{4(4)(4)\cdots(4)}^{N}$
4. *Multiplying two exponential quantities having the same base:* The product of two exponential quantities having the same base is the base raised to the sum of the exponents.	$(2)^2(2)^4 = (2)^{2+4} = (2)^6$ $a^N a^P = a^{N+P}$
5. *Dividing two exponential quantities having the same base:* The quotient of two exponential quantities having the same base is the base raised to an exponent equal to the exponent of the quantity in the numerator minus the exponent of the quantity in the denominator.	$\dfrac{(2)^4}{(2)^2} = (2)^2$ $\dfrac{a^N}{a^P} = a^{N-P}$
6. *Raising a base to a negative exponent:* A base raised to a negative exponent is equal to 1 divided by the base raised to the positive value of the exponent.	$(2)^{-3} = \dfrac{1}{(2)^3}$ $a^{-N} = \dfrac{1}{a^N}$

Factoring When factoring an algebraic expression, we try to reduce the expression to the simplest components that when multiplied together yield the original expression.

Example	Explanation
$ab + ac + ad = a(b + c + d)$	We factored out a from each item.
$abc - 2ab = ab(c - 2)$	We factored out ab from both terms.
$a^2 + 2ab + b^2 = (a + b)^2$	This expression can be reduced to $a + b$ times itself.

SOLVING EQUATIONS WITH ONE UNKNOWN

When solving equations with one unknown, the basic idea is to isolate the unknown on one side of the equation and reduce the other side to its smallest possible value. In so doing, we make use of the principle that the equation remains an equality if whatever we do to one side of the equation, we also do the same to the other side. Thus, for example, the equation remains an equality if we add the same number to both sides. In solving the equation, we alter the equation by adding, subtracting, multiplying dividing, squaring, and so forth, so as to isolate the unknown. This is permissible as long as we do the same operation to both sides of the equation, thus maintaining the equality. The following examples illustrate many of the operations commonly used to solve equations having one unknown. In each of the examples, we shall be solving the equation for Y.

Example	Explanation
$Y + 5 = 2$ $Y = 2 - 5$ $\quad = -3$	To isolate Y, subtract 5 from both sides of the equation.
$Y - 4 = 6$ $Y = 6 + 4$ $\quad = 10$	To isolate Y, add 4 to both sides of the equation.
$\dfrac{Y}{2} = 8$ $Y = 8(2)$ $\quad = 16$	To isolate Y, multiply both sides of the equation by 2.
$3Y = 7$ $Y = \frac{7}{3}$ $\quad = 2.33$	To isolate Y, divide both sides of the equation by 3.
$6 = \dfrac{Y - 3}{2}$ $12 = Y - 3$ $3 + 12 = Y$ $15 = Y$ $Y = 15$	To isolate Y, (1) multiply both sides by 2 and (2) add 3 to both sides.
$3 = \dfrac{2}{Y}$ $3Y = 2$ $Y = \frac{2}{3}$	To isolate Y, (1) multiply both sides by Y and (2) divide both sides by 3.
$4(Y + 1) = 3$ $Y + 1 = \frac{3}{4}$ $Y = \frac{3}{4} - 1$ $\quad = -\frac{1}{4}$	To isolate Y, (1) divide both sides by 4 and (2) subtract 1 from both sides.
$\dfrac{4}{Y + 2} = 8$ $\dfrac{Y + 2}{4} = \dfrac{1}{8}$ $Y + 2 = \frac{4}{8}$ $Y = \frac{4}{8} - 2$ $\quad = -1\frac{1}{2}$	To isolate Y, (1) take the reciprocal of both sides, (2) multiply both sides by 4, and (3) subtract 2 from both sides.
$2Y + 4 = 10$ $2Y = 10 - 4$ $Y = \dfrac{10 - 4}{2}$ $\quad = 3$	To isolate Y, (1) subtract 4 from both sides and (2) divide both sides by 2.

LINEAR INTERPOLATION

Linear interpolation is often necessary when looking up values in a table. For example, suppose we wanted to find the square root of 96.5 using a table that only has the square root of 96 and 97 but not 96.5, as shown here:

Number	Square root
96	9.7980
97	9.8489

Looking in the column headed by *Number,* we note that there is no value corresponding to 96.5. The closest values are 96 and 97. From the table, we can see that the square root of 96 is 9.7980 and that the square root of 97 is 9.8489. Obviously, the square root of 96.5 must lie between 9.7980 and 9.8489. Using *linear* interpolation, we assume there is a linear relationship between the number and its square root, and we use this linear relationship to approximate the square root of numbers not given in the table. Since 96.5 is halfway between 96 and 97, using linear interpolation, we would expect the square root of 96.5 to lie halfway between 9.7980 and 9.8489. If we let X equal the square root of 96.5, then

$$X = 9.7980 + 0.5(9.8489 - 9.7980) = 9.8234$$

Although it wasn't made explicit, the computed value for X was the result of setting up the following proportions and solving for X:

$$\frac{96.5 - 96}{97 - 96} = \frac{X - 9.7980}{9.8489 - 9.7980}$$

Number	Square Root
96	9.7980
96.5	X
97	9.8489

The relationship is shown graphically in Figure A.1.

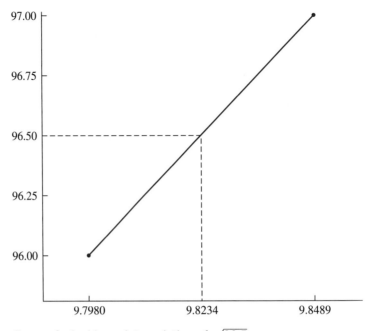

figure A. 1   Linear interpolation of $\sqrt{96.5}$.

Equations

Listed here are the computational equations used in this textbook. The page number refers to the page where the equation first appears.

Equation	Description	Equation First Occurs on Page:
$\sum_{i=1}^{N} X_i = X_1 + X_2 + X_3 + \cdots + X_N$	summation	27
$\text{cum }\% = \dfrac{\text{cum } f}{N} \times 100$	cumulative percentage	55
$\text{Percentile point} = X_L + (i/f_i)(\text{cum } f_p - \text{cum } f_L)$	equation for computing percentile point	58
$\text{Percentile rank} = \dfrac{\text{cum } f_L + (f_i/i)(X - X_L)}{N} \times 100$	equation for computing percentile rank	59
$\overline{X} = \dfrac{\Sigma X_i}{N}$	mean of a sample	81
$\mu = \dfrac{\Sigma X_i}{N}$	mean of a population of raw scores	81
$\overline{X}_{\text{overall}} = \dfrac{n_1\overline{X}_1 + n_2\overline{X}_2 + \cdots + n_k\overline{X}_k}{n_1 + n_2 + \cdots + n_k}$	overall mean of several groups	84
$\text{Mdn} = P_{50} = X_L + (i/f_i)(\text{cum } f_p - \text{cum } f_L)$	median of a distribution	85
$\text{Range} = \text{Highest score} - \text{Lowest score}$	range of a distribution	89
$X - \overline{X}$	deviation score for sample data	90
$X - \mu$	deviation score for population data	90
$\sigma = \sqrt{\dfrac{SS_{\text{pop}}}{N}} = \sqrt{\dfrac{\Sigma(X - \mu)^2}{N}}$	standard deviation of a population of raw scores	91
$s = \sqrt{\dfrac{SS}{N - 1}} = \sqrt{\dfrac{\Sigma(X - \overline{X})^2}{N - 1}}$	standard deviation of a sample of raw scores	91

Equation	Description	Equation First Occurs on Page:
$SS = \Sigma X^2 - \dfrac{(\Sigma X)^2}{N}$	sum of squares	93
$s^2 = \dfrac{SS}{N-1}$	variance of a sample of raw scores	95
$\sigma^2 = \dfrac{SS_{\text{pop}}}{N}$	variance of a population of raw scores	95
$z = \dfrac{X - \mu}{\sigma}$	z score for population data	106
$z = \dfrac{X - \overline{X}}{s}$	z score for sample data	106
$X = \mu + \sigma z$	equation for finding a population raw score from its z score	114
$Y = bX + a$	equation of a straight line	125
$b = \text{Slope} = \dfrac{\Delta Y}{\Delta X} = \dfrac{Y_2 - Y_1}{X_2 - X_1}$	slope of a straight line	125
$r = \dfrac{\Sigma z_X z_Y}{N-1}$	computational equation for Pearson r using z scores	133
$r = \dfrac{\Sigma XY - \dfrac{(\Sigma X)(\Sigma Y)}{N}}{\sqrt{\left[\Sigma X^2 - \dfrac{(\Sigma X)^2}{N}\right]\left[\Sigma Y^2 - \dfrac{(\Sigma Y)^2}{N}\right]}}$	computational equation for Pearson r	133
$r_s = 1 - \dfrac{6 \Sigma D_i^2}{N^3 - N}$	computational equation for Spearman rho	141
$Y' = b_Y X + a_Y$	linear regression equation for predicting Y given X	162
$b_Y = \dfrac{\Sigma XY - \dfrac{(\Sigma X)(\Sigma Y)}{N}}{\Sigma X^2 - \dfrac{(\Sigma X)^2}{N}}$	regression constant b for predicting Y given X, computational equation with raw scores	163
$a_Y = Y - b_Y \overline{X}$	regression constant a for predicting Y given X	163
$s_{Y|X} = \sqrt{\dfrac{SS_Y - \dfrac{[\Sigma XY - (\Sigma X)(\Sigma Y)/N]^2}{SS_X}}{N-2}}$	computational equation for the standard error of estimate when predicting Y given X	170
$b_Y = r\dfrac{s_Y}{s_X}$	equation relating r to the b_Y regression constant	173
$R^2 = \dfrac{r_{YX_1}^2 + r_{YX_2}^2 - 2r_{YX_1}r_{YX_2}r_{X_1X_2}}{1 - r_{X_1X_2}^2}$	equation for computing the squared multiple correlation	176

Equation	Description	Equation First Occurs on Page:
$p(A) = \dfrac{\text{Number of events classifiable as } A}{\text{Total number of possible events}}$	*a priori* probability	193
$p(A) = \dfrac{\text{Number of times } A \text{ has occurred}}{\text{Total number of occurrences}}$	*a posteriori* probability	194
$p(A \text{ or } B) = p(A) + p(B) - p(A \text{ and } B)$	addition rule for two events, general equation	196
$p(A \text{ or } B) = p(A) + p(B)$	addition rule when A and B are mutually exclusive	196
$p(A \text{ or } B \text{ or } C \text{ or } \dots \text{ or } Z) = p(A) + p(B) + p(C) + \dots + p(Z)$	addition rule with more than two mutually exclusive events	200
$p(A) + p(B) + p(C) + \dots + p(Z) = 1.00$	when events are exhaustive and mutually exclusive	200
$P + Q = 1.00$	when two events are exhaustive and mutually exclusive	201
$p(A \text{ and } B) = p(A)p(B\|A)$	multiplication rule with two events—general equation	201
$p(A \text{ and } B) = 0$	multiplication rule with mutually exclusive events	201
$p(A \text{ and } B) = p(A)p(B)$	multiplication rule with independent events	202
$p(A \text{ and } B \text{ and } C \text{ and } \dots \text{ and } Z) = p(A)p(B)p(C) \dots p(Z)$	multiplication rule with more than two independent events	206
$p(A \text{ and } B) = p(A)p(B\|A)$	multiplication rule with dependent events	207
$p(A \text{ and } B \text{ and } C \text{ and } \dots \text{ and } Z) = p(A)p(B\|A)p(C\|AB) \dots p(Z\|ABC \dots)$	multiplication rule with more than two dependent events	210
$p(A) = \dfrac{\text{Area under curve corresponding to } A}{\text{Total area under curve}}$	probability of A with a continuous variable	214
$(P + Q)^N$	binomial expansion	229
Number of Q events $= N - $ Number of P events	relationship between number of Q events, number of P events, and N	235
$\mu = NP$	mean of the normal distribution approximated by the binomial distribution	239
$\sigma = \sqrt{NPQ}$	standard deviation of the normal distribution approximated by the binomial distribution	239
Beta $= 1 - $ Power	relationship between beta and power	285
$\mu_{\overline{X}} = \mu$	mean of the sampling distribution of the mean	305
$\sigma_{\overline{X}} = \dfrac{\sigma}{\sqrt{N}}$	standard deviation of the sampling distribution of the mean or standard error of the mean	305

Equation	Description	Equation First Occurs on Page:
$z_{\text{obt}} = \dfrac{\overline{X}_{\text{obt}} - \mu}{\sigma_{\overline{X}}}$	z transformation for $\overline{X}_{\text{obt}}$	312
$z_{\text{obt}} = \dfrac{\overline{X}_{\text{obt}} - \mu}{\sigma/\sqrt{N}}$	z transformation for $\overline{X}_{\text{obt}}$	315
$N = \left[\dfrac{\sigma(z_{\text{crit}} - z_{\text{obt}})}{\mu_{\text{real}} - \mu_{\text{null}}} \right]^2$	determining N for a specified power	321
$t_{\text{obt}} = \dfrac{\overline{X}_{\text{obt}} - \mu}{s/\sqrt{N}}$	equation for calculating the t statistic	328
$t_{\text{obt}} = \dfrac{\overline{X}_{\text{obt}} - \mu}{s_{\overline{X}}}$	equation for calculating the t statistic	328
$s_{\overline{X}} = \dfrac{s}{\sqrt{N}}$	estimated standard error of the mean	328
$\text{df} = N - 1$	degrees of freedom for t test (single sample)	331
$t_{\text{obt}} = \dfrac{\overline{X}_{\text{obt}} - \mu}{\sqrt{\dfrac{SS}{N(N-1)}}}$	equation for calculating the t statistic from raw scores	334
$d = \dfrac{\lvert \text{mean difference} \rvert}{\text{population standard deviation}}$	general equation for size of effect	339
$d = \dfrac{\lvert \overline{X}_{\text{obt}} - \mu \rvert}{\sigma}$	conceptual equation for size of effect, single sample t test	339
$\hat{d} = \dfrac{\lvert \overline{X}_{\text{obt}} - \mu \rvert}{s}$	computational equation for size of effect, single sample t test	340
$\mu_{\text{lower}} = \overline{X}_{\text{obt}} - s_{\overline{X}} t_{0.025}$	lower limit for the 95% confidence interval	342
$\mu_{\text{upper}} = \overline{X}_{\text{obt}} + s_{\overline{X}} t_{0.025}$	upper limit for the 95% confidence interval	342
$\mu_{\text{lower}} = \overline{X}_{\text{obt}} - s_{\overline{X}} t_{\text{crit}}$	general equation for the lower limit of the confidence interval	343
$\mu_{\text{upper}} = \overline{X}_{\text{obt}} + s_{\overline{X}} t_{\text{crit}}$	general equation for the upper limit of the confidence interval	343
$\mu_{\text{lower}} = \overline{X}_{\text{obt}} - s_{\overline{X}} t_{0.005}$	lower limit for the 99% confidence interval	344
$\mu_{\text{upper}} = \overline{X}_{\text{obt}} + s_{\overline{X}} t_{0.005}$	upper limit for the 99% confidence interval	344
$t_{\text{obt}} = \dfrac{r_{\text{obt}} - p}{s_r}$	t test for testing the significance of r	346

Equation	Description	Equation First Occurs on Page:
$t_{obt} = \dfrac{r_{obt}}{\sqrt{\dfrac{1 - r_{obt}^2}{N - 2}}}$	t test for testing the significance of r	346
$t_{obt} = \dfrac{\overline{D}_{obt} - \mu_D}{s_D/\sqrt{N}}$	t test for correlated groups	360
$t_{obt} = \dfrac{\overline{D}_{obt} - \mu_D}{\sqrt{\dfrac{SS_D}{N(N - 1)}}}$	t test for correlated groups	360
$SS_D = \Sigma D^2 - \dfrac{(\Sigma D)^2}{N}$	sum of squares of the difference scores	360
$d = \dfrac{|\overline{D}_{obt}|}{\sigma_D}$	conceptual equation for size of effect, correlated groups t test	364
$\hat{d} = \dfrac{|\overline{D}_{obt}|}{s_D}$	computational equation for size of effect, correlated groups t test	364
$\mu_{\overline{X}_1 - \overline{X}_2} = \mu_1 - \mu_2$	mean of the difference between sample means	368
$\sigma_{\overline{X}_1 - \overline{X}_2} = \sqrt{\sigma^2\left(\dfrac{1}{n_1} + \dfrac{1}{n_2}\right)}$	standard deviation of the difference between sample means	368
$t_{obt} = \dfrac{(\overline{X}_1 - \overline{X}_2) - \mu_{\overline{X}_1 - \overline{X}_2}}{\sqrt{\left(\dfrac{SS_1 + SS_2}{n_1 + n_2 - 2}\right)\left(\dfrac{1}{n_1} + \dfrac{1}{n_2}\right)}}$	computational equation for t_{obt}, independent groups design	371
$t_{obt} = \dfrac{\overline{X}_1 - \overline{X}_2}{\sqrt{\left(\dfrac{SS_1 + SS_2}{n_1 + n_2 - 2}\right)\left(\dfrac{1}{n_1} + \dfrac{1}{n_2}\right)}}$	computational equation for t_{obt} assuming the independent variable has no effect	371
$SS_1 = \Sigma X_1^2 - \dfrac{(\Sigma X_1)^2}{n_1}$	sum of squares for group x	373
$SS_2 = \Sigma X_2^2 - \dfrac{(\Sigma X_2)^2}{n_2}$	sum of squares for group x	373
$t_{obt} = \dfrac{\overline{X}_1 - \overline{X}_2}{\sqrt{\dfrac{SS_1 + SS_2}{n(n - 1)}}}$	computational equation for t_{obt} when $n_1 = n_2$	373
$d = \dfrac{|\overline{X}_1 - \overline{X}_2|}{\sigma}$	conceptual equation for size of effect, independent groups t test	377
$\hat{d} = \dfrac{|\overline{X}_1 - \overline{X}_2|}{\sqrt{s_w^2}}$	computational equation for size of effect, independent groups t test	377

Equation	Description	Equation First Occurs on Page:
$\mu_{\text{lower}} = (\overline{X}_1 - \overline{X}_2) - s_{\overline{X}_1 - \overline{X}_2}t_{0.025}$	lower limit for the 95% confidence interval for $\mu_{\overline{X}_1 - \overline{X}_2}$	383
$\mu_{\text{upper}} = (\overline{X}_1 - \overline{X}_2) + s_{\overline{X}_1 - \overline{X}_2}t_{0.025}$	upper limit for the 95% confidence interval for $\mu_{\overline{X}_1 - \overline{X}_2}$	383
$\mu_{\text{lower}} = (\overline{X}_1 - \overline{X}_2) - s_{\overline{X}_1 - \overline{X}_2}t_{0.005}$	lower limit for the 99% confidence interval for $\mu_{\overline{X}_1 - \overline{X}_2}$	385
$\mu_{\text{upper}} = (\overline{X}_1 - \overline{X}_2) + s_{\overline{X}_1 - \overline{X}_2}t_{0.005}$	upper limit for the 99% confidence interval for $\mu_{\overline{X}_1 - \overline{X}_2}$	385
$F = \dfrac{\text{Variance estimate 1 of } \sigma^2}{\text{Variance estimate 2 of } \sigma^2}$	basic definition of F	402
$F_{\text{obt}} = \dfrac{\text{Between-groups variance estimate}}{\text{Within-groups variance estimate}} = \dfrac{MS_{between}}{MS_{within}}$	F equation for the analysis of variance	406
$s_W^2 = \dfrac{SS_1 + SS_2}{(n_1 - 1) + (n_2 - 1)}$	t test, only 2 groups	407
$MS_{within} = s_W^2$	both are estimates of σ^2	407
$MS_{within} = \dfrac{SS_1 + SS_2 + SS_3 + \cdots + SS_k}{(n_1 - 1) + (n_2 - 1) + (n_3 - 1) + \cdots + (n_k - 1)}$	conceptual equation for MS_{within}	407
$MS_{within} = \dfrac{SS_1 + SS_2 + SS_3 + \cdots + SS_k}{N - k}$	simplified equation for MS_{within}	407
$MS_{within} = \dfrac{SS_{within}}{\text{df}_{within}}$	within-groups variance estimate	407
$SS_{within} = SS_1 + SS_2 + SS_3 + \cdots + SS_k$	within-groups sum of squares	407
$\text{df}_{within} = N - k$	within-groups degrees of freedom	407
$SS_{within} = \overset{\text{all scores}}{\sum} X^2 - \left[\dfrac{(\Sigma X_1)^2}{n_1} + \dfrac{(\Sigma X_2)^2}{n_2} + \dfrac{(\Sigma X_3)^2}{n_3} + \cdots + \dfrac{(\Sigma X_k)^2}{n_k} \right]$	computational equation for within-groups sum of squares	408
$MS_{between} = \dfrac{n[(\overline{X}_1 - \overline{X}_G)^2 + (\overline{X}_2 - \overline{X}_G)^2 + (\overline{X}_3 - \overline{X}_G)^2 + \cdots + (\overline{X}_k - \overline{X}_G)^2]}{k - 1}$	conceptual equation for between-groups variance estimate	409
$MS_{between} = \dfrac{SS_{between}}{\text{df}_{between}}$	between-groups variance estimate	409
$SS_{between} = n[(\overline{X}_1 - \overline{X}_G)^2 + (\overline{X}_2 - \overline{X}_G)^2 + (\overline{X}_3 - \overline{X}_G)^2 + \cdots + (\overline{X}_3 - \overline{X}_G)^2$	between-groups sum of squares	409
$\text{df}_{between} = k - 1$	between-groups degrees of freedom	409
$SS_{between} = \left[\dfrac{(\Sigma X_1)^2}{n_1} + \dfrac{(\Sigma X_2)^2}{n_2} + \dfrac{(\Sigma X_3)^2}{n_3} + \cdots + \dfrac{(\Sigma X_k)^2}{n_k} \right] - \dfrac{\left(\overset{\text{all scores}}{\sum} X \right)^2}{N}$	computational equation for between-groups sum of squares	409
$SS_{total} = SS_{within} + SS_{between}$	equation for checking SS_{within} and $SS_{between}$	412

Equation	Description	Equation First Occurs on Page:
$SS_{total} = \overset{\text{all scores}}{\sum} X^2 - \dfrac{\left(\overset{\text{all scores}}{\sum} X\right)^2}{N}$	equation for calculating the total variability	412
$\hat{\omega}^2 = \dfrac{SS_{between} - (k-1)\,MS_{within}}{SS_{total} + MS_{within}}$	computational equation for estimating $\hat{\omega}^2$	419
$\eta^2 = \dfrac{SS_{between}}{SS_{total}}$	conceptual and computational equation for eta squared	420
$F_{obt} = \dfrac{MS_{between}}{MS_{within}} = \dfrac{n[(\overline{X}_1 - \overline{X}_G)^2 + (\overline{X}_2 - \overline{X}_G)^2 + (\overline{X}_3 - \overline{X}_G)^2]/2}{(SS_1 + SS_2 + SS_3)/(N-3)}$	F equation for three-group experiment	421
$t_{obt} = \dfrac{\overline{X}_1 - \overline{X}_2}{\sqrt{MS_{within}\left(\dfrac{1}{n_1} + \dfrac{1}{n_2}\right)}}$	general equation for t equation for *planned* comparisons,	422
$t_{obt} = \dfrac{\overline{X}_1 - \overline{X}_2}{\sqrt{2MS_{within}/n}}$	t equation for *planned* comparisons with equal n in the two groups	423
$Q_{obt} = \dfrac{\overline{X}_i - \overline{X}_j}{\sqrt{MS_{within}/n}}$	equation for calculating Q_{obt}	424
$SS_{between\,(groups\,i\,and\,j)} = \left[\dfrac{(\sum X_i)^2}{n_i} + \dfrac{(\sum X_j)^2}{n_j}\right] - \dfrac{\left(\overset{\text{groups }i\text{ and }j}{\sum} X\right)^2}{n_i + n_j}$	computational equation for $SS_{between\,(groups\,i\,and\,j)}$	426
$MS_{between\,(groups\,i\,and\,j)} = \dfrac{SS_{between\,(groups\,i\,and\,j)}}{df_{between\,(entire\,ANOVA)}}$	conceptual equation for $MS_{between\,(groups\,i\,and\,j)}$	426
$F_{Scheff\acute{e}} = \dfrac{MS_{between\,(groups\,i\,and\,j)}}{MS_{between\,(entire\,ANOVA)}}$	equation for $F_{Scheff\acute{e}}$	426
$F_{crit} = F_{crit\,(entire\,ANOVA)}$	F_{crit} for Scheffé test	426
$MS_{within-cells} = \dfrac{SS_{within-cells}}{df_{within-cells}}$	conceptual equation for equation for within-cells variance estimate	451
$SS_{within-cells} = SS_{11} + SS_{12} + \cdots + SS_{rc}$	conceptual equation for within-cells sum of squares	451
$SS_{within-cells} = \overset{\text{all scores}}{\sum} X^2 - \left[\dfrac{\left(\overset{\text{cell }11}{\sum} X\right)^2 + \left(\overset{\text{cell }12}{\sum} X\right)^2 + \cdots + \left(\overset{\text{cell }rc}{\sum} X\right)^2}{n_{cell}}\right]$	computational equation for within-cells sum of squares	452
$df_{within-cells} = rc(n-1)$	within-cells degrees of freedom	452
$MS_{rows} = \dfrac{SS_{rows}}{df_{rows}}$	conceptual equation for the row variance estimate	452
$SS_{rows} = n_{row}[(\overline{X}_{row\,1} - \overline{X}_G)^2 + (\overline{X}_{row\,2} - \overline{X}_G)^2 + \cdots + (\overline{X}_{row\,r} - \overline{X}_G)^2]$	conceptual equation for the row sum of squares	453

Equation	Description	Equation First Occurs on Page:
$df_{rows} = r - 1$	row degrees of freedom	453
$SS_{rows} = \left[\dfrac{\left(\overset{row}{\underset{1}{\sum}} X \right)^2 + \left(\overset{row}{\underset{2}{\sum}} X \right)^2 + \cdots + \left(\overset{row}{\underset{r}{\sum}} X \right)^2}{n_{row}} \right] - \dfrac{\left(\overset{all\ scores}{\sum} X \right)^2}{N}$	computational equation for the row sum of squares	453
$MS_{columns} = \dfrac{SS_{columns}}{df_{columns}}$	column variance estimate	454
$SS_{columns} = n_{column}\left[(\overline{X}_{column\ 1} - \overline{X}_G)^2 + (\overline{X}_{column\ 2} - \overline{X}_G)^2 + \cdots + (\overline{X}_{column\ c} - \overline{X}_G)^2 \right]$	conceptual equation for the column sum of squares	454
$df_{columns} = c - 1$	column degrees of freedom	454
$SS_{columns} = \left[\dfrac{\left(\overset{column}{\underset{1}{\sum}} X \right)^2 + \left(\overset{column}{\underset{2}{\sum}} X \right)^2 + \cdots + \left(\overset{column}{\underset{c}{\sum}} X \right)^2}{n_{column}} \right] - \dfrac{\left(\overset{all\ scores}{\sum} X \right)^2}{N}$	computational equation for the column sum of squares	454
$MS_{interaction} = \dfrac{SS_{interaction}}{df_{interaction}}$	interaction variance estimate	455
$SS_{interaction} = n_{cell}\left[(\overline{X}_{cell\ 11} - \overline{X}_G)^2 + (\overline{X}_{cell\ 12} - \overline{X}_G)^2 + \cdots + (\overline{X}_{cell\ rc} - \overline{X}_G)^2 \right] - SS_{rows} - SS_{columns}$	conceptual equation for the interaction sum of squares	455
$SS_{interaction} = \left[\dfrac{\left(\overset{cell}{\underset{11}{\sum}} X \right)^2 + \left(\overset{cell}{\underset{12}{\sum}} X \right)^2 + \cdots + \left(\overset{cell}{\underset{rc}{\sum}} X \right)^2}{n_{cell}} \right] - \dfrac{\left(\overset{all\ scores}{\sum} X \right)^2}{N} - SS_{rows} - SS_{columns}$	computational equation for the interaction sum of squares	455
$df_{interaction} = (r - 1)(c - 1)$	interaction degrees of freedom	455
$\chi^2_{obt} = \sum \dfrac{(f_o - f_e)^2}{f_e}$	equation for calculating χ^2_{obt}	485
$U_{obt} = n_1 n_2 + \dfrac{n_1(n_1 + 1)}{2} - R_1$	general equation for calculating U_{obt} or U'_{obt}	503
$U_{obt} = n_1 n_2 + \dfrac{n_2(n_2 + 1)}{2} - R_2$	general equation for calculating U_{obt} or U'_{obt}	503
$H_{obt} = \left[\dfrac{12}{N(N + 1)} \right]\left[\sum_{i=1}^{k} \dfrac{(R_i)^2}{n_i} \right] - 3(N + 1)$	*equation for computing H_{obt}*	509

Answers to End-of-Chapter Questions and Problems and SPSS Problems

■ CHAPTER 1

6. b. (1) functioning of the hypothalamus; (2) daily food intake; (3) the 30 rats selected for the experiment; (4) all rats living in the university vivarium at the time of the experiment; (5) the daily food intake of each animal during the 2-week period after recovery; (6) the mean daily food intake of each group **c.** (1) methods of treating depression; (2) degree or amount of depression; (3) the 60 depressed students; (4) the undergraduate body at a large university at the time of the experiment; (5) depression scores of the 60 depressed students; (6) mean of the depression scores of each treatment **d.** (1) and (2) since this is a study and not an experiment, there is no independent variable and no dependent variable. The two variables studied are the two levels of education and the annual salaries for each educational level; (3) the 200 individuals whose annual salaries were determined; (4) all individuals living in the city at the time of the experiment, having either of the educational levels; (5) the 200 annual salaries; (6) the mean annual salary for each educational level **e.** (1) spacing of practice sessions; (2) number of words correctly recalled; (3) the 30 seventh graders who participated in the experiment; (4) all seventh graders enrolled at the local junior high school at the time of the experiment; (5) the retention test scores of the 30 subjects; (6) mean values for each group

of the number of words correctly recalled in the test period **f.** (1) visualization versus visualization plus appropriate self-talk; (2) foul shooting accuracy; (3) the ten players participating in the experiment; (4) all players on the college basketball team at the time of the experiment; (5) foul shooting accuracy of the ten players, before and after 1 month of practicing the techniques; (6) the mean of the difference scores of each group **g.** (1) the arrangement of typing keys; (2) typing speed; (3) the 20 secretarial trainees who were in the experiment; (4) all secretarial trainees enrolled in the business school at the time of the experiment; (5) the typing speed scores of each trainee obtained at the end of training; (6) the mean typing speed of each group

7. a. constant   **b.** constant   **c.** variable
    d. variable   **e.** constant   **f.** variable
    g. variable   **h.** constant

8. a. descriptive statistics   **b.** descriptive statistics
    c. descriptive statistics   **d.** inferential statistics
    e. inferential statistics   **f.** descriptive statistics

9. a. The sample scores are the 20 scores given. The population scores are the 213 scores that would have resulted if the number of drinks during "happy hour" were measured from all of the bars. **c.** The sample scores are the 25 lengths measured. The

population scores are the 600 lengths that would be obtained if all 600 blanks were measured.
d. The sample scores are the 30 diastolic heart rates that were recorded. The population scores are the heart rate scores that would result from recording resting, diastolic heart rate from all the female students attending Tacoma University at the time of the experiment.

CHAPTER 2

2. **a.** continuous   **b.** discrete   **c.** discrete
   d. continuous   **e.** discrete   **f.** continuous
   g. continuous   **h.** continuous

3. **a.** ratio   **b.** nominal   **c.** interval
   d. ordinal   **e.** ordinal   **f.** ratio
   g. ratio   **h.** interval   **i.** ordinal

4. No, ratios are not legitimate on an interval scale. We need an absolute zero point to perform ratios. Since an ordinal scale does not have an absolute zero point, the ratio of the absolute values represented by 30 and 60 will not be $\dfrac{1}{2}$

5. **a.** 18   **b.** 21.1   **d.** 590

6. **a.** 14.54   **c.** 37.84   **d.** 46.50   **e.** 52.46   **f.** 25.49

7. **a.** 9.5–10.5   **b.** 2.45–2.55   **d.** 2.005–2.015
   e. 5.2315–5.2325

8. **a.** 25   **b.** 35   **d.** 101

9. **a.** $X_1 = 250$, $X_2 = 378$, $= X_3 = 451$, $X_4 = 275$, $X_5 = 225$, $X_6 = 430$, $X_7 = 325$, $X_8 = 334$
   b. 2668

10. **a.** $\displaystyle\sum_{i=1}^{N} X_i$ **b.** $\displaystyle\sum_{i=1}^{3} X_i$ **c.** $\displaystyle\sum_{i=2}^{4} X_i$ **d.** $\displaystyle\sum_{i=2}^{5} X_i^2$

11. **a.** 1.4   **b.** 23.2   **c.** 100.8   **d.** 41.7   **e.** 35.3

12. For 5b: $\sum X^2 = 104.45$ and $(\sum X)^2 = 445.21$; for 5c: $\sum X^2 = 3434$ and $(\sum X)^2 = 21{,}904$

13. **a.** 34   **b.** 14   **d.** 6.5

14. **a.** 4.1, 4.15   **b.** 4.2, 4.15   **c.** 4.2, 4.16
    d. 4.2, 4.20

SPSS

1. **a.** 176   **b.** 29.30   **c.** 3116.00

2. **a.** 42.70   **b.** 651.00

CHAPTER 3

5. **a.**

Score	f	Score	f	Score	f	Score	f
98	1	84	2	70	2	57	2
97	0	83	3	69	2	56	1
96	0	82	4	68	4	55	2
95	0	81	3	67	2	54	0
94	2	80	0	66	1	53	0
93	2	79	2	65	1	52	0
92	1	78	5	64	3	51	0
91	2	77	2	63	1	50	0
90	2	76	4	62	2	49	2
89	1	75	3	61	1	48	0
88	1	74	1	60	1	47	0
87	2	73	4	59	0	46	0
86	0	72	6	58	0	45	1
85	4	71	3				

b.

Class Interval	Real Limits	f
96–99	95.5–99.5	1
92–95	91.5–95.5	5
88–91	87.5–91.5	6
84–87	83.5–87.5	8
80–83	79.5–83.5	10
76–79	75.5–79.5	13
72–75	71.5–75.5	14
68–71	67.5–71.5	11
64–67	63.5–67.5	7
60–63	59.5–63.5	5
56–59	55.5–59.5	3
52–55	51.5–55.5	2
48–51	47.5–51.5	2
44–47	43.5–47.5	1
		88

6.

Class Interval	f	Relative f	Cumulative f	Cumulative %
96–99	1	0.01	88	100.00
92–95	5	0.06	87	98.86
88–91	6	0.07	82	93.18
84–87	8	0.09	76	86.36
80–83	10	0.11	68	77.27
76–79	13	0.15	58	65.91
72–75	14	0.16	45	51.14
68–71	11	0.12	31	35.23
64–67	7	0.08	20	22.73
60–63	5	0.06	13	14.77
56–59	3	0.03	8	9.09
52–55	2	0.02	5	5.68
48–51	2	0.02	3	3.41
44–47	1	0.01	1	1.14
	88	1.00		

7. a. 82.70   **b.** 72.70

8. a. 70.17   **c.** 85.23

10. a.

Class Interval	f
60–64	1
55–59	1
50–54	2
45–49	2
40–44	4
35–39	5
30–34	7
25–29	12
20–24	17
15–19	16
10–14	8
5–9	3
	78

11.

Class Interval	f	Relative f	Cumulative f
60–64	1	0.01	78
55–59	1	0.01	77
50–54	2	0.03	76
45–49	2	0.03	74
40–44	4	0.05	72
35–39	5	0.06	68
30–34	7	0.09	63
25–29	12	0.15	56
20–24	17	0.22	44
15–19	16	0.21	27
10–14	8	0.10	11
5–9	3	0.04	3
	78	1.00	

12. a. 23.03

13. a. 88.72   **b.** 67.18

16. a. 3.30   **b.** 2.09

17. 65.62

18. a. 34.38   **b.** 33.59   **c.** Some accuracy is lost when grouping scores because the grouped scores analysis assumes the scores are evenly distributed throughout the interval.

20. a.

Class Interval	f
360–369	1
350–359	2
340–349	5
330–339	3
320–329	3
310–319	9
300–309	4
290–299	8
280–289	5
270–279	5
260–269	4
250–259	1
	50

21.

Class Interval	f	Relative f	Cumulative f	Cumulative %
360–369	1	0.02	50	100.00
350–359	2	0.04	49	98.00
340–349	5	0.10	47	94.00
330–339	3	0.06	42	84.00
320–329	3	0.06	39	78.00
310–319	9	0.18	36	72.00
300–309	4	0.08	27	54.00
290–299	8	0.16	23	46.00
280–289	5	0.10	15	30.00
270–279	5	0.10	10	20.00
260–269	4	0.08	5	10.00
250–259	1	0.02	1	2.00
	50	1.00		

22. a. 304.50   **b.** 324.50

23. a. 15.50   **b.** 69.30

24. a.

Score	f
12	1
11	2
10	4
9	7
8	9
7	13
6	15
5	11
4	10
3	6
2	5
1	1
0	1

d. 27.06, 96.47

■ SPSS

1. b.

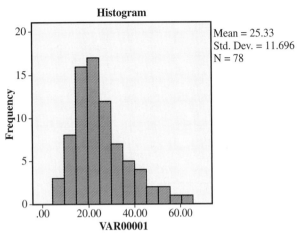

The distribution is not symmetrical; it is positively skewed

2.

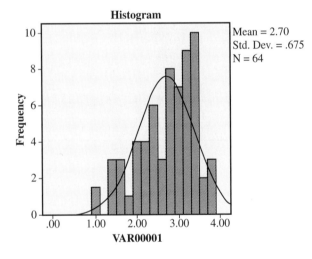

■ CHAPTER 4

13. All the scores must have the same value.

16. a. $\overline{X} = 3.56$, Mdn = 3, mode = 2
   c. $\overline{X} = 3.03$, Mdn = 2.70, no mode

17. a. $\overline{X}_{orig.} = 4.00$, $\overline{X}_{new} = 6.00$, $\overline{X}_{new} = \overline{X}_{orig.} + a$
   b. $\overline{X}_{orig.} = 4.00$, $\overline{X}_{new} = 2.00$, $\overline{X}_{new} = \overline{X}_{orig.} - a$
   c. $\overline{X}_{orig.} = 4.00$, $\overline{X}_{new} = 8.00$, $\overline{X}_{new} = a\overline{X}_{orig.}$
   d. $\overline{X}_{orig.} = 4.00$, $\overline{X}_{new} = 2.00$, $\overline{X}_{new} = \overline{X}_{orig.}/a$

18. a. 72.00   **b.** 72

19. a. 68.83   **b.** 64.5

20. a. 2.93   **b.** 2.8

21. a. the mean, because there are no extreme scores
   b. the mean, again because there are no extreme scores   **c.** the median, because the distribution contains an extreme score (25)

22. a. positively skewed   **b.** negatively skewed
   c. symmetrical

23. a. $\overline{X} = 2.54$ hours per day
   b. Mdn = 2.7 hours per day
   c. mode = 0 hours per day

25. a. $\overline{X} = 15.44$ **b.** Mdn = 14   **c.** There is no mode.

26. 197.44

27. a. range = 6, $s = 2.04$, $s^2 = 4.41$
   c. range = 9.1, $s = 3.64$, $s^2 = 13.24$

29. 4.00 minutes

30. 7.17 months

31. a. $s_{orig.} = 2.12$, $s_{new} = 2.12$, $s_{new} = s_{orig.}$
   b. $s_{orig.} = 2.12$, $s_{new} = 2.12$, $s_{new} = s_{orig.}$
   c. $s_{orig.} = 2.12$, $s_{new} = 4.24$, $s_{new} = as_{orig.}$
   d. $s_{orig.} = 2.12$, $s_{new} = 1.06$, $s_{new} = s_{orig.}/a$

32. a. 4.50   **b.** 5   **c.** 7   **d.** 7
   e. 2.67   **f.** 7.14

33. Distribution *b* is most variable, followed by distribution *a* and then distribution *c*. For distribution *b*, $s = 11.37$; for distribution *a*, $s = 3.16$; and for distribution *c*, $s = 0$.

34. a. $s = 1.86$ **b.** 11.96, Because the standard deviation is sensitive to extreme scores and 35 is an extreme score

35. a. $\overline{X} = 7.90$ **b.** 8   **c.** 8
   d. 15   **e.** 4.89   **f.** 23.88

36. a. 347.50   **b.** 335   **c.** There is no mode.
   d. 220   **e.** 87.28   **f.** 7617.50

37. a. 22.56   **b.** 21.50   **c.** There is no mode.
   d. 22   **e.** 7.80   **f.** 60.78

38. a. $\overline{X} + a$ **b.** $\overline{X} - a$ **c.** $a\overline{X}$ **d.** $\overline{X}/a$

39. a. *s* stays the same.   **b.** *s* stays the same.
   c. *s* is multiplied by *a*.   **d.** *s* is divided by *a*.

40. a. 2.67, 3.33, 6.33, 4.67, 6.67, 4.67, 3.00, 4.00, 7.67, 6.67   **b.** 3.00, 2.00, 8.00, 6.00, 6.00, 4.00,

2.00, 3.00, 8.00, 7.00   **c.** Expect more variability in the medians.   **d.** s(medians) = 2.38, s(means) = 1.76

■ SPSS

1. **a.**

Descriptive Statistics

	N	Range	Mean	Std. Deviation	Variance
Scores	13	13.00	8.3846	4.07305	16.590
Valid N (listwise)	13				

b.

Descriptive Statistics

	N	Range	Mean	Std. Deviation	Variance
Scores	10	6.90	5.1400	2.43867	5.947
Valid N (listwise)	10				

c.

Descriptive Statistics

	N	Range	Mean	Std. Deviation	Variance
Scores	13	63.00	50.2308	20.22850	409.192
Valid N (listwise)	13				

d.

Descriptive Statistics

	N	Range	Mean	Std. Deviation	Variance
Scores	11	388.00	413.9091	135.70000	18414.491
Valid N (listwise)	11				

■ CHAPTER 5

8. **a.**

Raw Score	z Score
10	−1.41
12	−0.94
16	0.00
18	0.47
19	0.71
21	1.18

b. mean = 0.00, standard deviation = 1.00

9. **a.**

Raw Score	z Score
10	−1.55
12	−1.03
16	0.00
18	0.52
19	0.77
21	1.29

b. mean = 0.00, standard deviation = 1.00

10. **a.** 1.14   **b.** −0.86   **c.** 1.86   **d.** 0.00   **e.** 1.00   **f.** −1.00

11. a. 50.00%   **b.** 15.87%   **c.** 6.18%
d. 2.02%   **e.** 0.07%   **f.** 32.64%

12. a. 34.13%   **b.** 34.13%   **c.** 49.04%
d. 49.87%   **e.** 0.00%   **f.** 25.17%
g. 26.73%

13. a. 0.00   **b.** 1.96   **c.** 1.64
d. 0.52   **e.** −0.84   **f.** −1.28

15. a. statistics   **b.** 92.07

16. a. 3.75%   **b.** 99.81%   **c.** 98.54%   **d.** 12.97%
e. 3.95 kilograms   **f.** 3.64   **g.** 13,623

17. a. 95.99%   **b.** 99.45%   **c.** 15.87%
d. 4.36%   **e.** 50.00%

18. a. 16.85%   **b.** 0.99%   **c.** 59.87%
d. 97.50%   **e.** 50.00%

19. a. 51.57%   **b.** 34.71%   **c.** 23.28%

20. a.

Distance	z Score
30	−1.88
31	−1.50
32	−1.13
33	−0.75
34	−0.38
35	0.00
36	0.38
37	0.75
38	1.13

b. and **c.**

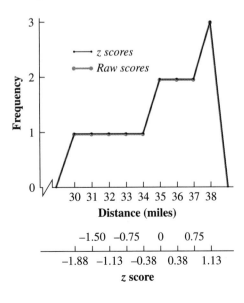

d. The z distribution is not normally shaped. The z distribution takes the same shape as the distribution of raw scores. In this problem, the raw scores are not normally shaped. Therefore, the z distribution will not be normally shaped either.   **e.** mean = 0.00, standard deviation = 1.00

21. a. 92.36%   **b.** 55.04%
c. 96.78%   **d.** $99.04

22. business

23. Both Rebecca and Maurice did better on Exam 2.

24. 93.64

■ SPSS

1.

	Y	ZY
1	10.00	-1.56616
2	13.00	-.90673
3	15.00	-.46710
4	16.00	-.24729
5	18.00	.19234
6	20.00	.63196
7	21.00	.85177
8	24.00	1.51121

2.

	Distance	ZDistance
1	32.00	-1.12771
2	35.00	.00000
3	30.00	-1.87952
4	38.00	1.12771
5	37.00	.75181
6	36.00	.37590
7	38.00	1.12771
8	36.00	.37590
9	38.00	1.12771
10	35.00	.00000
11	31.00	-1.50362
12	33.00	-.75181
13	34.00	-.37590
14	37.00	.75181

■ CHAPTER 6

3. **a.** linear, perfect positive
   b. curvilinear, perfect
   c. linear, imperfect negative
   d. curvilinear, imperfect
   e. linear, perfect negative
   f. linear, imperfect positive

14. **a.** For set A, $r = 1.00$; for set B, $r = 0.11$; for set C, $r = -1.00$ **b.** same value as in part **a**.   **c.** The r values are the same.   **d.** The r values remain the same.   **e.** The r values do not change if a constant is subtracted from the raw scores or if the raw scores are divided by a constant. The value of r does not change when the scale is altered by adding or subtracting a constant to it, nor does r change if the scale is transformed by multiplying or dividing by a constant.

16. **b.** $r = 0.79$

17. **b.** $r = 0.68$ **c.** 0.03. Decreasing the range produced a decrease in r.   **d.** $r^2 = 0.46$. If illness is causally related to smoking, r^2 allows us to evaluate how important a factor smoking is in producing illness.

18. **b.** $r = 0.98$ **c.** Yes, this is a reliable test because $r^2 = 0.95$. Almost all of the variability of the scores on the second administration can be accounted for by the scores on the first administration.

19. **a.** $r = 0.85$ **b.** $r_s = 0.86$

20. **b.** $r = -0.06$ **e.** $r = 0.93$

21. **b.** $r = 0.95$

22. **a.** negative   **b.** $r = -0.56$

23. **a.** $r_s = 0.85$ **b.** For the paper and pencil test and psychiatrist A, $r_s = 0.73$; for the paper and pencil test and psychiatrist B, $r_s = 0.79$.

24. **b.** $r = 0.59$ **d.** $r = 0.91$ **e.** Yes, test 2, because r^2 accounts for 82.4% of the variability in work performance. Although there are no doubt other factors operating, this test appears to offer a good adjunct to the interview. Test 1 does not do nearly as well.

■ SPSS

1. **a.**

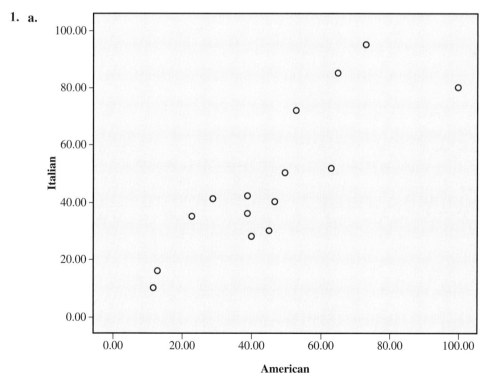

b. The relationship is linear, imperfect and positive.
c. The SPSS **Correlations** table shows that Pearson r for these data is **.852**.

2. **a.**

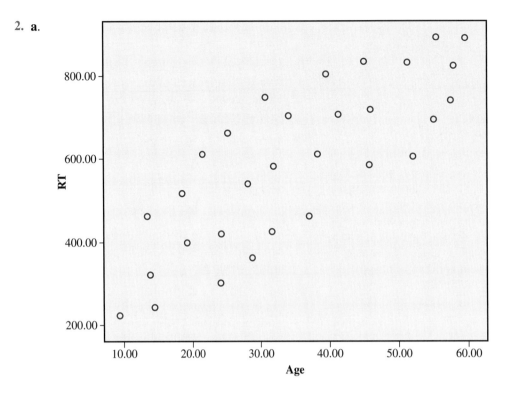

b. The relationship is linear, imperfect and positive.

c. The SPSS **Correlations** table shows that Pearson r for these data is **.807**.

d. 0.65 of the variance of **RT** is accounted for by **Age**. (SPSS table doesn't provide this information)

■ CHAPTER 7

10. **b.** The relationship is negative, imperfect, and linear. **c.** $r = -0.56$ **d.** $Y' = -0.513X + 24.964$; negative, because the relationship is negative **f.** 13.16

11. **a.** No. A scatter plot of the paired scores reveals that there is a perfect relationship between length of left index finger and weight. Thus, Mr. Clairvoyant can exactly predict my weight having measured the length of my left index finger.
b. $Y' = 7.5X + 37$ **c.** 79.75

12. **b.** The relationship is negative, imperfect, and linear. **c.** $r = -0.69$ **d.** $Y' = -1.429X + 125.883$; negative, because the relationship is negative **f.** 71.60 **g.** 10.87

13. **a.** $Y' = 10.828X + 11.660$ **b.** $196,000 **c.** Technically, the relationship holds only within the range of the base data. It may be that if a lot more money is spent, the relationship would change such that no additional profit or even loss is the result. Of course, the manager could experiment by "testing the waters" (e.g., by spending $25,000 on advertising to see whether the relationship still holds at that level).

14. **a.** Yes, $r = 0.85$ **b.** % games won = 5.557(tenure) + 34.592 **c.** 73.49%

15. **a.** $Y' = -1.212X + 131.77$ **b.** $130.96 **c.** 17.31

16. **a.** $Y' = 4.213X + 91.652$ **b.** 123.25

17. **a.** $Y' = 0.857X + 17.894$ **b.** 32.46

18. **a.** $Y' = -0.075X + 6.489$ **b.** 3.28

19. $R^2 = 85.3\%$, $r^2 = 82.4\%$; using test 1 doesn't seem worth the extra work.

■ SPSS

1. a.

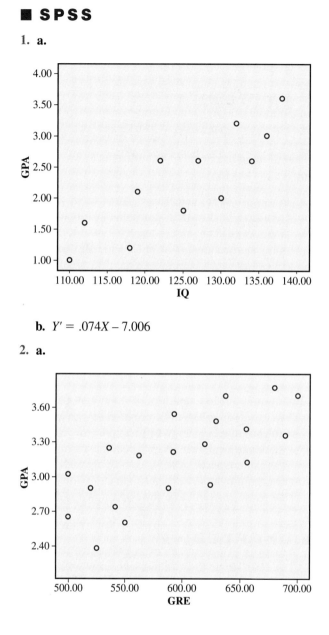

b. $Y' = .074X - 7.006$

2. a.

b. The relationship is linear, imperfect and positive. **c.** $Y' = .005X + .468$. **d.** Standard Error of Estimate = 0.27.

■ CHAPTER 8

9. a, c, e

10. a, c, d, e

11. a, c

12. a. 2 to 3   **b.** 0.6000   **c.** 0.4000

13. a. 0.0192   **b.** 0.0769   **c.** 0.3077
d. 0.5385

15. a. 0.4000   **b.** 0.0640   **c.** 0.0350
d. 0.2818

16. a. 0.4000   **b.** 0.0491   **c.** 0.0409
d. 0.2702

17. a. 0.0029   **b.** 0.0087   **c.** 0.0554

18. 0.0001

19. 0.0238

20. a. 0.0687   **b.** 0.5581   **c.** 0.2005

21. 0.1479

22. 0.00000001

23. a. 0.1429   **b.** 0.1648   **c.** 0.0116

25. a. 0.0301   **b.** 0.6268   **c.** 0.0301

26. a. 0.0192   **b.** 0.9525   **c.** 0.1949

28. a. 0.0146   **b.** 0.9233   **c.** 0.0344

29. a. 0.0764   **b.** 0.4983   **c.** 0.0021

30. a. 0.0400   **b.** 0.1200   **c.** 0.0400

■ CHAPTER 9

3. 0.90

4. a. 0.0369   **b.** $6P^5Q$
c. 0.0369 (The answers are the same.)

5. a. 0.0161   **b.** 0.0031   **c.** 0.0192
d. 0.0384

6. a. 0.0407   **b.** 0.0475   **c.** 0.0475

7. a. 0.1369   **b.** 0.2060   **c.** 0.2060

8. a. 0.0020   **b.** 0.0899   **c.** 0.1798

10. 0.0681

11. 0.3487

12. 0.0037

14. 0.0039

15. a. 0.0001   **b.** 0.0113

16. a. 32   **b.** 37

17. a. 0.0001   **b.** 0.4437   **c.** 0.1382
d. 0.5563

18. a. 0.0039   **b.** 0.3164   **c.** 0.6836

19. a. 0.0576   **b.** 0.0001   **c.** 0.0100
d. 0.8059

20. a. 0.0000   **b.** 0.0013

21. 0.8133

22. a. 0.0098   **b.** 1

■ CHAPTER 10

10. a. The alternative hypothesis states that the new teaching method increases the amount learned. **b.** The null hypothesis states that the two methods are equal in the amount of material learned or the old method does better.   **c.** $p(14$ or more pluses$) = 0.0577$. Since $0.0577 > 0.05$, you retain H_0. You cannot conclude that the new method is better.   **d.** You may be making a Type II error, retaining H_0 if it is false.   **e.** The results apply to the eighth-grade students in the school district at the time of the experiment.

11. a. The alternative hypothesis states that increases in the level of angiotensin II will produce change in thirst level.   **b.** The null hypothesis states that increases in the level of angiotensin II will not have any effect on thirst.   **c.** $p(0, 1, 2, 14, 15,$ or 16 pluses$) = 0.0040$. Since $0.0040 < 0.05$, you reject H_0. Increases in the level of angiotensin II appear to increase thirst.   **d.** You may be making a Type I error, rejecting H_0 if it is true.   **e.** The results apply to the rats living in the vivarium of the drug company at the time of the experiment.

12. a. The alternative hypothesis states that using Very Bright toothpaste instead of Brand X results in brighter teeth.   **b.** The null hypothesis states that Very Bright and Brand X toothpastes are equal in their brightening effects or Brand X is better. **c.** $p(7$ or more pluses$) = 0.1719$. Since $0.1719 > 0.05$, you retain H_0. You cannot conclude that Very Bright is better.   **d.** You may be making a Type II error, retaining H_0 if it is false.   **e.** The results apply to the employees of the Pasadena plant at the time of the experiment.

13. a. The alternative hypothesis states that acupuncture affects pain tolerance.   **b.** The null hypothesis states that acupuncture has no effect on pain tolerance.   **c.** $p(0, 1, 2, 3, 12, 13, 14,$ or 15 pluses$) = 0.0352$. Since $0.0352 < 0.05$, you reject H_0 and conclude that acupuncture affects pain tolerance.

It appears to increase pain tolerance.   **d.** You may have made a Type I error, rejecting H_0 if it is true.   **e.** The conclusion applies to the large pool of university undergraduate volunteers.

■ CHAPTER 11

8. For $P_{real} = 0.80$, power $= 0.8042$, and beta $= 0.1958$.

9. power $= 0.9559$, beta $= 0.0441$

10. power $= 0.1493$

11. Power $= 0.0955$, and beta $= 0.9045$. No, it is not legitimate to conclude that stimulus isolation had no effect on depression. That conclusion is the same thing as concluding that H_0 is true. Of course, we cannot prove H_0 is true from the data of an experiment. Particularly, in this case, the preceding analysis shows that this experiment has a low probability of detecting a real but small effect (power $= 0.0955$). This experiment is insensitive to small effects, and therefore, we cannot conclude stimulus isolation has no effect just because the results of the experiment were not significant.

12. Power $= 0.1268$, and beta $= 0.8732$. No, we cannot conclude that the TV program has no effect on violence in teenagers. We cannot prove H_0 is true. In this experiment, the power to detect a medium effect was quite low (0.1268). It could very well be true that the program really does increase violence, but due to lack of sufficient power, we failed to detect it.

■ CHAPTER 12

17. a. The sampling distribution of the mean is given here:

$(\bar{X})$	$p(\bar{X})$
7.0	0.04
6.5	0.08
6.0	0.12
5.5	0.16
5.0	0.20
4.5	0.16
4.0	0.12
3.5	0.08
3.0	0.04

b. From the population raw scores, $\mu = 5.00$, and from the 25 sample means, $\mu_{\bar{X}} = 5.00$. Therefore, $\mu_{\bar{X}} = \mu$.
c. From the 25 sample means, $\sigma_{\bar{X}} = 1.00$. From the population raw scores, $\sigma = 1.41$. Thus,

$$\sigma_{\bar{X}} = \sigma/\sqrt{N} = 1.41/\sqrt{2} = 1.41/1.41 = 1.00.$$

18. **a.** The distribution is normally shaped; $\mu_{\bar{X}} = 80$, $\sigma_{\bar{X}} = 2.00$. **b.** The distribution is normally shaped; $\mu_{\bar{X}} = 80$, $\sigma_{\bar{X}} = 1.35$. **c.** The distribution is normally shaped; $\mu_{\bar{X}} = 80$, $\sigma_{\bar{X}} = 1.13$ **d.** As N increases, $\mu_{\bar{X}}$ stays the same but $\sigma_{\bar{X}}$ decreases.

19. $z_{obt} = 3.16$, and $z_{crit} = \pm 1.96$. Since $|z_{obt}| > 1.96$, we reject H_0. It is not reasonable to consider the sample a random sample from a population with $\mu = 60$ and $\sigma = 10$.

20. **a.** $z_{obt} = -2.05$, and $z_{crit} = -2.33$. Since $|z_{obt}| < 2.33$, we can't reject the hypothesis that the sample is a random sample from a population with $\mu = 22$ and $\sigma = 8$. **b.** power $= 0.1685$ **c.** power $= 0.5675$ **d.** $N = 161$ (actually gives a power $= 0.7995$)

22. $z_{obt} = 4.03$, and $z_{crit} = 1.645$. Since $|z_{obt}| < 1.645$, reject H_0 and conclude that this year's class is superior to the previous ones.

23. $z_{obt} = 1.56$, and $z_{crit} = 1.645$. Since $|z_{obt}| < 1.645$, retain H_0. We cannot conclude that the new engine saves gas.

24. **a.** power $= 0.5871$ **b.** power $= 0.9750$
c. $N = 91$ (rounded to nearest integer)

25. $z_{obt} = 4.35$, and $z_{crit} = 1.645$. Since $|z_{obt}| > 1.645$, reject H_0 and conclude that exercise appears to slow down the "aging" process, at least as measured by maximum oxygen consumption.

■ CHAPTER 13

11. $t_{obt} = -1.37$, and t_{crit} with 29 df $= \pm 2.756$. Since $|t_{obt}| < 2.756$, we retain H_0. It is reasonable to consider the sample a random sample from a population with $\mu = 85$.

12. $t_{obt} = 3.08$, and t_{crit} with 28 df $= 2.467$. Since $|t_{obt}| > 2.467$, we can reject H_0, which specifies that the sample is a random sample from a population with a mean ≤ 72. Therefore, we can accept the hypothesis that the sample is a random sample from a population with a mean > 72.

13. $t_{obt} = 2.08$, and t_{crit} with 21 df $= 1.721$. Since $|t_{obt}| > 1.721$, we reject H_0. It is not reasonable to consider

the sample a random sample from a normal population with $\mu = 38$.

14.

	95%	**99%**
a.	21.68–28.32	20.39–29.61
c.	28.28–32.92	27.45–33.75
d.	22.76–27.24	21.98–28.02

Increasing N decreases the width of confidence interval.

15. $t_{obt} = -1.57$, and $t_{crit} = \pm 2.093$. Since $|t_{obt}| < 2.093$, you fail to reject H_0 and therefore cannot conclude that the student's technique shortens the duration of stay. The difference in conclusions is due to the greater sensitivity of the z test.

17. **a.** 18.34–21.66 **b.** 17.79–22.21

18. **a.** $z_{obt} = 3.96$, and $z_{crit} = 1.645$. Since $|z_{obt}| > 1.645$, we reject H_0 and conclude that the amount of smoking in women appears to have increased in recent years. The professor was correct. **b.** $t_{obt} = 3.67$, and $t_{crit} = 1.658$. Reject H_0. **c.** Same conclusion as in part **a**. **d.** $\hat{d} = 0.26$. This is a medium effect, according to Cohen's criteria.

19. **a.** $t_{obt} = 4.32$, and t_{crit} with 7 df $= \pm 2.365$. Since $|t_{obt}| > 2.365$, we reject H_0 and conclude that the drug affects short-term memory. It appears to improve it. **b.** $\hat{d} = 1.53$. This is a large effect, according to Cohen's criteria.

20. $t_{obt} = 1.65$, and $t_{crit} = 1.796$. Since $|t_{obt}| < 1.796$, we retain H_0. We cannot conclude that middle-age men employed by the corporation have become fatter.

21. $t_{obt} = 0.98$, and $t_{crit} = \pm 2.262$. Since $|t_{obt}| < 2.262$, we retain H_0. From these data, we cannot conclude that the graduates of the local business school get higher salaries for their first jobs than the national average.

22. **a.** 109.09–140.91 **b.** 103.14–146.86

23. **b.** $r = 0.675$. **c.** Yes, reject H_0 since $r_{crit} = \pm 0.5760$.

24. **a.** $r = 0.98$. **b.** Yes, reject H_0 since $r_{crit} = \pm 0.6319$.

25. **a.** $r_{obt} = 0.630$ **b.** $r_{crit} = \pm 0.7067$. Retain H_0; correlation is not significant. No, ρ may actually differ from 0, and power may be too low to detect it. **c.** $r_{crit} = \pm 0.4438$. Reject H_0; correlation is significant. Power is greater with $N = 20$.

26. $r_{crit} = \pm 0.5139$. Reject H_0; correlation is significant.

■ SPSS

1. $t = 1.387$ and **Sig. (2-tailed)** $= .208$. Since .208 > 0.05, we conclude by retaining H_0. Note that the SPSS t value (rounded to 2 decimal places) and the textbook t_{obt} value are the same. The conclusion reached in both cases is also the same.

2. **a.** $t = 1.649$ and **Sig. (2-tailed)/2** $= .064$. Since .064 > 0.05, we conclude by retaining H_0. . However, this result came very close to allowing us to reject H_0 and the experiment might be worth repeating with increased N. **b.** $t = 2.802$ and **Sig. (2-tailed)/2** $= .004$. Since .004 < 0.05, we conclude rejecting H_0 and affirming H_1. It appears that increasing N increased power enough to allow rejection of H_0.

■ CHAPTER 14

15. **a.** The alternative hypothesis states that memory for pictures is superior to memory for words. $\mu_1 > \mu_2$. **b.** The null hypothesis states that memory for pictures is not superior to memory for words. $\mu_1 \leq \mu_2$. **c.** $t_{obt} = 1.86$ and $t_{crit} = 1.761$. Since $|t_{obt}| > 1.761$, you reject H_0 and conclude that memory for pictures is superior to memory for words. **d.** $\hat{d} = 0.93$. This is a large effect, according to Cohen's criteria.

17. **a.** $t_{obt} = 4.10$, and $t_{crit} = \pm2.145$. Since $|t_{obt}| > 2.145$, you reject H_0 and conclude that newspaper advertising really does make a difference. It appears to increase cosmetics sales. **b.** $\hat{d} = 1.06$. According to Cohen's criteria, this is a large effect.

18. **a.** $t_{obt} = 4.09$, and $t_{crit} = \pm2.306$. Since $|t_{obt}| > 2.306$, you reject H_0 and conclude that biofeedback training reduces tension headaches. **b.** If the sampling distribution of D is not normally distributed, you cannot use the t test. However, you can use the sign test, because it does not assume anything about the shape of the scores. By using the sign test, $p(0, 1, 8, or 9 pluses) = 0.0392$. Since $0.0392 < 0.05$, you reject H_0, as before.

19. $t_{obt} = 3.50$, and $t_{crit} = \pm2.306$. Since $|t_{obt}| > 2.306$, we reject H_0 and conclude that other factors, such as attention, have an effect on tension headaches. They appear to decrease them.

20. $t_{obt} = 2.83$, and $t_{crit} = \pm2.120$. Since $|t_{obt}| > 2.120$, you reject H_0 and conclude that the decrease obtained

with biofeedback training cannot be attributed solely to other factors, such as attention. The biofeedback training itself has an effect on tension headaches. It appears to decrease them.

21. $t_{obt} = 0.60$, and $t_{crit} = \pm2.228$. Since $|t_{obt}| < 2.228$, we retain H_0. Based on these data, we cannot conclude that hiring part-time workers instead of full-time workers will affect productivity.

22. **a.** $t_{obt} = -2.83$, and $t_{crit} = \pm2.131$. Since $|t_{obt}| > 2.131$, you reject H_0 and conclude that the clinician was right. Depression interferes with sleep. **b.** $\hat{d} = 1.37$. Yes, this is a large effect, according to Cohen's criteria.

23. $t_{obt} = 2.11$ and $t_{crit} = \pm2.201$. Since $|t_{obt}| < 2.201$, you retain H_0. You cannot conclude that early exposure to schooling affects IQ.

24. **a.** $t_{obt} = 2.23$, and $t_{crit} = \pm2.074$. Since $|t_{obt}| > 2.074$, you reject H_0 and conclude that sleep has an effect on memory. It appears to improve it. **b.** 95% confidence interval $= 0.13–3.70$. Reject H_0. Size of effect $= 0.13–3.70$ more objects. **c.** 99% confidence interval $= -0.51–4.34$. We are 99% confident that the interval $-0.51–4.34$ contains the real effect. Since 0 is one of those values, we conclude by failing to reject H_0. We cannot affirm H_1. The results of the experiment are not significant at $\alpha = 0.01$. Check it out for yourself, using the null-hypothesis approach and $\alpha = 0.01$.

25. **a.** $t_{obt} = 5.36$, and $t_{crit} = \pm3.106$. Since $|t_{obt}| > 3.106$, you reject H_0 and conclude that high levels of curiosity in childhood appear to effect IQ. It seems to increase it. **b.** $\hat{d} = 1.55$. This is a large effect, according to Cohen's criteria.

26. $t_{obt} = 2.02$, and $t_{crit} = \pm2.160$. Since $|t_{obt}| < 2.160$, you retain H_0 and conclude that the data do not allow the conclusion that women and men differ in recalling emotional events. However, you also note that t_{obt} is very close to t_{crit}. With only 15 subjects, power is probably low and it may be premature to give up on H_1.

27. $t_{obt} = 3.28$, and $t_{crit} = \pm2.101$. Since $|t_{obt}| > 2.101$, you reject H_0 and conclude that women and men differ in recalling emotional events. Women appear to recall emotional events better than men. Increasing the power of the experiment allowed H_0 to be rejected.

28. $t_{obt} = 3.14$, and $t_{crit} = \pm2.179$. Since $|t_{obt}| > 2.179$, you reject H_0 and conclude that natural lighting affects student learning. It appears to improve it.

■ SPSS

Correlated Groups t test

1. **t** = 2.110 and **Sig. (2-tailed)** = .059. Since .059 > 0.05, H_0 is retained. We cannot conclude that early exposure to schooling affects IQ. Note, the textbook and SPSS analysis are in agreement.

2. **t** = −5.363 and **Sig. (2-tailed)** = .000. Since .000 < 0.01, we reject H_0 and conclude that high levels of curiosity in childhood appear to increase IQ. Note, the textbook and SPSS analysis are in agreement.

Independent Groups t test

1. **t** = 2.225 and **Sig. (2-tailed)** = .037. Since .037 < 0.05, we conclude by rejecting H_0. Sleep appears to facilitate memory consolidation. Again, the textbook and SPSS analysis are in agreement.

■ CHAPTER 15

10. **a.** $F_{crit} = 3.63$ **b.** $F_{crit} = 2.86$ **c.** $F_{crit} = 4.38$

11. $F_{obt} = 8.64$, and t_{obt} from Practice Problem 14.2 = −2.94. $8.64 = (-2.94)^2$. Therefore, $F = t^2$.

19. **a.**

Source	SS	df	MS	F_{obt}
Between groups	1253.68	3	417.89	4.00
Within groups	3762.72	36	104.52	
Total	5016.40	39		

b. 4   **c.** 10   **d.** 2.86   **e.** Yes, the effect is significant.

20. **a.** and **b.**

Source	SS	df	MS	F_{obt}
Between groups	30.333	2	15.167	5.21
Within groups	43.667	15	2.911	
Total	74.000	17		

$F_{crit} = 3.68$. Since $F_{obt} > 3.68$, you reject H_0 and conclude that at least one of the cereals differs in sugar content.

c. Tukey HSD

Comparison	Q_{obt}	Q_{crit}	Conclusion
Cereal A and *Cereal B*	3.83	3.67	Reject H_0
Cereal A and *Cereal C*	4.07	3.67	Reject H_0
Cereal B and *Cereal C*	0.24	3.67	Retain H_0

d. Scheffé

Groups (Cereals)	$SS_{between}$ (groups i and j)	$df_{between}$ from ANOVA	$MS_{between}$ (groups i and j)	MS_{within} from ANOVA	$F_{Scheffé}$	F_{crit} from ANOVA	Conclusion
A and B	21.333	2	10.667	2.911	3.66	3.68	Retain H_0
A and C	24.083	2	12.042	2.911	4.14	3.68	Reject H_0
B and C	0.083	2	0.042	2.911	0.01	3.68	Retain H_0

22. **a.**

Source	SS	df	MS	F_{obt}
Between groups	108.333	3	36.111	5.40
Within groups	133.667	20	6.683	
Total	242.000	23		

$F_{crit} = 3.10$. Since $F_{obt} > 3.10$, we reject H_0 and conclude that age affects memory.

b. $\hat{\omega}^2 = 0.355$, accounting for 35.5% of the variance.
c. $\eta^2 = 0.448$, accounting for 44.8% of the variance.
e. $t_{obt} = 3.13$, and $t_{crit} = \pm2.086$. Reject H_0 and conclude that the 60-year-old group is significantly different from the 30-year-old group.

f. Scheffé

Groups (Years Old)	$SS_{between\ (groups\ i\ and\ j)}$	$df_{between}$ from ANOVA	$MS_{between\ (groups\ i\ and\ j)}$	MS_{within} from ANOVA	$F_{Scheffé}$	F_{crit} from ANOVA	Conclusion
30-40	0.750	3	0.250	6.683	0.04	3.10	Retain H_0
30-50	0.083	3	0.028	6.683	0.004	3.10	Retain H_0
30-60	65.333	3	21.778	6.683	3.26	3.10	Reject H_0
40-50	0.333	3	0.111	6.683	0.02	3.10	Retain H_0
40-60	80.083	3	26.694	6.683	3.99	3.10	Reject H_0
50-60	70.083	3	23.361	6.683	3.50	3.10	Reject H_0

Reject H_0 for all comparisons involving the 60-year-old group. This age group is significantly different from each of the other age groups. Retain H_0 for all other comparisons. It appears that memory begins to deteriorate somewhere between the ages of 50 and 60.

23. a.

Source	SS	df	MS	F_{obt}
Between groups	100.167	2	50.084	4.87
Within groups	92.500	9	10.278	
Total	192.667	11		

$F_{crit} = 4.26$. Since $F_{obt} > 4.26$, we reject H_0 and conclude that the batteries of at least one manufacturer differ regarding useful life.

b. Tukey HSD

Comparison	Q_{obt}	Q_{crit}	Conclusion
Battery A-B	4.87	3.95	Reject H_0
Battery A-C	5.04	3.95	Reject H_0
Battery B-C	0.17	3.95	Retain H_0

Reject H_0 for the comparisons between the batteries of manufacturer A and the other two manufacturers. Retain H_0 for the comparison between the batteries of manufacturers B and C. The batteries of manufacturer A have significantly longer life. On this basis, you recommend them over the batteries made by manufacturers B and C.

24. a.

Source	SS	df	MS	F_{obt}
Between groups	221.6	3	73.867	22.77
Within groups	116.8	36	3.244	
Total	338.4	39		

$F_{crit} = 2.86$. Since $F_{obt} > 2.86$, we reject H_0 and conclude that hormone X affects sexual behavior.

b. $\hat{\omega}^2 = 0.620$, accounting for 62.0% of the variance.

c. $t_{obt} = 7.20$, and $t_{crit} = \pm2.029$. Reject H_0 and conclude that concentration 3 of hormone X significantly increases the number of matings.

d. Tukey HSD

Comparison	Q_{obt}	Q_{crit}	Conclusion
Concentration 0-1	0.35	3.79	Retain H_0
Concentration 0-2	4.92	3.79	Reject H_0
Concentration 0-3	10.18	3.79	Reject H_0
Concentration 1-2	4.56	3.79	Reject H_0
Concentration 1-3	9.83	3.79	Reject H_0
Concentration 2-3	5.27	3.79	Reject H_0

Reject H_0 for all comparisons except between the placebo and concentration 1. It appears that increasing the concentration of hormone X increases the number of matings. The failure to find a significant effect for concentration 1 was probably due to low power to detect a difference for this low level of concentration. Alternatively, there may be a threshold that must be exceeded before the hormone becomes effective.

25. a. The treatments are equally effective; $\mu_1 = \mu_2 = \mu_3 = \mu_4$

b.

Source	SS	df	MS	F_{obt}
Between groups	762.88	3	254.29	15.92
Within groups	574.90	36	15.97	
Total	1337.78	39		

$F_{obt} = 15.92$, $F_{crit} = 2.86$; reject H_0, affirm H_1.

c. Tukey HSD

Comparison	Q_{obt}	Q_{crit}	Conclusion
Placebo-Cognitive Restructuring	8.66	3.79	Reject H_0
Placebo-Assertiveness Training	5.38	3.79	Reject H_0
Placebo-Exercise/ Nutrition	1.50	3.79	Retain H_0
Cognitive Restructuring-Assertiveness Training	3.48	3.79	Retain H_0
Cognitive Restructuring-Exercise/Nutrition	7.36	3.79	Reject H_0
Assertiveness Training-Exercise/Nutrition	3.88	3.79	Reject H_0

Relative to the placebo, *Cognitive Restructuring* and *Assertiveness Training* were significantly more effective, whereas *Exercise/Nutrition was not.* Both *Cognitive Restructuring* and *Assertiveness Training* were significantly more effective than *Exercise/Nutrition.* There were no significant differences between *Cognitive Restructuring* and *Assertiveness Training.*

26. a.

Source	SS	df	MS	F_{obt}
Between groups	80.11	2	40.06	8.54
Within groups	70.33	15	4.69	
Total	150.44	17		

$F_{obt} = 8.54$, $F_{crit} = 3.68$, reject H_0 and conclude that acupuncture in combination with counseling affects cocaine addiction. They appear to help reduce cocaine addiction.

b. $\hat{\omega}^2 = 0.46$

c. $\eta^2 = 0.53$

27. a.

Source	SS	df	MS	F_{obt}
Between groups	16.53	2	8.27	0.12
Within groups	795.20	12	66.27	
Total	811.73	14		

$F_{obt} = 0.12$, $F_{crit} = 3.88$, retain H_0 and conclude that the data do not support the hypothesis that any of the tests are different in difficulty. Note, since $F_{obt} < 1.00$, you could have concluded to retain H_0 without determining F_{crit}.

■ SPSS

1. **F** = 14.062 and **Sig.** = .001. Since .001 < 0.05, our conclusion is to reject H_0. Sleep deprivation appears to affect sustained attention. Sure beats doing this by hand!!

2. **F** = 2.178 and **Sig.** = .101. Since .101 > 0.05, our conclusion is to retain H_0. We cannot conclude that there is a significant difference in acuity between any of the foreground and background color combinations.

■ CHAPTER 16

11. **a.** The different types of intervening material have the same effect on recall. $\mu_{a_1} = \mu_{a_2}$. The different amounts of repetition have the same effect on recall. $\mu_{b_1} = \mu_{b_2} = \mu_{b_3}$. There is no interaction between the number of repetitions and the type of intervening material in their effects on recall. With any main effects removed, $\mu_{a_1 b_1} = \mu_{a_1 b_2} = \mu_{a_1 b_3} = \mu_{a_2 b_1} = \mu_{a_2 b_2} = \mu_{a_2 b_3}$

b.

Source	SS	df	MS	F_{obt}	F_{crit}
Rows (intervening material)	121.000	1	121.000	41.41	4.17
Columns (number of repetitions)	176.167	2	88.084	30.14	3.32
Interaction	35.166	2	17.583	6.02	3.32
Within-cells	87.667	30	2.922		
Total	420.000	35			

Since in all cases $F_{obt} > F_{crit}$, H_0 is rejected for both main effects and the interaction effect. From the pattern of cell means, it is apparent that (1) increasing the number of repetitions increases recall; (2) using nonsense syllable pairs for intervening material decreases recall; and (3) there is an interaction such that the lower the number of repetitions, the greater the difference in effect between the two types of material.

13. **a.** For the concentrations administered, previous use of Drowson has no effect on its effectiveness. $\mu_{a_1} = \mu_{a_2}$. There is no difference between the placebo and the minimum recommended dosage of Drowson in their effects on insomnia. $\mu_{b_1} = \mu_{b_2}$. There is no interaction between the previous use of Drowson and the effect on insomnia of the two concentrations of

Drowson. With any main effects removed, $\mu_{a_1b_1} = \mu_{a_1b_2} = \mu_{a_2b_1} = \mu_{a_2b_2}$.

b.

Source	SS	df	MS	F_{obt}	F_{crit}
Rows (previous use)	639.031	1	639.031	11.63	4.20
Columns (concentration)	979.031	1	979.031	17.82	4.20
Interaction	472.782	1	472.782	8.61	4.20
Within-cells	1538.125	28	54.933		
Total	3628.969	31			

Since $F_{obt} > F_{crit}$ for each comparison, H_0 is rejected for both main effects and the interaction effect. From the pattern of cell means, it is apparent that Drowson promotes faster sleep onset in subjects who have had no previous use of the drug. However, the effect, if any, is much lower in chronic users, indicating that a tolerance to Drowson develops with chronic use.

■ SPSS

1. For the main effect of *Previous Use*, **F** = 11.633 and **Sig.** = .002. Since .002 < 0.05, we reject H_0. There is a significant main effect for *Previous Use*. For the main effect of *Drowson Concentration*, **F** = 17.822 and **Sig.** = .000. Since .000 < 0.05, we reject H_0. There is a significant main effect for *Drowson Concentration*. For the interaction between *Previous Use* and *Drowson Concentration*, **F** = 8.607 and **Sig.** = .007. Since .007 < 0.05, we reject H_0. There is a significant interaction between *Previous Use* and *Drowson Concentration*.

2. For the main effect of *Prior Usage*, **F** = 4.341 and **Sig.** = .022. Since .022 < 0.05, we reject H_0. There is a significant main effect for *Prior Usage*. For the main effect of *Marijuana Amount*, **F** = 16.896 and Sig. = .000. Since .000 < 0.05, we reject H_0. There is a significant main effect for *Marijuana Amount*. For the interaction between *Prior Use* and *Marijuana Amount*, **F** = 4.008 and **Sig.** = .029. Since .029 < 0.05, we reject H_0. There is a significant interaction between *Prior Use* and *Marijuana Amount*.

■ CHAPTER 17

16. $\chi^2_{obt} = 0.80$. Since $\chi^2_{crit} = 3.841$ we retain H_0. These data do not support the hypothesis that the prevailing view is fat people are more jolly.

17. $\chi^2_{obt} = 21.63$. Since $\chi^2_{crit} = 3.841$, we reject H_0 and conclude that big-city and small-town dwellers differ in their helpfulness to strangers. Converting the frequencies to proportions, we can see that the small-town dwellers were more helpful.

19. $\chi^2_{obt} = 13.28$ and $\chi^2_{crit} = 7.815$. Since $\chi^2_{obt} > 7.815$, we reject H_0 and conclude that the wrappings differ in their effect on sales. The manager should choose wrapping C.

20. $H_{obt} = 7.34$. Since $H_{crit} = 5.991$, we reject H_0 and conclude that at least one of the occupations differs from at least one of the others.

21. $\chi^2_{obt} = 3.00$. Since $\chi^2_{crit} = 3.841$, we retain H_0. Based on these data, we cannot conclude that church attendance and educational level are related.

22. $\chi^2_{obt} = 3.56$ and $\chi^2_{crit} = 9.488$. Since $\chi^2_{obt} < 9.488$, we retain H_0. Yes, the advertising is misleading because the data do not show a significant difference among brands.

23. $\chi^2_{obt} = 12.73$ and $\chi^2_{crit} = 5.991$. Since $\chi^2_{obt} > 5.991$, we reject H_0 and conclude that there is a relationship between the amount of contact white housewives have with blacks and changes in their attitudes toward blacks. The contact in the integrated housing projects appears to have had a positive effect on the attitude of the white housewives.

24. $H_{obt} = 0.69$. Since $H_{crit} = 9.210$, we must retain H_0. We cannot conclude that birth order affects assertiveness.

25. $\chi^2_{obt} = 29.57$ and $\chi^2_{crit} = 9.488$. Since $\chi^2_{obt} > 9.488$, we reject H_0. There is a relationship between gambling behavior and the different motives. Those high in power motivation appear to take the high risks more often. Most of the subjects with high power motivation placed high-risk bets, whereas the majority of those high in achievement motivation opted for medium-risk bets and the majority of those high in affiliation motivation chose the low-risk bets. These results are consistent with the views that (1) people with high power motivation will take high risks to achieve the attention and status that accompany such risk, (2) people high in achievement motivation will take medium risks to maximize the probability of having a sense of personal accomplishment, and (3) people with high affiliation motivation will take low risks to avoid competition and maximize the sense of belongingness.

26. $T_{obt} = 15$ and $T_{crit} = 17$. Since $T_{obt} < 17$, you reject H_0 and conclude that the film promotes more favorable attitudes toward major oil companies.

27. $T_{obt} = 1$ and $T_{crit} = 5$. Since $T_{obt} < 5$, you reject H_0 and conclude that biofeedback to relax frontalis muscle affects tension headaches. It appears to decrease them.

28. $T_{obt} = 14$ and $T_{crit} = 3$. Since $T_{obt} > 3$, you retain H_0. You cannot conclude that the pill affects blood pressure.

30. **a.** The alternative hypothesis states that FSH increases the singing rate in captive male cotingas.
b. The null hypothesis states that FSH does not increase the singing rate of captive male cotingas.
c. $U_{obt} = 15.5$ and $U'_{obt} = 64.5$. $U_{crit} = 20$. Since $U_{obt} < 20$, you reject H_0 and conclude that FSH appears to increase the singing rate of male cotingas.

31. **a.** The alternative hypothesis states that right-handed and left-handed people differ in spatial ability.
b. The null hypothesis states that right-handed people and left-handed people are equal in spatial ability.
c. $U_{obt} = 20.5$ and $U'_{obt} = 69.5$. $U_{crit} = 20$. Since $U_{obt} > 20$, you retain H_0. You cannot conclude that right-handed and left-handed people differ in spatial ability.

32. **a.** The alternative hypothesis states that hypnosis is more effective than the standard treatment in reducing test anxiety.
b. The null hypothesis states that hypnosis is not more effective than the standard treatment in reducing test anxiety.
c. $U_{obt} = 31$ and $U'_{obt} = 90$. $U_{crit} = 34$. Since $U_{obt} < 34$, you reject H_0 and conclude that hypnosis is more effective than the standard treatment in reducing test anxiety.

33. $H_{obt} = 10.12$ and $H_{crit} = 5.991$. Reject H_0; affirm H_1. Sleep deprivation has an effect on the ability to maintain sustained attention.

34. $\chi^2_{obt} = 6.454$ and $\chi^2_{crit} = 3.841$. Reject H_0; affirm H_1. There is a relationship between cohabitation and divorce. There is a significantly higher proportion of divorced couples among those that cohabited before marriage than among those that did not cohabit before marriage.

35. $\chi^2_{obt} = 19.82$. Since $\chi^2_{crit} = 5.991$, you reject H_0 and conclude that there is a relationship between gender and attitude regarding government involvement in citizen affairs. Men appear to favor a small role, whereas women seem to favor a large one.

36. $\chi^2_{obt} = 0.51$. Since $\chi^2_{crit} = 3.841$, retain H_0 and conclude that even though overall, black patients received fewer angiograms than white patients, physician racial bias does not appear to have contributed to this phenomenon.

37. $\chi^2_{obt} = 5.37$. Since $\chi^2_{crit} = 3.841$, you reject H_0 and conclude that the number of single-father homes has changed. It appears to have increased.

38. $\chi^2_{obt} = 8.333$. Since $\chi^2_{crit} = 5.991$, reject H_0 and conclude that cigarette smoking affects gender of offspring. It appears that when both parents smoke at least one pack of cigarettes a day, their offspring are more likely to be girls.

39. $\chi^2_{obt} = 80.00$. Since $\chi^2_{crit} = 5.991$, reject H_0 and conclude that the survey does reveal a reliable preference. Women undergraduates at the university seem to prefer soccer.

40. **a.** $\chi^2_{obt} = 79.23$. Since $\chi^2_{crit} = 3.841$, reject H_0 and conclude that the September 11, 2001, attacks affected religious sentiment. They appeared to increase it.
b. $\chi^2_{obt} = 0.69$. Since $\chi^2_{crit} = 3.841$, retain H_0; the data do not support the hypothesis that increased religious sentiment was still evident 1 year after the attacks. It appears that religious sentiment has returned to preattack levels.

■ **SPSS**

1. **Pearson Chi-Square** = 17.325 and **Asymp. Sig. (2-sided)** = .000. Since .000 < 0.05, we reject H_0 and conclude that preference for desktops or laptops and age of computer user are not independent; they appear to be related.

2. **Pearson Chi-Square** = 25.804 and **Asymp. Sig. (2-sided)** = .000. Since .000 < 0.05, we reject H_0 and conclude that preference for beer type and gender are not independent; they appear to be related.

■ **CHAPTER 18**

11. **a.** The alternative hypothesis states that the four brands of scotch whiskey are not equal in preference among the scotch drinkers in New York City.
b. The null hypothesis states that the four brands of scotch whiskey are equal in preference among the scotch drinkers in New York City. **c.** $\chi^2_{obt} = 2.72$ and $\chi^2_{crit} = 7.815$, Since $\chi^2_{obt} < 7.815$, retain H_0. You cannot conclude that the scotch drinkers

in New York City differ in their preference for the four brands of scotch whiskey.

12. **a.** The alternative hypothesis states that ACTH affects avoidance learning. $\mu_1 \neq \mu_2$ **b.** The null hypothesis states that ACTH has no effect on avoidance learning. $\mu_1 = \mu_2$ **c.** $t_{obt} = -3.24$, $t_{crit} = \pm2.878$. Since $|t_{obt}| > 2.878$, reject H_0 and conclude that ACTH has an effect on avoidance learning. It appears to facilitate avoidance learning. **d.** You may be making a Type I error. The null hypothesis may be true and it has been rejected. **e.** These results apply to the 100-day-old male rats living in the university vivarium at the time the sample was selected.   **f.** $\hat{d} = 1.45$, large effect.

14. **a.** Exogenous thyroxin has no effect on activity. Therefore, $\mu_1 = \mu_2 = \mu_3 = \mu_4$.

b.

Source	SS	df	MS	F_{obt}
Between groups	260.09	3	86.967	33.96
Within groups	92.2	36	2.561	
Total	353.1	39		

$F_{crit} = 2.86$. Since $F_{obt} > 2.86$, you reject H_0 and conclude that exogenous thyroxin affects activity level.

c. $\hat{\omega}^2 = 0.712$, accounting for 71.2% of the variability. **d.** $t_{obt} = 8.94$, and $t_{crit} = \pm2.029$. Reject H_0 and conclude that there is a significant difference between high amounts of exogenous thyroxin and saline on activity level. Exogenous thyroxin appears to increase activity level.
e. Tukey HSD

Comparison	Q_{obt}	Q_{crit}	Conclusion
Group 1-Group 2	1.78	3.79	Retain H_0
Group 1-Group 3	8.10	3.79	Reject H_0
Group 1-Group 4	12.65	3.79	Reject H_0
Group 2-Group 3	6.32	3.79	Reject H_0
Group 2-Group 4	10.87	3.79	Reject H_0
Group 3-Goup 4	4.54	3.79	Reject H_0

Reject H_0 for all comparisons except between groups 1 and 2. Increases in the amount of exogenous

thyroxin produce significantly higher levels of activity. The failure to find a significant difference between groups 1 and 2 is probably due to low power. Alternatively, there may be a threshold that must be exceeded before exogenous thyroxin becomes effective.

15. **a.** The alternative hypothesis states that dieting plus exercise is more effective in producing weight loss than dieting alone.   **b.** The null hypothesis states that dieting plus exercise is not more effective in producing weight loss than dieting alone. **c.** Since it is not valid to use the t test for correlated groups, the next most sensitive test is the Wilcoxon signed ranks test. $T_{obt} = 10.5$, and $T_{crit} = 17$. Since $T_{obt} < 17$, reject H_0 and conclude that dieting plus exercise is more effective than dieting alone in producing weight loss.

16. **a.** Sign test   **b.** $p(8, 9, 10, 11, \text{ or } 12 \text{ pluses}) = 0.1937$. Since the obtained probability is greater than alpha, you retain H_0.   **c.** The Wilcoxon signed ranks test is more powerful than the sign test. **d.** Power = 0.3907, beta = 0.6093

17. **a.** The null hypothesis states that there is no relationship between gender and time-of-day preference for having intercourse.   **b.** $\chi^2_{obt} = 2.14$ and $\chi^2_{crit} = 3.841$. Since $\chi^2_{obt} < 3.841$, retain H_0. You cannot conclude that there is a relationship between gender and time-of-day preference for having intercourse.

18. **a.** The null hypothesis states that food deprivation (hunger) has no effect on the number of food-related objects reported. Therefore, $\mu_1 = \mu_2 = \mu_3$

b.

Source	SS	df	MS	F_{obt}
Between groups	256.083	2	128.042	11.85
Within groups	226.875	21	10.804	
Total	482.958	23		

$F_{crit} = 3.47$. Since $F_{obt} > 3.47$, reject H_0 and conclude that food deprivation has an effect on the number of food-related objects reported.
c. $\hat{\omega}^2 = 0.47$
d. $\eta^2 = 0.53$

e. Scheffé

Groups	$SS_{between\ (groups\ i\ and\ j)}$	$df_{between}$ from ANOVA	$MS_{between\ (groups\ i\ and\ j)}$	MS_{within} from ANOVA	$F_{Scheffé}$	F_{crit} from ANOVA	Conclusion
1 and 4hr	60.062	2	30.031	10.084	2.78	3.47	Retain H_0
1 and 12hr	256.000	2	128.000	10.084	11.85	3.47	Reject H_0
4 and 12hr	68.062	2	34.031	10.084	3.15	3.47	Retain H_0

19. **a.** $t_{obt} = 10.78$, and $t_{crit} = 2.500$. Since $|t_{obt}| > 2.500$, reject H_0. Yes, the engineer is correct in her opinion. The new process results in significantly longer life for LCD TV's. **b.** $\hat{d} = 2.20$, large effect.

20. **a.** The alternative hypothesis states that alcohol has an effect on aggressiveness, $\mu_1 \neq \mu_2$. **b.** The null hypothesis states that alcohol has no effect on aggressiveness. $\mu_1 = \mu_2$. **c.** $t_{obt} = -3.93$, and $t_{crit} = \pm2.131$. Since $|t_{obt}| > 2.131$, reject H_0 and conclude that alcohol has an effect on aggressiveness. It appears to increase aggressiveness.

21. **a.** $r_{obt} = 0.70$. **b.** $r_{crit} = 0.5139$. Since $|r_{obt}| > 0.5139$, reject H_0. The correlation is significant. **c.** $r^2 = 0.48$. **d.** Yes, although it is clear that there is still a lot of variability unaccounted for.

22. **a.** The alternative hypothesis states that smoking affects heart rate. $\mu_D \neq 0$. **b.** The null hypothesis states that smoking has no effect on heart rate. $\mu_D = 0$. **c.** $t_{obt} = -3.40$, and $t_{crit} = \pm2.262$. Since $|t_{obt}| > 2.262$, reject H_0 and conclude that smoking affects heart rate. It appears to increase heart rate. **d.** $\hat{d} = 1.08$, large effect.

23. **a.** The null hypothesis states that there is no relationship between the occupations and the course of action that is favored. **b.** $\chi^2_{obt} = 13.79$ and $\chi^2_{crit} = 5.991$. Since $\chi^2_{obt} > 5.991$, reject H_0 and conclude that there is a relationship between the occupations and the course of action that is favored. From the proportions shown in the sample, business is in favor of letting the oil price rise, the homemakers favor gasoline rationing, and labor is fairly evenly divided.

24. **a.** The null hypothesis states that men and women do not differ in logical reasoning ability. **b.** $U_{obt} = 25.5$, $U'_{obt} = 37.5$, and $U_{crit} = 12$. Since $U_{obt} > 12$, retain H_0. You cannot conclude that men and women differ in logical reasoning ability. **c.** You may be making a Type II error. The null hypothesis may be false and you retained it.

25. 24.58–33.42

26. **a.** The null hypothesis states that Hispanics are not underrepresented in high school teachers in the part of the country the researcher lives. Therefore, the sample is a random sample from a population of high school teachers where the percentage of Hispanic teachers equals 22%. **b.** $\chi^2_{obt} = 12.59$ and $\chi^2_{crit} = 3.841$. Since $\chi^2_{obt} > 3.841$, reject H_0. It appears that high school teachers are underrepresented in the geographical locale studied.

27. $H_{obt} = 1.46$. Since $H_{crit} = 5.991$, we must retain H_0. We cannot conclude that physical science professors are more authoritarian than social science professors.

28.

Source	SS	df	MS	F_{obt}	F_{crit}
Rows (time)	17.633	1	17.633	11.76*	4.26
Columns (activity)	28.800	2	14.400	9.60*	3.40
Interaction	0.267	2	0.133	0.09	3.40
Within-cells	36.000	24	1.500		
Total	82.700	29			

*Since $F_{obt} > F_{crit}$ for the rows and columns effects, we reject H_0 for the main effects. We must retain H_0 for the interaction effect. It appears that performance is affected differently by at least one of the activity conditions and by the time of day when it is conducted. It appears that napping and afternoon produce superior performance.

Tables

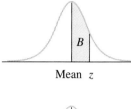

Mean *z*

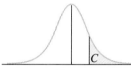

Mean *z*

Column A gives the positive *z* score.

Column B gives the area between the mean and *z*. Since the curve is symmetrical, areas for negative *z* scores are the same as for positive ones.

Column C gives the area that is beyond *z*.

table A Areas under the normal curve

z A	Area Between Mean and *z* B	Area Beyond *z* C	*z* A	Area Between Mean and *z* B	Area Beyond *z* C
0.00	.0000	.5000	0.45	.1736	.3264
0.01	.0040	.4960	0.46	.1772	.3228
0.02	.0080	.4920	0.47	.1808	.3192
0.03	.0120	.4880	0.48	.1844	.3156
0.04	.0160	.4840	0.49	.1879	.3121
0.05	.0199	.4801	0.50	.1915	.3085
0.06	.0239	.4761	0.51	.1950	.3050
0.07	.0279	.4721	0.52	.1985	.3015
0.08	.0319	.4681	0.53	.2019	.2981
0.09	.0359	.4641	0.54	.2054	.2946
0.10	.0398	.4602	0.55	.2088	.2912
0.11	.0438	.4562	0.56	.2123	.2877
0.12	.0478	.4522	0.57	.2157	.2843
0.13	.0517	.4483	0.58	.2190	.2810
0.14	.0557	.4443	0.59	.2224	.2776
0.15	.0596	.4404	0.60	.2257	.2743
0.16	.0636	.4364	0.61	.2291	.2709
0.17	.0675	.4325	0.62	.2324	.2676
0.18	.0714	.4286	0.63	.2357	.2643
0.19	.0753	.4247	0.64	.2389	.2611
0.20	.0793	.4207	0.65	.2422	.2578
0.21	.0832	.4168	0.66	.2454	.2546
0.22	.0871	.4129	0.67	.2486	.2514
0.23	.0910	.4090	0.68	.2517	.2483
0.24	.0948	.4052	0.69	.2549	.2451
0.25	.0987	.4013	0.70	.2580	.2420
0.26	.1026	.3974	0.71	.2611	.2389
0.27	.1064	.3936	0.72	.2642	.2358
0.28	.1103	.3897	0.73	.2673	.2327
0.29	.1141	.3859	0.74	.2704	.2296
0.30	.1179	.3821	0.75	.2734	.2266
0.31	.1217	.3783	0.76	.2764	.2236
0.32	.1255	.3745	0.77	.2794	.2206
0.33	.1293	.3707	0.78	.2823	.2177
0.34	.1331	.3669	0.79	.2852	.2148
0.35	.1368	.3632	0.80	.2881	.2119
0.36	.1406	.3594	0.81	.2910	.2090
0.37	.1443	.3557	0.82	.2939	.2061
0.38	.1480	.3520	0.83	.2967	.2033
0.39	.1517	.3483	0.84	.2995	.2005
0.40	.1554	.3446	0.85	.3023	.1977
0.41	.1591	.3409	0.86	.3051	.1949
0.42	.1628	.3372	0.87	.3078	.1922
0.43	.1664	.3336	0.88	.3106	.1894
0.44	.1700	.3300	0.89	.3133	.1867

(Continued)

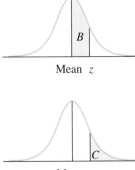

Mean z

Mean z

Column A gives the positive
z score.

Column B gives the area between
the mean and *z*. Since the curve
is symmetrical, areas for negative
z scores are the same as for positive ones.

Column C gives the area that is
beyond *z*.

table A Areas under the normal curve—*cont'd*

z A	Area Between Mean and z B	Area Beyond z C	z A	Area Between Mean and z B	Area Beyond z C
0.90	.3159	.1841	1.35	.4115	.0885
0.91	.3186	.1814	1.36	.4131	.0869
0.92	.3212	.1788	1.37	.4147	.0853
0.93	.3238	.1762	1.38	.4162	.0838
0.94	.3264	.1736	1.39	.4177	.0823
0.95	.3289	.1711	1.40	.4192	.0808
0.96	.3315	.1685	1.41	.4207	.0793
0.97	.3340	.1660	1.42	.4222	.0778
0.98	.3365	.1635	1.43	.4236	.0764
0.99	.3389	.1611	1.44	.4251	.0749
1.00	.3413	.1587	1.45	.4265	.0735
1.01	.3438	.1562	1.46	.4279	.0721
1.02	.3461	.1539	1.47	.4292	.0708
1.03	.3485	.1515	1.48	.4306	.0694
1.04	.3508	.1492	1.49	.4319	.0681
1.05	.3531	.1469	1.50	.4332	.0668
1.06	.3554	.1446	1.51	.4345	.0655
1.07	.3577	.1423	1.52	.4357	.0643
1.08	.3599	.1401	1.53	.4370	.0630
1.09	.3621	.1379	1.54	.4382	.0618
1.10	.3643	.1357	1.55	.4394	.0606
1.11	.3665	.1335	1.56	.4406	.0594
1.12	.3686	.1314	1.57	.4418	.0582
1.13	.3708	.1292	1.58	.4429	.0571
1.14	.3729	.1271	1.59	.4441	.0559
1.15	.3749	.1251	1.60	.4452	.0548
1.16	.3770	.1230	1.61	.4463	.0537
1.17	.3790	.1210	1.62	.4474	.0526
1.18	.3810	.1190	1.63	.4484	.0516
1.19	.3830	.1170	1.64	.4495	.0505
1.20	.3849	.1151	1.65	.4505	.0495
1.21	.3869	.1131	1.66	.4515	.0485
1.22	.3888	.1112	1.67	.4525	.0475
1.23	.3907	.1093	1.68	.4535	.0465
1.24	.3925	.1075	1.69	.4545	.0455
1.25	.3944	.1056	1.70	.4554	.0446
1.26	.3962	.1038	1.71	.4564	.0436
1.27	.3980	.1020	1.72	.4573	.0427
1.28	.3997	.1003	1.73	.4582	.0418
1.29	.4015	.0985	1.74	.4591	.0409
1.30	.4032	.0968	1.75	.4599	.0401
1.31	.4049	.0951	1.76	.4608	.0392
1.32	.4066	.0934	1.77	.4616	.0384
1.33	.4082	.0918	1.78	.4625	.0375
1.34	.4099	.0901	1.79	.4633	.0367

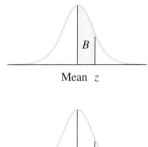

Mean z

Mean z

Column A gives the positive z score.

Column B gives the area between the mean and z. Since the curve is symmetrical, areas for negative z scores are the same as for positive ones.

Column C gives the area that is beyond z.

table A Areas under the normal curve—*cont'd*

z A	Area Between Mean and z B	Area Beyond z C	z A	Area Between Mean and z B	Area Beyond z C
1.80	.4641	.0359	2.25	.4878	.0122
1.81	.4649	.0351	2.26	.4881	.0119
1.82	.4656	.0344	2.27	.4884	.0116
1.83	.4664	.0336	2.28	.4887	.0113
1.84	.4671	.0329	2.29	.4890	.0110
1.85	.4678	.0322	2.30	.4893	.0107
1.86	.4686	.0314	2.31	.4896	.0104
1.87	.4693	.0307	2.32	.4898	.0102
1.88	.4699	.0301	2.33	.4901	.0099
1.89	.4706	.0294	2.34	.4904	.0096
1.90	.4713	.0287	2.35	.4906	.0094
1.91	.4719	.0281	2.36	.4909	.0091
1.92	.4726	.0274	2.37	.4911	.0089
1.93	.4732	.0268	2.38	.4913	.0087
1.94	.4738	.0262	2.39	.4916	.0084
1.95	.4744	.0256	2.40	.4918	.0082
1.96	.4750	.0250	2.41	.4920	.0080
1.97	.4756	.0244	2.42	.4922	.0078
1.98	.4761	.0239	2.43	.4925	.0075
1.99	.4767	.0233	2.44	.4927	.0073
2.00	.4772	.0228	2.45	.4929	.0071
2.01	.4778	.0222	2.46	.4931	.0069
2.02	.4783	.0217	2.47	.4932	.0068
2.03	.4788	.0212	2.48	.4934	.0066
2.04	.4793	.0207	2.49	.4936	.0064
2.05	.4798	.0202	2.50	.4938	.0062
2.06	.4803	.0197	2.51	.4940	.0060
2.07	.4808	.0192	2.52	.4941	.0059
2.08	.4812	.0188	2.53	.4943	.0057
2.09	.4817	.0183	2.54	.4945	.0055
2.10	.4821	.0179	2.55	.4946	.0054
2.11	.4826	.0174	2.56	.4948	.0052
2.12	.4830	.0170	2.57	.4949	.0051
2.13	.4834	.0166	2.58	.4951	.0049
2.14	.4838	.0162	2.59	.4952	.0048
2.15	.4842	.0158	2.60	.4953	.0047
2.16	.4846	.0154	2.61	.4955	.0045
2.17	.4850	.0150	2.62	.4956	.0044
2.18	.4854	.0146	2.63	.4957	.0043
2.19	.4857	.0143	2.64	.4959	.0041
2.20	.4861	.0139	2.65	.4960	.0040
2.21	.4864	.0136	2.66	.4961	.0039
2.22	.4868	.0132	2.67	.4962	.0038
2.23	.4871	.0129	2.68	.4963	.0037
2.24	.4875	.0125	2.69	.4964	.0036

(Continued)

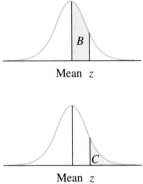

Mean z

Mean z

Column A gives the positive z score.

Column B gives the area between the mean and z. Since the curve is symmetrical, areas for negative z scores are the same as for positive ones.

Column C gives the area that is beyond z.

table A Areas under the normal curve—*cont'd*

z A	Area Between Mean and z B	Area Beyond z C	z A	Area Between Mean and z B	Area Beyond z C
2.70	.4965	.0035	3.00	.4987	.0013
2.71	.4966	.0034	3.01	.4987	.0013
2.72	.4967	.0033	3.02	.4987	.0013
2.73	.4968	.0032	3.03	.4988	.0012
2.74	.4969	.0031	3.04	.4988	.0012
2.75	.4970	.0030	3.05	.4989	.0011
2.76	.4971	.0029	3.06	.4989	.0011
2.77	.4972	.0028	3.07	.4989	.0011
2.78	.4973	.0027	3.08	.4990	.0010
2.79	.4974	.0026	3.09	.4990	.0010
2.80	.4974	.0026	3.10	.4990	.0010
2.81	.4975	.0025	3.11	.4991	.0009
2.82	.4976	.0024	3.12	.4991	.0009
2.83	.4977	.0023	3.13	.4991	.0009
2.84	.4977	.0023	3.14	.4992	.0008
2.85	.4978	.0022	3.15	.4992	.0008
2.86	.4979	.0021	3.16	.4992	.0008
2.87	.4979	.0021	3.17	.4992	.0008
2.88	.4980	.0020	3.18	.4993	.0007
2.89	.4981	.0019	3.19	.4993	.0007
2.90	.4981	.0019	3.20	.4993	.0007
2.91	.4982	.0018	3.21	.4993	.0007
2.92	.4982	.0018	3.22	.4994	.0006
2.93	.4983	.0017	3.23	.4994	.0006
2.94	.4984	.0016	3.24	.4994	.0006
2.95	.4984	.0016	3.30	.4995	.0005
2.96	.4985	.0015	3.40	.4997	.0003
2.97	.4985	.0015	3.50	.4998	.0002
2.98	.4986	.0014	3.60	.4998	.0002
2.99	.4986	.0014	3.70	.4999	.0001

table B Binomial distribution

N	No. of P or Q Events	P or Q .05	.10	.15	.20	.25	.30	.35	.40	.45	.50
1	0	.9500	.9000	.8500	.8000	.7500	.7000	.6500	.6000	.5500	.5000
	1	.0500	.1000	.1500	.2000	.2500	.3000	.3500	.4000	.4500	.5000
2	0	.9025	.8100	.7225	.6400	.5625	.4900	.4225	.3600	.3025	.2500
	1	.0950	.1800	.2550	.3200	.3750	.4200	.4550	.4800	.4950	.5000
	2	.0025	.0100	.0225	.0400	.0625	.0900	.1225	.1600	.2025	.2500
3	0	.8574	.7290	.6141	.5120	.4219	.3430	.2746	.2160	.1664	.1250
	1	.1354	.2430	.3251	.3840	.4219	.4410	.4436	.4320	.4084	.3750
	2	.0071	.0270	.0574	.0960	.1406	.1890	.2389	.2880	.3341	.3750
	3	.0001	.0010	.0034	.0080	.0156	.0270	.0429	.0640	.0911	.1250
4	0	.8145	.6561	.5220	.4096	.3164	.2401	.1785	.1296	.0915	.0625
	1	.1715	.2916	.3685	.4096	.4219	.4116	.3845	.3456	.2995	.2500
	2	.0135	.0486	.0975	.1536	.2109	.2646	.3105	.3456	.3675	.3750
	3	.0005	.0036	.0115	.0256	.0469	.0756	.1115	.1536	.2005	.2500
	4	.0000	.0001	.0005	.0016	.0039	.0081	.0150	.0256	.0410	.0625
5	0	.7738	.5905	.4437	.3277	.2373	.1681	.1160	.0778	.0503	.0312
	1	.2036	.3280	.3915	.4096	.3955	.3602	.3124	.2592	.2059	.1562
	2	.0214	.0729	.1382	.2048	.2637	.3087	.3364	.3456	.3369	.3125
	3	.0011	.0081	.0244	.0512	.0879	.1323	.1811	.2304	.2757	.3125
	4	.0000	.0004	.0022	.0064	.0146	.0284	.0488	.0768	.1128	.1562
	5	.0000	.0000	.0001	.0003	.0010	.0024	.0053	.0102	.0185	.0312
6	0	.7351	.5314	.3771	.2621	.1780	.1176	.0754	.0467	.0277	.0156
	1	.2321	.3543	.3993	.3932	.3560	.3025	.2437	.1866	.1359	.0938
	2	.0305	.0984	.1762	.2458	.2966	.3241	.3280	.3110	.2780	.2344
	3	.0021	.0146	.0415	.0819	.1318	.1852	.2355	.2765	.3032	.3125
	4	.0001	.0012	.0055	.0154	.0330	.0595	.0951	.1382	.1861	.2344
	5	.0000	.0001	.0004	.0015	.0044	.0102	.0205	.0369	.0609	.0938
	6	.0000	.0000	.0000	.0001	.0002	.0007	.0018	.0041	.0083	.0156
7	0	.6983	.4783	.3206	.2097	.1335	.0824	.0490	.0280	.0152	.0078
	1	.2573	.3720	.3960	.3670	.3115	.2471	.1848	.1306	.0872	.0547
	2	.0406	.1240	.2097	.2753	.3115	.3177	.2985	.2613	.2140	.1641
	3	.0036	.0230	.0617	.1147	.1730	.2269	.2679	.2903	.2918	.2734
	4	.0002	.0026	.0109	.0287	.0577	.0972	.1442	.1935	.2388	.2734
	5	.0000	.0002	.0012	.0043	.0115	.0250	.0466	.0774	.1172	.1641
	6	.0000	.0000	.0001	.0004	.0013	.0036	.0084	.0172	.0320	.0547
	7	.0000	.0000	.0000	.0000	.0001	.0002	.0006	.0016	.0037	.0078
8	0	.6634	.4305	.2725	.1678	.1001	.0576	.0319	.0168	.0084	.0039
	1	.2793	.3826	.3847	.3355	.2670	.1977	.1373	.0896	.0548	.0312
	2	.0515	.1488	.2376	.2936	.3115	.2965	.2587	.2090	.1569	.1094
	3	.0054	.0331	.0839	.1468	.2076	.2541	.2786	.2787	.2568	.2188
	4	.0004	.0046	.0185	.0459	.0865	.1361	.1875	.2322	.2627	.2734
	5	.0000	.0004	.0026	.0092	.0231	.0467	.0808	.1239	.1719	.2188
	6	.0000	.0000	.0002	.0011	.0038	.0100	.0217	.0413	.0703	.1094
	7	.0000	.0000	.0000	.0001	.0004	.0012	.0033	.0079	.0164	.0312
	8	.0000	.0000	.0000	.0000	.0000	.0001	.0002	.0007	.0017	.0039

(Continued)

table B Binomial distribution—*cont'd*

N	No. of P or Q Events	.05	.10	.15	.20	.25	.30	.35	.40	.45	.50
						P or Q					
9	0	.6302	.3874	.2316	.1342	.0751	.0404	.0277	.0101	.0046	.0020
	1	.2985	.3874	.3679	.3020	.2253	.1556	.1004	.0605	.0339	.0176
	2	.0629	.1722	.2597	.3020	.3003	.2668	.2162	.1612	.1110	.0703
	3	.0077	.0446	.1069	.1762	.2336	.2668	.2716	.2508	.2119	.1641
	4	.0006	.0074	.0283	.0661	.1168	.1715	.2194	.2508	.2600	.2461
	5	.0000	.0008	.0050	.0165	.0389	.0735	.1181	.1672	.2128	.2461
	6	.0000	.0001	.0006	.0028	.0087	.0210	.0424	.0743	.1160	.1641
	7	.0000	.0000	.0000	.0003	.0012	.0039	.0098	.0212	.0407	.0703
	8	.0000	.0000	.0000	.0000	.0001	.0004	.0013	.0035	.0083	.0176
	9	.0000	.0000	.0000	.0000	.0000	.0000	.0001	.0003	.0008	.0020
10	0	.5987	.3487	.1969	.1074	.0563	.0282	.0135	.0060	.0025	.0010
	1	.3151	.3874	.3474	.2684	.1877	.1211	.0725	.0403	.0207	.0098
	2	.0746	.1937	.2759	.3020	.2816	.2335	.1757	.1209	.0763	.0439
	3	.0105	.0574	.1298	.2013	.2503	.2668	.2522	.2150	.1665	.1172
	4	.0010	.0112	.0401	.0881	.1460	.2001	.2377	.2508	.2384	.2051
	5	.0001	.0015	.0085	.0264	.0584	.1029	.1536	.2007	.2340	.2461
	6	.0000	.0001	.0012	.0055	.0162	.0368	.0689	.1115	.1596	.2051
	7	.0000	.0000	.0001	.0008	.0031	.0090	.0212	.0425	.0746	.1172
	8	.0000	.0000	.0000	.0001	.0004	.0014	.0043	.0106	.0229	.0439
	9	.0000	.0000	.0000	.0000	.0000	.0001	.0005	.0016	.0042	.0098
	10	.0000	.0000	.0000	.0000	.0000	.0000	.0000	.0001	.0003	.0010
11	0	.5688	.3138	.1673	.0859	.0422	.0198	.0088	.0036	.0014	.0005
	1	.3293	.3835	.3248	.2362	.1549	.0932	.0518	.0266	.0125	.0054
	2	.0867	.2131	.2866	.2953	.2581	.1998	.1395	.0887	.0513	.0269
	3	.0137	.0710	.1517	.2215	.2581	.2568	.2254	.1774	.1259	.0806
	4	.0014	.0158	.0536	.1107	.1721	.2201	.2428	.2365	.2060	.1611
	5	.0001	.0025	.0132	.0388	.0803	.1231	.1830	.2207	.2360	.2256
	6	.0000	.0003	.0023	.0097	.0268	.0566	.0985	.1471	.1931	.2256
	7	.0000	.0000	.0003	.0017	.0064	.0173	.0379	.0701	.1128	.1611
	8	.0000	.0000	.0000	.0002	.0011	.0037	.0102	.0234	.0462	.0806
	9	.0000	.0000	.0000	.0000	.0001	.0005	.0018	.0052	.0126	.0269
	10	.0000	.0000	.0000	.0000	.0000	.0000	.0002	.0007	.0021	.0054
	11	.0000	.0000	.0000	.0000	.0000	.0000	.0000	.0000	.0002	.0005
12	0	.5404	.2824	.1422	.0687	.0317	.0138	.0057	.0022	.0008	.0002
	1	.3413	.3766	.3012	.2062	.1267	.0712	.0368	.0174	.0075	.0029
	2	.0988	.2301	.2924	.2835	.2323	.1678	.1088	.0639	.0339	.0161
	3	.0173	.0852	.1720	.2362	.2581	.2397	.1954	.1419	.0923	.0537
	4	.0021	.0213	.0683	.1329	.1936	.2311	.2367	.2128	.1700	.1208
	5	.0002	.0038	.0193	.0532	.1032	.1585	.2039	.2270	.2225	.1934
	6	.0000	.0005	.0040	.0155	.0401	.0792	.1281	.1766	.2124	.2256
	7	.0000	.0000	.0006	.0033	.0115	.0291	.0591	.1009	.1489	.1934
	8	.0000	.0000	.0001	.0005	.0024	.0078	.0199	.0420	.0762	.1208
	9	.0000	.0000	.0000	.0001	.0004	.0015	.0048	.0125	.0277	.0537
	10	.0000	.0000	.0000	.0000	.0000	.0002	.0008	.0025	.0068	.0161
	11	.0000	.0000	.0000	.0000	.0000	.0000	.0001	.0003	.0010	.0029
	12	.0000	.0000	.0000	.0000	.0000	.0000	.0000	.0000	.0001	.0002

table B Binomial distribution—*cont'd*

N	No. of P or Q Events	P or Q									
		.05	.10	.15	.20	.25	.30	.35	.40	.45	.50
13	0	.5133	.2542	.1209	.0550	.0238	.0097	.0037	.0013	.0004	.0001
	1	.3512	.3672	.2774	.1787	.1029	.0540	.0259	.0113	.0045	.0016
	2	.1109	.2448	.2937	.2680	.2059	.1388	.0836	.0453	.0220	.0095
	3	.0214	.0997	.1900	.2457	.2517	.2181	.1651	.1107	.0660	.0349
	4	.0028	.0277	.0838	.1535	.2097	.2337	.2222	.1845	.1350	.0873
	5	.0003	.0055	.0266	.0691	.1258	.1803	.2154	.2214	.1989	.1571
	6	.0000	.0008	.0063	.0230	.0559	.1030	.1546	.1968	.2169	.2095
	7	.0000	.0001	.0011	.0058	.0186	.0442	.0833	.1312	.1775	.2095
	8	.0000	.0000	.0001	.0011	.0047	.0142	.0336	.0656	.1089	.1571
	9	.0000	.0000	.0000	.0001	.0009	.0034	.0101	.0243	.0495	.0873
	10	.0000	.0000	.0000	.0000	.0001	.0006	.0022	.0065	.0162	.0349
	11	.0000	.0000	.0000	.0000	.0000	.0001	.0003	.0012	.0036	.0095
	12	.0000	.0000	.0000	.0000	.0000	.0000	.0000	.0001	.0005	.0016
	13	.0000	.0000	.0000	.0000	.0000	.0000	.0000	.0000	.0000	.0001
14	0	.4877	.2288	.1028	.0440	.0178	.0068	.0024	.0008	.0002	.0001
	1	.3593	.3559	.2539	.1539	.0832	.0407	.0181	.0073	.0027	.0009
	2	.1229	.2570	.2912	.2501	.1802	.1134	.0634	.0317	.0141	.0056
	3	.0259	.1142	.2056	.2501	.2402	.1943	.1366	.0845	.0462	.0222
	4	.0037	.0349	.0998	.1720	.2202	.2290	.2022	.1549	.1040	.0611
	5	.0004	.0078	.0352	.0860	.1468	.1963	.2178	.2066	.1701	.1222
	6	.0000	.0013	.0093	.0322	.0734	.1262	.1759	.2066	.2088	.1833
	7	.0000	.0002	.0019	.0092	.0280	.0618	.1082	.1574	.1952	.2095
	8	.0000	.0000	.0003	.0020	.0082	.0232	.0510	.0918	.1398	.1833
	9	.0000	.0000	.0000	.0003	.0018	.0066	.0183	.0408	.0762	.1222
	10	.0000	.0000	.0000	.0000	.0003	.0014	.0049	.0136	.0312	.0611
	11	.0000	.0000	.0000	.0000	.0000	.0002	.0010	.0033	.0093	.0222
	12	.0000	.0000	.0000	.0000	.0000	.0000	.0001	.0005	.0019	.0056
	13	.0000	.0000	.0000	.0000	.0000	.0000	.0000	.0001	.0002	.0009
	14	.0000	.0000	.0000	.0000	.0000	.0000	.0000	.0000	.0000	.0001
15	0	.4633	.2059	.0874	.0352	.0134	.0047	.0016	.0005	.0001	.0000
	1	.3658	.3432	.2312	.1319	.0668	.0305	.0126	.0047	.0016	.0005
	2	.1348	.2669	.2856	.2309	.1559	.0916	.0476	.0219	.0090	.0032
	3	.0307	.1285	.2184	.2501	.2252	.1700	.1110	.0634	.0318	.0139
	4	.0049	.0428	.1156	.1876	.2252	.2186	.1792	.1268	.0780	.0417
	5	.0006	.0105	.0449	.1032	.1651	.2061	.2123	.1859	.1404	.0916
	6	.0000	.0019	.0132	.0430	.0917	.1472	.1906	.2066	.1914	.1527
	7	.0000	.0003	.0030	.0138	.0393	.0811	.1319	.1771	.2013	.1964
	8	.0000	.0000	.0005	.0035	.0131	.0348	.0710	.1181	.1647	.1964
	9	.0000	.0000	.0001	.0007	.0034	.0116	.0298	.0612	.1048	.1527
	10	.0000	.0000	.0000	.0001	.0007	.0030	.0096	.0245	.0515	.0916
	11	.0000	.0000	.0000	.0000	.0001	.0006	.0024	.0074	.0191	.0417
	12	.0000	.0000	.0000	.0000	.0000	.0001	.0004	.0016	.0052	.0139
	13	.0000	.0000	.0000	.0000	.0000	.0000	.0001	.0003	.0010	.0032
	14	.0000	.0000	.0000	.0000	.0000	.0000	.0000	.0000	.0001	.0005
	15	.0000	.0000	.0000	.0000	.0000	.0000	.0000	.0000	.0000	.0000

(Continued)

table B Binomial distribution—*cont'd*

N	No. of P or Q Events	.05	.10	.15	.20	.25	.30	.35	.40	.45	.50
16	0	.4401	.1853	.0743	.0281	.0100	.0033	.0010	.0003	.0001	.0000
	1	.3706	.3294	.2097	.1126	.0535	.0228	.0087	.0030	.0009	.0002
	2	.1463	.2745	.2775	.2111	.1336	.0732	.0353	.0150	.0056	.0018
	3	.0359	.1423	.2285	.2463	.2079	.1465	.0888	.0468	.0215	.0085
	4	.0061	.0514	.1311	.2001	.2252	.2040	.1553	.1014	.0572	.0278
	5	.0008	.0137	.0555	.1201	.1802	.2099	.2008	.1623	.1123	.0667
	6	.0001	.0028	.0180	.0550	.1101	.1649	.1982	.1983	.1684	.1222
	7	.0000	.0004	.0045	.0197	.0524	.1010	.1524	.1889	.1969	.1746
	8	.0000	.0001	.0009	.0055	.0197	.0487	.0923	.1417	.1812	.1964
	9	.0000	.0000	.0001	.0012	.0058	.0185	.0442	.0840	.1318	.1746
	10	.0000	.0000	.0000	.0002	.0014	.0056	.0167	.0392	.0755	.1222
	11	.0000	.0000	.0000	.0000	.0002	.0013	.0049	.0142	.0337	.0667
	12	.0000	.0000	.0000	.0000	.0000	.0002	.0011	.0040	.0115	.0278
	13	.0000	.0000	.0000	.0000	.0000	.0000	.0002	.0008	.0029	.0085
	14	.0000	.0000	.0000	.0000	.0000	.0000	.0000	.0001	.0005	.0018
	15	.0000	.0000	.0000	.0000	.0000	.0000	.0000	.0000	.0001	.0002
	16	.0000	.0000	.0000	.0000	.0000	.0000	.0000	.0000	.0000	.0000
17	0	.4181	.1668	.0631	.0225	.0075	.0023	.0007	.0002	.0000	.0000
	1	.3741	.3150	.1893	.0957	.0426	.0169	.0060	.0019	.0005	.0001
	2	.1575	.2800	.2673	.1914	.1136	.0581	.0260	.0102	.0035	.0010
	3	.0415	.1556	.2359	.2393	.1893	.1245	.0701	.0341	.0144	.0052
	4	.0076	.0605	.1457	.2093	.2209	.1868	.1320	.0796	.0411	.0182
	5	.0010	.0175	.0668	.1361	.1914	.2081	.1849	.1379	.0875	.0472
	6	.0001	.0039	.0236	.0680	.1276	.1784	.1991	.1839	.1432	.0944
	7	.0000	.0007	.0065	.0267	.0668	.1201	.1685	.1927	.1841	.1484
	8	.0000	.0001	.0014	.0084	.0279	.0644	.1143	.1606	.1883	.1855
	9	.0000	.0000	.0003	.0021	.0093	.0276	.0611	.1070	.1540	.1855
	10	.0000	.0000	.0000	.0004	.0025	.0095	.0263	.0571	.1008	.1484
	11	.0000	.0000	.0000	.0001	.0005	.0026	.0090	.0242	.0525	.0944
	12	.0000	.0000	.0000	.0000	.0001	.0006	.0024	.0081	.0215	.0472
	13	.0000	.0000	.0000	.0000	.0000	.0001	.0005	.0021	.0068	.0182
	14	.0000	.0000	.0000	.0000	.0000	.0000	.0001	.0004	.0016	.0052
	15	.0000	.0000	.0000	.0000	.0000	.0000	.0000	.0001	.0003	.0010
	16	.0000	.0000	.0000	.0000	.0000	.0000	.0000	.0000	.0000	.0001
	17	.0000	.0000	.0000	.0000	.0000	.0000	.0000	.0000	.0000	.0000
18	0	.3972	.1501	.0536	.0180	.0056	.0016	.0004	.0001	.0000	.0000
	1	.3763	.3002	.1704	.0811	.0338	.0126	.0042	.0012	.0003	.0001
	2	.1683	.2835	.2556	.1723	.0958	.0458	.0190	.0069	.0022	.0006
	3	.0473	.1680	.2406	.2297	.1704	.1046	.0547	.0246	.0095	.0031
	4	.0093	.0070	.1592	.2153	.2130	.1681	.1104	.0614	.0291	.0117
	5	.0014	.0218	.0787	.1507	.1988	.2017	.1664	.1146	.0666	.0327
	6	.0002	.0052	.0310	.0816	.1436	.1873	.1941	.1655	.1181	.0708
	7	.0000	.0010	.0091	.0350	.0820	.1376	.1792	.1892	.1657	.1214
	8	.0000	.0002	.0022	.0120	.0376	.0811	.1327	.1734	.1864	.1669
	9	.0000	.0000	.0004	.0033	.0139	.0386	.0794	.1284	.1694	.1855
	10	.0000	.0000	.0001	.0008	.0042	.0149	.0385	.0771	.1248	.1669
	11	.0000	.0000	.0000	.0001	.0010	.0046	.0151	.0374	.0742	.1214

table B Binomial distribution—*cont'd*

N	No. of P or Q Events	.05	.10	.15	.20	.25	.30	.35	.40	.45	.50
18	12	.0000	.0000	.0000	.0000	.0002	.0012	.0047	.0145	.0354	.0708
	13	.0000	.0000	.0000	.0000	.0000	.0002	.0012	.0045	.0134	.0327
	14	.0000	.0000	.0000	.0000	.0000	.0000	.0002	.0011	.0039	.0117
	15	.0000	.0000	.0000	.0000	.0000	.0000	.0000	.0002	.0009	.0031
	16	.0000	.0000	.0000	.0000	.0000	.0000	.0000	.0000	.0001	.0006
	17	.0000	.0000	.0000	.0000	.0000	.0000	.0000	.0000	.0000	.0001
	18	.0000	.0000	.0000	.0000	.0000	.0000	.0000	.0000	.0000	.0000
19	0	.3774	.1351	.0456	.0144	.0042	.0011	.0003	.0001	.0000	.0000
	1	.3774	.2852	.1529	.0685	.0268	.0093	.0029	.0008	.0002	.0000
	2	.1787	.2852	.2428	.1540	.0803	.0358	.0138	.0046	.0013	.0003
	3	.0533	.1796	.2428	.2182	.1517	.0869	.0422	.0175	.0062	.0018
	4	.0112	.0798	.1714	.2182	.2023	.1491	.0909	.0467	.0203	.0074
	5	.0018	.0266	.0907	.1636	.2023	.1916	.1468	.0933	.0497	.0222
	6	.0002	.0069	.0374	.0955	.1574	.1916	.1844	.1451	.0949	.0518
	7	.0000	.0014	.0122	.0443	.0974	.1525	.1844	.1797	.1443	.0961
	8	.0000	.0002	.0032	.0166	.0487	.0981	.1489	.1797	.1771	.1442
	9	.0000	.0000	.0007	.0051	.0198	.0514	.0980	.1464	.1771	.1762
	10	.0000	.0000	.0001	.0013	.0066	.0220	.0528	.0976	.1449	.1762
	11	.0000	.0000	.0000	.0003	.0018	.0077	.0233	.0532	.0970	.1442
	12	.0000	.0000	.0000	.0000	.0004	.0022	.0083	.0237	.0529	.0961
	13	.0000	.0000	.0000	.0000	.0001	.0005	.0024	.0085	.0233	.0518
	14	.0000	.0000	.0000	.0000	.0000	.0001	.0006	.0024	.0082	.0222
	15	.0000	.0000	.0000	.0000	.0000	.0000	.0001	.0005	.0022	.0074
	16	.0000	.0000	.0000	.0000	.0000	.0000	.0000	.0001	.0005	.0018
	17	.0000	.0000	.0000	.0000	.0000	.0000	.0000	.0000	.0001	.0003
	18	.0000	.0000	.0000	.0000	.0000	.0000	.0000	.0000	.0000	.0000
	19	.0000	.0000	.0000	.0000	.0000	.0000	.0000	.0000	.0000	.0000
20	0	.3585	.1216	.0388	.0115	.0032	.0008	.0002	.0000	.0000	.0000
	1	.3774	.2702	.1368	.0576	.0211	.0068	.0020	.0005	.0001	.0000
	2	.1887	.2852	.2293	.1369	.0669	.0278	.0100	.0031	.0008	.0002
	3	.0596	.1901	.2428	.2054	.1339	.0716	.0323	.0123	.0040	.0011
	4	.0133	.0898	.1821	.2182	.1897	.1304	.0738	.0350	.0139	.0046
	5	.0022	.0319	.1028	.1746	.2023	.1789	.1272	.0746	.0365	.0148
	6	.0003	.0089	.0454	.1091	.1686	.1916	.1712	.1244	.0746	.0370
	7	.0000	.0020	.0160	.0545	.1124	.1643	.1844	.1659	.1221	.0739
	8	.0000	.0004	.0046	.0222	.0609	.1144	.1614	.1797	.1623	.1201
	9	.0000	.0001	.0011	.0074	.0271	.0654	.1158	.1597	.1771	.1602
	10	.0000	.0000	.0002	.0020	.0099	.0308	.0686	.1171	.1593	.1762
	11	.0000	.0000	.0000	.0005	.0030	.0120	.0336	.0710	.1185	.1602
	12	.0000	.0000	.0000	.0001	.0008	.0039	.0136	.0355	.0727	.1201
	13	.0000	.0000	.0000	.0000	.0002	.0010	.0045	.0146	.0366	.0739
	14	.0000	.0000	.0000	.0000	.0000	.0002	.0012	.0049	.0150	.0370
	15	.0000	.0000	.0000	.0000	.0000	.0000	.0003	.0013	.0049	.0148
	16	.0000	.0000	.0000	.0000	.0000	.0000	.0000	.0003	.0013	.0046
	17	.0000	.0000	.0000	.0000	.0000	.0000	.0000	.0000	.0002	.0011
	18	.0000	.0000	.0000	.0000	.0000	.0000	.0000	.0000	.0000	.0002
	19	.0000	.0000	.0000	.0000	.0000	.0000	.0000	.0000	.0000	.0000
	20	.0000	.0000	.0000	.0000	.0000	.0000	.0000	.0000	.0000	.0000

table C.1 Critical values of U and U' for a one-tailed test at $\alpha = 0.005$ or a two-tailed test at $\alpha = 0.01$

*To be significant for any given n_1 and n_2: U_{obt} must be equal to or **less than** the value shown in the table. U'_{obt} must be equal to or **greater than** the value shown in the table.*

Each cell shows U (top) over U' (underlined, bottom).

n_2 \ n_1	1	2	3	4	5	6	7	8	9	10	11	12	13	14	15	16	17	18	19	20
1	—	—	—	—	—	—	—	—	—	—	—	—	—	—	—	—	—	—	—	—
2	—	—	—	—	—	—	—	—	—	—	—	—	—	—	—	—	—	—	0 / 38	0 / 40
3	—	—	—	—	—	—	—	—	0 / 27	0 / 30	0 / 33	1 / 35	1 / 35	1 / 41	2 / 43	2 / 46	2 / 49	2 / 52	3 / 54	3 / 57
4	—	—	—	—	—	0 / 24	0 / 28	1 / 31	1 / 35	2 / 38	2 / 42	3 / 45	3 / 49	4 / 52	5 / 55	5 / 59	6 / 62	6 / 66	7 / 69	8 / 72
5	—	—	—	—	0 / 25	1 / 29	1 / 34	2 / 38	3 / 42	4 / 46	5 / 50	6 / 54	7 / 58	7 / 63	8 / 67	9 / 71	10 / 75	11 / 79	12 / 83	13 / 87
6	—	—	—	0 / 24	1 / 29	2 / 34	3 / 39	4 / 44	5 / 49	6 / 54	7 / 59	9 / 63	10 / 68	11 / 73	12 / 78	13 / 83	15 / 87	16 / 92	17 / 97	18 / 102
7	—	—	—	0 / 28	1 / 34	3 / 39	4 / 45	6 / 50	7 / 56	9 / 61	10 / 67	12 / 72	13 / 78	15 / 83	16 / 89	18 / 94	19 / 100	21 / 105	22 / 111	24 / 116
8	—	—	—	1 / 31	2 / 38	4 / 44	6 / 50	7 / 57	9 / 63	11 / 69	13 / 75	15 / 81	17 / 87	18 / 94	20 / 100	22 / 106	24 / 112	26 / 118	28 / 124	30 / 130
9	—	—	0 / 27	1 / 35	3 / 42	5 / 49	7 / 56	9 / 63	11 / 70	13 / 77	16 / 73	18 / 90	20 / 97	22 / 104	24 / 111	27 / 117	29 / 124	31 / 131	33 / 138	36 / 144
10	—	—	0 / 30	2 / 38	4 / 46	6 / 54	9 / 61	11 / 69	13 / 77	16 / 84	18 / 92	21 / 99	24 / 106	26 / 114	29 / 121	31 / 129	34 / 136	37 / 143	39 / 151	42 / 158
11	—	—	0 / 33	2 / 42	5 / 50	7 / 59	10 / 67	13 / 75	16 / 83	18 / 92	21 / 100	24 / 108	27 / 116	30 / 124	33 / 132	36 / 140	39 / 148	42 / 156	45 / 164	48 / 172
12	—	—	1 / 35	3 / 45	6 / 54	9 / 63	12 / 72	15 / 81	18 / 90	21 / 99	24 / 108	27 / 117	31 / 125	34 / 134	37 / 143	41 / 151	44 / 160	47 / 169	51 / 177	54 / 186
13	—	—	1 / 38	3 / 49	7 / 58	10 / 68	13 / 78	17 / 87	20 / 97	24 / 106	27 / 116	31 / 125	34 / 125	38 / 144	42 / 153	45 / 163	49 / 172	53 / 181	56 / 191	60 / 200
14	—	—	1 / 41	4 / 52	7 / 63	11 / 73	15 / 83	18 / 94	22 / 104	26 / 114	30 / 124	34 / 134	38 / 144	42 / 154	46 / 164	50 / 174	54 / 184	58 / 194	63 / 203	67 / 213
15	—	—	2 / 43	5 / 55	8 / 67	12 / 78	16 / 89	20 / 100	24 / 111	29 / 121	33 / 132	37 / 143	42 / 153	46 / 164	51 / 174	55 / 185	60 / 195	64 / 206	69 / 216	73 / 227
16	—	—	2 / 46	5 / 59	9 / 71	13 / 83	18 / 94	22 / 106	27 / 117	31 / 129	36 / 140	41 / 151	45 / 163	50 / 174	55 / 185	60 / 196	65 / 207	70 / 218	74 / 230	79 / 241
17	—	—	2 / 49	6 / 62	10 / 75	15 / 87	19 / 100	24 / 112	29 / 124	34 / 148	39 / 148	44 / 160	49 / 172	54 / 184	60 / 195	65 / 207	70 / 219	75 / 231	81 / 242	86 / 254
18	—	—	2 / 52	6 / 66	11 / 79	16 / 92	21 / 105	26 / 118	31 / 131	37 / 143	42 / 156	47 / 169	53 / 181	58 / 194	64 / 206	70 / 218	75 / 231	81 / 243	87 / 255	92 / 268
19	—	0 / 38	3 / 54	7 / 69	12 / 83	17 / 97	22 / 111	28 / 124	33 / 138	39 / 151	45 / 164	51 / 177	56 / 191	63 / 203	69 / 216	74 / 230	81 / 242	87 / 255	93 / 268	99 / 281
20	—	0 / 40	3 / 57	8 / 72	13 / 87	18 / 102	24 / 116	30 / 130	36 / 144	42 / 158	48 / 172	54 / 186	60 / 200	67 / 213	73 / 227	79 / 241	86 / 254	92 / 268	99 / 281	105 / 295

Dashes in the body of the table indicate that no decision is possible at the stated level of significance.

table C.2 Critical values of U and U' for a one-tailed test at $\alpha = 0.01$ or a two-tailed test at $\alpha = 0.02$

*To be significant for any given n_1 and n_2: U_{obt} must be equal to or **less than** the value shown in the table. U'_{obt} must be equal to or **greater than** the value shown in the table.*

n_2 \ n_1	1	2	3	4	5	6	7	8	9	10	11	12	13	14	15	16	17	18	19	20
1	—	—	—	—	—	—	—	—	—	—	—	—	—	—	—	—	—	—	—	—
2	—	—	—	—	—	—	—	—	—	—	—	—	0/26	0/28	0/30	0/32	0/34	0/36	1/37	1/39
3	—	—	—	—	—	—	0/21	0/24	1/26	1/29	1/32	2/34	2/37	2/40	3/42	3/45	4/47	4/50	4/52	5/55
4	—	—	—	—	0/20	1/23	1/27	2/30	3/33	3/37	4/40	5/43	5/47	6/50	7/53	7/57	8/60	9/63	9/67	10/70
5	—	—	—	0/20	1/24	2/28	3/32	4/36	5/40	6/44	7/48	8/52	9/56	10/60	11/64	12/68	13/72	14/76	15/80	16/84
6	—	—	—	1/23	2/28	3/33	4/38	6/42	7/47	8/52	9/57	11/61	12/66	13/71	15/75	16/80	18/84	19/89	20/94	22/93
7	—	—	0/21	1/27	3/32	4/38	6/43	7/49	9/54	11/59	12/65	14/70	16/75	17/81	19/86	21/91	23/96	24/102	26/107	28/112
8	—	—	0/24	2/30	4/36	6/42	7/49	9/55	11/61	13/67	15/73	17/79	20/84	22/90	24/96	26/102	28/108	30/114	32/120	34/126
9	—	—	1/26	3/33	5/40	7/47	9/54	11/61	14/67	16/74	18/81	21/87	23/94	26/100	28/107	31/113	33/120	36/126	38/133	40/140
10	—	—	1/29	3/37	6/44	8/52	11/59	13/67	16/74	19/81	22/88	24/96	27/103	30/110	33/117	36/124	38/132	41/139	44/146	47/153
11	—	—	1/32	4/40	7/48	9/57	12/65	15/73	18/81	22/88	25/96	28/104	31/112	34/120	37/128	41/135	44/143	47/151	50/159	53/167
12	—	—	2/34	5/43	8/52	11/61	14/70	17/79	21/87	24/96	28/104	31/113	35/121	38/130	42/138	46/146	49/155	53/163	56/172	60/180
13	—	0/26	2/37	5/47	9/56	12/66	16/75	20/84	23/94	27/103	31/112	35/121	39/130	43/139	47/148	51/157	55/166	59/175	63/184	67/193
14	—	0/28	2/40	6/50	10/60	13/71	17/81	22/90	26/100	30/110	34/120	38/130	43/139	47/149	51/159	56/168	60/178	65/187	69/197	73/207
15	—	0/30	3/42	7/53	11/64	15/75	19/86	24/96	28/107	33/117	37/128	42/138	47/148	51/159	56/169	61/179	66/189	70/200	75/210	80/220
16	—	0/32	3/45	7/57	12/68	16/80	21/91	26/102	31/113	36/124	41/135	46/146	51/157	56/168	61/179	66/190	71/201	76/212	82/222	87/233
17	—	0/34	4/47	8/60	13/72	18/84	23/96	28/108	33/120	38/132	44/143	49/155	55/166	60/178	66/189	71/201	72/212	82/224	88/234	93/247
18	—	0/36	4/50	9/63	14/76	19/89	24/102	30/114	36/126	41/139	47/151	53/163	59/175	65/187	70/200	76/212	82/224	88/236	94/248	100/260
19	—	1/37	4/53	9/67	15/80	20/94	26/107	32/120	38/133	44/146	50/159	56/172	63/184	69/197	75/210	82/222	88/235	94/248	101/260	107/273
20	—	1/39	5/55	10/70	16/84	22/98	28/112	34/126	40/140	47/153	53/167	60/180	67/193	73/207	80/220	87/233	93/247	100/260	107/273	114/286

Dashes in the body of the table indicate that no decision is possible at the stated level of significance.

table C.3   Critical values of U and U' for a one-tailed test at $\alpha = 0.025$ or a two-tailed test at $\alpha = 0.05$

*To be significant for any given n_1 and n_2: U_{obt} must be equal to or **less than** the value shown in the table. U'_{obt} must be equal to or* **greater than** *the value shown in the table.*

Each cell shows the U critical value (upper) over the U' critical value (lower), written here as U/U'.

$n_2 \backslash n_1$	1	2	3	4	5	6	7	8	9	10	11	12	13	14	15	16	17	18	19	20
1	—	—	—	—	—	—	—	—	—	—	—	—	—	—	—	—	—	—	—	—
2	—	—	—	—	—	—	—	0/16	0/18	0/20	0/22	1/23	1/25	1/27	1/29	1/31	2/32	2/34	2/36	2/38
3	—	—	—	—	0/15	1/17	1/20	2/22	2/25	3/27	3/30	4/32	4/35	5/37	5/40	6/42	6/45	7/47	7/50	8/52
4	—	—	—	0/16	1/19	2/22	3/25	4/28	4/32	5/35	6/38	7/41	8/44	9/47	10/50	11/53	11/57	12/60	13/63	13/67
5	—	—	0/15	1/19	2/23	3/27	5/30	6/34	7/38	8/42	9/46	11/49	12/53	13/57	14/61	15/65	17/68	18/72	19/76	20/80
6	—	—	1/17	2/22	3/27	5/31	6/36	8/40	10/44	11/49	13/53	14/58	16/62	17/67	19/71	21/75	22/80	24/84	25/89	27/93
7	—	—	1/20	3/25	5/30	6/36	8/41	10/46	12/51	14/56	16/61	18/66	20/71	22/76	24/81	26/86	28/91	30/96	32/101	34/106
8	—	0/16	2/22	4/28	6/34	8/40	10/46	13/51	15/57	17/63	19/69	22/74	24/78	26/86	29/91	31/97	34/102	36/108	38/111	41/119
9	—	0/18	2/25	4/32	7/38	10/44	12/51	15/57	17/64	20/70	23/76	26/82	28/89	31/95	34/101	37/107	39/114	42/120	45/126	48/132
10	—	0/20	3/27	5/35	8/42	11/49	14/56	17/63	20/70	23/77	26/84	29/91	33/97	36/104	39/111	42/118	45/125	48/132	52/138	55/145
11	—	0/22	3/30	6/38	9/46	13/53	16/61	19/69	23/76	26/84	30/91	33/99	37/106	40/114	44/121	47/129	51/136	55/143	58/151	62/158
12	—	1/23	4/32	7/41	11/49	14/58	18/58	22/74	26/82	29/91	33/99	37/107	41/115	45/123	49/131	53/139	57/147	61/155	65/163	69/171
13	—	1/25	4/35	8/44	12/53	16/62	20/71	24/80	28/89	33/97	37/106	41/115	45/124	50/132	54/141	59/149	63/158	67/167	72/175	76/184
14	—	1/27	5/37	9/47	13/51	17/67	22/76	26/86	31/95	36/104	40/114	45/123	50/132	55/141	59/151	64/160	67/171	74/178	78/188	83/197
15	—	1/29	5/40	10/50	14/61	19/71	24/81	29/91	34/101	39/111	44/121	49/131	54/141	59/151	64/161	70/170	75/180	80/190	85/200	90/210
16	—	1/31	6/42	11/53	15/65	21/75	26/86	31/97	37/107	42/118	47/129	53/139	59/149	64/160	70/170	75/181	81/191	86/202	92/212	98/222
17	—	2/32	6/45	11/57	17/68	22/80	28/91	34/102	39/114	45/125	51/136	57/147	63/158	67/171	75/180	81/191	87/202	93/213	99/224	105/235
18	—	2/34	7/47	12/60	18/72	24/84	30/96	36/108	42/120	48/132	55/143	61/155	67/167	74/168	80/190	86/202	93/213	99/225	106/236	112/248
19	—	2/36	7/50	13/63	19/76	25/89	32/101	38/114	45/126	52/138	58/151	65/163	72/175	78/188	85/200	92/212	99/224	106/236	113/248	119/261
20	—	2/38	8/52	13/67	20/80	27/93	34/106	41/119	48/132	55/145	62/158	69/171	76/184	83/197	90/210	98/222	105/235	112/248	119/261	127/273

Dashes in the body of the table indicate that no decision is possible at the stated level of significance.

table C.4   Critical values of U and U' for a one-tailed test at $\alpha = 0.05$ or a two-tailed test at $\alpha = 0.10$

*To be significant for any given n_1 and n_2: U_{obt} must be equal to or **less than** the value shown in the table. U'_{obt} must be equal to or **greater than** the value shown in the table.*

Each cell shows U (upper) / U' (lower).

n_2 \ n_1	1	2	3	4	5	6	7	8	9	10	11	12	13	14	15	16	17	18	19	20
1	—	—	—	—	—	—	—	—	—	—	—	—	—	—	—	—	—	—	0/19	0/20
2	—	—	—	—	0/10	0/12	0/14	1/15	1/17	1/19	1/21	2/22	2/24	2/26	3/27	3/29	3/31	4/32	4/34	4/36
3	—	—	0/9	0/12	1/14	2/16	2/19	3/21	3/24	4/26	5/28	5/31	6/33	7/35	7/38	8/40	9/42	9/45	10/47	11/49
4	—	—	0/12	1/15	2/18	3/21	4/24	5/27	6/30	7/36	8/36	9/39	10/42	11/45	12/48	14/50	15/53	16/56	17/59	18/62
5	—	0/10	1/14	2/18	4/21	5/25	6/29	8/32	9/36	11/39	12/43	13/47	15/50	16/54	18/57	19/61	20/65	22/68	23/72	25/75
6	—	0/12	2/16	3/21	5/25	7/29	8/34	10/38	12/42	14/46	16/50	17/55	19/59	21/63	23/67	25/71	26/76	28/80	30/84	32/88
7	—	0/14	2/19	4/24	6/29	8/34	11/38	13/43	15/48	17/53	19/58	21/63	24/67	26/72	28/77	30/82	33/86	35/91	37/96	39/101
8	—	1/15	3/21	5/27	8/32	10/38	13/43	15/49	18/54	20/60	23/65	26/70	28/76	31/81	33/87	36/92	39/97	41/103	44/108	47/113
9	—	1/17	3/24	6/30	9/36	12/42	15/48	18/54	21/60	24/66	27/72	30/78	33/84	36/90	39/96	42/102	45/108	48/114	51/120	54/126
10	—	1/19	4/26	7/33	11/39	14/46	17/53	20/60	24/66	27/73	31/79	34/86	37/93	41/99	44/106	48/112	51/119	55/125	58/132	62/138
11	—	1/21	5/28	8/36	12/43	16/50	19/58	23/65	27/72	31/79	34/87	38/94	42/101	46/108	50/115	54/122	57/130	61/137	65/144	69/151
12	—	2/22	5/31	9/39	13/47	17/55	21/63	26/70	30/78	34/86	38/94	42/102	47/109	51/117	55/125	60/132	64/140	68/148	72/156	77/163
13	—	2/24	6/33	10/42	15/50	19/59	24/67	28/76	33/84	37/93	42/101	47/109	51/118	56/126	61/134	65/143	70/151	75/159	80/167	84/176
14	—	2/26	7/35	11/45	16/54	21/63	26/72	31/81	36/90	41/99	46/108	51/117	56/126	61/135	66/144	71/153	77/161	82/170	87/179	92/188
15	—	3/27	7/38	12/48	18/57	23/67	28/77	33/87	39/96	44/106	50/115	55/125	61/134	66/144	72/153	77/163	83/172	88/182	94/191	100/200
16	—	3/29	8/40	14/50	19/61	25/71	30/82	36/92	42/102	48/112	54/122	60/132	65/143	71/153	77/163	83/173	89/183	95/193	101/203	107/213
17	—	3/31	9/42	15/53	20/65	26/76	33/86	39/97	45/108	51/119	57/130	64/140	70/151	77/161	83/172	89/183	96/193	102/204	109/214	115/225
18	—	4/32	9/45	16/56	22/68	28/80	35/91	41/103	48/114	55/125	61/137	68/148	75/159	82/170	88/182	95/193	102/204	109/215	116/226	123/237
19	0/19	4/34	10/47	17/59	23/72	30/84	37/96	44/108	51/120	58/132	65/144	72/156	80/167	87/179	94/191	101/203	109/214	116/226	123/238	130/250
20	0/20	4/36	11/49	18/62	25/75	32/88	39/101	47/113	54/126	62/138	69/151	77/163	84/176	92/188	100/200	107/213	115/225	123/237	130/250	138/262

Dashes in the body of the table indicate that no decision is possible at the stated level of significance.

table D Critical values of Student's *t* distribution

The values listed in the table are the critical values of t for the specified degrees of freedom (left column) and the alpha level (column heading). For two-tailed alpha levels, t_{crit} is both + and −. To be significant, $|t_{obt}| \geq |t_{crit}|$.

df	Level of Significance for One-Tailed Test, $\alpha_{1\ tail}$					
	.10	.05	.025	.01	.005	.0005
	Level of Significance for Two-Tailed Test, $\alpha_{2\ tail}$					
	.20	.10	.05	.02	.01	.001
1	3.078	6.314	12.706	31.821	63.657	636.619
2	1.886	2.920	4.303	6.965	9.925	31.598
3	1.638	2.353	3.182	4.541	5.841	12.941
4	1.533	2.132	2.776	3.747	4.604	8.610
5	1.476	2.015	2.571	3.365	4.032	6.859
6	1.440	1.943	2.447	3.143	3.707	5.959
7	1.415	1.895	2.365	2.998	3.499	5.405
8	1.397	1.860	2.306	2.986	3.355	5.041
9	1.383	1.833	2.262	2.821	3.250	4.781
10	1.372	1.812	2.228	2.764	3.169	4.587
11	1.363	1.796	2.201	2.718	3.106	4.437
12	1.356	1.782	2.179	2.681	3.055	4.318
13	1.350	1.771	2.160	2.650	3.012	4.221
14	1.345	1.761	2.145	2.624	2.977	4.140
15	1.341	1.753	2.131	2.602	2.947	4.073
16	1.337	1.746	2.120	2.583	2.921	4.015
17	1.333	1.740	2.110	2.567	2.898	3.965
18	1.330	1.734	2.101	2.552	2.878	3.922
19	1.328	1.729	2.093	2.539	2.861	3.883
20	1.325	1.725	2.086	2.528	2.845	3.850
21	1.323	1.721	2.080	2.518	2.831	3.819
22	1.321	1.717	2.074	2.508	2.819	3.792
23	1.319	1.714	2.069	2.500	2.807	3.767
24	1.318	1.711	2.064	2.492	2.797	3.745
25	1.316	1.708	2.060	2.485	2.787	3.725
26	1.315	1.706	2.056	2.479	2.779	3.707
27	1.314	1.703	2.052	2.473	2.771	3.690
28	1.313	1.701	2.048	2.467	2.763	3.674
29	1.311	1.699	2.045	2.462	2.756	3.659
30	1.310	1.697	2.042	2.457	2.750	3.646
40	1.303	1.684	2.021	2.423	2.704	3.551
60	1.296	1.671	2.000	2.390	2.660	3.460
120	1.289	1.658	1.980	2.358	2.617	3.373
∞	1.282	1.645	1.960	2.326	2.576	3.291

table E Critical values of Pearson r

The values listed in the table are the critical values of r for the specified degrees of freedom (left column) and the alpha level (column heading). For two-tailed alpha levels, r_{crit} is both + and —. To be significant, $|r_{obt}| \geq |r_{crit}|$.

df = N − 2	Level of Significance for One-Tailed Test, $\alpha_{1\ tail}$				
	.05	**.025**	**.01**	**.005**	**.0005**
	Level of Significance for Two-Tailed Test, $\alpha_{2\ tail}$				
	.10	**.05**	**.02**	**.01**	**.001**
1	.9877	.9969	.9995	.9999	1.0000
2	.9000	.9500	.9800	.9900	.9990
3	.8054	.8783	.9343	.9587	.9912
4	.7293	.8114	.8822	.9172	.9741
5	.6694	.7545	.8329	.8745	.9507
6	.6215	.7067	.7887	.8343	.9249
7	.5822	.6664	.7498	.7977	.8982
8	.5494	.6319	.7155	.7646	.8721
9	.5214	.6021	.6851	.7348	.8471
10	.4973	.5760	.6581	.7079	.8233
11	.4762	.5529	.6339	.6835	.8010
12	.4575	.5324	.6120	.6614	.7800
13	.4409	.5139	.5923	.6411	.7603
14	.4259	.4973	.5742	.6226	.7420
15	.4124	.4821	.5577	.6055	.7246
16	.4000	.4683	.5425	.5897	.7084
17	.3887	.4555	.5285	.5751	.6932
18	.3783	.4438	.5155	.5614	.6787
19	.3687	.4329	.5034	.5487	.6652
20	.3598	.4227	.4921	.5368	.6524
25	.3233	.3809	.4451	.4869	.5974
30	.2960	.3494	.4093	.4487	.5541
35	.2746	.3246	.3810	.4182	.5189
40	.2573	.3044	.3578	.3932	.4896
45	.2428	.2875	.3384	.3721	.4648
50	.2306	.2732	.3218	.3541	.4433
60	.2108	.2500	.2948	.3248	.4078
70	.1954	.2319	.2737	.3017	.3799
80	.1829	.2172	.2565	.2830	.3568
90	.1726	.2050	.2422	.2673	.3375
100	.1638	.1946	.2301	.2540	.3211

table **F** Critical values of the F distribution for α = 0.05 (Roman type) and α = 0.01 (boldface type)

The values listed in the table are the critical values of F for the degrees of freedom of the numerator of the F ratio (column headings) and the degrees of freedom of the denominator of the F ratio (row headings). To be significant, $F_{obt} \geq F_{crit}$.

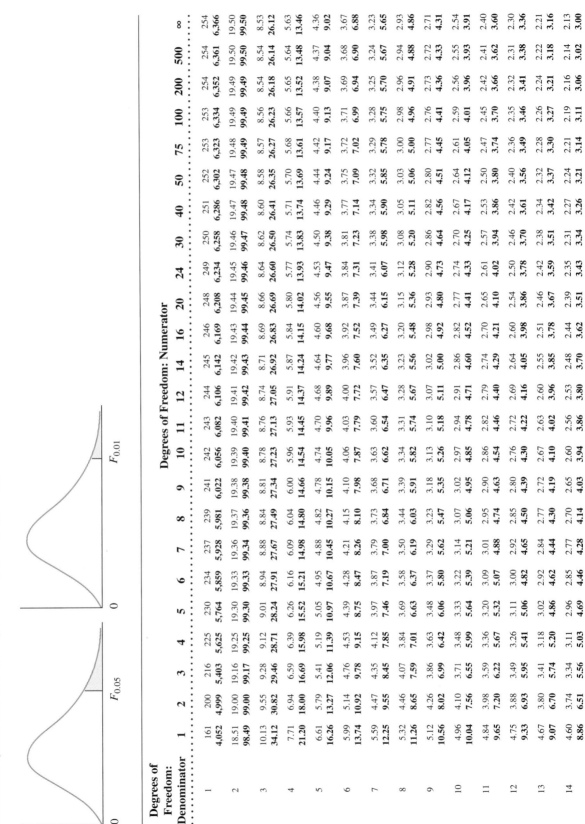

Degrees of Freedom: Numerator

Degrees of Freedom: Denominator	1	2	3	4	5	6	7	8	9	10	11	12	14	16	20	24	30	40	50	75	100	200	500	∞
1	161	200	216	225	230	234	237	239	241	242	243	244	245	246	248	249	250	251	252	253	253	254	254	254
	4,052	**4,999**	**5,403**	**5,625**	**5,764**	**5,859**	**5,928**	**5,981**	**6,022**	**6,056**	**6,082**	**6,106**	**6,142**	**6,169**	**6,208**	**6,234**	**6,258**	**6,286**	**6,302**	**6,323**	**6,334**	**6,352**	**6,361**	**6,366**
2	18.51	19.00	19.16	19.25	19.30	19.33	19.36	19.37	19.38	19.39	19.40	19.41	19.42	19.43	19.44	19.45	19.46	19.47	19.47	19.48	19.49	19.49	19.50	19.50
	98.49	**99.00**	**99.17**	**99.25**	**99.30**	**99.33**	**99.34**	**99.36**	**99.38**	**99.40**	**99.41**	**99.42**	**99.43**	**99.44**	**99.45**	**99.46**	**99.47**	**99.48**	**99.48**	**99.49**	**99.49**	**99.49**	**99.50**	**99.50**
3	10.13	9.55	9.28	9.12	9.01	8.94	8.88	8.84	8.81	8.78	8.76	8.74	8.71	8.69	8.66	8.64	8.62	8.60	8.58	8.57	8.56	8.54	8.54	8.53
	34.12	**30.82**	**29.46**	**28.71**	**28.24**	**27.91**	**27.67**	**27.49**	**27.34**	**27.23**	**27.13**	**27.05**	**26.92**	**26.83**	**26.69**	**26.60**	**26.50**	**26.41**	**26.35**	**26.27**	**26.23**	**26.18**	**26.14**	**26.12**
4	7.71	6.94	6.59	6.39	6.26	6.16	6.09	6.04	6.00	5.96	5.93	5.91	5.87	5.84	5.80	5.77	5.74	5.71	5.70	5.68	5.66	5.65	5.64	5.63
	21.20	**18.00**	**16.69**	**15.98**	**15.52**	**15.21**	**14.98**	**14.80**	**14.66**	**14.54**	**14.45**	**14.37**	**14.24**	**14.15**	**14.02**	**13.93**	**13.83**	**13.74**	**13.69**	**13.61**	**13.57**	**13.52**	**13.48**	**13.46**
5	6.61	5.79	5.41	5.19	5.05	4.95	4.88	4.82	4.78	4.74	4.70	4.68	4.64	4.60	4.56	4.53	4.50	4.46	4.44	4.42	4.40	4.38	4.37	4.36
	16.26	**13.27**	**12.06**	**11.39**	**10.97**	**10.67**	**10.45**	**10.27**	**10.15**	**10.05**	**9.96**	**9.89**	**9.77**	**9.68**	**9.55**	**9.47**	**9.38**	**9.29**	**9.24**	**9.17**	**9.13**	**9.07**	**9.04**	**9.02**
6	5.99	5.14	4.76	4.53	4.39	4.28	4.21	4.15	4.10	4.06	4.03	4.00	3.96	3.92	3.87	3.84	3.81	3.77	3.75	3.72	3.71	3.69	3.68	3.67
	13.74	**10.92**	**9.78**	**9.15**	**8.75**	**8.47**	**8.26**	**8.10**	**7.98**	**7.87**	**7.79**	**7.72**	**7.60**	**7.52**	**7.39**	**7.31**	**7.23**	**7.14**	**7.09**	**7.02**	**6.99**	**6.94**	**6.90**	**6.88**
7	5.59	4.47	4.35	4.12	3.97	3.87	3.79	3.73	3.68	3.63	3.60	3.57	3.52	3.49	3.44	3.41	3.38	3.34	3.32	3.29	3.28	3.25	3.24	3.23
	12.25	**9.55**	**8.45**	**7.85**	**7.46**	**7.19**	**7.00**	**6.84**	**6.71**	**6.62**	**6.54**	**6.47**	**6.35**	**6.27**	**6.15**	**6.07**	**5.98**	**5.90**	**5.85**	**5.78**	**5.75**	**5.70**	**5.67**	**5.65**
8	5.32	4.46	4.07	3.84	3.69	3.58	3.50	3.44	3.39	3.34	3.31	3.28	3.23	3.20	3.15	3.12	3.08	3.05	3.03	3.00	2.98	2.96	2.94	2.93
	11.26	**8.65**	**7.59**	**7.01**	**6.63**	**6.37**	**6.19**	**6.03**	**5.91**	**5.82**	**5.74**	**5.67**	**5.56**	**5.48**	**5.36**	**5.28**	**5.20**	**5.11**	**5.06**	**5.00**	**4.96**	**4.91**	**4.88**	**4.86**
9	5.12	4.26	3.86	3.63	3.48	3.37	3.29	3.23	3.18	3.13	3.10	3.07	3.02	2.98	2.93	2.90	2.86	2.82	2.80	2.77	2.76	2.73	2.72	2.71
	10.56	**8.02**	**6.99**	**6.42**	**6.06**	**5.80**	**5.62**	**5.47**	**5.35**	**5.26**	**5.18**	**5.11**	**5.00**	**4.92**	**4.80**	**4.73**	**4.64**	**4.56**	**4.51**	**4.45**	**4.41**	**4.36**	**4.33**	**4.31**
10	4.96	4.10	3.71	3.48	3.33	3.22	3.14	3.07	3.02	2.97	2.94	2.91	2.86	2.82	2.77	2.74	2.70	2.67	2.64	2.61	2.59	2.56	2.55	2.54
	10.04	**7.56**	**6.55**	**5.99**	**5.64**	**5.39**	**5.21**	**5.06**	**4.95**	**4.85**	**4.78**	**4.71**	**4.60**	**4.52**	**4.41**	**4.33**	**4.25**	**4.17**	**4.12**	**4.05**	**4.01**	**3.96**	**3.93**	**3.91**
11	4.84	3.98	3.59	3.36	3.20	3.09	3.01	2.95	2.90	2.86	2.82	2.79	2.74	2.70	2.65	2.61	2.57	2.53	2.50	2.47	2.45	2.42	2.41	2.40
	9.65	**7.20**	**6.22**	**5.67**	**5.32**	**5.07**	**4.88**	**4.74**	**4.63**	**4.54**	**4.46**	**4.40**	**4.29**	**4.21**	**4.10**	**4.02**	**3.94**	**3.86**	**3.80**	**3.74**	**3.70**	**3.66**	**3.62**	**3.60**
12	4.75	3.88	3.49	3.26	3.11	3.00	2.92	2.85	2.80	2.76	2.72	2.69	2.64	2.60	2.54	2.50	2.46	2.42	2.40	2.36	2.35	2.32	2.31	2.30
	9.33	**6.93**	**5.95**	**5.41**	**5.06**	**4.82**	**4.65**	**4.50**	**4.39**	**4.30**	**4.22**	**4.16**	**4.05**	**3.98**	**3.86**	**3.78**	**3.70**	**3.61**	**3.56**	**3.49**	**3.46**	**3.41**	**3.38**	**3.36**
13	4.67	3.80	3.41	3.18	3.02	2.92	2.84	2.77	2.72	2.67	2.63	2.60	2.55	2.51	2.46	2.42	2.38	2.34	2.32	2.28	2.26	2.24	2.22	2.21
	9.07	**6.70**	**5.74**	**5.20**	**4.86**	**4.62**	**4.44**	**4.30**	**4.19**	**4.10**	**4.02**	**3.96**	**3.85**	**3.78**	**3.67**	**3.59**	**3.51**	**3.42**	**3.37**	**3.30**	**3.27**	**3.21**	**3.18**	**3.16**
14	4.60	3.74	3.34	3.11	2.96	2.85	2.77	2.70	2.65	2.60	2.56	2.53	2.48	2.44	2.39	2.35	2.31	2.27	2.24	2.21	2.19	2.16	2.14	2.13
	8.86	**6.51**	**5.56**	**5.03**	**4.69**	**4.46**	**4.28**	**4.14**	**4.03**	**3.94**	**3.86**	**3.80**	**3.70**	**3.62**	**3.51**	**3.43**	**3.34**	**3.26**	**3.21**	**3.14**	**3.11**	**3.06**	**3.02**	**3.00**

Degrees of Freedom: Numerator

Each cell shows the upper value (top) and lower value (**bold**).

Degrees of Freedom: Denominator	1	2	3	4	5	6	7	8	9	10	11	12	14	16	20	24	30	40	50	75	100	200	500	∞
15	4.54 **8.68**	3.68 **6.36**	3.29 **5.42**	3.06 **4.89**	2.90 **4.56**	2.79 **4.32**	2.70 **4.14**	2.64 **4.00**	2.59 **3.89**	2.55 **3.80**	2.51 **3.73**	2.48 **3.67**	2.43 **3.56**	2.39 **3.48**	2.33 **3.36**	2.29 **3.29**	2.25 **3.20**	2.21 **3.12**	2.18 **3.07**	2.15 **3.00**	2.12 **2.97**	2.10 **2.92**	2.08 **2.89**	2.07 **2.87**
16	4.49 **8.53**	3.63 **6.23**	3.24 **5.29**	3.01 **4.77**	2.85 **4.44**	2.74 **4.20**	2.66 **4.03**	2.59 **3.89**	2.54 **3.78**	2.49 **3.69**	2.45 **3.61**	2.42 **3.55**	2.37 **3.45**	2.33 **3.37**	2.28 **3.25**	2.24 **3.18**	2.20 **3.10**	2.16 **3.01**	2.13 **2.96**	2.09 **2.89**	2.07 **2.86**	2.04 **2.80**	2.02 **2.77**	2.01 **2.75**
17	4.45 **8.40**	3.59 **6.11**	3.20 **5.18**	2.96 **4.67**	2.81 **4.34**	2.70 **4.10**	2.62 **3.93**	2.55 **3.79**	2.50 **3.68**	2.45 **3.59**	2.41 **3.52**	2.38 **3.45**	2.33 **3.35**	2.29 **3.27**	2.23 **3.16**	2.19 **3.08**	2.15 **3.00**	2.11 **2.92**	2.08 **2.86**	2.04 **2.79**	2.02 **2.76**	1.99 **2.70**	1.97 **2.67**	1.96 **2.65**
18	4.41 **8.28**	3.55 **6.01**	3.16 **5.09**	2.93 **4.58**	2.77 **4.25**	2.66 **4.01**	2.58 **3.85**	2.51 **3.71**	2.46 **3.60**	2.41 **3.51**	2.37 **3.44**	2.34 **3.37**	2.29 **3.27**	2.25 **3.19**	2.19 **3.07**	2.15 **3.00**	2.11 **2.91**	2.07 **2.83**	2.04 **2.78**	2.00 **2.71**	1.98 **2.68**	1.95 **2.62**	1.93 **2.59**	1.92 **2.57**
19	4.38 **8.18**	3.52 **5.93**	3.13 **5.01**	2.90 **4.50**	2.74 **4.17**	2.63 **3.94**	2.55 **3.77**	2.48 **3.63**	2.43 **3.52**	2.38 **3.43**	2.34 **3.36**	2.31 **3.30**	2.26 **3.19**	2.21 **3.12**	2.15 **3.00**	2.11 **2.92**	2.07 **2.84**	2.02 **2.76**	2.00 **2.70**	1.96 **2.63**	1.94 **2.60**	1.91 **2.54**	1.90 **2.51**	1.88 **2.49**
20	4.35 **8.10**	3.49 **5.85**	3.10 **4.94**	2.87 **4.43**	2.71 **4.10**	2.60 **3.87**	2.52 **3.71**	2.45 **3.56**	2.40 **3.45**	2.35 **3.37**	2.31 **3.30**	2.28 **3.23**	2.23 **3.13**	2.18 **3.05**	2.12 **2.94**	2.08 **2.86**	2.04 **2.77**	1.99 **2.69**	1.96 **2.63**	1.92 **2.56**	1.90 **2.53**	1.87 **2.47**	1.85 **2.44**	1.84 **2.42**
21	4.32 **8.02**	3.47 **5.78**	3.07 **4.87**	2.84 **4.37**	2.68 **4.04**	2.57 **3.81**	2.49 **3.65**	2.42 **3.51**	2.37 **3.40**	2.32 **3.31**	2.28 **3.24**	2.25 **3.17**	2.20 **3.07**	2.15 **2.99**	2.09 **2.88**	2.05 **2.80**	2.00 **2.72**	1.96 **2.63**	1.93 **2.58**	1.89 **2.51**	1.87 **2.47**	1.84 **2.42**	1.82 **2.38**	1.81 **2.36**
22	4.30 **7.94**	3.44 **5.72**	3.05 **4.82**	2.82 **4.31**	2.66 **3.99**	2.55 **3.76**	2.47 **3.59**	2.40 **3.45**	2.35 **3.35**	2.30 **3.26**	2.26 **3.18**	2.23 **3.12**	2.18 **3.02**	2.13 **2.94**	2.07 **2.83**	2.03 **2.75**	1.98 **2.67**	1.93 **2.58**	1.91 **2.53**	1.87 **2.46**	1.84 **2.42**	1.81 **2.37**	1.80 **2.33**	1.78 **2.31**
23	4.28 **7.88**	3.42 **5.66**	3.03 **4.76**	2.80 **4.26**	2.64 **3.94**	2.53 **3.71**	2.45 **3.54**	2.38 **3.41**	2.32 **3.30**	2.28 **3.21**	2.24 **3.14**	2.20 **3.07**	2.14 **2.97**	2.10 **2.89**	2.04 **2.78**	2.00 **2.70**	1.96 **2.62**	1.91 **2.53**	1.88 **2.48**	1.84 **2.41**	1.82 **2.37**	1.79 **2.32**	1.77 **2.28**	1.76 **2.26**
24	4.26 **7.82**	3.40 **5.61**	3.01 **4.72**	2.78 **4.22**	2.62 **3.90**	2.51 **3.67**	2.43 **3.50**	2.36 **3.36**	2.30 **3.25**	2.26 **3.17**	2.22 **3.09**	2.18 **3.03**	2.13 **2.93**	2.09 **2.85**	2.02 **2.74**	1.98 **2.66**	1.94 **2.58**	1.89 **2.49**	1.86 **2.44**	1.82 **2.36**	1.80 **2.33**	1.76 **2.27**	1.74 **2.23**	1.73 **2.21**
25	4.24 **7.77**	3.38 **5.57**	2.99 **4.68**	2.76 **4.18**	2.60 **3.86**	2.49 **3.63**	2.41 **3.46**	2.34 **3.32**	2.28 **3.21**	2.24 **3.13**	2.20 **3.05**	2.16 **2.99**	2.11 **2.89**	2.06 **2.81**	2.00 **2.70**	1.96 **2.62**	1.92 **2.54**	1.87 **2.45**	1.84 **2.40**	1.80 **2.32**	1.77 **2.29**	1.74 **2.23**	1.72 **2.19**	1.71 **2.17**
26	4.22 **7.72**	3.37 **5.53**	2.98 **4.64**	2.74 **4.14**	2.59 **3.82**	2.47 **3.59**	2.39 **3.42**	2.32 **3.29**	2.27 **3.17**	2.22 **3.09**	2.18 **3.02**	2.15 **2.96**	2.10 **2.86**	2.05 **2.77**	1.99 **2.66**	1.95 **2.58**	1.90 **2.50**	1.85 **2.41**	1.82 **2.36**	1.78 **2.28**	1.76 **2.25**	1.72 **2.19**	1.70 **2.15**	1.69 **2.13**
27	4.21 **7.68**	3.35 **5.49**	2.96 **4.60**	2.73 **4.11**	2.57 **3.79**	2.46 **3.56**	2.37 **3.39**	2.30 **3.26**	2.25 **3.14**	2.20 **3.06**	2.16 **2.98**	2.13 **2.93**	2.08 **2.83**	2.03 **2.74**	1.97 **2.63**	1.93 **2.55**	1.88 **2.47**	1.84 **2.38**	1.80 **2.33**	1.76 **2.25**	1.74 **2.21**	1.71 **2.16**	1.68 **2.12**	1.67 **2.10**
28	4.20 **7.64**	3.34 **5.45**	2.95 **4.57**	2.71 **4.07**	2.56 **3.76**	2.44 **3.53**	2.36 **3.36**	2.29 **3.23**	2.24 **3.11**	2.19 **3.03**	2.15 **2.95**	2.12 **2.90**	2.06 **2.80**	2.02 **2.71**	1.96 **2.60**	1.91 **2.52**	1.87 **2.44**	1.81 **2.35**	1.78 **2.30**	1.75 **2.22**	1.72 **2.18**	1.69 **2.13**	1.67 **2.09**	1.65 **2.06**
29	4.18 **7.60**	3.33 **5.42**	2.93 **4.54**	2.70 **4.04**	2.54 **3.73**	2.43 **3.50**	2.35 **3.33**	2.28 **3.20**	2.22 **3.08**	2.18 **3.00**	2.14 **2.92**	2.10 **2.87**	2.05 **2.77**	2.00 **2.68**	1.94 **2.57**	1.90 **2.49**	1.85 **2.41**	1.80	1.77 **2.32**	1.73 **2.27**	1.71 **2.19**	1.68 **2.15**	1.65 **2.10**	1.64 **2.06**
30	4.17 **7.56**	3.32 **5.39**	2.92 **4.51**	2.69 **4.02**	2.53 **3.70**	2.42 **3.47**	2.34 **3.30**	2.27 **3.17**	2.21 **3.06**	2.16 **2.98**	2.12 **2.90**	2.09 **2.84**	2.04 **2.74**	1.99 **2.66**	1.93 **2.55**	1.89 **2.47**	1.84 **2.38**	1.79 **2.29**	1.76 **2.24**	1.72 **2.16**	1.69 **2.13**	1.66 **2.07**	1.64 **2.03**	1.62 **2.01**
32	4.15 **7.50**	3.30 **5.34**	2.90 **4.46**	2.67 **3.97**	2.51 **3.66**	2.40 **3.42**	2.32 **3.25**	2.25 **3.12**	2.19 **3.01**	2.14 **2.94**	2.10 **2.86**	2.07 **2.80**	2.02 **2.70**	1.97 **2.62**	1.91 **2.51**	1.86 **2.42**	1.82 **2.34**	1.76 **2.25**	1.74 **2.20**	1.69 **2.12**	1.67 **2.08**	1.64 **2.02**	1.61 **1.98**	1.59 **1.96**
34	4.13 **7.44**	3.28 **5.29**	2.88 **4.42**	2.65 **3.93**	2.49 **3.61**	2.38 **3.38**	2.30 **3.21**	2.23 **3.08**	2.17 **2.97**	2.12 **2.89**	2.08 **2.82**	2.05 **2.76**	2.00 **2.66**	1.95 **2.58**	1.89 **2.47**	1.84 **2.38**	1.80 **2.30**	1.74 **2.21**	1.71 **2.15**	1.67 **2.08**	1.64 **2.04**	1.61 **1.98**	1.59 **1.94**	1.57 **1.91**
36	4.11 **7.39**	3.26 **5.25**	2.86 **4.38**	2.63 **3.89**	2.48 **3.58**	2.36 **3.35**	2.28 **3.18**	2.21 **3.04**	2.15 **2.94**	2.10 **2.86**	2.06 **2.78**	2.03 **2.72**	1.98 **2.62**	1.93 **2.54**	1.87 **2.43**	1.82 **2.35**	1.78 **2.26**	1.72 **2.17**	1.69 **2.12**	1.65 **2.04**	1.62 **2.00**	1.59 **1.94**	1.56 **1.90**	1.55 **1.87**
38	4.10 **7.35**	3.25 **5.21**	2.85 **4.34**	2.62 **3.86**	2.46 **3.54**	2.35 **3.32**	2.26 **3.15**	2.19 **3.02**	2.14 **2.91**	2.09 **2.82**	2.05 **2.75**	2.02 **2.69**	1.96 **2.59**	1.92 **2.51**	1.85 **2.40**	1.80 **2.32**	1.76 **2.22**	1.71 **2.14**	1.67 **2.08**	1.63 **2.00**	1.60 **1.97**	1.57 **1.90**	1.54 **1.86**	1.53 **1.84**

(Continued)

table F Critical values of the F distribution for $\alpha = 0.05$ (Roman type) and $\alpha = 0.01$ (boldface type)—cont'd

The values listed in the table are the critical values of F for the degrees of freedom of the numerator of the F ratio (column headings) and the degrees of freedom of the denominator of the F ratio (row headings). To be significant, $F_{obt} \geq F_{crit}$.

Degrees of Freedom: Numerator

Denominator	1	2	3	4	5	6	7	8	9	10	11	12	14	16	20	24	30	40	50	75	100	200	500	∞
40	4.08	3.23	2.84	2.61	2.45	2.34	2.25	2.18	2.12	2.07	2.04	2.00	1.95	1.90	1.84	1.79	1.74	1.69	1.66	1.61	1.59	1.55	1.53	1.51
	7.31	**5.18**	**4.31**	**3.83**	**3.51**	**3.29**	**3.12**	**2.99**	**2.88**	**2.80**	**2.73**	**2.66**	**2.56**	**2.49**	**2.37**	**2.29**	**2.20**	**2.11**	**2.05**	**1.97**	**1.94**	**1.88**	**1.84**	**1.81**
42	4.07	3.22	2.83	2.59	2.44	2.32	2.24	2.17	2.11	2.06	2.02	1.99	1.94	1.89	1.82	1.78	1.73	1.68	1.64	1.60	1.57	1.54	1.51	1.49
	7.27	**5.15**	**4.29**	**3.80**	**3.49**	**3.26**	**3.10**	**2.96**	**2.86**	**2.77**	**2.70**	**2.64**	**2.54**	**2.46**	**2.35**	**2.26**	**2.17**	**2.08**	**2.02**	**1.94**	**1.91**	**1.85**	**1.80**	**1.78**
44	4.06	3.21	2.82	2.58	2.43	2.31	2.23	2.16	2.10	2.05	2.01	1.98	1.92	1.88	1.81	1.76	1.72	1.66	1.63	1.58	1.56	1.52	1.50	1.48
	7.24	**5.12**	**4.26**	**3.78**	**3.46**	**3.24**	**3.07**	**2.94**	**2.84**	**2.75**	**2.68**	**2.62**	**2.52**	**2.44**	**2.32**	**2.24**	**2.15**	**2.06**	**2.00**	**1.92**	**1.88**	**1.82**	**1.78**	**1.75**
46	4.05	3.20	2.81	2.57	2.42	2.30	2.22	2.14	2.09	2.04	2.00	1.97	1.91	1.87	1.80	1.75	1.71	1.65	1.62	1.57	1.54	1.51	1.48	1.46
	7.21	**5.10**	**4.24**	**3.76**	**3.44**	**3.22**	**3.05**	**2.92**	**2.82**	**2.73**	**2.66**	**2.60**	**2.50**	**2.42**	**2.30**	**2.22**	**2.13**	**2.04**	**1.98**	**1.90**	**1.86**	**1.80**	**1.76**	**1.72**
48	4.04	3.19	2.80	2.56	2.41	2.30	2.21	2.14	2.08	2.03	1.99	1.96	1.90	1.86	1.79	1.74	1.70	1.64	1.61	1.56	1.53	1.50	1.47	1.45
	7.19	**5.08**	**4.22**	**3.74**	**3.42**	**3.20**	**3.04**	**2.90**	**2.80**	**2.71**	**2.64**	**2.58**	**2.48**	**2.40**	**2.28**	**2.20**	**2.11**	**2.02**	**1.96**	**1.88**	**1.84**	**1.78**	**1.73**	**1.70**
50	4.03	3.18	2.79	2.56	2.40	2.29	2.20	2.13	2.07	2.02	1.98	1.95	1.90	1.85	1.78	1.74	1.69	1.63	1.60	1.55	1.52	1.48	1.46	1.44
	7.17	**5.06**	**4.20**	**3.72**	**3.41**	**3.18**	**3.02**	**2.88**	**2.78**	**2.70**	**2.62**	**2.56**	**2.46**	**2.39**	**2.26**	**2.18**	**2.10**	**2.00**	**1.94**	**1.86**	**1.82**	**1.76**	**1.71**	**1.68**
55	4.02	3.17	2.78	2.54	2.38	2.27	2.18	2.11	2.05	2.00	1.97	1.93	1.88	1.83	1.76	1.72	1.67	1.61	1.58	1.52	1.50	1.46	1.43	1.41
	7.12	**5.01**	**4.16**	**3.68**	**3.37**	**3.15**	**2.98**	**2.85**	**2.75**	**2.66**	**2.59**	**2.53**	**2.43**	**2.35**	**2.23**	**2.15**	**2.06**	**1.96**	**1.90**	**1.82**	**1.78**	**1.71**	**1.66**	**1.64**
60	4.00	3.15	2.76	2.52	2.37	2.25	2.17	2.10	2.04	1.99	1.95	1.92	1.86	1.81	1.75	1.70	1.65	1.59	1.56	1.50	1.48	1.44	1.41	1.39
	7.08	**4.98**	**4.13**	**3.65**	**3.34**	**3.12**	**2.95**	**2.82**	**2.72**	**2.63**	**2.56**	**2.50**	**2.40**	**2.32**	**2.20**	**2.12**	**2.03**	**1.93**	**1.87**	**1.79**	**1.74**	**1.68**	**1.63**	**1.60**
65	3.99	3.14	2.75	2.51	2.36	2.24	2.15	2.08	2.02	1.98	1.94	1.90	1.85	1.80	1.73	1.68	1.63	1.57	1.54	1.49	1.46	1.42	1.39	1.37
	7.04	**4.95**	**4.10**	**3.62**	**3.31**	**3.09**	**2.93**	**2.79**	**2.70**	**2.61**	**2.54**	**2.47**	**2.37**	**2.30**	**2.18**	**2.09**	**2.00**	**1.90**	**1.84**	**1.76**	**1.71**	**1.64**	**1.60**	**1.56**
70	3.98	3.13	2.74	2.50	2.35	2.23	2.14	2.07	2.01	1.97	1.93	1.89	1.84	1.79	1.72	1.67	1.62	1.56	1.53	1.47	1.45	1.40	1.37	1.35
	7.01	**4.92**	**4.08**	**3.60**	**3.29**	**3.07**	**2.91**	**2.77**	**2.67**	**2.59**	**2.51**	**2.45**	**2.35**	**2.28**	**2.15**	**2.07**	**1.98**	**1.88**	**1.82**	**1.74**	**1.69**	**1.62**	**1.56**	**1.53**
80	3.96	3.11	2.72	2.48	2.33	2.21	2.12	2.05	1.99	1.95	1.91	1.88	1.82	1.77	1.70	1.65	1.60	1.54	1.51	1.45	1.42	1.38	1.35	1.32
	6.96	**4.88**	**4.04**	**3.56**	**3.25**	**3.04**	**2.87**	**2.74**	**2.64**	**2.55**	**2.48**	**2.41**	**2.32**	**2.24**	**2.11**	**2.03**	**1.94**	**1.84**	**1.78**	**1.70**	**1.65**	**1.57**	**1.52**	**1.49**
100	3.94	3.09	2.70	2.46	2.30	2.19	2.10	2.03	1.97	1.92	1.88	1.85	1.79	1.75	1.68	1.63	1.57	1.51	1.48	1.42	1.39	1.34	1.30	1.28
	6.90	**4.82**	**3.98**	**3.51**	**3.20**	**2.99**	**2.82**	**2.69**	**2.59**	**2.51**	**2.43**	**2.36**	**2.26**	**2.19**	**2.06**	**1.98**	**1.89**	**1.79**	**1.73**	**1.64**	**1.59**	**1.51**	**1.46**	**1.43**
125	3.92	3.07	2.68	2.44	2.29	2.17	2.08	2.01	1.95	1.90	1.86	1.83	1.77	1.72	1.65	1.60	1.55	1.49	1.45	1.39	1.36	1.31	1.27	1.25
	6.84	**4.78**	**3.94**	**3.47**	**3.17**	**2.95**	**2.79**	**2.65**	**2.56**	**2.47**	**2.40**	**2.33**	**2.23**	**2.15**	**2.03**	**1.94**	**1.85**	**1.75**	**1.68**	**1.59**	**1.54**	**1.46**	**1.40**	**1.37**
150	3.91	3.06	2.67	2.43	2.27	2.16	2.07	2.00	1.94	1.89	1.85	1.82	1.76	1.71	1.64	1.59	1.54	1.47	1.44	1.37	1.34	1.29	1.25	1.22
	6.81	**4.75**	**3.91**	**3.44**	**3.14**	**2.92**	**2.76**	**2.62**	**2.53**	**2.44**	**2.37**	**2.30**	**2.20**	**2.12**	**2.00**	**1.91**	**1.83**	**1.72**	**1.66**	**1.56**	**1.51**	**1.43**	**1.37**	**1.33**
200	3.89	3.04	2.65	2.41	2.26	2.14	2.05	1.98	1.92	1.87	1.83	1.80	1.74	1.69	1.62	1.57	1.52	1.45	1.42	1.35	1.32	1.26	1.22	1.19
	6.76	**4.71**	**3.88**	**3.41**	**3.11**	**2.90**	**2.73**	**2.60**	**2.50**	**2.41**	**2.34**	**2.28**	**2.17**	**2.09**	**1.97**	**1.88**	**1.79**	**1.69**	**1.62**	**1.53**	**1.48**	**1.39**	**1.33**	**1.28**
400	3.86	3.02	2.62	2.39	2.23	2.12	2.03	1.96	1.90	1.85	1.81	1.78	1.72	1.67	1.60	1.54	1.49	1.42	1.38	1.32	1.28	1.22	1.16	1.13
	6.70	**4.66**	**3.83**	**3.36**	**3.06**	**2.85**	**2.69**	**2.55**	**2.46**	**2.37**	**2.29**	**2.23**	**2.12**	**2.04**	**1.92**	**1.84**	**1.74**	**1.64**	**1.57**	**1.47**	**1.42**	**1.32**	**1.24**	**1.19**
1000	3.85	3.00	2.61	2.38	2.22	2.10	2.02	1.95	1.89	1.84	1.80	1.76	1.70	1.65	1.58	1.53	1.47	1.41	1.36	1.30	1.26	1.19	1.13	1.08
	6.66	**4.62**	**3.80**	**3.34**	**3.04**	**2.82**	**2.66**	**2.53**	**2.43**	**2.34**	**2.26**	**2.20**	**2.09**	**2.01**	**1.89**	**1.81**	**1.71**	**1.61**	**1.54**	**1.44**	**1.38**	**1.28**	**1.19**	**1.11**
∞	3.84	2.99	2.60	2.37	2.21	2.09	2.01	1.94	1.88	1.83	1.79	1.75	1.69	1.64	1.57	1.52	1.46	1.40	1.35	1.28	1.24	1.17	1.11	1.00
	6.64	**4.60**	**3.78**	**3.32**	**3.02**	**2.80**	**2.64**	**2.51**	**2.41**	**2.32**	**2.24**	**2.18**	**2.07**	**1.99**	**1.87**	**1.79**	**1.69**	**1.59**	**1.52**	**1.41**	**1.36**	**1.25**	**1.15**	**1.00**

table G Critical values of the studentized range (Q) distribution

The values listed in the table are the critical values of Q for α = 0.05 and 0.01, as a function of degrees of freedom of MS_{within} and k (the number of means). To be significant, $Q_{obt} \geq Q_{crit}$.

MS_{within} df	α	\multicolumn{10}{c}{k (Number of Means)}
		2	3	4	5	6	7	8	9	10	11
5	.05	3.64	4.60	5.22	5.67	6.03	6.33	6.58	6.80	6.99	7.17
	.01	5.70	6.98	7.80	8.42	8.91	9.32	9.67	9.97	10.24	10.48
6	.05	3.46	4.34	4.90	5.30	5.63	5.90	6.12	6.32	6.49	6.65
	.01	5.24	6.33	7.03	7.56	7.97	8.32	8.61	8.87	9.10	9.30
7	.05	3.34	4.16	4.68	5.06	5.36	5.61	5.82	6.00	6.16	6.30
	.01	4.95	5.92	6.54	7.01	7.37	7.68	7.94	8.17	8.37	8.55
8	.05	3.26	4.04	4.53	4.89	5.17	5.40	5.60	5.77	5.92	6.05
	.01	4.75	5.64	6.20	6.62	6.96	7.24	7.47	7.68	7.86	8.03
9	.05	3.20	3.95	4.41	4.76	5.02	5.24	5.43	5.59	5.74	5.87
	.01	4.60	5.43	5.96	6.35	6.66	6.91	7.13	7.33	7.49	7.65
10	.05	3.15	3.88	4.33	4.65	4.91	5.12	5.30	5.46	5.60	5.72
	.01	4.48	5.27	5.77	6.14	6.43	6.67	6.87	7.05	7.21	7.36
11	.05	3.11	3.82	4.26	4.57	4.82	5.03	5.20	5.35	5.49	5.61
	.01	4.39	5.15	5.62	5.97	6.25	6.48	6.67	6.84	6.99	7.13
12	.05	3.08	3.77	4.20	4.51	4.75	4.95	5.12	5.27	5.39	5.51
	.01	4.32	5.05	5.50	5.84	6.10	6.32	6.51	6.67	6.81	6.94
13	.05	3.06	3.73	4.15	4.45	4.69	4.88	5.05	5.19	5.32	5.43
	.01	4.26	4.96	5.40	5.73	5.98	6.19	6.37	6.53	6.67	6.79
14	.05	3.03	3.70	4.11	4.41	4.64	4.83	4.99	5.13	5.25	5.36
	.01	4.21	4.89	5.32	5.63	5.88	6.08	6.26	6.41	6.54	6.66
15	.05	3.01	3.67	4.08	4.37	4.59	4.78	4.94	5.08	5.20	5.31
	.01	4.17	4.84	5.25	5.56	5.80	5.99	6.16	6.31	6.44	6.55
16	.05	3.00	3.65	4.05	4.33	4.56	4.74	4.90	5.03	5.15	5.26
	.01	4.13	4.79	5.19	5.49	5.72	5.92	6.08	6.22	6.35	6.46
17	.05	2.98	3.63	4.02	4.30	4.52	4.70	4.86	4.99	5.11	5.21
	.01	4.10	4.74	5.14	5.43	5.66	5.85	6.01	6.15	6.27	6.38
18	.05	2.97	3.61	4.00	4.28	4.49	4.67	4.82	4.96	5.07	5.17
	.01	4.07	4.70	5.09	5.38	5.60	5.79	5.94	6.08	6.20	6.31
19	.05	2.96	3.59	3.98	4.25	4.47	4.65	4.79	4.92	5.04	5.14
	.01	4.05	4.67	5.05	5.33	5.55	5.73	4.89	6.02	6.14	6.25
20	.05	2.95	3.58	3.96	4.23	4.45	4.62	4.77	4.90	5.01	5.11
	.01	4.02	4.64	5.02	5.29	5.51	5.69	5.84	5.97	6.09	6.19
24	.05	2.92	3.53	3.90	4.17	4.37	4.54	4.68	4.81	4.92	5.01
	.01	3.96	4.55	4.91	5.17	5.37	5.54	5.69	5.81	5.92	6.02
30	.05	2.89	3.49	3.85	4.10	4.30	4.46	4.60	4.72	4.82	4.92
	.01	3.89	4.45	4.80	5.05	5.24	5.40	5.54	5.65	5.76	5.85
40	.05	2.86	3.44	3.79	4.04	4.23	4.39	4.52	4.63	4.73	4.82
	.01	3.82	4.37	4.70	4.93	5.11	5.26	5.39	5.50	5.60	5.69
60	.05	2.83	3.40	3.74	3.98	4.16	4.31	4.44	4.55	4.65	4.73
	.01	3.76	4.28	4.59	4.82	4.99	5.13	5.25	5.36	5.45	5.53
120	.05	2.80	3.36	3.68	3.92	4.10	4.24	4.36	4.47	4.56	4.64
	.01	3.70	4.20	4.50	4.71	4.87	5.01	5.12	5.21	5.30	5.37
∞	.05	2.77	3.31	3.63	3.86	4.03	4.17	4.29	4.39	4.47	4.55
	.01	3.64	4.12	4.40	4.60	4.76	4.88	4.99	5.08	5.16	5.23

table H Chi-square (χ^2) distribution

The first column (df) locates each χ^2 distribution. The other columns give the proportion of area under the χ^2 distribution that is above the tabled value of χ^2. The χ^2 values under the column headings of .05 and .01 are the critical values of χ^2 for $\alpha = 0.05$ and 0.01. To be significant, $\chi^2_{obt} \geq \chi^2_{crit}$.

Degrees of Freedom df	P = .99	.98	.95	.90	.80	.70	.50	.30	.20	.10	.05	.02	.01
1	.000157	.000628	.00393	.0158	.0642	.148	.455	1.074	1.642	2.706	3.841	5.412	6.635
2	.0201	.0404	.103	.211	.446	.713	1.386	2.408	3.219	4.605	5.991	7.824	9.210
3	.115	.185	.352	.584	1.005	1.424	2.366	3.665	4.642	6.251	7.815	9.837	11.341
4	.297	.429	.711	1.064	1.649	2.195	3.357	4.878	5.989	7.779	9.488	11.668	13.277
5	.554	.752	1.145	1.610	2.343	3.000	4.351	6.064	7.289	9.236	11.070	13.388	15.086
6	.872	1.134	1.635	2.204	3.070	3.828	5.348	7.231	8.558	10.645	12.592	15.033	16.812
7	1.239	1.564	2.167	2.833	3.822	4.671	6.346	8.383	9.803	12.017	14.067	16.622	18.475
8	1.646	2.032	2.733	3.490	4.594	5.527	7.344	9.524	11.030	13.362	15.507	18.168	20.090
9	2.088	2.532	3.325	4.168	5.380	6.393	8.343	10.656	12.242	14.684	16.919	19.679	21.666
10	2.558	3.059	3.940	4.865	6.179	7.267	9.342	11.781	13.442	15.987	18.307	21.161	23.209
11	3.053	3.609	4.575	5.578	6.989	8.148	10.341	12.899	14.631	17.275	19.675	22.618	24.725
12	3.571	4.178	5.226	6.304	7.807	9.034	11.340	14.011	15.812	18.549	21.026	24.054	26.217
13	4.107	4.765	5.892	7.042	8.634	9.926	12.340	15.119	16.985	19.812	22.362	25.472	27.688
14	4.660	5.368	6.571	7.790	9.467	10.821	13.339	16.222	18.151	21.064	23.685	26.873	29.141
15	5.229	5.985	7.261	8.547	10.307	11.721	14.339	17.322	19.311	22.307	24.996	28.259	30.578
16	5.812	6.614	7.962	9.312	11.152	12.624	15.338	18.418	20.465	23.542	26.296	29.633	32.000
17	6.408	7.255	8.672	10.085	12.002	13.531	16.338	19.511	21.615	24.769	27.587	30.995	33.409
18	7.015	7.906	9.390	10.865	12.857	14.440	17.338	20.601	22.760	25.989	28.869	32.346	34.805
19	7.633	8.567	10.117	11.651	13.716	15.352	18.338	21.689	23.900	27.204	30.144	33.687	36.191
20	8.260	9.237	10.851	12.443	14.578	16.266	19.337	22.775	25.038	28.412	31.410	35.020	37.566
21	8.897	9.915	11.591	13.240	15.445	17.182	20.337	23.858	26.171	29.615	32.671	36.343	38.932
22	9.542	10.600	12.338	14.041	16.314	18.101	21.337	24.939	27.301	30.813	33.924	37.659	40.289
23	10.196	11.293	13.091	14.848	17.187	19.021	22.337	26.018	28.429	32.007	35.172	38.968	41.638
24	10.856	11.992	13.848	15.659	18.062	19.943	23.337	27.096	29.553	33.196	36.415	40.270	42.980
25	11.524	12.697	14.611	16.473	18.940	20.867	24.337	28.172	30.675	34.382	37.652	41.566	44.314
26	12.198	13.409	15.379	17.292	19.820	21.792	25.336	29.246	31.795	35.563	38.885	42.856	45.642
27	12.879	14.125	16.151	18.114	20.703	22.719	26.336	30.319	32.912	36.741	40.113	44.140	46.963
28	13.565	14.847	16.928	18.939	21.588	23.647	27.336	31.391	34.027	37.916	41.337	45.419	48.278
29	14.256	15.574	17.708	19.768	22.475	24.577	28.336	32.461	35.139	39.087	42.557	46.693	49.588
30	14.953	16.306	18.493	20.599	23.364	25.508	29.336	33.530	36.250	40.256	43.773	47.962	50.892

table I   Critical values of *T* for Wilcoxon signed ranks test

The values listed in the table are the critical values of T for the specified N (left column) and alpha level (column heading). To be significant, $T_{obt} \leq T_{crit}$.

	Level of Significance for One-Tailed Test					Level of Significance for One-Tailed Test			
	.05	.025	.01	.005		.05	.025	.01	.005
	Level of Significance for Two-Tailed Test					Level of Significance for Two-Tailed Test			
N	.10	.05	.02	.01	N	.10	.05	.02	.01
5	0	—	—	—	28	130	116	101	91
6	2	0	—	—	29	140	126	110	100
7	3	2	0	—	30	151	137	120	109
8	5	3	1	0	31	163	147	130	118
9	8	5	3	1	32	175	159	140	128
10	10	8	5	3	33	187	170	151	138
11	13	10	7	5	34	200	182	162	148
12	17	13	9	7	35	213	195	173	159
13	21	17	12	9	36	227	208	185	171
14	25	21	15	12	37	241	221	198	182
15	30	25	19	15	38	256	235	211	194
16	35	29	23	19	39	271	249	224	207
17	41	34	27	23	40	286	264	238	220
18	47	40	32	27	41	302	279	252	233
19	53	46	37	32	42	319	294	266	247
20	60	52	43	37	43	336	310	281	261
21	67	58	49	42	44	353	327	296	276
22	75	65	55	48	45	371	343	312	291
23	83	73	62	54	46	389	361	328	307
24	91	81	69	61	47	407	378	345	322
25	100	89	76	68	48	426	396	362	339
26	110	98	84	75	49	446	415	379	355
27	119	107	92	83	50	466	434	397	373

table J Random numbers

	1	2	3	4	5	6	7	8	9
1	32942	95416	42339	59045	26693	49057	87496	20624	14819
2	07410	99859	83828	21409	29094	65114	36701	25762	12827
3	59981	68155	45673	76210	58219	45738	29550	24736	09574
4	46251	25437	69654	99716	11563	08803	86027	51867	12116
5	65558	51904	93123	27887	53138	21488	09095	78777	71240
6	99187	19258	86421	16401	19397	83297	40111	49326	81686
7	35641	00301	16096	34775	21562	97983	45040	19200	16383
8	14031	00936	81518	48440	02218	04756	19506	60695	88494
9	60677	15076	92554	26042	23472	69869	62877	19584	39576
10	66314	05212	67859	89356	20056	30648	87349	20389	53805
11	20416	87410	75646	64176	82752	63606	37011	57346	69512
12	28701	56992	70423	62415	40807	98086	58850	28968	45297
13	74579	33844	33426	07570	00728	07079	19322	56325	84819
14	62615	52342	82968	75540	80045	53069	20665	21282	07768
15	93945	06293	22879	08161	01442	75071	21427	94842	26210
16	75689	76131	96837	67450	44511	50424	82848	41975	71663
17	02921	16919	35424	93209	52133	87327	95897	65171	20376
18	14295	34969	14216	03191	61647	30296	66667	10101	63203
19	05303	91109	82403	40312	62191	67023	90073	83205	71344
20	57071	90357	12901	08899	91039	67251	28701	03846	94589
21	78471	57741	13599	84390	32146	00871	09354	22745	65806
22	89242	79337	59293	47481	07740	43345	25716	70020	54005
23	14955	59592	97035	80430	87220	06392	79028	57123	52872
24	42446	41880	37415	47472	04513	49494	08860	08038	43624
25	18534	22346	54556	17558	73689	14894	05030	19561	56517
26	39284	33737	42512	86411	23753	29690	26096	81361	93099
27	33922	37329	89911	55876	28379	81031	22058	21487	54613
28	78355	54013	50774	30666	61205	42574	47773	36027	27174
29	08845	99145	94316	88974	29828	97069	90327	61842	29604
30	01769	71825	55957	98271	02784	66731	40311	88495	18821
31	17639	38284	59478	90409	21997	56199	30068	82800	69692
32	05851	58653	99949	63505	40409	85551	90729	64938	52403
33	42396	40112	11469	03476	03328	84238	26570	51790	42122
34	13318	14192	98167	75631	74141	22369	36757	89117	54998
35	60571	54786	26281	01855	30706	66578	32019	65884	58485
36	09531	81853	59334	70929	03544	18510	89541	13555	21168
37	72865	16829	86542	00396	20363	13010	69645	49608	54738
38	56324	31093	77924	28622	83543	28912	15059	80192	83964
39	78192	21626	91399	07235	07104	73652	64425	85149	75409
40	64666	34767	97298	92708	01994	53188	78476	07804	62404
41	82201	75694	02808	65983	74373	66693	13094	74183	73020
42	15360	73776	40914	85190	54278	99054	62944	47351	89098
43	68142	67957	70896	37983	20487	95350	16371	03426	13895
44	19138	31200	30616	14639	44406	44236	57360	81644	94761
45	28155	03521	36415	78452	92359	81091	56513	88321	97910
46	87971	29031	51780	27376	81056	86155	55488	50590	74514
47	58147	68841	53625	02059	75223	16783	19272	61994	71090
48	18875	52809	70594	41649	32935	26430	82096	01605	65846
49	75109	56474	74111	31966	29969	70093	98901	84550	25769
50	35983	03742	76822	12073	59463	84420	15868	99505	11426

table J   Random numbers—*cont'd*

	1	2	3	4	5	6	7	8	9
51	12651	61646	11769	75109	86996	97669	25757	32535	07122
52	81769	74436	02630	72310	45049	18029	07469	42341	98173
53	36737	98863	77240	76251	00654	64688	09343	70278	67331
54	82861	54371	76610	94934	72748	44124	05610	53750	95938
55	21325	15732	24127	37431	09723	63529	73977	95218	96074
56	74146	47887	62463	23045	41490	07954	22597	60012	98866
57	90759	64410	54179	66075	61051	75385	51378	08360	95946
58	55683	98078	02238	91540	21219	17720	87817	41705	95785
59	79686	17969	76061	83748	55920	83612	41540	86492	06447
60	70333	00201	86201	69716	78185	62154	77930	67663	29529
61	14042	53536	07779	04157	41172	36473	42123	43929	50533
62	59911	08256	06596	48416	69770	68797	56080	14223	59199
63	62368	62623	62742	14891	39247	52242	98832	69533	91174
64	57529	97751	54976	48957	74599	08759	78494	52785	68526
65	15469	90574	78033	66885	13936	42117	71831	22961	94225
66	18625	23674	53850	32827	81647	80820	00420	63555	74489
67	74626	68394	88562	70745	23701	45630	65891	58220	35442
68	11119	16519	27384	90199	79210	76965	99546	30323	31664
69	41101	17336	48951	53674	17880	45260	08575	49321	36191
70	32123	91576	84221	78902	82010	30847	62329	63898	23268
71	26091	68409	69704	82267	14751	13151	93115	01437	56945
72	67680	79790	48462	59278	44185	29616	76531	19589	83139
73	15184	19260	14073	07026	25264	08388	27182	22557	61501
74	58010	45039	57181	10238	36874	28546	37444	80824	63981
75	56425	53996	86245	32623	78858	08143	60377	42925	42815
76	82630	84066	13592	60642	17904	99718	63432	88642	37858
77	14927	40909	23900	48761	44860	92467	31742	87142	03607
78	23740	22505	07489	85986	74420	21744	97711	36648	35620
79	32990	97446	03711	63824	07953	85965	87089	11687	92414
80	05310	24058	91946	78437	34365	82469	12430	84754	19354
81	21839	39937	27534	88913	49055	19218	47712	67677	51889
82	08833	42549	93981	94051	28382	83725	72643	64233	97252
83	58336	11139	47479	00931	91560	95372	97642	33856	54825
84	62032	91144	75478	47431	52726	30289	42411	91886	51818
85	45171	30557	53116	04118	58301	24375	65609	85810	18620
86	91611	62656	60128	35609	63698	78356	50682	22505	01692
87	55472	63819	86314	49174	93582	73604	78614	78849	23096
88	18573	09729	74091	53994	10970	86557	65661	41854	26037
89	60866	02955	90288	82136	83644	94455	06560	78029	98768
90	45043	55608	82767	60890	74646	79485	13619	98868	40857
91	17831	09737	79473	75945	28394	79334	70577	38048	03607
92	40137	03981	07585	18128	11178	32601	27994	05641	22600
93	77776	31343	14576	97706	16039	47517	43300	59080	80392
94	69605	44104	40103	95635	05635	81673	68657	09559	23510
95	19916	52934	26499	09821	87331	80993	61299	36979	73599
96	02606	58552	07678	56619	65325	30705	99582	53390	46357
97	65183	73160	87131	35530	47946	09854	18080	02321	05809
98	10740	98914	44916	11322	89717	88189	30143	52687	19420
99	98642	89822	71691	51573	83666	61642	46683	33761	47542
100	60139	25601	93663	25547	02654	94829	48672	28736	84994

ACKNOWLEDGMENTS

The tables contained in this appendix have been adapted with permission from the following sources:

Table A   R. Clarke, A. Coladarch, and J. Caffrey, *Statistical Reasoning and Procedures,* Charles E. Merrill Publishers, Columbus, Ohio, 1965, Appendix 2.

Table B   R. S. Burington and D. C. May, *Handbook of Probability and Statistics with Tables,* 2nd ed., McGraw-Hill Book Company, New York, 1970.

Table C   H. B. Mann and D. R. Whitney, "On a Test of Whether One of Two Random Variables Is Stochastically Larger Than the Other," *Annals of Mathematical Statistics,* 18 (1947), 50–60, and D. Auble, "Extended Tables for the Mann–Whitney Statistic," *Bulletin of the Institute of Educational Research at Indiana University, 1,* No. 2 (1953), as used in Runyon and Haber, *Fundamentals of Behavioral Statistics,* 3rd ed., Addison-Wesley Publishing Company, Inc., Reading, Mass., 1976.

Table D   Fisher and Yates, *Statistical Tables for Biological, Agricultural, and Medical Research,* Longman Group Ltd., London (previously published by Oliver & Boyd Ltd., Edinburgh), 1974, Table III.

Table E   Fisher and Yates, *Statistical Tables for Biological, Agricultural, and Medical Research,* Longman Group Ltd., London (previously published by Oliver & Boyd Ltd., Edinburgh), 1974, Table VII.

Table F   G. W. Snedecor, *Statistical Methods,* 5th ed., Iowa State University Press, Ames, 1956.

Table G   E. S. Pearson and H. O. Hartley, eds., *Biometrika Tables for Statisticians,* Vol. 1, 3rd ed., Cambridge University Press, New York, 1966, Table 29.

Table H   Fisher and Yates, *Statistical Tables for Biological, Agricultural, and Medical Research,* Longman Group Ltd., London (previously published by Oliver & Boyd Ltd., Edinburgh), 1974, Table IV.

Table I   F. Wilcoxon, S. Katte, and R. A. Wilcox, *Critical Values and Probability Levels for the Wilcoxon Rank Sum Test and the Wilcoxon Signed Ranks Test,* American Cyanamid Co., New York, 1963, and F. Wilcoxon and R. A. Wilcox, *Some Rapid Approximate Statistical Procedures,* Lederle Laboratories, New York, 1964, as used in Runyon and Haber, *Fundamentals of Behavioral Statistics,* 3rd ed., Addison-Wesley Publishing Company, Inc., Reading, Mass., 1976.

Table J   RAND Corporation, *A Million Random Digits,* Free Press of Glencoe, Glencoe, Ill., 1955.

Introduction to SPSS

Introduction   The Statistical Package for the Social Sciences is usually referred to as *SPSS*. SPSS is a statistical software package that runs on PCs and Macs. It is widely used within psychology and is available in colleges and universities throughout the United States. This material is written for *SPSS, an IBM company,* Windows Version 19.

Your textbook contains one or two SPSS illustrative examples and at least two additional problems at the end of each relevant chapter, placed just before the *Notes* section. The illustrative examples provide detailed instruction on how to analyze the data that is appropriate for the chapter. The additional problems are intended to give you practice with what you have just learned.

Before moving ahead to analyze data using SPSS, it is useful to consider some general features that you will probably use, regardless of the data you are analyzing. If you have a question concerning something I have not covered, it is also worth consulting the SPSS *Help* function that SPSS offers located on the menu bar at the top right of the windows it displays. If it turns out that you want to learn about SPSS in more detail than I have presented here, I recommend reading L. A. Kirkpatrick and B. C. Feeney, *A Simple Guide to SPSS for Windows for Version 18*, Wadsworth/Cengage, Belmont, CA, 2011.

Basic Steps in Entering and Analyzing Data   Using SPSS makes data analysis very easy. All that you need to do is:

1. **Enter the data into SPSS.** The two most common methods are to enter the data by typing it directly into the *SPSS Data Editor* (I will discuss the SPSS Data Editor in a moment), or to open a saved SPSS data file residing on your computer into the Data Editor. Since it is highly unlikely that you have any saved data files residing on your computer for the examples in our textbook, I have assumed that you will be entering the data by typing it into the Data Editor.

2. **Select a procedure.** Select a procedure (doing a statistical analysis or producing a graph), from the menu bar or tool bar at the top of the Data Editor.
3. **Interact with one or more dialog box(es).** Selecting a procedure produces one of more dialog boxes. Input from you to the dialog box(es) gives SPSS the information it needs to carry out the procedure you requested.
4. **Give SPSS the OK command to run the procedure and output the results.** Once the information is entered in the dialog box(es), clicking the **OK** button located on the appropriate dialog box gives SPSS the go-ahead and causes the procedure to be carried out. The results are then displayed in a window called the *Viewer* (more about the Viewer in a moment).

SPSS Windows SPSS has several windows it can present. We will discuss two of them, the *Data Editor Window*, and the *Viewer Window*. The Data Editor window displays the Data Editor. The Viewer window displays the output resulting from procedures usually initiated from the Data Editor. You can move from one to the other by clicking **Window** on the menu bar at the top of either window and then clicking either _____**SPSS Data Editor** or _____**SPSS Viewer** as appropriate. (The "_____" indicates variable material that precedes **SPSS Data Editor or SPSS Viewer** depending on several factors, like the name of the file, how many different data sets you have analyzed in a session, etc.

SPSS Data Editor When you first open SPSS, you will see the Data Editor displayed, or a smaller screen, asking, "What would you like to do?" If you encounter the small screen, you can switch to the Data Editor by **clicking Type in data**, and then **clicking OK. Clicking** the **X** button in the upper right corner of the small screen will also put you in the Data Editor.

The Data Editor is a large table of rows and columns (similar to an Excel spreadsheet) where you enter, edit, save, analyze/graph, and print the data. The results of analyzing or graphing the data are output from the Data Editor to the Viewer. The Data Editor is presented when you first open SPSS because the first thing you will do is enter the data and specify variable information, such as the variable name. Once the data are entered, analysis or graphing can be carried out and the results sent to the Viewer for your consideration.

The Data Editor has two possible views, the **Data View** and the **Variable View.** Let's discuss the Data View first.

Data View Figure E.1 shows the Data Editor displaying the Data View. The Data View is composed of a data table, a menu bar and a tool bar for performing various procedures and for getting help if desired. The tool bar is located under the menu bar; it allows procedures to be selected by clicking icons instead of using menus. Our discussion will proceed via menus rather than using the tool bar. When you open SPSS or obtain a new data table, the Data Editor table will be blank as shown in Figure E.1.

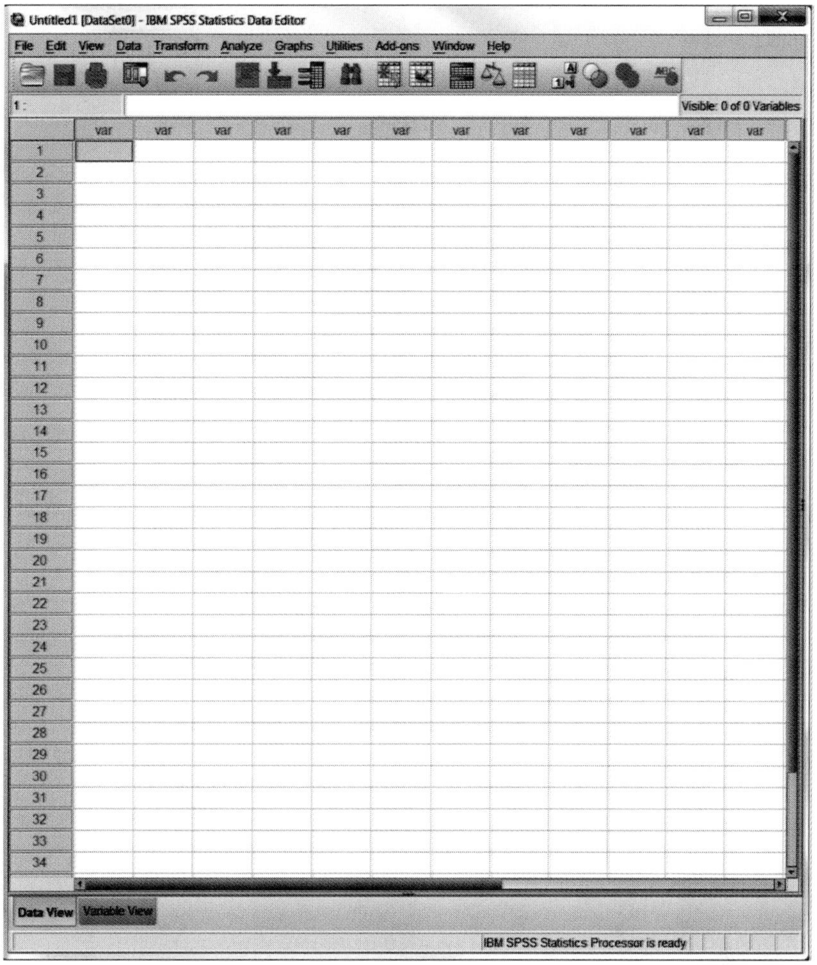

Figure E.1 Data Editor-Data View at start-up.

Figure E.2 shows the Data View of an untitled Data Editor in which I have entered the scores of **10**, **12**, **15**, **23**, **18**, **31**, **40**, **16**, **28**, and **36** for a variable named **X**, and **7, 9, 4, 14, 21, 15, 10, 13, 18**, and **5** for the variable named **Y**. SPSS automatically adds the **.00** after each score. You will get plenty of practice entering scores if you actually use SPSS as you work through the SPSS illustrative examples and problems contained in the textbook.

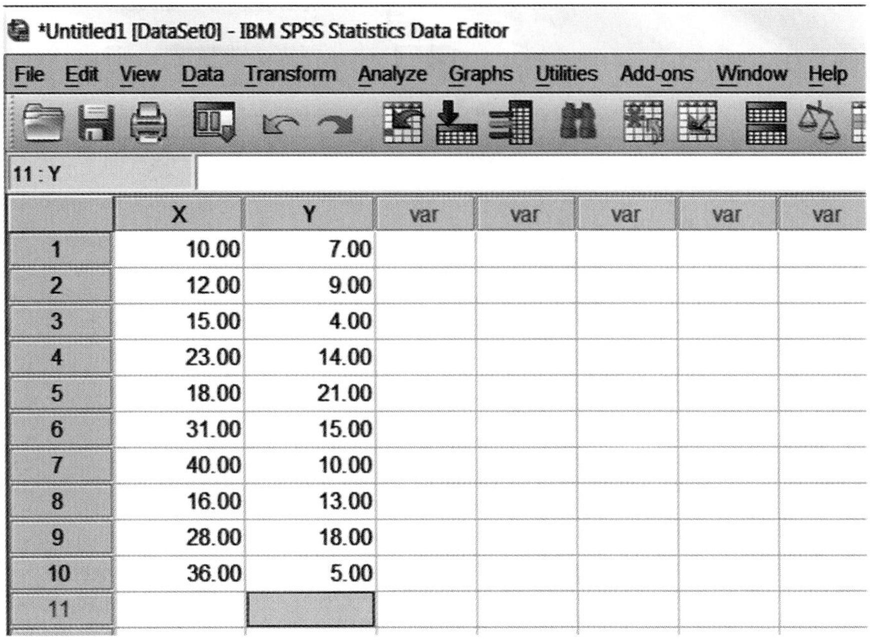

	X	Y	var	var	var	var	var
1	10.00	7.00					
2	12.00	9.00					
3	15.00	4.00					
4	23.00	14.00					
5	18.00	21.00					
6	31.00	15.00					
7	40.00	10.00					
8	16.00	13.00					
9	28.00	18.00					
10	36.00	5.00					
11							

Figure E.2 The Data View showing two variables, ***X*** and ***Y***, and the scores for each variable

From Figure E.2, you can see that the Data Editor-Data View displays a table where each column pertains to a variable. In columns that contain data, each column displays a variable name that is specific to the column, and the scores for that variable. At the top of the screen, there is a menu bar that permits data entry, obtaining new (blank) Data Editor screens, editing, saving and printing, procedure selection (analyzing the data, graphing the data, etc), and getting help should you have questions. We will discuss some of these functions here, and the rest as appropriate in the textbook chapters, in conjunction with data from specific experiments or problems. I hope you will like and be excited by how easily, quickly, accurately and esthetically you can accomplish these functions using SPSS.

Variable View The Variable View of the Data Editor is shown in Figure E.3. It is obtained by clicking the **Variable View** tab at the bottom left of the Data View. Conversely, clicking the **Data View** tab when the screen is displaying the Variable View will produce the Data View. Like the Data View, the Variable View displays a table; only in this table each row represents a variable, giving its name and other important information about the variable. When you open SPSS or obtain a new data table, the table presented in the Variable View is blank. The default heading for each column is **VAR**. When you enter data into the Data Editor, SPSS gives the data a variable name, changing **VAR** to the new variable name. If the data are entered in the first column, the new column heading name is **VAR00001.** If the data are entered in the second column, the new heading is **VAR00002**, and so forth. You can also give the variables names of your own choosing. Generally, I believe it is better to name the variables yourself, because it avoids confusion when interacting with dialog boxes and when interpreting the results of an analysis. In the table displayed in Figure E.3, I have entered the names **X**, and **Y**. The other table entries are the default entries that SPSS gives to each numeric variable when it is entered into the table.

Figure E.3 Data Editor-Variable View showing the names of two variables, **X** and **Y**, along with other information concerning each variable.

Entering Data by Typing Directly into the Data Editor   Data are entered via the Data Editor-Data View. Let's assume that you have the following *IQ* scores that you desire to enter.

IQ: 100, 105, 118, 120, 123.

I will assume SPSS is running and that a blank Data Editor-Data View screen is displayed, as shown in Figure E.1. If you have just opened SPSS and encounter the small screen mentioned above on p. 616, you can switch to the Data Editor-Data View by **clicking Type in data**, then **pressing** OK, or by **Clicking** the **X** button in the upper right corner of the screen. Assuming a blank Data Editor-Data View is displayed, the cell located at row 1 of the first column of the Data View table should be highlighted. If not, **clicking** the cell will highlight it. I will assume the cell is highlighted in the ensuing discussion.

Next, let's see how to enter the data into the Data Editor

---

1. **Type 100** in the highlighted cell, then **press Enter**.

The value **100.00** is entered in the first cell of the first column. SPSS automatically gives the variable the name **VAR00001**, because the score is located in the first column, and the cursor moves down one cell. The SPSS default for numeric variables is 2 decimal places; so when the score of **100** was entered, SPSS automatically added **.00** to the score of **100**, resulting in the value **100.00**. To correct a score that was entered incorrectly, move the cursor to the cell containing the incorrect score, and type in the correct value.

2. **Type 105**; then **press Enter**.

The value of **105.00** is entered in the cell directly under **100.00**.

3. **Type 118**; then **press Enter**.

The value of **118.00** is entered in the cell directly under **105.00**.

4. **Type 120**; then **press Enter**.

The value of **120.00** is entered in the cell directly under **118.00**.

5. **Type 123**; then **press Enter**.

The value of **123.00** is entered cell directly under **120.00**.

Figure E.4 shows the Data Editor after the scores have been entered.

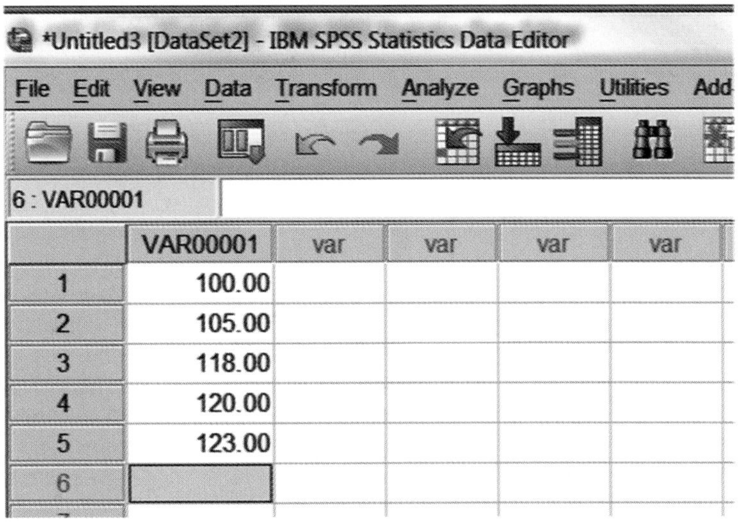

Figure E.4   Five scores entered into the Data Editor

Let's now see how to assign our own name by changing **VAR00001** to **IQ**. To do so:

1. Click Variable View, next to *Data View* in the lower left corner of the screen.	This causes the Data Editor-Variable View to be displayed, with the cell containing the name **VAR00001** highlighted.
2. Type IQ in the highlighted cell; then **press Enter**.	**IQ** is entered as the variable name, replacing **VAR00001**. Note that when you change the name of a variable in the Variable View screen, that name change is carried through in the Data View table as well. Figure E.5 shows the Data View.

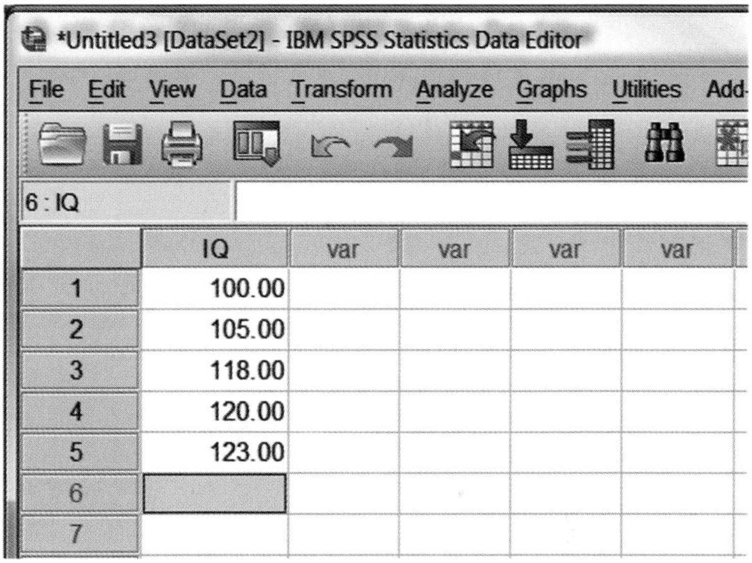

Figure E.5 Data View showing the scores with **IQ** as the variable name.

Saving Data Files   It is a good idea when you have finished entering the data and naming the variable to save the data file. This is because any changes to data files made in a session, including initial data input, only last as long as the session, unless you save the file. Let's assume you have just entered the IQ data, and named the variable, IQ, and are in the Data View. Next, you want to save the file. You decide to name the file *IQexp*. To save this file on your computer with the name **IQexp**,

1. Click File on the menu bar at the top of the screen.	This produces a drop-down menu.
2. Click Save on the drop-down menu.	This produces **the Save Data As** dialog box shown below, with the cursor located in the **File name:** box. Note that the folders and files that are displayed in the large box contain material already saved and can vary widely from computer to computer.

The **Look in**: box at the top shows the directory in which the file will be saved. You can browse to another directory if you choose.

3. Type IQexp in the **File name**: box, replacing **Untitled3**; then **click** the **Save** button.

SPSS saves the file, adding the extension **.sav**. You are then returned to the Data View with the name of the data file, **IQexp. sav**, entered in the title bar at the top left of the screen.

Obtaining a New (Blank) Data Editor   When you have finished analyzing the data for one problem or experiment, and you want to move on to another problem, it will be necessary for you to enter the data of the new problem. If the Data Editor already contains a data file and you are going to type the new data directly into the Data Editor, it is useful to obtain first a new or blank Data Editor into which you can enter the new data. To illustrate how to do this, we will assume that you are in the Data Editor, displaying the Data View, and you have the scores from a saved data file currently entered in the data table. To obtain a new (blank) Data Editor,

1. Click File, then **select New**; then **click** on **Data**.

SPSS displays a new Data Editor. Since this is a new Data Editor, the table is blank.

Analyzing Data   Before presenting a specific example, it is worthwhile to discuss a general procedure that SPSS uses to analyze data. For any data set, you must tell SPSS the name of the variable you want analyzed. SPSS accomplishes this by displaying a dialog box. When you tell SPSS that you want a particular procedure done, such as computing the mean of a set of scores currently entered in the Data Editor, SPSS will display the appropriate dialog box to tell it the name of the variable to be analyzed. In our examples so far, there has been only one variable. However, often the Data Editor contains more than one variable. Whichever is the case for a specific data set, the dialog box will list all of the variables contained in the Data Editor for that set in a large box on the left. You must then move the variable(s) that you want analyzed into the designated blank box, which is usually on the right. This is accomplished by clicking the variable(s) and then clicking the arrow that is located next to the designated box. When this is done, the variable(s) moves from the box on the left into the designated box. Please note, **SPSS only analyzes variables that are contained in the designated box**.

Let's now do an example to illustrate how SPSS analyzes data. You will see the above general procedure at work in this example. Assume that you want to compute the mean, standard deviation, and range of the **IQ** data shown in Figure E.5, and that these data are currently entered into the Data Editor. To compute the **mean**, **standard deviation** and **range** of the **IQ** scores,

1. **Click** **Analyze** on the menu bar at the top of the screen; then **select Descriptive Statistics**; then; **click Descriptives**….

This produces the **Descriptives** dialog box, shown below, that SPSS uses to do descriptive statistics. It also is the dialog box in which you tell SPSS the name of the variable(s) you want analyzed. Notice that **IQ** is located and highlighted in the large box on the left.

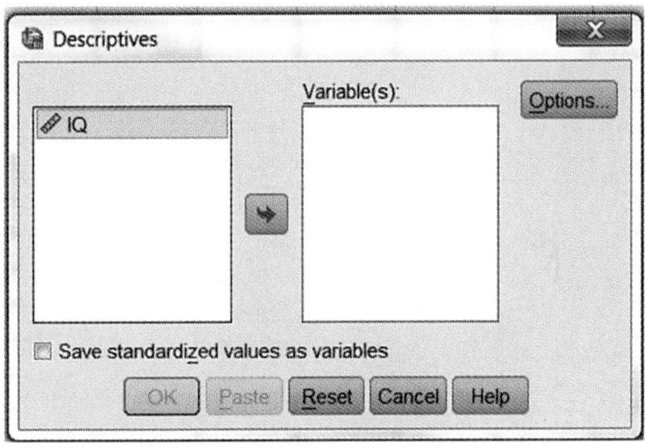

2. **Click** the **arrow** in the middle of the dialog box.

This moves **IQ** from the large box on the left into the designated **Variable(s):** box on the right, telling SPSS that you want to analyze the **IQ** scores.

3. **Click** **Options…** at the top right of the dialog box.

This produces the **Descriptives: Options** dialog box shown below. This dialog box allows you to tell SPSS which statistics you want to compute. Checked boxes indicate the default statistics that SPSS computes.

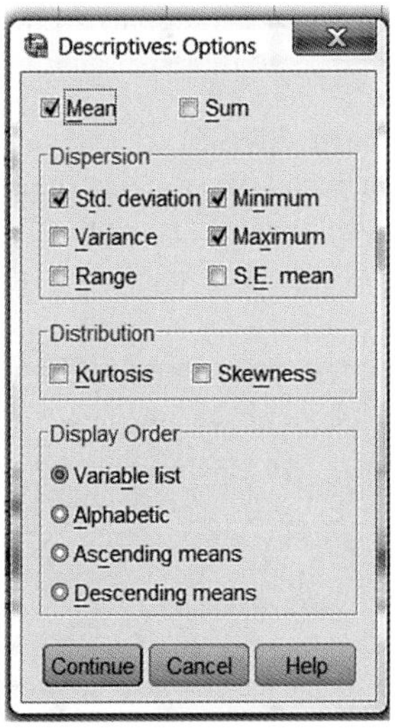

4. Click Mi_n_imum and Ma_x_imum; then click _R_ange.

This removes the default **checked** entries for **Mi_n_imum** and **Ma_x_imum**, and produces a **check** in the **Range** box. Since the **Mean** and **Std. deviation** boxes were already checked, the boxes for **Mean**, **Std. deviation**, and **Range** should now be the only checked boxes. SPSS will compute these statistics when given the **OK** command from the **Descriptions** dialog box.

5. Click Continue.

This returns you to the **Descriptions** dialog box where you can give the **OK** command.

6. Click OK.

SPSS then analyzes the data and displays the results shown below.

Analysis Results

The results are displayed in the SPSS Viewer window as shown in E.6 below. From the **Descriptive Statistics** table, we note that **Range = 23.00**, **Mean = 113.2000**, and **Std. Deviation = 10.08464**. The **Descriptive Statistics** table also shows that *N* = 5.

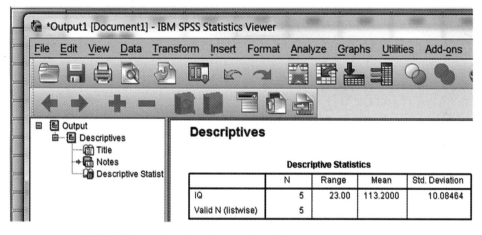

Figure E.6 SPSS Viewer showing the results of the analysis.

Exiting from SPSS There are several ways to exit from SPSS. One way is to click on **File** on the menu bar of either the Data Editor or Viewer. Then **click Exit** on the drop-down menu. Another way is to **click** the ✕ (close) button in the upper right corner of either screen. If this is the last Data Editor window, SPSS will display the following screen

If you **click Yes**, and the data have been saved, SPSS will close. If you haven't saved your work (data, analyses, graphs, or output) you will be prompted about whether you want to save it. Answer the dialog box(es) as appropriate. After doing so, SPSS will close. If you have saved your work before exiting, when you **click Exit** or **click** the ✕ button, SPSS will immediately close.

GLOSSARY

Alpha level A probability level set by an investigator at the beginning of an experiment to limit the probability of making a Type I error. *(p. 252,255)*

***A posteriori* comparisons** Comparisons that are not planned before doing the experiment. They usually arise after the experimenter sees the data and chooses groups with mean values that are far apart, or else they arise from doing all the possible comparisons with no theoretical *a priori* basis. *(p. 423)*

***A posteriori* probability** Probability determined after the fact, after some data have been collected. In equation form,

$$p(A) = \frac{\text{Number of times } A \text{ has occurred}}{\text{Total number of occurrences}}$$

(p. 194)

***A priori* comparisons** Comparisons that are planned in advance of the experiment. They often arise from predictions that are based on theory and prior research. *(p. 422)*

***A priori* probability** Probability determined without collecting any data; deduced from reason alone. In equation form,

$$p(A) = \frac{\text{Number of events classifiable as } A}{\text{Total number of possible events}}$$

(p. 193)

Addition rule Gives the probability of occurrence of one of several events. If there are only two events, A and B, the addition rule gives the probability of occurrence of A or B. In equation form,

$$p(A \text{ or } B) = p(A) + p(B) - p(A \text{ and } B)$$

(p. 196)

Alternative hypothesis Symbolized by H_1. The hypothesis that claims the differences in results between conditions is due to the independent variable. *(p. 252)*

Analysis of variance Abbreviated ANOVA. Statistical technique used to analyze multigroup experiments. Uses the F test as the basis of the analysis(es). *(p. 405)*

Arithmetic mean The sum of the scores divided by the number of scores. In equation form,

$$\overline{X} = \frac{\sum X_i}{N} = \frac{X_1 + X_2 + X_3 + \cdots + X_N}{N}$$

mean of a sample

or

$$\mu = \frac{\sum X_i}{N} = \frac{X_1 + X_2 + X_3 + \cdots + X_N}{N}$$

mean of a population set of scores

where
$$\begin{aligned}
X_1,\ldots, X_N &= \text{raw scores} \\
\overline{X} \text{ (read "X bar")} &= \text{mean of a sample set of scores} \\
\mu \text{ (read "mew")} &= \text{mean of a population set of scores} \\
\Sigma \text{ (read "sigma")} &= \text{summation sign} \\
N &= \text{number of scores}
\end{aligned}$$

(p. 80)

Asymptotic Approaching a given value as a function extends to infinity. For the normal curve, it refers to how the Y value of the normal curve approaches 0 (the X axis) as X extends to + and – infinity. Y gets closer and closer to 0, but never quite reaches it. *(p. 103)*.

Bar graph Graph of nominal or ordinal data, where a bar is drawn for each category and the height of the bar represents the frequency or number of members of that category. *(p. 63)*

Bell-shaped curve Frequency graph named "bell-shaped" because it looks like a bell. *(p. 67)*

Beta The probability of making a Type II error. *(p. 255)*

Between-groups degrees of freedom Symbolized by $df_{between}$. Statistic computed in the one-way ANOVA. The denominator for the between-groups variance estimate, $MS_{between}$. *(p. 409)*

Between-groups sum of squares Symbolized by $SS_{between}$. Statistic computed in the one-way ANOVA. The numerator of the equation for the between-groups variance estimate, $MS_{between}$. *(p. 406, 409)*

Between-groups variance estimate Symbolized by $MS_{between}$. Statistic computed in the one-way ANOVA. Estimate of the null-hypothesis population variance that is based on the variability between the groups. *(p. 406, 408)*

Biased coins Coins for which p(head) $\neq$ p(tail) for any coin when flipped. Expressed in terms of P and Q, $P \neq Q \neq 0.50$. *(p. 200)*

Binomial distribution A probability distribution that results when five preconditions are met: (1) There is a series of N trials; (2) on each trial there are only two possible outcomes; (3) on each trial, the two possible outcomes are mutually exclusive; (4) there is independence between the outcomes of each trial; and (5) the probability of each possible outcome on any trial stays the same from trial to trial. The binomial distribution gives each possible outcome of the N trials and the probability of getting each of these outcomes. *(p. 226)*

Binomial expansion Mathematical expression used to generate the binomial distribution. The expression is given by $(P + Q)^N$. *(p. 229)*

Binomial table Table that contains binomial distribution probabilities for many values of N and P. *(p. 230)*

Biserial coefficient A correlation coefficient, symbolized by r_b. It is used when one of the variables is at least of interval scaling and the other is dichotomous. *(p. 140)*

Central tendency The average, middle, or most frequent value of a set of scores. *(p. 80)*

Chi-square Nonparametric inference test that is used with nominal scaling. Statistic computed is χ^2. *(p. 484)*

Coefficient of determination Symbolized by r^2. Tells us the proportion of the total variability that is accounted for by X. *(p. 139)*

Cohen's *d* Statistic, associated with J. Cohen, that is used to measure the size of effect. *(p. 339)*

Column degrees of freedom Symbolized by $df_{columns}$. Statistic computed in two-way ANOVA. The denominator of the equation for computing the column variance estimate, $MS_{columns}$. *(p. 454)*

Column sum of squares Symbolized by $SS_{columns}$. Statistic computed in two-way ANOVA. The numerator

of the equation for computing the column variance estimate, $MS_{columns}$. *(p. 454)*

Column variance estimate Symbolized by $MS_{columns}$. Statistic computed in two-way ANOVA. Estimate of the null-hypothesis population variance that is based on the between columns variability. *(p. 449, 454)*

Confidence interval A range of values that probably contains the population value. *(p. 341)*

Confidence limits The values that state the boundaries of the confidence interval. *(p. 341)*

Confidence-interval approach Alternative approach to null-hypothesis approach. Uses confidence intervals as a method that allows conclusions with regard both to whether there is a real effect and to the size of the effect. *(p. 382)*

Constant A quantity whose value doesn't change. Pi (π) is an example; it has a value that never changes. The value of Pi to 5 decimal place accuracy is 3.14159. *(p. 6)*

Contingency table A two-way table showing the contingency between two variables where the variables have been classified into mutually exclusive categories and the cell entries are frequencies. *(p. 489)*

Continuous variable A variable that theoretically can have an infinite number of values between adjacent units on the scale. *(p. 35)*

Correct decision Rejecting H_0 when H_0 is false; retaining H_0 when H_0 is true. *(p. 255)*

Correlated groups design There are paired scores in the conditions, and the differences between paired scores are analyzed. *(p. 251)*

Correlation The association or relationship between two variables. It focuses on the direction and degree of the relationship. *(p. 130)*

Correlation coefficient A quantitative expression of the magnitude and direction of a relationship. *(p. 130)*

Critical region Short for "critical region for rejection of the null hypothesis." Region that contains values of the statistic that allow rejection of the null hypothesis. *(p. 312)*

Critical region for rejection of the null hypothesis The area under the curve that contains all the values of the statistic that allow rejection of the null hypothesis. *(p. 312)*

Critical value of a statistic The value of the statistic that bounds the critical region. *(p. 312)*

Critical value of *F* Symbolized F_{crit}. The value of F that bounds the critical region. *(p. 403)*

Critical value of *r* Symbolized by r_{crit}. The value of r that bounds the critical region. *(p. 347)*

Critical value of *t* Symbolized by t_{crit}. The value of *t* that bounds the critical region. *(p. 332)*

Critical value of $\overline{X}$ Symbolized by $\overline{X}_{crit}$. The value of $\overline{X}$ that bounds the critical region. *(p. 318)*

Critical value of *z* Symbolized by z_{crit}. The value of z that bounds the critical region. *(p. 312)*

Cumulative frequency distribution The number of scores that fall below the upper real limit of each interval. *(p. 54)*

Cumulative percentage distribution The percentage of scores that fall below the upper real limit of each interval. *(p. 54)*

Curvilinear relationship The relationship between two variables is curved, rather than linear. In this case, a curved line fits the data better than a straight line. *(p. 127)*

Data The measurements that are made on the subjects of an experiment. *(p. 7)*

Degree of separation Used in conjunction with the Mann–Whitney *U* test. Refers to the lack of overlap between the sample scores of the two groups. *(p. 502)*

Degrees of freedom (df) The number of scores that are free to vary in calculating a statistic. *(p. 330, 371)*

Dependent variable The variable in an experiment that an investigator measures to determine the effect of the independent variable. *(p. 7)*

Descriptive statistics Techniques that are used to describe or characterize the obtained sample data. *(p. 10)*

Deviation score The distance of the raw score from the mean of its distribution. *(p. 89)*

Direct relationship As *X* increases, *Y* increases. As *X* decreases, *Y* decreases. The slope of the relationship is positive. Higher values of *X* are associated with higher values of *Y*. Lower values of *X* are associated with lower values of *Y*. Also called a *positive relationship*. *(p. 127)*

Directional hypothesis An hypothesis that specifies the direction of the effect of the independent variable on the dependent variable. *(p. 252)*

Discrete variable A variable for which no values are possible between adjacent units on the scale. *(p. 35)*

Dispersion The spread of a set of scores. *(p. 89)*

Estimated standard error of the difference between sample means Symbolized by $s_{\overline{X}_1 - \overline{X}_2}$. Estimate of $\sigma_{\overline{X}_1 - \overline{X}_2}$. *(p. 370)*

Eta squared Biased estimate of the size of effect of the independent variable. *(p. 420)*

Exhaustive set of events A set that includes all of the possible events. *(p. 200)*

Expected frequency Symbolized by f_e. Statistic computed for the chi-square test. The expected frequency under the assumption sampling is random from the null-hypothesis population. *(p. 485)*

Exploratory data analysis A recently developed technique that employs easily constructed diagrams that are useful in summarizing and describing sample data. *(p. 67)*

F test Inference test based on the ratio of two independent estimates of the same population variance, σ^2. Used in conjunction with the analysis of variance. *(p. 402)*

Factorial experiment An experiment in which the effects of two or more factors are assessed and the treatments used are combinations of the levels of the factors. *(p. 446)*

Fail to reject null hypothesis Conclusion when analyzing the data of an experiment that retains the null hypothesis as a reasonable explanation of the data. *(p. 253)*

Fair coins Coins for which, when flipped, $p(\text{head}) = p(\text{tail})$ for any coin. Expressed in terms of *P* and *Q*, $P = Q = 0.50$. *(p. 200)*.

Frequency distribution A listing of score values and their frequency of occurrence. *(p. 48)*

Frequency polygon Graph that is used with interval or ratio data. Identical to a histogram, except that instead of using bars, the midpoints of each interval are plotted and joined together with straight lines, and the lines extended to meet the horizontal axis at the midpoint of the intervals that are immediately beyond the lowest and highest intervals. *(p. 64)*

Grand mean Symbolized $\overline{X}_G$. Statistic computed in one-way and two-way ANOVA. The overall mean of all the scores combined. *(p. 408)*

Histogram Similar to a bar graph, except that it is used with interval or ratio data. Class intervals are plotted on the horizontal axis, a bar is drawn over each class interval such that each class bar begins and ends at the real limits of the interval. The height of each bar corresponds to the frequency of the interval and the vertical bars touch each other rather than spaced apart as with the bar graph. *(p. 63)*

Homogeneity of variance Assumption underlying the independent groups *t* test and ANOVA. If there are *k* groups, the assumption is that the variances of the populations from which the *k* samples are drawn, are equal. In equation form, $\sigma_1^2 = \sigma_2^2 = \cdots = \sigma_k^2$. *(p. 375)*

Homoscedasticity Assumption used in conjunction with the standard error of estimate. The assumption is that the variability of Y remains constant for all values of X. *(p. 170)*

Imperfect relationship A positive or negative relationship for which all of the points do not fall on the line. *(p. 128)*

Importance of an effect A real effect that in addition to being statistically significant, is of practical or theoretical importance. *(p. 265)*

Independence of two events The occurrence of one event has no effect on the probability of occurrence of the other. *(p. 201)*

Independent groups design Involves experiments using two or more conditions. Each condition employs a different level of the independent variable. The most basic experiment has two conditions. Subjects are randomly selected from the subject population and then randomly assigned to the two conditions. Since subjects are randomly assigned to the conditions, there is no basis for pairing of scores between conditions. Rather, a statistic is computed for the scores of each group separately, and the two group statistics are compared to determine if chance alone is a reasonable explanation of the data. *(p. 366)*

Independent variable The variable in an experiment that is systematically manipulated by an investigator. *(p. 6)*

Inferential statistics Techniques that use the obtained sample data to infer to populations. *(p. 10)*

Interaction degrees of freedom Symbolized by df $_{interaction}$. Statistic computed in two-way ANOVA. The denominator of the equation for computing the interaction variance estimate, $MS_{interaction}$. *(p. 455)*

Interaction effect The result observed when the effect of one factor is not the same at all levels of the other factor. *(p. 447)*

Interaction sum of squares Symbolized by $SS_{interaction}$. Statistic computed in two-way ANOVA. The numerator of the equation for computing the interaction variance estimate, $MS_{interaction}$. *(p. 455)*

Interaction variance estimate Symbolized by $MS_{interaction}$. Statistic computed in two-way ANOVA. Estimate of the null-hypothesis population variance that is based on the variability of the cell means. *(p. 449, 455)*

Interval scale A measuring scale that possesses the properties of magnitude and equal interval between adjacent units on the scale, but doesn't have an absolute zero point. Celsius scale of temperature measurement is a good example of an interval scale. *(p. 32)*

Inverse relationship As X increases, Y decreases; as X decreases, Y increases. The slope of the relationship is negative. Higher values of X are associated with lower values of Y. Lower values of X are associated with higher values of Y. Also called a *negative* relationship. *(p. 127)*

J-shaped curve Frequency graph named *J-shaped* because it has the shape of the letter "J." *(p. 67)*

Kruskal–Wallis test Nonparametric inference test used as a substitute for the parametric, one-way, independent groups ANOVA when the assumptions of that test are seriously violated. Statistic computed is H. *(p. 507)*

Least-squares regression line The prediction line that minimizes the total error of prediction according to the least-squares criterion of $\Sigma (Y - Y')^2$. *(p. 161)*

Linear relationship A relationship between two variables that can be most accurately represented by a straight line. *(p. 124)*

Main effect The effect of factor A (averaged over the levels of factor B) and the effect of factor B (averaged over the levels of factor A). *(p. 447)*

Mann–Whitney U test Nonparametric inference test used as a substitute for the independent groups t test when the assumptions of that test are seriously violated. Statistics computed are U and U'. *(p. 501)*

Marginals Used in conjunction with contingency tables. Marginals are the row and column totals lying outside the contingency table. *(p. 491)*

Mean of the population of difference scores Symbolized by μ_D. Mean of a hypothetical population of difference scores from which the sample difference scores are assumed to have been drawn. If the independent variable has no effect, then $\mu_D = 0$. *(p. 360)*

Mean of the sampling distribution of the difference between sample means Symbolized by $\mu_{\bar{X}_1 - \bar{X}_2}$. Mean of the complete population distribution of $(\bar{X}_1 - \bar{X}_2)$ scores. *(p. 368)*

Mean of the sampling distribution of the mean Symbolized by $\mu_{\bar{X}}$. This is the mean of the full set of sample means. Also called the standard error of the mean. *(p. 305)*

Median (Mdn) The scale value below which 50% of the scores fall. *(p. 85)*

Method of authority Something is considered true because of tradition or because some person of distinction says it is true. *(p. 4)*

Method of intuition Sudden insight, or clarifying idea that springs into consciousness, all at once as a whole. *(p. 5)*

Method of rationalism Uses reason alone to arrive at knowledge. It assumes that if the premises are sound and the reasoning is carried out correctly according to the rules of logic, then the conclusions will yield truth. *(p. 4)*

Mode The most frequent score in the distribution. *(p. 87)*

Multiple coefficient of determination Symbolized by R^2. Gives the proportion of the total variance in Y accounted for by the multiple X variables. Also called *squared multiple correlation. (p. 176)*

Multiple regression Technique used for predicting Y from multiple associated X variables. *(p. 174)*

Multiplication rule Gives the probability of joint or successive occurrence of several events. If there are only two events, the multiplication rule gives the probability of occurrence of A and B. In equation form,

$$p(A \text{ and } B) = p(A)p(B|A)$$

(p. 201)

Mutually exclusive events Two events that cannot occur together; that is, the occurrence of one precludes the occurrence of the other. *(p. 196)*

Naturalistic observation research A type of observational study in which the subjects of interest are observed in their natural setting. A goal of this research is to obtain an accurate description of behaviors of interest occurring in the natural setting. *(p. 9)*

Negative relationship An inverse relationship between two variables. *(p. 127)*

Negatively skewed curve A curve on which most of the scores occur at the higher values, and the curve tails off toward the lower end of the horizontal axis. *(p. 65)*

Nominal scale The scale is composed of categories, and the object is "measured" by determining to which category the object belongs. The categories comprise the units of the scale. An example would be brands of computers; the units would be Apple, Dell, HP, etc. *(p. 31)*

Nondirectional hypothesis An hypothesis that doesn't specify the direction of the effect of the independent variable on the dependent variable. *(p. 252)*

Normal approximation Technique used to solve binomial problems when $N > 20$. *(p. 239)*

Normal curve A symmetrical, bell-shaped curve with mean, median, and mode equal to each other, and specified kurtosis. Kurtosis refers to the sharpness or flatness of a curve as it reaches its peak. In equation form, the normal curve equals

$$Y = \frac{N}{\sqrt{2\pi}\sigma} e^{-(X-\mu)^2/2\sigma^2}$$

where e = a constant of 2.7183
π = a constant of 3.1416

(p. 103)

Null hypothesis Symbolized by H_0. Logical counterpart to the alternative hypothesis. It either specifies that there is no effect, or that there is a real effect in the direction opposite to that specified by the alternative hypothesis. *(p. 252)*

Null-hypothesis approach Main approach used in this textbook for analyzing data to determine if the independent variable has a real effect. In this approach, we assume that chance alone is responsible for the difference between the scores in each group, calculate the obtained probability, and determine if the obtained probability is low enough to rule out chance as a reasonable explanation of the score differences between groups. *(p. 382)*

Null-hypothesis population An actual or theoretical set of population scores that would result if the experiment were done on the entire population and the independent variable had no effect; it is used to test the validity of the null hypothesis. *(p. 300)*

Number of P events A P event is one of the two possible outcomes of any trial. The number of P events is the number of such outcomes. *(p. 229)*

Number of Q events A Q event is one of the two possible outcomes of any trial. The number of Q events is the number of such outcomes. *(p. 229)*

Observational studies A type of research in which no variables are actively manipulated. The researcher observes and records the data of interest. *(p. 9)*

Observed frequency Symbolized by f_o. Statistic computed for the chi-square test. Observed frequency in the sample. *(p. 485)*

Omega squared Symbolized $\hat{\omega}^2$. Unbiased estimate of the size of the effect of the independent variable. *(p. 419)*

One-tailed probability Probability that results when all of the outcomes being evaluated are under one tail of the distribution. *(p. 259)*

One-way ANOVA, independent groups design Statistical technique used to analyze multigroup experiments in which the experimental design is an independent groups design and only one independent variable is studied. *(p. 405)*

Ordinal scale This is a rank-ordered scale in which the objects being measured are rank-ordered according to whether they possess more, less, or the same amount of the variable being measured. An example is ranking Division I NCAA college football teams according to which college or university football team is considered the best, the next best, the next next best, and so on. *(p. 32)*

Overall mean Sometimes called *weighted mean*. The average value of several sets or groups of scores. It takes into account the number of scores in each group and in effect, weights the mean of each group by the number of scores in the group. In equation form,

$$\overline{X}_{\text{overall}} = \frac{n_1\overline{X}_1 + n_2\overline{X}_2 + \cdots + n_k\overline{X}_k}{n_1 + n_2 + \cdots + n_k}$$

(p. 83)

Parameter A number calculated on population data that quantifies a characteristic of the population. *(p. 7)*

Parameter estimation research A type of observational study in which the goal is to determine a characteristic of a population. An example might be the mean age of all psychology majors at your university. *(p. 9)*

Pearson r A measure of the extent to which paired scores occupy the same or opposite positions within their own distributions. *(p. 131)*

Percentile The value on the measurement scale below which a specified percentage of the scores in the distribution falls. *(p. 56)*

Percentile point See Percentile.

Percentile rank (of a score) The percentage of scores with values lower than the score in question *(p. 59)*

Perfect relationship A positive or negative relationship for which all of the points fall on the line. *(p. 128)*

Phi coefficient A correlation coefficient, symbolized by ϕ. Used when each of the variables is dichotomous. *(p. 140)*

Planned comparisons See *a posteriori* comparisons.

Population The complete set of individuals, objects, or scores that an investigator is interested in studying. *(p. 6)*

Positive relationship A direct relationship between two variables. *(p. 127)*

Positively skewed curve A curve on which most of the scores occur at the lower values, and the curve tails off toward the higher end of the horizontal axis. *(p. 65)*

Post hoc comparisons See *a posteriori* comparisons.

Power The probability that the results of an experiment will allow rejection of the null hypothesis if the independent variable has a real effect. *(p. 278)*

Probability Expressed as a fraction or decimal number, probability is fundamentally a proportion; it gives the chances that an event will or will not occur. *(p. 193)*

Probability of occurrence of A or B The probability of occurrence of A plus the probability of occurrence of B minus the probability of occurrence of both A and B. *(p. 196)*

Probability of occurrence of both A and B The probability of occurrence of A times the probability of occurrence of B given that A has occurred. *(p. 201)*

Q_{crit} The value of Q that bounds the critical region. *(p. 424)*

Q_{obt} The obtained value of Q. *(p. 424)*

Random sample A sample selected from the population by a process that ensures that (1) each possible sample of a given size has an equal chance of being selected and (2) all the members of the population have an equal chance of being selected into the sample. *(p. 190)*

Range The difference between the highest and lowest scores in the distribution. *(p. 89)*

Ratio scale A measuring scale that possesses the properties of magnitude, equal intervals between adjacent units on the scale, and also possesses an absolute zero point. The Kelvin scale of temperature measurement is an example of a ratio scale. *(p. 33)*

Real effect An effect of the independent variable that produces a change in the dependent variable. *(p. 278)*

Real limits of a continuous variable Those values that are above and below the recorded value by one-half of the smallest measuring unit of the scale. *(p. 36)*

Regression A topic that considers using the relationship between two or more variables for prediction. *(p. 160)*

Regression constant The a_Y and b_Y terms in the equation, $Y' = b_Y X + a_Y$. *(p. 162)*

Regression line A best fitting line used for prediction. *(p. 160)*

Regression of *Y* on *X* Technique used to derive the regression line for predicting *Y* given *X*. *(p. 162)*

Reject null hypothesis Conclusion when analyzing the data of an experiment that rejects the null hypothesis as a reasonable explanation of the data. *(p. 254)*

Relative frequency distribution The proportion of the total number of scores that occur in each interval. *(p. 54)*

Repeated measures design A form of the correlated groups design. There are paired scores in the conditions, and the differences between paired scores are analyzed. *(p. 251)*

Replicated measures design Same as the repeated measures design. There are paired scores in the conditions, and the differences between paired scores are analyzed. *(p. 251)*

Retain null hypothesis Same as fail to reject null hypothesis. Conclusion when analyzing the data of an experiment that fails to reject the null hypothesis as a reasonable explanation of the data. *(p. 252)*

Row degrees of freedom Symbolized by df_{rows}. Statistic computed in two-way ANOVA. Degrees of freedom in forming the row variance estimate, MS_{rows}. *(p. 452)*

Row sum of squares Symbolized by SS_{rows}. Statistic computed in two-way ANOVA. The numerator of the equation for computing the row variance estimate, MS_{rows}. *(p. 452)*

Row variance estimate Symbolized by MS_{rows}. Estimate of the null-hypothesis population variance that is based on the between rows variability. *(p. 449, 452)*

Sample A subset of the population. *(p. 6)*

Sampling distribution of a statistic A listing of (1) all the values that the statistic can take and (2) the probability of getting each value under the assumption that it results from chance alone, or if sampling is random from the null-hypothesis population. *(p. 299)*

Sampling distribution of *F* Gives all the possible *F* values along with the $p(F)$ for each value, assuming sampling is random from the population. *(p. 402)*

Sampling distribution of *t* A probability distribution of the *t* values that would occur if all possible different samples of a fixed size *N* were drawn from the null-hypothesis population. It gives (1) all the possible different *t* values for samples of size *N* and (2) the probability of getting each value if sampling is random from the null-hypothesis population. *(p. 329)*

Sampling distribution of the difference between sample means Hypothetical population distribution of $(\overline{X}_1 - \overline{X}_2)$ scores obtained from taking all possible samples of size n_1 and n_2 from populations of means μ_1 and μ_2, and standard deviations σ_1 and σ_2. *(p. 368)*

Sampling distribution of the mean A listing of all the values the mean can take, along with the probability of getting each value if sampling is random from the null-hypothesis population. *(p. 303)*

Sampling with replacement A method of sampling in which each member of the population selected for the sample is returned to the population before the next member is selected. *(p. 193)*

Sampling without replacement A method of sampling in which the members of the sample are not returned to the population before selecting subsequent members. *(p. 193)*

Scatter plot A graph of paired *X* and *Y* values. *(p. 124)*

Scientific method The scientist has a hypothesis about some feature of realty that he or she wishes to test. An objective, observational study or experiment is carried out. The data is analyzed statistically, and conclusions are drawn either supporting or rejecting the hypothesis. *(p. 6)*

Scheffé test *Post hoc*, multiple comparisons test for doing all possible *post hoc* comparisons, not just pair-wise mean comparisons. The most conservative of all the possible *post hoc* tests. *(p. 425)*

Sign test Statistical inference test, appropriate for the repeated measures or correlated groups design, involving only two groups, that ignores the magnitude of the difference scores and considers only their direction or sign. *(p. 250)*

Significant The result of an experiment that is statistically reliable. *(p. 253, 265)*

Simple randomized-group design See one-way ANOVA, independent groups design. *(p. 406)*

Single factor experiment, independent groups design See one-way ANOVA, independent groups design. *(p. 406)*

Size of effect Magnitude of the real effect of the independent variable on the dependent variable. *(p. 265, 363, 376)*

Skewed curve A curve whose two sides do not coincide if the curve is folded in half; that is, a curve that is not symmetrical. *(p. 65)*

Slope Rate of change. For a straight line,

$$\text{Slope} = \frac{\Delta Y}{\Delta X} = \frac{Y_2 - Y_1}{X_2 - X_1}$$

(p. 125)

Spearman rho A correlation coefficient, symbolized by r_s. Used when one or both of the variables are of ordinal scaling. *(p. 141)*

Standard deviation A measure of variability that gives the average deviation of a set of scores about the mean. In equation form,

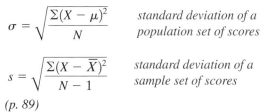

$$\sigma = \sqrt{\frac{\Sigma(X - \mu)^2}{N}} \qquad \text{standard deviation of a population set of scores}$$

$$s = \sqrt{\frac{\Sigma(X - \overline{X})^2}{N - 1}} \qquad \text{standard deviation of a sample set of scores}$$

(p. 89)

Standard deviation of the sampling distribution of the difference between sample means Symbolized by $\sigma_{\overline{X}_1 - \overline{X}_2}$. Standard deviation of the complete population distribution of $(\overline{X}_1 - \overline{X}_2)$ scores. *(p. 368)*

Standard error of estimate Symbolized by $s_{Y|X}$. Gives us a measure of the average deviation of prediction errors about the regression line. *(p. 169)*

Standard error of the mean Symbolized by $\mu_{\overline{X}}$. The mean of the sampling distribution of the mean. *(p. 305)*

Standard score See z score. *(p. 105)*

State of reality Truth regarding H_0 and H_1. *(p. 255)*

Statistic A number calculated on sample data that quantifies a characteristic of the sample. *(p. 7)*

Statistical Package for the Social Sciences Abbreviated SPSS. Statistical software package widely used in the social sciences. *(p. 11)*

Stem-and-leaf diagram An alternative to the histogram, which is used in exploratory data analysis. A picture is shown of each score divided into a stem and leaf, separated by a vertical line. The leaf for each score is usually the last digit, and the stem is the remaining digits. Occasionally, the leaf is the last two digits depending on the range of the scores. The stem is placed to the left of the vertical line, and the leaf to the right of the line. Stems are placed vertically down the page, and leafs are placed in order horizontally across the page. *(p. 67)*

Sum of squares The sum of $(X - \mu)^2$ or $(X - \overline{X})^2$ is called the sum of squares. It is symbolized by SS_{pop} for population data or just SS for sample data. In equation form,

$$SS_{pop} = \Sigma(X - \mu)^2 = \Sigma X^2 - \frac{(\Sigma X)^2}{N}$$

sum of squares for population data

$$SS = \Sigma(X - \overline{X})^2 = \Sigma X^2 - \frac{(\Sigma X)^2}{N}$$

sum of squares for sample data

(p. 91, 92)

Summation Operation very often performed in statistics in which all or parts of a set (or sets) of scores are added. *(p. 27)*

Symmetrical curve A curve whose two sides coincide if the curve is folded in half. *(p. 65)*

***t* test for correlated groups** Inference test using Student's t statistic. Employed with correlated groups, replicated measures, and repeated measures designs. *(p. 358)*

***t* test for independent groups** Inference test using Student's t statistic. Employed with independent groups design. *(p. 366, 370)*

***t* test for single samples** Inference test using Student's t statistic. Employed with single sample design. *(p. 328)*

Total sum of squares Symbolized by SS_{total}. Statistic computed in the analysis of variance. The variability of all the scores about the grand mean. *(p. 406, 414)*

True experiment In a true experiment, an independent variable is manipulated and its effect on some dependent variable is studied. Has the potential to determine causality. *(p. 9)*

Tukey HSD test *Post hoc*, multiple comparisons test that makes all possible pairwise comparisons among the sample means. *(p. 424)*

Two-tailed probability Probability that results when the outcomes being evaluated are under both tails of the distribution. *(p. 258)*

Two-way analysis of variance Statistical technique for assessing the effects of two variables that are manipulated in one experiment. *(p. 446, 450)*

Type I error A decision to reject the null hypothesis when the null hypothesis is true. *(p. 254)*

Type II error A decision to retain the null hypothesis when the null hypothesis is false. *(p. 254)*

U-shaped curve Frequency graph named *U-shaped* because it has the shape of the letter "U." *(p. 67)*

Variability Refers to the spread of a set of scores. *(p. 80)*

Variability accounted for by *X* The change in Y that is explained by the change in X. Used in measuring the strength of a relationship. *(p. 138)*

Variable Any property or characteristic of some event, object, or person that may have different values at different times depending on the conditions. *(p. 6)*

Variance The standard deviation squared. In equation form,

$$\sigma^2 = \frac{\Sigma(X - \mu)^2}{N} \qquad \text{variance of a population set of scores}$$

$$s^2 = \frac{\Sigma(X - \bar{X})^2}{N - 1} \quad \textit{variance of a sample set of scores}$$

(p. 95)

Weighted mean See overall mean

Weighted variance estimate Symbolized s_W^2. Used in the *t* test for independent groups to estimate the population variance. *(p. 370)*

Wilcoxon matched-pairs signed ranks test Nonparametric inference test used as a substitute for the correlated groups *t* test when the assumptions of that test are seriously violated. Statistic computed is *T*. *(p. 498)*

Within-cells degrees of freedom Symbolized by $df_{within\text{-}cells}$. Statistic computed in two-way ANOVA. The denominator of the equation for computing the within-cells variance estimate, $MS_{within\text{-}cells}$. *(p. 451)*

Within-cells sum of squares Symbolized by $SS_{within\text{-}cells}$. Statistic computed in two-way ANOVA. The numerator of the equation for computing the within-cells variance estimate, $MS_{within\text{-}cells}$. *(p. 451)*

Within-cells variance estimate Symbolized by $MS_{within\text{-}cells}$. Statistic computed in two-way ANOVA. Estimate of the null-hypothesis population variance that is based on the within-cells variability. *(p. 449, 451)*

Within-groups degrees of freedom Symbolized by df_{within}. Statistic computed in the one-way ANOVA. The denominator of the equation for computing the within-groups variance estimate, MS_{within}. *(p. 407)*

Within-groups sum of squares Symbolized by SS_{within}. Statistic computed in the one-way ANOVA. The total of the sum of squares for each group. *(p. 406, 407)*

Within-groups variance estimate Symbolized by MS_{within}. Statistic computed in the one-way ANOVA. Estimate of the null-hypothesis population variance that is based on the within groups variability. *(p. 406)*

X axis The horizontal axis of a graph. *(p. 61)*

Y axis The vertical axis of a graph. *(p. 61)*

Y intercept The *Y* value of a function where the function intersects the *Y* axis. For the linear relationship $Y = bX + a$, *a* is the *Y* intercept. *(p. 125)*

z score A transformed score that designates how many standard deviation units the corresponding raw score is above or below the mean. *(p. 105)*

z test for single samples Inference test using the *z* statistic. Employed with single sample designs. Also called the Normal Deviate test. *(p. 303)*

Page numbers followed by "n" refer to notes at the bottom of the page or at the end of the chapter.

Symbols

Listed below are the symbols we have used in this textbook. The meaning of each symbol is given to the right of the symbol. The last column gives the page number where the symbol first appears.

Symbol	Meaning	Symbol First Occurs on Page:
α	threshold probability level for rejecting H_0	252
	the probability of a Type I error	255
β	probability of a Type II error	255
χ^2	chi-square	484
ϕ	correlation coefficient for dichotomous variables	140
η	curvilinear correlation coefficient	140
η^2	estimate of size of effect	420
μ	mean of a population	81
μ_D	mean of the population of difference scores	359
μ_{null}	mean of the null-hypothesis population	318
μ_{real}	mean of population when there is a real effect	318
$\mu_{\bar{X}}$	mean of the sampling distribution of the mean	305
$\mu_{\bar{X}_1 - \bar{X}_2}$	mean of the sampling distribution of the difference between sample means	368
ρ	population linear correlation coefficient	346
Σ	the sum of	27
σ	standard deviation of a population	91
σ^2	variance of a population	95
$\sigma_{\bar{X}}$	standard deviation of the sampling distribution of the mean; standard error of the mean	305
$\sigma_{\bar{X}}^2$	variance of the sampling distribution of the mean	305
$\sigma_{\bar{X}_1 - \bar{X}_2}$	standard deviation of the sampling distribution of the difference between sample means; standard error of the difference beween sample means	368
$\hat{\omega}^2$	estimate of size of effect	419
a_Y	Y-axis intercept for minimizing errors in predicting Y	162
b_Y	slope of the line for minimizing errors in predicting Y given X	162

(Continued)

Symbol	Meaning	Symbol First Occurs on Page:
c	number of columns in a contingency table	492
	number of columns in a two-way ANOVA data table	452
cum f	cumulative frequency	55
cum f_L	frequency of scores below the lower real limit of the interval containing the percentile point	57
cum f_P	frequency of scores below the percentile point	57
cum %	cumulative percentage	55
d	size of effect	339
$\hat{d}$	estimated size of effect	340
D	difference between paired scores	360
$\overline{D}_{obt}$	mean of the sample difference scores	360
df	degrees of freedom	330
$df_{between}$	between-group degrees of freedom	409
$df_{columns}$	column degrees of freedom	454
$df_{interaction}$	interaction degrees of freedom	455
df_{rows}	row degrees of freedom	453
df_{within}	within-groups degrees of freedom	407
$df_{within\text{-}cells}$	within-cells degrees of freedom	452
F	ratio of two variance estimates	402
F_{crit}	critical value of f	403
F_{obt}	statistic computed in one-way ANOVA	402
	statistic computed in two-way ANOVA	449
$F_{Scheffé}$	statistic computed in Scheffé test	426
f	frequency	48
f_e	expected frequency	485
f_i	frequency of the interval containing the percentile point	58
f_o	observed frequency	485
H_0	null hypothesis	252
H_1	alternative hypothesis	252
H_{obt}	statistic calculated with Kruskal–Wallis	508
i	width of the interval	51
k	number of groups or means	407
Mdn	median	85
$MS_{between}$	between-groups variance estimate	409
$MS_{between\ (groups\ i\ and\ j)}$	between-groups variance estimate, Scheffé test	426
$MS_{columns}$	column variance estimate	449
MS_{rows}	row variance estimate	449
$MS_{interaction}$	interaction variance estimate	449

Symbol	Meaning	Symbol First Occurs on Page:
MS_{within}	within-groups variance estimate	406
$MS_{within\text{-}cells}$	within-cells variance estimate	449
N	total number of scores	27
	number of paired scores	163
n_k	number of scores in the kth or last group	84
P	in a two-event situation, the probability of one of the events	200
p	probability	194
$p(A)$	probability of event A	193
$p(B|A)$	probability of B, given A has occurred	201
P_{null}	the proportion of pluses in the population if the independent variable has no effect	279
P_{real}	the proportion of pluses in the population if the independent variable has a real effect	279
Q	in a two-event situation, the probability of one of the events	201
	Studentized range statistic	424
Q_{crit}	the critical value of Q	424
Q_{obt}	statistic computed in the Tukey HSD test	424
r	Pearson product moment correlation coefficient	130
	number of rows in a contingency table	492
	number of rows in a two-way ANOVA data table	452
r^2	coefficient of determination	139
R^2	multiple coefficient of determination	176
	squared multiple correlation	176
r_b	biserial correlation coefficient	140
r_s	Spearman rank order correlation coefficient, rho	140
s	standard deviation of a sample	91
	estimate of a population standard deviation	91
s_D	standard deviation of sample difference scores	360
s_X	standard deviation of the X variable	173
s_Y	standard deviation of the Y variable	173
$s_{Y|X}$	standard error of estimate when predicting Y given X	170
$s_{\overline{X}}$	estimated standard error of the mean	328
$s_{\overline{x}_1-\overline{x}_2}$	estimated standard error of the difference between sample means	370
s^2	variance of a sample	95
s_w^2	weighted estimate of the population variance	370
SS	sum of squares of a sample	91
$SS_{between}$	between-groups sum of squares	406
$SS_{between\ (groups\ i\ and\ j)}$	between-groups sum of squares, Scheffé test	426
$SS_{columns}$	column sum of squares	450

(Continued)